Grundlehren der mathematischen Wissenschaften 346

A Series of Comprehensive Studies in Mathematics

For further volumes:
www.springer.com/series/138

Jean-Louis Loday · Bruno Vallette

Algebraic Operads

Springer

Jean-Louis Loday (1946–2012)
IRMA, CNRS et Université de Strasbourg
Strasbourg, France

Bruno Vallette
Lab. de Mathématiques J.A. Dieudonné
Université de Nice-Sophia Antipolis
Nice, France

ISSN 0072-7830 Grundlehren der mathematischen Wissenschaften
ISBN 978-3-642-30361-6 ISBN 978-3-642-30362-3 (eBook)
DOI 10.1007/978-3-642-30362-3
Springer Heidelberg New York Dordrecht London

Library of Congress Control Number: 2012944723

Mathematics Subject Classification: 18D50, 17AXX, 18G50, 55P48, 57T30

Printed on acid-free paper

Springer is part of Springer Science+Business Media (www.springer.com)

À la mémoire de JLL

Preface

An operad is an algebraic device which encodes a type of algebras. Instead of studying the properties of a particular algebra, we focus on the universal operations that can be performed on the elements of any algebra of a given type. The information contained in an operad consists in these operations and all the ways of composing them. The classical types of algebras, that is associative algebras, commutative algebras and Lie algebras, give the first examples of algebraic operads. Recently, there has been much interest in other types of algebras, to name a few: Poisson algebras, Gerstenhaber algebras, Jordan algebras, pre-Lie algebras, Batalin–Vilkovisky algebras, Leibniz algebras, dendriform algebras and the various types of algebras up to homotopy. The notion of operad permits us to study them conceptually and to compare them.

The operadic point of view has several advantages. First, many results known for classical types of algebras, when written in the operadic language, can be applied to other types of algebras. Second, the operadic language simplifies both the statements and the proofs. So, it clarifies the global understanding and allows one to go further. Third, even for classical algebras, the operad theory provides new results that had not been unraveled before. Operadic theorems have been applied to prove results in other fields, like the deformation-quantization of Poisson manifolds by Maxim Kontsevich and Dmitry Tamarkin for instance. Nowadays, operads appear in many different themes: algebraic topology, differential geometry, noncommutative geometry, C^*-algebras, symplectic geometry, deformation theory, quantum field theory, string topology, renormalization theory, combinatorial algebra, category theory, universal algebra and computer science.

Historically, the theoretical study of compositions of operations appeared in the 1950s in the work of Michel Lazard as "analyseurs". Operad theory emerged as an efficient tool in algebraic topology in the 1960s in the work of Frank Adams, J. Michael Boardmann, André Joyal, Gregory Kelly, Peter May, Saunders McLane, Jim Stasheff, Rainer Vogt and other topologists and category theorists. In the 1990s, there was a "renaissance" of the theory in the development of deformation theory and quantum field theory, with a shift from topology to algebra, that can be found in the work of Ezra Getzler, Victor Ginzburg, Vladimir Hinich, John Jones, Mikhail

Kapranov, Maxim Kontsevich, Yuri I. Manin, Martin Markl, Vadim Schechtman, Vladimir Smirnov and Dmitry Tamarkin for instance. Ten years later, a first monograph [MSS02] on this subject was written by Martin Markl, Steve Shnider and Jim Stasheff in which one can find more details on the history of operad theory.

Now, 20 years after the renaissance of the operad theory, most of the basic aspects of it have been settled and it seems to be the right time to provide a comprehensive account of algebraic operad theory. This is the purpose of this book.

One of the main fruitful problems in the study of a given type of algebras is its relationship with algebraic homotopy theory. For instance, starting with a chain complex equipped with some compatible algebraic structure, can this structure be transferred to any homotopy equivalent chain complex? In general, the answer is negative. However, one can prove the existence of higher operations on the homotopy equivalent chain complex, which endow it with a richer algebraic structure. In the particular case of associative algebras, this higher structure is encoded into the notion of associative algebra up to homotopy, alias A-infinity algebra, unearthed by Stasheff in the 1960s. In the particular case of Lie algebras, it gives rise to the notion of L-infinity algebras, which was successfully used in the proof of the Kontsevich formality theorem. It is exactly the problem of governing these higher structures that prompted the introduction of the notion of operad.

Operad theory provides an explicit answer to this transfer problem for a large family of types of algebras, for instance those encoded by Koszul operads. Koszul duality was first developed at the level of associative algebras by Stewart Priddy in the 1970s. It was then extended to algebraic operads by Ginzburg and Kapranov, and also Getzler and Jones in the 1990s (part of the renaissance period). The duality between Lie algebras and commutative algebras in rational homotopy theory was recognized to coincide with the Koszul duality theory between the operad encoding Lie algebras and the operad encoding commutative algebras. The application of Koszul duality theory for operads to homotopical algebra is a far-reaching generalization of the ideas of Dan Quillen and Dennis Sullivan.

The aim of this book is, first, to provide an introduction to algebraic operads, second, to give a conceptual treatment of Koszul duality, and, third, to give applications to homotopical algebra.

We begin by developing the general theory of twisting morphisms, whose main application here is the Koszul duality theory for associative algebras. We do it in such a way that this pattern can be adapted to the operad setting. After giving the definition and the main properties of the notion of operad, we develop the operadic homological algebra. Finally, Koszul duality theory of operads permits us to study the homotopy properties of algebras over an operad.

We are very grateful to the many friends and colleagues who have helped us and in particular to pioneers of the subject Jim Stasheff, Dennis Sullivan, and Yuri I. Manin. We owe thanks to Olivia Bellier, Alexander Berglund, Emily Burgunder, Damien Calaque, Yongshan Chen, Pierre-Louis Curien, Vladimir Dotsenko, Gabriel Drummond-Cole, Clément Dupont, Yaël Frégier, Benoit Fresse, Hidekazu Furusho, Ezra Getzler, Darij Grinberg, Moritz Groth, Li Guo, Kathryn Hess, Joseph Hirsh, Laurent Hofer, Eric Hoffbeck, Ralf Holkamp, Magdalena Kędziorek, Muriel Livernet, Joan Millès, Nikolay Nikolov, Todor Popov, Maria Ronco, Henrik Strohmayer,

Antoine Touzé, Christine Vespa, Yong Zhang, and to the referees for their helpful and critical comments.

We wish to express our appreciation to the Centre National de Recherche Scientifique, the Eidgenössische Technische Hochschule (Zürich), and the Max-Planck Institut für Mathematik (Bonn) for their support.

Last but not least, nous sommes heureux de remercier tout particulièrement Eliane et Catherine pour avoir su créer autour de nous l'environnement idéal à la rédaction d'un tel ouvrage.

Strasbourg, France
18th January 2012

Jean-Louis Loday
Bruno Vallette

Contents

Leitfaden

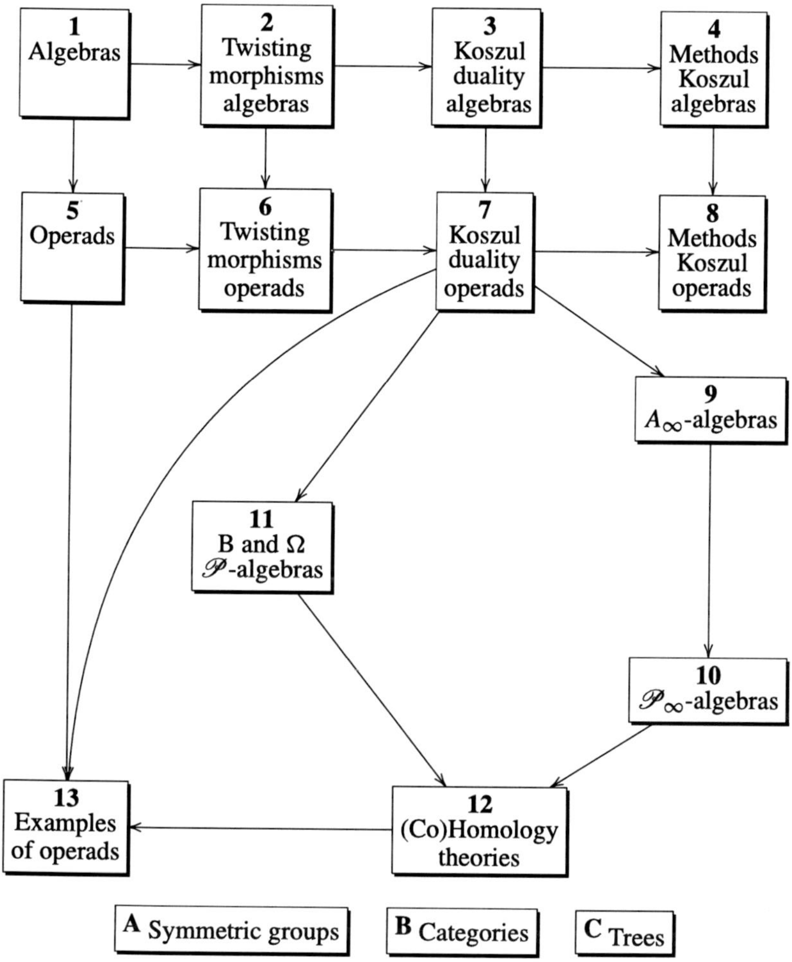
1 Algebras
2 Twisting morphisms algebras
3 Koszul duality algebras
4 Methods Koszul algebras
5 Operads
6 Twisting morphisms operads
7 Koszul duality operads
8 Methods Koszul operads
9 A_∞-algebras
11 B and Ω $\mathscr{P}$-algebras
10 $\mathscr{P}_\infty$-algebras
13 Examples of operads
12 (Co)Homology theories
A Symmetric groups
B Categories
C Trees

Introduction

The broad scheme of this book is as follows:

- Koszul duality of associative algebras: Chaps. 1, 2, 3, 4.
- Algebraic operads and their Koszul duality: Chaps. 5, 6, 7, 8.
- Homotopy theory of algebras over an operad: Chaps. 9, 10, 11, 12.
- Examples of operads: Chaps. 9, 13.

Koszul Duality of Associative Algebras

Koszul duality theory is a homological method whose aim is to provide an explicit quasi-free resolution (in fact the minimal model) for quadratic algebras. The algorithm for such a construction splits into two steps. First, from the data defining a quadratic algebra A, we construct a quadratic coalgebra $A^{¡}$. Second, we apply to this coalgebra the cobar construction Ω in order to get a differential graded algebra $\Omega A^{¡}$. When some condition is fulfilled, $\Omega A^{¡}$ is the expected quasi-free resolution. In that case A is called a *Koszul algebra*. The main feature of our treatment of Koszul duality is to keep algebras and coalgebras on the same footing.

Before treating Koszul duality of quadratic algebras, we fully develop the homological algebra properties of *twisting morphisms*. By definition, a twisting morphism is a linear map $\alpha : C \to A$ from a differential graded associative (dga) coalgebra C to a dga algebra A, satisfying the Maurer–Cartan equation:

$$\partial(\alpha) + \alpha \star \alpha = 0.$$

It gives rise to a chain complex structure on $C \otimes A$, called the *twisted tensor product*. The set $\mathrm{Tw}(C, A)$, made up of the twisting morphisms, is a bifunctor which is representable both in C and in A by the cobar construction Ω and the bar construction B respectively. The key point of this general theory is the characterization of the twisted morphisms whose associated twisted tensor product is acyclic. They are called *Koszul morphisms* and they form the set $\mathrm{Kos}(C, A)$. Under the bijections of the adjunction, they correspond to quasi-isomorphisms:

$$\begin{array}{ccccc} \mathrm{Hom}_{\mathsf{dga\ alg}}(\Omega C, A) & \cong & \mathrm{Tw}(C, A) & \cong & \mathrm{Hom}_{\mathsf{dga\ coalg}}(C, \mathrm{B}A) \\ \cup & & \cup & & \cup \\ \text{q-Iso}_{\mathsf{dga\ alg}}(\Omega C, A) & \cong & \mathrm{Kos}(C, A) & \cong & \text{q-Iso}_{\mathsf{dga\ coalg}}(C, \mathrm{B}A). \end{array}$$

Then, Koszul duality of quadratic algebras consists in applying this general theory to the following situation. Let $(V, R \subset V^{\otimes 2})$ be a quadratic data, where V is a vector space. It gives rise to a quadratic algebra $A = T(V)/(R)$, which is a quotient of the free algebra over V. It also gives rise to a quadratic coalgebra $A^{¡}$, which is a subcoalgebra of the cofree coalgebra over V. It is called the *Koszul dual coalgebra*. There is a natural twisting morphism $\kappa : A^{¡} \to A$ to which we apply the preceding theory: if the twisting morphism κ is a Koszul morphism, then A is a Koszul algebra and $\Omega A^{¡}$ is its minimal model.

We first treat the homogeneous case $R \subset V^{\otimes 2}$ and then we show how to extend the results to the inhomogeneous case $R \subset V \oplus V^{\otimes 2}$. The toy model is the universal enveloping algebra $A = U(\mathfrak{g})$ of a Lie algebra $\mathfrak{g}$, which was studied originally by Jean-Louis Koszul.

In the last chapter of this first part, we provide methods (Poincaré–Birkhoff–Witt and Gröbner bases, rewriting method) to prove the Koszulity for quadratic algebras and we introduce the Manin products of quadratic algebras. Since this presentation of the Koszul duality theory of associative algebras relies only on the universal properties of the various objects, it paves the way to other settings, in particular to operads as treated in the next part.

Algebraic Operads and Their Koszul Duality

Let us consider a "type of algebras" $\mathcal{P}$ for which there is a notion of free algebra $\mathcal{P}(V)$ over a generic vector space V. From the universal properties of a free object, one can deduce immediately that $\mathcal{P}$, considered as an endofunctor of the category of vector spaces Vect, inherits a monoid structure $\gamma : \mathcal{P} \circ \mathcal{P} \to \mathcal{P}$. This is the notion of algebraic operad. Since the underlying object of this monoid is an endofunctor, which is therefore a monad in Vect, there is a notion of algebra A over $\mathcal{P}$ determined by a map $\gamma_A : \mathcal{P}(A) \to A$. The notion of cooperad $\mathcal{C}$ is obtained similarly by replacing monoid by comonoid in the definition of an operad.

In fact, in this book, we work with a more specific notion. We suppose that, as an endofunctor of Vect, $\mathcal{P}$ is a Schur functor. It means that $\mathcal{P}$ is supposed to be built out of a collection of modules $\mathcal{P}(n)$ over the symmetric group $\mathbb{S}_n$, for $n \geq 0$, as

$$\mathcal{P}(V) := \bigoplus_{n \geq 0} \mathcal{P}(n) \otimes_{\mathbb{S}_n} V^{\otimes n}.$$

The elements of $\mathcal{P}(n)$ are the n-ary operations acting on the algebras of type $\mathcal{P}$:

$$(\mu; a_1, \ldots, a_n) \in \mathcal{P}(A) \mapsto \gamma_A(\mu; a_1, \ldots, a_n) \in A.$$

The action of the group $\mathbb{S}_n$ on $\mathcal{P}(n)$ encodes their symmetries.

From the specific form of the underlying endofunctor that we have assumed, it follows that the definition of an operad can be formalized into several guises:

- ◇ MONOIDAL DEFINITION. It is a monoid $(\mathscr{P}, \gamma, \eta)$ in the monoidal category of $\mathbb{S}$-modules.
- ◇ CLASSICAL DEFINITION. It consists in making the monoid structure on $\mathscr{P}$ explicit on the spaces $\mathscr{P}(n)$ of n-ary operations.
- ◇ PARTIAL DEFINITION. A presentation of an operad using the so-called partial operations $\circ_i$ involving only two operations.
- ◇ COMBINATORIAL DEFINITION. It consists in viewing an operad as an algebra over some monad. This monad is built on "rooted trees" and "substitution". It allows for several variations of the notion of operad.

The rooted trees are naturally involved in the theory of operads since they describe the free operad over an $\mathbb{S}$-module.

We also study, along the same lines, algebraic operads such that the functor $\mathscr{P}$ is built out of a collection of vector spaces $\mathscr{P}_n$, for $n \geq 0$, as

$$\mathscr{P}(V) := \bigoplus_{n\geq 0} \mathscr{P}_n \otimes_{\mathbb{K}} V^{\otimes n}.$$

In this case, there is no involvement of the symmetric groups. They are called *nonsymmetric operads*. They encode types of algebras for which the generating operations do not satisfy any symmetry properties, and, in the relations, the variables stay in the same order in each monomial. The combinatorial objects involved in the description of a free nonsymmetric operad are the planar rooted trees. Associative algebras and dendriform algebras are examples giving rise to nonsymmetric operads.

Taking advantage of the fact that an operad (symmetric or nonsymmetric) is a monoid in a certain linear category, we can extend the notions of twisting morphism and Koszul morphism to the operad setting. This generalization of the Koszul duality theory is not straightforward because of the following phenomenon. In the monoidal category of vector spaces, in which the definition of an associative algebra takes place, the tensor product $V \otimes W$ is linear in both variables, for instance $V \otimes (W \oplus W') = (V \otimes W) \oplus (V \otimes W')$. However, in the category of endofunctors, in which the definition of an operad takes place, the composite $\mathscr{P} \circ \mathscr{Q}$ is linear in the left variable, but not in the right variable. So several constructions, like the convolution product $\star$ and the notion of derivation, have to be adapted for the final results to hold true.

Once this generalization has been made, we develop homological algebra at the operadic level: there are notions of twisting morphisms, twisted composite product, Koszul morphisms, bar and cobar constructions in the operadic context. The main theorem gives rise to the following bijections:

$$\begin{array}{ccccc} \mathrm{Hom}_{\mathsf{dg\ Op}}(\Omega\mathscr{C}, \mathscr{P}) & \cong & \mathrm{Tw}(\mathscr{C}, \mathscr{P}) & \cong & \mathrm{Hom}_{\mathsf{dg\ coOp}}(\mathscr{C}, \mathrm{B}\mathscr{P}) \\ \cup & & \cup & & \cup \\ \text{q-Iso}_{\mathsf{dg\ Op}}(\Omega\mathscr{C}, \mathscr{P}) & \cong & \mathrm{Kos}(\mathscr{C}, \mathscr{P}) & \cong & \text{q-Iso}_{\mathsf{dg\ coOp}}(\mathscr{C}, \mathrm{B}\mathscr{P}), \end{array}$$

where the set of Koszul morphisms $\mathrm{Kos}(\mathscr{C}, \mathscr{P})$ is made up of the twisting morphisms whose associated twisted composite product is acyclic.

Koszul duality theory consists in applying the preceding general theory to a specific twisting morphism $\kappa : \mathscr{P}^{¡} \to \mathscr{P}$. Here $\mathscr{P}$ is the quadratic operad associated to a given operadic quadratic data (generating operations and quadratic relations). A similar construction, in the cooperad setting, gives rise to the quadratic cooperad $\mathscr{P}^{¡}$, called the *Koszul dual cooperad*. We first treat the homogeneous case and then we extend it to the inhomogeneous case. The paradigm of this last case is the operad encoding Batalin–Vilkovisky algebras.

The ultimate consequence and useful result of this Koszul duality theory is to provide an explicit algorithm to construct a quasi-free resolution (in fact the minimal model) of the quadratic operad $\mathscr{P}$, whenever it is Koszul. This quasi-free resolution is the differential graded operad $\Omega\,\mathscr{P}^{¡}$. So the algorithm consists in

◇ constructing the cooperad $\mathscr{P}^{¡}$ out of the quadratic data,
◇ taking the free operad on the underlying $\mathbb{S}$-module of $\mathscr{P}^{¡}$,
◇ constructing the differential on the free operad out of the cooperad structure of $\mathscr{P}^{¡}$.

We adopt the notation $\mathscr{P}_\infty := \Omega\,\mathscr{P}^{¡}$ and the terminology *$\mathscr{P}_\infty$-algebras*, also called *homotopy $\mathscr{P}$-algebras*. It provides a generalization up to homotopy of the notion of $\mathscr{P}$-algebra.

We give several methods for proving Koszulity of operads. They rely either on rewriting systems, PBW and Gröbner bases, distributive laws (Diamond Lemma), or combinatorics (partition poset method). The notion of shuffle operad (Vladimir Dotsenko, Eric Hoffbeck, Anton Khoroshkin) plays a key role in this respect. We also introduce the Manin products constructions for operads.

Homotopy Theory of Algebras over an Operad

Knowing about the Koszul resolution of a given operad enables us to answer many questions. For instance, it permits us to construct a "small" chain complex for computing the homology and the cohomology of an algebra over an operad. Of course, in the classical cases of associative algebras, commutative algebras and Lie algebras, that is for the operads *As*, *Com*, and *Lie*, one recovers the complex of Hochschild, Harrison and Chevalley–Eilenberg respectively. In the Leibniz case, we recover the complex constructed by the first author. Surprisingly, in the Poisson case it gives a complex different from the one which was used by the specialists, as shown by Benoit Fresse.

The Koszul resolution permits us to give a precise meaning to the notion of homotopy algebra of a given type. Again in the classical cases *As*, *Com*, and *Lie*, we recover the known structures of A_∞-algebra, C_∞-algebra and L_∞-algebra:

$$As_\infty = A_\infty, \qquad Com_\infty = C_\infty, \qquad Lie_\infty = L_\infty.$$

We devote an independent and almost self-contained chapter to the properties of A_∞-algebras.

The advantage of this treatment of operad theory is that we can extend known results on some specific operads to all Koszul operads at once. This is the case for the *Homotopy Transfer Theorem*, long known for associative algebras. It takes the following form.

Let

$$h \circlearrowright (A, d_A) \underset{i}{\overset{p}{\rightleftarrows}} (V, d_V),$$

$$\mathrm{Id}_A - ip = d_A h + h d_A, \qquad i = \text{quasi-isomorphism},$$

be a *homotopy retract* of chain complexes. Let $\mathscr{P}$ be a Koszul operad and suppose that (A, d_A) is a differential graded $\mathscr{P}$-algebra. It is natural to ask oneself whether we can transfer the $\mathscr{P}$-algebra structure of (A, d_A) to (V, d_V). In general we cannot, however we can transfer it into a $\mathscr{P}_\infty$-algebra structure. In fact, the answer to the first question would be positive if the operad $\mathscr{P}$ were quasi-free, for instance by replacing $\mathscr{P}$ by $\mathscr{P}_\infty$ (recall that a $\mathscr{P}$-algebra is a particular case of $\mathscr{P}_\infty$-algebra). This transferred $\mathscr{P}_\infty$-algebra structure is homotopically equivalent to the first one.

An example of a homotopy retract is given by $(V, d_V) := (H(A), 0)$, that is the homology of the underlying chain complex (A, d_A), called the *homotopy of the differential graded $\mathscr{P}$-algebra A*. As a result, we obtain higher operations on the homotopy of A, which encompass and generalize the notion of Massey products.

Examples of Operads and of Types of Operads

Throughout the aforementioned theoretical chapters, we illustrate the various results with the three classical operads *As*, *Com* and, *Lie* (the "three graces") encoding respectively the associative algebras, the commutative algebras and the Lie algebras.

We treat in detail the case of the operad *As* in a separate chapter (Chap. 9), not only because it is the most common type of algebras, but also because it serves as a paradigm for the other Koszul operads. The notion of A_∞-algebra is related to the Stasheff polytope, also known as the associahedron. We give a detailed treatment of the Homotopy Transfer Theorem in this case.

In Chap. 13, on top of the *Com* and *Lie* case, we treat many examples of operads together with their inter-relationship:

- ◇ POISSON ALGEBRA and GERSTENHABER ALGEBRA: it mixes both *Lie* and *Com*. It comes from Poisson geometry and deformation theory.
- ◇ PRE-LIE ALGEBRA: it permits us to analyze the properties of Hochschild homology and also the properties of derivations. It is closely related to the notion of operad itself, since the convolution algebra of maps from a cooperad to an operad is a pre-Lie algebra.

◇ LEIBNIZ ALGEBRA and ZINBIEL ALGEBRA: it comes as a natural generalization of Lie algebras with nonantisymmetric bracket. Zinbiel algebras (the Koszul dual) play a salient role in the analysis of divided power algebras.
◇ DENDRIFORM ALGEBRA: it arises naturally when the product of an associative algebra splits into the sum of two binary operations. It models a sort of "noncommutative shuffle". Many combinatorial Hopf algebras bear a dendriform structure.
◇ BATALIN–VILKOVISKY ALGEBRA: it is a particular class of Gerstenhaber algebras endowed with a extra square-zero unary operator. It plays a crucial role in mathematical physics (Batalin–Vilkovisky formalism in quantum field theory), in string topology (homology of the free loop spaces) and in the study of double loop spaces (cyclic Deligne conjecture).
◇ MAGMATIC ALGEBRA: it is an algebra with no relation. This type of algebras is important for operad theory because the operad encoding magmatic algebras is free and any operad is a quotient of a free operad. Though their Koszul duality is obvious, it gives a nice explanation of the inversion of power series formula.
◇ JORDAN ALGEBRA: coming from the properties of the symmetrized product in an associative algebra, they play a key role in differential geometry (symmetric spaces). The analysis of the associated operad, which is cubic, is still to be done.
◇ MULTI-ARY ALGEBRA: there are various types of algebras with generating operations of higher arity. The higher structure encoded in the Koszul resolution $\mathscr{P}_\infty$ is of this form. Several other examples appear in Deligne conjecture (brace operations) and Gromov–Witten invariants (moduli space of curves).

We have seen that the term *algebra*, which, most often, means "associative algebra" has been broadened to encompass many other types of algebras. So we have "commutative algebras", "Lie algebras", "Poisson algebras", "Leibniz algebras" and so forth. The same fact occurs for the term "operad". A priori it means what is more accurately called "symmetric operad". But its meaning has been broadened. We have already mentioned the notions of "nonsymmetric operad" and "shuffle operad". Replacing the rooted trees by other combinatorial objects gives rise to other types of operads: "colored operad", "cyclic operad", "permutad", "modular operad", "properad". We only briefly mention the definition of these structures in the last section of the last chapter. The conceptual presentation of Koszul duality given in this book has the advantage of being applicable to some of these other types of operads.

Recent Literature on Operads

The operad theory is connected to many subjects in mathematics and, so, can be treated in different guises as in the following books, which appeared since 2000:

◇ in [Smi01] Vladimir Smirnov aims at applications to algebraic topology,
◇ in [MSS02] Martin Markl, Steve Shnider and Jim Stasheff gave applications in algebra, topology and geometry,

◇ in [Lei04] Tom Leinster treats operad theory from the higher category point of view,
◇ in [Fre09a] Benoit Fresse studies the modules over algebraic operads.

Various Ways of Reading This Book

◇ The reader interested in *operad theory* can begin right away with Chap. 5: "Algebraic operad". We treat both symmetric operads and nonsymmetric operads. We give five equivalent definitions of an operad, we introduce the notion of algebra over an operad and we study the notion of free operad in detail. The reader may refer to Chap. 1 for classical notions, notations and results in elementary algebra, if necessary. Then, in Chap. 13: "Examples of algebraic operads" he/she will find a lot of examples. After this reading, the reader should be fluent in "operadic calculus".
◇ The study of *homology and homotopy theory of algebras* over an operad is first done in Chap. 9 in the associative algebra case, and in Chaps. 10, 11, 12 in the general case. A priori, it requires the knowledge of Koszul duality of operads. However, one need not know all the details of this theory but only the statement of the main results. In particular, we treat the Homotopy Transfer Theorem, which reveals hidden structures in homological algebra.
◇ *Koszul duality of operads* is done in Chaps. 6 and 7. Chapter 6 is devoted to twisting morphisms and the characterization of Koszul morphisms in general. Chapter 7 is devoted to the Koszul morphism associated to an operadic quadratic data. As said before, the overall presentation of results and proofs is analogous to the algebra case. So the reader not familiar with Koszul duality is advised to read Chaps. 2 and 3 first. Chapter 8 provides effective methods to prove that an operad is Koszul.
◇ The reader interested in *Koszul duality of algebras* should read Chaps. 2, 3 and 4, which give a new point of view of this classical theory. This part is a self-contained study of Koszul duality of associative algebras. For examples and applications, we refer the reader to the monograph [PP05] by Alexander Polishchuk and Leonid Positselski.

There are three appendices dealing respectively with the representations of the symmetric groups, notions of category theory, and constructions on trees, including the marvelous Stasheff polytope.

Notation and Conventions

The ground field (resp. commutative ring) is denoted by $\mathbb{K}$. The category of vector spaces over $\mathbb{K}$ is denoted by $\mathsf{Vect}_{\mathbb{K}}$ or by Vect. We often say *space* instead of vector space, and *map* instead of linear map. We also say $\mathbb{K}$-module, whenever the constructions and results are valid for $\mathbb{K}$ a commutative ring, not necessarily a field.

The linear dual of the $\mathbb{K}$-vector space V is $V^* := \mathrm{Hom}(V, \mathbb{K})$. When V is finite dimensional, it is canonically isomorphic to the dual of its dual. In the sign-graded case, this isomorphism $\Phi : V \to V^{**}$ is given by $\Phi(x)(f) := (-1)^{|x||f|} f(x)$.

The tensor product of two $\mathbb{K}$-vector spaces V and W is denoted by $V \otimes_{\mathbb{K}} W$ or, more often, by $V \otimes W$. The tensor product of n copies of V is denoted by

$$V^{\otimes n} := \underbrace{V \otimes \cdots \otimes V}_{n}.$$

The monomial $v_1 \otimes \cdots \otimes v_n \in V^{\otimes n}$ is also written $(v_1, \ldots, v_n)$ or, more simply, $v_1 \cdots v_n$ (concatenation of the vectors) when there is no ambiguity. The homogeneous element $v_1 \cdots v_n$ is said to be of *weight* n. The *tensor module* over V is by definition the direct sum

$$T(V) := \mathbb{K} \oplus V \oplus V^{\otimes 2} \oplus \cdots \oplus V^{\otimes n} \oplus \cdots .$$

The *reduced tensor module* is defined as

$$\overline{T}(V) := V \oplus V^{\otimes 2} \oplus \cdots \oplus V^{\otimes n} \oplus \cdots .$$

It can be viewed either as a subspace or as a quotient of $T(V)$. The terminology "tensor module" will prove helpful, because this module can be equipped with many different kinds of algebraic structures. The tensor algebra structure is only one of them.

The symmetric group $\mathbb{S}_n$ is the automorphism group of the set $\{1, \ldots, n\}$. An element of $\mathbb{S}_n$ is called a *permutation*. A permutation $\sigma \in \mathbb{S}_n$ is denoted by

$$\big[\sigma(1)\sigma(2)\cdots\sigma(n)\big].$$

We also sometimes adopt the classical cycle notation with parentheses, so, for instance, $(12) = [2\,1]$. The action on the right of $\mathbb{S}_n$ on $V^{\otimes n}$ is given by

$$(v_1 \cdots v_n)^\sigma := v_{\sigma(1)} \cdots v_{\sigma(n)}.$$

The action on the left is given by $\sigma \cdot (v_1 \cdots v_n) := v_{\sigma^{-1}(1)} \cdots v_{\sigma^{-1}(n)}$.

We freely use the language of categories and functors, see for instance [ML98] and Appendix B.1. While we review some elementary homological algebra in the first chapter, the reader is expected to be familiar with the theory; see for instance [ML95].

The exercises labeled ★ are more advanced.

Chapter 1
Algebras, Coalgebras, Homology

C'est de l'algèbre, se dit d'une chose à laquelle on ne comprend rien.

Petit Littré

In this chapter we recall elementary facts about algebras and homological algebra, essentially to establish the terminology and the notation. We first review the notions of associative, commutative and Lie algebra. Then we deal with the notion of coalgebra, which is going to play a key role in this book. This leads to the notion of convolution. The last sections cover bialgebras, pre-Lie algebras, differential graded objects and convolution algebra.

1.1 Classical Algebras (Associative, Commutative, Lie)

We review the classical notions of associative algebra, commutative algebra (meaning commutative and associative), and Lie algebra.

1.1.1 Associative Algebras

An *associative algebra* over $\mathbb{K}$ is a vector space A equipped with a binary operation (linear map)

$$\mu : A \otimes A \to A$$

which is associative, i.e. $\mu \circ (\mu \otimes \mathrm{id}) = \mu \circ (\mathrm{id} \otimes \mu)$. Here id is the identity map from A to A (sometimes denoted by id_A), and the operation μ is called the *product*. Denoting $ab := \mu(a \otimes b)$, associativity reads:

$$(ab)c = a(bc).$$

J.-L. Loday, B. Vallette, *Algebraic Operads*,
Grundlehren der mathematischen Wissenschaften 346,
DOI 10.1007/978-3-642-30362-3_1, © Springer-Verlag Berlin Heidelberg 2012

An associative algebra A is said to be *unital* if there is a map $u : \mathbb{K} \to A$ such that $\mu \circ (u \otimes \mathrm{id}) = \mathrm{id} = \mu \circ (\mathrm{id} \otimes u)$. We denote by 1_A or simply by 1 the image of $1_{\mathbb{K}}$ in A under u. Under this notation unitality reads:

$$1a = a = a1.$$

Associativity and unitality can be viewed as commutative diagrams:

$$\begin{array}{ccc} A \otimes A \otimes A & \xrightarrow{\mathrm{id} \otimes \mu} & A \otimes A \\ {\scriptstyle \mu \otimes \mathrm{id}} \downarrow & & \downarrow {\scriptstyle \mu} \\ A \otimes A & \xrightarrow{\mu} & A \end{array}$$

$$\begin{array}{ccccc} \mathbb{K} \otimes A & \xrightarrow{u \otimes \mathrm{id}} & A \otimes A & \xleftarrow{\mathrm{id} \otimes u} & A \otimes \mathbb{K} \\ & {\scriptstyle \simeq} \searrow & \downarrow {\scriptstyle \mu} & \swarrow {\scriptstyle \simeq} & \\ & & A. & & \end{array}$$

It is sometimes helpful to picture the binary operation μ as a graph

or simply as

if there is no ambiguity on μ.

The associativity relation becomes:

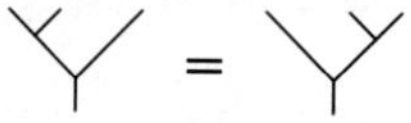

and the unitality relations become:

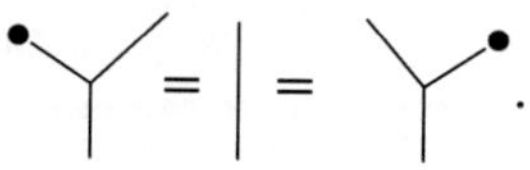

As usual we sometimes abbreviate (unital) associative algebra as algebra if no confusion can arise, and we abbreviate (A, μ, u) as (A, μ) or even just A.

Equivalently a unital associative algebra is a monoid in the tensor category $(\mathsf{Vect}, \otimes, \mathbb{K})$ (cf. Appendix B.3).

An *algebra morphism* (or simply morphism) is a linear map $f : A \to A'$ such that $f(ab) = f(a)f(b)$. If A is also unital, then we further assume $f(1) = 1$.

The category of nonunital associative algebras is denoted by As-alg, and the category of unital associative algebras by uAs-alg (or by As-alg if there is no risk of confusion).

An algebra is *augmented* when there is given a morphism of algebras $\varepsilon : A \to \mathbb{K}$, called the *augmentation map*. In particular $\varepsilon(1_A) = 1_{\mathbb{K}}$. If A is augmented, then A

is canonically isomorphic, as a vector space, to $\mathbb{K}1 \oplus \operatorname{Ker} \varepsilon$. The ideal $\operatorname{Ker} \varepsilon$ is called the *augmentation ideal*, which we denote by $\bar{A}$, so

$$A = \mathbb{K}1_A \oplus \bar{A}.$$

On the other hand, starting with a (not necessarily unital) algebra I, one can construct an augmented algebra $I_+ := \mathbb{K}1 \oplus I$. The product is given by

$$(\lambda 1 + a)(\lambda' 1 + a') = \lambda\lambda' 1 + (\lambda a' + \lambda' a + aa').$$

Therefore the category of nonunital associative algebras is equivalent to the category of unital augmented associative algebras.

For an augmented algebra I_+ the space of *indecomposables* is

$$\operatorname{Indec}(I_+) := I/I^2.$$

1.1.2 Free Associative Algebra

The *free associative algebra* over the vector space V is an associative algebra $\mathscr{F}(V)$ equipped with a linear map $i : V \to \mathscr{F}(V)$ which satisfies the following universal condition:

Any map $f : V \to A$, where A is an associative algebra, extends uniquely into an algebra morphism $\tilde{f} : \mathscr{F}(V) \to A$ such that the following diagram commutes:

$$\begin{array}{ccc} V & \xrightarrow{i} & \mathscr{F}(V) \\ & {}_{f}\searrow & \downarrow{\scriptstyle \tilde{f}} \\ & & A. \end{array}$$

Observe that the free algebra over V is well-defined up to a unique isomorphism. Categorically, $\mathscr{F}$ is a functor from the category of vector spaces to the category of associative algebras (or unital associative algebras depending on the context we choose), which is left adjoint to the forgetful functor assigning to A its underlying vector space:

$$\operatorname{Hom}_{As\text{-}\mathsf{alg}}(\mathscr{F}(V), A) \cong \operatorname{Hom}_{\mathsf{Vect}}(V, A).$$

There are forgetful functors As-alg $\to$ Vect $\to$ Set and sometimes it is useful to consider the free associative algebra over a set X. This is the same as the associative algebra over the space $\mathbb{K}X$ spanned by X since the functor given by $X \mapsto \mathbb{K}X$ is left adjoint to the functor Vect $\to$ Set.

1.1.3 Tensor Module, Tensor Algebra

By definition the *tensor algebra* over the vector space V is the *tensor module*

$$T(V) := \mathbb{K}1 \oplus V \oplus \cdots \oplus V^{\otimes n} \oplus \cdots$$

equipped with the *concatenation* product $T(V) \otimes T(V) \to T(V)$ given by

$$v_1 \cdots v_p \otimes v_{p+1} \cdots v_{p+q} \mapsto v_1 \cdots v_p v_{p+1} \cdots v_{p+q}.$$

This operation is clearly associative and $1 \in V^{\otimes 0} = \mathbb{K}1$ is taken as a unit. Observe that $T(V)$ is augmented by $\varepsilon(v_1 \cdots v_n) = 0$ for $n \geq 1$ and $\varepsilon(1) = 1$. The map $\varepsilon : T(V) \to \mathbb{K}$ is called the *augmentation*. For a homogeneous element $x \in V^{\otimes n}$, the integer n is called the *weight* of x. We say that $T(V)$ is *weight graded*.

The *reduced tensor algebra* $\overline{T}(V)$ is the *reduced tensor module*

$$\overline{T}(V) := V \oplus \cdots \oplus V^{\otimes n} \oplus \cdots$$

equipped with the concatenation product. It is a nonunital associative algebra (augmentation ideal of $T(V)$).

Proposition 1.1.1. *The tensor algebra (resp. reduced tensor algebra) is free in the category of unital associative algebras (resp. nonunital associative algebras).*

Proof. Let $f : V \to A$ be a map. If the extension $\tilde{f}$ exists, then we should have

$\tilde{f}(1) = 1$ by unitality,
$\tilde{f}(v) = f(v)$ by compatibility with f,
$\tilde{f}(v_1 \cdots v_n) = f(v_1) \cdots f(v_n)$, since $\tilde{f}$ is a morphism.

These equalities define a map $\tilde{f} : T(V) \to A$. We immediately check that it is a morphism. Since it is unique we have proved that $T(V)$, equipped with the inclusion $i : V \rightarrowtail T(V)$, is free over V. □

1.1.4 Noncommutative Polynomial Algebra

Let $V = \mathbb{K}x_1 \oplus \cdots \oplus \mathbb{K}x_n$ be a finite dimensional vector space with basis $\{x_1, \ldots, x_n\}$. The tensor algebra $T(V)$ is simply the algebra of noncommutative polynomials in n variables and is denoted by $\mathbb{K}\langle x_1, \ldots, x_n\rangle$.

1.1.5 Module and Bimodule

A *left module* M over an algebra A is a vector space equipped with a linear map

$$\lambda : A \otimes M \to M, \qquad \lambda(a, m) = am,$$

called the *left action*, which is compatible with the product and the unit of A, in the sense that the following diagrams commute

$$\begin{array}{ccc} A \otimes A \otimes M & \xrightarrow{\mathrm{Id} \otimes \lambda} & A \otimes M \\ {\scriptstyle \mu \otimes \mathrm{Id}} \downarrow & & \downarrow {\scriptstyle \lambda} \\ A \otimes M & \xrightarrow{\lambda} & A \end{array} \qquad \begin{array}{ccc} \mathbb{K} \otimes M & \xrightarrow{u \otimes \mathrm{Id}} & A \otimes M \\ & \searrow {\scriptstyle \simeq} & \downarrow {\scriptstyle \lambda} \\ & & M. \end{array}$$

There is a similar notion of *right module* involving a right action $\lambda' : M \otimes A \to M$, $\lambda'(m, a') = ma'$. Finally, a *bimodule* M over the algebra A is a vector space which is both a left module and a right module and which satisfies $(am)a' = a(ma')$ for any $a, a' \in A$ and $m \in M$.

If A is unital, then 1 is required to act by the identity.

For any vector space V the *free left A-module* over V is $M := A \otimes V$ equipped with the obvious left operation. Similarly the *free A-bimodule* over V is $M := A \otimes V \otimes A$.

Let

$$0 \to M \to A' \to A \to 0$$

be an exact sequence of associative algebras such that the product in M is 0. Then it is easy to check that M is a bimodule over A.

1.1.6 Derivations

Let M be a bimodule over an algebra A. A linear map $d : A \to M$ is a *derivation* if the Leibniz rule holds:

$$d(ab) = d(a)b + ad(b) \quad \text{for any } a, b \in A.$$

Graphically we get:

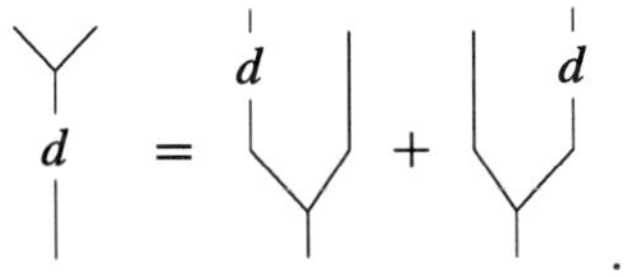

Any element $m \in M$ defines a derivation $d_m(a) = [a, m] := am - ma$ called an *inner derivation*. We denote by $\mathrm{Der}(A, M)$ the space of derivations of A with values in M. We remark that $d(1) = 0$. In the case where $M = A$ with its usual bimodule structure we denote the space of derivations simply by $\mathrm{Der}(A)$.

Proposition 1.1.2. *Let M be a bimodule over the free algebra $T(V)$. Any linear map $f : V \to M$ can be extended uniquely into a derivation $d_f : T(V) \to M$. More* *precisely, there is an isomorphism:*

$$\mathrm{Hom}(V, M) \cong \mathrm{Der}\big(T(V), M\big), \qquad f \mapsto d_f,$$

given by $d_f(v_1, \ldots, v_n) = \sum_{i=1}^n v_1 \cdots f(v_i) \cdots v_n$.

Proof. Since $T(V)$ is free, the extension of f into a derivation from $T(V)$ to M is unique. The above formula is obtained by induction on n. On the other hand, any derivation from $T(V)$ to M gives, by restriction, a linear map from V to M. These two constructions are inverse to each other. □

1.1.7 Universality of the Free Algebra

We have seen that any linear map $f : V \to A$ extends uniquely into a map $f : \overline{T}(V) \to A$ which satisfies $f(xy) = f(x)f(y)$ (morphism of associative algebras). We have also seen that any linear map $d : V \to A$ extends uniquely into a map $d : \overline{T}(V) \to A$ which satisfies $d(xy) = d(x)y + xd(y)$. More generally, if we are given some formula expressing $f(xy)$ in terms of $f(x)$, $f(y)$, the sum and the product, then the extension to $\overline{T}(V)$, when it exists, is going to be unique.

1.1.8 Commutative Algebra

By definition a *commutative algebra* is a vector space A over $\mathbb{K}$ equipped with a binary operation $\mu : A \otimes A \to A$, $\mu(a, b) = ab$, which is both associative and commutative (i.e. symmetric):

$$ab = ba.$$

In terms of the *switching map*

$$\tau : A \otimes A \to A \otimes A$$

defined by $\tau(x, y) := (y, x)$ (in the nongraded context; see Sect. 1.5.3 for the definition in the graded context), the commutation condition reads $\mu \circ \tau = \mu$. Sometimes one needs to work with algebras whose binary operation satisfies this symmetry condition but is not associative. We propose calling them *commutative magmatic algebras* (cf. Sect. 13.8.4), since we call magmatic algebra an algebra equipped with a binary operation.

The free unital commutative algebra over the vector space V (cf. Sect. 5.2.5 for the general definition of a free algebra) is the *symmetric algebra*

$$S(V) = \bigoplus_{n \geq 0} S^n(V) := \bigoplus_{n \geq 0} \big(V^{\otimes n}\big)_{\mathbb{S}_n}.$$

Recall that if M is a right $\mathbb{S}_n$-module the space $M_{\mathbb{S}_n}$ is the quotient of M by the action of $\mathbb{S}_n$. It is also called the space of *coinvariants* of M since

$$M_{\mathbb{S}_n} = M \otimes_{\mathbb{K}[\mathbb{S}_n]} \mathbb{K}$$

where $\mathbb{K}$ is equipped with the trivial action of $\mathbb{S}_n$. The right action of the symmetric group on $V^{\otimes n}$ is by permutation of the variables $(v_1, \dots, v_n)^\sigma := (v_{\sigma(1)}, \dots, v_{\sigma(n)})$. If $V = \mathbb{K}x_1 \oplus \cdots \oplus \mathbb{K}x_k$, then $S(V)$ is isomorphic to the algebra of polynomials $\mathbb{K}[x_1, \dots, x_k]$.

The category of nonunital commutative algebras is denoted by *Com*-alg, and the category of unital commutative algebras is denoted by *uCom*-alg. Since a commutative algebra is associative, there is a forgetful functor

$$Com\text{-alg} \longrightarrow As\text{-alg}.$$

The free object in the category of nonunital commutative algebras is the *reduced symmetric algebra*:

$$\overline{S}(V) := \bigoplus_{n\geq 1} S^n(V).$$

1.1.9 Lie Algebras

A *Lie algebra* is a vector space $\mathfrak{g}$ over $\mathbb{K}$ equipped with a binary operation $c : \mathfrak{g} \otimes \mathfrak{g} \to \mathfrak{g}$, $c(x, y) := [x, y]$, (c for "crochet" in French), called *bracket*, which is antisymmetric:

$$[x, y] = -[y, x], \quad \text{equivalently} \quad c \circ \tau = -c,$$

and satisfies the *Leibniz identity*

$$\big[[x, y], z\big] = \big[x, [y, z]\big] + \big[[x, z], y\big],$$
$$\text{equivalently} \quad c \circ (c \otimes \mathrm{id}) = c \circ (\mathrm{id} \otimes c) + c \circ (c \otimes \mathrm{id}) \circ (\mathrm{id} \otimes \tau).$$

In the literature this relation is, most of the time, replaced by the *Jacobi identity*

$$\big[[x, y], z\big] + \big[[y, z], x\big] + \big[[z, x], y\big] = 0.$$

The Jacobi relation and the Leibniz relation are equivalent under the anti-symmetry condition, but not otherwise. Observe that the Leibniz identity is equivalent to saying that the map $\mathrm{Ad}_z := [-, z] : \mathfrak{g} \to \mathfrak{g}$ is a derivation for the bracket, since it can be written

$$\mathrm{Ad}_z\big([x, y]\big) = \big[x, \mathrm{Ad}_z(y)\big] + \big[\mathrm{Ad}_z(x), y\big].$$

Sometimes one needs to work with algebras whose binary operation is antisymmetric but is not a Lie bracket. We propose calling them *antisymmetric magmatic algebras*.

If the bracket operation satisfies the Leibniz identity, but not necessarily the antisymmetry condition, then the algebra is called a *Leibniz algebra* (this notion was first introduced and studied in [Lod93]), cf. Sect. 13.5.1.

Any associative algebra A can be made into a Lie algebra, denoted by A_{Lie}, under the formula

$$[x, y] := xy - yx,$$

since an inner derivation satisfies the Leibniz relation. It gives a forgetful functor

$$(-)_{Lie} : As\text{-}\mathsf{alg} \longrightarrow Lie\text{-}\mathsf{alg}.$$

See Proposition 9.1.1 for an explanation of the "forgetfulness". The free Lie algebra over the vector space V (cf. Sect. 5.2.5) is denoted by $Lie(V)$. It will be studied in Sect. 1.3.

1.1.10 Universal Enveloping Algebra

We construct a functor $U : Lie\text{-}\mathsf{alg} \longrightarrow As\text{-}\mathsf{alg}$ as follows. Let $\mathfrak{g}$ be a Lie algebra and let $T(\mathfrak{g})$ be the tensor algebra over the vector space $\mathfrak{g}$. By definition, the *universal enveloping algebra* $U(\mathfrak{g})$ is the quotient of $T(\mathfrak{g})$ by the two-sided ideal generated by the elements

$$x \otimes y - y \otimes x - [x, y], \quad \text{for all } x, y \in \mathfrak{g}.$$

Proposition 1.1.3. *The functor* $U : Lie\text{-}\mathsf{alg} \longrightarrow As\text{-}\mathsf{alg}$ *is left adjoint to the functor* $(-)_{Lie} : As\text{-}\mathsf{alg} \longrightarrow Lie\text{-}\mathsf{alg}$.

Proof. We prove that for any Lie algebra $\mathfrak{g}$ and any associative algebra A there is an isomorphism

$$\mathrm{Hom}_{\mathsf{As}}\big(U(\mathfrak{g}), A\big) \cong \mathrm{Hom}_{\mathsf{Lie}}(\mathfrak{g}, A_{Lie}).$$

To any morphism $f : U(\mathfrak{g}) \to A$ we associate its restriction to $\mathfrak{g}$, which is a Lie algebra morphism $\mathfrak{g} \to A_{Lie}$. To any Lie algebra morphism $g : \mathfrak{g} \to A_{Lie}$ we associate the unique extension as associative algebra morphism $\tilde{g} : T(\mathfrak{g}) \to A$. Since $\tilde{g}(x \otimes y - y \otimes x) = g(x)g(y) - g(y)g(x) = [g(x), g(y)] = g([x, y])$, it follows that $\tilde{g}$ is trivial on the two-sided ideal generated by the elements $x \otimes y - y \otimes x - [x, y]$. So it gives a map $\tilde{g} : U(\mathfrak{g}) \to A$. It is immediate to check that these two constructions are inverse to each other. □

1.1.11 Lie Module

A *Lie module*, that is a module over a Lie algebra $\mathfrak{g}$, is a vector space M equipped with a linear map $\mathfrak{g} \otimes M \to M$, $(x, m) \mapsto [x, m]$, which satisfies the Jacobi identity whenever one of the variables is in M and the other two are in $\mathfrak{g}$ (convention $[m, x] := -[x, m]$). It is well known that a module over $\mathfrak{g}$ is equivalent to a left module over the associative algebra $U(\mathfrak{g})$.

1.2 Coassociative Coalgebras

Formally the notion of coalgebra is obtained by linearly dualizing the notion of algebra. But these two notions are not equivalent, because, though a coalgebra gives an algebra by dualization, an algebra gives a coalgebra only under a finiteness hypothesis. In analysis this phenomenon gives rise to the notion of "distribution".

1.2.1 Definition

A *coassociative coalgebra* (or associative coalgebra) over $\mathbb{K}$ is a vector space C equipped with a binary cooperation (linear map):

$$\Delta : C \to C \otimes C$$

which is coassociative, i.e. $(\Delta \otimes \mathrm{id}) \circ \Delta = (\mathrm{id} \otimes \Delta) \circ \Delta$. The map Δ is called a *coproduct*. Sometimes we use *Sweedler's notation*:

$$\Delta(x) = \sum x_{(1)} \otimes x_{(2)}, \qquad (\Delta \otimes \mathrm{id}) \circ \Delta(x) = \sum x_{(1)} \otimes x_{(2)} \otimes x_{(3)}.$$

More generally, we define the iterated coproduct $\Delta^n : C \to C^{\otimes n+1}$ by $\Delta^0 = \mathrm{id}$, $\Delta^1 = \Delta$ and

$$\Delta^n := (\Delta \otimes \mathrm{id} \otimes \cdots \otimes \mathrm{id}) \circ \Delta^{n-1}.$$

We write $\Delta^n(x) = \sum x_{(1)} \otimes \cdots \otimes x_{(n+1)} \in C^{\otimes n+1}$. Since Δ is coassociative, we have $\Delta^n = (\mathrm{id} \otimes \cdots \otimes \mathrm{id} \otimes \Delta \otimes \mathrm{id} \otimes \cdots \otimes \mathrm{id}) \circ \Delta^{n-1}$.

A coassociative coalgebra is said to be *counital* if it is equipped with a map $\varepsilon : C \to \mathbb{K}$ called the *counit* (or the augmentation map) such that $(\varepsilon \otimes \mathrm{id}) \circ \Delta = \mathrm{id} = (\mathrm{id} \otimes \varepsilon) \circ \Delta$. Coassociativity and counitality can be viewed as commutative diagrams:

$$\begin{array}{ccc} C & \xrightarrow{\Delta} & C \otimes C \\ {\scriptstyle \Delta}\downarrow & & \downarrow{\scriptstyle \mathrm{id} \otimes \Delta} \\ C \otimes C & \xrightarrow{\Delta \otimes \mathrm{id}} & C \otimes C \otimes C \end{array}$$

$$\begin{array}{ccccc} & & C & & \\ & \swarrow^{\simeq} & \downarrow \Delta & \searrow^{\simeq} & \\ C\otimes\mathbb{K} & \xleftarrow{\mathrm{id}\otimes\varepsilon} & C\otimes C & \xrightarrow{\varepsilon\otimes\mathrm{id}} & \mathbb{K}\otimes C. \end{array}$$

Observe that $\mathbb{K}$ itself is a counital coassociative coalgebra for $\Delta(1)=1\otimes 1$. We often abbreviate (counital) coassociative coalgebra into coalgebra if no confusion can arise, and we abbreviate the notation (C,Δ,ε) into (C,Δ) or even just C.

Pictorially the coproduct is represented as

and the coassociativity by

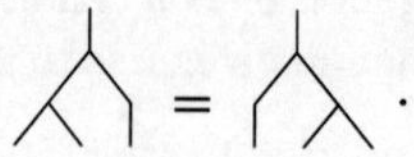

A *morphism of coalgebras* $f : C \to C'$ is a linear map which commutes with the coproducts, i.e. $(f\otimes f)\circ\Delta_C=\Delta_{C'}\circ f$, and with the counits. The category of associative coalgebras is denoted by *As*-coalg (counital or not, depending on the context).

A coalgebra is *coaugmented* if there is given a morphism of coalgebras $u:\mathbb{K}\to C$; in particular $\varepsilon\circ u=\mathrm{Id}_{\mathbb{K}}$. If C is coaugmented, then C is canonically isomorphic to $\mathrm{Ker}\,\varepsilon\oplus\mathbb{K}1$. The kernel $\mathrm{Ker}\,\varepsilon$ is often denoted by $\overline{C}$, so

$$C=\overline{C}\oplus\mathbb{K}1.$$

The *reduced coproduct* $\bar{\Delta}:\overline{C}\to\overline{C}\otimes\overline{C}$ is the map given by

$$\bar{\Delta}(x):=\Delta(x)-x\otimes 1-1\otimes x.$$

The iterated reduced coproduct is denoted by $\bar{\Delta}^n:\overline{C}\to\overline{C}^{\otimes n+1}$. So $\bar{\Delta}^n$ is an $(n+1)$-ary cooperation.

The coproduct Δ is said to be cocommutative (or simply commutative) if it satisfies the relation $\Delta=\tau\circ\Delta$, where τ is the switching map.

1.2.2 From Algebra to Coalgebra and Vice Versa

Let $V^*:=\mathrm{Hom}(V,\mathbb{K})$ be the linear dual of the space V. There is a canonical map $\omega:V^*\otimes V^*\to(V\otimes V)^*$ given by $\omega(f\otimes g)(x\otimes y)=f(x)g(y)$ (we work in the nongraded framework). When V is finite dimensional, ω is an isomorphism.

If (C,Δ) is a coalgebra, then $(C^*,\Delta^*\circ\omega)$ is an algebra (no need of finiteness hypothesis).

If (A,μ) is an algebra which is finite dimensional, then $(A^*,\omega^{-1}\circ\mu^*)$ is a coalgebra.

1.2.3 Structure Constants

Let us suppose that A is a finite dimensional algebra with basis $x_1, x_2, \ldots$. The product is completely determined by constants $a_{xy}^z \in \mathbb{K}$ such that $xy = \sum_z a_{xy}^z z$, where x, y, z run over all basis vectors. For $C := A^*$ let us choose the dual basis $x_1^*, x_2^*, \ldots$. Then the coproduct in C is given by the following formula on this basis:

$$\Delta(z^*) = \sum_{x,y} a_{xy}^z x^* \otimes y^*.$$

1.2.4 Conilpotency, Primitive Elements

Let $C = \mathbb{K}1 \oplus \overline{C}$ be a coaugmented coalgebra. The *coradical filtration* on C is defined as follows:

$$F_0C := \mathbb{K}1,$$
$$F_rC := \mathbb{K}1 \oplus \{x \in \overline{C} \mid \bar{\Delta}^n(x) = 0, \text{ for any } n \geq r\} \text{ for a given } r \geq 1.$$

A coalgebra C is said to be *conilpotent* (or sometimes *connected* in the literature) if it is coaugmented and if the filtration is exhaustive, that is $C = \bigcup_r F_rC$.

Observe that, under this hypothesis, any element in $\overline{C}$ is *conilpotent* i.e. for any $x \in \overline{C}$ there exists n such that $\bar{\Delta}^m(x) = 0$ for any $m \geq n$. By definition an element $x \in C$ is said to be *primitive* if

$$\Delta(x) = x \otimes 1 + 1 \otimes x,$$

or, equivalently, $x \in \overline{C}$ and $\bar{\Delta}(x) = 0$. The space of primitive elements of C is denoted by $\operatorname{Prim} C$. It is clear that $F_1(C) = \mathbb{K}1 \oplus \operatorname{Prim} C$.

Under finite dimensional assumption the dual of the primitives are the indecomposables:

$$(\operatorname{Prim} C)^* = \operatorname{Indec} C^*.$$

1.2.5 Cofree Associative Coalgebra

By definition the *cofree associative coalgebra* over the vector space V is a conilpotent associative coalgebra $\mathscr{F}^c(V)$ equipped with a linear map $p : \mathscr{F}^c(V) \to V$, so that $1 \mapsto 0$ and which satisfies the following universal condition:

Any linear map $\varphi : C \to V$, where C is a conilpotent associative coalgebra, satisfying $\varphi(1) = 0$, extends uniquely into a coaugmented coalgebra morphism: $\tilde{\varphi} : C \to \mathscr{F}^c(V)$:

$$\begin{array}{ccc} C & & \\ {\scriptstyle \tilde{\varphi}}\big\downarrow & \searrow^{\varphi} & \\ \mathscr{F}^c(V) & \xrightarrow{p} & V. \end{array}$$

Observe that the cofree coalgebra over V is well-defined up to a unique isomorphism. Categorically, $\mathscr{F}^c$ is a functor from the category of vector spaces to the category of conilpotent coalgebras. It is right adjoint to the forgetful functor which assigns to C the underlying vector space of $\overline{C}$:

$$\mathrm{Hom}_{\mathsf{Vect}}(\overline{C}, V) \cong \mathrm{Hom}_{\mathsf{conil}\text{-}\mathit{As}\text{-}\mathsf{coalg}}\big(C, \mathscr{F}^c(V)\big).$$

It is important to notice that here the conilpotency condition is part of the definition of cofree. The cofree object in the category of not necessarily conilpotent coalgebras is completely different.

1.2.6 Tensor Coalgebra, Cofree Coalgebra

By definition the *tensor coalgebra* over the vector space V, denoted by $T^c(V)$, is the tensor module

$$T^c(V) := \mathbb{K}1 \oplus V \oplus \cdots \oplus V^{\otimes n} \oplus \cdots$$

equipped with the *deconcatenation* coproduct $T^c(V) \to T^c(V) \otimes T^c(V)$ given by

$$\Delta(v_1 \cdots v_n) := \sum_{i=0}^{n} v_1 \cdots v_i \otimes v_{i+1} \cdots v_n \quad \text{and} \quad \Delta(1) = 1 \otimes 1,$$

and the counit $T^c(V) \to \mathbb{K}$ which is the identity on $\mathbb{K}$ and 0 otherwise. The cooperation Δ is clearly coassociative and counital. Observe that $T^c(V)$ is coaugmented by the inclusion $i : \mathbb{K} \to T(V)$. It is conilpotent and the filtration is $F_r(T^c(V)) := \bigoplus_{n \leq r} V^{\otimes n}$.

Observe that in the noncounital context the deconcatenation is given by the formula:

$$\bar{\Delta}(v_1 \cdots v_n) := \sum_{i=1}^{n-1} v_1 \cdots v_i \otimes v_{i+1} \cdots v_n.$$

We denote by $proj_V : T^c(V) \to V$ the projection map which is the identity on V and 0 otherwise.

Proposition 1.2.1. *The tensor coalgebra is cofree in the category of conilpotent coalgebras.*

Proof. We equip $T^c(V)$ with the map $proj_V : T^c(V) \to V$. For $\omega \in T^c(V)$ we denote by ω_n its component in $V^{\otimes n}$. Observe that only finitely many components are nonzero since $T^c(V)$ is a "sum".

Let $\varphi : C \to V$ be a map such that $\varphi(1) = 0$. If there exists $\tilde{\varphi} : C \to T^c(V)$ which is a morphism of coaugmented coalgebras and which lifts φ, then, for any $x \in \overline{C}$, we should have

$\tilde{\varphi}(1) = 1$ by coaugmentation,
$\tilde{\varphi}(x)_0 = 0$ by counitality,
$\tilde{\varphi}(x)_1 = \varphi(x)$ by compatibility with φ,
$\tilde{\varphi}(x)_n = \sum \varphi(x_{(1)}) \otimes \cdots \otimes \varphi(x_{(n)})$, since $\tilde{\varphi}$ is a coalgebra morphism.

Here we use Sweedler's notation for $\bar{\Delta}$, that is $\bar{\Delta}^{n-1}(x) = \sum x_{(1)} \otimes \cdots \otimes x_{(n)}$. This argument proves the uniqueness of $\tilde{\varphi}$. Let us define $\tilde{\varphi}(x)_n$ by the aforementioned formula. Since C is supposed to be conilpotent, there is only a finite number of nontrivial elements $\tilde{\varphi}(x)_n$. Therefore $\tilde{\varphi}(x) := \sum_n \tilde{\varphi}(x)_n$ belongs to $T^c(V)$, and so we have defined a map $\tilde{\varphi} : C \to T^c(V)$.

It is immediate to check that $\tilde{\varphi}$ is a coalgebra map whose projection onto V coincides with φ. So, we have proved the universal condition. □

1.2.7 Coderivation

Let $C = (C, \Delta)$ be a conilpotent coalgebra. By definition a *coderivation* is a linear map $d : C \to C$ such that $d(1) = 0$ and

$$\Delta \circ d = (d \otimes \mathrm{id}) \circ \Delta + (\mathrm{id} \otimes d) \circ \Delta .$$

We denote by $\mathrm{Coder}(C)$ the space of coderivations of C.

Proposition 1.2.2. *If C is cofree, i.e. $C = T^c(V)$ for some vector space V, then a coderivation $d \in \mathrm{Coder}(T^c(V))$ is completely determined by its weight 1 component*

$$T^c(V) \xrightarrow{d} T^c(V) \xrightarrow{proj_V} V.$$

Proof. Let us denote by $f(x) := d(x)^{(1)} = proj_V(d(x))$ the weight-one component of $d(x)$. Since $T^c(V)$ is cofree, the weight n component of $d(x)$, denoted by $d(x)^{(n)} \in V^{\otimes n}$, is given by

$$d(x)^{(n)} = \sum_{i=1}^{n} \sum_{(x)} proj_V(x_{(1)}) \otimes \cdots \otimes f(x_{(i)}) \otimes \cdots \otimes proj_V(x_{(n)}),$$

where $\sum_{(x)} x_{(1)} \otimes \cdots \otimes x_{(n)} := \bar{\Delta}^{n-1}(x)$ and $d(x)^{(0)} = 0$. So the coderivation d is completely determined by $f := proj_V \circ d : T^c(V) \to V$. □

1.2.8 Comodule

A *left comodule* N over a coalgebra C is a vector space endowed with a linear map

$$\Delta^l : N \to C \otimes N,$$

which is compatible with the coproduct and the counit

$$\begin{CD} N @>\Delta^l>> C\otimes N \\ @V\Delta^lVV @VV\mathrm{Id}\otimes\Delta^lV \\ C\otimes N @>\Delta\otimes\mathrm{Id}>> C\otimes C\otimes N \end{CD} \qquad \begin{array}{ccc} N & & \\ {\scriptstyle\Delta^l}\downarrow & \searrow^{\cong} & \\ C\otimes N & \xrightarrow{\varepsilon\otimes\mathrm{id}} & \mathbb{K}\otimes N. \end{array}$$

It is called a *coaction map*. The notion of a *right comodule* using $\Delta^r : N \to N \otimes C$ is analogous. A *co-bimodule* is a left and right comodule such that the two coaction maps satisfy the coassociativity condition:

$$(\mathrm{id}_C \otimes \Delta^r) \circ \Delta^l = (\Delta^l \otimes \mathrm{id}_C) \circ \Delta^r.$$

For instance, any coalgebra is a co-bimodule over itself.

1.2.9 Cocommutative Coalgebra

An associative coalgebra (C, Δ) is said to be *cocommutative* (or sometimes simply commutative) if the coproduct Δ satisfies the following symmetry condition:

$$\Delta = \tau \circ \Delta.$$

In other words we assume that the image of Δ lies in the invariant space $(C \otimes C)^{\mathbb{S}_2}$ where the generator of $\mathbb{S}_2$ acts via the switching map τ. The cofree commutative coalgebra over the space V (taken in the category of conilpotent commutative coalgebras of course) can be identified, in characteristic zero, to the symmetric module $S(V)$ equipped with the following coproduct:

$$\Delta'(v_1 \cdots v_n) = \sum v_{i_1} \cdots v_{i_p} \otimes v_{j_1} \cdots v_{j_q},$$

where the sum is extended to all the (p, q)-shuffles $(i_1 \cdots i_p, j_1 \cdots j_q)$, see Sect. 1.3.2 below for details. It is denoted by $S^c(V)$.

1.3 Bialgebra

We introduce the classical notions of bialgebra and Hopf algebra, which are characterized by the Hopf compatibility relation. We make explicit the coproduct in the tensor bialgebra in terms of shuffles and we characterize the space of primitive elements (CMM theorem and PBW theorem). We introduce the notion of convolution and the Eulerian idempotents.

1.3.1 Definition

A *bialgebra* $\mathcal{H} = (\mathcal{H}, \mu, \Delta, u, \varepsilon)$ over $\mathbb{K}$ is a vector space $\mathcal{H}$ equipped with an algebra structure $\mathcal{H} = (\mathcal{H}, \mu, u)$ and a coalgebra structure $\mathcal{H} = (\mathcal{H}, \Delta, \varepsilon)$ related by the *Hopf compatibility relation*

$$\Delta(xy) = \Delta(x)\Delta(y),$$

where $xy := \mu(x \otimes y)$ and the product on $\mathcal{H} \otimes \mathcal{H}$ is given (in the nongraded case) by $(x \otimes y)(x' \otimes y') = xx' \otimes yy'$. It is often better to write the compatibility relation in terms of Δ and μ. Then one has to introduce the switching map

$$\tau : \mathcal{H} \otimes \mathcal{H} \to \mathcal{H} \otimes \mathcal{H}, \qquad \tau(x \otimes y) := y \otimes x.$$

With this notation the compatibility relation reads:

$$\Delta \circ \mu = \underbrace{(\mu \otimes \mu) \circ (\mathrm{id} \otimes \tau \otimes \mathrm{id})}_{\mu_{\mathcal{H} \otimes \mathcal{H}}} \circ (\Delta \otimes \Delta) : \mathcal{H} \otimes \mathcal{H} \longrightarrow \mathcal{H} \otimes \mathcal{H}.$$

It can be represented by the following picture:

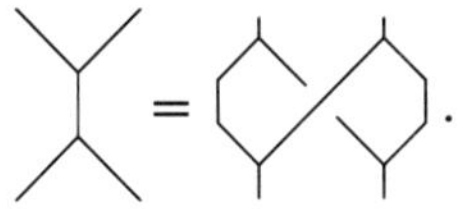

It is also assumed that ε is a morphism of algebras and u is a morphism of coalgebras.

1.3.2 The Tensor Bialgebra, Shuffles

Consider the free associative algebra $(T(V), \mu)$ over the space V. Since $T(V)$ is free, there is only one algebra morphism $\Delta' : T(V) \to T(V) \otimes T(V)$ such that

$$\Delta'(v) = v \otimes 1 + 1 \otimes v \quad \text{for } v \in V.$$

It is immediate to check (again from the freeness property) that Δ' is coassociative and counital. Hence $(T(V), \mu, \Delta')$ is a conilpotent bialgebra. Observe that it is cocommutative, that is $\Delta' = \tau \circ \Delta'$. It is called the *shuffle coproduct* for reasons explained below.

Dually, on the cofree associative coalgebra $(T^c(V), \Delta)$, there is a unique coaugmented coalgebra morphism

$$\mu' : T^c(V) \otimes T^c(V) \to T^c(V)$$

whose projection onto V is 0 except on $(V \otimes \mathbb{K}) \oplus (\mathbb{K} \otimes V)$ where it is given by:

$$\mu'(v \otimes 1) = v, \qquad \mu'(1 \otimes v) = v \quad \text{for } v \in V.$$

It is immediate to check (again from the cofreeness property) that μ' is associative and unital. Hence $(T^c(V), \mu', \Delta)$ is a conilpotent bialgebra. Observe that it is commutative, that is $\mu' = \mu' \circ \tau$. It is called the *shuffle product* for reasons explained below.

One can make the coproduct Δ' of $T(V)$ and the product μ' of $T^c(V)$ explicit as follows.

By definition a (p, q)-*shuffle* is a sequence of integers

$$[i_1 \cdots i_p \mid j_1 \ \cdots \ j_q]$$

which is a permutation of $\{1, \ldots, p+q\}$ such that

$$i_1 < \cdots < i_p \quad \text{and} \quad j_1 < \cdots < j_q.$$

The associated permutation $\sigma \in \mathbb{S}_{p+q}$ given by

$$\sigma(1) = i_1, \ldots, \sigma(p) = i_p \quad \text{and} \quad \sigma(p+1) = j_1, \ldots, \sigma(p+q) = j_q$$

is also called a (p, q)-shuffle by abuse of terminology. We denote by $Sh(p, q)$ the subset of (p, q)-shuffles in $\mathbb{S}_{p+q}$. For instance the three $(1, 2)$-shuffles are $[1 \mid 2\ 3]$, $[2 \mid 1\ 3]$ and $[3 \mid 1\ 2]$. Observe that the identity permutation $[1\ 2\ 3]$ is both a $(1, 2)$-shuffle and a $(2, 1)$-shuffle. The notion of $(i_1, \ldots, i_k)$-shuffle and the set $Sh(i_1, \ldots, i_k)$ are defined analogously.

Following Jim Stasheff, we call *unshuffle* the inverse of a shuffle. For instance the three $(1, 2)$-unshuffles are $[1\ 2\ 3]$, $[2\ 1\ 3]$ and $[2\ 3\ 1]$. We denote the set of (p, q)-unshuffles by $Sh^{-1}_{p,q}$.

Lemma 1.3.1. *For any integers p and q and any $\sigma \in \mathbb{S}_{p+q}$ there exist unique permutations $\alpha \in \mathbb{S}_p$, $\beta \in \mathbb{S}_q$ and $\omega \in Sh(p, q)$ such that:*

$$\sigma = \omega \cdot (\alpha \times \beta).$$

Proof. The permutation α is the unique element of $\mathrm{Aut}\{1, \ldots, p\}$ such that $\sigma(\alpha^{-1}(i)) < \sigma(\alpha^{-1}(i+1))$ for any $i = 1, \ldots, p-1$. The permutation β is the unique element of $\mathrm{Aut}\{p+1, \ldots, p+q\}$ such that $\sigma(\beta^{-1}(i)) < \sigma(\beta^{-1}(i+1))$ for any $i = p+1, \ldots, p+q-1$. Since $\omega := \sigma \cdot (\alpha \times \beta)^{-1}$ is a (p,q)-shuffle, we are done. □

The composite

$$Sh(p,q) \rightarrowtail \mathbb{S}_{p+q} \twoheadrightarrow \mathbb{S}_{p+q}/(\mathbb{S}_p \times \mathbb{S}_q)$$

is a bijection by the preceding lemma. Hence $Sh(p,q)$ gives a preferred splitting to the surjective map $\mathbb{S}_{p+q} \twoheadrightarrow \mathbb{S}_{p+q}/(\mathbb{S}_p \times \mathbb{S}_q)$. In other words, for any $\sigma \in \mathbb{S}_{p+q}$ the class $[\sigma]$ (modulo $\mathbb{S}_p \times \mathbb{S}_q$) contains one and only one (p,q)-shuffle.

Proposition 1.3.2. *The coproduct Δ' of the bialgebra $T(V)$ is given by*

$$\begin{aligned}\Delta'(v_1 \cdots v_n) &= \sum v_{i_1} \cdots v_{i_p} \otimes v_{j_1} \cdots v_{j_q}, \\ &= \sum_{\substack{p+q=n \\ \sigma \in Sh(p,q)}} v_{\sigma(1)} \cdots v_{\sigma(p)} \otimes v_{\sigma(p+1)} \cdots v_{\sigma(p+q)}.\end{aligned}$$

The product $\mu'(a,b)$ of the bialgebra $T^c(V)$ is given by

$$\begin{aligned}\mu'(v_1 \cdots v_p, v_{p+1} \cdots v_{p+q}) &= \sum_{\sigma \in Sh(p,q)} v_{\sigma^{-1}(1)} \cdots v_{\sigma^{-1}(p+q)} \\ &= \sum_{\sigma \in Sh(p,q)} \sigma \cdot (v_1 \cdots v_{p+q}).\end{aligned}$$

Proof. The proof is by direct inspection. □

The tensor module equipped with the product μ' constructed out of the shuffles in the above proposition is called the *shuffle algebra* and often denoted by $T^{sh}(V)$.

Proposition 1.3.3. [Qui69, Wig89] *Let $\mathbb{K}$ be a characteristic zero field and let $(T(V), \mu, \Delta')$ be the tensor bialgebra. Denote by $L(V)$ the Lie subalgebra of $T(V)$ generated by V under the bracket operation. Then, for any $x \in V^{\otimes n}$, the following are equivalent*:

(a) $x \in L(V)$,
(b) x *is primitive*,
(c) $\gamma(x) = nx$,

where $\gamma : T(V) \to T(V)$ is given by right bracketing:

$$\gamma(v_1 \cdots v_n) := \big[v_1, \big[v_2, \ldots, [v_{n-1}, v_n]\ldots\big]\big].$$

Proof. For this proof we use the convolution product which is recalled in Sect. 1.6.1. Let $E : T(V) \to T(V)$ be the *Euler operator* defined by $x \mapsto nx$ for $x \in V^{\otimes n}$. By induction it is easy to check that $\gamma \star \mathrm{id} = E$.

(a) $\Rightarrow$ (b). If $n = 1$, that is $x \in V$, then $\Delta'(x) = x \otimes 1 + 1 \otimes x$, so x is primitive. If x and y are primitive, then so is $[x, y]$. Therefore we are done by induction.

(b) $\Rightarrow$ (c). From the definition of the convolution we have, for any primitive element x,

$$\begin{aligned} nx = E(x) &= \big(\mu \circ (\gamma \otimes \mathrm{id}) \circ \Delta'\big)(x) \\ &= \mu \circ (\gamma \otimes \mathrm{id})(x \otimes 1 + 1 \otimes x) \\ &= \mu\big(\gamma(x), 1\big) = \gamma(x). \end{aligned}$$

(c) $\Rightarrow$ (a). It is immediate since we are in characteristic 0. □

Let us mention that the map γ (suitably normalized) is called the *Dynkin idempotent*.

Theorem 1.3.4 (Structure Theorem for cocommutative bialgebras). *Let $\mathbb{K}$ be a characteristic zero field. For any cocommutative bialgebra $\mathcal{H}$ over $\mathbb{K}$ the following are equivalent:*

(a) *$\mathcal{H}$ is conilpotent,*
(b) *$\mathcal{H} \cong U(\mathrm{Prim}\,\mathcal{H})$ as a bialgebra,*
(c) *$\mathcal{H} \cong S^c(\mathrm{Prim}\,\mathcal{H})$ as a conilpotent coalgebra.*

Proof. This is a classical result for which we refer to the "classics": [Car56, MM65, Qui69]. □

This statement contains several classical results, namely:

(a) $\Rightarrow$ (b) is essentially the Cartier–Milnor–Moore theorem.

(b) $\Rightarrow$ (c) is the Poincaré–Birkhoff–Witt theorem. A nice history of it can be found in [Gri04].

So we call it the CMM-PBW theorem. A far reaching generalization of this structure theorem can be found in [Lod08].

Corollary 1.3.5. *In the tensor algebra $T(V)$ we have the following identification:*

$$Lie(V) = L(V) = \mathrm{Prim}\,T(V)$$

where $Lie(V)$ is the free Lie algebra on V.

Proof. Recall that the subspace $L(V)$ generated by V under the bracket is a Lie algebra which contains V. So there is a natural map $Lie(V) \to L(V)$. Proposition 1.3.3 shows that $L(V) = \mathrm{Prim}\,T(V)$. Applying the structure theorem to $T(V)$ we get $Lie(V) = \mathrm{Prim}\,T(V)$ since $U(Lie(V)) = T(V)$ (the composite of two left adjoint functors is a left adjoint functor). Therefore we obtain the expected identifications. □

1.3.3 The Module *Lie*(*n*)

Let V_n be the space spanned by the variables $\{x_1, \ldots, x_n\}$. We denote by $Lie(n)$ the subspace of $Lie(V_n) \subset T(V_n)$ which is linear in each variable x_i. For instance $Lie(2)$ is one-dimensional spanned by $[x_1, x_2] = x_1x_2 - x_2x_1$. The action of $\mathbb{S}_n$ makes it into an $\mathbb{S}_n$-module and we have

$$Lie(V) = \bigoplus_n Lie(n) \otimes_{\mathbb{S}_n} V^{\otimes n}.$$

We introduce the *space of nontrivial shuffles* $\overline{Sh}_n \subset \mathbb{K}[\mathbb{S}_n]$ as follows. Let $\overline{T}^{sh}(V)$ be the augmentation ideal of the shuffle algebra $T^{sh}(V)$, cf. Proposition 1.3.2. We consider the quotient $T^{sh}(V)/(\overline{T}^{sh}(V))^2$. The space $\overline{Sh}_n$ is the $\mathbb{S}_n$-submodule of $\mathbb{K}[\mathbb{S}_n]$ such that $T^{sh}(V)/(\overline{T}^{sh}(V))^2 = \oplus_{n\geq 0}\overline{Sh}_n \otimes_{\mathbb{S}_n} V^{\otimes n}$. For instance $\overline{Sh}_1$ is one-dimensional spanned by [1 2] − [2 1].

Theorem 1.3.6. [Ree58] *Let* $Lie(n) \rightarrowtail \mathbb{K}[\mathbb{S}_n]$ *be the inclusion deduced from the functorial inclusion*

$$Lie(V) = \bigoplus_n Lie(n) \otimes_{\mathbb{S}_n} V^{\otimes n} \rightarrowtail T(V) = \bigoplus_n \mathbb{K}[\mathbb{S}_n] \otimes_{\mathbb{S}_n} V^{\otimes n}.$$

Under the isomorphism $\mathbb{K}[\mathbb{S}_n] \cong \mathbb{K}[\mathbb{S}_n]^*$ *obtained by taking the dual basis, the kernel* $\mathrm{Ker}(\mathbb{K}[\mathbb{S}_n]^* \to Lie(n)^*)$ *is the subspace* $\overline{Sh}_n$ *of* $\mathbb{K}[\mathbb{S}_n]$ *spanned by the nontrivial shuffles*:

$$\mathrm{Ker}\left(\mathbb{K}[\mathbb{S}_n]^* \to Lie(n)^*\right) = \overline{Sh}_n.$$

Proof. Since the graded dual of $T(V)$ is $T(V)^* = T^c(V^*)$, and since the dual of the space of primitives is the space of indecomposables (cf. Sect. 1.2.2), the map $T^c(V^*) \to Lie(V)^*$ gets identified with the map $T^c(V^*) \to T^c(V^*)/(\overline{T}^c(V^*))^2$. By Proposition 1.3.2 it follows that this kernel is spanned by the nontrivial shuffles. □

1.3.4 Hopf Algebra

Let $(\mathscr{H}, \mu, \Delta)$ be a bialgebra. If f and g are two linear maps from $\mathscr{H}$ to itself, then one can construct a third one, called the *convolution* of f and g, as

$$f \star g := \mu \circ (f \otimes g) \circ \Delta.$$

For the properties of the convolution product, see Sect. 1.6.

A *Hopf algebra* is a bialgebra $\mathscr{H}$ equipped with a linear map $S : \mathscr{H} \to \mathscr{H}$ which is an inverse of the identity under the convolution product:

$$S \star \mathrm{id} = u\varepsilon = \mathrm{id} \star S.$$

It is called an *antipode*.

Observe that a conilpotent bialgebra has automatically an antipode S given by

$$S(x) := -x + \sum_{n\geq 1} (-1)^{n+1} \mu^n \circ \bar{\Delta}^{n-1}(x),$$

where $\bar{\Delta}^{n-1}$ is the iterated reduced coproduct. Hence any conilpotent bialgebra is a Hopf algebra.

1.3.5 Eulerian Idempotents

Let $T(V)$ be the tensor algebra considered as a bialgebra, cf. Sect. 1.3.2. Since it is both an algebra and a coalgebra we can consider its convolution algebra $(\mathrm{Hom}(T(V), T(V)), \star, u\varepsilon)$. Let us write $\mathrm{Id} = u\varepsilon + J$ so that J is the identity on $V^{\otimes n}$ except for $n = 0$ on which it is 0. We define

$$e^{(1)} := \log^\star(\mathrm{Id}) = \log^\star(u\varepsilon + J) = \sum_{n\geq 1} (-1)^{n+1} \frac{J^{\star n}}{n}.$$

In weight n we get $e^{(1)} : V^{\otimes n} \to V^{\otimes n}$. Since it is functorial in V, by the Schur Lemma, cf. Sect. A.2.3, it is given by

$$e^{(1)}(v_1 \cdots v_n) = e_n^{(1)} \cdot (v_1 \cdots v_n)$$

for some uniquely defined element $e_n^{(1)} \in \mathbb{Q}[\mathbb{S}_n]$. These elements are called the (first) *Eulerian idempotents*.

It can be shown that $\mathrm{Im}\, e^{(1)}(V) = Lie(V) \subset T(V)$.

For any $i \geq 1$ let us define

$$e^{(i)} := \frac{(e^{(1)})^{\star i}}{i!}.$$

It can be shown that the elements $e_n^{(i)} \in \mathbb{Q}[\mathbb{S}_n], i = 1, \ldots, n$, are orthogonal idempotents. Hence the tensor product $V^{\otimes n}$ splits as

$$V^{\otimes n} = \bigoplus_{i=1}^{n} \mathrm{Im}\, e_n^{(i)}(V).$$

We refer to [Reu93] and [Lod94] for proofs and details.

1.4 Pre-Lie Algebras

The notion of pre-Lie algebra appeared in a work of Cayley on trees and then, later on, in differential geometry (Hochschild homology, flat affine connections on a given manifold). Nowadays it is also present in many topics, including algebraic topology, algebraic combinatorics and theoretical physics (renormalization).

1.4.1 Pre-Lie Algebra

By definition a (right) *pre-Lie algebra* is a vector space A equipped with a binary operation $\{x, y\}$ which satisfies the following relation, called *pre-Lie relation*:

$$\big\{\{x, y\}, z\big\} - \big\{x, \{y, z\}\big\} = \big\{\{x, z\}, y\big\} - \big\{x, \{z, y\}\big\}.$$

In plain words, the associator (left-hand side part of the equality) is right-symmetric. For the opposite type (i.e. $\langle x, y\rangle := \{y, x\}$) the associator is left-symmetric. It appeared in the work of Gerstenhaber [Ger63] and Vinberg [Vin63] in differential geometry and in several other papers subsequently. So it appears in the literature under various names, for instance: Vinberg algebras, right-symmetric algebras. The paper [Bur06] is a survey on this notion.

The notion of morphism of pre-Lie algebras is obvious. The category of pre-Lie algebras is denoted by *preLie*-alg.

Lemma 1.4.1. *The antisymmetrization of a pre-Lie operation is a Lie bracket. As a consequence there is a forgetful functor*

$$(-)_{Lie} : preLie\text{-}\mathsf{alg} \to Lie\text{-}\mathsf{alg}.$$

Proof. The product $[x, y] := \{x, y\} - \{y, x\}$ is antisymmetric by definition. The jacobiator is the sum of 12 elements that can be grouped into 3×4. The pre-Lie relators for (x, y, z), (y, z, x), (z, x, y) account for these 3 packets. □

1.4.2 Remark

The aforementioned lemma is valid for both the right pre-Lie product and the left pre-Lie product. Observe that for left pre-Lie algebras the linear map $L(x)$ defined by $L(x)(y) := \{x, y\}$ satisfies the following functional equation:

$$L\big([x, y]\big) = \big[L(x), L(y)\big].$$

1.4.3 Examples

(a) An associative algebra is an example of a pre-Lie algebra since the associator is trivial.

(b) *Faà di Bruno*. On $L_{-1} = \bigoplus_{n\geq -1} \mathbb{K}x_n$ the operation:

$$\{x_p, x_q\} := (p+1)x_{p+q}$$

is a pre-Lie product since

$$\big\{\{x_p, x_q\}, x_r\big\} - \big\{x_p, \{x_q, x_r\}\big\} = p(p+1)x_{p+q+r}$$

is symmetric in q and r. The associated Lie bracket is given by

$$[x_p, x_q] = (p-q)x_{p+q},$$

hence L_{-1} (also denoted W_1 in the literature) is the Lie algebra of polynomial vector fields on the affine line $\mathbb{K}^1$. Taking the universal algebra of the associated Lie algebra, we get a Hopf algebra $U(L_{-1})$, which turns out to be the dual of the celebrated Faà di Bruno Hopf algebra [JR82].

When n ranges over $\mathbb{Z}$ we get a pre-Lie algebra whose associated Lie algebra is well known. For $\mathbb{K} = \mathbb{R}$ it is the Lie algebra of polynomial vector fields over the circle (Virasoro algebra without center). For $\mathbb{K}$ being a finite field it is called the Witt algebra.

(c) *Derivations*. Let $D_i, i = 1, \ldots, k$, be commuting derivations of a commutative algebra A. On the free right A-module spanned by the D_i one defines

$$\{D_i a, D_j b\} := D_i D_j(a) b.$$

Since we assumed that the derivations are commuting, it is immediate to verify that this is a pre-Lie product. The previous case is a particular example.

1.4.4 Module over a Pre-Lie Algebra

For any pre-Lie algebra A, there is an obvious notion of module M over A. It is given by two operations $\{-,-\} : M \otimes A \to M$ and $\{-,-\} : A \otimes M \to M$ such that the pre-Lie relation holds whenever one of the variables is in M and the other two are in A.

But, due to the peculiar form of the pre-Lie relation, there is also a notion of *left pre-Lie module*. It is given by one operation $\{-,-\} : M \otimes A \to M$ such that the pre-Lie relation holds whenever $x \in M$ and $y, z \in A$.

1.5 Differential Graded Algebra

We present the notions of algebra and coalgebra in the differential graded framework.

1.5.1 Graded Vector Space

A *graded vector space* V is a family of vector spaces $\{V_n\}_{n\in\mathbb{Z}}$. The direct sum is denoted by

$$V_\bullet := \cdots \oplus V_{-n} \oplus \cdots \oplus V_0 \oplus V_1 \oplus \cdots \oplus V_n \oplus \cdots,$$

and the product is denoted by

$$\widehat{V}_\bullet := \prod_{n\in\mathbb{Z}} V_n.$$

By abuse of notation we often write V in place of $V_\bullet$ (resp. $\widehat{V}$ in place of $\widehat{V}_\bullet$). The degree of $v \in V_n$ is denoted by $|v|$, so here $|v| = n$. Most of the time the index will run over $\mathbb{N}$ only. A *morphism of degree r*, say $f : V \to W$, of graded vector spaces is a family of maps $f_n : V_n \to W_{n+r}$ for all n. The integer r is called the degree of f and denoted by $|f|$. So we have $|f(v)| = |f| + |v|$. If the nonnegative components are all zero, then it is helpful to write $V^n := V_{-n}$. We will say that $V_\bullet = \bigoplus_n V_n$ is *homologically graded*, and that $V^\bullet = \bigoplus_n V^n$ is *cohomologically graded.*

There are obvious notions of subvector space and quotient vector space in the graded framework. The grading of the tensor product $V \otimes W$ of two graded spaces V and W is described explicitly as:

$$(V \otimes W)_n := \bigoplus_{i+j=n} V_i \otimes W_j.$$

Hence $V^{\otimes n}$ is graded and so is the tensor module $T(V)$. Any element $\underline{v} = v_1 \cdots v_n \in V^{\otimes n}$ admits a *degree* and a *weight*:

$$|\underline{v}| := |v_1| + \cdots + |v_n|, \qquad \text{weight}(\underline{v}) := n.$$

Let $\mathbb{K}s$ be the one-dimensional graded vector space spanned by s with $|s| = 1$. By definition the *suspension* of the graded space V is

$$sV := \mathbb{K}s \otimes V.$$

In particular $(sV)_i = V_{i-1}$. Any $v \in V_n$ determines an element $sv \in (sV)_{n+1}$ of degree $n+1$. Alternative notations for sV used in the literature are $V[1]$ and $\uparrow V$.

Similarly let $\mathbb{K}s^{-1}$ be the graded vector space spanned by s^{-1} put in degree -1. By definition the *desuspension* of the graded space V is $s^{-1}V := \mathbb{K}s^{-1} \otimes V$. In particular $(s^{-1}V)_i = V_{i+1}$. Any $v \in V_n$ determines an element $s^{-1}v \in (sV)_{n-1}$ of degree $n-1$.

If $V = \{V_n\}_{n\geq 0}$ is a graded vector space, then its dual is the graded vector space $V^* = \{V^*_{-n}\}_{n\leq 0} = \{V^{*n}\}_{n\leq 0} = \{\text{Hom}(V_{-n}, \mathbb{K})\}_{n\leq 0}$. Observe that the direct sum $\oplus_n V^{*n}$ is not in general the dual of the direct sum $\oplus_n V_n$ (unless there is only finitely many nonzero summands). By abuse of terminology we say that the infinite sum $\oplus_{i\geq 0} V_i^*$ is the *graded dual* of $\oplus_{i\geq 0} V_i$.

1.5.2 Switching Map

The category of graded vector spaces (gVect, $\otimes$, $\mathbb{K}$) is a monoidal category. It is usually equipped with a symmetric structure given by the *switching map*

$$\tau : V \otimes W \to W \otimes V, \qquad \tau(v \otimes w) := (-1)^{|v||w|} w \otimes v.$$

In this formula the elements v and w are supposed to be homogeneous. Then τ is extended by linearity. The category gVect, equipped with this symmetry, is called the category of *sign-graded* vector spaces.

Observe that there is another choice of symmetry: $\tau(v \otimes w) := w \otimes v$. Equipped with this symmetry gVect is simply called the category of graded vector spaces.

1.5.3 Koszul Convention

When working in the symmetric monoidal category of sign-graded vector spaces with switching map τ, there are signs involved in the formulas. Indeed the isomorphism $\tau : V \otimes W \to W \otimes V$ is given by $v \otimes w \mapsto (-1)^{|v||w|} w \otimes v$ on homogeneous elements. In plain words, the rule (called sign rule) is the following: when two variables switch in a formula, they create a sign which is minus one to the power the product of the degrees.

For any maps $f : V \to V'$ and $g : W \to W'$ between graded spaces the map $V \otimes W \to V' \otimes W'$ coming from the symmetric monoidal structure is given by

$$v \otimes w \mapsto (-1)^{|g||v|} f(v) \otimes g(w).$$

Hence it is convenient to introduce the notation $f \otimes g : V \otimes W \to V' \otimes W'$ defined as

$$(f \otimes g)(v \otimes w) := (-1)^{|g||v|} f(v) \otimes g(w).$$

This trick is the *Koszul convention*. It permits us to avoid complicated signs in the formulas provided one works with the maps (or functions) *without* evaluating them on the elements. When all the involved operations are of degree 0, the formulas in the nongraded case apply mutatis mutandis to the graded case.

As a consequence the sign rule is also valid for morphisms:

$$(f \otimes g) \circ (f' \otimes g') = (-1)^{|g||f'|} f \circ f' \otimes g \circ g'.$$

1.5.4 Differential Graded Vector Space (Chain Complex)

A *differential graded vector space* (V, d), abbreviated into dg vector space and also called a *chain complex*, is a graded space $V_\bullet$ equipped with a linear map $d = d_V :$

$V_\bullet \to V_{\bullet-1}$ of degree -1, called the *differential*, satisfying $d^2 = 0$,

$$\cdots \xleftarrow{d} V_{-1} \xleftarrow{d} V_0 \xleftarrow{d} V_1 \xleftarrow{d} V_2 \xleftarrow{d} \cdots \xleftarrow{d} V_n \xleftarrow{d} \cdots .$$

Observe that, if we write V_{n-1} in place of $(sV)_n$ in the suspended space, then the new differential is $-d$ since $d(sv) = (-1)^{|d|} sd(v) = -sd(v)$. If $V_n = 0$ for $n < 0$ we say that the complex is nonnegatively graded. If the complex is negatively graded, then we sometimes adopt the cohomological grading and write $V^n := V_{-n}$:

$$\cdots \to V^{-1} \to V^0 \to V^1 \to V^2 \to \cdots \to V^n \to \cdots ,$$

and $(V^\bullet, d)$ is called a *cochain complex*. By definition the dual of the chain complex $C = (V, d)$ is the cochain complex $C^\bullet$ whose module of n-cochains is $C^n := \mathrm{Hom}(V_n, \mathbb{K})$ and the boundary map $d^n : C^n \to C^{n+1}$ is given by $d^n(f) := (-1)^{n+1} f d_{n+1}$, for $f : V_n \to \mathbb{K}$ and $d_{n+1} : V_{n+1} \to V_n$. This is a particular case of the derivative of a graded linear map, see below. A degree r *morphism of chain complexes*, denoted by $f : V \to W$, is a morphism of graded vector spaces of degree r such that $d_W \circ f = (-1)^r f \circ d_V$.

A *bicomplex* is a bigraded vector space $V = \{V_{pq}\}_{p\geq 0, q\geq 0}$ equipped with a horizontal differential $d^h : V_{pq} \to V_{p-1q}$ and a vertical differential $d^v : V_{pq} \to V_{pq-1}$ satisfying

$$d^h \circ d^v + d^v \circ d^h = 0.$$

The *total complex* associated to a bicomplex (V, d^h, d^v) is defined by

$$(\mathrm{Tot}\, V)_n := \bigoplus_{p+q=n} V_{pq} \quad \text{and} \quad d = d^h + d^v.$$

It is immediate to verify that $d^2 = 0$.

Let (V, d_V) and (W, d_W) be two differential graded vector spaces. Their *tensor product* $(V \otimes W)_n := \bigoplus_{p+q=n} V_p \otimes W_q$ is equipped with the differential

$$d_{V\otimes W} := d_V \otimes \mathrm{Id}_W + \mathrm{Id}_V \otimes d_W,$$

that is

$$d_{V\otimes W}(v \otimes w) := d_V(v) \otimes w + (-1)^p v \otimes d_W(w),$$

for $v \otimes w \in V_p \otimes W_q$.

The suspension sV of the chain complex V is by definition the tensor product of $s\mathbb{K}$ with V. Here $s\mathbb{K}$ is considered as a chain complex concentrated in degree 1 with 0 differential: $(sV)_n = V_{n-1}$ and $d_{sV} = -d_V$.

The *derivative* (or boundary) of a graded linear map $f : V_\bullet \to W_{\bullet+r}$ of degree r is the graded map

$$\partial(f) = [d, f] := d_W \circ f - (-1)^r f \circ d_V,$$

which is of degree $r-1$. So f is a morphism of chain complexes if and only if $\partial(f)=0$. The derivative is a differential on the space of graded morphisms from V to W, that is $\partial^2=0$. Under this definition, a morphism f of chain complexes of degree r is an element of $\mathrm{Hom}(V,W)_r := \prod_{p\in\mathbb{Z}} \mathrm{Hom}(V_p, W_{p+r})$ such that $\partial(f)=0$. Remark that both $(V\otimes W, d)$ and $(\mathrm{Hom}(V,W), \partial)$ are total complexes associated to bicomplexes.

The Hom complex bifunctor and the tensor complex bifunctor satisfy the following adjunction property:

$$\mathrm{Hom}\big(U, \mathrm{Hom}(V,W)\big) \cong \mathrm{Hom}(U\otimes V, W),$$

for any complexes U, V, W.

1.5.5 Homology and Cohomology

Given a chain complex (V,d) its nth *homology group* is by definition

$$H_n(V,d) := \mathrm{Ker}(d: V_n \to V_{n-1})/\,\mathrm{Im}(d: V_{n+1}\to V_n).$$

The inclusion $\mathrm{Im}\, d \subset \mathrm{Ker}\, d$ is a consequence of $d^2=0$. We also adopt the notation $H_\bullet(V,d) := \bigoplus_{n\in\mathbb{Z}} H_n(V,d)$, or $H_\bullet(V)$ for short. A morphism $f: V\to W$ of chain complexes induces a morphism on homology denoted either by $f_\bullet$ or by $H_\bullet(f)$.

For a "cohomological chain complex" (V,d) (that is the differential map d is of degree $+1$), the nth *cohomology group* is by definition

$$H^n(V,d) := \big(\mathrm{Ker}\, d: V^n \to V^{n+1}\big)/\big(\mathrm{Im}\, d: V^{n-1}\to V^n\big).$$

We also adopt the notation $H^\bullet(V,d) := \bigoplus_{n\in\mathbb{Z}} H^n(V,d)$, or $H^\bullet(V)$ for short.

When $\mathbb{K}$ is a field, we recall that the *Künneth formula* asserts that the homology of the tensor product of two chain complexes is the tensor product of their homology: $H_\bullet(V\otimes W) \cong H_\bullet(V)\otimes H_\bullet(W)$, cf. [ML95, Chapter V].

By definition a *quasi-isomorphism* is a morphism of chain complexes which induces an isomorphism on homology (or cohomology). We denote quasi-isomorphisms by the symbol $\xrightarrow{\sim}$.

A chain complex is said to be *acyclic* if its homology is 0 everywhere. A nonnegatively graded chain complex is said to be *augmented* if there is given a map to the chain complex which is 0 everywhere except in degree 0 where it is $\mathbb{K}$. By abuse of terminology we say that this augmented chain complex is acyclic when the augmentation map is a quasi-isomorphism. In other words the homology is 0 everywhere except in degree 0 where it is $\mathbb{K}$.

Let f and g be two chain maps between the two graded chain complexes (V,d) and (V',d'). A *homotopy* between f and g is a map $h: V\to V'$ of degree $+1$ such that

$$f-g = d'h + hd =: \partial(h).$$

It is easy to check that the induced maps on homology are equal: $f_\bullet = g_\bullet$. Applied to Id and 0, it provides a way to prove that a chain complex is acyclic.

A *homotopy equivalence* between two chain complexes (W, d_W) and (V, d_V) is a chain map $i : W \to V$ such that there exists a chain map $p : V \to W$, with $i \circ p$ homotopic to Id_V and $p \circ i$ homotopic to Id_W.

$$h' \circlearrowright (W, d_W) \underset{p}{\overset{i}{\rightleftarrows}} (V, d_V) \circlearrowleft h$$

$$\mathrm{Id}_V - i \circ p = d_V \circ h + h \circ d_V, \qquad \mathrm{Id}_W - p \circ i = d_W \circ h' + h' \circ d_W.$$

In this case, the chain complexes W and V are *homotopy equivalent.* If $h' = 0$ (i.e. $\mathrm{Id}_W = p \circ i$), then i is injective, p is surjective and the chain complex W is called a *deformation retract of* V.

There is an intermediate notion called *homotopy retract* consisting in

$$(W, d_W) \underset{p}{\overset{i}{\rightleftarrows}} (V, d_V) \circlearrowleft h$$

where it is assumed that i and p are chain maps, that h is a homotopy and that i is a quasi-isomorphism. Obviously a deformation retract is a particular case of homotopy retract.

1.5.6 Spectral Sequence

Let

$$C : \quad \cdots \to C_n \xrightarrow{d} C_{n-1} \to \cdots$$

be a chain complex. We suppose that it is equipped with a filtration:

$$\cdots \subset F_{-1}C_n \subset F_0 C_n \subset F_1 C_n \subset \cdots \subset C_n,$$

compatible with the differential, that is, F_pC is a chain sub-complex of C, $d : F_pC_n \to F_pC_{n-1}$.

Given such a filtered complex, we can either take the homology and then take the graded associated module, or, take the graded module and then the homology of the induced differential. It turns out that, in this second case, there is a hidden structure. It takes the form of a new differential. Therefore we can again take the homology. But, again, there is a hidden structure: a new differential, and so on and so forth. This family of complexes, usually denoted by $(E^r, d^r)_{r \geq 0}$, are related by the formula $H(E^r, d^r) = E^{r+1}$. It is called a *spectral sequence*. The relationship between the two ways of handling the filtration and the differential is, under some mild assumptions, the following:

$$\mathrm{gr}\, H(C_\bullet, d) = E^\infty.$$

This remarkable discovery is due to Jean Leray [Ler46] and was made available through the seminal paper [Kos47] by Jean-Louis Koszul.

More precisely, starting with a filtered chain complex $F_\bullet C_\bullet$, we put $E^0_{pq} := F_pC_{p+q}/F_{p-1}C_{p+q}$. The boundary map of C induces a boundary map on E^0 that we denote by d^0. It is of bidegree $(0, -1)$. We denote by E^1 the homology of (E^0, d^0). The first key point is the existence of a boundary map d^1 on E^1 of bidegree $(-1, 0)$. It is constructed by the so-called *staircase argument* (diagram chasing), see for instance [Kos47], [ML95, Chapter XI], [Lod98, Appendix D], [BT82, Sect. 14], [Wei94, Chapter 5]. Since d^1 is a boundary map, we can take its homology to get $E^2 := H(E^1, d^1)$. The theory of *spectral sequences*, cf. loc. cit., asserts that there is an infinite sequence of bigraded chain complexes $(E^r_{pq}, d^r)_{r\geq 0}$, where d^r is of bidegree $(-r, r-1)$ satisfying $E^{r+1} = H(E^r, d^r)$. For any pair (p, q), the spaces E^r_{pq} give rise to a limit E^∞_{pq}.

The filtration F is said to be *bounded below* whenever, for each n, there exists k such that $F_pC_n = 0$, for any $p < k$. The filtration F is said to be *exhaustive* if $C_n = \cup_p F_pC_n$.

Theorem 1.5.1 (Classical convergence theorem of spectral sequences). *If the filtration $F_\bullet C$ of the chain complex $C = (C_\bullet, d)$ is exhaustive and bounded below, then the spectral sequence converges. This means that there is an isomorphism*

$$F_pH_{p+q}(C)/F_{p-1}H_{p+q}(C) \cong E^\infty_{pq}.$$

When the differential maps d^r are 0 for $r \geq k$, the spectral sequence is said to *degenerate* at page k (or to collapse at rank k). In this case, we get

$$E^k_{pq} = E^{k+1}_{pq} = \cdots = E^\infty_{pq}.$$

COMMENT. The aforementioned theorem will be the main tool to prove the various homological results throughout this book. The spectral sequence argument is used as follows. We will encounter many chain complexes with differential maps made up of the sum of several terms. In these cases, we will introduce a suitable filtration to investigate their homology via its associated spectral sequence.

1.5.7 Differential Graded Algebra

A *graded algebra* A is a graded vector space $\{A_n\}_{n\geq 0}$ equipped with a unital product μ of degree 0. Hence it sends $A_p \otimes A_q$ into A_{p+q}. For instance, if we put V in degree 1, then the tensor algebra $T(V)$ is a graded algebra (cf. Sect. 1.1.3). In this case the degree coincides with the weight. Throughout the book, we will mainly consider nonnegatively graded algebras.

A *differential graded associative algebra* (A, d) (dga algebra for short) is a graded algebra equipped with a differential map $d : A \to A$ of degree -1 which is a *derivation* for the product, that is, satisfies the identity:

$$d(ab) = (da)b + (-1)^{|a|}a(db).$$

This identity is better written as follows:

$$d \circ \mu = \mu \circ (d \otimes \mathrm{id} + \mathrm{id} \otimes d)$$

where μ is the product in A. It means that the product $\mu : A \otimes A \to A$ is a morphism of chain complexes. Observe that the unit 1 belongs to A_0. The dga algebra A is said to be *connected* if $A_0 = \mathbb{K}1$.

A dga algebra is said to be a *quasi-free* algebra if, as a graded algebra, it is free over some graded vector space V. Observe that, in the notation "dga", the letter "a" stands for "associative" not for algebra in this book.

1.5.8 Minimal Models

Let $p : M \twoheadrightarrow A$ be a surjective map of dga algebras. If p is a quasi-isomorphism, then (M, p) is called a *model* of A.

A dga algebra (M, d) is called *quasi-free* if M is free as a graded algebra, that is after forgetting the differential: $M \cong T(V)$. By definition, a *minimal dga algebra* is a quasi-free dga algebra $(T(V), d)$

1. whose differential is *decomposable*, that is $d : V \to T(V)^{(\geq 2)}$, and
2. such that the generating graded module V admits a decomposition into

$$V = \bigoplus_{k \geq 1} V^{(k)}$$

satisfying

$$d\big(V^{(k+1)}\big) \subset T\left(\bigoplus_{i=1}^{k} V^{(i)}\right).$$

Finally, a *minimal model* of a dga algebra A is the data of a minimal algebra $(T(V), d)$ together with a quasi-isomorphism of dga algebras

$$\big(T(V), d\big) \xrightarrow{\sim} A\ ,$$

which is an epimorphism. Notice that this last condition is always satisfied when the differential of A is trivial.

Theorem 1.5.2 (Fundamental theorem of minimal models [DGMS75]). *Let A be a dga algebra. When A admits a minimal model, it is unique up to a (non-unique) isomorphism.*

Proof. Let $M = (T(V), d) \xrightarrow{\sim} A$ and $M' = (T(V'), d') \xrightarrow{\sim} A$ be two minimal models of a dga algebra A. Using the decompositions of the spaces of generators, one develops an obstruction theory to prove that there exists a quasi-isomorphism $f : M \xrightarrow{\sim} M'$ of dga algebras. It induces a quasi-isomorphism of chain complexes between the space of generators $(V, d_V) \xrightarrow{\sim} (V', d_{V'})$. Since the differentials d and d' are decomposable, we get $d_V = 0$ and $d_{V'} = 0$. So the aforementioned quasi-isomorphism is actually an isomorphism of graded modules $V \cong V'$. Therefore, the map f is an isomorphism of dga algebras. □

The proof shows that the generating space V of minimal models is uniquely determined up to isomorphism. The study of the automorphism group of minimal models is an interesting domain of study. The Koszul duality theory, developed in this book, produces minimal models with differential maps satisfying $d(V) \subset T(V)^{(2)}$. In this case, the differential is called *quadratic* and the minimal model is called a *quadratic model.*

1.5.9 Differential Graded Coalgebra

A *graded coalgebra* C is a graded vector space $\{C_n\}_{n\in\mathbb{Z}}$ equipped with a counital coproduct Δ of degree 0, that is sending C_n into $\bigoplus_{p+q=n} C_p \otimes C_q$. For instance, if we put V in homological degree 1, then the tensor coalgebra $T^c(V)$ is a graded coalgebra. We will almost always work with nonnegatively graded coalgebras.

A *differential graded associative coalgebra* (C, d) (dga coalgebra for short) is a graded coalgebra equipped with a differential map $d : C \to C$ (of degree -1) which is a *coderivation* for the coproduct, that is, satisfies the identity:

$$\Delta \circ d = (d \otimes \mathrm{id} + \mathrm{id} \otimes d) \circ \Delta .$$

Equivalently, a dga coalgebra is a chain complex endowed with a coproduct which is a morphism of chain complexes. Observe that the counit sends C_n to 0 for $n > 0$ and C_0 to $\mathbb{K}$.

Since the differential d is a coderivation, it preserves the coradical filtration of Sect. 1.2.4: $d : F_r \to F_r$. The dga coalgebra is said to be *conilpotent* if the underlying coalgebra C is conilpotent in the sense of Sect. 1.2.4, that is when the coradical filtration is exhaustive. A nonnegatively graded dga coalgebra C is called *connected* if $C_0 = \mathbb{K}1$. A connected dga coalgebra is conilpotent since $C_r \subset F_r$ in this case.

A dga coalgebra is said to be a *quasi-cofree* coalgebra if, as a graded coalgebra, it is cofree over some graded vector space V.

1.5.10 Weight Graded Framework

We will often need the assumption that dg modules, algebras and coalgebras have an extra grading, which we call the *weight* to avoid confusion with the homological degree. This means that a weighted dg module M is a direct sum of sub-dg modules $M^{(n)}$ indexed by the weight n. A *weight-graded dga algebra A*, *wdga algebra* for short, is an associative algebra structure on a weight-graded dg module A. Its product is supposed to preserve the homological degree as well as the weight grading. We denote $A_d^{(n)}$ the sub-module of degree d and weight n of A. Similarly, there is the notion of *weight-graded dg associative coalgebra C*, *wdga coalgebra* for short, which is an associative coalgebra structure on a weight graded dg module C. In this context the morphisms are supposed to respect the weight grading.

EXAMPLE. Let V be a graded vector space and let $T(V)$ be the tensor module. An element in $V^{\otimes n}$ is said to be of weight n. Therefore the element $\omega = v_1 \dots v_n \in V^{\otimes n}$ has weight n (the number of factors) and degree $|\omega| = |v_1| + \cdots + |v_n|$.

In this book, the homological degree and the weight grading are supposed to be nonnegative gradings. A wdga algebra A is called *connected* if it decomposes as

$$A := \mathbb{K}1 \oplus A^{(1)} \oplus \cdots \oplus A^{(n)} \oplus \cdots$$

with $A^{(0)} = \mathbb{K}1$ concentrated in degree 0. In particular we assume $A_d^{(0)} = 0$ for $d \neq 0$. Dually, a wdga coalgebra is called *connected* if it satisfies the same decomposition. Notice that in the weight graded case, the word "connected" refers to the weight and not to the homological degree. A connected wdga coalgebra is conilpotent with coradical filtration $F_r C = \bigoplus_{n=0}^{r} C^{(r)}$.

1.5.11 Differential Graded Module and Comodule

Let A be an associative algebra equipped with a derivation d_A (for instance a graded differential algebra). An *A-derivation* (or simply a derivation by abuse of language) on a right A-module M is a linear map $d_M : M \to M$ which satisfies:

$$d_M(ma) = d_M(m)a \pm m d_A(a)$$

for any $m \in M$ and any $a \in A$. If A is a dga algebra and M is a chain complex such that d_M is an A-derivation, then M is called a differential graded A-module.

Analogously one can define the notion of C-coderivation on a left comodule (for C a coalgebra equipped with a coderivation) and the notion of differential graded C-comodule (for C a dga coalgebra).

Proposition 1.5.3. *Let A be an associative algebra and let N be a vector space. There is a one-to-one correspondence between A-derivations on the free A-module $N \otimes A$ and linear maps from N to $N \otimes A$:*

$$\mathrm{Der}(N \otimes A) \cong \mathrm{Hom}(N, N \otimes A), \qquad d_f = (\mathrm{Id} \otimes \mu) \circ (f \otimes \mathrm{Id}) \leftrightarrow f.$$

Dually, let C be a coassociative coalgebra and let N be a vector space. There is a one-to-one correspondence between coderivations on the cofree C-module $C \otimes N$ and linear maps from $C \otimes N$ to N:

$$\mathrm{Coder}(C \otimes N) \cong \mathrm{Hom}(C \otimes N, N), \qquad d_f = (\mathrm{Id} \otimes f) \circ (\Delta \otimes \mathrm{Id}) \leftrightarrow f.$$

Proof. First statement. In one direction it is simply the restriction to N. In the other direction, the unique derivation which extends a map $f : N \to N \otimes A$ is equal to the following composite

$$d_f : N \otimes A \xrightarrow{f \otimes \mathrm{Id}} N \otimes A \otimes A \xrightarrow{\mathrm{Id} \otimes \mu} N \otimes A.$$

Second statement. In one direction it is simply the projection onto N. In the other direction, the unique coderivation which extends a map $f : C \otimes N \to N$ is equal to the following composite

$$d_f : C \otimes N \xrightarrow{\Delta \otimes \mathrm{Id}} C \otimes C \otimes N \xrightarrow{\mathrm{Id} \otimes f} C \otimes N.$$ □

This proposition extends to differential graded objects by adding the underlying differential of the tensor modules: $d_f + d_{N \otimes A}$ and $d_f + d_{C \otimes N}$ respectively.

1.6 Convolution

1.6.1 Convolution Algebra

Let (C, Δ, ε) be a coalgebra and (A, μ, u) be an algebra. Let $f, g : C \to A$ be two linear maps. The composite

$$f \star g := \mu \circ (f \otimes g) \circ \Delta : C \longrightarrow A$$

is called the *convolution* of f and g.

Proposition 1.6.1. *The convolution product $\star$ is associative. The composite $u \circ \varepsilon$ is the unit for $\star$.*

Proof. Associativity of μ and coassociativity of Δ imply that the convolution operation $\star$ is an associative operation on $\mathrm{Hom}(C, A)$. □

The associative algebra $(\mathrm{Hom}(C, A), \star, u \circ \varepsilon)$ is called a *convolution algebra*.

$d_\alpha \circ d_\beta =$

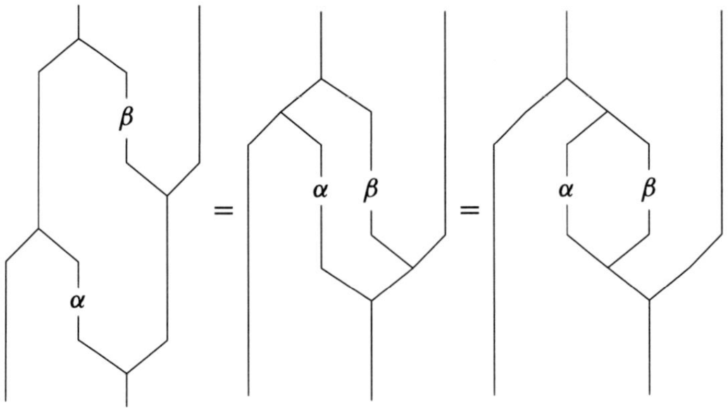

$= d_{\alpha\star\beta}$.

Fig. 1.1 Electronic proof

1.6.2 From Convolution Algebra to Tensor Product

Any linear map $\alpha : C \to A$ from a graded coalgebra C to a graded algebra A, defines a morphism

$$C \xrightarrow{\Delta} C \otimes C \xrightarrow{\mathrm{Id}\otimes\alpha} C \otimes A$$

which induces a unique derivation on $C \otimes A$ and a morphism

$$C \otimes A \xrightarrow{\alpha\otimes\mathrm{Id}} A \otimes A \xrightarrow{\mu} A$$

which induces a unique coderivation on $C \otimes A$ by Proposition 1.5.3. Both extended maps are equal to

$$d_\alpha := (\mathrm{Id}_C \otimes \mu) \circ (\mathrm{Id}_C \otimes \alpha \otimes \mathrm{Id}_A) \circ (\Delta \otimes \mathrm{Id}_A).$$

The following result gives a condition under which d_α is a boundary map.

Proposition 1.6.2. *For any* $\alpha, \beta \in \mathrm{Hom}(C, A)$ *one has*

$$d_{\alpha\star\beta} = d_\alpha \circ d_\beta \quad \textit{and} \quad d_{u\varepsilon} = \mathrm{Id}_{C\otimes A}\,.$$

So, $d_- : (\mathrm{Hom}(C, A), \star) \to (\mathrm{End}(C \otimes A), \circ)$ *is a morphism of associative algebras.*
If $\alpha \star \alpha = 0$, *then* $(d_\alpha)^2 = 0$.

Proof. The last assertion follows immediately from the first. Under some obvious convention (see Sect. 1.3.1), Fig. 1.1 (to be read from top to bottom) is a proof of the first assertion.

Observe that we use only associativity and coassociativity to prove these equalities. The second picture is a proof of the second assertion.

$d_{u\varepsilon} =$ 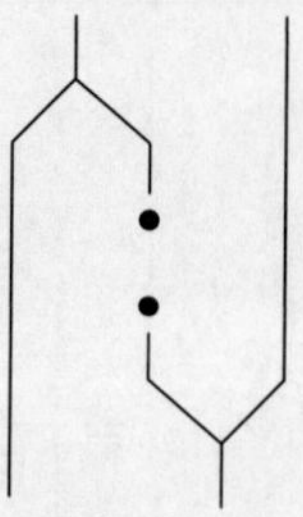 $=$  $= \mathrm{Id}_{C\otimes A}.$

□

As a consequence we get a chain complex $(C \otimes A, d_\alpha)$. We will begin the next chapter with a differential graded version of this result.

1.7 Résumé

Unital associative algebra: $A = (A, \mu, u)$, augmentation $\varepsilon : A \to \mathbb{K}$.

Counital coassociative coalgebra: $C = (C, \Delta, \varepsilon)$, coaugmentation $u : \mathbb{K} \to C$.

Linear duality: C^* is an algebra, A^* is a coalgebra if A is finite-dimensional.

Free unital associative algebra over V*:* $T(V)$ with concatenation product.

Cofree counital coassociative coalgebra over V*:* $T^c(V)$ with deconcatenation coproduct.

Bialgebra: $\mathscr{H} = (\mathscr{H}, \mu, \Delta, u, \varepsilon)$ algebra + coalgebra + Hopf compatibility condition:

$$\Delta \circ \mu = (\mu \otimes \mu) \circ (\mathrm{id} \otimes \tau \otimes \mathrm{id}) \circ (\Delta \otimes \Delta).$$

Pre-Lie algebra: $(A, \{-,-\})$ where the associator of $\{-,-\}$ is left-symmetric.

Derivative of a linear map: $\partial(f) = [d, f] := d \circ f - (-1)^{|f|} f \circ d$.

Convolution: $f, g \in \mathrm{Hom}(C, A)$, $f \star g := \mu \circ (f \otimes g) \circ \Delta$.

Twisted tensor product: $C \otimes_\alpha A := (C \otimes A, d_\alpha)$, for $\alpha : C \to A$ where

$$d_\alpha := (\mathrm{Id}_C \otimes \mu) \circ (\mathrm{Id}_C \otimes \alpha \otimes \mathrm{Id}_A) \circ (\Delta \otimes \mathrm{Id}_A).$$

Theorem. *If* $\alpha \star \alpha = 0$, *then* $(d_\alpha)^2 = 0$.

1.8 Exercises

Exercise 1.8.1 (Action of the symmetric group). Prove that the formula

$$\sigma \cdot (v_1 \cdots v_n) := v_{\sigma^{-1}(1)} \cdots v_{\sigma^{-1}(n)}$$

defines a left action of the symmetric group $\mathbb{S}_n$ on the tensor product $V^{\otimes n}$.

Show that applying σ on the left to $v_1 \cdots v_n$ means putting v_i at the place number $\sigma(i)$.

Exercise 1.8.2 (Uniqueness of free algebra). Let V be a vector space. Let A and A' be two associative algebras which are free over V. Show that there exists a unique isomorphism $A \to A'$ under V.

Exercise 1.8.3 (Coproduct on a basis). Prove the assertion of Sect. 1.2.3.

Exercise 1.8.4 (Small coalgebras). Make explicit the coalgebra which is the linear dual of the dual numbers algebra $\mathbb{K}[t]/(t^2 = 0)$, resp. the group algebra $\mathbb{K}[t]/(t^2 = 1)$. Are they conilpotent?

Exercise 1.8.5 (Polynomial algebra). Let $\mathbb{K}[x] = \mathbb{K}1 \oplus \mathbb{K}x \oplus \cdots \oplus \mathbb{K}x^n \oplus \cdots$. It is a unital commutative algebra for the product $x^n x^m = x^{n+m}$. Show that the product

$$x^n * x^m := \binom{n+m}{n} x^{n+m}$$

makes it also a unital commutative algebra, that we denote by $\Gamma(\mathbb{K}x)$. Compute explicitly:

a) the dual coalgebra of $\mathbb{K}[x]$ and of $\Gamma(\mathbb{K}x)$,
b) the coalgebra structure of $\mathbb{K}[x]$, resp. $\Gamma(\mathbb{K}x)$, which makes it a bialgebra and which is uniquely determined by $\Delta(x) = x \otimes 1 + 1 \otimes x$.
c) Compare the results of a) and b).

Exercise 1.8.6 (Polynomial algebra continued). Same as in the preceding exercise but with several variables, i.e. for the algebra of coinvariants $S(V) = \bigoplus_n ((V^{\otimes n})_{\mathbb{S}_n})$ and the algebra of invariants $\Gamma(V) = \bigoplus_n ((V^{\otimes n})^{\mathbb{S}_n})$ over the vector space V.

Exercise 1.8.7 (Symmetric algebra as bialgebra). Show that $S(V)$ is a Hopf algebra for Δ uniquely determined by $\Delta(v) = v \otimes 1 + 1 \otimes v$. Show that the linear dual of the coalgebra $(S(V), \Delta)$ is $\Gamma(V^*)$.

Exercise 1.8.8 (Universal enveloping algebra). Show that the universal algebra $U(\mathfrak{g})$ of a Lie algebra $\mathfrak{g}$ is a conilpotent Hopf algebra.

Exercise 1.8.9 (Group algebra). Show that for any group G the group algebra $\mathbb{K}[G]$ is a Hopf algebra. Show that it is not conilpotent in general.

Exercise 1.8.10 (Shuffles). Show that for $\sigma \in \mathbb{S}_{p+q}$ the class $[\sigma] \in \mathbb{S}_{p+q}/\mathbb{S}_p \times \mathbb{S}_q$, contains one and only one (p, q)-shuffle.

Exercise 1.8.11 (Nonunital Hopf relation). Show that the restriction of the product and the reduced coproduct on the augmentation ideal $\overline{\mathscr{H}}$ of a bialgebra $\mathscr{H}$ satisfy the following relation:

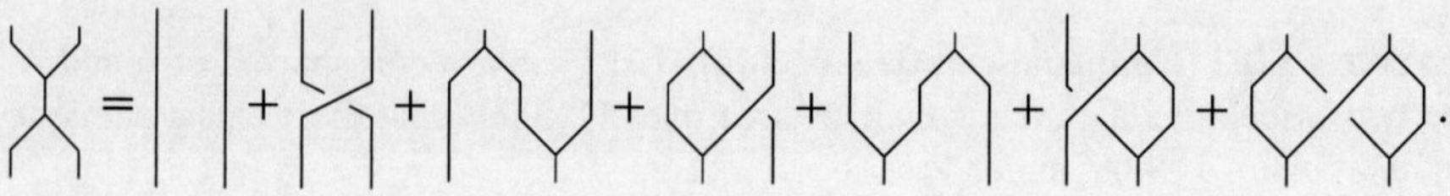

Exercise 1.8.12 (Unital infinitesimal bialgebra). Let $T(V)$ be the tensor algebra over V. Its product (concatenation) is denoted by μ. Let us denote by Δ the deconcatenation coproduct. Show that they satisfy the following compatibility relation:

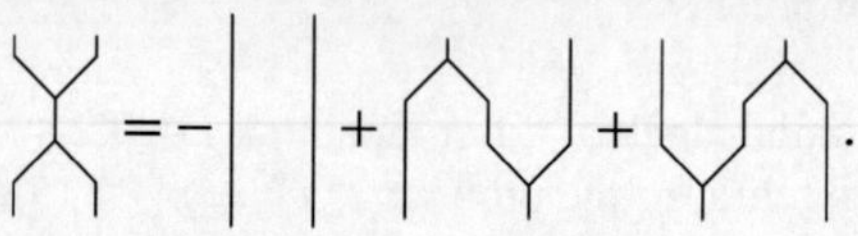

Let $\overline{T}(V)$ be the augmentation ideal and let $\bar{\Delta}$ be the reduced diagonal. What is the compatibility relation between μ and $\bar{\Delta}$ on $\overline{T}(V)$?

Exercise 1.8.13 (Baker–Campbell–Hausdorff ★). Show that the polynomials $H_n(x, y)$ appearing in the BCH formula

$$\exp(x)\exp(y) = \exp\bigl(x + y + \cdots + H_n(x, y) + \cdots\bigr)$$

can be computed out of the Eulerian idempotent $e_n^{(1)}$ (see for instance [Lod94]).

Exercise 1.8.14 (Relative tensor product ★). Let A be a $\mathbb{K}$-algebra, M be a right A-module and N be a left A-module. Show that the surjection map $\pi : M \otimes_{\mathbb{K}} N \to M \otimes_A N$ is the coequalizer (cokernel of the difference map):

$$M \otimes_{\mathbb{K}} A \otimes_{\mathbb{K}} N \rightrightarrows M \otimes_{\mathbb{K}} N \xrightarrow{\pi} M \otimes_A N$$

where the two maps on the left-hand side are using the right A-module structure of M and the left A-module structure of N respectively.

Chapter 2
Twisting Morphisms

... remember young fellow, Ω is left adjoint ...
Dale Husemöller, MPIM (Bonn),
private communication

In this chapter, we introduce the bar construction and the cobar construction as follows. A *twisting morphism* is a linear map $f : C \to A$, from a dga coalgebra C to a dga algebra A, which satisfies the Maurer–Cartan equation:

$$\partial(f) + f \star f = 0.$$

The set of twisting morphisms $\mathrm{Tw}(C, A)$ is shown to be representable both in C and in A. More precisely, the cobar construction is a functor Ω from dga coalgebras to dga algebras and the bar construction is a functor B from dga algebras to dga coalgebras which satisfy the following properties: there are natural isomorphisms

$$\mathrm{Hom}_{\mathsf{dga\ alg}}(\Omega C, A) \cong \mathrm{Tw}(C, A) \cong \mathrm{Hom}_{\mathsf{dga\ coalg}}(C, \mathrm{B}A).$$

As an immediate consequence the functors cobar and bar are adjoint to each other. Then we investigate the twisting morphisms which give rise to quasi-isomorphisms under the aforementioned identifications. We call them *Koszul morphisms*.

The main point is the following characterization of the Koszul morphisms. Any linear map $\alpha : C \to A$ gives rise to a map $d_\alpha : C \otimes A \to C \otimes A$, which is a differential if and only if α is a twisting morphism. Moreover, α is a Koszul morphism if and only if the chain complex $(C \otimes A, d_\alpha)$ is acyclic. This is the first step of Koszul duality theory, which will be treated in the next chapter.

As a corollary, it is shown that the unit and the counit of the bar–cobar adjunction

$$C \to \mathrm{B}\Omega C \quad \text{and} \quad \Omega \mathrm{B}A \to A,$$

are quasi-isomorphisms. Hence, the latter provides a canonical free resolution of A.

This chapter is inspired by H. Cartan [Car55], E. Brown [Bro59], J.C. Moore [Moo71], Husemoller–Moore–Stasheff [HMS74], A. Prouté [Pro86] and K. Lefèvre-Hasegawa [LH03].

J.-L. Loday, B. Vallette, *Algebraic Operads*,
Grundlehren der mathematischen Wissenschaften 346,
DOI 10.1007/978-3-642-30362-3_2,

2.1 Twisting Morphisms

We introduce the Maurer–Cartan equation in the convolution algebra. Its solutions are called twisting morphisms (sometimes called twisting cochains in the literature). To such a twisting morphism we associate a twisted structure on the convolution algebra and on the tensor product, thereby introducing the notion of twisted tensor product of chain complexes.

In this section (C, d_C) is a differential graded coaugmented coalgebra and (A, d_A) is a differential graded augmented algebra, where the differentials are both of degree -1.

2.1.1 Convolution in the DG Framework

We extend the result of Sect. 1.6.1 to graded vector spaces, that is $\mathrm{Hom}(C, A)$ is a graded associative algebra under the convolution product $\star$ (called *cup-product* in [HMS74]). The derivative ∂ of graded linear maps defined in p. 25 makes $\mathrm{Hom}(C, A)$ into a dg vector space.

Proposition 2.1.1. *The convolution algebra* $(\mathrm{Hom}(C, A), \star, \partial)$ *is a dga algebra.*

Proof. It suffices to prove that the derivative ∂ is a derivation for the convolution product $\star$. Let f and g be two maps of degree p and q respectively. We have

$$\begin{aligned}
\partial(f \star g) &= d_A \circ (f \star g) - (-1)^{p+q}(f \star g) \circ d_C \\
&= d_A \circ \mu \circ (f \otimes g) \circ \Delta - (-1)^{p+q} \mu \circ (f \otimes g) \circ \Delta \circ d_C \\
&= \mu \circ (d_A \otimes \mathrm{id} + \mathrm{id} \otimes d_A) \circ (f \otimes g) \circ \Delta \\
&\quad - (-1)^{p+q} \mu \circ (f \otimes g) \circ (d_C \otimes \mathrm{id} + \mathrm{id} \otimes d_C) \circ \Delta \\
&= \mu \circ \big((d_A \circ f) \otimes g + (-1)^p f \otimes (d_A \circ g) \\
&\quad - (-1)^p (f \circ d_C) \otimes g - (-1)^{p+q} f \otimes (g \circ d_C)\big) \circ \Delta \\
&= \mu \circ \big(\partial(f) \otimes g + (-1)^p f \otimes \partial(g)\big) \circ \Delta \\
&= \partial(f) \star g + (-1)^p f \star \partial(g).
\end{aligned}$$

□

2.1.2 Maurer–Cartan Equation, Twisting Morphism

In the dga algebra $\mathrm{Hom}(C, A)$ we consider the *Maurer–Cartan equation*

$$\partial(\alpha) + \alpha \star \alpha = 0.$$

By definition a *twisting morphism* (terminology of John Moore [Moo71], "fonctions tordantes" in H. Cartan [Car58]) is a solution $\alpha : C \to A$ of degree -1 of the

Maurer–Cartan equation, which is null when composed with the augmentation of A and also when composed with the coaugmentation of C.

We denote by $\mathrm{Tw}(C, A)$ the set of twisting morphisms from C to A. Recall from Sect. 1.1.9 that a graded associative algebra is a graded Lie algebra, with the graded bracket defined by $[a, b] := a \star b - (-1)^{|a|.|b|} b \star a$. When 2 is invertible in the ground ring $\mathbb{K}$, we have $\alpha \star \alpha = \frac{1}{2}[\alpha, \alpha]$, when α has degree -1. Therefore, the "associative" Maurer–Cartan equation, written above, is equivalent to the "classical" Maurer–Cartan equation $\partial(\alpha) + \frac{1}{2}[\alpha, \alpha] = 0$ in the Lie convolution algebra $(\mathrm{Hom}(C, A), [-, -])$.

Until the end of next section, we assume that the characteristic of the ground field is not equal to 2.

2.1.3 Twisted Structure on the Hom Space

Let $\alpha \in \mathrm{Hom}(C, A)$ be a map of degree -1. We define a *twisted derivation* ∂_α on $\mathrm{Hom}(C, A)$ by the formula

$$\partial_\alpha(f) := \partial(f) + [\alpha, f].$$

Lemma 2.1.2. *Let* $(\mathrm{Hom}(C, A), [,], \partial)$ *be the dg Lie convolution algebra. For any map* $\alpha \in \mathrm{Hom}(C, A)$ *of degree* -1 *the twisted derivation* $\partial_\alpha(x) := \partial(x) + [\alpha, x]$ *satisfies*

$$\partial_\alpha^2(x) = \big[\partial(\alpha) + \alpha \star \alpha, x\big].$$

Proof. We have

$$\begin{aligned}
\partial_\alpha^2(x) &= \partial_\alpha\big(\partial(x) + [\alpha, x]\big) \\
&= \partial^2(x) + \partial\big([\alpha, x]\big) + \big[\alpha, \partial(x)\big] + \big[\alpha, [\alpha, x]\big] \\
&= \big[\partial(\alpha), x\big] + \big[\alpha, [\alpha, x]\big] \quad (\partial \text{ is a derivation for } [,]) \\
&= \big[\partial(\alpha), x\big] + [\alpha \star \alpha, x] \quad (\text{graded Jacobi relation}) \\
&= \big[\partial(\alpha) + \alpha \star \alpha, x\big].
\end{aligned}$$

□

As a consequence, when α is a twisting morphism in $\mathrm{Hom}(C, A)$, the map ∂_α is a differential. We denote by $\mathrm{Hom}^\alpha(C, A) := (\mathrm{Hom}(C, A), \partial_\alpha)$ this chain complex.

Proposition 2.1.3. *Let* α *be a twisting morphism. The convolution algebra* $(\mathrm{Hom}^\alpha(C, A), \star, \partial_\alpha)$ *is a dga algebra.*

Proof. The twisted derivation ∂_α is the sum of a derivation ∂ with $[\alpha, -]$. Therefore, it is enough to prove that the latter is a derivation with respect to the convolution product $\star$:

$$\begin{aligned}&[\alpha, f]\star g+(-1)^p f\star[\alpha, g]\\&\quad=\alpha\star f\star g-(-1)^p f\star\alpha\star g+(-1)^p f\star\alpha\star g-(-1)^{p+q} f\star g\star\alpha\\&\quad=[\alpha, f\star g],\end{aligned}$$

for f of degree p and g of degree q. □

The dga algebras of the form $(\mathrm{Hom}^\alpha(C,A),\star,\partial_\alpha)$ are called *twisted convolution algebras*. We leave it to the reader to prove that $(\mathrm{Hom}^\alpha(C,A),[,],\partial_\alpha)$ is a dg Lie algebra *twisted* by the twisting morphism α.

2.1.4 Twisted Tensor Product

We saw in Sect. 1.6.2 that the differential on the free A-module (resp. cofree C-comodule) $C\otimes A$ is a derivation (resp. coderivation). Any map $\alpha: C\to A$ induces a unique (co)derivation on $C\otimes A$, which we denote by d^r_α here. Since C and A are dga (co)algebras, we consider the total (co)derivation

$$d_\alpha := d_{C\otimes A}+d^r_\alpha = d_C\otimes \mathrm{Id}_A+\mathrm{Id}_C\otimes d_A+d^r_\alpha.$$

So d_α is a perturbation of the differential of the tensor product.

Lemma 2.1.4. *The (co)derivation d_α satisfies*

$$d_\alpha{}^2 = d^r_{\partial(\alpha)+\alpha\star\alpha}.$$

Therefore, α satisfies the Maurer–Cartan equation if and only if the (co)derivation d_α satisfies $d_\alpha{}^2=0$.

Proof. The first relation comes from $d_\alpha{}^2=(d_{C\otimes A}+d^r_\alpha)^2=d_{C\otimes A}\circ d^r_\alpha+d^r_\alpha\circ d_{C\otimes A}+d^{r\,2}_\alpha$. We saw in Proposition 1.6.2 that $d^{r\,2}_\alpha=d^r_{\alpha\star\alpha}$. And we have $d_{C\otimes A}\circ d^r_\alpha+d^r_\alpha\circ d_{C\otimes A}=d^r_{d_A\circ\alpha+\alpha\circ d_C}=d^r_{\partial(\alpha)}$.

Hence, if $\alpha\in \mathrm{Tw}(C,A)$, then $d_\alpha{}^2=d^r_0=0$. Conversely, we notice that the restriction of d^r_f on $C\otimes \mathbb{K}1_A\to\mathbb{K}1_C\otimes A$ is equal to f. So if $d_\alpha{}^2=0$, then $\partial(\alpha)+\alpha\star\alpha=0$. □

From the preceding lemma, it follows that, when $\alpha: C\to A$ is a twisting morphism, there exists a chain complex

$$C\otimes_\alpha A := (C\otimes A, d_\alpha)$$

which is called the *(right) twisted tensor product* (or twisted tensor complex). Since the tensor product is symmetric, this construction is also symmetric in A and C. So we can define a left twisted tensor product $A\otimes_\alpha C$. Warning: even if the two underlying modules $C \otimes A$ and $A \otimes C$ are isomorphic, the left and the right twisted tensor products are not isomorphic as chain complexes in general. The twisting term of the differential is not symmetric; it uses one particular side of the coproduct of the coalgebra and one particular side of the product of the algebra but not the same ones. If C were cocommutative and if A were commutative, then they would be isomorphic. Since the two constructions are symmetric, we will only state the related properties for the right twisted tensor product in the rest of this chapter.

This construction is functorial both in C and in A. Let $g : A \to A'$ be a morphism of dga algebras and $f : C \to C'$ be a morphism of dga coalgebras. Consider $C \otimes_\alpha A$ and $C' \otimes_{\alpha'} A'$ two twisted tensor products. We say that the morphisms f and g are *compatible* with the twisting morphisms α and α' if $\alpha' \circ f = g \circ \alpha$. One can show that $f \otimes g : C \otimes_\alpha A \to C' \otimes_{\alpha'} A'$ is then a morphism of chain complexes.

In the weight-graded context, we require that the twisting morphisms preserve the weight. In this case, the following lemma states that if two among these three morphisms are quasi-isomorphisms, then so is the third one. This result first appeared in the Cartan seminar [Car55].

Lemma 2.1.5 (Comparison Lemma for twisted tensor product). *Let $g : A \to A'$ be a morphism of wdga connected algebras and $f : C \to C'$ be a morphism of wdga connected coalgebras. Let $\alpha : C \to A$ and $\alpha' : C' \to A'$ be two twisting morphisms, such that f and g are compatible with α and α'.*

If two morphisms among f, g and $f \otimes g : C \otimes_\alpha A \to C' \otimes_{\alpha'} A'$ (or $g \otimes f : A \otimes_\alpha C \to A' \otimes_{\alpha'} C'$) are quasi-isomorphisms, then so is the third one.

Proof. We postpone the proof to the end of the chapter (see Sect. 2.5). □

2.2 Bar and Cobar Construction

We construct the cobar and bar functors and we prove that they give representing objects for the twisting morphisms bifunctor $\mathrm{Tw}(-,-)$. As a consequence the bar and cobar functors form a pair of adjoint functors. The bar construction goes back to Samuel Eilenberg and Saunders Mac Lane [EML53] and the cobar construction goes back to Franck Adams [Ada56].

2.2.1 Bar Construction

We are going to construct a functor from the category of augmented dga algebras to the category of conilpotent dga coalgebras:

$$\mathrm{B} : \{\text{aug. dga algebras}\} \longrightarrow \{\text{con. dga coalgebras}\}$$

called the *bar construction*.

Let A be an augmented algebra: $A = \mathbb{K}1 \oplus \bar{A}$ (concentrated in degree 0) with product μ. The bar construction of A is a differential graded coalgebra defined on the cofree coalgebra $T^c(s\bar{A})$ over the suspension $s\bar{A} = \mathbb{K}s \otimes \bar{A}$ as follows. We denote it by $\mathrm{B}A$, using a slight but usual abuse of notation.

Consider the map $\Pi_s : \mathbb{K}s \otimes \mathbb{K}s \to \mathbb{K}s$ of degree -1 induced by $\Pi_s(s \otimes s) := s$. The restriction $\mu_{\bar{A}}$ of the product of the algebra A to $\bar{A}$ induces the following map

$$f : T^c(s\bar{A}) \twoheadrightarrow \mathbb{K}s \otimes \bar{A} \otimes \mathbb{K}s \otimes \bar{A} \xrightarrow{\mathrm{Id}\otimes\tau\otimes\mathrm{Id}} \mathbb{K}s \otimes \mathbb{K}s \otimes \bar{A} \otimes \bar{A} \xrightarrow{\Pi_s \otimes \mu_{\bar{A}}} \mathbb{K}s \otimes \bar{A}.$$

Since $T^c(s\bar{A})$ is cofree, by Proposition 1.2.2 there is a unique coderivation $d_2 : T^c(s\bar{A}) \to T^c(s\bar{A})$ which extends the map $f : T^c(s\bar{A}) \to s\bar{A}$:

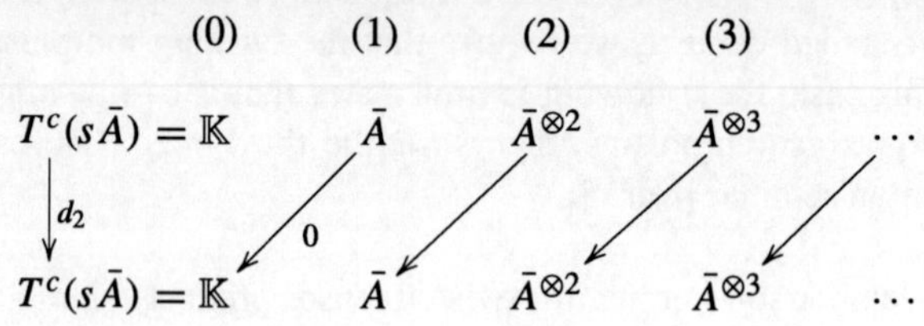

Proposition 2.2.1. *The associativity of μ implies that $(d_2)^2 = 0$, hence $(T^c(s\bar{A}), d_2)$ is a chain complex.*

Proof. We will give the proof in the dual case in Proposition 2.2.4. It is also a direct consequence of the next lemma. □

The complex $\mathrm{B}A := (T^c(s\bar{A}), d_2)$ is a conilpotent differential graded coalgebra, called the *bar construction* of the augmented graded algebra A. It is obviously a functor from the category of augmented graded algebras to the category of conilpotent differential graded coalgebras.

Lemma 2.2.2. *For any augmented associative algebra A, concentrated in degree* 0, *the bar complex of A can be identified with the nonunital Hochschild complex of $\bar{A}$:*

$$\cdots \to \bar{A}^{\otimes n} \xrightarrow{b'} \bar{A}^{\otimes n-1} \to \cdots \to \bar{A} \to \mathbb{K},$$

where $b'[a_1 \mid \ldots \mid a_n] = \sum_{i=1}^{n-1} (-1)^{i-1} [a_1 \mid \ldots \mid \mu(a_i, a_{i+1}) \mid \ldots \mid a_n]$.

Proof. Here we have adopted Mac Lane's notation $[a_1 \mid \ldots \mid a_n] \in \bar{A}^{\otimes n}$. Since $\bar{A}$ is in degree 0, the space $s\bar{A}$ is in degree 1 and $(s\bar{A})^{\otimes n}$ is in degree n. So the module of n-chains can be identified with $\bar{A}^{\otimes n}$. Let us identify the boundary map. Since d_2 is induced by the product and is a derivation, it has the form indicated in the statement. The signs come from the presence of the shift s. For instance:

$$\begin{aligned}[][a_1 \mid a_2 \mid a_3] = (sa_1, sa_2, sa_3) \mapsto & \big(d_2(sa_1, sa_2), sa_3\big) - \big(sa_1, d_2(sa_2, sa_3)\big) \\ &= \big[\mu(a_1, a_2) \mid a_3\big] - \big[a_1 \mid \mu(a_2, a_3)\big].\end{aligned}$$

The minus sign appears because d_2 "jumps" over sa_1 which is of degree one. □

In general, the formula is the same with only the following change of sign:

$$d_2(sa_1 \otimes \cdots \otimes sa_n) = \sum_{i=1}^{n-1} (-1)^{i-1+|a_1|+\cdots+|a_i|} sa_1 \otimes \cdots \otimes s\mu(a_i, a_{i+1}) \otimes \cdots \otimes sa_n.$$

One can extend this functor to the case where (A, d_A) is an augmented differential graded algebra. Indeed, the differential $d_A : A \to A$ induces a differential on $A^{\otimes n}$ by

$$d_1 := \sum_{i=1}^{n} (\mathrm{id}, \ldots, \mathrm{id}, d_A, \mathrm{id}, \ldots, \mathrm{id}).$$

We denote by d_1 the differential on $T^c(s\bar{A})$. Since μ_A is a morphism of differential graded vector spaces, one can check that d_1 and d_2 anticommute: $d_1 \circ d_2 + d_2 \circ d_1 = 0$. The chain complex associated to the total differential $d_1 + d_2$ is called the *bar construction* of the augmented differential graded algebra

$$\mathrm{B}A := \big(T^c(s\bar{A}), d_{\mathrm{B}A} = d_1 + d_2\big).$$

The analogous construction in algebraic topology (classifying space of a topological group) is also called bar construction and denoted by B.

Proposition 2.2.3. *For any quasi-isomorphism $f : A \to A'$ of augmented dga algebras, the induced morphism $\mathrm{B}f : \mathrm{B}A \to \mathrm{B}A'$ is a quasi-isomorphism.*

Proof. We consider the filtration on $\mathrm{B}A$ defined by

$$F_p\mathrm{B}A := \Big\{\sum sa_1 \otimes \cdots \otimes sa_n \mid n \leq p\Big\}.$$

It is stable under $d_{\mathrm{B}A}$, $d_1 : F_p \to F_p$ and $d_2 : F_p \to F_{p-1}$. This filtration is increasing, bounded below and exhaustive. Hence, the classical convergence theorem of spectral sequences (Theorem 1.5.1) applies. The first page is equal to

$$E^0_{pq}\mathrm{B}A = (F_p\mathrm{B}A)_{p+q}/(F_{p-1}\mathrm{B}A)_{p+q} \cong \big\{sa_1 \otimes \cdots \otimes sa_p \mid |a_1| + \cdots + |a_p| = q\big\}.$$

Finally $E^0_{p\bullet}(f) = (sf)^{\otimes p}$ is a quasi-isomorphism by Künneth formula. □

2.2.2 Cobar Construction

Analogously one can construct a functor from the category of coaugmented dga coalgebras to the category of augmented dga algebras:

$$\Omega : \{\text{coaug. dga coalgebras}\} \longrightarrow \{\text{aug. dga algebras}\}$$

called the *cobar construction*, as follows.

Let C be a coaugmented graded coalgebra: $C = \overline{C} \oplus \mathbb{K}1$ with coproduct Δ. The reduced coproduct $\bar{\Delta} : \overline{C} \to \overline{C} \otimes \overline{C}$ is defined by the equality $\Delta(x) = x \otimes 1 + 1 \otimes x + \bar{\Delta}(x)$ for any $x \in \overline{C}$. It is obviously coassociative and of degree 0.

Consider now $\mathbb{K}s^{-1}$ equipped with the diagonal map $\Delta_s(s^{-1}) := -s^{-1} \otimes s^{-1}$ of degree -1, see Exercise 2.7.3. Then, one defines a map f on $s^{-1}\overline{C} = \mathbb{K}s^{-1} \otimes \overline{C}$ as the composite

$$f : \mathbb{K}s^{-1} \otimes \overline{C} \xrightarrow{\Delta_s \otimes \bar{\Delta}} \mathbb{K}s^{-1} \otimes \mathbb{K}s^{-1} \otimes \overline{C} \otimes \overline{C} \xrightarrow{\mathrm{Id} \otimes \tau \otimes \mathrm{Id}} \mathbb{K}s^{-1} \otimes \overline{C} \otimes \mathbb{K}s^{-1} \otimes \overline{C}.$$

Consider the free algebra $T(s^{-1}\overline{C})$ over the desuspension $s^{-1}\overline{C}$. Since it is free, the degree -1 map $f : s^{-1}\overline{C} \to s^{-1}\overline{C} \otimes s^{-1}\overline{C}$ has a unique extension to $T(s^{-1}\overline{C})$ as a derivation by Proposition 1.1.2. We denote it by

$$d_2 : T\left(s^{-1}\overline{C}\right) \to T\left(s^{-1}\overline{C}\right),$$

$$\begin{array}{ccccccc}
 & (0) & (1) & (2) & (3) & \\
T(s^{-1}\overline{C}) = & \mathbb{K} & \overline{C} & \overline{C}^{\otimes 2} & \overline{C}^{\otimes 3} & \cdots \\
\downarrow d_2 & \searrow^{0} & \searrow & \searrow & \searrow & \\
T(s^{-1}\overline{C}) = & \mathbb{K} & \overline{C} & \overline{C}^{\otimes 2} & \overline{C}^{\otimes 3} & \cdots
\end{array}$$

Proposition 2.2.4. *The coassociativity of $\bar{\Delta}$ implies that $d_2 \circ d_2 = 0$ on $s^{-1}\overline{C}$. Therefore d_2 is a differential and $(T(s^{-1}\overline{C}), d_2)$ is a chain complex.*

Proof. For any $x \in \overline{C}$, let us write $\bar{\Delta}(x) = \sum x_{(1)} \otimes x_{(2)}$. We also adopt the notation

$$(\bar{\Delta} \otimes \mathrm{id})\bar{\Delta}(x) = \sum x_{(1)} \otimes x_{(2)} \otimes x_{(3)} = (\mathrm{id} \otimes \bar{\Delta})\bar{\Delta}(x).$$

We have defined

$$d_2\left(s^{-1}x\right) := -\sum (-1)^{|x_{(1)}|} s^{-1}x_{(1)} \otimes s^{-1}x_{(2)} \in \overline{C}^{\otimes 2}.$$

Let us prove that $d_2 \circ d_2 = 0$. Let $p := |x_{(1)}|$, $q := |x_{(2)}|$, $r := |x_{(3)}|$. The term $s^{-1}x_{(1)} \otimes s^{-1}x_{(2)} \otimes s^{-1}x_{(3)}$ coming from $(\bar{\Delta} \otimes \mathrm{id})\bar{\Delta}$ under $d_2 \circ d_2$ comes with the sign $(-1)^{p+q}(-1)^p$. Indeed, the first one comes from the application of the first copy of d_2, the second one comes from the application of the second copy of d_2. The term $s^{-1}x_{(1)} \otimes s^{-1}x_{(2)} \otimes s^{-1}x_{(3)}$ coming from $(\mathrm{id} \otimes \bar{\Delta})\bar{\Delta}$ under $d_2 \circ d_2$ comes with the sign $(-1)^{p+q}(-1)^{1+p+q}(-1)^q$. Indeed, the first one comes from the application of the first copy of d_2, the second one comes from the fact that d_2, which is of degree -1, jumps over a variable of degree $p-1$, the third one comes from the application of the second copy of d_2.

Adding up these two elements we get 0 as expected. □

By definition the *cobar construction* of the coaugmented graded coalgebra C is the augmented dga algebra

$$\Omega C := \big(T\big(s^{-1}\overline{C}\big), d_2\big).$$

It obviously gives a functor Ω from the category of coaugmented graded coalgebras to the category of augmented differential graded algebras.

One easily extends this functor to coaugmented differential graded coalgebras (C, Δ, d_C) by adding to d_2 the differential d_1 induced by the differential d_C. Since Δ is a morphism of chain complexes, d_1 and d_2 anticommute and one has a well-defined bicomplex. The chain complex associated to this total differential is called the *cobar construction* of the coaugmented coalgebra

$$\Omega C := \big(T\big(s^{-1}\overline{C}\big), d_{\Omega C} = d_1 + d_2\big).$$

The notation Ω is by analogy with the loop space construction in algebraic topology.

A nonnegatively graded dga coalgebra C is called 2-*connected* if $C_0 = \mathbb{K}1$ and $C_1 = 0$.

Proposition 2.2.5. *Let $f : C \to C'$ be a quasi-isomorphism between two 2-connected dga coalgebras. The induced morphism $\Omega f : \Omega C \to \Omega C'$ between the cobar constructions is a quasi-isomorphism.*

Proof. We consider the following filtration on the cobar construction

$$F_p \Omega C := \Big\{\sum s^{-1}c_1 \otimes \cdots \otimes s^{-1}c_n \mid n \geq -p\Big\}.$$

This increasing filtration is preserved by the differential of the cobar construction, $d_1 : F_p \to F_p$ and $d_2 : F_p \to F_{p-1}$. So the first term of the associated spectral sequence is equal to

$$\begin{aligned} E^0_{pq} &= (F_p\Omega C)_{p+q}/(F_{p-1}\Omega C)_{p+q} \\ &\cong \Big\{\sum s^{-1}c_1 \otimes \cdots \otimes s^{-1}c_p \big| |c_1| + \cdots + |c_p| = 2p + q\Big\}, \end{aligned}$$

with $d^0 = d_1$. Since $E^0_{p\bullet}(\Omega f) = (s^{-1}f)^{\otimes p}$, it is a quasi-isomorphism by Künneth formula. Since C (respectively C') is 2-connected, the degree of an element $s^{-1}c \in s^{-1}\overline{C}$ is at least 1 and $(F_p\Omega C)_n = 0$ for $p < -n$. The filtration being exhaustive and bounded below, this spectral sequence converges to the homology of the cobar construction by the classical convergence theorem of spectral sequences (Theorem 1.5.1), which concludes the proof. □

This result does not hold when the dga coalgebras are not 2-connected. We give a counterexample in Proposition 2.4.3. Beyond the 2-connected case, the relationship between the cobar construction and quasi-isomorphisms is more subtle. This question is fully studied in Sect. 2.4.

2.2.3 Bar–Cobar Adjunction

We show that the bar and cobar constructions form a pair of adjoint functors

$$\Omega : \{\text{con. dga coalgebras}\} \rightleftharpoons \{\text{aug. dga algebras}\} : \mathrm{B}.$$

More precisely, this adjunction is given by the space of twisting morphisms. When A is augmented and C coaugmented, a twisting morphism between C and A is supposed to send $\mathbb{K}$ to 0 and $\overline{C}$ to $\bar{A}$.

Theorem 2.2.6. *For every augmented dga algebra A and every conilpotent dga coalgebra C there exist natural bijections*

$$\mathrm{Hom}_{\mathsf{dga\ alg}}(\Omega C, A) \cong \mathrm{Tw}(C, A) \cong \mathrm{Hom}_{\mathsf{dga\ coalg}}(C, \mathrm{B}A).$$

Proof. Let us make the first bijection explicit. Since $\Omega C = T(s^{-1}\overline{C})$ is a free algebra, any morphism of algebras from ΩC to A is characterized by its restriction to $\overline{C}$ (cf. Proposition 1.1.1). Let φ be a map from $\overline{C}$ to A of degree -1. Define the map $\bar{\varphi} : s^{-1}\overline{C} \to A$ of degree 0 by the formula $\bar{\varphi}(s^{-1}c) := \varphi(c)$. Similarly, $\bar{\varphi}$ induces a unique morphism Φ of algebras from ΩC to A. The map Φ commutes with the differentials, meaning $d_A \circ \Phi = \Phi \circ (d_1 + d_2)$, or equivalently to $d_A \circ \varphi = -\varphi \circ d_C - \varphi \star \varphi$. Finally, we get $\partial(\varphi) + \varphi \star \varphi = 0$. Notice that the map φ lands in $\bar{A}$ since the map Φ is a morphism of augmented algebras.

The second bijection is given by the same method, so the rest of the proof is left to the reader as an exercise. Notice that we need the coalgebra C to be conilpotent in order to be able to extend a map $C \to sA$ into a morphism of coalgebras $C \to \mathrm{B}A = T^c(s\bar{A})$ (see Sect. 1.2.6). □

As a consequence of this proposition Ω and B form a pair of adjoint functors (Ω is left adjoint and B is right adjoint), which represent the bifunctor Tw.

2.2.4 Universal Twisting Morphisms

Several universal morphisms appear from this pair of adjoint functors. Applying Theorem 2.2.6 to $C = \mathrm{B}A$ we get the *counit* of the adjunction $\varepsilon : \Omega \mathrm{B}A \to A$ (see Appendix B.2.1) and the universal twisting morphism $\pi : \mathrm{B}A \to A$. Then applying Theorem 2.2.6 to $A = \Omega C$ we get the *unit* of the adjunction $\upsilon : C \to \mathrm{B}\Omega C$ (this is upsilon not v) and the universal twisting morphism $\iota : C \to \Omega C$.

By Theorem 2.2.6 the twisting morphisms π and ι have the following property.

Proposition 2.2.7. *Any twisting morphism $\alpha : C \to A$ factorizes uniquely through π and ι:*

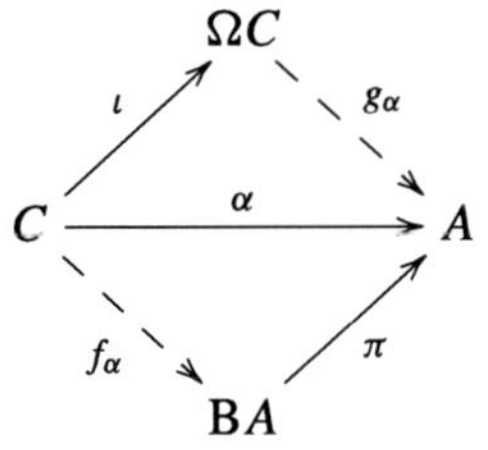

where g_α *is a dga algebra morphism and* f_α *is a dga coalgebra morphism.*

2.2.5 Augmented Bar and Cobar Construction

The *universal twisting morphism* $\pi : \mathrm{B}A = T^c(s\bar{A}) \twoheadrightarrow s\bar{A} \cong \bar{A} \rightarrowtail A$ gives rise to the twisted tensor product $\mathrm{B}A \otimes_\pi A$ (cf. Sect. 2.1.4). It is called the *augmented bar construction of* A.

Dually, the universal twisting morphism $\iota : C \twoheadrightarrow \overline{C} \cong s^{-1}\overline{C} \rightarrowtail T(s^{-1}\overline{C}) = \Omega C$ gives rise to the *coaugmented cobar construction of* C denoted $C \otimes_\iota \Omega C = (C \otimes \Omega C, d_\iota)$.

Proposition 2.2.8. *The chain complexes* $\mathrm{B}A \otimes_\pi A$ *(resp.* $A \otimes_\pi \mathrm{B}A$*) and* $C \otimes_\iota \Omega C$ *(resp.* $\Omega C \otimes_\iota C$*) are acyclic.*

Proof. Once made explicit, the chain complex is the nonunital Hochschild complex with coefficients in A whose module of n-chains is $\bar{A}^{\otimes n} \otimes A$ and whose boundary map is b' given by

$$b'\big([a_1 \mid \ldots \mid a_n]a_{n+1}\big) = \sum_{i=1}^{n-1} (-1)^{i-1}[a_1 \mid \ldots \mid a_i a_{i+1} \mid \ldots \mid a_n]a_{n+1} + (-1)^{n-1}[a_1 \mid \ldots \mid a_{n-1}]a_n a_{n+1}.$$

We consider the kernel K of the augmentation map

$$K \rightarrowtail \mathrm{B}A \otimes_\pi A \twoheadrightarrow \mathbb{K}.$$

It is immediate to check that the map $h : K \to K$ given by

$$[a_1 \mid \ldots \mid a_n]a_{n+1} \mapsto (-1)^n\big[a_1 \mid \ldots \mid a_n \mid a_{n+1} - \varepsilon(a_{n+1})\big]$$

is a homotopy from id_K to 0:

$$b'h + hb' = \mathrm{id}_K.$$

Hence the twisted tensor complex $\mathrm{B}A \otimes_\pi A$ is acyclic.

The proof for the other case is similar. □

2.3 Koszul Morphisms

We have just seen that the twisted tensor products associated to the two universal twisting morphisms π and ι are acyclic. When the twisted complex $C \otimes_\alpha A$, or equivalently $A \otimes_\alpha C$, happens to be acyclic, the twisting morphism α is called a *Koszul morphism.* We denote the set of Koszul morphisms by $\mathrm{Kos}(C, A)$.

In this section, we give the main theorem of this chapter which relates Koszul morphisms with bar and cobar resolutions. As a corollary, we prove that the unit and the counit of the bar–cobar adjunction are quasi-isomorphisms.

2.3.1 Koszul Criterion

Here we give the main result of this section, which is a criterion about Koszul morphisms. It comes from E. Brown's paper [Bro59].

Theorem 2.3.1 (Twisting morphism fundamental theorem). *Let A be a connected wdga algebra and let C be a connected wdga coalgebra. For any twisting morphism $\alpha : C \to A$ the following assertions are equivalent*:

1. *the right twisted tensor product $C \otimes_\alpha A$ is acyclic,*
2. *the left twisted tensor product $A \otimes_\alpha C$ is acyclic,*
3. *the dga coalgebra morphism $f_\alpha : C \xrightarrow{\sim} \mathrm{B}A$ is a quasi-isomorphism,*
4. *the dga algebra morphism $g_\alpha : \Omega C \xrightarrow{\sim} A$ is a quasi-isomorphism.*

Proof. Since we require A to be connected, we have $A = \bar{A} \oplus \mathbb{K}1$, where the elements of the augmentation ideal $\bar{A}$ have positive degree and positive weight. There is a similar statement for C. Recall that wdga (co)algebras were introduced in Sect. 1.5.10.

We first notice that the bar construction of a wgda connected algebra is a wgda connected coalgebra. And dually, the cobar construction of a wgda connected coalgebra is a wgda connected algebra. The weight of an element of $\mathrm{B}A$ is equal to the total weight $\omega(sa_1, \ldots, sa_k) = \omega(a_1) + \cdots + \omega(a_k)$.

We consider the commutative diagram of Sect. 2.2.4, where $f_\alpha : C \to \mathrm{B}A$, resp. $g_\alpha : \Omega C \to A$, is the morphism of wdga coalgebras, resp. algebras, associated to the twisting morphism α and respecting the weight grading. Notice that the universal twisting morphisms π and ι also preserve the weight.

(1) $\Leftrightarrow$ (3). Consider the tensor map $f_\alpha \otimes \mathrm{Id}_A : C \otimes A \to \mathrm{B}A \otimes A$. Since $\pi \circ f_\alpha = \alpha = \mathrm{Id}_A \circ \alpha$, the map $f_\alpha \otimes \mathrm{Id}_A$ is a morphism of chain complexes from $C \otimes_\alpha A$ to $\mathrm{B}A \otimes_\pi A$. We have seen in Proposition 2.2.8 that the augmented bar construction is always acyclic. Therefore, the twisted complex $C \otimes_\alpha A$ is acyclic if and only if $f_\alpha \otimes \mathrm{Id}_A$ is a quasi-isomorphism. The Comparison Lemma 2.1.5 implies that $C \otimes_\alpha A$ is acyclic if and only if f_α is a quasi-isomorphism.

(1) ⇔ (4). We use the same method with the tensor map $\mathrm{Id}_C \otimes g_\alpha : C \otimes_\iota \Omega(C) \to C \otimes_\alpha A$. Since $g_\alpha \circ \iota = \alpha = \alpha \circ \mathrm{Id}_C$, the map $\mathrm{Id}_C \otimes g_\alpha$ is a morphism of chain complexes. The acyclicity of the coaugmented cobar construction (Proposition 2.2.8) and the Comparison Lemma 2.1.5 imply that the twisted chain complex $C \otimes_\alpha A$ is acyclic if and only if g_α is a quasi-isomorphism.

The proof of the equivalence (2) ⇔ (3) ⇔ (4) is similar and uses the two other cases of Proposition 2.2.8 and Lemma 2.1.5. □

2.3.2 Bar–Cobar Resolution

We consider the counit $\varepsilon : \Omega \mathrm{B} A \to A$ and the unit $\upsilon : C \to \mathrm{B}\Omega C$ of the bar–cobar adjunction. The counit is a canonical resolution of A which is called the *bar–cobar resolution*. The following statement shows that it provides a quasi-free model for A, which is not minimal in general.

Corollary 2.3.2. *Let A be an augmented dga algebra and let C be a conilpotent dga coalgebra.*

The counit $\varepsilon : \Omega \mathrm{B} A \xrightarrow{\sim} A$ is a quasi-isomorphism of dga algebras. Dually, the unit $\upsilon : C \xrightarrow{\sim} \mathrm{B}\Omega C$ is a quasi-isomorphism of dga coalgebras.

Proof. We give a proof under the hypothesis that A (resp. C) is a connected wdga algebra (resp. connected wdga coalgebra). However the result holds in full generality (see [HMS74]). We apply Theorem 2.3.1 to the following diagram

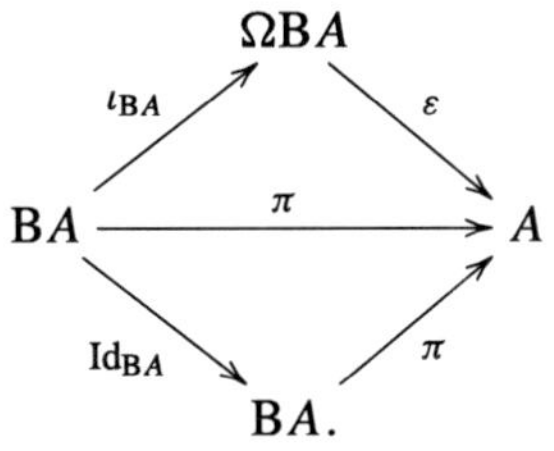

Since $\mathrm{Id}_{\mathrm{B}A}$ is an isomorphism, it follows that the counit ε is a quasi-isomorphism.

Following the same method, since $\mathrm{Id}_{\Omega C}$ is an isomorphism, the unit υ is a quasi-isomorphism. □

2.4 Cobar Construction and Quasi-isomorphisms

Using the previous results, we study the relationship between the cobar construction and quasi-isomorphisms. The main source of inspiration for this section is Lefèvre-Hasegawa's thesis [LH03].

To any dga coalgebra C, we consider the graded modules associated to the coradical filtration: $\mathrm{gr}_r\, C := F_r C / F_{r-1} C$. Let $f : C \to C'$ be a morphism of conilpotent dga coalgebras. Since the map f and the differentials preserve the coradical filtrations, f induces a morphism of chain complexes $[f] : \mathrm{gr}\, C \to \mathrm{gr}\, C'$ between the associated graded modules. If $[f]$ is a quasi-isomorphism, then f is called a *graded quasi-isomorphism.*

Proposition 2.4.1. *For any morphism $f : C \to C'$ of conilpotent dga coalgebras which is a graded quasi-isomorphism the induced morphism $\Omega f : \Omega C \xrightarrow{\sim} \Omega C'$ is a quasi-isomorphism.*

Proof. We consider the following grading for any element c in a conilpotent coalgebra C, $\mathrm{gr}\, c := \min\{r \mid c \in F_r C\}$. We consider the filtration of the cobar construction ΩC defined by

$$F_p \Omega C := \{s^{-1}c_1 \otimes \cdots \otimes s^{-1}c_n \mid \mathrm{gr}\, c_1 + \cdots + \mathrm{gr}\, c_n \leq p\}.$$

The increasing filtration is bounded below and exhaustive so the associated spectral sequence converges to the homology of ΩC. Its first term is equal to

$$E^0_{pq} \Omega C = (F_p \Omega C)_{p+q} / (F_{p-1} \Omega C)_{p+q} \cong (\Omega\, \mathrm{gr}\, C)^{(p)}_{p+q},$$

where

$$(\Omega\, \mathrm{gr}\, C)^{(p)} = \{s^{-1}c_1 \otimes \cdots \otimes s^{-1}c_n \mid \mathrm{gr}\, c_1 + \cdots + \mathrm{gr}\, c_n = p\}.$$

Hence $E^0(\Omega f) = \Omega[f]$, under the preceding notation. For any fixed p, we now prove that $E^0_{p\bullet}(\Omega C) \to E^0_{p\bullet}(\Omega C')$ is quasi-isomorphism. On $E^0_{p\bullet}(\Omega C)$, we define the filtration F_k as follows: an element $s^{-1}c_1 \otimes \cdots \otimes s^{-1}c_n$ is in F_k if and only if $n \geq -k$. This filtration is increasing. Since C is conilpotent the grading gr of the elements of $\overline{C}$ is strictly greater than 0, and we have $\mathsf{F}_{-p-1} = 0$. Since it is bounded below and exhaustive, the associated spectral sequence converges by Theorem 1.5.1. The first term $\mathsf{E}^0_{k\bullet}$ is isomorphic to the sub-module of $(s^{-1}\, \mathrm{gr}\, C)^{\otimes k}$ of grading p and degree $k + \bullet$ with differential d^0 induced by the differential of $\mathrm{gr}\, C$. The morphism f being a graded quasi-isomorphism, $\mathsf{E}^0(\Omega[f])$ is also a quasi-isomorphism by Künneth formula, which concludes the proof. □

2.4.1 Weak Equivalence

Any morphism $f : C \to C'$ of dga coalgebras, such that the induced morphism $\Omega f : \Omega C \to \Omega C'$ is a quasi-isomorphism, is called a *weak equivalence.*

Proposition 2.4.2. *Any* weak equivalence $f : C \to C'$ *of conilpotent dga coalgebras is a quasi-isomorphism.*

Proof. Since f is a weak equivalence, Ωf is a quasi-isomorphism and by Proposition 2.2.3, the morphism of dga coalgebras $\mathrm{B}\Omega f : \mathrm{B}\Omega C \to \mathrm{B}\Omega C'$ is a quasi-isomorphism. We conclude with the following commutative diagram, where all the maps are quasi-isomorphisms by Proposition 2.3.2

$$\begin{array}{ccc} C & \xrightarrow{\upsilon_C} & \mathrm{B}\Omega C \\ \downarrow f & & \downarrow \mathrm{B}\Omega f \\ C' & \xrightarrow{\upsilon_{C'}} & \mathrm{B}\Omega C'. \end{array}$$

□

In conclusion, the exact relationship between these notions is the following:

graded quasi-isomorphisms $\subseteq$ weak equivalences $\subsetneq$ quasi-isomorphisms.

Proposition 2.4.3. *There exist quasi-isomorphisms of dga coalgebras which are not weak equivalences.*

Proof. Let A be a unital dga algebra A, which is not acyclic. Consider its augmentation $A_+ := A \oplus \mathbb{K}1$, where 1 acts as a unit. The dga coalgebra $C := \mathrm{B}A_+ \cong T^c(sA)$ is isomorphic to $\mathbb{K} \oplus \mathrm{B}A \otimes_\pi A$. So it is quasi-isomorphic to the trivial dga coalgebra $\mathbb{K}$ by Proposition 2.2.8. But the cobar construction of $\mathbb{K}$ is acyclic, whereas the cobar construction $\Omega \mathrm{B}A_+$ is quasi-isomorphic to A_+ by Corollary 2.3.2, which is not acyclic. □

Notice that $C = \mathrm{B}A_+$ is connected but not 2-connected since C_1 contains $s1_A$, the suspension of the unit of A. So Proposition 2.2.5 does not hold for connected dga coalgebras in general. For 2-connected dga coalgebras, a quasi-isomorphism is a weak equivalence and vice versa.

2.5 Proof of the Comparison Lemma

In this section, we prove the Comparison Lemma 2.1.5 used in the proof of the fundamental theorem of twisting morphisms (Theorem 2.3.1). We assume here that the reader is familiar with the following notions of homological algebra: long exact sequences, cones, filtrations and spectral sequences. We refer the reader to any textbook on homological algebra, for instance [ML95] by Saunders MacLane.

Lemma 2.5.1 (Comparison Lemma for twisted tensor product, Cartan [Car55]). *Let $g : A \to A'$ be a morphism of wdga connected algebras and $f : C \to C'$ be a morphism of wdga connected coalgebras. Let $\alpha : C \to A$ and $\alpha' : C' \to A'$ be two twisting morphisms, such that f and g are compatible with α and α'.*

If two morphisms among f, g and $f \otimes g : C \otimes_\alpha A \to C' \otimes_{\alpha'} A'$ (or $g \otimes f : A \otimes_\alpha C \to A' \otimes_{\alpha'} C'$) are quasi-isomorphisms, then so is the third one.

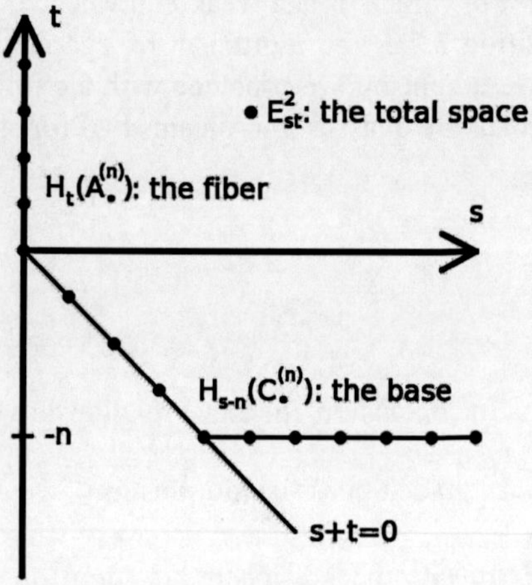

Fig. 2.1 The page E^2_{st} of the spectral sequence

Proof. Recall that the notion of weight-graded dga algebra was defined in Sect. 1.5.10. We denote by $M = \oplus_{n\geq 0} M^{(n)}$ (resp. $M' = \oplus_{n\geq 0} M'^{(n)}$) the weight-graded chain complex $C \otimes_\alpha A$ (resp. $C' \otimes_{\alpha'} A'$). We define a filtration F_s on $M^{(n)}$, where $n \in \mathbb{N}$ is the weight, by the formula

$$F_s(M^{(n)}) := \bigoplus_{d+m\leq s} (C_d^{(m)} \otimes A)^{(n)} = \bigoplus_{d+m\leq s} C_d^{(m)} \otimes A^{(n-m)}.$$

The differential d_α on $M = C \otimes_\alpha A$ is the sum of three terms $\mathrm{Id}_C \otimes d_A$, $d_C \otimes \mathrm{Id}_A$ and d^r_α. One has $\mathrm{Id}_C \otimes d_A : F_s \to F_s$, $d_C \otimes \mathrm{Id}_A : F_s \to F_{s-1}$ and $d^r_\alpha : F_s \to F_{s-2}$. Therefore, F_s is a filtration on the chain complex $M^{(n)}$. We consider the associated spectral sequence $\{E^\bullet_{st}\}_{s,t}$. One has

$$E^0_{st} = F_s(M^{(n)})_{s+t} / F_{s-1}(M^{(n)})_{s+t} = \bigoplus_{m=0}^{n} C^{(m)}_{s-m} \otimes A^{(n-m)}_{t+m}.$$

The study of the differential d_α on the filtration F_s of M shows that $d_0 = \mathrm{Id}_C \otimes d_A$ and that $d_1 = d_C \otimes \mathrm{Id}_A$. It follows that

$$E^2_{st} = \bigoplus_{m=0}^{n} H_{s-m}(C^{(m)}_\bullet) \otimes H_{t+m}(A^{(n-m)}_\bullet).$$

Since A and C are weight graded and connected, the part $m = 0$ is concentrated in $s = 0$ and $t \geq 0$, where it is equal to $E^2_{0t} = H_t(A^{(n)}_\bullet)$. The part $m = n$ is concentrated in $t = -n$ and $s \geq n$, where it is equal to $E^2_{s-n} = H_{s-n}(C^{(n)}_\bullet)$. For any $0 < m < n$, the nonvanishing part of $H_{s-m}(C^{(m)}_\bullet) \otimes H_{t+m}(A^{(n-m)}_\bullet)$ is in $s \geq 1$ and $t \geq -n+1$. See Fig. 2.1.

The filtration F_s is exhaustive $M^{(n)} = \bigcup_{s\geq 0} F_s(M^{(n)})$ and bounded below $F_{-1}(M^{(n)}) = \{0\}$, so the spectral sequence converges to the homology of $M^{(n)}$ by the classical convergence theorem 1.5.1:

$$E^{\infty}_{st}\big(M^{(n)}\big) \cong F_s\big(H_{s+t}\big(M^{(n)}\big)\big)/F_{s-1}\big(H_{s+t}\big(M^{(n)}\big)\big).$$

We consider the same filtration on M' and we denote by Φ the morphism of chain complexes $\Phi := f \otimes g$. We treat the three cases one after the other.

(1) *If f and g are quasi-isomorphisms, then $\Phi = f \otimes g$ is a quasi-isomorphism.*

For every s, t and n, the maps $E^2_{st}(M^{(n)}) \xrightarrow{H_\bullet(f)\otimes H_\bullet(g)} E^2_{st}(M'^{(n)})$ are isomorphisms. By the convergence of the two spectral sequences, the maps

$$E^{\infty}_{st}\big(M^{(n)}\big) \xrightarrow{\sim} E^{\infty}_{st}\big(M'^{(n)}\big)$$

are again isomorphisms. So the map Φ is a quasi-isomorphism.

(2) *If $\Phi = f \otimes g$ and g are quasi-isomorphisms, then f is a quasi-isomorphism.*

Let us work by induction on the weight n. When $n = 0$, the map $f^{(0)} : \mathbb{K} \to \mathbb{K}$, which is the identity, is a quasi-isomorphism. Suppose now that the result is true up to weight $n-1$. We consider the mapping cone of $\Phi^{(n)}$: $cone(\Phi^{(n)}) := s^{-1}M^{(n)} \oplus M'^{(n)}$ and the associated filtration $F_s(cone(\Phi^{(n)})) := F_{s-1}(M^{(n)}) \oplus F_s(M'^{(n)})$, which satisfies $E^1_{\bullet t}(cone(\Phi^{(n)})) = cone(E^1_{\bullet t}(\Phi^{(n)}))$. The long exact sequence of the mapping cone reads

$$\cdots \to H_{s+1}\big(cone\big(E^1_{\bullet t}\big(\Phi^{(n)}\big)\big)\big) \to H_s\big(E^1_{\bullet t}\big(M^{(n)}\big)\big)$$
$$\xrightarrow{H_s(E^1_{\bullet t}(\Phi^{(n)}))} H_s\big(E^1_{\bullet t}\big(M'^{(n)}\big)\big) \to H_s\big(cone\big(E^1_{\bullet t}\big(\Phi^{(n)}\big)\big)\big) \to \cdots.$$

Therefore there is a long exact sequence (ξ_t)

$$(\xi_t) \qquad \cdots \to E^2_{s+1t}\big(cone\big(\Phi^{(n)}\big)\big) \to E^2_{st}\big(M^{(n)}\big)$$
$$\xrightarrow{E^2_{st}(\Phi^{(n)})} E^2_{st}\big(M'^{(n)}\big) \to E^2_{st}\big(cone\big(\Phi^{(n)}\big)\big) \to \cdots$$

where $E^2_{st}(\Phi^{(n)})$ is given by $H_\bullet(f) \otimes H_\bullet(g)$.

When $t > -n$, we have seen that only $C^{(m)}$ (and $C'^{(m)}$) with $m < n$ are involved in E^2_{st}. In that case, since $E^2_{st}(M^{(n)}) = \bigoplus_{m=0}^{n-1} H_{s-m}(C^{(m)}_\bullet) \otimes H_{t+m}(A^{(n-m)}_\bullet)$, the induction hypothesis implies that

$$E^2_{st}\big(M^{(n)}\big) \xrightarrow{H_\bullet(f)\otimes H_\bullet(g)} E^2_{st}\big(M'^{(n)}\big)$$

is an isomorphism for every s and every $t > -n$. Using the long exact sequence (ξ_t) for $t > -n$, it gives $E^2_{st}\big(cone\big(\Phi^{(n)}\big)\big) = 0$ for every s and every $t \neq -n$. The collapsing of the spectral sequence $E^\bullet_{st}(cone(\Phi^{(n)}))$ at rank 2 implies the equality $E^\infty_{st}(cone(\Phi^{(n)})) = E^2_{st}(cone(\Phi^{(n)}))$. The convergence of the spectral sequence $E^\bullet_{st}(cone(\Phi^{(n)}))$ shows that

$$E^2_{st}\big(cone\big(\Phi^{(n)}\big)\big) = F_s\big(H_{s+t}\big(cone\big(\Phi^{(n)}\big)\big)\big)/F_{s-1}\big(H_{s+t}\big(cone\big(\Phi^{(n)}\big)\big)\big) = 0$$

since $\Phi^{(n)}$ is a quasi-isomorphism. Since $E^2_{s-n}(cone(\Phi^{(n)})) = 0$, the long exact sequence (ξ_{-n}) gives the isomorphism

$$H_{s-n}(C^{(n)}_\bullet) = E^2_{s-n}(M^{(n)}) \xrightarrow{H_\bullet(f)} E^2_{s-n}(M'^{(n)}) = H_{s-n}(C'^{(n)}_\bullet),$$

for every s. So f is a quasi-isomorphism as expected.

(3) *If $\Phi = f \otimes g$ and f are quasi-isomorphisms, then g is a quasi-isomorphism.*

Once again, we work by induction on the weight n. For $n = 0$, the map $g^{(0)} : \mathbb{K} \to \mathbb{K}$ is an isomorphism. Suppose that the result is true up to weight $n - 1$. When $s \geq 1$, we have seen that only $A^{(n-m)}$ (and $A'^{(n-m)}$) with $m > 0$ are involved in E^2_{st},

$$E^2_{st}(M^{(n)}) = \bigoplus_{m=1}^{n} H_{s-m}(C^{(m)}_\bullet) \otimes H_{t+m}(A^{(n-m)}_\bullet).$$

In this case, the induction hypothesis implies that $E^2_{st}(M^{(n)}) \xrightarrow{H_\bullet(f)\otimes H_\bullet(g)} E^2_{st}(M'^{(n)})$ is an isomorphism for every $s \geq 1$ and every t. The long exact sequence (ξ_t) shows that $E^2_{st}(cone(\Phi^{(n)})) = 0$ for $s \geq 2$ and every t. The spectral sequence of the cone of $\Phi^{(n)}$ converges to its homology, which is null since $\Phi^{(n)}$ is a quasi-isomorphism. Therefore, $E^2_{1,t-1}(cone(\Phi^{(n)})) = E^2_{0,t}(cone(\Phi^{(n)})) = 0$ for every t. This implies $E^2_{st}(cone(\Phi^{(n)})) = 0$ for every t and s. Finally, the beginning $(s = 0)$ of the exact sequence (ξ_t) gives the isomorphism

$$H_t(A^{(n)}_\bullet) = E^2_{0t}(M^{(n)}) \xrightarrow{H_\bullet(g)} E^2_{0t}(M'^{(n)}) = H_t(A'^{(n)}_\bullet).$$

So g is a quasi-isomorphism as expected. □

2.5.1 Comparison with Algebraic Topology

The Comparison Lemma is the algebraic avatar of the following result in algebraic topology. Let

$$\begin{array}{ccccc} F & \longrightarrow & X & \longrightarrow & B \\ \downarrow{\scriptstyle f_F} & & \downarrow{\scriptstyle f_X} & & \downarrow{\scriptstyle f_B} \\ F' & \longrightarrow & X' & \longrightarrow & B' \end{array}$$

be a morphism between two fibrations of simply-connected spaces. If two of the morphisms f_F, f_X, f_B are isomorphisms in homology, then so is the third. Using the Whitehead theorem it can be proved as follows: homology isomorphism is equivalent to homotopy isomorphism for simply-connected CW-complexes. When two of the morphisms are homotopy isomorphisms, then so is the third by the long

Serre exact homotopy sequence. Since a homotopy isomorphism is also a homology isomorphism, we are done.

In order to translate this result into homological algebra one needs some extra idea since the trick of passing to homotopy is not available anymore. The idea goes back to the Cartan seminar [Car55] and was later generalized to any first quadrant spectral sequence by Zeeman [Zee57]. This latter one applies to Leray–Serre spectral sequence of fiber spaces, whence the name *base* for the x-axis terms E^2_{s0} (E^2_{s-n} in the present proof) and *fiber* for the y-axis terms E^2_{0t}. More precisely, there is a twisting morphism between the singular chain complex $S(B)$ of the base space, which is a dg coalgebra, and the singular chain complex $S(F)$ of the fiber space which is a module over the algebra of the singular chain complex $S(\Omega B)$ of the loops of B. The induced twisted tensor product is shown to be quasi-isomorphic to the singular chain complex $S(X)$ of the total space, under certain hypotheses, by E.H. Brown in [Bro59]. The spectral sequence introduced in the core of this proof is an algebraic analog of the Leray–Serre spectral sequence.

2.6 Résumé

2.6.1 Twisting Morphism and Twisted Tensor Products

Convolution dga algebra: C dga coalgebra and A dga algebra:

$$\big(\operatorname{Hom}(C, A), \star, \partial\big),$$
$$f \star g = \mu \circ (f \otimes g) \circ \Delta, \qquad \partial(f) = d^A f - (-1)^{|f|} f d^C.$$

Twisting morphism, $\mathrm{Tw}(C, A)$*:* Solution of degree -1 to the Maurer–Cartan equation

$$\partial(\alpha) + \alpha \star \alpha \equiv \partial(\alpha) + \frac{1}{2}[\alpha, \alpha] = 0.$$

Any $\alpha \in \mathrm{Tw}(C, A)$ induces

▷ a twisted differential $\partial_\alpha := \partial + [\alpha, -]$ in $\operatorname{Hom}(C, A)$,
▷ a differential $d_\alpha := d_{C\otimes A} + d^r_\alpha$ on the tensor product $C \otimes A$ defining the right twisted tensor product $C \otimes_\alpha A$,
▷ a differential $d_\alpha := d_{A\otimes C} + d^l_\alpha$ on the tensor product $A \otimes C$ defining the left twisted tensor product $A \otimes_\alpha C$.

Table 2.1 summarizes this hierarchy of notions.

2.6.2 Bar and Cobar Constructions

Bar construction:

$$\mathrm{B}A := \big(T^c(s\bar{A}), d_1 + d_2\big), \qquad d_2(sx \otimes sy) = (-1)^{|x|} s(xy).$$

Table 2.1 Hierarchy of notions

$\alpha \in$	Determines:
$\mathrm{Hom}(C, A)_{-1}$	$d_\alpha : C \otimes A \to C \otimes A$
$\cup$	
$\mathrm{Tw}(C, A)$	chain complex $C \otimes_\alpha A$, $d_\alpha^2 = 0$
$\cup$	
$\mathrm{Kos}(C, A)$	acyclicity of $C \otimes_\alpha A$

Cobar construction:

$$\Omega C := \left(T\left(s^{-1}\overline{C}\right), d_1 + d_2\right), \qquad d_2\left(s^{-1}x\right) = -\sum (-1)^{|x_{(1)}|} s^{-1}x_{(1)} \otimes s^{-1}x_{(2)}.$$

Summary of Theorem 2.2.6 (second row) and Theorem 2.3.1 (third row):

$$\begin{array}{ccccc}
\mathrm{Hom}_{\mathsf{ga\ alg}}\left(T\left(s^{-1}\overline{C}\right), A\right) & \cong & \mathrm{Hom}(\overline{C}, \bar{A})_{-1} & \cong & \mathrm{Hom}_{\mathsf{ga\ coalg}}\left(C, T^c(s\bar{A})\right) \\
\cup & & \cup & & \cup \\
\mathrm{Hom}_{\mathsf{dga\ alg}}(\Omega C, A) & \cong & \mathrm{Tw}(C, A) & \cong & \mathrm{Hom}_{\mathsf{dga\ coalg}}(C, \mathrm{B}A) \\
\cup & & \cup & & \cup \\
\text{q-Iso}_{\mathsf{dga\ alg}}(\Omega C, A) & \cong & \mathrm{Kos}(C, A) & \cong & \text{q-Iso}_{\mathsf{dga\ coalg}}(C, \mathrm{B}A).
\end{array}$$

With $C = \mathrm{B}A$, we get

$$\Omega \mathrm{B}A \xrightarrow{\varepsilon} A \leftrightarrow \mathrm{B}A \xrightarrow{\pi} A \leftrightarrow \mathrm{B}A \xrightarrow{\mathrm{Id}} \mathrm{B}A,$$

and with $A = \Omega C$, we get

$$\Omega C \xrightarrow{\mathrm{Id}} \Omega C \leftrightarrow C \xrightarrow{\iota} \Omega C \leftrightarrow C \xrightarrow{\upsilon} \mathrm{B}\Omega C.$$

2.6.3 Universal Twisting Morphisms and Fundamental Theorem

Universal twisting morphisms: $\iota : C \to \Omega C$ and $\pi : \mathrm{B}A \to A$, which are Koszul.

Factorization of any twisting morphism $\alpha : C \to A$:

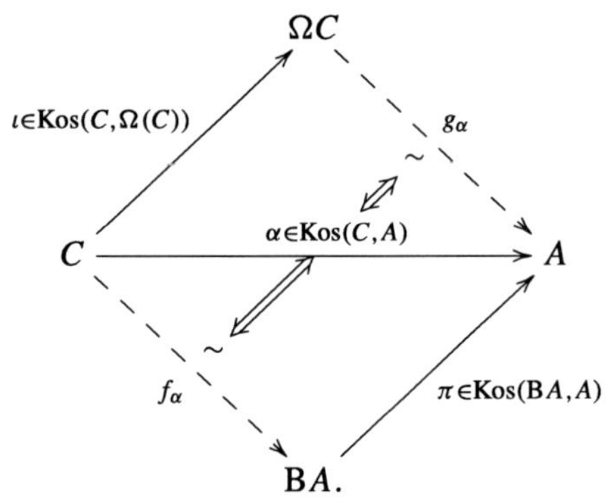

- ▷ $g_\alpha : \Omega C \to A$ morphism of dg algebras,
- ▷ $f_\alpha : C \to \mathrm{B}A$ morphism of dg coalgebras.

Twisting Morphisms Fundamental Theorem. *The following assertions are equivalent*

- ▷ *a twisting morphism $\alpha : C \to A$ is Koszul,*
- ▷ *the morphism of dg algebras $g_\alpha : \Omega C \xrightarrow{\sim} A$ is a quasi-isomorphism,*
- ▷ *the morphism of dg coalgebras $f_\alpha : C \xrightarrow{\sim} \mathrm{B}A$ is a quasi-isomorphism.*

Corollary. $\varepsilon : \Omega \mathrm{B}A \xrightarrow{\sim} A$ *and* $\upsilon : C \xrightarrow{\sim} \mathrm{B}\Omega C$.

2.6.4 Quasi-isomorphisms Under Bar and Cobar Constructions

Proposition. *The bar construction* B *preserves quasi-isomorphisms between dga algebras.*

Proposition. *The cobar construction* Ω *preserves quasi-isomorphisms between* 2-*connected dga coalgebras.*

Weak equivalence: $f : C \to C'$ such that $\Omega f : \Omega C \xrightarrow{\sim} \Omega C'$.

graded quasi-isomorphisms $\subseteq$ weak equivalences $\subsetneq$ quasi-isomorphisms.

2.7 Exercises

Exercise 2.7.1 (Convolution dga algebra). Draw a picture proof of Proposition 2.1.1, as in Proposition 1.6.2.

Exercise 2.7.2 (Bar construction as an algebra). We know that the cofree coalgebra can be endowed with a commutative algebra structure through the shuffle product, cf. Sect. 1.3.2. Show that the bar construction of a dga algebra is a dg commutative Hopf algebra.

Exercise 2.7.3 (Application of the sign rule). Let $\mathbb{K}s$ be the vector space of degree 1 equipped with the degree -1 product $s \otimes s \mapsto s$. Show that the transpose of the product on the linear dual $\mathbb{K}s^*$ is given by $\Delta(s^*) = -s^* \otimes s^*$.

Exercise 2.7.4 (Universal twisting morphism). Verify directly that $\iota : C \to \Omega C$ is a twisting morphism.

Exercise 2.7.5 (Functoriality). Prove that $\mathrm{Tw} : \textsf{dga coalg}^{\textsf{op}} \times \textsf{dga alg} \to \textsf{Set}$ is a bifunctor.

Exercise 2.7.6 (Cotangent complex). Let A be a dga algebra, C a dga coalgebra and let $\alpha : C \to A$ be a twisting morphism. We consider the following twisted differential on $A \otimes C \otimes A$, the free A-bimodule on C:

$$d_\alpha := d_{A\otimes C\otimes A} + \mathrm{Id}_A \otimes d^r_\alpha - d^l_\alpha \otimes \mathrm{Id}_A,$$

where

$$d^r_\alpha := (\mathrm{Id}_C \otimes \mu) \circ (\mathrm{Id}_C \otimes \alpha \otimes \mathrm{Id}_A) \circ (\Delta \otimes \mathrm{Id}_A),$$

and

$$d^l_\alpha := (\mu \otimes \mathrm{Id}_C) \circ (\mathrm{Id}_A \otimes \alpha \otimes \mathrm{Id}_C) \circ (\mathrm{Id}_A \otimes \Delta).$$

◇ Prove that ${d_\alpha}^2 = 0$.
We denote this chain complex by

$$A \otimes_\alpha C \otimes_\alpha A := (A \otimes C \otimes A, d_\alpha).$$

◇ Show that there is an isomorphism of chain complexes

$$\big(\mathrm{Hom}^\alpha(C, A), \partial_\alpha\big) \cong \big(\mathrm{Hom}_{A-\mathsf{biMod}}(A \otimes_\alpha C \otimes_\alpha A, A), \partial\big).$$

◇ Show that the following composite

$$\xi : A \otimes C \otimes A \xrightarrow{\mathrm{Id} \otimes \varepsilon \otimes \mathrm{Id}} A \otimes \mathbb{K} \otimes A \cong A \otimes A \xrightarrow{\mu} A$$

is a morphism of dg A-bimodules.
◇ Under the same weight grading assumptions as in Theorem 2.3.1, prove that $\xi : A \otimes_\alpha C \otimes_\alpha A \xrightarrow{\sim} A$ is a quasi-isomorphism if and only if α is a Koszul morphism.

Exercise 2.7.7 (Naturality). Prove that the bijections given in Theorem 2.2.6 are functorial in A and C.

Exercise 2.7.8 (Fundamental Theorem). Using the Comparison Lemma 2.1.5, prove directly the equivalence (2) $\Longleftrightarrow$ (3) of Theorem 2.3.1.

Exercise 2.7.9 (Unit of adjunction). Use the same kind of filtrations as in the proof of Proposition 2.4.1 to prove that the unit of adjunction $\upsilon : C \to \mathrm{B}\Omega C$ is a quasi-isomorphism, when C is a conilpotent dga coalgebra.

Chapter 3
Koszul Duality for Associative Algebras

In the process of its internal development and prompted by its inner logic, mathematics, too, creates virtual worlds of great complexity and internal beauty which defy any attempt to describe them in natural language but challenge the imagination of a handful of professionals in many successive generations.

Yuri I. Manin in "Mathematics as metaphor"

A minimal model for the associative algebra A is a quasi-free resolution $(T(W), d)$ such that the differential map d maps W into $\oplus_{n\geq 2} W^{\otimes n}$. We would like to find a method to construct this minimal model when A is quadratic, that is $A = T(V)/(R)$ where the ideal (R) is generated by $R \subset V^{\otimes 2}$ (this is the quadratic hypothesis). We will see that the quadratic data (V, R) permits us to construct explicitly a coalgebra $A^{¡}$ and a twisting morphism $\kappa : A^{¡} \to A$. Then, applying the theory of Koszul morphisms given in the previous chapter, we obtain a simple condition which ensures that the cobar construction on the Koszul dual coalgebra, that is $\Omega A^{¡}$, is the minimal model of A.

If one tries to construct by hand the space W, then one is led to take $W = V \oplus R \oplus (R \otimes V \cap V \otimes R) \oplus \cdots$. In fact, $\mathbb{K} \oplus V \oplus R \oplus (R \otimes V \cap V \otimes R)$ is the beginning of a certain sub-coalgebra of the cofree coalgebra over V, which is uniquely determined by V and R. This is precisely the expected coalgebra $A^{¡}$, up to suspension. The twisting morphism κ is simply the composite $A^{¡} \twoheadrightarrow V \rightarrowtail A$. The expected condition is the acyclicity of the Koszul complex $A^{¡} \otimes_\kappa A$. This is the Koszul duality theory for homogeneous quadratic algebras as introduced by Stewart Priddy in [Pri70]. In practice it is easier to work with algebras instead of coalgebras. When V is finite dimensional we consider the "graded linear dual" of $A^{¡}$ which is, up to suspension, a quadratic algebra $A^!$, usually called the *Koszul dual algebra* of A.

The quadratic hypothesis $R \subset V^{\otimes 2}$ can be weakened by only requiring $R \subset V^{\otimes 2} \oplus V$. In this case, we say that the algebra is *inhomogeneous quadratic*. We show how to modify the preceding method to handle the inhomogeneous quadratic case, also done in [Pri70]. Two examples are: the universal enveloping algebra $U(\mathfrak{g})$ of a Lie algebra $\mathfrak{g}$ (original example due to J.-L. Koszul) and the Steenrod algebra.

J.-L. Loday, B. Vallette, *Algebraic Operads*,
Grundlehren der mathematischen Wissenschaften 346,
DOI 10.1007/978-3-642-30362-3_3, © Springer-Verlag Berlin Heidelberg 2012

Inhomogeneous Koszul duality theory gives a proof of a general *Poincaré–Birkhoff–Witt theorem*, which, applied to $U(\mathfrak{g})$, gives the classical one.

In our treatment of Koszul duality of associative algebras, we keep algebras and coalgebras on the same footing. Working with coalgebras allows us to avoid the finite dimensional hypothesis. Moreover we give conceptual proofs so that they can be generalized to other monoidal categories. Our interest for Koszul duality of associative algebras is to serve as a paradigm for Koszul duality of algebraic operads.

Koszul algebras have applications in many fields of mathematics, which will not be discussed at all here (see the introduction of [PP05]). Classical references on Koszul duality of associative algebras include: S. Priddy [Pri70], Yu.I. Manin [Man87, Man88], R. Fröberg [Frö99], A. Polishchuk and L. Positselski [PP05].

3.1 Quadratic Data, Quadratic Algebra, Quadratic Coalgebra

We start with a quadratic data (V, R) to which we associate an algebra and a coalgebra

$$\begin{array}{ccc} & (V, R) & \\ \swarrow & & \searrow \\ A(V, R) & & C(V, R). \end{array}$$

In this chapter we suppose that $\mathbb{K}$ is a field, though most of the definitions and constructions are valid over a commutative ring.

3.1.1 Quadratic Data

By definition a *quadratic data* (V, R) is a graded vector space V and a graded subspace $R \subseteq V \otimes V$. A *morphism* of quadratic data $f : (V, R) \to (W, S)$ is a graded linear map $f : V \to W$ such that $(f \otimes f)(R) \subseteq S$.

3.1.2 Quadratic Algebra

The *quadratic algebra* $A(V, R) := T(V)/(R)$ is, by definition, the quotient of the free associative algebra over V by the two-sided ideal (R) generated by $R \subseteq V^{\otimes 2}$. In other words, $A(V, R)$ is the quotient of $T(V)$ which is *universal* among the quotient algebras A of $T(V)$ such that the composite

$$R \rightarrowtail T(V) \twoheadrightarrow A$$

is 0. It means that, for any such algebra A, there is a unique algebra morphism $A(V,R) \to A$ which makes the following diagram commutative:

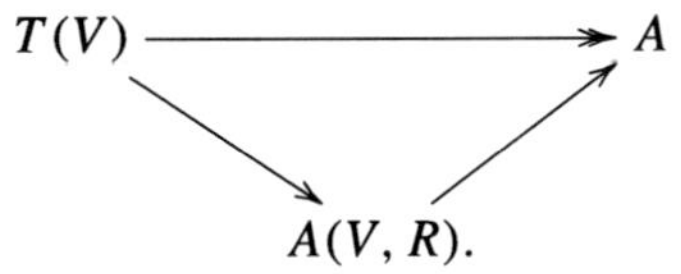

Since (R) is a homogeneous ideal, it follows that $A(V,R)$ is graded and augmented. This degree is called the *weight* and denoted as a superscript in parentheses. Explicitly it is given by:

$$A = \bigoplus_{n\in\mathbb{N}} A^{(n)} = \mathbb{K}1 \oplus V \oplus \left(V^{\otimes 2}/R\right) \oplus \cdots \oplus \left(V^{\otimes n}/ \sum_{i+2+j=n} V^{\otimes i} \otimes R \otimes V^{\otimes j}\right) \oplus \cdots.$$

Any basis of V is called a set of *generators* of A. Any basis $\{r_i\}$ of R determines a set of *relations* $r_i = 0$ in A. By abuse of terminology r_i, which should be called a *relator*, is often called a relation.

A morphism of quadratic data, $f : (V,R) \to (W,S)$ induces a natural morphism of weight graded algebras $A(V,R) \to A(W,S)$. Any morphism of algebras which respects the weight grading is of this form. But it is not the case for every morphism of algebras.

3.1.3 Quadratic Coalgebra

The *quadratic coalgebra* $C(V,R)$ is, by definition, the sub-coalgebra of the cofree coassociative coalgebra $T^c(V)$ which is *universal* among the sub-coalgebras C of $T^c(V)$ such that the composite

$$C \rightarrowtail T^c(V) \twoheadrightarrow V^{\otimes 2}/R$$

is 0. It means that, for any such coalgebra C, there is a unique coalgebra morphism $C \to C(V,R)$ which makes the following diagram commutative:

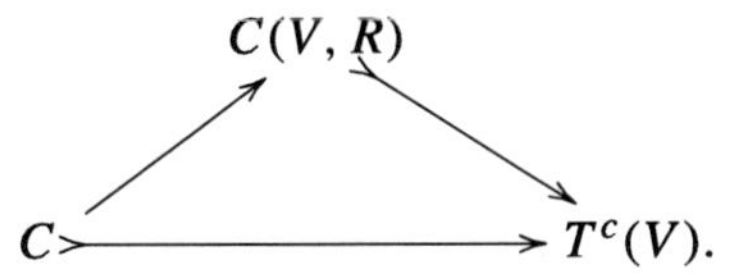

The coalgebra $C(V,R)$ is weight graded. Explicitly it is given by:

$$C = \bigoplus_{n\in\mathbb{N}} C^{(n)} = \mathbb{K}1 \oplus V \oplus R \oplus \cdots \oplus \left(\bigcap_{i+2+j=n} V^{\otimes i} \otimes R \otimes V^{\otimes j}\right) \oplus \cdots.$$

Observe that the restriction of the coproduct of C (that is the deconcatenation) to the weight 2 component $C^{(2)} = R$ is given by

$$r = r_1 \otimes r_2 \mapsto r \otimes 1 + r_1 \otimes r_2 + 1 \otimes r \in \left(V^{\otimes 2} \otimes \mathbb{K}\right) \oplus (V \otimes V) \oplus \left(\mathbb{K} \otimes V^{\otimes 2}\right).$$

We will say that $C(V, R)$ is *cogenerated* by V with *corelations* R in $T^c(V)$. Observe that the coalgebra $C(V, R)$ is conilpotent, cf. Sect. 1.2.4.

A morphism of quadratic data, $f : (V, R) \to (W, S)$ induces a natural morphism of weight graded coalgebras $C(V, R) \to C(W, S)$. Any morphism of coalgebras which respects the weight grading is of this form. But it is not the case for every morphism of coalgebras.

3.1.4 The Graded Framework

Both constructions $A(V, R)$ and $C(V, R)$ can be extended to the category of graded vector spaces. In this framework, V is a graded module and R is a graded submodule of the graded module $V^{\otimes 2}$. Then the algebra $A(V, R)$, resp. the coalgebra $C(V, R)$, is bigraded by degree and weight (cf. Sect. 1.5.1). Both $A(V, R)$ and $C(V, R)$ are connected weight graded in the sense of Sect. 1.5.10, with trivial differential.

3.2 Koszul Dual of a Quadratic Algebra

We construct the Koszul dual coalgebra and the Koszul dual algebra of a quadratic algebra. We work here in the homogeneous framework. The inhomogeneous framework, where it is only supposed that $R \subset V \oplus V^{\otimes 2}$, is treated in Sect. 3.6.

3.2.1 Koszul Dual Coalgebra of a Quadratic Algebra

Let (V, R) be a graded quadratic data. By definition the *Koszul dual coalgebra* of the quadratic algebra $A(V, R)$ is the coalgebra

$$A^{¡} := C\left(sV, s^2R\right),$$

where s^2R is the image of R in $(sV)^{\otimes 2}$ under the map $V^{\otimes 2} \to (sV)^{\otimes 2}$, $vw \mapsto svsw$. The upside down exclamation point ¡ (left exclamation point in the Spanish language) is usually pronounced "anti-shriek". If V is a graded space concentrated in degree 0, then sV is concentrated in degree 1. Observe that $C(sV, s^2R)$ is equal to $C(V, R)$ as a coalgebra. The decoration "s" is modifying the degree of the objects. It plays a role when we apply the Koszul sign rule to morphisms. We can omit it in the notation at the expense of changing the signs of the maps accordingly.

3.2.2 Koszul Dual Algebra of a Quadratic Algebra

The algebra obtained as the linear dual of the coalgebra $A^{¡}$ carries a desuspension sign. In the literature, one sometimes finds its unsuspended analog, denoted by $A^!$ and called the *Koszul dual algebra of the quadratic algebra* A. Explicitly it is defined by

$$(A^!)^{(n)} := s^n (A^{¡*})^{(n)}$$

and carries the obvious associative algebra structure.

Dualizing linearly the exact sequence

$$0 \to R \rightarrowtail V^{\otimes 2} \twoheadrightarrow V^{\otimes 2}/R \to 0,$$

provides the exact sequence

$$0 \leftarrow R^* \twoheadleftarrow (V^*)^{\otimes 2} \leftarrowtail R^{\perp} \leftarrow 0.$$

In other words the *orthogonal space* $R^{\perp}$ is defined as the image of $(V^{\otimes 2}/R)^*$ in $(V^*)^{\otimes 2}$ under the isomorphism $(V^{\otimes 2})^* \cong V^* \otimes V^*$, cf. Sect. 1.2.2.

Proposition 3.2.1. *The Koszul dual algebra* $A^!$ *admits the following quadratic presentation*

$$A^! = A(V^*, R^{\perp}).$$

Proof. First notice that the linear dual of the quadratic coalgebra $A^{¡} = C(sV, s^2R)$ is the quadratic algebra $A^{¡*} = A(s^{-1}V^*, s^{-2}R^{\perp})$. The last step can be proved either directly or by using the notion of Manin products of Sect. 4.5.1: the Koszul dual algebra is equal to $A^! = (A^{¡*}) \bigcirc T(s\mathbb{K}) = A(V^*, R^{\perp})$. □

3.2.3 Koszul Dual Algebra of a Coalgebra

It is also useful to introduce the *Koszul dual algebra* of a quadratic coalgebra

$$C^{¡} := A(s^{-1}V, s^{-2}R) \quad \text{for } C = C(V, R).$$

It follows immediately that

$$(A^{¡})^{¡} = A \quad \text{and} \quad (C^{¡})^{¡} = C.$$

As an immediate consequence we have, under finite dimensionality assumption:

$$(A^!)^! = A.$$

Observe that the coalgebra $A^{¡}$ is well-defined even in the graded framework and without any finiteness hypothesis.

3.2.4 Examples

1. Let V be a finite dimensional vector space and let $R = 0$. Then we have $A = T(V)$. Its Koszul dual algebra is the *algebra of dual numbers* $A^! = D(V^*) := \mathbb{K}1 \oplus V^*$, with trivial multiplication.
2. The symmetric algebra $S(V)$ is the quadratic algebra $T(V)/(R)$, where the space of relations R is the subvector space of $V^{\otimes 2}$ spanned by the elements $x \otimes y - y \otimes x$ for $x, y \in V$. The coalgebra $\Lambda^c(sV)$ is the subcoalgebra of $T^c(sV)$ satisfying the universal property of Sect. 3.1.3 with the subspace $s^2R = \langle sx \otimes sy - sy \otimes sx \mid x, y \in V\rangle$. Therefore, its component of weight n is equal to

$$\Lambda^c(sV)^{(n)} = \left\langle \sum_{\sigma \in \mathbb{S}_n} \operatorname{sgn}(\sigma) s^n x_{\sigma(1)} \otimes \cdots \otimes x_{\sigma(n)} \;\middle|\; x_1, \ldots, x_n \in V \right\rangle.$$

   The coalgebra structure is given by the deconcatenation coproduct and is cocommutative. When V is an n-dimensional vector space with basis $\{x_1, \ldots, x_n\}$ in degree 0, one gets the polynomial algebra $S(V) = \mathbb{K}[x_1, \ldots, x_n]$. In this case, its Koszul dual algebra is the exterior algebra $S(V)^! = \Lambda(V^*)$, since $R^\perp$ is spanned by the elements $x_i^* x_j^* + x_j^* x_i^*$, where $\{x_1^*, \ldots, x_n^*\}$ is the dual basis.
3. We refer to [Pri70, Man87, Man88, Frö99, PP05] for many more examples.

3.3 Bar and Cobar Construction on a Quadratic Data

We make explicit the dga coalgebra BA and the dga algebra ΩC in the quadratic case. The Koszul dual objects are shown to be equal to the syzygy degree 0 homology group in both cases.

3.3.1 Bar Construction on a Quadratic Algebra

The bar construction $\mathrm{B}A := T^c(s\bar{A})$ over the quadratic dga algebra $A = A(V, R)$ (whose differential is trivial) is equipped with a homological degree and a weight grading. We now introduce the *syzygy degree*.

The *weight grading* on BA is defined by the sum of the weight of each element: $\omega(sa_1, \ldots, sa_k) := \omega(a_1) + \cdots + \omega(a_k)$. Since A is a connected wgda algebra, the augmentation ideal $\bar{A}$ is concentrated in weight grading ≥ 1. We define another degree on $\bar{A}$ by the weight grading of A minus 1. It induces a new nonnegative degree on the bar construction, called the *syzygy degree* which is equal to $\omega(a_1) + \cdots + \omega(a_k) - k$. The component of syzygy degree d of BA is denoted by $\mathrm{B}^d A$, whereas the homological degree r component is denoted by $(\mathrm{B}A)_r$.

Since A has trivial internal differential, the differential on BA reduces to d_2, which raises the syzygy degree by 1 and preserves the weight grading. So it forms a cochain complex with respect to the syzygy degree, which splits with respect to the weight grading. Hence the associated cohomology groups will be bigraded, by the syzygy degree and by the weight grading.

The following diagram depicts this decomposition. The syzygy degree is indicated on the last row, so we delete the notation s for simplicity. We write $\bar{A} = V \oplus V^2/R \oplus V^3/(RV + VR) \oplus \cdots$ for more clarity, the tensor product notation being reserved for the one in $\mathrm{B}A$ only.

$$\begin{array}{ccccccccc}
\cdots & & \cdots & & \cdots & & \cdots & & (4) \\
0 & \leftarrow & V^3/(VR+RV) & \leftarrow & \big(V^2/R \otimes V\big) \oplus \big(V \otimes V^2/R\big) & \leftarrow & V \otimes V \otimes V & & (3) \\
 & & 0 & \leftarrow & V^2/R & \leftarrow & V \otimes V & & (2) \\
 & & & & 0 & \leftarrow & V & & (1) \\
 & & & & & & \mathbb{K} & & (0) \\
3 & & 2 & & 1 & & 0 & &
\end{array}$$

On the weight (3) row the map from $V \otimes V \otimes V$ is

$$u \otimes v \otimes w \mapsto [uv] \otimes w - u \otimes [vw]$$

where $[-]$ denotes the class in the quotient. The other map on this row is

$$\big([uv] \otimes w, u' \otimes [v'w']\big) \mapsto [uvw] + [u'v'w'].$$

From this description we see immediately that the syzygy degree 0 column forms the cofree coalgebra $T^c(sV)$. Hence the Koszul dual coalgebra $A^{¡} = \mathbb{K} \oplus sV \oplus s^2R \oplus \cdots$ is a subspace of this column.

Let $x \in A^{¡}$ and $\bar{\Delta}(x) = \sum x_{(1)} \otimes x_{(2)}$. The boundary map of $\Omega A^{¡}$ is given explicitly by the formula

$$d_2\big(s^{-1}x\big) = \sum (-1)^{|x_{(1)}|} s^{-1}x_{(1)} \otimes s^{-1}x_{(2)}.$$

The next proposition shows that $A^{¡}$ is equal to the kernel of the boundary map.

Proposition 3.3.1. *Let (V, R) be a quadratic data, $A = A(V, R)$ the quadratic algebra and $A^{¡} = C(sV, s^2R)$ its Koszul dual coalgebra. The natural coalgebra inclusion $i : A^{¡} \rightarrowtail \mathrm{B}A$ induces an isomorphism of graded coalgebras:*

$$i : A^{¡} \xrightarrow{\cong} H^0\big(\mathrm{B}^\bullet A\big), \quad \textit{i.e.} \quad A^{¡(n)} \cong H^0\big(\mathrm{B}^\bullet A\big)^{(n)} \quad \textit{for any } n.$$

Proof. We claim that, for each n, the inclusion $A^{¡(n)} \to (sV)^{\otimes n}$ is exactly the kernel of the horizontal differential, that is $H^0(\mathrm{B}^\bullet A)^{(n)}$. It is obvious for $n = 0$ and $n = 1$. For $n = 2$ the boundary map in $(\mathrm{B}A)^{(2)}$ is the quotient map $V^{\otimes 2} \to V^{\otimes 2}/R$, hence its kernel is R. More generally, since the boundary map is a derivation, it is given in degree 0 by the sum of the maps $(sV)^{\otimes n} \to (sV)^{\otimes i} \otimes sV^{\otimes 2}/R \otimes (sV)^{\otimes j}$. So the kernel is

$$\bigcap_{i+2+j=n} (sV)^{\otimes i} \otimes s^2 R \otimes (sV)^{\otimes j} = A^{¡(n)}. \qquad \square$$

3.3.2 Cobar Construction on a Quadratic Coalgebra

Like the bar construction, the cobar construction $\Omega C = T(s^{-1}\overline{C})$ over the quadratic dga coalgebra $C = C(V, R)$ (whose differential is trivial) has several gradings.

We introduce the same definitions as for the bar construction. We consider the *weight grading* $(\Omega C)^{(n)}$, which is the sum of the weights of the elements of $\overline{C}$. The *syzygy degree* of ΩC is induced by the weight of elements of $\overline{C}$ minus 1 in the same way. We denote it by $\Omega_d C$.

Since the internal differential of the coalgebra C is trivial, the differential of the cobar construction ΩC reduces to d_2, which lowers the syzygy degree by 1. Hence, $(\Omega_\bullet C, d_2)$ becomes a chain complex. Since the differential d_2 preserves the weight of the elements of C, this chain complex splits with respect to the weight: it is isomorphic to the following direct sum of sub-chain complexes $\Omega C \cong \bigoplus_{n\geq 0} (\Omega C)^{(n)}$.

The diagram below represents this weight decomposition. The syzygy degree is indicated on the last row, so we delete the notation s^{-1} for simplicity.

$$\begin{array}{ccccccccc}
\cdots & & \cdots & & \cdots & & \cdots & & (4) \\
0 \to & & VR \cap RV & \to & (V \otimes R) \oplus (R \otimes V) & \to & V \otimes V \otimes V & & (3) \\
 & & 0 & \to & R & \to & V \otimes V & & (2) \\
 & & & & 0 & \to & V & & (1) \\
 & & & & & & \mathbb{K} & & (0) \\
3 & & 2 & & 1 & & 0 & &
\end{array}$$

In degrees 0 and 1, the maps $R \to V^{\otimes 2}$ and $(V \otimes R) \oplus (R \otimes V) \to V^{\otimes 3}$ are simply the inclusions. The map $VR \cap RV \to (V \otimes R) \oplus (R \otimes V)$ is $\mathrm{inc}_1 - \mathrm{inc}_2$ where inc_1, resp. inc_2 is the inclusion of the first, resp. second, summand. From this description we see immediately that the syzygy degree 0 column forms the free algebra $T(s^{-1}V)$ and that the algebra

$$C^{¡} = \mathbb{K} \oplus s^{-1}V \oplus \left(s^{-1}V\right)^{\otimes 2}/s^{-2}R \oplus \cdots$$

is a quotient of it.

Proposition 3.3.2. *Let $C = C(V, R)$ be the quadratic coalgebra associated to the quadratic data (V, R), and let $C^{¡} := A(s^{-1}V, s^{-2}R)$ be its Koszul dual algebra.*

The natural algebra projection $p : \Omega C \twoheadrightarrow C^{¡}$ induces an isomorphism of graded algebras:

$$p : H_0(\Omega_\bullet C) \xrightarrow{\cong} C^{¡}, \quad \text{i.e.} \quad H_0(\Omega_\bullet C)^{(n)} \cong C^{¡(n)} \quad \text{for any } n.$$

Proof. The proof is analogous to the proof of Proposition 3.3.1. □

3.4 Koszul Algebras

For any quadratic data, we define a twisting morphism from the Koszul dual coalgebra to the quadratic algebra. This gives a twisted tensor product, called the *Koszul complex*, which plays a central role in the Koszul duality theory. We state and prove the main theorem of this chapter which says that the Koszul complex is acyclic if and only if the cobar construction over the Koszul dual coalgebra gives the minimal model of the algebra. The other definitions of a Koszul algebra appearing in the literature are given and we conclude with examples.

3.4.1 The Koszul Complex of a Quadratic Data

Starting with a quadratic data (V, R) we define $\kappa : C(sV, s^2R) \to A(V, R)$ as the linear map of degree -1 which is 0 everywhere except on V where it identifies sV to V:

$$\kappa : C\big(sV, s^2R\big) \twoheadrightarrow sV \xrightarrow{s^{-1}} V \rightarrowtail A(V, R).$$

Observe that the shift s in the definition of $A^{¡}$ makes κ a degree -1 map. The following result shows that κ is a twisting morphism.

Lemma 3.4.1. *We have $\kappa \star \kappa = 0$, and therefore $\kappa \in \mathrm{Tw}(A^{¡}, A)$.*

Proof. Since κ is 0 almost everywhere, the convolution product $\kappa \star \kappa$ is 0 except maybe on $V^{\otimes 2}$. Computing $\kappa \star \kappa$ explicitly on $V^{\otimes 2}$ we find that it is equal to the composite

$$C^{(2)} = R \to V \otimes V \to V^{\otimes 2}/R = A^{(2)},$$

hence it is 0 as expected.

So the map κ is a twisting morphism by Sect. 2.1.2. □

Proposition 3.4.2. *The twisting morphism $\kappa : C(sV, s^2R) \twoheadrightarrow V \rightarrowtail A(V, R)$ induces a map d_κ which makes*

$$A^{¡}c \otimes_\kappa A := \big(C\big(sV, s^2R\big) \otimes A(V, R), d_\kappa\big)$$

(*respectively $A \otimes_\kappa A^{¡}$*) *into a weight graded chain complex.*

Proof. The differential d_κ was constructed out of κ in Sect. 1.6.1. It is a differential by Lemmas 3.4.1 and 1.6.2. Since κ has degree -1 and weight 0, it is the same for the differential d_κ. Hence this chain complex splits with respects to the total weight. □

The chain complex $A^{¡} \otimes_\kappa A$ (resp. $A \otimes_\kappa A^{¡}$) is called the *Koszul complex*, or left Koszul complex (resp. right Koszul complex) of the quadratic algebra $A(V, R)$. Its summand $(A^{¡} \otimes_\kappa A)^{(n)}$ of weight (n) is equal to:

$$0 \to A^{¡(n)} \to A^{¡(n-1)} \otimes A^{(1)} \to \cdots \to A^{¡(1)} \otimes A^{(n-1)} \to A^{¡(n)} \to 0.$$

3.4.2 Koszul Criterion

In this section, we derive the main theorem of Koszul duality theory for associative algebras from the preceding chapter.

Proposition 3.4.3. *The maps corresponding to the twisting morphism $\kappa : A^{¡} \to A$ under the isomorphisms of Theorem 2.2.6 are exactly $i = f_\kappa : A^{¡} \rightarrowtail \mathrm{B}A$ and $p = g_\kappa : \Omega A^{¡} \twoheadrightarrow A$.*

Proof. By direct inspection. □

Theorem 3.4.4 (Koszul criterion). *Let (V, R) be a quadratic data. Let $A := A(V, R)$ be the associated quadratic algebra and let $A^{¡} := C(sV, s^2R)$ be the associated quadratic coalgebra. Then the following assertions are equivalent:*

1. *the right Koszul complex $A^{¡} \otimes_\kappa A$ is acyclic,*
2. *the left Koszul complex $A \otimes_\kappa A^{¡}$ is acyclic,*
3. *the inclusion $i : A^{¡} \rightarrowtail \mathrm{B}A$ is a quasi-isomorphism,*
4. *the projection $p : \Omega A^{¡} \twoheadrightarrow A$ is a quasi-isomorphism.*

When these assertions hold, the cobar construction on $A^{¡}$ gives a minimal resolution of A.

Proof. Theorem 2.3.1 can be applied to $A := A(V, R)$, $C := A^{¡} = C(sV, s^2R)$ and to $\alpha = \kappa$ since by Lemma 3.4.1 κ is a twisting morphism and since the connectivity and weight grading assumptions are satisfied.

Let us verify that $\Omega A^{¡}$ is the minimal model of A when the Koszul complex is acyclic. First, the dga algebra $\Omega A^{¡}$ is free as a graded algebra by construction (but not as a dga algebra). Second, its differential $d_{\Omega A^{¡}} = d_2$ satisfies the minimal hypothesis $d(W) \subset \bigoplus_{n \geq 2} W^{\otimes n}$ also by construction. Third, by Proposition 3.3.2 we have $H_0(\Omega_\bullet A^{¡}) = A$ and by (4) the resulting map $p : \Omega A^{¡} \twoheadrightarrow A$ is a quasi-isomorphism. □

Observe that starting with $C(V,R)$ instead of $A(V,R)$ with the following twisting morphism

$$C = C(V,R) \twoheadrightarrow V \xrightarrow{s^{-1}} s^{-1}V \rightarrowtail C^{¡} = A\left(s^{-1}V, s^{-2}R\right)$$

gives the same result up to a shift of grading. So we get Koszul duality theory for coalgebras.

3.4.3 Definition of a Koszul Algebra

A quadratic data (resp. quadratic algebra, resp. quadratic coalgebra) is said to be *Koszul* if its Koszul complex is acyclic.

By Theorem 3.4.4 we see that A is Koszul if and only if there is an isomorphism $A^{¡} \cong H^{\bullet}(\mathrm{B}A)$ (resp. $H_{\bullet}(\Omega A^{¡}) \cong A$). By Propositions 3.3.1 and 3.3.2, this is equivalent to the vanishing of the (co)homology groups: $H^{d}(\mathrm{B}^{\bullet}A)$ and $H_{d}(\Omega_{\bullet}A^{¡}) = 0$ for $d > 0$. More generally, a connected weight graded algebra A is said to be Koszul if the cohomology $H^{d}(\mathrm{B}^{\bullet}A) = 0$ of its bar construction is concentrated in syzygy degree $d = 0$. In this case, Exercise 3.8.1 shows that A admits a quadratic presentation. Therefore, there is no restriction to treat only the quadratic case.

The bar–cobar construction $\Omega\mathrm{B}A$ is always a resolution of A. To simplify it, one idea is to apply the cobar construction to the homology $H^{\bullet}(\mathrm{B}A)$ rather than to $\mathrm{B}A$. When A is Koszul, the homology of $\mathrm{B}A$ is exactly $A^{¡}$ and one gets the resolution $\Omega A^{¡}$ of A. For any quadratic algebra we have the following commutative diagrams:

$$\Omega A^{¡} \rightarrowtail \Omega\mathrm{B}A \xrightarrow{\sim} A$$

$$A^{¡} \overset{\sim}{\rightarrowtail} \mathrm{B}\Omega A^{¡} \twoheadrightarrow \mathrm{B}A$$

The algebra A is Koszul if and only if all these maps are quasi-isomorphisms by Corollary 2.3.2 and Theorem 3.4.4. Both $\Omega A^{¡}$ and $\Omega\mathrm{B}A$ are models of A and $\Omega A^{¡}$ is the minimal model.

With the aforementioned definitions, a quadratic algebra A is Koszul if and only if its Koszul dual coalgebra $A^{¡}$ is Koszul. The following proposition states the same property with the Koszul dual algebra.

Proposition 3.4.5. *Let (V,R) be a finite dimensional quadratic data. The quadratic algebra $A = A(V,R)$ is Koszul if and only if its Koszul dual algebra $A^{!} = A(V^{*}, R^{\perp})$ is Koszul.*

Proof. The left Koszul complex $A^{!} \otimes_{\kappa'} A^{!¡}$ associated to the twisting morphism $\kappa' : A^{!¡} \to A^{!}$ is made up of finite dimensional vector spaces in each degree and weight. Its linear dual is equal to the right Koszul complex $A^{¡} \otimes_{\kappa} A$, up to suspension.

Therefore one is acyclic if and only if the other one is acyclic and we conclude by Theorem 3.4.4. □

3.4.4 Other Equivalent Definitions

In the literature [Löf86, Frö99], one encounters the following equivalent definitions of a Koszul algebra.

Lemma 3.4.6. *Let $A = A(V, R)$ be a quadratic algebra. It is a Koszul algebra if and only if the homology of its bar construction $H^\bullet(\mathrm{B}A)$ is a sub-coalgebra of $T^c(sV)$.*

Proof. It is a direct consequence of Proposition 3.3.1 and Theorem 3.4.4. □

Let $A = A(V, R)$ be a finitely generated quadratic algebra. Recall that the derived Ext-functor $\mathrm{Ext}^\bullet_A(\mathbb{K}, \mathbb{K})$ is defined as the homology $H_\bullet(\mathrm{Hom}_A(R, \mathbb{K}))$, where $R \xrightarrow{\sim} \mathbb{K}$ is any projective resolution of $\mathbb{K}$ in the category of A-modules. It can be endowed with an associative algebra structure called the *Yoneda algebra*. Considering the quasi-free resolution $A \otimes_\iota \mathrm{B}A \xrightarrow{\sim} \mathbb{K}$, the Ext-functor can be computed by $\mathrm{Ext}^\bullet_A(\mathbb{K}, \mathbb{K}) = H_\bullet((\mathrm{B}A)^*)$, where $(\mathrm{B}A)^*$ is the degreewise and weightwise dual of $\mathrm{B}A$. Since it is the homology of the linear dual of a dga coalgebra, the Yoneda algebra structure is easily described.

Proposition 3.4.7. *A finitely generated quadratic algebra $A(V, R)$ is Koszul if and only if its Yoneda algebra $\mathrm{Ext}^\bullet_A(\mathbb{K}, \mathbb{K})$ is generated by its weight 1 elements.*

Proof. This proposition is linear dual to the previous lemma. □

Another equivalent definition of a Koszul algebra amounts to saying that the ground field $\mathbb{K}$ has a "linear minimal graded resolution of $\mathbb{K}$ with free A-modules". Such a resolution is provided by the Koszul complex $A \otimes_\kappa A^{¡}$. For the definitions of these terms and a proof of the equivalence, we refer the reader to [Frö99].

3.4.5 Examples

The symmetric algebra $S(V)$ and the exterior coalgebra $\Lambda^c(sV)$ are Koszul dual to each other. The tensor algebra and the dual numbers coalgebra are also Koszul dual to each other. Here are the proofs of the acyclicity of the associated Koszul complexes, which proves that they are Koszul.

Proposition 3.4.8. *The Koszul complex $(\Lambda^c(sV) \otimes S(V), d_\kappa)$ is acyclic.*

Proof. Though this statement is true over $\mathbb{Z}$, we will prove it only over a characteristic zero field. We represent any element $\sum_{\sigma\in\mathbb{S}_p} \operatorname{sgn}(\sigma)s^p x_{\sigma(1)} \otimes \cdots \otimes x_{\sigma(p)}$ of $\Lambda^c(sV)^{(p)}$ simply by $x_1 \wedge \cdots \wedge x_p$, keeping in mind that $x_1 \wedge \cdots \wedge x_p = \operatorname{sgn}(\sigma) x_{\sigma(1)} \wedge \cdots \wedge x_{\sigma(p)}$ holds for any $\sigma \in \mathbb{S}_p$, like in the Koszul dual algebra $\Lambda(V^*)$. (This identification is nothing but the isomorphism between $(\Lambda^c(sV))^*$ and $\Lambda(V^*)$, up to suspension.)

The boundary map

$$d = d_\kappa : \Lambda^c(sV)^{(p)} \otimes S(V)^{(q)} \longrightarrow \Lambda^c(sV)^{(p-1)} \otimes S(V)^{(q+1)}$$

is given by

$$d(x_1 \wedge \cdots \wedge x_p \otimes y_1 \cdots y_q) = \sum_{j=1}^{p} (-1)^{p-j} x_1 \wedge \cdots \wedge \widehat{x_j} \wedge \cdots \wedge x_p \otimes x_j y_1 \cdots y_q.$$

Define

$$h : \Lambda^c(sV)^{(p)} \otimes S(V)^{(q)} \longrightarrow \Lambda^c(sV)^{(p+1)} \otimes S(V)^{(q-1)}$$

by the formula

$$h(x_1 \wedge \cdots \wedge x_p \otimes y_1 \cdots y_q) := \sum_{i=1}^{q} x_1 \wedge \cdots \wedge x_p \wedge y_i \otimes y_1 \cdots \widehat{y_i} \cdots y_q.$$

One checks that $hd + dh = (p+q)\,\mathrm{id}$. Since we work in characteristic zero it shows that id is homotopic to 0 and therefore the complex is acyclic. □

Proposition 3.4.9. *For any graded vector space V, the Koszul complex $((\mathbb{K} \oplus V) \otimes T(V), d_\kappa)$ of the quadratic algebra $T(V)$ is acyclic.*

Proof. Since $T(V) = A(V, 0)$, we get $R = 0$ and therefore $C(V, R) \cong \mathbb{K} \oplus V$, where $\Delta(1) = 1 \otimes 1$, $\Delta(v) = v \otimes 1 + 1 \otimes v$.

The boundary map $d = d_\kappa$ of the Koszul complex $(\mathbb{K} \oplus V) \otimes T(V)$ is zero on the component $\mathbb{K} \otimes T(V)$ and is the identification of $V \otimes T(V)$ with $\mathbb{K} \otimes T(V)^{\geq 1}$ on the other component. Indeed, it is a consequence of the formulas for Δ and of $\kappa(1) = 0$, $\kappa(v) = v$.

So, the homology of the Koszul complex is $\operatorname{Ker} d / \operatorname{Im} d = T(V)/T(V)^{\geq 1} = \mathbb{K}$ concentrated in bidegree $(0, 0)$. Hence the Koszul complex is acyclic. □

3.5 Generating Series

Let (V, R) be a quadratic data such that V is finite dimensional. The weight-graded algebra $A(V, R) = \oplus_{n\geq 0} A^{(n)}$ is such that $A_0 = \mathbb{K}1$ and $A^{(n)}$ is finite dimensional.

By definition the *generating series* or *Hilbert–Poincaré series* of A is

$$f^A(x) := \sum_{n\geq 0} \dim A^{(n)} x^n.$$

Theorem 3.5.1. *If (V, R) is a finite dimensional quadratic data which is Koszul, then the following identity holds between the generating series of A and $A^!$:*

$$f^{A^!}(x) f^A(-x) = 1.$$

Proof. The Euler–Poincaré characteristic of the sub-chain complex of weight (n) of the Koszul complex of A is equal to $\sum_{k=0}^{n} (-1)^k \dim A^{(k)} \dim A^{¡(n-k)}$. By definition, it is equal to the coefficient of x^n of $f^{A^!}(x) f^A(-x)$. When the quadratic data (V, R) is Koszul, the Koszul complex is acyclic. It implies that the Euler–Poincaré characteristic is equal to 0, for $n > 0$, and it is equal to 1, for $n = 0$, which concludes the proof. □

Notice that one can also define the generating series of a quadratic coalgebra. In that case, we have $f^{A^¡} = f^{A^!}$.

Let us apply this theorem to the examples of Sect. 3.4.5. When the dimension of V is equal to k, we have

$$f^{T(V)}(x) = \frac{1}{1-kx} \quad \text{and} \quad f^{D(V^*)}(x) = 1 + kx,$$

which satisfy $f^{\mathbb{K}\oplus V}(x) f^{T(V)}(-x) = 1$. In the case of the symmetric algebra, we have

$$f^{S(V)}(x) = \frac{1}{(1-x)^k} \quad \text{and} \quad f^{\Lambda(V^*)}(x) = (1+x)^k,$$

which satisfy $f^{\Lambda(V^*)}(x) f^{S(V)}(-x) = 1$.

Theorem 3.5.1 provides a method to prove that an algebra is not Koszul. One first computes the Hilbert–Poincaré series $f^A(x)$ of the quadratic algebra A and then its inverse series $f^A(-x)^{-1}$. If this last one has at least one strictly negative coefficient, then it cannot be the series associated to a quadratic algebra. Therefore, the algebra A is not a Koszul algebra. (See [PP05, Sect. 2.2] for an example.) For a more exhaustive treatment of generating series, we refer the reader to [PP05, Ufn95].

If a chain complex is acyclic, then its Euler–Poincaré characteristic is equal to zero; but the converse is not true. This motivates us to look for quadratic algebras satisfying the functional equation of Theorem 3.5.1 but which fail to be Koszul. Such examples are given in [Pos95, Roo95, Pio01]. In the next section, we give a necessary and sufficient combinatorial condition for an algebra to be Koszul and in Sect. 4.3 we give a sufficient algebraic condition for an algebra to be Koszul.

3.6 Koszul Duality Theory for Inhomogeneous Quadratic Algebras

In the preceding sections, we dealt with Koszul duality of homogeneous quadratic algebras. In [Pri70] Priddy considered more general objects: inhomogeneous quadratic algebras with quadratic and linear relations. They are algebras whose relators contain not only quadratic terms but also possibly linear terms. The main example is the universal enveloping algebra of a Lie algebra: $U(\mathfrak{g}) = T(\mathfrak{g})/(R)$ where the relator is $[x, y] - x \otimes y + y \otimes x$. The purpose of this section is to adapt our treatment of Koszul duality theory to this more general framework. The modification consists in adding a suitable internal differential in the construction of the Koszul dual coalgebra.

There exists an even more general case allowing also constant terms in the space of relations, cf. [PP05].

3.6.1 Quadratic-Linear Algebra

A *quadratic-linear* data (V, R) is a graded vector space V together with a degree homogeneous subspace

$$R \subset V \oplus V^{\otimes 2}.$$

So, there may be linear terms in the space of relations. We still denote by $A = A(V, R) = T(V)/(R)$ the associated quotient. We consider $q : T(V) \twoheadrightarrow V^{\otimes 2}$ the projection onto the quadratic part of the tensor algebra. The image of R under q, denoted qR, is homogeneous quadratic, so (V, qR) is a quadratic data in the sense of Sect. 3.1. We denote by qA its associated algebra: $qA := A(V, qR)$. We assume that R satisfies the property

$$(ql_1) : R \cap V = \{0\}.$$

If it is not the case, by removing some elements of V one can choose another presentation of A which does satisfy (ql_1). This condition amounts to the minimality of the space of generators of A. Under this assumption, there exists a map $\varphi : qR \to V$ such that R is the graph of φ:

$$R = \{X - \varphi(X) \mid X \in qR\}.$$

For instance, if $A = U(\mathfrak{g})$, then $\varphi(x \otimes y - y \otimes x) = [x, y]$ and $qA = S\mathfrak{g}$. The weight grading on $T(V)$ induces a filtration which is compatible with the ideal (R). Hence the quotient algebra A is filtered by $\mathrm{F}_n A := \mathrm{Im}(\bigoplus_{k \leq n} V^{\otimes k})$. The assumption $R \cap V = \{0\}$ implies $\mathrm{F}_1 A = \mathbb{K} \oplus V$. We denote by $\mathrm{gr}\, A$ the graded algebra associated to the filtration of A, $\mathrm{gr}_n A := \mathrm{F}_n A/\mathrm{F}_{n-1} A$. We denote by

$$p : qA \twoheadrightarrow \mathrm{gr}\, A$$

the resulting epimorphism. It is obviously an isomorphism in weight 0 and 1, but not necessarily in weight 2. A corollary of the present theory shows that p is an isomorphism provided that qA is Koszul, see Theorem 3.6.4. In the example $A = U(\mathfrak{g})$ the map $p : S(\mathfrak{g}) \to \operatorname{gr} U(\mathfrak{g})$ is the PBW isomorphism.

3.6.2 Koszul Dual Coalgebra

The map φ permits us to construct the composite map

$$\tilde{\varphi} : (qA)^{\text{i}} = C\big(sV, s^2 qR\big) \twoheadrightarrow s^2 qR \xrightarrow{s^{-1}\varphi} sV.$$

By Proposition 1.2.2 there exists a unique coderivation, $d_{\tilde{\varphi}} : (qA)^{\text{i}} \to T^c(sV)$, which extends this composite.

Lemma 3.6.1.

(a) *If* $\{R \otimes V + V \otimes R\} \cap V^{\otimes 2} \subset qR$, *then the image of the coderivation* $d_{\tilde{\varphi}}$ *lives in* $(qA)^{i} = C(sV, s^2 qR) \subset T^c(sV)$, *thereby defining a coderivation* d_φ *of the coalgebra* $(qA)^{i}$.

(b) *If the condition*

$$(ql_2) : \{R \otimes V + V \otimes R\} \cap V^{\otimes 2} \subset R \cap V^{\otimes 2}$$

is satisfied, then the coderivation d_φ *squares to* 0.

Proof. If $\{R\otimes V + V\otimes R\} \cap V^{\otimes 2} \subset qR$, then we prove that $d_{\tilde{\varphi}}(C(sV, s^2qR)^{(3)}) \subset C(sV, s^2qR)^{(2)} = s^2qR$. The proof of the general case is done in the same way with the formula $(qA)^{\text{i}(n)} = \bigcap_{i+2+j=n} (sV)^{\otimes i} \otimes s^2qR \otimes (sV)^{\otimes j}$. Since $C(sV, s^2qR)^{(3)}$ is equal to $s^2qR \otimes sV \cap sV \otimes s^2qR$, any of its elements can be written $Y = \sum s^2X \otimes sv = \sum sv' \otimes s^2X'$, with $v, v' \in V$ and $X, X' \in qR$. The formula for the unique coderivation on the cofree coalgebra $T^c(sV)$ of Proposition 1.2.2 gives

$$\begin{aligned} d_{\tilde{\varphi}}(Y) &= \sum \tilde{\varphi}\big(s^2X\big) \otimes sv - \sum (-1)^{|v'|} sv' \otimes \tilde{\varphi}\big(s^2X'\big) \\ &= \sum \big(s\varphi(X) - s^2X\big) \otimes sv + \sum sv' \otimes \big(s^2X' - (-1)^{|v'|} s\varphi\big(X'\big)\big). \end{aligned}$$

Hence, forgetting the suspension for simplicity, we have

$$\begin{aligned} d_{\tilde{\varphi}}(Y) &= \sum \big(\varphi(X) - X\big) \otimes v + \sum v' \otimes \big(X' - \varphi\big(X'\big)\big) \\ &\in \{R \otimes V + V \otimes R\} \cap V^{\otimes 2} \subset qR = C(V, qR)^{(2)}. \end{aligned}$$

Since $\ker\varphi = R \cap V^{\otimes 2}$, we get $d_\varphi{}^2(C(V, qR)^{(3)}) = \{0\}$ if $\{R \otimes V + V \otimes R\} \cap V^{\otimes 2} \subset R \cap V^{\otimes 2}$. Once again, the proof of the general case follows from the same pattern using the explicit formula of the coalgebra $(qA)^{\text{i}}$. □

Since $R \cap V^{\otimes 2} \subset qR$, condition (ql_2) implies $\{R \otimes V + V \otimes R\} \cap V^{\otimes 2} \subset qR$. Condition (ql_2) amounts to saying that one cannot create new quadratic relations in R by adding an element to the relations of the presentation.

Let (V, R) be a quadratic-linear data satisfying the conditions (ql_1) and (ql_2). By definition the *Koszul dual dga coalgebra* of $A = A(V, R)$ is the dga coalgebra

$$A^{¡} := \left((qA)^{¡}, d_\varphi\right) = \left(C\left(sV, s^2 qR\right), d_\varphi\right).$$

3.6.3 Koszulity in the Inhomogeneous Quadratic Framework

A quadratic-linear data (resp. a quadratic-linear algebra) is said to be *Koszul* if it satisfies conditions (ql_1), (ql_2) and if the quadratic data (V, qR), or equivalently the quadratic algebra qA, is Koszul in the sense of Sect. 3.4.

Notice that for a homogeneous quadratic data, Koszul in the classical sense is Koszul in this sense. In this case, the conditions (ql_1), (ql_2) are trivially satisfied and the inner coderivation d_φ vanishes.

3.6.4 Cobar Construction in the Inhomogeneous Quadratic Framework

Under the hypotheses (ql_1) and (ql_2), we have constructed a conilpotent dga coalgebra $A^{¡}$. Applying the cobar construction of Sect. 2.2.2, we get a dga algebra $\Omega A^{¡}$, whose differential is of the form $d_1 + d_2$. The internal derivation d_1 is the unique derivation which extends d_φ. The derivation d_2 is induced by the coalgebra structure of $A^{¡}$.

We consider the same map κ in this context

$$\kappa : A^{¡} = C\left(sV, s^2 qR\right) \twoheadrightarrow sV \xrightarrow{s^{-1}} V \rightarrowtail A.$$

Lemma 3.6.2. *The map κ is a twisting morphism in* $\mathrm{Hom}(A^{¡}, A)$, *that is* $\partial(\kappa) + \kappa \star \kappa = 0$.

Proof. We refine the proof of Lemma 3.4.1, taking care of the internal differential d_φ of $A^{¡}$. The Maurer–Cartan equation becomes $-\kappa \circ d_\varphi + \kappa \star \kappa = 0$. The map $-\kappa \circ d_\varphi + \kappa \star \kappa$ is equal to 0 everywhere except on $(A^{¡})^{(2)} = s^2 qR$ where its image is $\{-\varphi(X) + X \mid X \in qR\} = R$, which vanishes in A. □

The twisting morphism κ induces a morphism of dga algebras $g_\kappa : \Omega A^{¡} \to A$ by Theorem 2.2.6.

Theorem 3.6.3. *Let A be an inhomogeneous quadratic Koszul algebra satisfying the conditions (ql_1) and (ql_2). Let $A^{¡} = ((qA)^{¡}, d_\varphi)$ be its Koszul dual dga coalgebra. The morphism of dga coalgebras $g_\kappa : \Omega A^{¡} \xrightarrow{\sim} A$ is a quasi-isomorphism.*

Proof. In this proof, we consider the cobar construction as a chain complex graded by the syzygy degree as in Sect. 3.3.2: both the internal differential d_1 of $\Omega A^{¡}$ induced by d_φ and the differential d_2 induced by the coproduct of the coalgebra $(qA)^{¡}$ lower the syzygy degree by 1. So we have a well-defined nonnegatively graded chain complex.

Since $(qA)^{¡}$ is a weight graded coalgebra, the underlying module $\Omega A^{¡} = T(s^{-1}\overline{(qA)^{¡}})$ of the bar construction is weight-graded. We consider the filtration F_r of $\Omega A^{¡}$ defined by its weight: the elements of F_r are the elements of weight less than r. The two components of the differential map $d = d_1 + d_2$ satisfy

$$d_2 : F_r \to F_r \quad \text{and} \quad d_1 : F_r \to F_{r-1}.$$

The filtration F_r is therefore stable under the boundary map d. Since it is bounded below and exhaustive, the associated spectral sequence $E^{\bullet}_{rs}$ converges to the homology of $\Omega A^{¡}$ by the classical convergence theorem of spectral sequences [ML95, Proposition 3.2, Chap. 11]. Hence, F_r induces a filtration F_r on the homology of $\Omega A^{¡}$ such that

$$E^{\infty}_{rs} \cong \mathrm{F}_r\big(H_{r+s}(\Omega A^{¡})\big)/\mathrm{F}_{r-1}\big(H_{r+s}(\Omega A^{¡})\big) =: \mathrm{gr}_r\big(H_{r+s}(\Omega A^{¡})\big).$$

The first term of this spectral sequence is equal to $E^0_{rs} = T(s^{-1}\overline{(qA)^{¡}})^{(r)}_{r+s}$, which is made up of the elements of syzygy degree equal to $r+s$ and grading equal to (r). The differential map d^0 is given by d_2. Since the algebra qA is Koszul, the spectral sequence is equal to $E^1_{rs} = qA^{(r)}$ at rank 1. More precisely E^1_{rs} is concentrated in the line $r+s=0$: $E^1_{rs} \cong qA^{(r)}$, for $r+s=0$ and $E^1_{rs}=0$, for $r+s \neq 0$. Therefore, the spectral sequence collapses at rank 1.

In conclusion, the convergence theorem gives

$$E^1_{r-r} \cong qA^{(r)} \cong E^{\infty}_{r-r} \cong \mathrm{gr}_r\big(H_0(\Omega A^{¡})\big),$$
$$E^1_{rs} \cong 0 \cong E^{\infty}_{rs} \cong \mathrm{gr}_r\big(H_{r+s}(\Omega A^{¡})\big), \quad \text{for } r+s \neq 0.$$

The result of Proposition 3.3.2 still holds in the inhomogeneous case, that is $H_0(\Omega A^{¡}) \cong A$, with the syzygy degree. Hence the quotient $\mathrm{gr}_r(H_0(\Omega A^{¡}))$ is equal to $\mathrm{gr}_r A$ and the morphism $\Omega A^{¡} \xrightarrow{\sim} A$ is a quasi-isomorphism. □

Notice that, in the inhomogeneous case, this resolution is not minimal because of the internal differential d_1.

3.6.5 Poincaré–Birkhoff–Witt Theorem

Theorem 3.6.4 (Poincaré–Birkhoff–Witt Theorem). *When a quadratic-linear algebra A is Koszul, then the epimorphism $p : qA \twoheadrightarrow \mathrm{gr}\, A$ is an isomorphism of graded* *algebras*

$$qA \cong \mathrm{gr}\, A.$$

Proof. This theorem was already proved in the proof of the previous theorem, where the convergence of the spectral sequence gave

$$E^1_{r-r} \cong qA^{(r)} \cong E^\infty_{r-r} \cong \mathrm{gr}_r A. \qquad \square$$

Another proof of this theorem, based on deformation theory, can be found in [BG96]. Even if the Poincaré–Birkhoff–Witt theorem is a direct consequence of the proof of Proposition 3.6.3, it has the following two nontrivial consequences: Corollary 3.6.5 and Proposition 3.6.6.

Corollary 3.6.5. *Let $A(V, R)$ be an algebra with the quadratic-linear presentation (V, R). If the quadratic algebra $qA = A(V, qR)$ is Koszul, then conditions (ql_1) and (ql_2) are equivalent to conditions*

$$(ql_1{}') : (R) \cap V = \{0\} \quad \textit{and} \quad (ql_2{}') : R = (R) \cap \{V \oplus V^{\otimes 2}\}.$$

Proof. Condition $(ql_1{}')$ is the generalization of condition (ql_1) from R to (R). Condition $(ql_2{}')$ is the generalization of condition (ql_2) from $R \otimes V + V \otimes R$ to (R). In the other way round, if conditions (ql_1) and (ql_2) are satisfied and if the algebra $A(V, R)$ is Koszul, then we get the Poincaré–Birkhoff–Witt isomorphism $qA \cong \mathrm{gr}\, A$ of Theorem 3.6.4. In weight 1, it implies condition $(ql_1{}')$. In weight 2, it implies $qR = q((R) \cap \{V \oplus V^{\otimes 2}\})$, which is equivalent to condition $(ql_2{}')$ by condition $(ql_1{}')$. $\square$

Conditions $(ql_1{}')$ and $(ql_2{}')$ amount to say that the ideal generated by R does not create any new quadratic-linear relation. It is equivalent to the maximality of the space of relations in the presentation of the inhomogeneous quadratic algebra. Such conditions can be hard to check in practice because one would have to compute the full ideal generated by R. But this proposition shows that if one finds a quadratic-linear presentation of an algebra satisfying conditions (ql_1), (ql_2) and whose homogeneous quadratic data is Koszul, then the space of relations R is maximal.

REMARK. This result is "Koszul dual" to the Diamond Lemma 4.2.4, since we work with the cobar construction Ω instead of the bar construction B in Sect. 4.2.4. Here it gives, under condition (ql_1),

$$(qA)^{¡} \text{ Koszul } \& \ (ql_2) \quad \Rightarrow \quad A^{¡} \text{ Koszul } \& \ (ql_2{}'),$$

where condition (ql_2) has to be seen as the particular case of condition $(ql_2{}')$ in weight 3. These two conditions refer to the ideal generated by R, whereas the condition of the Diamond Lemma refers to the quotient by some ideal associated to R. Also, in a similar way, we get the following isomorphism between the Koszul dual algebras (of the aforementioned coalgebras): $qA \cong \mathrm{gr}\, A \cong A$ as a direct byproduct. This result is better seen as a Diamond Lemma for Gröbner bases, see Sect. 4.3.7.

3.6.6 Acyclicity of the Koszul Complex

As in the quadratic case, the Koszul complex associated to an inhomogeneous Koszul algebra is acyclic.

Proposition 3.6.6. *When $A(V,R)$ is a quadratic-linear Koszul algebra, its Koszul complexes $A^{¡} \otimes_\kappa A$ and $A \otimes_\kappa A^{¡}$ are acyclic.*

Proof. We consider the Koszul complex as a chain complex graded by the weight of the elements of $A^{¡}$. The two parts $d_\varphi \otimes \mathrm{id}_A$ and d^r_κ of the differential map lower this degree by -1, so it is a well-defined chain complex.

The natural filtration on A plus the weight grading on $(qA)^{¡}$ induce an exhaustive and bounded below filtration F_r on $A^{¡} \otimes_\kappa A$. The differential maps satisfy $d^r_\kappa : F_r \to F_r$ and $d_\varphi \otimes \mathrm{id}_A : F_r \to F_{r-1}$. Therefore, E^0 is equal to $A^{¡} \otimes_{\tilde\kappa} \mathrm{gr}A$ where $\tilde\kappa : A^{¡} \to \mathrm{gr}A$ is the associated twisting morphism and where $d^0 = d^r_{\tilde\kappa}$.

By the Poincaré–Birkhoff–Witt Theorem, E^0 is equal to the twisted tensor product $(qA)^{¡} \otimes_{\tilde\kappa} qA$ of the Koszul quadratic algebra qA, with $\tilde\kappa : (qA)^{¡} \to qA$ being the Koszul twisting morphism. Therefore, it is acyclic and we conclude by the convergence theorem for spectral sequences [ML95, Proposition 3.2, Chap. 11]). □

In [Pri70], Priddy called *Koszul resolutions*, the resolution $A \otimes_\kappa A^{¡}$ (resp. $A^{¡} \otimes_\kappa A$) of $\mathbb{K}$ by free A-modules. They provide chain complexes, smaller than the augmented bar construction $A \otimes_\pi BA$, which allow one to compute the Tor functors $\mathrm{Tor}^A_\bullet(\mathbb{K}, M)$ for any A-module M (see [CE56, ML95] for the definition of Tor functors). In the example of the universal enveloping algebra of a Lie algebra, Priddy recovers the original Koszul resolution [CE56], which computes the Chevalley–Eilenberg homology of Lie algebras, see Sect. 3.6.7. Applied to restricted Lie algebras, this gives May resolutions [May66]. For the Steenrod algebra, it provides resolutions based on the Λ (co)algebra of [BCK+66], see Sect. 3.6.8 for more details.

Dually, the twisted convolution algebra $\mathrm{Hom}^\kappa(A^{¡}, A)$ computes the homology functors $\mathrm{Ext}^\bullet_A(\mathbb{K}, A)$ as in [BCK+66] (see Exercise 3.8.11).

3.6.7 The Example of the Universal Enveloping Algebra

The universal enveloping algebra of a Lie algebra $\mathfrak{g}$ is $U(\mathfrak{g}) := T(\mathfrak{g})/(x \otimes y - y \otimes x - [x,y])$. So it is defined as a quadratic-linear algebra with $V = \mathfrak{g}$. Its associated quadratic algebra is the symmetric algebra on $\mathfrak{g}$: $q(U(\mathfrak{g})) \cong S(\mathfrak{g})$.

Proposition 3.6.7. *When the characteristic of the ground field is not* 2, *the universal enveloping algebra $U(\mathfrak{g})$ of a Lie algebra $\mathfrak{g}$ is a Koszul algebra.*

Proof. A direct inspection shows that condition (ql_1) is satisfied. Let us prove that condition (ql_2) is also satisfied. The subspace $R \cap V^{\otimes 2}$ of $V^{\otimes 2}$ is equal to $\{\sum x \otimes y \mid$

$\sum[x,y]=0\}$. Let $\xi=\sum(x\otimes y\otimes z-y\otimes x\otimes z-[x,y]\otimes z)+\sum(t\otimes u\otimes v-t\otimes v\otimes u-t\otimes[u,v])$ be an element of $(R\otimes V+V\otimes R)$. It belongs to $V^{\otimes 2}$ if and only if $\sum(x\otimes y\otimes z-y\otimes x\otimes z)+\sum(t\otimes u\otimes v-t\otimes v\otimes u)=0$. In this case, applying $[[-,-],-]$ to this element, we get $2\sum[[x,y],z]+2\sum[[t,u],v]=0$. This proves that $\xi\in R\cap V^{\otimes 2}$ and that (ql_2) holds, when the characteristic of $\mathbb{K}$ is not 2. Finally, Proposition 3.4.8 shows that $S(\mathfrak{g})$ is a Koszul algebra, therefore $U(\mathfrak{g})$ is a Koszul algebra. □

Among other consequences, Theorem 3.6.4 can be applied and gives the "classical" Poincaré–Birkhoff–Witt theorem: there is an isomorphism of graded algebras

$$S(\mathfrak{g})\cong \operatorname{gr} U(\mathfrak{g}),$$

which is sometimes stated in terms of the monomial basis of the symmetric algebra.

Proposition 3.6.8. *The Koszul dual dga coalgebra of the universal enveloping algebra $U(\mathfrak{g})$ is the following dga coalgebra*

$$U(\mathfrak{g})^{¡}\cong\left(\Lambda^c(s\mathfrak{g}),d_\varphi\right),$$

where d_φ is the Chevalley–Eilenberg boundary map defining the homology of the Lie algebra $\mathfrak{g}$.

Proof. First, we have $q(U(\mathfrak{g}))^{¡}=S(\mathfrak{g})^{¡}=\Lambda^c(s\mathfrak{g})$. Recall that $\Lambda^c(s\mathfrak{g})$ is linearly spanned by the elements $\sum_{\sigma\in\mathbb{S}_n}\operatorname{sgn}(\sigma)s^n x_{\sigma(1)}\otimes\cdots\otimes x_{\sigma(n)}$, which we denote by $x_1\wedge\cdots\wedge x_n$. The internal differential d_φ is the unique coderivation which extends $\varphi: x\otimes y-y\otimes x\mapsto[x,y]$. Therefore it is equal to

$$d_\varphi(x_1\wedge\cdots\wedge x_n)=\sum_{i<j}(-1)^{i+j-1}[x_i,x_j]\wedge x_1\wedge\cdots\wedge\widehat{x_i}\wedge\cdots\wedge\widehat{x_j}\wedge\cdots\wedge x_n,$$

which is the Chevalley–Eilenberg differential [CE48, Kos50], see Sect. 13.2.7. □

Corollary 3.6.9. *The twisted tensor product $U(\mathfrak{g})\otimes_\kappa\Lambda^c(s\mathfrak{g})$ is a resolution of $\mathbb{K}$ by free $U(\mathfrak{g})$-modules.*

Proof. Direct corollary of Proposition 3.6.6 and Proposition 3.6.7. □

This is the original Koszul resolution which computes Chevalley–Eilenberg homology of Lie algebras [CE56].

3.6.8 The Example of the Steenrod Algebra

The *Steenrod algebra* $\mathscr{A}_2$ is the quadratic-linear algebra

$$\mathscr{A}_2:=A\left(\left\{Sq^i\right\}_{i\geq 1},R_{Adem}\right)$$

over the characteristic 2 field $\mathbb{K} = \mathbb{F}_2$, where $|Sq^i| = i$ and where R_{Adem} stands for the *Adem relations*

$$Sq^i Sq^j = \binom{j-1}{i} Sq^{i+j} + \sum_{k=1}^{[\frac{i}{2}]} \binom{j-k-1}{i-2k} Sq^{i+j-k} Sq^k, \quad \forall i, j > 0 \text{ with } i < 2j.$$

The quadratic analog $q\mathscr{A}_2$ is obtained by omitting the linear term $\binom{j-1}{i} Sq^{i+j}$. The images of the elements $\{Sq^{i_1} \cdots Sq^{i_k};\, i_l \geq 2i_{l+1}\}$ form a basis of $q\mathscr{A}_2$ and $\mathscr{A}_2$, called the *Cartan–Serre basis* of *admissible monomials*.

The degree-wise linear dual of the Koszul dual dga coalgebra $\mathscr{A}_2^{¡}$ is a dga algebra, which is anti-isomorphic to the *Λ algebra* of [BCK+66]. Notice that its homology gives the second page of the Adams spectral sequence which computes homotopy groups of spheres. The dga algebra Λ is generated by the elements $\{\lambda_i\}_{i \geq 0}$ of degree $|\lambda_i| = i$ and satisfies the relations

$$\lambda_i \lambda_{2i+1+j} = \sum_{k \geq 0} \binom{j-k-1}{k} \lambda_{i+j-k} \lambda_{2i+1+k}.$$

Its differential is the unique derivation extending

$$\lambda_j \mapsto \sum_{k \geq 0} \binom{j-k-1}{k+1} \lambda_{j-k-1} \lambda_k.$$

The mod-p Steenrod algebra can be treated in the same way. For more details, we refer the reader to [Wan67, Pri70].

3.7 Résumé

3.7.1 Quadratic Data and Koszul Dual Constructions

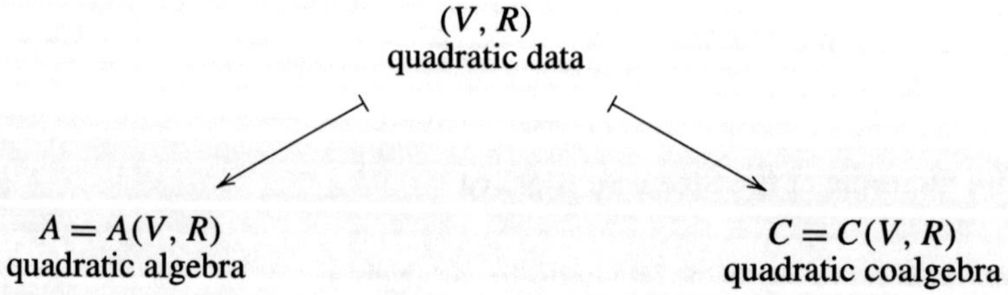

The quadratic algebra:

$$A = A(V,R) = T(V)/(R) \cong \bigoplus_{n\in\mathbb{N}} A^{(n)}$$
$$= \mathbb{K}1 \oplus V \oplus \left(V^{\otimes 2}/R\right) \oplus \cdots \oplus \left(V^{\otimes n}/\sum_{i+2+j=n} V^{\otimes i}\otimes R\otimes V^{\otimes j}\right) \oplus \cdots.$$

The quadratic coalgebra:

$$C = C(V,R) \subset T^c(V) \cong \bigoplus_{n\in\mathbb{N}} C^{(n)}$$
$$= \mathbb{K}1 \oplus V \oplus R \oplus \cdots \oplus \left(\bigcap_{i+2+j=n} V^{\otimes i}\otimes R\otimes V^{\otimes j}\right) \oplus \cdots.$$

Koszul dual coalgebra of an algebra:

$$A(V,R)^{¡} := C\left(sV, s^2R\right).$$

Koszul dual algebra of a coalgebra:

$$C(V,R)^{¡} := A\left(s^{-1}V, s^{-2}R\right), \qquad \left(A^{¡}\right)^{¡} \cong A.$$

Koszul dual algebra of an algebra: When V is finite dimensional, the linear dual of the desuspension of $A^{¡}$ is the quadratic algebra $A^! \cong A(V^*, R^\perp)$.

EXAMPLES: $T(V)^! \cong D(V^*)$ and $S(V)^! \cong \Lambda(V^*)$.

3.7.2 Koszul Duality Theory

Twisting morphism:

$$\kappa : A^{¡} = C\left(sV, s^2R\right) \twoheadrightarrow sV \xrightarrow{s^{-1}} V \rightarrowtail A(V,R) = A.$$

Koszul complexes: $A \otimes_\kappa A^{¡}$ and $A^{¡} \otimes_\kappa A$,

$$A^{¡} \rightarrowtail \mathrm{B}A \quad \text{and} \quad \Omega A^{¡} \twoheadrightarrow A,$$

with the *syzygy degree*: $H^0(\mathrm{B}^\bullet A) \cong A^{¡}$ and $H_0(\Omega_\bullet A^{¡}) \cong A$.

The quadratic data (V,R) is *Koszul* when one of the following equivalent assertions is satisfied.

1. the right Koszul complex $A^{¡} \otimes_\kappa A$ is acyclic,
2. the left Koszul complex $A \otimes_\kappa A^{¡}$ is acyclic,
3. the inclusion $i : A^{¡} \rightarrowtail \mathrm{B}A$ is a quasi-isomorphism,

4. the projection $p: \Omega A^{¡} \twoheadrightarrow A$ is a quasi-isomorphism,
5. $H^n(\mathrm{B}^\bullet A) = 0$ for $n \geq 1$,
6. $H_n(\Omega_\bullet A^{¡}) = 0$ for $n \geq 1$,
7. $H^\bullet(\mathrm{B}^\bullet A)$ is a sub-coalgebra of $T^c(sV)$,
8. the Yoneda algebra $\mathrm{Ext}_A(\mathbb{K}, \mathbb{K})$ is generated by its weight 1 elements [when V is finite dimensional].

EXAMPLES: $T(V)$, $D(V)$, $S(V)$, $\Lambda(V)$.

3.7.3 Generating Series or Hilbert–Poincaré Series

$$f^A(t) := \sum_{n\geq 0} \dim A^{(n)} t^n,$$

$$A \text{ Koszul} \quad \Longrightarrow \quad f^{A^!}(t) f^A(-t) = 1.$$

3.7.4 Inhomogeneous Koszul Duality Theory

Quadratic-linear data: (V, R), with $R \subset V \oplus V^{\otimes 2}$.

Quadratic analog: $qR := \mathrm{proj}_{V^{\otimes 2}}(R)$ and $qA := A(V, qR)$.

$$(ql_1): R \cap V = \{0\} \quad \Rightarrow \quad R = \mathrm{Graph}(\varphi : qR \to V),$$
$$(ql_2): \{R \otimes V + V \otimes R\} \cap V^{\otimes 2} \subset R \cap V^{\otimes 2}.$$

$(qA)^{¡} \twoheadrightarrow qR \xrightarrow{\varphi} V$ induces a coderivation $d_\varphi (qA)^{¡} \to T^c(V)$, (ql_1) and (ql_2) imply d_φ well-defined and $(d_\varphi)^2 = 0$.

Koszul dual dga coalgebra: $A^{¡} := ((qA)^{¡}, d_\varphi)$.

$A(V, R)$ *Koszul algebra* when (ql_1), (ql_2) and qA quadratic Koszul algebra. In this case:

- quasi-free resolution: $\Omega A^{¡} \xrightarrow{\sim} A$,
- Poincaré–Birkhoff–Witt theorem: $qA \cong \mathrm{gr}\, A$,
- Koszul complex: $A \otimes_\kappa A^{¡}$ acyclic.

EXAMPLE: $A = U(\mathfrak{g})$, universal enveloping algebra of a Lie algebra $\mathfrak{g}$,

- $U(\mathfrak{g})^{¡} = (\Lambda^c(s\mathfrak{g}), \text{Chevalley–Eilenberg differential})$,
- Original Poincaré–Birkhoff–Witt theorem: $S(V) \cong \mathrm{gr}\, U(\mathfrak{g})$,
- Original Koszul complex: $U(\mathfrak{g}) \otimes_\kappa \Lambda^c(s\mathfrak{g})$ acyclic.

EXAMPLE: $A = \mathscr{A}_2$, the mod-2 Steenrod algebra,

- Cartan–Serre basis,
- the dga algebra $(\mathscr{A}_2{}^{¡})^*$ is the Λ algebra.

3.8 Exercises

Exercise 3.8.1 (Koszul implies quadratic). Let A be a connected weight graded algebra (see Sect. 1.5.10). Its bar construction $\mathrm{B}A$ splits with respect to the weight and we consider the same syzygy degree as in Sect. 3.3.1. Show that if the homology of $\mathrm{B}A$ is concentrated in syzygy degree 0, then the algebra has a quadratic presentation.

Exercise 3.8.2 (Two-sided Koszul complex). Let (V, R) be a quadratic data. Under the notation of Sect. 2.1.4, we define the *two-sided Koszul complex* on $A \otimes A^{\text{¡}} \otimes A$ by the differential $d_\kappa^l \otimes \mathrm{Id}_A + \mathrm{Id}_A \otimes d_\kappa^r$ and we denote it by $A \otimes_\kappa A^{\text{¡}} \otimes_\kappa A$. Show that the quadratic data is Koszul if and only if the morphism of dg A-bimodules

$$A \otimes A^{\text{¡}} \otimes A \xrightarrow{\mathrm{Id}_A \otimes \varepsilon \otimes \mathrm{Id}_A} A \otimes \mathbb{K} \otimes A \cong A \otimes A \xrightarrow{\mu} A$$

is a resolution of A.

(★) When (V, R) is a quadratic-linear data satisfying conditions (ql_1) and (ql_2), we add the term $\mathrm{Id}_A \otimes d_\varphi \otimes \mathrm{Id}_A$ to the differential defining the two-sided Koszul complex. Prove the same result in this case.

Exercise 3.8.3 (Dual numbers algebra). Show that, for the quadratic algebra of dual numbers $A = \mathbb{K}[\varepsilon]/(\varepsilon^2 = 0)$, with ε of degree 0, the cobar construction of $A^{\text{¡}}$ is isomorphic to the dga algebra $\Omega A^{\text{¡}} = \mathbb{K}\langle t_1, t_2, \ldots, t_n, \ldots\rangle$, where $|t_n| = n - 1$ and $d(t_n) = -\sum_{i+j=n}(-1)^i t_i t_j$.

Exercise 3.8.4 (Inhomogeneous algebra ★). Let A be an inhomogeneous quadratic algebra. Show that if A is Koszul, then $f_\kappa : A^! \to \mathrm{B}A$ is a quasi-isomorphism of dga coalgebras.

Exercise 3.8.5 (Koszul complex of the symmetric algebra). Prove that $(\Lambda^c(sV) \otimes S(V), d_\kappa)$ is acyclic over $\mathbb{Z}$.

HINT. Use a suitable filtration.

Exercise 3.8.6 (Koszul complexes ★). Consider the three functors S, Λ and Γ (cf. Exercise 1.8.6). Show that there are acyclic complexes $\Lambda \otimes S$, $\Gamma \otimes \Lambda$. Show that $S \otimes \Lambda$ is not acyclic in characteristic p and defines the Cartier homomorphism, see for instance [Pir02b].

Exercise 3.8.7 (Koszul complex in local cohomology ★). Let A be a commutative algebra concentrated in degree 0. Let x be an element of A. We define the "Koszul complex" by

$$K_\bullet^A(x) : 0 \to A \to A \to 0,$$

concentrated in degrees 0 and 1, where the boundary map defined by $d(a) := ax$. More generally, for n elements $\{x_1, \ldots, x_n\}$ of A, the "Koszul complex" is defined by the tensor product

$$K_\bullet^A(x_1,\dots,x_n) := K_\bullet^A(x_1)\otimes\cdots\otimes K_\bullet^A(x_n)$$

of chain complexes.

Show that the degree -1 map

$$\tau : \Lambda^c(sx_1,\dots,sx_n) \twoheadrightarrow \mathbb{K}sx_1\oplus\cdots\oplus\mathbb{K}sx_n \xrightarrow{s^{-1}} \mathbb{K}x_1\oplus\cdots\oplus\mathbb{K}x_n \to A$$

is a twisting morphism from the symmetric cofree coalgebra on the suspension of the basis $\{x_1,\dots,x_n\}$ to the algebra A.

Prove that the "Koszul complex" $K_\bullet^A(x_1,\dots,x_n)$ is isomorphic to the twisted tensor product $\Lambda^c(sx_1,\dots,sx_n\mathbb{K})\otimes_\tau A$.

Considering the canonical twisting morphism $\kappa : \Lambda^c(sx_1,\dots,sx_n)\to S(x_1,\dots,x_n)$, show that the "Koszul complex" $K_\bullet^A(x_1,\dots,x_n)$ is isomorphic to the relative tensor product

$$\big(\Lambda^c(sx_1,\dots,sx_n)\otimes_\kappa S(x_1,\dots,x_n)\big)\otimes_{S(x_1,\dots,x_n)} A,$$

where A is considered a left $S(x_1,\dots,x_n)$-module.

We say that $\{x_1,\dots,x_n\}$ is a *regular sequence* when the image of x_i in $A/(x_1,\dots,x_{i-1})A$ has no nonzero divisor, for $1\le i\le n$. When it is the case, prove that the Koszul complex $K_\bullet^A(x_1,\dots,x_n)$ is a resolution of $A/(x_1,\dots,x_n)A$ by free A-modules.

This chain complex is used to compute local cohomology (see [Wei94, Sects. 4.5–4.6]).

Exercise 3.8.8 (Homological degree). Let (V,R) be a quadratic data such that V is concentrated in degree 0. We consider the bar construction $\mathrm{B}_\bullet A$ of the quadratic algebra $A = A(V,R)$ as a chain complex with the homological degree.

Show that this chain complex splits with respect to the weight grading: $\mathrm{B}_\bullet A = \bigoplus_{n\in\mathbb{N}}(\mathrm{B}_\bullet A)^{(n)}$. For $n\ge 1$, prove that the sub-chain complex $(\mathrm{B}_\bullet A)^{(n)}$ is finite, concentrated in degrees $1\le\bullet\le n$ and that $H_n((\mathrm{B}_\bullet A)^{(n)})\cong A^{¡(n)}$.

Show that the quadratic data (V,R) is Koszul if and only if the homology of the bar construction $\mathrm{B}_\bullet A$ is concentrated on the diagonal $\bigoplus_{n\in\mathbb{N}} H_n((\mathrm{B}_\bullet A)^{(n)})$.

Exercise 3.8.9 (Double Hilbert–Poincaré series). Pursuing the preceding exercise, we require here the vector space V to be finite dimensional. In this case, show that all the components $(\mathrm{B}_m A)^{(n)}$ of the bar construction of A are finite dimensional, for any $m,n\in\mathbb{N}$.

We define the *double Hilbert–Poincaré series* of A by

$$F^A(x,t) := \sum_{m,n\ge 0}\dim H_m\big((\mathrm{B}_\bullet A)^{(n)}\big)x^m t^n.$$

Show that the quadratic data is Koszul if and only if the double Hilbert–Poincaré series has only nontrivial coefficients in powers $x^m t^n$ for $m = n$. Prove that it is also equivalent to $F^A(x,t) = f^{A^!}(xt)$.

Prove the functional equation $f^A(t)F^A(-1,t) = 1$ and recover the equation of Theorem 3.5.1.

Exercise 3.8.10 (Every augmented algebra is inhomogeneous Koszul). Let A be an augmented associative algebra.

1. Show that $V := \bar{A}$ and $R := \{a \otimes b - ab \mid a, b \in \bar{A}^{\otimes 2}\}$ is a quadratic-linear presentation of A satisfying conditions (ql_1) and (ql_2).
2. Prove that qA is nilpotent and that the Koszul dual dg coalgebra $A^¡ \cong BA$ is isomorphic to the bar construction of the algebra A.
3. Finally, show that this quadratic-linear presentation is Koszul and that the Koszul resolution is nothing but the bar–cobar resolution.

Exercise 3.8.11 (BCKQRS spectral sequence as twisted convolution algebra ★). Show that the first page E^1X of the spectral sequence of [BCK+66] for any spectrum X is equal to the convolution algebra $\mathrm{Hom}^\kappa(\mathscr{A}_2{}^¡, \mathscr{A}_2)$ as follows: $H_\bullet(X)$ and $\mathrm{Hom}^\kappa(\mathscr{A}_2{}^¡, \mathscr{A}_2)$ are $\mathscr{A}_2$-modules and

$$E^1X \cong \mathrm{Hom}^\kappa\left(\mathscr{A}_2{}^¡, \mathscr{A}_2\right) \otimes_{\mathscr{A}_2} H_\bullet(X).$$

Show that $E^2X \cong \mathrm{Ext}^\bullet_{\mathscr{A}_2}(\mathbb{K}, H_\bullet(X))$.

Chapter 4
Methods to Prove Koszulity of an Algebra

"Là, tout n'est qu'ordre et beauté, Luxe, calme et volupté.
Charles Baudelaire

After having introduced the notion of Koszul algebra in the preceding chapter, we give here methods to prove that an algebra is Koszul together with constructions to produce new Koszul algebras.

We begin by describing a short algorithmic method, called *rewriting method*. It amounts to choosing first an ordered basis of the generating space. Then, we interpret the relations as rewriting rules, replacing each leading term by a sum of lower terms, with respect to a suitable ordering on monomials. If applying the rewriting rules to the critical monomials leads to the same element (confluence property), then the algebra is Koszul.

This method is the simplest case of a general one, which relies on an extra data: a decomposition of the generating space $V \cong V_1 \oplus \cdots \oplus V_k$ of a quadratic algebra $A = A(V, R)$ and a suitable order on the set of tuples in $\{1, \ldots, k\}$. Such a data induces a filtration on the algebra A. When the associated graded algebra $\operatorname{gr} A$ is Koszul, the algebra A itself is also Koszul. So the problem reduces to the graded algebra $\operatorname{gr} A$, whose product is simpler than the product of A. What about its underlying module? We have a tentative quadratic presentation $\mathring{A} := A(V, R_{\text{lead}}) \twoheadrightarrow \operatorname{gr} A$, where the module R_{lead} is made up of the leading terms of the relations. The *Diamond Lemma* asserts that it is enough to prove the injectivity of this map in weight 3 and that the algebra $\mathring{A}$ is Koszul, to get the isomorphism of algebras $\mathring{A} \cong \operatorname{gr} A$. It implies that A is Koszul. This method reduces the problem to proving the Koszulity of the simpler quadratic algebra $\mathring{A}$.

The particular case where each component V_i is one-dimensional gives rise to the notion of *Poincaré–Birkhoff–Witt (PBW) basis* of a quadratic algebra. Here the quadratic algebra $\mathring{A}$ is a *quadratic monomial algebra*, which is always a Koszul algebra, thereby simplifying the theory. For instance, any quadratic algebra admitting a PBW basis is a Koszul algebra. In this case, we refine even further the Diamond Lemma to give a simple way to check whether a quadratic algebra admits a PBW

J.-L. Loday, B. Vallette, *Algebraic Operads*,
Grundlehren der mathematischen Wissenschaften 346,
DOI 10.1007/978-3-642-30362-3_4, © Springer-Verlag Berlin Heidelberg 2012

basis. This is the aforementioned rewriting method. We also introduce the notion of quadratic *Gröbner basis* for the ideal (R) and prove that it is equivalent to a PBW basis for the quotient algebra $T(V)/(R)$.

The last method uses a family of lattices associated to any quadratic data. The *Backelin criterion* states that these lattices are distributive if and only if the quadratic data is Koszul.

Finally, we introduce the two Manin products, white $\bigcirc$ and black $\bullet$, in the category of quadratic data. They are sent to one another under the Koszul dual functor and they preserve the Koszul property by Backelin's criterion. This allows us to construct a new chain complex, called the *Manin complex*, on the white product $A \bigcirc A^!$ of a quadratic algebra and its Koszul dual algebra (not coalgebra). Dually, the black product $A \bullet A^!$ is endowed with a Hopf algebra structure.

This chapter is essentially extracted from Priddy [Pri70], Bergman [Ber78], Backelin [Bac83], Manin [Man87, Man88] and Polishchuk–Positselski [PP05].

4.1 Rewriting Method

In this section, we give a short algorithmic method, based on the rewriting rules given by the relations, to prove that an algebra is Koszul. We give no proof here since this method is a particular case of a more general theory explained in detail in the next two sections.

Let $A(V, R)$ be a quadratic algebra, for instance

$$A\big(v_1, v_2, v_3; v_1^2 - v_1v_2, v_2v_3 + v_2v_2, v_1v_3 + 2v_1v_2 - v_1^2\big).$$

Step 1. We choose a basis $\{v_i\}_{i=1,\dots,k}$ for the space of generators V. We consider the ordering $v_1 < v_2 < \cdots < v_k$.

Step 2. We consider the induced basis of $V^{\otimes 2}$, which we order lexicographically:

$$v_1v_1 < v_1v_2 < \cdots < v_1v_k < v_2v_1 < \cdots.$$

(One can choose other suitable orders, like

$$v_1v_1 < v_1v_2 < v_2v_1 < v_1v_3 < v_2v_2 < v_3v_1 < v_1v_4 < \cdots$$

see the discussion at the end of Sect. 4.2.1.)

We choose a basis of R. Any one of its elements is of the form

$$r = \lambda v_iv_j - \sum_{(k,l)<(i,j)} \lambda_{k,l}^{i,j} v_kv_l, \quad \lambda \neq 0.$$

The monomial v_iv_j is called the *leading term* of r. We can always change this basis for one with the following *normalized form*. First, the coefficient of the leading term

Fig. 4.1 Pentagonal diamond

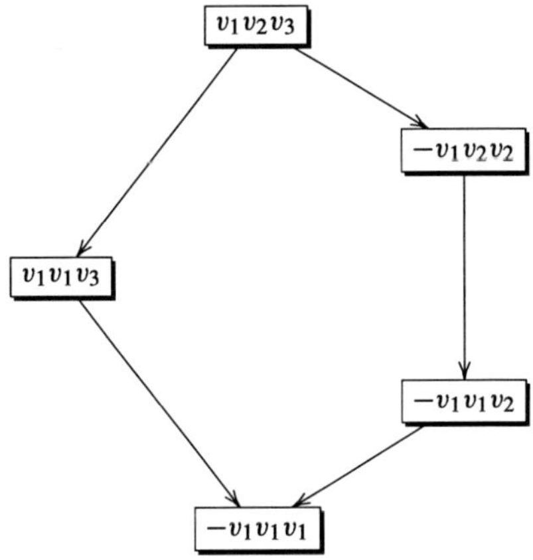

can always be supposed to be 1 since $\mathbb{K}$ is a field. Then, we can always suppose that two different relators in the basis have different leading terms and that the sum in the right-hand side of any relator contains no leading term of any other relator.

In the example at hand, the space of relations R admits the following normalized basis:

$$\{v_1v_2 - (v_1^2), v_2v_3 - (-v_2v_2), v_1v_3 - (-v_1^2)\}.$$

The three leadings terms are v_1v_2, v_2v_3 and v_1v_3.

Step 3. These choices provide rewriting rules of the form

$$v_iv_j \mapsto \sum_{(k,l)<(i,j)} \lambda_{k,l}^{i,j} v_kv_l,$$

$$\text{leading term} \mapsto \text{sum of lower terms},$$

for any relator r in the normalized basis of R. A monomial $v_iv_jv_k$ is called *critical* if both v_iv_j and v_jv_k are leading terms. Any critical monomial gives rise to a graph made up of the successive application of the aforementioned rewriting rule.

In the example at hand, we have the following rewriting rules

$$v_1v_2 \mapsto v_1^2, \qquad v_2v_3 \mapsto -v_2v_2, \qquad v_1v_3 \mapsto -v_1^2.$$

There is only one critical monomial: $v_1v_2v_3$.

Step 4. Any critical monomial $v_iv_jv_k$ gives a graph under the rewriting rules. It is *confluent*, if it has only one terminal vertex.

In the example at hand, the only critical monomial induces the confluent graph shown in Fig. 4.1.

Conclusion. If each critical monomial is confluent, then the algebra A is Koszul.

This assertion is a consequence of the following result.

Theorem 4.1.1 (Rewriting method). *Let $A = A(V, R)$ be a quadratic algebra. If its generating space V admits an ordered basis, for which there exists a suitable order on the set of tuples, such that every critical monomial is confluent, then the algebra A is Koszul.*

Proof. This result is Theorem 4.3.7. □

In this case, the algebra A is equipped with an induced basis sharing nice properties, called a PBW basis, see Sect. 4.3. For other examples, like the symmetric algebra, and for more details, we refer the reader to Sect. 4.3.5.

4.2 Reduction by Filtration

The idea of the "reduction by filtration" method can be shortened as follows: when a quadratic algebra $A = A(V, R)$ admits a filtration with nice properties, there exists a morphism of algebras

$$\mathring{A} := A(V, R_{\text{lead}}) \twoheadrightarrow \operatorname{gr} A := \operatorname{gr}\big(A(V, R)\big)$$

from the quadratic algebra defined by the associated graded presentation to the associated graded algebra. If the quadratic algebra $\mathring{A}$ is Koszul and if this map is an isomorphism (in weight 3), then the algebra A itself is Koszul. This reduces the problem of the Koszulity of the algebra A to the algebra $\mathring{A}$. Koszulity of $\mathring{A}$ is easier to check in general.

4.2.1 Extra Ordered Grading

The aim of this section is to endow the free algebra $T(V)$ with an extra grading, which refines the weight grading, such that the product is strictly increasing: $a < a'$, $b < b'$ implies $ab < a'b'$.

Let $A = A(V, R) = T(V)/(R)$ be a quadratic algebra, i.e. $R \subset V^{\otimes 2}$. We suppose here that V is equipped with an extra grading $V \cong V_1 \oplus \cdots \oplus V_k$, which is finite. This induces the following grading on $T(V)$:

$$V^{\otimes n} \cong \bigoplus_{(i_1,\ldots,i_n)\in\{1,\ldots,k\}^n} V_{i_1} \otimes \cdots \otimes V_{i_n},$$

under the lexicographical order

$$0 < 1 < \cdots < k < (1,1) < (1,2) < \cdots < (k,k) < (1,1,1) < (1,1,2) < \cdots,$$

where $\mathbb{K}1$ is in degree 0. This lexicographical order induces a bijection of totally ordered sets between the set of tuples in $\{1,\ldots,k\}$ and the set of integers $\mathbb{N}$. For any tuple $(i_1, \ldots, i_n)$ sent to $p \in \mathbb{N}$, we will denote the sub-space $V_{i_1} \otimes \cdots \otimes V_{i_n}$ of $V^{\otimes n}$ simply by $T(V)_p$. Under this bijection, there exists a map χ which corresponds to the concatenation of tuples:

$$\begin{array}{ccc} ((i_1, \ldots, i_n), (j_1, \ldots, j_m)) & \longmapsto & (i_1, \ldots, i_n, j_1, \ldots, j_m) \\ \downarrow \cong & & \downarrow \cong \\ (p, q) & \overset{\chi}{\longmapsto} & \chi(p, q). \end{array}$$

Under the lexicographical order on $\mathbb{N} \times \mathbb{N}$, the map $\chi : \mathbb{N} \times \mathbb{N} \to \mathbb{N}$ is strictly increasing. The concatenation product on the free associative algebra $T(V)$ satisfies

$$\mu : T(V)_p \otimes T(V)_q \to T(V)_{\chi(p,q)}.$$

Hence, this grading $(\mathbb{N}, \chi)$ refines the weight grading of $T(V)$. Notice that the map χ defines a monoid structure $(\mathbb{N}, \chi, 0)$.

Associated to this grading, we consider the increasing and exhaustive filtration $F_pT(V) := \bigoplus_{q=0}^{p} T(V)_q$ on $T(V)$. The image of this filtration under the canonical projection $T(V) \twoheadrightarrow A$ defines an increasing filtration

$$\mathrm{F}_0A \subset \mathrm{F}_1A \subset \mathrm{F}_2A \subset \cdots \subset \mathrm{F}_pA \subset \mathrm{F}_{p+1}A \subset \cdots$$

of the underlying module of A. The strictly increasing map χ allows us to define a χ-graded product on the associated graded module $\mathrm{gr}_p A := \mathrm{F}_pA/\mathrm{F}_{p-1}A$:

$$\bar{\mu} : \mathrm{gr}_p A \otimes \mathrm{gr}_q A \to \mathrm{gr}_{\chi(p,q)} A.$$

This algebra is denoted by $\mathrm{gr}_\chi A$, or simply by $\mathrm{gr}\, A$, when there is no possible confusion. Since the extra grading refines the weight grading, the algebra $\mathrm{gr}\, A$ is also weight graded.

There are two ways of generalizing the aforementioned arguments. First, one can allow k to be infinite, that is V can admit an extra grading labeled by $\mathbb{N}$: $V \cong \bigoplus_{i \in \mathbb{N}} V_i$. Then, one need not work only with the lexicographical order. Let $I := \{1, \ldots, k\}$ denote the labeling set of the extra grading on V. We consider any bijection $\bigsqcup_{n \geq 0} I^n \cong \mathbb{N}$. This endows the set of tuples $\bigsqcup_{n \geq 0} I^n$ with a total order isomorphic to $\mathbb{N}$. To define the graded algebra $\mathrm{gr}\, A$, it is enough to require that the map χ, or equivalently the concatenation product, be strictly increasing. In this case, we call the total order on the set of tuples a *suitable order*. For instance, when $k = \infty$, we can consider the following suitable total order

$$0 < 1 < 2 < (1,1) < 3 < (1,2) < (2,1) < (1,1,1) < 4 < (1,3) < (2,2) < \cdots,$$

isomorphic to $\mathbb{N}$.

Such a data, the decomposition of V and the suitable order on tuples, is called an *extra ordered grading*.

4.2.2 The Koszul Property

Recall from Sect. 3.4.3 that a connected weight graded algebra is called Koszul if the cohomology of its bar construction is concentrated in syzygy degree 0.

Proposition 4.2.1. *Let $A = A(V, R)$ be a quadratic algebra equipped with an extra ordered grading. If the algebra* $\mathrm{gr}\, A$ *is Koszul, then the algebra A is also Koszul.*

Proof. We consider the bar construction $\mathrm{B}^{-\bullet}A$ as a chain complex with the opposite of the syzygy degree (see Sect. 3.3.1). We extend the filtration on the free algebra $T(V)$ to its bar construction $\mathrm{B}T(V)$ as follows:

$$F_p\mathrm{B}T(V) := \big\{sx_1 \otimes \cdots \otimes sx_m \mid x_1x_2\ldots x_m \in F_pT(V)\big\}.$$

This filtration is stable under the differential map. The canonical projection $T(V) \twoheadrightarrow A$ induces an epimorphism of dg coalgebras $\mathrm{B}T(V) \twoheadrightarrow \mathrm{B}A$ between the bar constructions. The image under this map of the preceding filtration defines a filtration $\mathrm{F}_p\mathrm{B}A$ of the bar construction of A. The first page of the associated spectral sequence $\mathrm{E}^0_{pq} \cong \mathrm{F}_p\mathrm{B}^{-p-q}A/\mathrm{F}_{p-1}\mathrm{B}^{-p-q}A$ is isomorphic to the bar construction of the associated graded algebra $\mathrm{gr}\, A$:

$$\big(\mathrm{E}^0_{pq}, d^0\big) \cong \mathrm{B}_p^{-p-q}\, \mathrm{gr}\, A,$$

where the index p refers to the total grading induced by the finer grading on the bar construction, see Fig. 4.2.

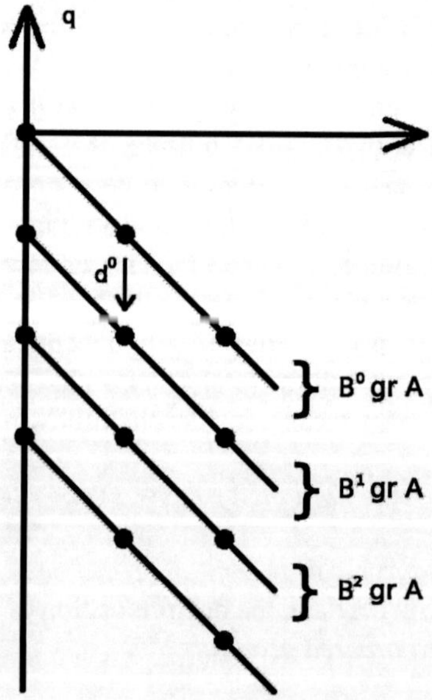

Fig. 4.2 The page E^0 of the spectral sequence

This latter algebra being Koszul, the homology of its bar construction is concentrated in syzygy degree $p+q=0$. This implies the collapsing of the spectral sequence at rank 1. The filtration being bounded below and exhaustive, it converges by Theorem 1.5.1:

$$\mathrm{E}^{\infty}_{pq} \cong \mathrm{E}^{1}_{pq} \cong \mathrm{gr}_p\, H^{-p-q}\big(\mathrm{B}^{\bullet}A\big) = 0, \quad \text{for } p+q \neq 0.$$

Since the homology of the bar construction of A is concentrated in syzygy degree 0, the algebra A is a Koszul algebra. □

Thanks to the finer filtration, we have reduced the Koszul problem for the algebra A to the algebra $\mathrm{gr}\, A$, whose product is simpler. But the underlying module of $\mathrm{gr}\, A$ might be difficult to describe. However, Exercise 3.8.1 implies that if the algebra $\mathrm{gr}\, A$ is Koszul, then it admits a quadratic presentation.

4.2.3 Quadratic Analog and Leading Space of Relations

By the universal property of the free algebra, there exists a morphism of algebras $T(V) \twoheadrightarrow \mathrm{gr}\, A$, which is an epimorphism of χ-graded algebras. Hence, it also preserves the weight grading. It is obviously an isomorphism in weights 0 and 1. Let us denote by R_{lead} the kernel of its restriction to $V^{\otimes 2}$. We consider the quadratic algebra defined by

$$\mathring{A} := T(V)/(R_{\mathrm{lead}}).$$

Let us now make R_{lead} explicit. Any element $r \in R$ decomposes according to the finer grading as $r = X_1 + \cdots + X_p$, with $X_i \in V^{\otimes 2}$ and where $X_p \neq 0$ is the term of greatest grading. We call X_p the *leading term* of r. The space R_{lead} is spanned by the leading terms of all the elements of R. Hence, we call it the *leading space of relations*.

Proposition 4.2.2. *Let $A = A(V,R)$ be a quadratic algebra equipped with an extra ordered grading. We have the following commutative diagram of epimorphisms of χ-graded, thus weight graded, algebras*

$$\begin{array}{ccc} T(V) & & \\ \downarrow & \searrow & \\ \psi : \mathring{A} = T(V)/(R_{\mathrm{lead}}) & \twoheadrightarrow & \mathrm{gr}\, A, \end{array}$$

where the space of relations R_{lead} is equal to

$$R_{\mathrm{lead}} = \langle X_p, r = X_1 + \cdots + \underbrace{X_p}_{\neq 0} \in R\rangle = \big\langle \textit{Leading Term } (r), r \in R\big\rangle.$$

By definition, the canonical projection of weight graded algebras $\psi : \mathring{A} \twoheadrightarrow \mathrm{gr}\, A$ is bijective in weights 0, 1 and 2. Therefore the algebra $\mathring{A}$ is the best candidate for a quadratic presentation of the weight graded algebra $\mathrm{gr}\, A$. Before studying this situation, let us give the following straightforward but useful result.

Proposition 4.2.3. *Let $A = A(V, R)$ be a quadratic algebra equipped with an extra ordered grading. If the algebra $\mathring{A} := T(V)/(R_{\mathrm{lead}})$ is Koszul and if the canonical projection $\mathring{A} \cong \mathrm{gr}\, A$ is an isomorphism of algebras, then the algebra A is Koszul.*

Proof. In this case, the algebra $\mathrm{gr}\, A$ is Koszul and we conclude with Theorem 4.2.1. □

4.2.4 The Diamond Lemma

Under the assumption that the quadratic algebra $\mathring{A}$ is Koszul, the Diamond Lemma asserts that it is enough for the canonical projection $\psi : \mathring{A} \twoheadrightarrow \mathrm{gr}\, A$ to be injective in weight 3, to ensure that it is an isomorphism.

Theorem 4.2.4 (Diamond Lemma for quadratic algebras). *Let $A = A(V, R)$ be a quadratic algebra equipped with an extra ordered grading. Suppose that the quadratic algebra $\mathring{A} := T(V)/(R_{\mathrm{lead}})$ is Koszul.*

If the canonical projection $\mathring{A} \twoheadrightarrow \mathrm{gr}\, A$ is injective in weight 3, then it is an isomorphism. In this case, the algebra A is Koszul.

Proof. We prove this theorem in two steps. We first filter the bar construction to get an isomorphism on the level of the Koszul dual coalgebras. Then we filter the cobar construction, in the same way, to get the final isomorphism.

Step 1. We consider the same filtration of the bar construction $\mathrm{B}^{-\bullet}A$ as in the proof of Theorem 4.2.1. Since the canonical projection ψ is an isomorphism in weight less than 3, the first page of the associated spectral sequence is equal to

$$\left(\mathrm{E}^0_{pq}, d^0\right) \cong \mathrm{B}_p^{-p-q}\, \mathrm{gr}\, A \cong \mathrm{B}_p^{-p-q}\, \mathring{A},$$

for $p + q \geq -2$ (the syzygy degree being defined by the weight grading minus 1). The algebra $\mathring{A}$ being Koszul, we get $\mathrm{E}^1_{pq} = 0$ for $p + q = -1$ and $\mathrm{E}^1_{p-p} = \mathring{A}^{¡}_{p}$, see Fig. 4.3.

We conclude by the same argument as in the proof of Theorem 4.2.1: the convergence of the spectral sequence shows that there is an isomorphism $\mathrm{gr}_p A^{¡} \cong \mathring{A}^{¡}_p$.

Step 2. Dually, we apply the same method and consider the same kind of filtration on $T^c(sV)$ as on $T(V)$ and on $\Omega T^c(sV)$ as on $BT(V)$:

$$F_p \Omega T^c(sV) := \left\{ s^{-1}x_1 \otimes \cdots \otimes s^{-1}x_m \mid x_1 x_2 \ldots x_m \in F_p T(V) \right\}.$$

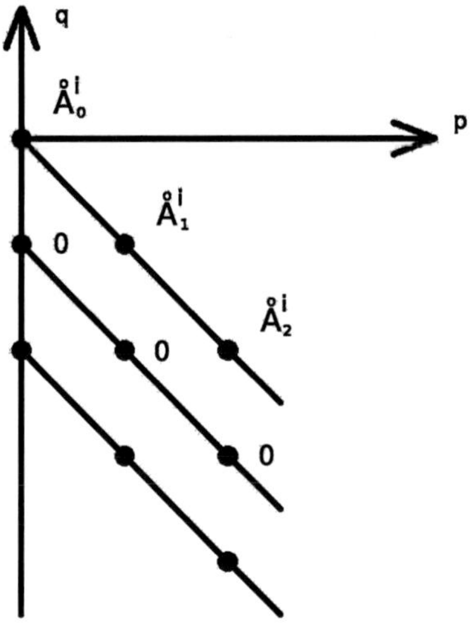

Fig. 4.3 The page E^1 of the spectral sequence

Since $A^{¡}$ is a sub-coalgebra of $T^c(sV)$, this filtration restricts to a filtration $\mathsf{F}_p\Omega A^{¡}$ on the cobar construction of $A^{¡}$. We consider the cobar construction $\Omega_\bullet A^{¡}$ with its syzygy degree, see Sect. 3.3.2. The first page of the associated spectral sequence $\mathsf{E}^0_{pq} \cong \mathsf{F}_p\Omega_{p+q}A^{¡}/\mathsf{F}_{p-1}\Omega_{p+q}A^{¡}$ is isomorphic to the cobar construction of the Koszul dual coalgebra $\mathring{A}^{¡}$:

$$\left(\mathsf{E}^0_{pq}, d^0\right) \cong \left(\Omega_{p+q}\operatorname{gr} A^{¡}\right)_p \cong \left(\Omega_{p+q}\mathring{A}^{¡}\right)_p.$$

The isomorphism between the underlying modules is induced by the aforementioned isomorphism $\operatorname{gr}_p A^{¡} \cong \mathring{A}^{¡}_p$. The part d^0 of the boundary map of $\Omega A^{¡}$ is the part which preserves the extra grading. Hence it is given by the deconcatenation of the leading terms of the elements of $A^{¡}$ or equivalently by the coproduct of the coalgebra $\mathring{A}^{¡}$ under the above isomorphism. This proves that d^0 is in one-to-one correspondence with the differential of $\Omega\mathring{A}^{¡}$. Since the quadratic algebra $\mathring{A}$ is Koszul, the homology of $\Omega\mathring{A}^{¡}$ is concentrated in syzygy degree $p+q=0$. Therefore, $\mathsf{E}^1_{pq}=0$ for $p+q\neq 0$ and $\mathsf{E}^1_{p-p}=\mathring{A}_p$. The convergence theorem for spectral sequences (Theorem 1.5.1) finally gives the desired isomorphism:

$$\psi : \mathring{A}_p \cong \mathsf{E}^1_{p-p} \cong \mathsf{E}^\infty_{p-p} \cong \operatorname{gr}_p H_0\left(\Omega_\bullet A^{¡}\right) = \operatorname{gr}_p A. \qquad \square$$

Notice that in the proof, we have proved the same isomorphism of χ-graded modules: $\operatorname{gr}_p A^{¡} \cong \mathring{A}^{¡}_p$, but on the level of the Koszul dual coalgebras.

4.2.5 The Inhomogeneous Case

When the associative algebra A is inhomogeneous quadratic, we choose, if possible, a presentation $A = A(V, R)$ with $R \subset V \oplus V^{\otimes 2}$ satisfying conditions (ql_1) and

(ql_2) of Sect. 3.6. Applying the previous propositions to the quadratic algebra $qA := A(V, qR)$, we get the same kind of results.

We start with the same refining data. Let the generating module V be endowed with an extra grading $V \cong V_1 \oplus \cdots \oplus V_k$ together with a suitable order on tuples. As above, this defines a filtration of the algebra A. It induces a χ-graded algebra, denoted $\mathrm{gr}_\chi A$, whose underlying module refines that of the weight graded algebra $\mathrm{gr}\, A$ of Sect. 3.6. This extra grading also induces a filtration on the quadratic algebra qA and the canonical projection $qA \twoheadrightarrow \mathrm{gr}\, A$ refines as follows

$$\begin{array}{ccccc}
\mathring{q}A & & & & \\
\downarrow & & & & \\
\mathrm{gr}_\chi\, qA & \xrightarrow{\cong} & qA & & \\
\downarrow & & \downarrow & & \\
\mathrm{gr}_\chi\, A & \xrightarrow{\cong} & \mathrm{gr}\, A & \xrightarrow{\cong} & A,
\end{array}$$

where the vertical maps are epimorphisms of algebras and where the horizontal maps are linear isomorphisms. The first column is made up of χ-graded algebras and the second column is made up of weight graded algebras. We only state the last theorem in the inhomogeneous case, leaving the details of the other ones to the reader.

Theorem 4.2.5 (Diamond Lemma for inhomogeneous quadratic algebras). *Let $A = A(V, R)$ be a quadratic-linear algebra with a presentation satisfying conditions (ql_1) and (ql_2). We suppose that $T(V)$ comes equipped with an extra ordered grading.*

If the quadratic algebra $\mathring{q}A$ is Koszul and if the canonical projection $\mathring{q}A \twoheadrightarrow \mathrm{gr}_\chi\, qA$ is injective in weight 3, then the algebra A is Koszul and all the maps of the above diagram are isomorphisms, in particular:

$$\mathring{q}A \cong \mathrm{gr}_\chi\, qA \cong qA \cong \mathrm{gr}\, A \cong A.$$

Proof. The Diamond Lemma 4.2.4, applied to the quadratic algebra qA, gives that the algebra qA is Koszul and the isomorphism $\mathring{q}A \cong \mathrm{gr}_\chi\, qA$. It implies that the inhomogeneous algebra A is Koszul. The last isomorphism is given by the PBW Theorem 3.6.4. □

In this case, we can compute the Koszul dual dg coalgebra of A from the isomorphism of weight graded modules $(\mathring{q}A)^{¡} \cong qA^{¡} \cong A^{¡}$.

4.2.6 Koszul Dual Algebra

In this section, we suppose that the generating space V is finite dimensional to be able to consider the Koszul dual algebra $A^! = T(V^*)/(R^\perp)$, see Sect. 3.2.2. The extra grading on $V \cong V_1 \oplus \cdots \oplus V_k$ induces a dual grading $V^* \cong V_k^* \oplus \cdots \oplus V_1^*$, that we choose to order as indicated. To any suitable order on the tuples indexing $T(V)$, we consider the locally reversed order for $T(V^*)$: we keep the global ordering but completely reverse the total order on the subset of n-tuples, for any n. For example, with the lexicographic order, it gives the following grading on $T(V^*)$:

$$T(V^*) = \mathbb{K} \bigoplus V_k^* \oplus \cdots \oplus V_1^* \bigoplus V_k^* \otimes V_k^* \oplus \cdots \oplus V_1^* \otimes V_1^* \bigoplus V_k^* \otimes V_k^* \otimes V_k^* \oplus \cdots.$$

This order allows us to consider the graded algebra $\mathrm{gr}(A^!)$, as in Sect. 4.2.1, which comes with its quadratic analog $(\mathring{A}^!) := A(V^*, (R^\perp)_{\mathrm{lead}}) \twoheadrightarrow \mathrm{gr}(A^!)$.

Lemma 4.2.6. *Let $A = A(V, R)$ be a finitely generated quadratic algebra equipped with an extra ordered grading. The isomorphism $(R^\perp)_{\mathrm{lead}} \cong (R_{\mathrm{lead}})^\perp$ of sub-modules of $(V^*)^{\otimes 2}$ induces the isomorphism of quadratic algebras $(\mathring{A}^!) \cong (\mathring{A})^!$.*

Proof. We write the proof for the lexicographic order, the general case being similar. Let us denote $\mathsf{R}_i := R_{\mathrm{lead}} \cap T(V)_{k+i}$, for $1 \le i \le k^2$. Therefore, $R_{\mathrm{lead}} = \mathsf{R}_1 \oplus \cdots \oplus \mathsf{R}_{k^2}$. Since the vector space V is finite dimensional, each $T(V)_{k+i}$ is finite dimensional and we can consider a direct summand $\breve{\mathsf{R}}_i$ such that $T(V)_{k+i} \cong \mathsf{R}_i \oplus \breve{\mathsf{R}}_i$. Hence the linear dual is equal to $(T(V)_{k+i})^* \cong \mathsf{R}_i^\perp \oplus \breve{\mathsf{R}}_i^\perp$.

By definition of R_{lead}, the space of relations R is linearly spanned by elements of the form $r = X_1 + \cdots + X_p$ with $X_i \in \breve{\mathsf{R}}_i$, for $1 \le i < p \le k^2$ and $X_p \neq 0$, $X_p \in \mathsf{R}_p$. Dually, any element $\rho \in R^\perp$ decomposes as $\rho = Y_q - Y_{q+1} - \cdots - Y_{k^2}$ with $Y_q \neq 0$, $Y_q \in \mathsf{R}_q^\perp$ and $Y_i \in \breve{\mathsf{R}}_i^\perp$, for $q < i \le k^2$. This implies finally the isomorphism $(R^\perp)_{\mathrm{lead}} \cong \mathsf{R}_1^\perp \oplus \cdots \oplus \mathsf{R}_{k^2}^\perp \cong (R_{\mathrm{lead}})^\perp$. □

Proposition 4.2.7. *Let $A = A(V, R)$ be a finitely generated quadratic algebra equipped with an extra ordered grading. Suppose that the quadratic algebra $\mathring{A} := T(V)/(R_{\mathrm{lead}})$ is Koszul. If the canonical projection $\mathring{A} \twoheadrightarrow \mathrm{gr}\, A$ is injective in weight 3, then the dual canonical projection $(\mathring{A}^!) \cong \mathrm{gr}(A^!)$ is an isomorphism.*

Proof. We pursue the proof of Theorem 4.2.4. The proper desuspension of the linear dual of the isomorphism $\mathrm{gr}_p\, A^{\text{¡}} \cong \mathring{A}^{\text{¡}}_p$ gives the isomorphism $\mathrm{gr}_p\, A^! \cong (\mathring{A})^!_p$. We conclude with the isomorphism of Lemma 4.2.6. □

4.3 Poincaré–Birkhoff–Witt Bases and Gröbner Bases

In this section, we study the particular case of the preceding section when the generating space V is equipped with an extra grading $V \cong V_1 \oplus \cdots \oplus V_k$ such that each sub-space V_i is one-dimensional. This gives rise to the notion of Poincaré–Birkhoff–Witt basis, or PBW basis for short. Quadratic algebras which admit such a basis share nice properties. For instance, they are Koszul algebras.

We introduce the equivalent notion of quadratic Gröbner basis (also called Gröbner–Shirshov basis), which is to the ideal (R) what PBW basis is to the quotient algebra $T(V)/(R)$.

The notion of PBW basis comes from Sect. 5 of the original paper of S. Priddy [Pri70]. We refer the reader to Chap. 4 of the book [PP05] for more details on the subject.

4.3.1 Ordered Bases

We now restrict ourself to the case where the generating space V of a quadratic algebra $A(V,R)$ is equipped with an extra grading $V \cong V_1 \oplus \cdots \oplus V_k$ such that each V_i is one-dimensional. This datum is equivalent to a *totally ordered basis* $\{v_i\}_{i\in\{1,\dots,k\}}$ of V, which is a basis labeled by a totally ordered set, by definition. Let us denote $I := \{1,\dots,k\}$ and let us use the convention $I^0 := \{0\}$. As in Sect. 4.2.1, we consider the set $J := \bigsqcup_{n\geq 0} I^n$ of tuples $\bar{i} = (i_1,\dots,i_n)$ in $\{1,\dots,k\}$ equipped with a suitable order, for instance the lexicographic order. With this definition, the set $\{v_{\bar{i}} = v_{i_1}v_{i_2}\cdots v_{i_n}\}_{\bar{i}\in J}$ becomes a totally ordered basis of $T(V)$. In this case, we say that $T(V)$ is equipped with a *suitable ordered basis.*

Written in this basis, the space of relations R is equal to

$$R = \left\{ \lambda v_i v_j - \sum_{(k,l)<(i,j)} \lambda_{k,l}^{i,j} v_k v_l;\ \lambda \neq 0 \right\}.$$

Proposition 4.3.1. *Let A be a quadratic algebra $A(V,R)$, with $T(V)$ equipped with a suitable ordered basis $\{v_{\bar{i}}\}_{\bar{i}\in J}$. The associated quadratic algebra $\mathring{A}$ is equal to $A(V, R_{\text{lead}})$, with $R_{\text{lead}} \cong \langle v_i v_j, (i,j) \in \overline{L}^{(2)}\rangle$, where the set $\overline{L}^{(2)}$ is the set of labels of the leading terms of the relations of R.*

Proof. It is a direct corollary of Proposition 4.2.2. □

Notice that the space of relations admits a normalized basis of the form

$$R = \left\langle v_i v_j - \sum_{(k,l)\notin \overline{L}^{(2)},\,(k,l)<(i,j)} \lambda_{k,l}^{i,j} v_k v_l;\ (i,j) \in \overline{L}^{(2)} \right\rangle.$$

The algebra $\mathring{A}$ depends only on the ordered basis of V and on the suitable order on tuples, but it is always a quadratic algebra whose ideal is generated by monomial elements.

4.3.2 Quadratic Monomial Algebras

We introduce the notion of quadratic monomial algebra, which is the structure carried by the quadratic algebra $\mathring{A}$ in the PBW bases theory.

A *quadratic monomial algebra* is a quadratic algebra $\mathring{A} = A(V,R) = T(V)/(R)$ with a (non-necessarily ordered) basis $\{v_i\}_{i\in I}$ of V such that the space of relations R is linearly spanned by a set $\{v_i v_j\}_{(i,j)\in \overline{L}^{(2)}}$, where $\overline{L}^{(2)} \subset I^2$. We denote by $L^{(2)}$ the complement of $\overline{L}^{(2)}$ in I^2, $L^{(2)} := I^2 \backslash \overline{L}^{(2)}$, which labels a basis of the quotient $\mathring{A}^{(2)} = V^{\otimes 2}/R$. We set $L^{(0)} := \{0\}$ and $L^{(1)} := I$. We will prove that a quadratic monomial algebra is Koszul by making its Koszul complex explicit and by computing its homology.

Proposition 4.3.2. *For any quadratic monomial algebra $\mathring{A} = A(V,R)$, the subset $L = \bigsqcup_{n\in\mathbb{N}} L^{(n)} \subset J$ defined by*

$$\bar{i} = (i_1,\dots,i_n) \in L^{(n)} \quad \Longleftrightarrow \quad (i_m, i_{m+1}) \in L^{(2)}, \quad \forall 1 \le m < n$$

labels a basis of the monomial algebra $\mathring{A}$.

Though this statement is obvious, it will play a key role in the definition of PBW bases. Dually, we make explicit a monomial basis for the Koszul dual coalgebra.

Proposition 4.3.3. *For any quadratic monomial algebra $\mathring{A} = A(V,R)$, the subset $\overline{L} = \bigsqcup_{n\in\mathbb{N}} \overline{L}^{(n)} \subset J$ defined by*

$$\bar{i} = (i_1,\dots,i_n) \in \overline{L}^{(n)} \quad \Longleftrightarrow \quad (i_m, i_{m+1}) \in \overline{L}^{(2)}, \quad \forall 1 \le m < n$$

labels a basis of its Koszul dual coalgebra $\mathring{A}^{¡}$.

When the generating space V is finite dimensional, the Koszul dual algebra $\mathring{A}^!$ is also a monomial algebra, with presentation

$$\mathring{A}^! \cong A\big(V^*, R^\perp\big), \quad \text{with } R^\perp = \big\langle v_i^* v_j^*, (i,j) \in L^{(2)}\big\rangle$$

and basis labeled by $\overline{L}$.

Proof. The elements $s^n v_{\bar{i}}$ for $\bar{i} \in \overline{L}^{(n)}$ form a basis of $(\mathring{A}^{¡})^{(n)}$ by the intersection formula $(\mathring{A}^{¡})^{(n)} = \bigcap_{i+2+j=n} (sV)^{\otimes i} \otimes s^2 R \otimes (sV)^{\otimes j}$ of Sect. 3.1.3. □

Theorem 4.3.4. *Any quadratic monomial algebra is a Koszul algebra.*

Proof. By the two preceding propositions, the Koszul complex $\mathring{A}^{¡} \otimes_\kappa \mathring{A}$ admits a basis of the form $s^k v_{\bar{i}} \otimes v_{\bar{j}}$ with $\bar{i} \in \overline{L}^{(k)}$ and with $\bar{j} \in L^{(l)}$. In this basis, its boundary map is equal to

$$d_\kappa\big(s^k v_{\bar{i}} \otimes v_{\bar{j}}\big) = \pm s^{k-1} v_{i_1} \dots v_{i_{k-1}} \otimes v_{i_k} v_{j_1} \dots v_{j_l}, \quad \text{when } (i_k, j_1) \in L^{(2)}, \quad \text{and}$$
$$d_\kappa\big(s^k v_{\bar{i}} \otimes v_{\bar{j}}\big) = 0, \quad \text{when } (i_k, j_1) \in \overline{L}^{(2)}.$$

In the latter case, such a cycle is a boundary since $(i_1,\dots,i_k,j_1) \in \overline{L}^{(k+1)}$ and $d_\kappa(s^{k+1} v_{i_1} \dots v_{i_k} v_{j_1} \otimes v_{j_2} \dots v_{j_l}) = \pm s^k v_{\bar{i}} \otimes v_{\bar{j}}$. Finally the Koszul complex is acyclic and the algebra $\mathring{A}$ is Koszul. □

This result is a key point in the PBW basis theory. It says that when the decomposition $V \cong V_1 \oplus \cdots \oplus V_k$ is made up of one-dimensional sub-spaces, the quadratic analog $\mathring{A}$ is always a Koszul algebra.

4.3.3 PBW Basis

The image of the monomial basis $\{v_{\bar{i}}\}_{\bar{i}\in L}$ of $\mathring{A}$, given in Proposition 4.3.2, under the successive morphisms of graded modules $\mathring{A} \twoheadrightarrow \operatorname{gr} A \cong A$ provides a family of elements $\{a_{\bar{i}}\}_{\bar{i}\in L}$, which linearly span the algebra A. When these elements are linearly independent, they form a basis of the algebra A, called a *Poincaré–Birkhoff–Witt basis*, or *PBW basis* for short. This condition corresponds to the bijection of the canonical projection $\psi : \mathring{A} \twoheadrightarrow \operatorname{gr} A$. We say that an algebra $A = A(V, R)$ admits a PBW basis if there exists a totally ordered basis of V and a suitable order on tuples such that the associated elements $\{a_{\bar{i}}\}_{\bar{i}\in L}$ form a basis of the algebra A.

EXAMPLE. The symmetric algebra $S(v_1, \ldots, v_k)$ admits the following PBW basis, with the lexicographic order: $\{v_1^{\nu_1} \ldots v_k^{\nu_k}\}$ with $\nu_1, \ldots, \nu_k \in \mathbb{N}$.

The main property of PBW bases lies in the following result.

Theorem 4.3.5. *Any quadratic algebra endowed with a PBW basis is Koszul.*

Proof. Since the monomial algebra $\mathring{A}$ is always Koszul by Proposition 4.3.4, it is a direct corollary of Proposition 4.2.3. □

The existence of a PBW basis gives a purely algebraic condition to prove that an algebra is Koszul, without having to compute any homology group.

There are Koszul algebras which do not admit any PBW basis. The quadratic algebra $A(V, R)$ generated by $V := \mathbb{K}x \oplus \mathbb{K}y \oplus \mathbb{K}z$ with the two relations $x^2 - yz$ and $x^2 + 2zy$ is Koszul but does not admit a PBW basis. This example comes from Sect. 4.3 of [PP05] and is due to J. Backelin.

4.3.4 Diamond Lemma for PBW Bases

Since the canonical projection $\psi : \mathring{A} \twoheadrightarrow \operatorname{gr} A$ is bijective in weights 0, 1 and 2, the elements $\{a_{\bar{i}}\}_{\bar{i}\in L^{(n)}}$ form a basis of $A^{(n)}$, for $n \le 2$. It is enough to check only the next case, $n = 3$, as the following theorem shows.

Theorem 4.3.6 (Diamond Lemma for PBW bases). *Let $A = A(V, R)$ be a quadratic algebra, with $T(V)$ equipped with a suitable ordered basis $\{v_{\bar{i}}\}_{\bar{i}\in J}$. If the elements $\{a_{\bar{i}}\}_{\bar{i}\in L^{(3)}}$ are linearly independent in $A^{(3)}$, then the elements $\{a_{\bar{i}}\}_{\bar{i}\in L}$ form a PBW basis of A. In that case, the algebra A is Koszul.*

Proof. It is a direct corollary of Theorems 4.3.4 and 4.2.4. □

EXAMPLE. Let us consider the quadratic algebra $A := A(V, R)$ generated by a two-dimensional vector space $V := \mathbb{K}v_1 \oplus \mathbb{K}v_2$ with relation $R := \mathbb{K}(v_1v_2 - v_1^2)$. With the lexicographic order $1 < 2 < (1,1) < \cdots$, we have $R_{\text{lead}} = \mathbb{K}v_1v_2$, $\overline{L}^{(2)} = \{(1,2)\}$ and $L^{(2)} = \{(1,1), (2,1), (2,2)\}$. One easily verifies that the elements $a_2a_2a_2$, $a_2a_2a_1$, $a_2a_1a_1$ and $a_1a_1a_1$ are linearly independent in A. Therefore, the family of monomial elements $\{a_{\bar{i}}\}_{\bar{i}\in L}$ indexed by $L^{(n)} = \{(2, 2, \ldots, 2),$ $(2, 2, \ldots, 2, 1), \ldots, (1, 1, \ldots, 1)\}$, for $n \in \mathbb{N}$, form a PBW basis of the algebra A.

COUNTER-EXAMPLE. Let us consider the same quadratic algebra with the extra relation $v_2^2 - v_1^2$. In this case, $R_{\text{lead}} = \mathbb{K}v_1v_2 \oplus \mathbb{K}v_2^2$, $\overline{L}^{(2)} = \{(1,2), (2,2)\}$ and $L^{(2)} = \{(1,1), (2,1)\}$. Therefore, the monomial basis of the quadratic algebra $\mathring{A}$ is indexed by $L^{(n)} = \{(2, 1, \ldots, 1), (1, 1, \ldots, 1)\}$. In weight 3, the relation $a_1a_1a_1 = a_2a_1a_1$, obtained by calculating $a_2a_2a_2$ by two different methods, shows that it does not form a PBW basis.

4.3.5 Recollection with the Classical Diamond Lemma

The aforementioned result can also be seen as a direct consequence of the classical Diamond Lemma of G.M. Bergman [Ber78], which comes from graph theory [New42].

Let us first recall the statement. Starting from a vertex in an oriented graph, one might have the choice of two outgoing edges. Such a configuration is called an *ambiguity* in rewriting systems. An ambiguity is called *solvable* or *confluent* if, starting from each of these two edges, there exists one path ending at a common vertex. In this case, we get a diamond shape graph like in Fig. 4.4. Under the condition that any path has an end (*termination hypothesis*), the classical Diamond Lemma asserts that, if every ambiguity is confluent, then any connected component of a graph has a unique terminal vertex.

The relationship with ring theory comes from the following graphical representation. We depict the elements of $T(V)$ by vertices of a graph with edges labeled by the relations of R oriented from the leading term to the rest. In the case of PBW bases, we restrict ourself to the generating relations given at the end of Sect. 4.3.1:

$$v_iv_j \longmapsto \sum_{(k,l)\in L^{(2)},(k,l)<(i,j)} \lambda_{k,l}^{i,j} v_kv_l, \quad (i,j) \in \overline{L}^{(2)}.$$

EXAMPLE. For instance, in the symmetric algebra

$$S(v_1, v_2, v_3) := A\big(\{v_1, v_2, v_3\}, \{r_{12} = v_2v_1 - v_1v_2, r_{23} = v_3v_2 - v_2v_3, r_{13} = v_3v_1 - v_1v_3\}\big)$$

(see Fig. 4.4).

Fig. 4.4 Hexagonal diamond

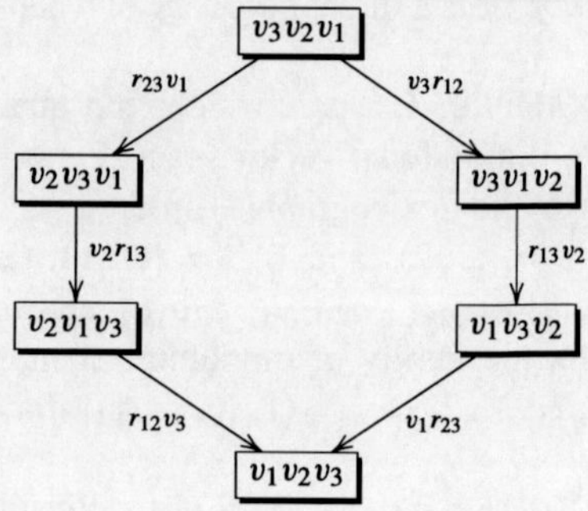

In this example, there are two ways (ambiguity) of rewriting the monomial $v_3v_2v_1$, which finally end up on the same element (confluence). Notice that this diamond is exactly the Yang–Baxter equation.

Therefore, the connected graphs depict the successive relations applied to elements in $A^{(n)}$. The terminal vertices correspond to linear combinations of monomials labeled by elements of $L^{(n)}$; there is no more leading term of any relation inside them. Having two different terminal vertices would imply that the two associated elements are equal in A. Therefore, the set L labels a PBW basis of the algebra A if and only if every connected graph has a unique terminal vertex. The PBW basis is then made up of the labels of these terminal vertices, like the element $v_1v_2v_3$ in the above example.

To prove the Diamond Lemma for PBW bases from the classical Diamond Lemma, one has just to prove that any path is finite, which is given by the bounded below suitable order, and that any ambiguity is confluent. There are here only two types of ambiguities: the square type ambiguities where one applies two relations to two distinct sub-monomials, of the same monomial or of two different monomials of a sum, and the ones starting from a sub-monomial of length 3, where one either applies a relation to the two first elements or to the two last elements. The first type is obviously confluent. The confluence of the second type of ambiguities is precisely given by the assumption that the elements labeled by $L^{(3)}$ are linearly independent. These weight 3 monomials $v_iv_jv_k$ are called *critical*. They are such that v_iv_j and v_jv_k are both leading terms of some relator.

Bergman in [Ber78] extended the Diamond Lemma beyond the set theoretic case of graph theory and showed that it is enough to check the confluence condition on monomials as the next proposition shows.

Theorem 4.3.7. *Let $A = A(V,R)$ be a quadratic algebra, with $T(V)$ equipped with an extra ordered basis $\{v_{\bar{i}}\}_{\bar{i}\in J}$. If the ambiguities coming from the critical monomials are confluent, then the elements $\{a_{\bar{i}}\}_{\bar{i}\in L}$ form a PBW basis of A and A is Koszul.*

Proof. The monomial elements with the second type of ambiguities are the elements labeled $\overline{L}^{(3)}$. Suppose that there exists a nontrivial linear combination between elements labeled by $L^{(3)}$. In terms of graph, it corresponds to a zig-zag like

$$X \leftarrow \bullet \leftarrow \bullet \rightarrow \bullet \leftarrow \cdots \leftarrow \bullet \rightarrow \bullet \rightarrow Y,$$

where $X = Y$ is the relation, with X and Y two sums of elements labeled by $L^{(3)}$. Since all the ambiguities are confluent by hypothesis, we can find another zig-zag, where the distance between X and the first $\leftarrow \bullet \rightarrow$ is strictly less than in the first zig-zag. By iteration, we prove the existence of a zig-zag of the shape $X \leftarrow \bullet \rightarrow \cdots$. Finally, by confluence, there exists an edge leaving X, which is impossible. We conclude with the Diamond Lemma 4.3.6. □

Finally, to prove that one has a PBW basis, it is enough to draw the graphs generated by elements of $\overline{L}^{(3)}$ only and to show that each of them has only one terminal vertex. We refer to the above figure for the example of the symmetric algebra. In the counterexample of Sect. 4.3.4, the element $v_2v_2v_2$ gives the following graph

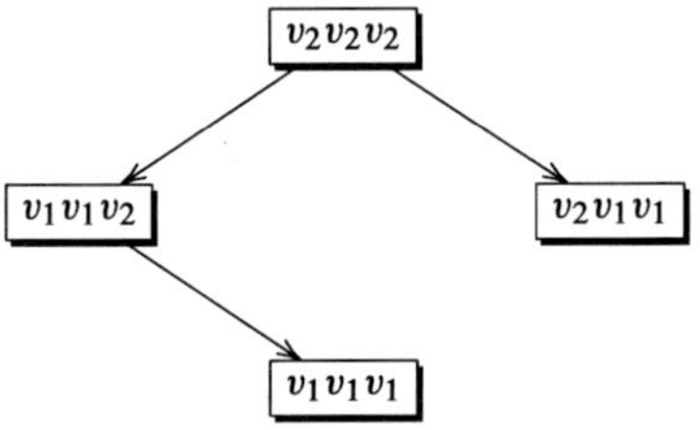

We have finally proved here the rewriting method Theorem 4.1.1 given in Sect. 4.1.

4.3.6 Product of Elements of a PBW Basis

The canonical projection $\psi : \mathring{A} \twoheadrightarrow A$ is an epimorphism of graded modules, but not of algebras in general. Therefore, the product of two elements of the generating family $\{a_{\bar{i}}\}_{\bar{i}\in L}$ is not always equal to an element of this family, but to a sum of lower terms, as the following proposition shows.

Proposition 4.3.8. *The elements $\{a_{\bar{i}}\}_{\bar{i}\in L}$ satisfy the following property: for any pair $\bar{i}, \bar{j} \in L$, if $(\bar{i}, \bar{j}) \notin L$ the product $a_{\bar{i}}a_{\bar{j}}$ in A can be written as a linear combination of strictly lower terms labeled by L:*

$$a_{\bar{i}}a_{\bar{j}} = \sum_{\bar{l}\in L, \bar{l}<(\bar{i},\bar{j})} \lambda_{\bar{l}}^{\bar{i},\bar{j}}\, a_{\bar{l}},$$

with $\lambda_{\bar{l}}^{\bar{i},\bar{j}} \in \mathbb{K}$.

Proof. The proof is done by a simple induction argument with the suitable order on tuples. □

In the example of Sect. 4.3.4, we have $(a_2a_1)(a_2a_1) = a_2a_1a_1a_1$. The original definition of a PBW basis given by Priddy in [Pri70] is a basis made up of a family of monomial elements $\{a_{\bar{i}}\}_{\bar{i}\in L}$ of A labeled by a set $L \subset J$, which satisfies the property of Proposition 4.3.8 and: any $\bar{i} = (i_1, \ldots, i_n) \in L$ if and only if $(i_1, \ldots, i_m) \in L$ and $(i_{m+1}, \ldots, i_n) \in L$ for any $1 \le m < n$.

We leave it to the reader to check that this definition of a PBW basis is equivalent to the one given in Sect. 4.3.3.

4.3.7 Gröbner Bases

In this section, we introduce the notion of (noncommutative) Gröbner basis for an ideal I of the free algebra, see [Buc06]. In the quadratic case, when $I = (R)$, it is equivalent to a PBW basis for the quotient algebra $A = T(V)/(R)$.

Any element P in $T(V)$ is a linear combination of monomials. When $T(V)$ is equipped with a suitable ordered basis, we denote by P_{lead} the leading term of P. For any subset $M \subset T(V)$, we consider the set made up of the leading terms of any element of M and we denote it by $\text{Lead}(M)$. Under this notation, the space of relations R_{lead} of Proposition 4.3.1 is equal to the linear span of $\text{Lead}(R)$: $R_{\text{lead}} = \langle \text{Lead}(R) \rangle$.

A *(noncommutative) Gröbner basis of an ideal I in $T(V)$* is a set $G \subset I$ which generates the ideal I, i.e. $(G) = I$, such that the leading terms of G and the leading terms of the elements of I generate the same ideal: $(\text{Lead}(G)) = (\text{Lead}(I))$.

Proposition 4.3.9. *Let $A = A(V, R)$ be a quadratic algebra such that $T(V)$ is equipped with a suitable ordered basis $\{v_{\bar{i}}\}_{\bar{i}\in J}$. The elements $\{a_{\bar{i}}\}_{\bar{i}\in L}$ form a PBW basis of A if and only if the elements*

$$\left\{ v_i v_j - \sum_{(k,l)\in L^{(2)},\,(k,l)<(i,j)} \lambda_{k,l}^{i,j} v_k v_l \right\}_{(i,j)\in \overline{L}^{(2)}},$$

spanning R, form a Gröbner basis of the ideal (R) in $T(V)$.

Proof. (⇒) When L labels a PBW basis, the elements

$$\left\{ v_{\bar{i}} - \sum_{\bar{l}\in L,\,\bar{l}<\bar{i}} \lambda_{\bar{l}}^{\bar{i}} v_{\bar{l}} \right\}_{\bar{i}\in J\backslash L},$$

form a linear basis of (R). The leading terms of (R) are $\text{Lead}(R) = \{\mathbb{K} v_{\bar{i}}\}_{\bar{i}\in J\backslash L}$ and $\text{Lead}(v_i v_j - \sum_{(k,l)\in L^{(2)},(k,l)<(i,j)} \lambda_{k,l}^{i,j} v_k v_l) = v_i v_j$. Condition (2) implies that $\bar{i} =$

$(i_1, \ldots, i_n) \in J \backslash L$ if and only if there exists $1 \leq m < n$ such that $(i_m, i_{m+1}) \in \overline{L}^{(2)}$. Therefore, the two following ideals are equal $(v_i v_j, (i, j) \in \overline{L}^{(2)}) = (v_{\bar{i}}, \bar{i} \in J \backslash L)$.

($\Leftarrow$) We show that any $\bar{l} \in L$, $a_{\bar{l}}$ is not equal in A to a linear combination of strictly lower terms labeled by L. Suppose that there is an $\bar{l} \in L$ such that $a_{\bar{l}} = \sum_{\bar{k} \in L, \bar{k} < \bar{l}} \lambda_{\bar{k}}^{\bar{l}} a_{\bar{k}}$, with $\lambda_{\bar{l}}^{\bar{i}} \in \mathbb{K}$. This is equivalent to $v_{\bar{l}} - \sum_{\bar{k} \in L, \bar{k} < \bar{l}} \lambda_{\bar{k}}^{\bar{l}} v_{\bar{k}} \in (R)$, whose leading term is $v_{\bar{l}}$. By definition, this element belongs to the ideal generated by the elements $v_i v_j$ with $(i, j) \in \overline{L}^{(2)}$, which is impossible by condition (2). □

In the quadratic case, the two notions of PBW basis and Gröbner basis are equivalent dual notions. The terminology "PBW basis" refers to the basis of the quotient algebra while the terminology "noncommutative Gröbner basis" refers to the ideal (R).

We refer to Sect. 2.12 of [Ufn95] for the history of the Gröbner–Shirshov bases.

4.3.8 PBW Bases for Inhomogeneous Quadratic Algebras

Following Sect. 4.2.5, we say that *an inhomogeneous quadratic algebra A admits a PBW basis* if there exists a presentation $A = A(V, R)$, satisfying conditions (ql_1) and (ql_2), such that the associated quadratic algebra $qA = A(V, qR)$ admits a PBW basis. In this case, the image $\{a_{\bar{i}}\}_{\bar{i} \in L} \subset A$ of the basis elements $\{v_{\bar{i}}\}_{\bar{i} \in L}$ of the quadratic monomial algebra $q\mathring{A}$ gives a basis of the inhomogeneous quadratic algebra A. Such a result is once again proved using the following version of the Diamond Lemma.

Theorem 4.3.10. *Let $A = A(V, R)$ be an inhomogeneous quadratic algebra with a quadratic-linear presentation satisfying conditions (ql_1) and (ql_2) and such that $T(V)$ is equipped with a suitable ordered basis $\{v_{\bar{i}}\}_{\bar{i} \in J}$.*

If the images of the elements $\{v_{\bar{i}}\}_{\bar{i} \in L^{(3)}}$ in qA are linearly independent, then the images $\{a_{\bar{i}}\}_{\bar{i} \in L}$ of the elements $\{v_{\bar{i}}\}_{\bar{i} \in L}$ form a basis of A and the algebra A is Koszul.

Proof. It is a particular case of Theorem 4.2.5. □

In the example of the universal enveloping algebra $U(\mathfrak{g})$ of a Lie algebra $\mathfrak{g}$ (cf. Sect. 3.6.7), the symmetric monomials basis of $S(\mathfrak{g})$ induces a PBW basis of $U(\mathfrak{g})$. With the suitable order

$$0 < 1 < 2 < (1, 1) < 3 < (2, 1) < (1, 2) < (1, 1, 1) < 4 < (3, 1) < (2, 2) < \cdots,$$

the Cartan–Serre basis of Sect. 3.6.8 is a PBW basis of the Steenrod algebra.

In the inhomogeneous case too, the notion of PBW basis is equivalent and dual to that of Gröbner basis.

Proposition 4.3.11. *Let A be an inhomogeneous quadratic algebra with a quadratic-linear presentation $A = A(V, R)$ satisfying conditions (ql_1) and (ql_2) and such that $T(V)$ is equipped with a suitable ordered basis $\{v_{\bar{i}}\}_{\bar{i}\in J}$. Let $\varphi : qR \to V$ be the linear map whose graph gives R. The elements $\{a_{\bar{l}}\}_{\bar{l}\in L}$ form a PBW basis of A if and only if the elements*

$$\left\{(\mathrm{Id}-\varphi)\left(v_i v_j - \sum_{(k,l)\in L^{(2)},(k,l)<(i,j)} \lambda_{k,l}^{i,j} v_k v_l\right)\right\}_{(i,j)\in\overline{L}^{(2)}}$$

spanning R, form a Gröbner basis of the ideal (R) in $T(V)$.

Proof. The proof of the inhomogeneous case is similar but uses the PBW Theorem 3.6.4. □

4.3.9 PBW Basis of the Koszul Dual Algebra

Proposition 4.2.7 shows that any PBW basis of a quadratic algebra induces a dual PBW basis on the Koszul dual algebra. In this section, we provide further details.

Let $\{v_i\}_{i\in I}$ be a finite ordered basis of the vector space V and consider a suitable order on tuples. The elements $\{a_k a_l\}_{(k,l)\in L^{(2)}}$ form a basis of $V^{\otimes 2}/R$. In this case, there are elements

$$\left\{v_i v_j - \sum_{(k,l)\in L^{(2)},(k,l)<(i,j)} \lambda_{k,l}^{i,j} v_k v_l\right\}_{(i,j)\in\overline{L}^{(2)}},$$

which form a basis of R. So the complement set $\overline{L}^{(2)} = I^2 \setminus L^{(2)}$ labels a basis of R, which is not $\{v_i v_j\}_{(i,j)\in\overline{L}^{(2)}}$ itself in general. The dual elements $\{v_i^*\}_{i\in I}$ provide a dual basis of V^* and the elements

$$\left\{v_k^* v_l^* + \sum_{(i,j)\in\overline{L}^{(2)},(i,j)>(k,l)} \lambda_{k,l}^{i,j} v_i^* v_j^*\right\}_{(k,l)\in L^{(2)}}$$

provide a basis of $R^\perp$. Therefore the image of the elements $\{v_i^* v_j^*\}_{(i,j)\in\overline{L}^{(2)}}$ in $A^! = A(V^*, R^\perp)$, denoted $a_i^* a_j^*$, form a basis of $V^{*\otimes 2}/R^\perp$. We consider the opposite order $i \leqslant^{\mathrm{op}} j$, defined by $i \geqslant j$, on the labeling set I of the dual basis of V^*.

Theorem 4.3.12. *Let $A = A(V, R)$ be a quadratic algebra endowed with a PBW basis $\{a_{\bar{i}}\}_{\bar{i}\in L}$. Its Koszul dual algebra $A^!$ admits the PBW basis $\{a_{\bar{j}}^*\}_{\bar{j}\in\overline{L}}$, with opposite order.*

Proof. It is a direct corollary of Proposition 4.2.7. □

When the algebra $A = A(V, R)$ admits a PBW basis labeled by a set L, the Koszul dual coalgebra $A^{¡}$ admits a basis indexed by the set $\overline{L}$ by Proposition 4.2.3. More precisely, the Koszul dual coalgebra admits a basis of the form

$$\left\{ s^n v_{\bar{i}} + \sum_{\bar{j}<\bar{i},\ \bar{j}\in L^{(n)}} s^n \lambda_{\bar{j}}^{\bar{i}} v_{\bar{j}},\ \bar{i} \in \overline{L}^{(n)} \right\}_{n\in\mathbb{N}}.$$

4.4 Koszul Duality Theory and Lattices

We introduce a combinatorial criterion for Koszulity. It states that a certain family of lattices associated to a quadratic data is distributive if and only if the quadratic data is Koszul (Backelin criterion). It will allow us to prove that the Koszul property is stable under Manin products in the next section.

4.4.1 Poset and Lattice

This section recalls the basic properties of posets (partially ordered sets) and lattices. It mainly comes from R.P. Stanley's book [Sta97a].

In a poset with a partial order denoted $\leqslant$, a *least upper bound* z for two elements x and y, when it exists, is an upper bound, meaning $x \leqslant z$ and $y \leqslant z$, which is less than any other upper bound. When it exists, it is unique. It is denoted by $x \vee y$ and is called the *join*. Dually, there is the notion of *greatest lower bound* which is denoted by $x \wedge y$ and called the *meet*.

A *lattice* is a poset where the join and the meet exist for every pair of elements. These two operations are associative, commutative and idempotent, that is $x \vee x = x = x \wedge x$. They satisfy the *absorption law* $x \wedge (x \vee y) = x = x \vee (x \wedge y)$ and the partial order can be recovered by $x \leqslant y \iff x \wedge y = x \iff x \vee y = y$. A *sublattice generated by a subset* of a lattice L is the smallest sublattice of L stable for the operations join and meet. It is explicitly composed by the elements obtained by composing the generating elements with the operations join and meet.

A lattice is *distributive* if it satisfies the equivalent *distributivity relations*

$$x \vee (y \wedge z) = (x \vee y) \wedge (x \vee z) \iff x \wedge (y \vee z) = (x \wedge y) \vee (x \wedge z).$$

The subsets of a set form a distributive lattice where the partial order is defined by the inclusion $\subset$ and where the join and meet are given by the union $\cup$ and the intersection $\cap$ respectively. Actually, any finite distributive lattice is of this form (fundamental theorem for finite distributive lattices). Notice also that a distributive sublattice generated by a finite number of elements is finite.

In the linear context, we will consider the lattice of sub-spaces of a vector space. The order is given by the inclusion $\subset$ and the join and meet are given by the sum

$+$ and the intersection $\cap$ respectively. In the sequel, we will study finitely generated sublattices of such a lattice. Our main tool will be the following result, which is the analog, in the linear setting, of the fundamental theorem for finite distributive lattices.

Lemma 4.4.1. *Let U be a vector space and let L be a finitely generated sublattice of the lattice of sub-spaces of U. The lattice L is distributive if and only if there exists a basis $\mathscr{B}$ of U such that, $\mathscr{B} \cap X$ is a basis of X, for any $X \in L$.*

In this case, we say that the basis $\mathscr{B}$ *distributes* the sublattice L.

4.4.2 Lattice Associated to a Quadratic Data

Let (V, R) be a quadratic data. For every $n \in \mathbb{N}$, we consider the lattice of sub-spaces of $V^{\otimes n}$, where the order is given by the inclusion: $X \leqslant Y$ if $X \subset Y$. The join of two sub-spaces X and Y is their sum $X \vee Y := X + Y$ and their meet is the intersection $X \wedge Y := X \cap Y$.

For every $n \in \mathbb{N}$, we denote by $L(V, R)_{(n)}$ the sublattice of the lattice of sub-spaces of $V^{\otimes n}$ generated by the finite family $\{V^{\otimes i} \otimes R \otimes V^{\otimes n-2-i}\}_{i=0,\dots,n-2}$.

The example of $L(V, R)_{(3)}$ is depicted below.

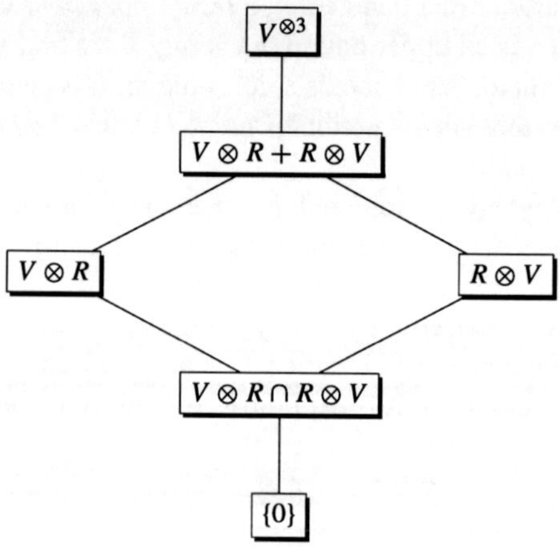

4.4.3 Backelin's Criterion

The following result belongs to the long list of properties between the Koszul duality theory and the poset theory.

Theorem 4.4.2 (Backelin [Bac83]). *A quadratic data (V, R) is Koszul if and only if the lattices $L(V, R)_{(n)}$ are distributive, for every $n \in \mathbb{N}$.*

In this case, the lattices $L(V,R)_{(n)}$, $n \geq 0$ are finite. We refer the reader to the Ph.D. thesis of J. Backelin [Bac83] and to the book of A. Polishchuk and L. Positselski [PP05] for the proof of this result and for more details about this subject.

4.5 Manin Products for Quadratic Algebras

In this section, extracted from Yu.I. Manin [Man87, Man88], we define two products $\bigcirc$ and $\bullet$ for quadratic data. They share nice properties with respect to Koszul duality theory. The black product is shown to produce Hopf algebras, some of which appear in quantum group theory.

4.5.1 Black and White Manin Products

Let (V,R) and (W,S) be two quadratic data. We denote by τ_{23} the isomorphism induced by the switching of the two middle terms:

$$\tau_{23} := \mathrm{Id}_V \otimes \tau \otimes \mathrm{Id}_W : V \otimes V \otimes W \otimes W \xrightarrow{\sim} V \otimes W \otimes V \otimes W.$$

By definition *Manin's white product of* (V,R) *and* (W,S) is the quadratic data given by

$$(V,R) \bigcirc (W,S) := \big(V \otimes W, \tau_{23}\big(R \otimes W^{\otimes 2} + V^{\otimes 2} \otimes S\big)\big).$$

By definition *Manin's black product of* (V,R) *and* (W,S) is the quadratic data given by

$$(V,R) \bullet (W,S) := \big(V \otimes W, \tau_{23}(R \otimes S)\big).$$

The quadratic data $(\mathbb{K}x, 0)$ is the unit object for the white product $\bigcirc$, where $\mathbb{K}x$ stands for a one-dimensional vector space spanned by x. Dually, the quadratic data $(\mathbb{K}x, (\mathbb{K}x)^{\otimes 2})$ is the unit object for the black product $\bullet$. The associated algebras are the free associative algebra on one generator $\mathbb{K}[x]$ and the algebra of dual numbers $D(\mathbb{K}x) = \mathbb{K}[x]/(x^2)$ on one generator respectively.

We denote by $A(V,R) \bigcirc A(W,S)$ and by $A(V,R) \bullet A(W,S)$ the algebras associated to the quadratic data obtained by white and black products. Notice that there is a morphism of quadratic algebras

$$A(V,R) \bullet A(W,S) \to A(V,R) \bigcirc A(W,S),$$

for any pair of quadratic data. The algebra associated to the white product is isomorphic to the Hadamard (or Segre) product

$$A(V,R) \bigcirc A(W,S) \cong A(V,R) \underset{\mathrm{H}}{\otimes} A(W,S) := \bigoplus_{n \in \mathbb{N}} A(V,R)_{(n)} \otimes A(W,S)_{(n)},$$

which is the weight-wise tensor product.

4.5.2 Manin Products and Koszul Duality

Manin's black and white products behave well with respect to Koszul duality theory.

Proposition 4.5.1. *Let (V,R) and (W,S) be two quadratic data, where V and W are finite dimensional. Black and white products are sent one to the other under the Koszul duality functor*

$$\big(A(V,R)\circ A(W,S)\big)^! = A(V,R)^! \bullet A(W,S)^!.$$

Proof. The quadratic algebra on the left-hand side is equal to

$$\begin{aligned} &A\big((V\otimes W)^*, \big(\tau_{23}\big(R\otimes W^{\otimes 2}+V^{\otimes 2}\otimes S\big)\big)^{\perp}\big) \\ &\cong A\big(V^*\otimes W^*, \tau_{23}\big(R^{\perp}\otimes W^{*\otimes 2}\cap V^{*\otimes 2}\otimes S^{\perp}\big)\big) \\ &\cong A\big(V^*\otimes W^*, \tau_{23}\big(R^{\perp}\otimes S^{\perp}\big)\big). \end{aligned}$$

□

Theorem 4.5.2. [BF85] *If two quadratic data are Koszul, then their white product and their black product are Koszul.*

Proof. First we prove the white product property.

Let (V,R) and (W,S) denote two Koszul quadratic data. By Theorem 4.4.2, the sublattices $L(V,R)_{(n)}$ of $V^{\otimes n}$ and $L(W,S)_{(n)}$ of $W^{\otimes n}$ are distributive, for any $n\in\mathbb{N}$. By Lemma 4.4.1, there exist bases $\mathscr{B}'_{(n)}$ and $\mathscr{B}''_{(n)}$ of $V^{\otimes n}$ and $W^{\otimes n}$ respectively that distribute $L(V,R)_{(n)}$ and $L(W,S)_{(n)}$. For any $n\in\mathbb{N}$, the sublattice $L(V\otimes W, \tau_{23}(R\otimes W^{\otimes 2}+V^{\otimes 2}\otimes S))_{(n)}$ of $(V\otimes W)^{\otimes n}$ is isomorphic to the sublattice of $V^{\otimes n}\otimes W^{\otimes n}$ generated by the finite family

$$\big\{V^{\otimes i}\otimes R\otimes V^{\otimes n-2-i}\otimes W^{\otimes n}, V^{\otimes n}\otimes W^{\otimes i}\otimes S\otimes W^{\otimes n-2-i}\big\}_{i=0,\dots,n-2}.$$

Therefore, the basis $\mathscr{B}_{(n)} := \{\beta'\otimes\beta'' \mid \beta'\in\mathscr{B}'_{(n)}, \beta''\in\mathscr{B}''_{(n)}\}$ distributes $L(V\otimes W, \tau_{23}(R\otimes W^{\otimes 2}+V^{\otimes 2}\otimes S))_{(n)}$. We conclude by using Theorem 4.4.2 in the other way round.

To prove the same result for the black product, we consider the Koszul dual algebras and apply Proposition 4.5.1 and Proposition 3.4.5. □

4.5.3 Adjunction and Internal (Co)Homomorphism

The white and black products satisfy the following adjunction formula.

Proposition 4.5.3. *There is a natural bijection in the category* Quad-alg *of quadratic algebras (or equivalently quadratic data):*

$$\mathrm{Hom}_{\textsf{quad alg}}(A\bullet B, C)\cong \mathrm{Hom}_{\textsf{quad alg}}\big(A, B^!\circ C\big),$$

when B is a finitely generated algebra.

Proof. Let A, B and C be the three algebras associated to the three quadratic data (V, R), (W, S) and (X, T) respectively. There is a one-to-one correspondence between the maps $f : V \otimes W \to X$ and the maps $\tilde{f} : V \to W^* \otimes X$. Such a map satisfies $f^{\otimes 2} : \tau_{23}(R \otimes S) \to T$ if and only if $\tilde{f}^{\otimes 2} : R \to \tau_{23}(S^\perp \otimes X + (W^*)^{\otimes 2} \otimes T)$. □

In other words, $\mathsf{Hom}(B, C) := B^! \bigcirc C$ is the internal 'Hom' functor in the monoidal category of finitely generated quadratic algebras with the black product as tensor product. Dually, $\mathsf{CoHom}(A, B) := A \bullet B^!$ is the internal 'coHom' (or inner) functor in the monoidal category of finitely generated quadratic algebras with the white product as tensor product.

4.5.4 Manin Complexes

Let us apply this adjunction to the three quadratic algebras $(\mathbb{K}x, (\mathbb{K}x)^{\otimes 2})$, $A(V, R)$, $A(V, R)$, where V is finite dimensional. As usual we write $A = A(V, R)$. Since the first one is the unit object for the black product, we get the bijection

$$\mathrm{Hom}_{\mathsf{quad\ alg}}(A, A) \cong \mathrm{Hom}_{\mathsf{quad\ alg}}\big(\mathbb{K}[x]/(x^2), A^! \bigcirc A\big).$$

To the identity of A on the left-hand side corresponds a natural morphism of quadratic algebras $\mathbb{K}[x]/(x^2) \to A^! \bigcirc A = A^! \underset{\mathrm{H}}{\otimes} A$, which is equivalent to a square zero element ξ in $A^! \bigcirc A$. We define a differential d_ξ by multiplying elements of $A^! \underset{\mathrm{H}}{\otimes} A$ by ξ, that is $d_\xi(\alpha) := \alpha\xi$. The chain complex $(A^! \underset{\mathrm{H}}{\otimes} A, d_\xi)$ thereby obtained is called the *first Manin complex* denoted by $L(A)$ in [Man88, Chap. 9]. If we choose a basis $\{v_i\}_{i=1,\dots,n}$ for V and denote by $\{v_i^*\}_{i=1,\dots,n}$ the dual basis of V^*, the square zero element ξ is equal to $\sum_{i=1}^n v_i^* \otimes v_i$.

The *second Manin complex* $\widetilde{L}(A)$ is defined on the tensor product $A^! \otimes A$ by the differential $\tilde{d}_\xi(\alpha) := \alpha \mapsto \xi\alpha - (-1)^{|\alpha|}\alpha\xi$, where the cohomological degree is given by the weight of $A^!$. Under this degree convention, there is a isomorphism of graded modules

$$A^! \otimes A \cong \mathrm{Hom}\big(A^{\text{¡}}, A\big),$$

which sends the squarezero element ξ to the twisting morphism κ and the aforementioned differential to $\partial_\kappa(f) = [f, \kappa]$. This induces an isomorphism of cochain complexes

$$\widetilde{L}(A) = \big(A^! \otimes A, \tilde{d}_\xi\big) \cong \mathrm{Hom}^\kappa\big(A^{\text{¡}}, A\big).$$

So, when the algebra A is Koszul, the second Manin complex computes the homology functor $\mathrm{Ext}_A^\bullet(\mathbb{K}, A)$ and likewise the Hochschild cohomology of A with coefficients into itself, see Sect. 9.1.7.

4.5.5 Hopf Algebras

We show that the black product construction gives rise to Hopf algebras.

Proposition 4.5.4. *Let (V, R) be a quadratic data with V finite dimensional and let $A = A(V, R)$. The algebra $A^! \bullet A$ is a Hopf algebra.*

Proof. By general properties of adjunction (see Appendix B.2.1), $A^! \bullet A$ is a comonoid in the monoidal category (quad alg, $\bigcirc = \underset{H}{\otimes}, \mathbb{K}[\varepsilon]$). Since $A^! \underset{H}{\otimes} A$ embeds into $A^! \otimes A$, the space $A^! \bullet A$ is a comonoid in the monoidal category of graded algebras with the classical tensor product, which makes it into a bialgebra. Since it is conilpotent, the antipode comes for free. □

This method was used by Yu.I. Manin to study quantum groups in [Man87, Man88].

4.6 Résumé

4.6.1 Rewriting Method

Let $A(V, R)$ be a quadratic algebra such that $V = \oplus_{i=1}^{n} \mathbb{K} v_i$ is a vector space equipped with a finite ordered basis. We order $V^{\otimes 2}$ by using, for instance, the lexicographical order:

$$v_1 v_1 < v_1 v_2 < \cdots < v_1 v_n < v_2 v_1 < \cdots < v_n v_n.$$

Typical relation:

$$v_i v_j = \sum_{(k,l)<(i,j)} \lambda_{k,l}^{i,j} v_k v_l, \quad \lambda_{k,l}^{i,j} \in \mathbb{K}.$$

The element $v_i v_j$ is called a *leading term*. The monomial $v_i v_j v_k$ is called *critical* if both $v_i v_j$ and $v_j v_k$ are leading terms.

Theorem. *Confluence for all the critical monomials $\Rightarrow$ Koszulity of the algebra.*

4.6.2 Reduction by Filtration and Diamond Lemma

Let $A = A(V, R)$ be a quadratic algebra. Any grading on $V \cong V_1 \oplus \cdots \oplus V_k$ together with a suitable order on tuples induce a filtration on the algebra A and

$$\psi : \mathring{A} := A(V, R_{\text{lead}}) \twoheadrightarrow \operatorname{gr} A,$$

with $R_{\text{lead}} = \langle \text{Leading Term}(r),\ r \in R \rangle$.

DIAMOND LEMMA FOR QUADRATIC ALGEBRAS.

$$\mathring{A} \quad \textit{Koszul} \quad \textit{and} \quad \mathring{A}^{(3)} \rightarrowtail (\operatorname{gr} A)^{(3)} \quad \Longrightarrow \quad A \quad \textit{Koszul} \quad \textit{and} \quad \mathring{A} \cong \operatorname{gr} A$$

INHOMOGENEOUS CASE.

$$\begin{array}{lcl} q\mathring{A} \quad \textit{Koszul} \quad \textit{and} & & A \quad \textit{Koszul} \quad \textit{and} \\ q\mathring{A}^{(3)} \rightarrowtail (\operatorname{gr}_\chi qA)^{(3)} & \Longrightarrow & q\mathring{A} \cong \operatorname{gr}_\chi qA \cong qA \cong \operatorname{gr} A \cong A. \end{array}$$

4.6.3 PBW Basis, Gröbner Basis and Diamond Lemma

Particular case:

$$\forall i \in I = \{1, \dots, k\}, \quad \dim(V_i) = 1 \quad \Leftrightarrow \quad \{v_i\}_{i \in I} \text{ basis of } V,$$
$$\mathring{A} \text{ monomial algebra} \Rightarrow \mathring{A} \text{ Koszul and basis } \{v_{\bar{i}}\}_{\bar{i} \in L}.$$

PBW basis of $A(V, R)$: basis $\{a_{\bar{i}}\}_{\bar{i} \in L} :=$ image of $\{v_{\bar{i}}\}_{\bar{i} \in L}$ under $\mathring{A} \twoheadrightarrow \operatorname{gr} A$.

MAIN PROPERTIES OF PBW BASES.

$$A(V, R) \textit{ PBW basis} \quad \Rightarrow \quad A(V, R) \textit{ Koszul algebra.}$$

DIAMOND LEMMA.

$$\{a_{\bar{i}}\}_{\bar{i} \in L^{(3)}} \textit{ linearly independent} \quad \Longrightarrow \quad \{a_{\bar{i}}\}_{\bar{i} \in L} \textit{ PBW basis.}$$

GRÖBNER BASIS.

$$\textit{Gröbner basis of } (R) \subset T(V) \quad \Longleftrightarrow \quad \textit{PBW basis of } T(V)/(R).$$

PBW bases for inhomogeneous quadratic algebras:

$$qA = A(V, qR) \text{ PBW basis} \quad \Rightarrow \quad A(V, R) \text{ PBW basis.}$$

PBW bases on Koszul dual algebra:

$$A = A(V, R) \text{ PBW basis} \quad \Longleftrightarrow \quad A^! = A(V^*, R^\perp) \text{ PBW basis.}$$

4.6.4 Backelin Criterion

$L(V, R)_{(n)}$: lattice of sub-spaces of $V^{\otimes n}$ generated by

$\{V^{\otimes i} \otimes R \otimes V^{\otimes n-2-i}\}_{i=0,\dots,n-2}$,

(V, R) Koszul quadratic data $\Longleftrightarrow$ $L(V, R)_{(n)}$ distributive lattice, $\forall n \in \mathbb{N}$.

4.6.5 Manin Black and White Products

$$(V,R)\bigcirc(W,S) := \big(V\otimes W, \tau_{23}\big(R\otimes W^{\otimes 2}+V^{\otimes 2}\otimes S\big)\big),$$
$$(V,R)\bullet(W,S) := \big(V\otimes W, \tau_{23}(R\otimes S)\big).$$

Unit for the white product: $\mathbb{K}[x]$. Unit for the black product: $\mathbb{K}[x]/(x^2)$.

$$(A\bigcirc B)^! = A^!\bullet B^!,$$

Theorem. *Manin products preserve the Koszul property.*

$$\mathrm{Hom}_{\text{quad alg}}(A\bullet B, C)\cong \mathrm{Hom}_{\text{quad alg}}\big(A, B^!\bigcirc C\big),$$
$$A^!\bigcirc A : \textit{Manin chain complex}; \quad A^!\bullet A : \textit{Hopf algebra}.$$

4.7 Exercises

Exercise 4.7.1 (An example). Show that the quadratic algebra presented by the generators x, y, z and the relators $xy-yz, zy-yx, xz-zx, y^2-zx$ is Koszul by the rewriting method.

Exercise 4.7.2 (Distributive law ★). Apply the method of Sect. 4.2 to the following case. Let $A(V\oplus W, R\oplus D_\lambda\oplus S)$ be a quadratic algebra, where $R\subset V^{\otimes 2}$, $S\subset W^{\otimes 2}$ and where $D_\lambda\subset V\otimes W\bigoplus W\otimes V$ is the graph of a linear morphism $\lambda: W\otimes V\to V\otimes W$. Let us use the following notations $A:=A(V,R)$, $B:=A(W,S)$ and $A\vee_\lambda B:=A(V\oplus W, R\oplus D_\lambda\oplus S)$.

We consider the following ordered grading $V_1:=V$ and $V_2:=W$ together with the lexicographic order. In this case, prove that $R_{\text{lead}}=R\bigoplus W\otimes V\bigoplus S$ and that $\mathring{A}=A\vee_0 B$. Show that the underlying module satisfies $\mathring{A}\cong A(V,R)\otimes A(W,S)$ and make the product explicit. Dually, show that the underlying module satisfies $\mathring{A}^{\text{¡}}\cong A(W,S)^{\text{¡}}\otimes A(V,R)^{\text{¡}}$ and make the coproduct explicit. We now suppose that the two quadratic data (V,R) and (W,S) are Koszul. Show that the quadratic data $(V\oplus W, R\bigoplus W\otimes V\bigoplus S)$ is also Koszul.

Finally, show that if the maps $V^{\otimes 2}/R\otimes W\to A$ and $V\otimes W^{\otimes 2}/S\to A$ are injective, then the algebra A is Koszul and its underlying graded module is isomorphic to $A\cong A(V,R)\otimes A(W,S)$.

Extra question: when the generating spaces V and W are finite dimensional, prove that the Koszul dual algebra has the same form:

$$A^! = A\big(V^*\oplus W^*, R^\perp\oplus D_{{}^t\lambda}\oplus S^\perp\big) = A^!\vee_{{}^t\lambda} B^!,$$

where ${}^t\lambda: V^*\otimes W^*\to W^*\otimes V^*$ is the transpose map.

Exercise 4.7.3 (Equivalent definitions of PBW bases ★). Let $A=(V,R)$ be a quadratic algebra endowed with a family of elements $\{a_{\bar{i}}\}_{\bar{i}\in L}$ of A labeled by a set $L\subset J$. Prove that, under condition (2) of Sect. 4.3.6, condition (1) is equivalent to

(1′) for any pair $(i, j) \in I^2$, if $(i, j) \notin L^{(2)}$, the product $a_i a_j$ in A can be written as a linear combination of strictly lower terms labeled by $L^{(2)}$:

$$a_i a_j = \sum_{(k,l)\in L^{(2)}, (k,l)<(i,j)} \lambda_{k,l}^{i,j} a_k a_l,$$

with $\lambda_{k,l}^{i,j} \in \mathbb{K}$.

In the same way, prove that, under condition (2), condition (1) and (1′) are equivalent to

(1″) for any $\bar{i} \in J$, if $\bar{i} \notin L$, then the element $a_{\bar{i}} \in A$ can be written as a linear combination of strictly lower terms labeled by L:

$$a_{\bar{i}} = \sum_{\bar{l}\in L, \bar{l}<\bar{i}} \lambda_{\bar{l}}^{\bar{i}} a_{\bar{l}},$$

with $\lambda_{\bar{l}}^{\bar{i}} \in \mathbb{K}$.

Exercise 4.7.4 (Hilbert–Poincaré series and PBW bases ★). Compute the Hilbert–Poincaré series of a quadratic algebra endowed with a PBW basis, see [PP05, Sect. 4.6].

Exercise 4.7.5 (From PBW to Koszul ★). A quadratic algebra $A(V, R)$ is called *n-PBW* if it admits an extra ordered grading $V \cong V_1 \oplus \cdots \oplus V_k$ such that $\dim V_i \leq n$, for any i, if the algebra $\mathring{A}$ is Koszul and if the isomorphism $\mathring{A} \cong \operatorname{gr} A$ holds. Notice that 1-PBW algebra are the algebras having a PBW basis.

Prove the following inclusions of categories

$$\mathsf{PBW} = 1\text{-}\mathsf{PBW} \subset 2\text{-}\mathsf{PBW} \subset \cdots \subset \mathsf{fg\ Koszul},$$

where the last category is the category of finitely generated Koszul algebras.

Show that the quadratic algebra $A(x, y, z; x^2 - yz, x^2 + 2zy)$ is 2-PBW but not 1-PBW.

Exercise 4.7.6 (Inhomogeneous Koszul duality theory with PBW-bases ★). Let $A(V, R)$ be an inhomogeneous Koszul algebra endowed with a PBW basis. Make the constructions of Sect. 3.6 explicit with this basis.

For instance, the degree-wise linear dual of the Koszul dual dg coalgebra $A^{¡} = ((qA)^{¡}, d_\varphi)$ is a dga algebra. Make it explicit with its differential. In the case of the Steenrod algebra, show that this gives the Λ algebra (see [BCK+66, Wan67, Pri70]).

Exercise 4.7.7 (PBW bases and Manin products ★). Let $A := A(V, R)$ and $B := A(W, S)$ be two quadratic algebras with ordered bases of V and W labeled by I_A and I_B respectively. Suppose that we have a PBW basis on A and on B labeled respectively by L_A and L_B. We denote by $\bar{i} = (i_1, \ldots, i_n)$ the elements of I_A^n and by $\bar{j} = (j_1, \ldots, j_n)$ the elements of I_B^n.

Show that the following set labels a PBW basis of the white product $A \bigcirc B$:

$$L_{A \bigcirc B} = \bigcup_{n \in \mathbb{N}} \{(i_1, j_1, i_2, j_2, \ldots, i_n, j_n) \mid \bar{i} \in L_A^{(n)} \text{ and } \bar{j} \in L_B^{(n)}\}$$
$$\cong \bigcup_{n \in \mathbb{N}} L_A^{(n)} \times L_B^{(n)}.$$

Dually, show that the following set labels a PBW basis of the black product $A \bullet B$:

$$L_{A \bullet B} = \{(i_1, j_1, i_2, j_2, \ldots, i_n, j_n) \mid \forall 1 \leq k < n, (i_k, i_{k+1}) \in L_A^{(2)} \text{ or } (j_k, j_{k+1}) \in L_B^{(2)}\}.$$

Exercise 4.7.8 (Coproduct ★). Let A and B be two nonunital associative algebras. Show that their coproduct $A \vee B$ in the category of nonunital associative algebras is given by a suitable product on

$$A \vee B = A \bigoplus B \bigoplus A \otimes B \bigoplus B \otimes A \bigoplus A \otimes B \otimes A \bigoplus B \otimes A \otimes B \bigoplus \cdots.$$

Let (V, R) and (W, S) be two quadratic data. Show that they admit a coproduct in the category of quadratic data, which is given by $(V \oplus W, R \oplus S)$. Prove that the quadratic algebra $A(V \oplus W, R \oplus S)$ is the coproduct of the quadratic algebras $A(V, R)$ and $A(W, S)$ in the category of unital associative algebras. When V and W are finite dimensional, compute its Koszul dual algebra.

Consider now the example of Exercise 4.7.2 and show that $A(V \oplus W, R \oplus D_\lambda \oplus S) \cong A \vee B/(D_\lambda)$, whence the notation $A \vee_\lambda B$.

Prove that if (V, R) and (W, S) are Koszul quadratic data, then so is their coproduct. [Give several different proofs, using the Koszul complex and the distributive lattices for instance.]

Prove that if $A := A(V, R)$ and $B := A(W, S)$ admit a PBW basis, then they can be used to construct a PBW basis on $A \vee B$.

Chapter 5
Algebraic Operad

The name 'operad' is a word that I coined myself, spending a week thinking about nothing else.

J.P. May in "Operads, algebras and modules"

An algebra of a certain type is usually defined by generating operations and relations, see for instance the classical definition of associative algebras, commutative algebras, Lie algebras. Given a type of algebras there is a notion of "free" algebra over a generic vector space V. Let us denote it by $\mathscr{P}(V)$. Viewed as a functor from the category Vect of vector spaces to itself, $\mathscr{P}$ is equipped with a monoid structure, that is a transformation of functors $\gamma : \mathscr{P} \circ \mathscr{P} \to \mathscr{P}$, which is associative, and another one $\eta : \mathrm{I} \to \mathscr{P}$ which is a unit. The existence of this structure follows readily from the universal properties of free algebras. Such a data $(\mathscr{P}, \gamma, \eta)$ is called an *algebraic operad.*

On the other hand, any algebraic operad $\mathscr{P}$ determines a type of algebras: the $\mathscr{P}$-algebras. The main advantage of this point of view on types of algebras is that operads look like associative algebras but in a different monoidal category (see Table 5.1).

So most of the constructions for associative algebras can be translated into this new context. This is exactly what we intend to do with Koszul duality theory in the following chapters.

Depending on further properties of the type of algebras, the associated operad might be of a special kind. For instance, if the generating operations are not supposed to satisfy any symmetry, if the relations are multilinear and if, in these relations, the variables stay in the same order, then the functor $\mathscr{P}$ is of the form

$$\mathscr{P}(V) = \bigoplus_n \mathscr{P}_n \otimes V^{\otimes n}$$

and the composition map γ is completely determined by $\mathbb{K}$-linear maps

$$\gamma : \mathscr{P}_k \otimes \mathscr{P}_{i_1} \otimes \cdots \otimes \mathscr{P}_{i_k} \longrightarrow \mathscr{P}_{i_1+\cdots+i_k}.$$

Then $\mathscr{P}$ is called a *nonsymmetric operad.*

J.-L. Loday, B. Vallette, *Algebraic Operads*,
Grundlehren der mathematischen Wissenschaften 346,
DOI 10.1007/978-3-642-30362-3_5,

Table 5.1 Operads as monoids

	Category	Product	Unit
monoid	Set	$\times$	$\{*\}$
algebra	Vect	$\otimes$	$\mathbb{K}$
operad	$\mathsf{EndoFunct}_{\mathsf{Vect}}$	$\circ$	I

More generally, if the relations are multilinear, without any further hypothesis, then the functor $\mathscr{P}$ is completely determined by a family of $\mathbb{S}_n$-modules $\{\mathscr{P}(n)\}_{n\geq 0}$,

$$\mathscr{P}(V) := \bigoplus_n \mathscr{P}(n) \otimes_{\mathbb{S}_n} V^{\otimes n}$$

and the composition map γ is completely determined by $\mathbb{K}$-linear maps

$$\gamma : \mathscr{P}(k) \otimes \mathscr{P}(i_1) \otimes \cdots \otimes \mathscr{P}(i_k) \longrightarrow \mathscr{P}(i_1 + \cdots + i_k).$$

Then $\mathscr{P}$ is called a *symmetric operad.*

Another interesting case, leading to the study of algebras with divided powers, consists in taking

$$\Gamma\mathscr{P}(V) := \bigoplus_n \big(\mathscr{P}(n) \otimes V^{\otimes n}\big)^{\mathbb{S}_n}.$$

Of course, taking invariants instead of coinvariants leads to a different type of algebras only in positive characteristic.

In this book we are going to work mainly with symmetric operads, that we simply call operads. Since $\mathbb{S}$-modules (family of representations over all the finite symmetric groups) play a prominent role in this case, we devote Sect. 5.1 of this chapter to their study and to the Schur functors that they determine. For symmetric operads the *monoidal* definition can be made explicit in several ways.

The *classical* definition consists in describing an operad in terms of the spaces $\mathscr{P}(n)$ of n-ary operations. This family of spaces forms the $\mathbb{S}$-module, which is equipped with "compositions of operations". They satisfy some properties which reflect functoriality, associativity and unitality of the monoidal definition.

The *partial* definition is a variation of the classical definition which takes advantage of the fact that we only need to know how to compose two operations to describe the whole structure. It is a description by generators and relations.

There is also a *combinatorial* way of describing an operad. It is based on the combinatorial objects which crop up in the description of a free operad, namely the rooted trees. One can construct a monad in the monoidal category of $\mathbb{S}$-modules out of the rooted trees, and an operad is simply a representation of this monad (i.e. an algebra over the monad). It has the advantage of deserving many variations by changing these combinatorial objects. For instance, nonsymmetric operads, shuffle operads, cyclic operads, modular operads, properads, permutads and several others can be described analogously by replacing the rooted trees by some other combinatorial objects, see Sect. 13.14.

One should keep in mind that the most economical way of defining a concrete operad is, most often, by describing the type of algebras that it determines. The relationship with the monoidal definition is via the notion of "free algebra" as mentioned above. Another way is the following. A type of algebras is determined by generating operations (possibly with symmetries) and relations (supposed to be multilinear). The generating operations and their symmetries determine the $\mathbb{S}$-module. Taking all the formal compositions of operations gives the free operad on the generating operations. The relations can be translated as relators which are operations in the free operad. The relators determine an operadic ideal and the expected operad is the quotient of the free operad by this ideal. The algebras over the quotient operad are exactly the algebras of the starting type.

Historically one can say that operad theory began with composition of functions. Let us mention the seminal paper of Michel Lazard "Lois de groupes et analyseurs" [Laz55] where a system of compositions was called an "analyseur" (in French). It gave rise to the notion of formal groups. For more about the history of "operads" we refer to the first chapter of [MSS02].

Here is the content of this chapter. In Sect. 5.1 we introduce the notions of $\mathbb{S}$-module and of Schur functor, and various constructions on them. In Sect. 5.2 we give the monoidal definition of an operad, and we define the notion of algebra (and also of coalgebra) over an operad. Then we restrict ourselves for the rest of the book to symmetric operads and nonsymmetric operads. In Sect. 5.3 we give the classical and the partial definitions of a symmetric operad. In Sects. 5.4 and 5.5 we describe in detail the free operad over an $\mathbb{S}$-module. In Sect. 5.6 we give the combinatorial definition of an operad. Then we make explicit the relationship between "types of algebras" and algebraic operads in Sect. 5.7.

In Sect. 5.8 we introduce the notion of cooperad which will play a prominent role in Koszul duality theory of quadratic operads.

In Sect. 5.9 we treat the notion of nonsymmetric operad. It can be read independently of the first eight sections of this chapter. It consists in replacing the starting $\mathbb{S}$-modules by graded vector spaces. So, it is a simpler object and it can be considered as a toy-model in the operad theory.

Then we give a brief résumé of all the definitions and we end this chapter with a list of exercises.

Though we work over a ground field $\mathbb{K}$, many of the notions presented in this chapter are valid when $\mathbb{K}$ is a commutative ring.

5.1 $\mathbb{S}$-Module and Schur Functor

We introduce $\mathbb{S}$-modules upon which the notion of algebraic operad is based in this book. Composition of $\mathbb{S}$-modules is the core of operad theory. To any $\mathbb{S}$-module is associated an endofunctor of Vect, called the Schur functor, and vice versa. The interplay between both structures is a fruitful game.

5.1.1 $\mathbb{S}$-Module

By definition an $\mathbb{S}$-*module* over $\mathbb{K}$ is a family

$$M = \big(M(0), M(1), \dots, M(n), \dots\big),$$

of right $\mathbb{K}[\mathbb{S}_n]$-modules $M(n)$ (cf. Appendix A). It is sometimes called a "collection" in the literature. An $\mathbb{S}$-module is said to be *finite dimensional* if $M(n)$ is a finite dimensional vector space for any n. For $\mu \in M(n)$ the integer n is called the *arity* of μ. A morphism of $\mathbb{S}$-modules $f : M \to N$ is a family of $\mathbb{S}_n$-equivariant maps $f_n : M(n) \to N(n)$. When all the maps f_n are injective the $\mathbb{S}$-module M is said to be a *sub-$\mathbb{S}$-module* of N.

When $M(0) = 0$, the $\mathbb{S}$-module is called *reduced*.

5.1.2 Schur Functor

To any $\mathbb{S}$-module M we associate its *Schur functor* $\widetilde{M} : \mathsf{Vect} \to \mathsf{Vect}$ defined by

$$\widetilde{M}(V) := \bigoplus_{n\geq 0} M(n) \otimes_{\mathbb{S}_n} V^{\otimes n}.$$

Here $V^{\otimes n} = \underbrace{V \otimes \cdots \otimes V}_{n}$ is viewed as a left $\mathbb{S}_n$-module under the left action

$$\sigma \cdot (v_1, \dots, v_n) := (v_{\sigma^{-1}(1)}, \dots, v_{\sigma^{-1}(n)}).$$

So the tensor product over $\mathbb{S}_n$ (i.e. over the ring $\mathbb{K}[\mathbb{S}_n]$) used in the definition of $\widetilde{M}$ is well-defined. Equivalently $\widetilde{M}(V)$ is the sum over n of the coinvariants of $M(n) \otimes V^{\otimes n}$ by the diagonal right action of $\mathbb{S}_n$.

Any morphism of $\mathbb{S}$-modules $\alpha : M \to N$ gives rise to a transformation of functors $\tilde{\alpha} : \widetilde{M} \to \widetilde{N}$.

Sometimes we need to work with the product instead of the sum of the components. We call *complete Schur functor* the infinite product:

$$\widehat{M}(V) := \prod_{n\geq 0} M(n) \otimes_{\mathbb{S}_n} V^{\otimes n}.$$

If the $\mathbb{S}$-module M is concentrated in arity 0 (resp. 1, resp. n), then the functor $\widetilde{M}$ is constant (resp. linear, resp. homogeneous polynomial of degree n). Observe that we get the *identity functor*, denoted $\widetilde{\mathrm{I}}$, by taking the Schur functor of $\mathrm{I} := (0, \mathbb{K}, 0, 0, \dots)$, so $\widetilde{\mathrm{I}}(V) = \mathrm{Id}_{\mathsf{Vect}}(V) = V$.

In this subsection we use the two notations M and $\widetilde{M}$, but later on we will use only M for both notions.

There are three important constructions on endofunctors of Vect: the direct sum, the tensor product and the composition.

The *direct sum* of two functors $F, G : \mathsf{Vect} \to \mathsf{Vect}$ is given by

$$(F \oplus G)(V) := F(V) \oplus G(V).$$

The *tensor product* $(F \otimes G)$ is given by

$$(F \otimes G)(V) := F(V) \otimes G(V).$$

The *composition* of functors, denoted $F \circ G$, is given by

$$(F \circ G)(V) := F\big(G(V)\big).$$

We are going to show that in each case, if the functors F and G are Schur functors, then the resulting functor is also a Schur functor. We also unravel the $\mathbb{S}$-module from which it comes.

For the direct sum, it is immediate: for any $\mathbb{S}$-modules M and N their *direct sum* is the $\mathbb{S}$-module $M \oplus N$ defined by

$$(M \oplus N)(n) := M(n) \oplus N(n).$$

It is obvious that

$$(\widetilde{M \oplus N}) = \widetilde{M} \oplus \widetilde{N}.$$

Lemma 5.1.1. *Let M be an $\mathbb{S}$-module. For any $n \geq 0$ the multilinear part of $\widetilde{M}(\mathbb{K}x_1 \oplus \cdots \oplus \mathbb{K}x_n)$ is isomorphic, as an $\mathbb{S}_n$-module, to $M(n)$.*

Proof. First, it is clear that the multilinear part of $\widetilde{M}(\mathbb{K}x_1 \oplus \cdots \oplus \mathbb{K}x_n)$ inherits a structure of $\mathbb{S}_n$-module from the action of the symmetric group on the set of variables $\{x_1, \ldots, x_n\}$. Second, the identification of these two $\mathbb{S}_n$-modules is given by $\mu \mapsto (\mu \otimes x_1 \cdots x_n)$ for $\mu \in M(n)$. □

5.1.3 Tensor Product of $\mathbb{S}$-Modules

For any $\mathbb{S}$-modules M and N their *tensor product* is the $\mathbb{S}$-module $M \otimes N$ defined by

$$(M \otimes N)(n) := \bigoplus_{i+j=n} \operatorname{Ind}_{\mathbb{S}_i \times \mathbb{S}_j}^{\mathbb{S}_n} M(i) \otimes N(j).$$

In this formula we use the notion of induced representation, cf. Appendix A.1. Since the subset $Sh(i, j)$ of (i, j)-shuffles of $\mathbb{S}_n$ is a convenient set of representatives of the quotient $\mathbb{S}_i \times \mathbb{S}_j \backslash \mathbb{S}_n$ (cf. 1.3.2), we have:

$$(M \otimes N)(n) \cong \bigoplus_{i+j=n} M(i) \otimes N(j) \otimes \mathbb{K}\big[Sh(i, j)\big].$$

This tensor product of $\mathbb{S}$-modules is sometimes called in the literature the *Cauchy product*.

Proposition 5.1.2. *The tensor product of $\mathbb{S}$-modules is associative with unit the $\mathbb{S}$-module $(\mathbb{K}, 0, 0, \ldots)$. There is an equality of Schur functors:*

$$\widetilde{(M \otimes N)} = \widetilde{M} \otimes \widetilde{N}.$$

Proof. The first part is straightforward. The proof of the equality follows from the identities (where $i + j = n$):

$$\begin{aligned}
&\left(\operatorname{Ind}_{\mathbb{S}_i \times \mathbb{S}_j}^{\mathbb{S}_n} M(i) \otimes N(j)\right) \otimes_{\mathbb{S}_n} V^{\otimes n} \\
&= \left(\left(M(i) \otimes N(j)\right) \otimes_{\mathbb{S}_i \times \mathbb{S}_j} \mathbb{K}[\mathbb{S}_n]\right) \otimes_{\mathbb{S}_n} \left(V^{\otimes i} \otimes V^{\otimes j}\right) \\
&= \left(M(i) \otimes N(j)\right) \otimes_{\mathbb{S}_i \times \mathbb{S}_j} \left(V^{\otimes i} \otimes V^{\otimes j}\right) \\
&= \left(M(i) \otimes_{\mathbb{S}_i} V^{\otimes i}\right) \otimes \left(N(j) \otimes_{\mathbb{S}_j} V^{\otimes j}\right).
\end{aligned}$$

□

5.1.4 Composite of $\mathbb{S}$-Modules

By definition the *composite* of the two $\mathbb{S}$-modules M and N is the $\mathbb{S}$-module

$$M \circ N := \bigoplus_{k \geq 0} M(k) \otimes_{\mathbb{S}_k} N^{\otimes k}.$$

The notation $N^{\otimes k}$ stands for the tensor product of k copies of the $\mathbb{S}$-module N. Observe that $\mathbb{S}_k$ is acting on $N^{\otimes k}$, that is $\mathbb{S}_k$ is acting on $N^{\otimes k}(n)$ for all n and this action commutes with the action of $\mathbb{S}_n$.

For instance, let $k = 2$. Then $N^{\otimes 2}(n) = (N \otimes N)(n) = \oplus_{i+j=n} N(i) \otimes N(j) \otimes \mathbb{K}[Sh(i, j)]$. The transposition $[2\ 1] \in \mathbb{S}_2$ is acting on the direct sum by sending

$$(\mu, \nu, \sigma) \in N(i) \otimes N(j) \otimes \mathbb{K}\left[Sh(i, j)\right]$$

to

$$(\nu, \mu, \sigma') \in N(j) \otimes N(i) \otimes \mathbb{K}\left[Sh(j, i)\right] \quad \text{where } \sigma' = \sigma[j+1\ \cdots\ i+j\ 1\ \cdots\ j].$$

When M and N are determined by only one representation, the operation $\circ$ is called the *plethysm* in representation theory.

Proposition 5.1.3. *The composite of the two $\mathbb{S}$-modules M and N satisfies the formula*

$$\widetilde{(M \circ N)} = \widetilde{M} \circ \widetilde{N},$$

where, on the right-hand side, the symbol $\circ$ stands for the composition of functors.

Proof. We want to prove $\widetilde{M}(\widetilde{N}(V)) = \widetilde{M \circ N}(V)$. We get:

$$\begin{aligned}
\widetilde{M} \circ \widetilde{N}(V) &= \bigoplus_k M(k) \otimes_{\mathbb{S}_k} \widetilde{N}(V)^{\otimes k} & \\
&= \bigoplus_k M(k) \otimes_{\mathbb{S}_k} \widetilde{N^{\otimes k}}(V) & \text{by compatibility of } \otimes \\
&= \bigoplus_{k,p} M(k) \otimes_{\mathbb{S}_k} \big(N^{\otimes k}(p) \otimes_{\mathbb{S}_p} V^{\otimes p}\big) & \text{by inspection} \\
&= \bigoplus_{k,p} \big(M(k) \otimes_{\mathbb{S}_k} N^{\otimes k}(p)\big) \otimes_{\mathbb{S}_p} V^{\otimes p} & \text{by associativity} \\
&= \bigoplus_p (M \circ N)(p) \otimes_{\mathbb{S}_p} V^{\otimes p} & \text{by definition of } \circ \\
&= \widetilde{M \circ N}(V). &
\end{aligned}$$

□

Corollary 5.1.4. *For any two $\mathbb{S}$-modules M and N one has*

$$(M \circ N)(n) = \bigoplus_{k \geq 0} M(k) \otimes_{\mathbb{S}_k} \Big(\bigoplus \operatorname{Ind}^{\mathbb{S}_n}_{\mathbb{S}_{i_1} \times \cdots \times \mathbb{S}_{i_k}} \big(N(i_1) \otimes \cdots \otimes N(i_k)\big)\Big)$$

where the second sum is extended, for fixed k and n, to all the nonnegative k-tuples $(i_1, \ldots, i_k)$ satisfying $i_1 + \cdots + i_k = n$.

Recall that a positive k-tuple $(i_1, \ldots, i_k)$ such that $i_1 + \cdots + i_k = n$ is called a *k-composition* of n.

The action of $\mathbb{S}_k$ on the right-hand side factor is on the set of k-tuples $\{(i_1, \ldots, i_k)\}$. This action is well-defined since the tensor product of vector spaces is associative and commutative.

Proof. It follows from the preceding propositions. □

5.1.5 Example

Suppose that $M(0) = N(0) = 0$ and $M(1) = N(1) = \mathbb{K}$. Then we get

$$\begin{aligned}
(M \circ N)(2) &\cong M(2) \oplus N(2), \\
(M \circ N)(3) &\cong M(3) \oplus \big(M(2) \otimes \operatorname{Ind}^{\mathbb{S}_3}_{\mathbb{S}_2}(N(2))\big) \oplus N(3),
\end{aligned}$$

where, as a vector space, $\operatorname{Ind}^{\mathbb{S}_3}_{\mathbb{S}_2}(N(2))$ is the sum of three copies of $N(2)$. Indeed, for $k = 3$, we get the component $M(3) \otimes_{\mathbb{S}_3} N(1)^{\otimes 3}$ which is isomorphic to $M(3)$. For $k = 1$, we get the component $M(1) \otimes_{\mathbb{S}_1} N(3)$ which is isomorphic to $N(3)$. For $k = 2$, we get the component

$$M(2) \otimes_{\mathbb{S}_2} \big(\operatorname{Ind}^{\mathbb{S}_3}_{\mathbb{S}_1 \times \mathbb{S}_2}\big(N(1) \otimes N(2)\big) \oplus \operatorname{Ind}^{\mathbb{S}_3}_{\mathbb{S}_2 \times \mathbb{S}_1}\big(N(2) \otimes N(1)\big)\big).$$

Since $\mathbb{S}_2$ is exchanging the two summands, we get the expected result.

5.1.6 Notation

From now on we abandon the notation $\tilde{}$ and so we denote by the same symbol the $\mathbb{S}$-module and its associated Schur functor. Hence we freely treat an $\mathbb{S}$-module as an endofunctor of Vect. Accordingly a morphism of $\mathbb{S}$-modules $\alpha : M \to N$ is sometimes called a transformation of functors (meaning: transformation of Schur functors).

If $f : F \to F'$ and $g : G \to G'$ are two morphisms of $\mathbb{S}$-modules (equivalently natural transformations of Schur functors), then we denote sometimes the morphisms

$$f \oplus g : F \bigoplus G \to F' \bigoplus G', \tag{5.1}$$

$$f \otimes g : F \otimes G \to F' \otimes G', \tag{5.2}$$

$$f \circ g : F \circ G \to F' \circ G' \tag{5.3}$$

by (f, g) when there is no confusion.

5.1.7 On the Notation of Elements in a Composite $\mathbb{S}$-Module

As a consequence of Corollary 5.1.4 the space $(M \circ N)(n)$ is spanned by the equivalence classes of the elements

$$(\mu; \nu_1, \dots, \nu_k; \sigma)$$

(under the action of $\mathbb{S}_k$) where $\mu \in M(k)$, $\nu_1 \in N(i_1), \dots, \nu_k \in N(i_k)$, $\sigma \in Sh(i_1, \dots, i_k)$. When $\sigma = \mathrm{id}_n \in \mathbb{S}_n$ (the identity permutation), we denote the relevant class either by

$$\mu \circ (\nu_1, \dots, \nu_k)$$

or by

$$(\mu; \nu_1, \dots, \nu_k).$$

5.1.8 Associativity Isomorphism of the Composite [Sign Warning]

The composition of Schur functors is associative. It implies that the composition of $\mathbb{S}$-modules is associative too. We would like to insist on the following phenomenon which does not happen in the algebra case (versus the operad case): in the associativity isomorphism $(M \circ N) \circ P \cong M \circ (N \circ P)$ of $\mathbb{S}$-modules, the switching map τ (see Sect. 1.5.2) plays a role. Indeed, in the identification of the component

$$\big(M(a) \otimes N(b) \otimes N(c)\big) \otimes P(d) \otimes P(e) \otimes P(f) \otimes P(g)$$

in $(M \circ N) \circ P$ with the component

$$M(a) \otimes \big((N(b) \otimes P(d) \otimes P(e)\big) \otimes \big(N(c) \otimes P(f) \otimes P(g)\big)$$

in $M \circ (N \circ P)$ we need to use the switching map to carry $N(c)$ over $P(d) \otimes P(e)$. As said above this phenomenon[1] does not occur in the algebra case and on the left-hand side because the product $\otimes$ is bilinear and the product $\circ$ is linear on the left-hand side. This phenomenon is important to notice in the sign-graded case since the occurrence of τ may result in signs in the formulas.

5.1.9 Composite of Morphisms

For any pair $f : M \to M'$, $g : N \to N'$ of morphisms of $\mathbb{S}$-modules, their *composite product* $f \circ g : M \circ N \to M' \circ N'$ is given explicitly by the formula

$$(f \circ g)(\mu; \nu_1, \ldots, \nu_k) := \big(f(\mu); g(\nu_1), \ldots, g(\nu_k)\big),$$

where $(\mu; \nu_1, \ldots, \nu_k)$ represents an element of

$$M(k) \otimes_{\mathbb{S}_k} \Big(\bigoplus \mathrm{Ind}_{\mathbb{S}_{i_1} \times \cdots \times \mathbb{S}_{i_k}}^{\mathbb{S}_n} \big(N(i_1) \otimes \cdots \otimes N(i_k)\big)\Big).$$

Beware: $f \circ g$ does not mean the composite of g and f in the sense of composition in a category, which has no meaning here anyway.

Observe that this composite is not linear in the right-hand side variable.

Proposition 5.1.5. *The category of* $\mathbb{S}$*-modules* ($\mathbb{S}$-Mod, $\circ$, I) *is a monoidal category.*

Proof. It follows from the comparison to Schur functors. □

In a first reading, the rest of this section can be bypassed and the reader can move to the beginning of Sect. 5.2.

---

[1]If the grandfather J wants to make a picture of his family, then he has two choices. He can put his children E, H and S on his right side, and then the grandchildren Y, B, A further right. Or, he can put the grandchildren on the right side of their parent: E, Y, B and H, A, and then put these subfamilies on his right. That gives two different pictures since JEHSYBA≠JEYBHAS. If J had only one child, the pictures would have been the same.

5.1.10 Generating Series

To any finite dimensional $\mathbb{S}$-module $M = \{M(n)\}_{n\geq 0}$ we associate its *generating series* (also called *Hilbert–Poincaré series*) defined by

$$f^M(x) := \sum_{n\geq 0} \frac{\dim M(n)}{n!} x^n.$$

Proposition 5.1.6. *Let M and N be two finite dimensional* $\mathbb{S}$*-modules. The following equalities hold:*

$$f^{M\oplus N}(x) = f^M(x) + f^N(x),$$
$$f^{M\otimes N}(x) = f^M(x) f^N(x),$$
$$f^{M\circ N}(x) = f^M\big(f^N(x)\big),$$

assuming $N(0) = 0$ *in the last equality.*

Proof. The first equality is immediate. The second one follows from Sect. 5.1.3. The third one (with $N(0) = 0$) follows from Corollary 5.1.4. □

5.1.11 Symmetric Function Indicator

There is a finer invariant than the generating series: the Frobenius characteristic. Starting with an $\mathbb{S}$-module M, it consists in taking the isomorphism class of the $\mathbb{S}_n$-representation $M(n)$ in the Grothendieck group of the representations of $\mathbb{S}_n$. The sum over n gives an element in $\prod_n Rep(\mathbb{S}_n)$ which is known to be isomorphic to the algebra of symmetric functions. The image of this element, denoted by F^M, is called the *Frobenius characteristic*. The operations $\oplus$, $\otimes$, $\circ$ on $\mathbb{S}$-modules commute with their counterpart in the algebra of symmetric functions, cf. [Mac95].

5.1.12 Hadamard Product of $\mathbb{S}$-Modules

By definition the *Hadamard product* of the two $\mathbb{S}$-modules $\mathscr{P}$ and $\mathscr{Q}$ is the $\mathbb{S}$-module $\mathscr{P} \underset{\mathrm{H}}{\otimes} \mathscr{Q}$ given by

$$\big(\mathscr{P} \underset{\mathrm{H}}{\otimes} \mathscr{Q}\big)(n) := \mathscr{P}(n) \otimes \mathscr{Q}(n),$$

where the action of $\mathbb{S}_n$ is the diagonal action.

The generating series of $\mathscr{P} \underset{\mathrm{H}}{\otimes} \mathscr{Q}$ is not the Hadamard product of the generating series of $\mathscr{P}$ and $\mathscr{Q}$, but it will be it in the nonsymmetric framework, see Sect. 5.9.10 for a discussion on this matter.

5.1.13 Linear Species

A right $\mathbb{S}$-module M can be viewed as a functor from the groupoid $\mathbb{S}$ of symmetric groups to the category Vect (or the category of $\mathbb{K}$-modules if $\mathbb{K}$ is a commutative ring). As a consequence it can be extended as a contravariant functor from the groupoid Bij of finite sets and all bijections to the category of vector spaces. We suppose that the empty set is an object of Bij. If X is a finite set, then the extended functor, still denoted by M, is given by the coinvariant space:

$$M(X) := \Big(\bigoplus_{f:\underline{n}\to X} M(n) \Big)_{\mathbb{S}_n}$$

where the sum is over all the bijections from $\underline{n} := \{1, \dots, n\}$ to X. The right action of $\sigma \in \mathbb{S}_n$ on $(f; \mu)$ for $\mu \in M(n)$ is given by $(f; \mu)^\sigma = (f\sigma; \mu^\sigma)$.

Such a functor is sometimes called a *linear species*, cf. [Joy86], [AM10, Sect. B.1.1]. Here is the translation of the above constructions into this language. Let M and N be functors from Bij to Vect. For any set X we have:

$$(M \oplus N)(X) = M(X) \oplus N(X),$$

$$(M \otimes N)(X) = \bigoplus_{X=Y\sqcup Z} M(Y) \otimes N(Z),$$

where the sum is over all the ordered disjoint unions$Y \sqcup Z$ of X,

$$(M \circ N)(X) = \bigoplus_{B\in \mathrm{PART}(X)} M(B) \otimes \bigotimes_{b\in \mathrm{B}} N(X_b),$$

where $\mathrm{PART}(X) =$ set of decompositions of X (see below)

$$\big(M \underset{\mathrm{H}}{\otimes} N\big)(X) = M(X) \otimes N(X).$$

See for instance [AM10, Appendix B].

5.1.14 On the Notation $\bigotimes_{b\in \mathbf{B}} N(X_b)$

A *decomposition* of the finite set X is a family of subsets $\{X_b\}_{b\in B}$ of X such that their disjoint union is X. We let n be the number of elements in B. For any contravariant functor $N : \mathrm{Bij} \to \mathrm{Vect}$ we define

$$\bigotimes_{b\in B} N(X_b) := \Big(\bigoplus_{f:\underline{n}\to B} N(X_{f(1)}) \otimes \cdots \otimes N(X_{f(n)}) \Big)_{\mathbb{S}_n}$$

where the sum is over all the bijections from $\underline{n}$ to B. As usual, the right action of $\mathbb{S}_n$ on the direct sum is given by

$$(f; \mu_1, \dots, \mu_n)^\sigma = (f\sigma; \mu_{\sigma(1)}, \dots, \mu_{\sigma(n)}),$$

where $\sigma \in \mathbb{S}_n$ and $\mu_i \in N(X_{f(i)})$.

5.1.15 Invariants Versus Coinvariants

The Schur functor $\mathscr{P}(V)$ can be written as a sum of coinvariants spaces

$$\mathscr{P}(V)=\bigoplus_n (\mathscr{P}(n)\otimes V^{\otimes n})_{\mathbb{S}_n}$$

where the symmetric group is acting diagonally on the tensor product. Here we use the fact that $V^{\otimes n}$ is a right module over $\mathbb{S}_n$.

Instead of working with coinvariants we could choose to work with invariants, that is to define

$$\Gamma\mathscr{P}(V):=\bigoplus_n (\mathscr{P}(n)\otimes V^{\otimes n})^{\mathbb{S}_n}.$$

Everything would work, because the direct sum, the tensor product and the composition of such functors are of the same type. In particular, there exists a composition $\bar{\circ}$ of $\mathbb{S}$-modules such that, for any two $\mathbb{S}$-modules $\mathscr{P}$ and $\mathscr{Q}$, one has

$$\Gamma\mathscr{P}\circ\Gamma\mathscr{Q}=\Gamma(\mathscr{P}\,\bar{\circ}\,\mathscr{Q}).$$

This composite is given by

$$(\mathscr{P}\,\bar{\circ}\,\mathscr{Q})(n):=\bigoplus_r (\mathscr{P}(r)\otimes \mathscr{Q}^{\otimes r})^{\mathbb{S}_r}(n).$$

Recall that the norm map of an $\mathbb{S}_n$-module M is given by

$$M_{\mathbb{S}_n}\to M^{\mathbb{S}_n},\qquad x\mapsto \sum_{\sigma\in\mathbb{S}_n} x^{\sigma}.$$

The norm map induces an $\mathbb{S}$-module map $\mathscr{P}\circ\mathscr{Q}\to\mathscr{P}\,\bar{\circ}\,\mathscr{Q}$ since we took coinvariants on the left-hand side and invariants on the right-hand side. Whenever $\mathscr{Q}(0)=0$ the induced transformation of functors

$$\Phi:\Gamma(\mathscr{P}\circ\mathscr{Q})\to\Gamma(\mathscr{P}\,\bar{\circ}\,\mathscr{Q})$$

is an isomorphism since $\mathbb{S}_r$ is acting freely on $\mathscr{Q}^{\otimes r}$ (cf. [Sto93, Fre00]).

In characteristic zero the norm map is an isomorphism, so $\mathscr{P}(V)\to\Gamma\mathscr{P}(V)$ is an isomorphism (see Appendix A). However in positive characteristic we get two different functors.

5.2 Algebraic Operad and Algebra over an Operad

We define a symmetric operad as a monoid in the monoidal category of symmetric modules. Since this is a monad, that is a monoid in the category of endofunctors of

Vect, one can define the notion of an algebra over an operad. Replacing $\mathbb{S}$-modules by arity graded vector spaces we get the notion of nonsymmetric operad. Taking invariants in place of coinvariants we get the notion of divided power operad. We call them collectively "algebraic operads".

5.2.1 Monoidal Definition of an Operad

By definition a *symmetric operad* $\mathscr{P} = (\mathscr{P}, \gamma, \eta)$ is an $\mathbb{S}$-module $\mathscr{P} = \{\mathscr{P}(n)\}_{n\geq 0}$ endowed with morphisms of $\mathbb{S}$-modules

$$\gamma : \mathscr{P} \circ \mathscr{P} \to \mathscr{P}$$

called *composition map*, and

$$\eta : \mathrm{I} \to \mathscr{P}$$

called the *unit map*, which make $\mathscr{P}$ into a monoid.

Explicitly, the morphisms γ and η satisfy the classical axioms for monoids, that is associativity:

$$\begin{array}{ccccc}
 & & \mathscr{P} \circ (\mathscr{P} \circ \mathscr{P}) & \xrightarrow{\mathrm{Id}\circ\gamma} & \mathscr{P} \circ \mathscr{P} \\
 & \overset{\cong}{\nearrow} & & & \\
(\mathscr{P} \circ \mathscr{P}) \circ \mathscr{P} & & & & \Big\downarrow \gamma \\
\Big\downarrow \gamma\circ\mathrm{Id} & & & & \\
\mathscr{P} \circ \mathscr{P} & & \xrightarrow{\qquad\gamma\qquad} & & \mathscr{P}
\end{array}$$

and unitality:

$$\begin{array}{ccccc}
\mathrm{I} \circ \mathscr{P} & \xrightarrow{\eta\circ\mathrm{Id}} & \mathscr{P} \circ \mathscr{P} & \xleftarrow{\mathrm{Id}\circ\eta} & \mathscr{P} \circ \mathrm{I} \\
 & \underset{=}{\searrow} & \Big\downarrow \gamma & \underset{=}{\swarrow} & \\
 & & \mathscr{P}. & &
\end{array}$$

Hence for any vector space V one has linear maps

$$\gamma(V) : \mathscr{P}\big(\mathscr{P}(V)\big) \to \mathscr{P}(V) \quad \text{and} \quad \eta(V) : V \to \mathscr{P}(V).$$

In the literature a monoid structure on an endofunctor is often called a monad, cf. Appendix B.4. Interpreting the $\mathbb{S}$-module $\mathscr{P}$ as a Schur functor, the monoid structure on $\mathscr{P}$ is nothing but a monad in the category of vector spaces.

Let $\mathscr{Q}$ be another symmetric operad. A *morphism of operads* from $\mathscr{P}$ to $\mathscr{Q}$ is a morphism of $\mathbb{S}$-modules $\alpha : \mathscr{P} \to \mathscr{Q}$, which is compatible with the composition maps. In other words, the following diagrams are supposed to be commutative:

$$\begin{array}{ccc} \mathscr{P} \circ \mathscr{P} & \xrightarrow{(\alpha,\alpha)} & \mathscr{Q} \circ \mathscr{Q} \\ {\scriptstyle \gamma_{\mathscr{P}}} \downarrow & & \downarrow {\scriptstyle \gamma_{\mathscr{Q}}} \\ \mathscr{P} & \xrightarrow{\alpha} & \mathscr{Q} \end{array} \qquad \begin{array}{ccc} & \mathrm{I} & \\ {\scriptstyle \eta_{\mathscr{P}}} \swarrow & & \searrow {\scriptstyle \eta_{\mathscr{Q}}} \\ \mathscr{P} & \xrightarrow{\alpha} & \mathscr{Q}. \end{array}$$

The category of operads over $\mathsf{Vect}_{\mathbb{K}}$ is denoted by $\mathsf{Op}_{\mathbb{K}}$ or Op.

In order to differentiate between the notion of composition in the operadic framework (the map γ) and the classical notion of composition of functors in category theory (denoted by $\circ$), we will sometimes say "operadic composition" for the first one.

Here we are mainly interested in the notion of operads in the category of vector spaces, or modules over a commutative ring, or in the category of chain complexes (dg spaces), but it is immediate to verify that it makes sense in any symmetric monoidal category with infinite sums such that finite sums commute with the monoidal structure.

When $\mathscr{P}(0) = 0$, the operad is called *reduced*.

5.2.2 Operadic Module

A *left module* over the symmetric operad $\mathscr{P}$ is an $\mathbb{S}$-module M together with an $\mathbb{S}$-module morphism $\mathscr{P} \circ M \to M$ satisfying associativity and unitality with respect to the operad structure on $\mathscr{P}$. The terminology varies a lot in the literature. It is sometimes called a "twisted $\mathscr{P}$-algebra", or sometimes simply a "$\mathscr{P}$-algebra" ; see [Fre09a] for a thorough study of this structure. There is a particular case which is important: when M is constant, that is concentrated in arity 0. This gives rise to the notion of algebra over an operad, see below.

Observe that the notion of *right module*, which is obvious to define, gives rise to a completely different structure.

A bimodule (or two-sided module) over $\mathscr{P}$ is an $\mathbb{S}$-module $\mathscr{M}$ together with morphisms of $\mathbb{S}$-modules $\mathscr{P} \circ \mathscr{M} \to \mathscr{M}$ and $\mathscr{M} \circ \mathscr{P} \to \mathscr{M}$ satisfying the classical axioms of associativity and unitality for two-sided modules.

5.2.3 Algebra over an Operad

By definition an *algebra over the operad* $\mathscr{P}$, or a $\mathscr{P}$*-algebra* for short, is a vector space A equipped with a linear map $\gamma_A : \mathscr{P}(A) \to A$ such that the following diagrams commute:

$$
\begin{array}{ccc}
 & \mathscr{P}(\mathscr{P}(A)) \xrightarrow{\mathscr{P}(\gamma_A)} \mathscr{P}(A) & \\
 \overset{=}{\nearrow} & & \\
(\mathscr{P}\circ\mathscr{P})(A) & & \downarrow{\scriptstyle \gamma_A} \\
{\scriptstyle \gamma(A)}\downarrow & & \\
\mathscr{P}(A) & \xrightarrow{\qquad\quad \gamma_A \quad\qquad} & A
\end{array}
\qquad\qquad
\begin{array}{ccc}
\mathrm{I}(A) & \xrightarrow{\eta(A)} & \mathscr{P}(A) \\
 & \underset{=}{\searrow} & \downarrow{\scriptstyle \gamma_A} \\
 & & A.
\end{array}
$$

The transformation of functors γ applied to A, that is $\gamma(A)$, is not to be confused with $\mathscr{P}(\gamma_A)$ which is the functor $\mathscr{P}$ applied to the map γ_A. They have the same source and the same target, but they are different.

Let A' be another $\mathscr{P}$-algebra. A *morphism of $\mathscr{P}$-algebras* is a linear map $f : A \to A'$ which makes the following diagram commutative:

$$
\begin{array}{ccc}
\mathscr{P}(A) & \xrightarrow{\gamma_A} & A \\
{\scriptstyle \mathscr{P}(f)}\downarrow & & \downarrow{\scriptstyle f} \\
\mathscr{P}(A') & \xrightarrow{\gamma_{A'}} & A'.
\end{array}
$$

We denote by $\mathscr{P}$-alg the category of $\mathscr{P}$-algebras.

Observe that if $\mathscr{P}$ is interpreted as a monad, then this is the classical notion of an algebra over a monad, see Appendix B.4.

5.2.4 Functors Between Categories of Algebras

Let $\alpha : \mathscr{P} \to \mathscr{Q}$ be a morphism of operads. Then there is a well-defined functor

$$\alpha^* : \mathscr{Q}\text{-alg} \longrightarrow \mathscr{P}\text{-alg}.$$

Indeed, the $\mathscr{P}$-algebra associated to the $\mathscr{Q}$-algebra A has the same underlying vector space structure and the composition map is the composite

$$\mathscr{P}(A) \xrightarrow{\alpha(A)} \mathscr{Q}(A) \to A.$$

Observe that the functor which assigns to an operad the category of algebras over this operad is contravariant.

We give in Proposition 5.2.2 an interpretation of a $\mathscr{P}$-algebra as a morphism of operads.

5.2.5 Free $\mathscr{P}$-Algebra

In the category of $\mathscr{P}$-algebras, a $\mathscr{P}$-algebra $\mathscr{F}(V)$, equipped with a linear map $\eta : V \to \mathscr{F}(V)$ is said to be *free* over the vector space V if it satisfies the following universal condition:

For any $\mathscr{P}$-algebra A and any linear map $f : V \to A$ there is a unique $\mathscr{P}$-algebra extension $\tilde{f} : \mathscr{F}(V) \to A$ of f:

$$\begin{array}{ccc} V & \xrightarrow{\eta} & \mathscr{F}(V) \\ & \searrow^{f} & \downarrow{\scriptstyle \tilde{f}} \\ & & A. \end{array}$$

Observe that a free algebra is unique up to a unique isomorphism, cf. Appendix B.2.2.

In other words, $\mathscr{F}$ is a functor $\mathsf{Vect} \to \mathscr{P}\text{-}\mathsf{alg}$ which is left adjoint to the forgetful functor:

$$\mathrm{Hom}_{\mathscr{P}\text{-}\mathsf{alg}}\big(\mathscr{F}(V), A\big) \cong \mathrm{Hom}_{\mathsf{Vect}}(V, A).$$

For any vector space V one can equip $\mathscr{P}(V)$ with a $\mathscr{P}$-algebra structure as follows. Define

$$\gamma_{\mathscr{P}(V)} := \gamma(V) : \mathscr{P}\big(\mathscr{P}(V)\big) \to \mathscr{P}(V).$$

The axioms defining the operad $\mathscr{P}$ show that $(\mathscr{P}(V), \gamma(V))$ is a $\mathscr{P}$-algebra.

Proposition 5.2.1. *The $\mathscr{P}$-algebra $(\mathscr{P}(V), \gamma(V))$ equipped with $\eta(V) : V \to \mathscr{P}(V)$ is the free $\mathscr{P}$-algebra over V.*

Proof. For any linear map $f : V \to A$, where A is a $\mathscr{P}$-algebra, we consider the composition $\tilde{f} : \mathscr{P}(V) \xrightarrow{\mathscr{P}(f)} \mathscr{P}(A) \xrightarrow{\gamma_A} A$. It extends f since the composite

$$V \xrightarrow{\eta(V)} \mathscr{P}(V) \xrightarrow{\mathscr{P}(f)} \mathscr{P}(A) \xrightarrow{\gamma_A} A$$

is f by $\mathscr{P}(f) \circ \eta(V) = \eta(A) \circ f$ and $\gamma_A \circ \eta(A) = \mathrm{I}_A$.

The following diagram is commutative by functoriality and the fact that A is a $\mathscr{P}$-algebra:

$$\begin{array}{ccc} \mathscr{P}(\mathscr{P}(V)) & \longrightarrow & \mathscr{P}(V) \\ \downarrow & & \downarrow \\ \mathscr{P}(\mathscr{P}(A)) & \longrightarrow & \mathscr{P}(A) \quad \tilde{f} \\ \downarrow & & \downarrow \\ \mathscr{P}(A) & \longrightarrow & A. \end{array}$$

It implies that the map $\tilde{f}$ is a $\mathscr{P}$-algebra morphism.

Let us show that the map $\tilde{f}$ is unique. Since we want $\tilde{f}$ to coincide with f on V and we want $\tilde{f}$ to be an algebra morphism, there is no other choice by $\tilde{f}$. $\square$

5.2.6 Endofunctors of Vect

In Sect. 5.1 we showed that any $\mathbb{S}$-module gives rise to an endofunctor of Vect, called the associated Schur functor. Similarly any graded vector space $\mathscr{P} = \{\mathscr{P}_n\}_{n\geq 0}$ gives rise to an endofunctor of Vect by the formula

$$\mathscr{P}(V) := \bigoplus_{n\geq 0} \mathscr{P}_n \otimes V^{\otimes n}.$$

We remark immediately that this endofunctor is the Schur functor associated to the $\mathbb{S}$-module, still denoted by $\mathscr{P}$, given by

$$\mathscr{P}(n) := \mathscr{P}_n \otimes \mathbb{K}[\mathbb{S}_n].$$

Writing $\mathscr{P}(n)$ or $\mathscr{P}_n$ suffices to indicate which framework we are working in. Moreover, most of the time, it is only the endofunctor which is relevant.

A third interesting case consists in starting with an $\mathbb{S}$-module and taking the invariants instead of the coinvariants when forming the endofunctor:

$$\Gamma\mathscr{P}(V) := \bigoplus_{n\geq 0} \big(\mathscr{P}(n) \otimes V^{\otimes n}\big)^{\mathbb{S}_n}.$$

5.2.7 Symmetric Operads

In a symmetric operad $(\mathscr{P}, \gamma, \eta)$ the composition map γ is made up of linear maps

$$\gamma(i_1, \ldots, i_k) : \mathscr{P}(k) \otimes \mathscr{P}(i_1) \otimes \cdots \otimes \mathscr{P}(i_k) \longrightarrow \mathscr{P}(i_1 + \cdots + i_k)$$

that will be studied in the next section.

If A is a $\mathscr{P}$-algebra, then the structure map γ_A determines maps

$$\mathscr{P}(n) \otimes A^{\otimes n} \twoheadrightarrow \mathscr{P}(n) \otimes_{\mathbb{S}_n} A^{\otimes n} \xrightarrow{\gamma_A(n)} A.$$

Therefore, any element $\mu \in \mathscr{P}(n)$ and any n-tuple $a_1 \ldots a_n \in A^{\otimes n}$ give rise to an element

$$\gamma_A(n)(\mu; a_1 \ldots a_n) \in A.$$

Such an element μ is called an *n-ary operation* and $\mathscr{P}(n)$ is called the space of n-ary operations. By abuse of notation we write

$$\mu(a_1 \ldots a_n) := \gamma_A(n)(\mu; a_1 \ldots a_n),$$

and we call μ an operation on A.

The unit functor $\eta : \mathrm{I} \to \mathscr{P}$ defines a particular element in $\mathscr{P}(1)$, namely the image of $1 \in \mathbb{K} = \mathrm{I}(1)$, which we denote by $\mathrm{id} \in \mathscr{P}(1)$ and call the *identity operation*.

Indeed we have $\mathrm{id}(a) = a$ for any $a \in A$. For any symmetric operad $\mathscr{P}$ the space $\mathscr{P}(1)$ inherits the structure of a unital associative algebra over $\mathbb{K}$. It is given by the map $\gamma(1) : \mathscr{P}(1) \otimes \mathscr{P}(1) \to \mathscr{P}(1)$ and id is the unit of $\mathscr{P}(1)$.

5.2.8 Nonsymmetric Operads

A *nonsymmetric operad* (ns operad for short) is an arity graded vector space $\mathscr{P} = \{\mathscr{P}_n\}_{n\geq 0}$ endowed with morphisms $\gamma : \mathscr{P} \circ \mathscr{P} \to \mathscr{P}$ and $\eta : \mathrm{I} \to \mathscr{P}$ which make it into a monoid in the monoidal category of arity graded vector spaces. The composition map γ is completely determined by maps

$$\gamma_{i_1,\ldots,i_k} : \mathscr{P}_k \otimes \mathscr{P}_{i_1} \otimes \cdots \otimes \mathscr{P}_{i_k} \longrightarrow \mathscr{P}_{i_1+\cdots+i_k}$$

for $n = i_1 + \cdots + i_k$ and η is determined by an element $\mathrm{id} \in \mathscr{P}_1$. See Sect. 5.9 for more.

A nonsymmetric operad $\mathscr{P}$ gives rise to an operad, usually still denoted by $\mathscr{P}$, such that $\mathscr{P}(n) = \mathscr{P}_n \otimes \mathbb{K}[\mathbb{S}_n]$. The action of the symmetric group is induced by the $\mathbb{S}_n$-module structure of the regular representation $\mathbb{K}[\mathbb{S}_n]$. The composition map is the tensor product

$$\gamma(i_1, \ldots, i_k) = \gamma_{i_1,\ldots,i_k} \otimes \gamma^{Ass}(i_1, \ldots, i_k)$$

where γ^{Ass} is the composition map of the operad *Ass* that will be described below.

Such a symmetric operad is sometimes called a *regular operad*.

5.2.9 Operads with Divided Powers

Definitions and results of this subsection come from Benoit Fresse's paper [Fre00] in which the reader will find the details. Let $\{\mathscr{P}(n)\}_{n\geq 0}$ be an $\mathbb{S}$-module with $\mathscr{P}(0) = 0$. Recall that there is defined an endofunctor $\Gamma\mathscr{P}$ by using invariants instead of coinvariants:

$$\Gamma\mathscr{P}(V) := \bigoplus_{n\geq 1} \big(\mathscr{P}(n) \otimes V^{\otimes n}\big)^{\mathbb{S}_n}.$$

An *operad with divided powers* is a monoid structure on $\Gamma\mathscr{P}$, that is a composition map $\bar{\gamma} : \Gamma\mathscr{P} \circ \Gamma\mathscr{P} \to \Gamma\mathscr{P}$ which is associative and unital.

If $\mathscr{P} = (\mathscr{P}, \gamma, \eta)$ is a symmetric operad, then it determines an operad with divided powers as follows. First, recall from Sect. 5.1.15 that the norm map permits us to construct a map $\Phi : \Gamma(\mathscr{P} \circ \mathscr{P}) \to \Gamma(\mathscr{P} \mathbin{\bar{\circ}} \mathscr{P})$ which happens to be an isomorphism. The composition map $\bar{\gamma}$ is defined as the composite

$$\Gamma\mathscr{P} \circ \Gamma\mathscr{P} = \Gamma(\mathscr{P} \mathbin{\bar{\circ}} \mathscr{P}) \xrightarrow{\Phi^{-1}} \Gamma(\mathscr{P} \circ \mathscr{P}) \xrightarrow{\Gamma(\gamma)} \Gamma\mathscr{P}.$$

An algebra over $\Gamma\mathscr{P}$ is called a *$\mathscr{P}$-algebra with divided powers*. It can be shown that if $\mathscr{P}(n)$ is $\mathbb{S}_n$-projective, e.g. $\mathscr{P}(n)$ is the regular representation, then a $\mathscr{P}$-algebra with divided powers is the same as a $\mathscr{P}$-algebra. It is also the case in characteristic zero since the norm map is then an isomorphism.

From this construction it follows that there is a forgetful functor from the category of $\mathscr{P}$-algebras with divided powers to the category of $\mathscr{P}$-algebras. It is often a challenge to find a presentation of the first out of a presentation of the second.

One of the interests of the notion of algebras with divided powers is the following result, proved in [Fre00]. Let $A_\bullet$ be a simplicial $\mathscr{P}$-algebra. If $A_\bullet$ is 2-reduced (that is $A_0 = A_1 = 0$), then its homotopy $\pi_*(A_\bullet)$ is a graded $\Gamma\mathscr{P}$-algebra. For $\mathscr{P} = Com$ it is a result of Henri Cartan (cf. loc.cit.).

5.2.10 First Examples of Operads

We show that a unital associative algebra can be interpreted as an operad. Then we introduce the "three graces", the operads *Ass*, *Com* and *Lie*. In the next section we treat the endomorphism operad which can be seen as a toy-model for the operad structure.

EXAMPLE 0. A unital associative algebra is an example of an operad. Indeed, let R be a unital associative algebra and consider the $\mathbb{S}$-module M given by $M(1) = R$ and $M(n) = 0$ otherwise. Then we have $M(V) = R \otimes V$, and an operad structure on M is equivalent to a unital associative algebra structure on R. The composition map γ is induced by the product on R:

$$\gamma(V) : M \circ M(V) \to M(V),$$
$$R \otimes R \otimes V \to R \otimes V, \qquad (r, s, v) \mapsto (rs, v).$$

The unit map η is induced by the unit of R. An algebra over this operad is simply a left R-module. So any unital associative algebra is an example of an algebraic operad.

In particular, if $R = \mathbb{K}$, then we get the *identity operad* I, that we sometimes denote by $Vect$ to emphasize the fact that its category of algebras is simply the category of vector spaces Vect.

EXAMPLE 1. Let $Ass : \mathsf{Vect} \to \mathsf{Vect}$ be the Schur functor given by $Ass(V) := \overline{T}(V) = \oplus_{n\geq 1} V^{\otimes n}$ (reduced tensor module). As an $\mathbb{S}$-module we have $Ass(n) = \mathbb{K}[\mathbb{S}_n]$ (regular representation), since $\mathbb{K}[\mathbb{S}_n] \otimes_{\mathbb{S}_n} V^{\otimes n} = V^{\otimes n}$ for $n \geq 1$, and $Ass(0) = 0$. The map $\gamma(V) : Ass(Ass(V)) \to Ass(V)$ is given by "composition of noncommutative polynomials". This is the symmetric operad encoding associative algebras since an algebra over Ass is a nonunital associative algebra. So the free Ass-algebra over the vector space V is nothing but the reduced tensor algebra $\overline{T}(V)$ (cf. Sect. 1.1.3).

The symmetric operad Ass comes from a nonsymmetric operad, denoted As, for which $As_n = \mathbb{K}\mu_n$ (one-dimensional space) for $n \geq 1$. On associative algebras, μ_n

is the n-ary operation $\mu_n(x_1, \ldots, x_n) = x_1 \cdots x_n$. This basic example is treated in more detail in Chap. 9.

The operad of unital associative algebras, denoted *uAss*, is the same except that $uAss(0) = \mathbb{K}$. The image of $1 \in uAss(0)$, in the unital associative algebra A, is the unit of A. The free algebra is the tensor algebra: $uAss(V) = T(V)$.

In the process which associates a symmetric operad to a nonsymmetric operad the composition map is given by

$$\gamma(i_1, \ldots, i_k) = \gamma^{\mathcal{P}}_{i_1,\ldots,i_k} \otimes \gamma^{Ass}(i_1, \ldots, i_k)$$

up to a reordering of the factors on the source space.

Since $Ass(n)$ is the regular representation, there is no difference between associative algebras with divided powers and associative algebras.

EXAMPLE 2. Let $Com : \mathsf{Vect} \to \mathsf{Vect}$ be the Schur functor given by

$$Com(V) := \overline{S}(V) = \bigoplus_{n\geq 1} S^n V = \bigoplus_{n\geq 1} (V^{\otimes n})_{\mathbb{S}_n}.$$

As an $\mathbb{S}$-module we have $Com(n) = \mathbb{K}$ with trivial action, since

$$\mathbb{K} \otimes_{\mathbb{S}_n} V^{\otimes n} = (V^{\otimes n})_{\mathbb{S}_n} = S^n V \quad \text{for } n \geq 1,$$

and $Com(0) = 0$. The map $\gamma(V) : Com(Com(V)) \to Com(V)$ is given by "composition of polynomials". This is the symmetric operad encoding commutative algebras since an algebra over *Com* is a nonunital commutative algebra (in the sense commutative and associative). So the free *Com*-algebra over the vector space V is nothing but the (nonunital) symmetric algebra $\overline{S}(V)$ (cf. Sect. 1.1.8).

Since any commutative algebra is an associative algebra, there is a functor $Com\text{-}\mathsf{alg} \longrightarrow Ass\text{-}\mathsf{alg}$. It is induced by the morphism of operads $Ass \to Com$, which, in degree n, is the augmentation map $\mathbb{K}[\mathbb{S}_n] \to \mathbb{K}$, $\sigma \mapsto 1$ (projection onto the trivial representation). This case is treated in more details in Chap. 13.

It is proved in [Fre00] that the notion of "divided power commutative algebras" is the classical one, see Sect. 13.1.12 for the precise presentation.

EXAMPLE 3. Let $Lie : \mathsf{Vect} \to \mathsf{Vect}$ be the functor such that the space $Lie(V) \subset \overline{T}(V)$ is generated by V under the *bracket* operation $[x, y] := xy - yx$. We know by Corollary 1.3.5 that this is the free Lie algebra on V. Let $Lie(n)$ be the multilinear part of degree n in the free Lie algebra $Lie(\mathbb{K}x_1 \oplus \cdots \oplus \mathbb{K}x_n)$. One can show that there is an operad structure on the Schur functor *Lie* induced by the operad structure on *Ass* (Lie polynomials of Lie polynomials are again Lie polynomials). An algebra over the operad *Lie* is a Lie algebra.

Any associative algebra is a Lie algebra under the antisymmetrization of the product $[x, y] = xy - yx$. This functor $Ass\text{-}\mathsf{alg} \longrightarrow Lie\text{-}\mathsf{alg}$ is induced by the morphism of operads $Lie \to Ass$, which, in arity n, is the inclusion $Lie(n) \rightarrowtail Ass(n) = \mathbb{K}[\mathbb{S}_n]$ mentioned above. This case is treated in more detail in Sect. 13.2.

It is proved in [Fre00] that the notion of "divided power Lie algebras" over a field with positive characteristic coincides with the notion of restricted Lie algebras introduced by Nathan Jacobson [Jac62], see Sect. 13.2.15 for the precise presentation.

5.2.11 Endomorphism Operad

For any vector space V the endomorphism operad End_V is given by

$$\mathrm{End}_V(n) := \mathrm{Hom}\big(V^{\otimes n}, V\big),$$

where, by convention, $V^{\otimes 0} = \mathbb{K}$. The right action of $\mathbb{S}_n$ on $\mathrm{End}_V(n)$ is induced by the left action on $V^{\otimes n}$. The composition map γ is given by composition of endomorphisms:

$$\begin{array}{ccccccc}
V^{\otimes i_1}\otimes & & \cdots & & \otimes V^{\otimes i_k} & = & V^{\otimes n} \\
\downarrow{\scriptstyle f_1} & & & & \downarrow{\scriptstyle f_k} & & \downarrow{\scriptstyle f_1\otimes\cdots\otimes f_k} \\
V\otimes & & \cdots & & \otimes V & = & V^{\otimes k} \\
 & & \downarrow{\scriptstyle f} & & & & \downarrow{\scriptstyle f} \\
 & & V & & & = & V
\end{array}$$

$$\gamma(f; f_1, \ldots, f_k) := f\big(f_1 \otimes \cdots \otimes f_k\big).$$

It is immediate to verify that End_V is an algebraic operad.

Proposition 5.2.2. *A $\mathscr{P}$-algebra structure on the vector space A is equivalent to a morphism of operads $\mathscr{P} \to \mathrm{End}_A$.*

Proof. This statement follows from the natural isomorphism

$$\mathrm{Hom}_{\mathbb{S}_n}\big(\mathscr{P}(n), \mathrm{Hom}\big(A^{\otimes n}, A\big)\big) = \mathrm{Hom}\big(\mathscr{P}(n) \otimes_{\mathbb{S}_n} A^{\otimes n}, A\big). \qquad \square$$

By definition a *graded $\mathscr{P}$-algebra* over the graded operad $\mathscr{P}$ is a graded vector space A (i.e. an object in the sign-graded category gVect) and a morphism of graded operads $\mathscr{P} \to \mathrm{End}_A$. We leave it to the reader to write down the compatibility conditions in terms of the map $\gamma_A : \mathscr{P}(A) \to A$.

5.2.12 Algebras over an Operad: Functorial Properties

By abuse of notation, we often denote by $\mu : A^{\otimes n} \to A$ the image of $\mu \in \mathscr{P}(n)$ under γ_A in $\mathrm{End}_A(n)$. It follows immediately from the interpretation of an algebra over an operad given in Proposition 5.2.2 that if $\alpha : \mathscr{P} \to \mathscr{Q}$ is a morphism of operads, then any $\mathscr{Q}$-algebra A has a $\mathscr{P}$-algebra structure via the composition of operad morphisms

$$\mathscr{P} \to \mathscr{Q} \to \operatorname{End}_A.$$

Hence we get the functor

$$\alpha^* : \mathscr{Q}\text{-alg} \longrightarrow \mathscr{P}\text{-alg}.$$

This functor, which is analogous to the restriction functor for modules, admits a left adjoint, analogous to the induced functor for modules. It is denoted by

$$\alpha_! : \mathscr{P}\text{-alg} \to \mathscr{Q}\text{-alg}$$

and constructed as follows. For any $\mathscr{P}$-algebra A, the $\mathscr{Q}$-algebra $\alpha_!(A)$ is the quotient of the free $\mathscr{Q}$-algebra $\mathscr{Q}(A)$ which identifies the two different $\mathscr{P}$-algebra structures. It is a particular case the relative composite product which will appear in Sect. 11.2.1. Explicitly it is given by the coequalizer:

$$\mathscr{Q} \circ \mathscr{P} \circ A \underset{\operatorname{Id}_{\mathscr{Q}} \circ \gamma_A}{\overset{\rho \circ \operatorname{Id}_A}{\rightrightarrows}} \mathscr{Q} \circ A \longrightarrow\!\!\!\!\!\rightarrow \mathscr{Q} \circ_{\mathscr{P}} A =: \alpha_!(A),$$

where the right $\mathscr{P}$-action ρ on $\mathscr{Q}$ is the composition $\mathscr{Q} \circ \mathscr{P} \xrightarrow{\operatorname{Id}_{\mathscr{Q}} \circ \alpha} \mathscr{Q} \circ \mathscr{Q} \xrightarrow{\gamma^{\mathscr{Q}}} \mathscr{Q}$. In the particular case of the morphism $\alpha : Lie \to Ass$ we obtain the universal algebra of a Lie algebra: $\alpha_!(\mathfrak{g}) = U(\mathfrak{g})$.

5.2.13 Ubiquity of the Elements of $\mathscr{P}(n)$

Let $V_n = \mathbb{K}x_1 \oplus \cdots \oplus \mathbb{K}x_n$ be an n-dimensional vector space with preferred basis. The element

$$x_1 \otimes \cdots \otimes x_n \in V_n^{\otimes n} \subset \mathscr{P}(V_n)^{\otimes n}$$

is called the *generic element*. Applying the n-ary operation $\mu \in \mathscr{P}(n)$ to the generic element gives an element of the free $\mathscr{P}$-algebra over V_n:

$$\gamma_{\mathscr{P}(V_n)} : \mathscr{P}(n) \otimes \mathscr{P}(V_n)^{\otimes n} \to \mathscr{P}(V_n),$$
$$\mu \otimes (x_1 \otimes \cdots \otimes x_n) \mapsto \mu(x_1, \ldots, x_n)$$

(this is a slight abuse of notation since we do not mention γ). The resulting map

$$\mathscr{P}(n) \to \mathscr{P}(V_n),$$
$$\mu \mapsto \mu(x_1, \ldots, x_n)$$

is one-to-one onto the multilinear part of degree n of $\mathscr{P}(V_n)$. The relationship with the action of the symmetric group is as follows. For $\sigma \in \mathbb{S}_n$, we have

$$\mu^{\sigma}(x_1, \ldots, x_n) = \mu\big(\sigma \cdot (x_1, \ldots, x_n)\big) = \mu(x_{\sigma^{-1}(1)}, \ldots, x_{\sigma^{-1}(n)})$$

in $\mathscr{P}(V_n)$.

So, any such μ can be viewed either as an n-ary operation or as an element of some specific free $\mathscr{P}$-algebra.

In practice we will often talk about "the operation $x * y$" to mean "the operation $\mu \in \mathscr{P}(2)$ determined by $\mu(x, y) := x * y$". Similarly we will allow ourselves to say "the relation $(x * y) * z = x * (y * z)$", when we really mean "the relator $\mu \circ (\mu, \mathrm{id}) - \mu \circ (\mathrm{id}, \mu) \in \mathscr{T}(E)(3)$", see Sect. 5.5 for the notation $\mathscr{T}$.

5.2.14 Operadic Ideal and Quotient Operad

An *operadic ideal* (or simply ideal) of an operad $\mathscr{P}$ is a sub-$\mathbb{S}$-module $\mathscr{I}$ of $\mathscr{P}$ such that the operad structure of $\mathscr{P}$ passes to the quotient $\mathscr{P}/\mathscr{I}$. Explicitly it is equivalent to the following conditions. For any family of operations $\{\mu; \nu_1, \ldots, \nu_k\}$, if one of them is in $\mathscr{I}$, then we require that the composite $\gamma(\mu; \nu_1, \ldots, \nu_k)$ is also in $\mathscr{I}$.

5.2.15 Coalgebra over an Operad

Let V be a vector space. By definition the *co-endomorphism operad* over V, denoted coEnd_V, is given by

$$\mathrm{coEnd}_V(n) := \mathrm{Hom}\big(V, V^{\otimes n}\big).$$

The right action of $\mathbb{S}_n$ on $\mathrm{coEnd}_V(n)$ is induced by the right action on $V^{\otimes n}$. The composition map is given by composition of morphisms:

$$\begin{array}{ccccc}
 & V & & = & V \\
 & \downarrow & & & \downarrow \\
V\otimes & \cdots & \otimes V & = & V^{\otimes k} \\
\downarrow & & \downarrow & & \downarrow \\
V^{\otimes i_1}\otimes & \cdots & \otimes V^{\otimes i_k} & = & V^{\otimes n}.
\end{array}$$

By definition a *coalgebra C over the operad $\mathscr{P}$* is a vector space C and a morphism of operads

$$\mathscr{P} \longrightarrow \mathrm{coEnd}_C.$$

Explicitly, for any n, the data is an $\mathbb{S}_n$-equivariant map

$$\mathscr{P}(n) \otimes C \longrightarrow C^{\otimes n}.$$

The image of $\mu \in \mathscr{P}(n)$, that is the map $C \to C^{\otimes n}$, is called an *n-ary cooperation* and, often, still denoted by μ by abuse of notation. In order to simplify the terminology we allow ourselves to call C a $\mathscr{P}$-coalgebra.

When $\mathscr{P}$ is the associative operad *Ass*, an *Ass*-coalgebra is a coassociative coalgebra (also called associative coalgebra) as defined in Sect. 1.2.1. In the case $\mathscr{P} = Com$ it is a cocommutative (and coassociative) coalgebra. We simply say commutative coalgebra. When $\mathscr{P} = Lie$ we get the notion of *Lie coalgebra* (sometimes referred to as coLie coalgebra). Explicitly a Lie coalgebra is a vector space L equipped with a linear map $\Delta : L \to L \otimes L$ which is antisymmetric, i.e. $\tau\Delta = -\Delta$, and satisfies the co-Leibniz rule:

$$(\Delta \otimes \mathrm{Id})\Delta = (\mathrm{Id} \otimes \Delta)\Delta + (\mathrm{Id} \otimes \tau)(\Delta \otimes \mathrm{Id})\Delta.$$

5.3 Classical and Partial Definition of an Operad

From now on, by "operad" we mean symmetric operad. So we suppose that the endofunctor is in fact a Schur functor induced by an $\mathbb{S}$-module.

The classical definition of an operad is quite technical, but it is the most common form appearing in algebraic topology papers on operads. For the third definition one takes advantage of the fact that the operadic structure can be determined out of some elementary compositions called "partial compositions". It is very helpful in some frameworks because it has the minimal number of generators.

5.3.1 Classical Definition of an Operad (J.P. May [May72])

Let us now describe explicitly the operad structure on $\mathbb{S}$-modules. By Corollary 5.1.4 the vector space $(\mathscr{P} \circ \mathscr{P})(n)$ is a quotient of the direct sum of all the possible tensor products $\mathscr{P}(k) \otimes \mathscr{P}(i_1) \otimes \cdots \otimes \mathscr{P}(i_k)$ for $i_1 + \cdots + i_k = n$. So, the composition map γ of the operad $\mathscr{P}$ defines linear maps

$$\gamma(i_1, \ldots, i_k) : \mathscr{P}(k) \otimes \mathscr{P}(i_1) \otimes \cdots \otimes \mathscr{P}(i_k) \longrightarrow \mathscr{P}(i_1 + \cdots + i_k).$$

Pictorially this composition looks as follows:

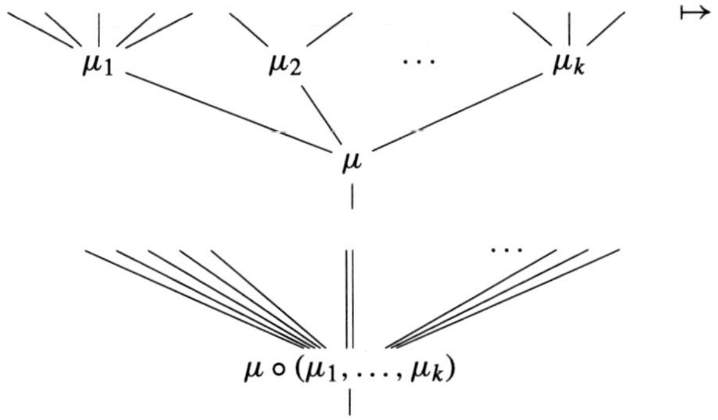

The next proposition gives the conditions under which a family of linear maps $\gamma(i_1, \ldots, i_k)$ gives rise to an operad.

Proposition 5.3.1. *Let* $\mathcal{P} = \{\mathcal{P}(n)\}_{n \geq 0}$ *be an* $\mathbb{S}$*-module. The maps*

$$\gamma(i_1, \ldots, i_k) : \mathcal{P}(k) \otimes \mathcal{P}(i_1) \otimes \cdots \otimes \mathcal{P}(i_k) \longrightarrow \mathcal{P}(i_1 + \cdots + i_k)$$

define an operad structure on $\mathcal{P}$ *if and only if they satisfy the following conditions*:

(a) *for any integers* k *and* n, *the map*

$$\sum \gamma(i_1, \ldots, i_k) : \mathcal{P}(k) \otimes \Big(\bigoplus \mathcal{P}(i_1) \otimes \cdots \otimes \mathcal{P}(i_k) \Big) \longrightarrow \mathcal{P}(n),$$

where the direct sum is over all k*-tuples* $(i_1, \ldots, i_k)$ *such that* $i_1 + \cdots + i_k = n$, *factors through the tensor product over* $\mathbb{S}_k$. *Moreover it is equivariant with respect to the action of* $\mathbb{S}_{i_1} \times \cdots \times \mathbb{S}_{i_k}$ *(we use the natural embedding of this group product into* $\mathbb{S}_n$*)*,

(b) *for any set of indices* $(j_{1,1}, \ldots, j_{1,i_1}, j_{2,1}, \ldots, j_{2,i_2}, \ldots, j_{n,1}, \ldots, j_{n,i_n})$ *the following square is commutative (we leave out the* $\otimes$ *signs)*:

$$\begin{array}{ccc}
\mathcal{P}(n)\mathcal{P}(r_1)\cdots\mathcal{P}(r_n) & \longrightarrow & \\
\uparrow & & \\
\mathcal{P}(n)\mathcal{P}(i_1)\mathcal{P}(j_{1,1})\cdots\mathcal{P}(j_{1,i_1})\mathcal{P}(i_2)\mathcal{P}(j_{2,1})\cdots\cdots\mathcal{P}(j_{n,i_n}) & & \\
\downarrow \cong & & \\
\mathcal{P}(n)\mathcal{P}(i_1)\cdots\mathcal{P}(i_n)\mathcal{P}(j_{1,1})\cdots\mathcal{P}(j_{1,i_1})\mathcal{P}(j_{2,1})\cdots\cdots\mathcal{P}(j_{n,i_n}) & & \downarrow \\
\downarrow & & \\
\mathcal{P}(m)\mathcal{P}(j_{1,1})\cdots\mathcal{P}(j_{1,i_1})\mathcal{P}(j_{2,1})\cdots\cdots\mathcal{P}(j_{n,1})\cdots\mathcal{P}(j_{n,i_n}) & \longrightarrow & \mathcal{P}(\ell)
\end{array}$$

where $r_k = j_{k,1} + \cdots + j_{k,i_k}$ *for* $k = 1$ *to* n, $m = i_1 + \cdots + i_n$ *and* $\ell = r_1 + \cdots + r_n$,

(c) *there is an element* id *in* $\mathscr{P}(1)$ *such that the evaluation of* $\gamma(n) : \mathscr{P}(1) \otimes \mathscr{P}(n) \to \mathscr{P}(n)$ *on* (id, μ) *is equal to* μ, *and the evaluation of* γ *on* (μ; id..., id) *is equal to* μ.

Proof. Starting with an algebraic operad $(\mathscr{P}, \gamma, \eta)$, we get the maps $\gamma(i_1, \ldots, i_k)$ by restriction to the identity shuffle. The unit map defines an inclusion $\eta : \mathbb{K} \to \mathscr{P}(1)$, whose image of $1 = 1_{\mathbb{K}}$ is the identity operation id. Then, it is clear that the axioms of functoriality, associativity and unitality of the operad data imply the properties (a), (b) and (c).

On the other hand, starting with an $\mathbb{S}$-module $\mathscr{P}$ and maps $\gamma(i_1, \ldots, i_k)$, we construct a monoid structure on the Schur functor as follows. Condition (a) provides a transformation of functors $\gamma : \mathscr{P} \circ \mathscr{P} \to \mathscr{P}$. Condition (b) ensures associativity of γ. Condition (c) ensures unitality. □

As a consequence of Proposition 5.3.1 one can define an operad as an $\mathbb{S}$-module $\{\mathscr{P}(n)\}_{n \geq 0}$ equipped with maps $\gamma(i_1, \ldots, i_k)$ for all k-tuples $(i_1, \ldots, i_k)$ satisfying the equivariance conditions (a), the associativity condition (b), and the unitality condition (c). This is what we call the *classical definition of an operad.*

5.3.2 Hadamard Product of Operads

Let $\mathscr{P}$ and $\mathscr{Q}$ be two operads. The Hadamard tensor product $\mathscr{P} \underset{\mathrm{H}}{\otimes} \mathscr{Q}$ of the underlying $\mathbb{S}$-modules (cf. Sect. 5.1.12) has a natural operad structure:

$$\begin{aligned}
&(\mathscr{P} \underset{\mathrm{H}}{\otimes} \mathscr{Q})(k) \otimes (\mathscr{P} \underset{\mathrm{H}}{\otimes} \mathscr{Q})(n_1) \otimes \cdots \otimes (\mathscr{P} \underset{\mathrm{H}}{\otimes} \mathscr{Q})(n_k) \\
&\quad = \mathscr{P}(k) \otimes \mathscr{Q}(k) \otimes \mathscr{P}(n_1) \otimes \mathscr{Q}(n_1) \otimes \cdots \otimes \mathscr{P}(n_k) \otimes \mathscr{Q}(n_k) \\
&\quad \cong \mathscr{P}(k) \otimes \mathscr{P}(n_1) \otimes \cdots \otimes \mathscr{P}(n_k) \otimes \mathscr{Q}(k) \otimes \mathscr{Q}(n_1) \otimes \cdots \otimes \mathscr{Q}(n_k) \\
&\quad \longrightarrow \mathscr{P}(n) \otimes \mathscr{Q}(n) = (\mathscr{P} \underset{\mathrm{H}}{\otimes} \mathscr{Q})(n)
\end{aligned}$$

for $n = n_1 + \cdots + n_k$. Observe that we use the switching map in the category Vect to put the factors $\mathscr{Q}(i)$ in the correct position. Therefore, when Vect is replaced by another symmetric monoidal category (cf. Appendix B.3) signs might be involved. The operad $uCom$ is obviously a unit for this operation.

The operad $\mathscr{P} \underset{\mathrm{H}}{\otimes} \mathscr{Q}$ is the *Hadamard product* of the operads $\mathscr{P}$ and $\mathscr{Q}$.

Proposition 5.3.2. *Let A be a $\mathscr{P}$-algebra and let B be a $\mathscr{Q}$-algebra. The tensor product $A \otimes B$ is a $\mathscr{P} \underset{\mathrm{H}}{\otimes} \mathscr{Q}$-algebra.*

Proof. Let us denote by $\gamma_A : \mathscr{P} \to \mathrm{End}_A$ and by $\gamma_B : \mathscr{Q} \to \mathrm{End}_B$ the respective actions of $\mathscr{P}$ on A and of $\mathscr{Q}$ on B. Then the action of $\mathscr{P} \underset{\mathrm{H}}{\otimes} \mathscr{Q}$ on $A \otimes B$ is given by the following composite

$$\mathscr{P} \underset{\mathrm{H}}{\otimes} \mathscr{Q} \xrightarrow{\gamma_A \underset{\mathrm{H}}{\otimes} \gamma_B} \mathrm{End}_A \underset{\mathrm{H}}{\otimes} \mathrm{End}_B \to \mathrm{End}_{A\otimes B},$$

where the last map is defined by

$$\begin{aligned}\mathrm{Hom}\big(A^{\otimes n}, A\big) \otimes \mathrm{Hom}\big(B^{\otimes n}, B\big) \to \mathrm{Hom}\big(A^{\otimes n} \otimes B^{\otimes n}, A \otimes B\big)\\ \cong \mathrm{Hom}\big((A \otimes B)^{\otimes n}, A \otimes B\big).\end{aligned}$$

We leave it to the reader to verify that this map is a morphism of operads. □

5.3.3 Hopf Operads

A *Hopf operad* is a reduced operad $\mathscr{P}$ with a morphism of operads $\Delta_{\mathscr{P}} : \mathscr{P} \to \mathscr{P} \underset{\mathrm{H}}{\otimes} \mathscr{P}$ called the coproduct of $\mathscr{P}$ and a morphism of operads $\varepsilon_{\mathscr{P}} : \mathscr{P} \to \mathit{Com}$ called the counit. This structure is supposed to be coassociative and counital. Since $\Delta_{\mathscr{P}}$ and $\varepsilon_{\mathscr{P}}$ are determined by their arity n components

$$\Delta_{\mathscr{P}}(n) : \mathscr{P}(n) \to (\mathscr{P} \underset{\mathrm{H}}{\otimes} \mathscr{P})(n) = \mathscr{P}(n) \otimes \mathscr{P}(n), \qquad \varepsilon_{\mathscr{P}} : \mathscr{P}(n) \to \mathit{Com}(n) = \mathbb{K},$$

a Hopf operad is equivalently defined as an operad in the category of counital coalgebras. The main purpose of this definition lies in the following result.

Proposition 5.3.3. *When $\mathscr{P}$ is a Hopf operad, the tensor product $A \otimes B$ of two $\mathscr{P}$-algebras A and B is again a $\mathscr{P}$-algebra, and there is a natural isomorphism*

$$(A \otimes B) \otimes C \cong A \otimes (B \otimes C)$$

where C is also a $\mathscr{P}$-algebra.

Proof. Proposition 5.3.2 asserts that $A \otimes B$ is a $\mathscr{P} \underset{\mathrm{H}}{\otimes} \mathscr{P}$-algebra. Then the following composite

$$\mathscr{P} \xrightarrow{\Delta_{\mathscr{P}}} \mathscr{P} \underset{\mathrm{H}}{\otimes} \mathscr{P} \to \mathrm{End}_{A\otimes B}$$

defines a $\mathscr{P}$-algebra structure on $A \otimes B$ as explained above in Sect. 5.2.11. Coassociativity of $\Delta_{\mathscr{P}}$ ensures the validity of the last assertion. □

The operads *Ass* and *Com* are Hopf operads, their diagonals are given by $\Delta_{\mathit{Ass}} : \sigma \in \mathbb{S}_n \mapsto \sigma \otimes \sigma \in \mathbb{K}[\mathbb{S}_n] \otimes \mathbb{K}[\mathbb{S}_n]$ and by $\Delta_{\mathit{Com}} : \mathit{Com}(n) = \mathbb{K} \xrightarrow{\sim} \mathbb{K} \otimes \mathbb{K} = \mathit{Com}(n) \otimes \mathit{Com}(n)$ respectively. (It is a good basic exercise to prove that they are morphisms of operads.) With these definitions in mind, we get an operadic interpretation of the fact that the tensor product of two associative (resp. commutative) algebras is again an associative (resp. commutative) algebra with the product given by $\mu(a \otimes b, a' \otimes b') = \mu(a, a') \otimes \mu(b, b')$.

It is also a good exercise to show directly that the operad *Lie* has no nontrivial diagonal, that is *Lie* is not a Hopf operad.

5.3.4 Partial Definition of an Operad

Let $\mathscr{P}$ be an operad and let $\mu \in \mathscr{P}(m)$, $\nu \in \mathscr{P}(n)$ be two operations. By definition the *partial composition* $(\mu, \nu) \mapsto \mu \circ_i \nu \in \mathscr{P}(m-1+n)$ is defined, for $1 \le i \le m$, by "substitution":

$$- \circ_i - : \mathscr{P}(m) \otimes \mathscr{P}(n) \longrightarrow \mathscr{P}(m-1+n),$$
$$\mu \circ_i \nu := \gamma(\mu; \mathrm{id}, \dots, \mathrm{id}, \nu, \mathrm{id}, \dots, \mathrm{id}).$$

Pictorially it is represented by the following grafting of trees, where the root of ν is grafted onto the ith leaf of μ:

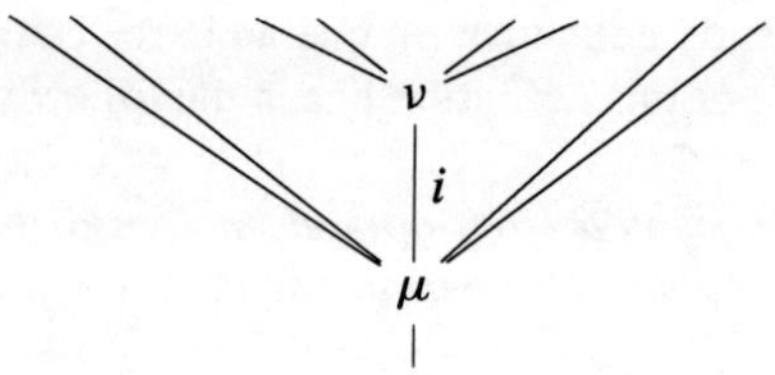

The relationship between this partial composition and the action of the symmetric groups is given by the following two relations. First, for any $\sigma \in \mathbb{S}_n$ we have:

$$\mu \circ_i \nu^\sigma = (\mu \circ_i \nu)^{\sigma'}$$

where $\sigma' \in \mathbb{S}_{m-1+n}$ is the permutation which acts by the identity, except on the block $\{i, \dots, i-1+n\}$ on which it acts via σ. Second, for any $\sigma \in \mathbb{S}_m$ we have:

$$\mu^\sigma \circ_i \nu = (\mu \circ_{\sigma(i)} \nu)^{\sigma''}$$

where $\sigma'' \in \mathbb{S}_{m-1+n}$ is acting like σ on the block $\{1, \dots, m-1+n\} \backslash \{i, \dots, i-1+n\}$ with values in $\{1, \dots, m-1+n\} \backslash \{\sigma(i), \dots, \sigma(i)-1+n\}$ and identically on the block $\{i, \dots, i-1+n\}$ with values in $\{\sigma(i), \dots, \sigma(i)-1+n\}$.

There are two different cases for two-stage partial compositions, depending on the relative positions of the two graftings:

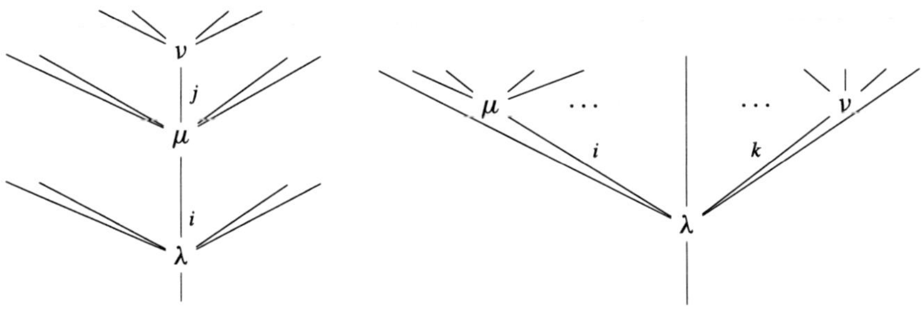

(I) (II)

In both cases associativity of the composition in an operad leads to some relation for the partial composition:

$$\begin{cases} \text{(I)} \quad (\lambda \circ_i \mu) \circ_{i-1+j} \nu = \lambda \circ_i (\mu \circ_j \nu), & \text{for } 1 \le i \le l,\ 1 \le j \le m, \\ \text{(II)} \quad (\lambda \circ_i \mu) \circ_{k-1+m} \nu = (\lambda \circ_k \nu) \circ_i \mu, & \text{for } 1 \le i < k \le l, \end{cases}$$

for any $\lambda \in \mathscr{P}(l)$, $\mu \in \mathscr{P}(m)$, $\nu \in \mathscr{P}(n)$. Relation (I) is called the *sequential composition* axiom and relation (II) is called the *parallel composition* axiom.

Conversely an operad can be defined as being an $\mathbb{S}$-module $\mathscr{P}$ equipped with partial compositions $\circ_i$ satisfying the compatibility with the action of the symmetric groups, and the two associativity relations (I) and (II) described above. It is also assumed that there is an element id in $\mathscr{P}(1)$ satisfying $\mathrm{id} \circ_1 \nu = \nu$ and $\mu \circ_i \mathrm{id} = \mu$. This gives the *partial definition of an operad.*

In case where the $\mathbb{S}$-modules are graded, there is a sign in formula (II) (because μ and ν are exchanged):

$$(\lambda \circ_i \mu) \circ_{k-1+m} \nu = (-1)^{|\mu||\nu|} (\lambda \circ_k \nu) \circ_i \mu.$$

Proposition 5.3.4. *The partial definition of an operad is equivalent to the classical definition of an operad, and so to all the other definitions.*

Proof. We already remarked that, starting with an operad $\mathscr{P}$, we get the partial compositions which satisfy the aforementioned properties. In the other direction, starting with partial compositions $- \circ_i -$ one constructs maps

$$\gamma(i_1, \ldots, i_n) : \mathscr{P}(n) \otimes \mathscr{P}(i_1) \otimes \cdots \otimes \mathscr{P}(i_n) \longrightarrow \mathscr{P}(i_1 + \cdots + i_n)$$

as

$$\gamma(i_1, \ldots, i_n) = \big(- \circ_1 \big(\cdots \big(- \circ_{n-1} (- \circ_n -)\big) \cdot \big)\big).$$

It is a tedious, but straightforward, task to verify that the axioms of the classical definition of an operad are fulfilled. □

5.3.5 Set Operads and Other Types of Operads

In this monograph we are mainly concerned with algebraic operads, that is operads in the category of vector spaces and dg vector spaces (or dg modules). The properties of the category Vect which are used here are: the tensor product is associative, commutative, unital (symmetric monoidal category) and distributive with respect to the direct sum. Notice that the classical and the partial definitions of an operad do not require that, in the underlying symmetric monoidal category, the monoidal product commutes with the coproduct.

One can define operads with values in other symmetric monoidal categories (for instance tensor categories). For instance, the category of sets (resp. simplicial sets, resp. topological sets) equipped with the cartesian product is a symmetric monoidal category. Here are some details for the category Set.

By definition a *set operad* (sometimes called *set-theoretic operad*) is a family of $\mathbb{S}_n$-sets $P(n)$ such that the functor

$$P : \mathsf{Set} \to \mathsf{Set}, \qquad X \mapsto \coprod_n P(n) \times_{\mathbb{S}_n} X^n$$

is equipped with a monoid structure. Here X^n denotes the cartesian product of n copies of the set X. The composition map gives rise to maps

$$\gamma(i_1, \ldots, i_k) : P(k) \times P(i_1) \times \cdots \times P(i_k) \to P(i_1 + \cdots + i_k),$$

which satisfy properties analogous to those of the linear case (cf. Sect. 5.3.1).

To any set X we can associate the vector space $\mathbb{K}[X]$ based on X. This functor is the left adjoint to the forgetful functor from vector spaces to sets. Any set operad P gives rise to an algebraic operad $\mathscr{P}$ under this functor: $\mathscr{P}(n) := \mathbb{K}[P(n)]$. We will meet some algebraic operads coming from set operads in the sequel. For instance the operad *Ass* comes from the set operad $P(n) = \mathbb{S}_n$ and the operad *Com* comes from the set operad $P(n) = \{*\}$.

The category of $\mathbb{S}$-modules can be equipped with an associative and commutative tensor product $M \otimes N$, cf. Sect. 5.1.3. So we can define an operad in this symmetric monoidal category. Such an object is sometimes ambiguously called a *twisted operad*.

Starting with the category of topological spaces (resp. simplicial spaces) equipped with the cartesian product, one can define analogously the notion of topological operads (resp. simplicial operads). There is a large amount of literature on these objects, see for instance [May72, MSS02]. In the next section we give an example, which has the advantage of exposing the main feature of the operadic calculus.

5.3.6 The Little Discs Operad

The *little discs operad* $\mathscr{D}$ is a topological symmetric operad defined as follows. The topological space $\mathscr{D}(n)$ is made up of the unit disc (in $\mathbb{C}$) with n non-intersecting

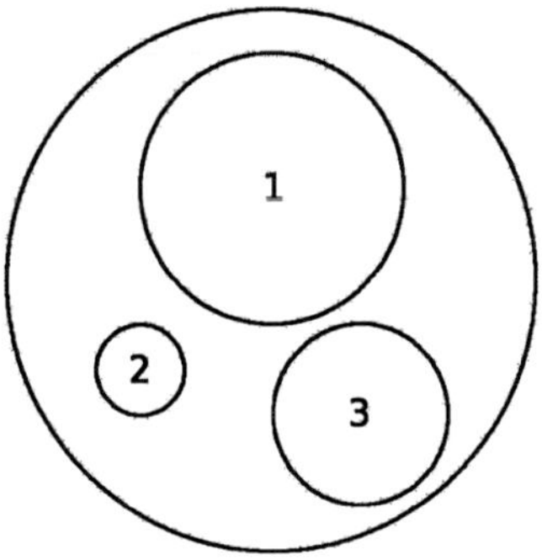

Fig. 5.1 Little discs configuration in $\mathscr{D}(3)$

subdiscs in its interior. So, an element on $\mathscr{D}(n)$ is completely determined by a family of n continuous maps $f_i : S^1 \to D^2$, $i = 1, \ldots, n$, satisfying the non-intersecting condition, see Fig. 5.1.

The enumeration of the interior discs is part of the structure. The operadic composition is given by insertion of a disc in an interior disc. The symmetric group action is given by permuting the labels. Figure 5.2 gives an example of a partial composition.

It is clear how to proceed to define the *little k-discs operad* or the *little k-cubes operad*. For $k = 1$ it is called the *little interval operad.*

The main property of the little k-discs operad is the following "recognition principle" proved by Boardman and Vogt in [BV73] and May in [May72]:

Claim. *If the connected topological space X is an algebra over the little discs operad, then it is homotopy equivalent to the k-fold loop space of some other pointed space Y:*

$$X \sim \Omega^k(Y).$$

5.4 Various Constructions Associated to an Operad

From an operad, one can construct a symmetric monoidal category, a group, a preLie algebra and a Hopf algebra.

5.4.1 Category Associated to an Operad [BV73]

Let $\mathscr{P}$ be an algebraic operad. We associate to it a symmetric monoidal category denoted $\mathsf{cat}\mathscr{P}$ as follows. The objects of $\mathsf{cat}\mathscr{P}$ are the natural numbers: $\underline{0}, \underline{1}, \ldots, \underline{n}, \ldots$. It will prove helpful to consider $\underline{n}$ as the set $\{1, 2, \ldots, n\}$, so $\underline{0} = \emptyset$.

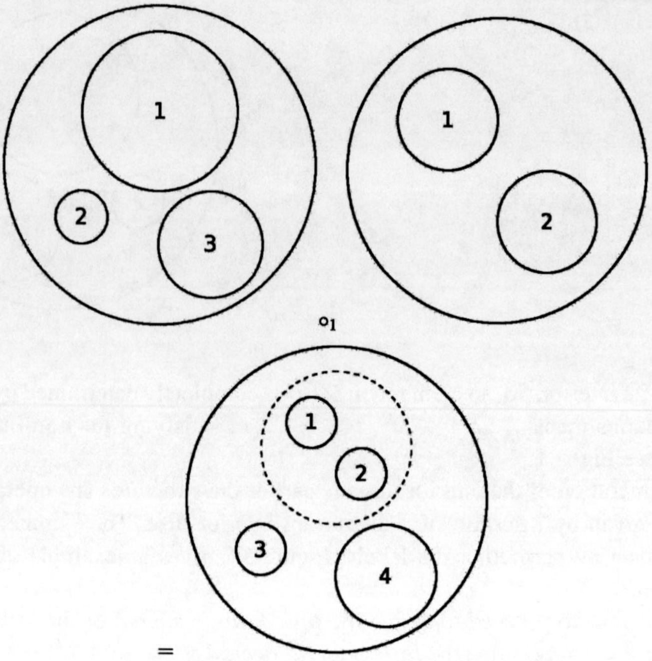

Fig. 5.2 Example of partial composition in the little discs operad

The morphisms of $\mathrm{cat}\mathscr{P}$ are defined as

$$\mathrm{cat}\mathscr{P}(\underline{m},\underline{n}) := \bigoplus_{f:\underline{m}\to\underline{n}} \bigotimes_{i=1}^{n} \mathscr{P}(f^{-1}(i))$$

where f is a set map from $\{1,2,\ldots,m\}$ to $\{1,2,\ldots,n\}$. Here we use the extension of the functor $\underline{n} \mapsto \mathscr{P}(n)$ to the category of finite sets, cf. Sect. 5.1.13. Observe that for $n = 1$ we get $\mathrm{cat}\mathscr{P}(\underline{m},\underline{1}) = \mathscr{P}(m)$.

Here is an example of a morphism in $\mathrm{cat}\mathscr{P}$:

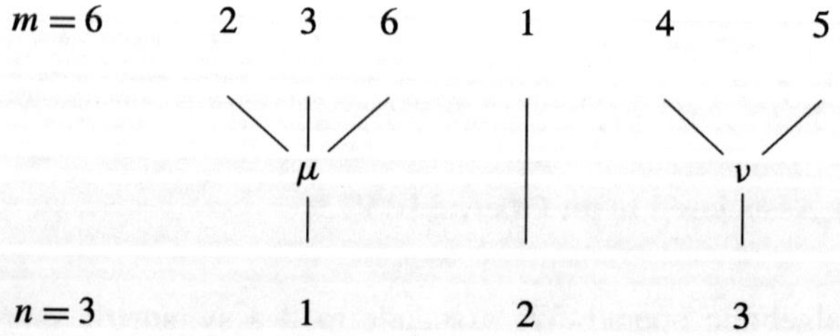

where $\mu \in \mathscr{P}(3)$, $\nu \in \mathscr{P}(2)$.

Observe that there is no harm in taking the finite sets as objects.

Proposition 5.4.1. *The operad structure of* $\mathcal{P}$ *induces on* $\mathsf{cat}\,\mathcal{P}$ *a structure of symmetric monoidal category which is the addition of integers on objects.*

Proof. The composition of morphisms in the category $\mathsf{cat}\mathcal{P}$ is obtained through the compositions in $\mathcal{P}$, see [MT78, Sect. 4] for an explicit formula.

Associativity of the composition in $\mathsf{cat}\mathcal{P}$ follows readily from associativity of the composition in $\mathcal{P}$.

The $\mathbb{S}_n$-module structure of $\mathcal{P}(n)$ accounts for the action of the automorphism group of $\underline{n}$.

The symmetric monoidal structure of $\mathsf{cat}\mathcal{P}$ is given by the addition of integers on objects, and therefore by concatenation of morphisms. We see by direct inspection that it is compatible with composition. □

Observe that the symmetric monoidal category $\mathsf{cat}\mathcal{P}$ is completely determined by the Hom-spaces $\mathrm{Hom}_{\mathsf{cat}\mathcal{P}}(\underline{n},\underline{1}) = \mathcal{P}(n)$ under the composition product (and concatenation). So even among the symmetric monoidal categories based on $\mathbb{N}$ they are very special categories. There are examples of more general symmetric monoidal categories called "props", which code for the "generalized bialgebras", cf. [Lod08].

Proposition 5.4.2. *Let* $\mathcal{P}$ *be an algebraic operad. A* $\mathcal{P}$*-algebra* A *is equivalent to a symmetric monoidal functor* $\mathsf{cat}\mathcal{P} \to \mathsf{Vect}$ *of the form* $\underline{n} \mapsto A^{\otimes n}$.

Proof. The functor in the other direction is simply given by evaluation at $\underline{1}$. The properties are verified straightforwardly. □

EXAMPLE 1 ($\mathsf{cat}\,uCom$). Let $\mathcal{P} = uCom$ be the operad of unital commutative algebras. The category $\mathsf{cat}\,uCom$ is the linearization of the category of finite sets, denoted Fin:

$$\mathsf{cat}\,uCom = \mathbb{K}[\mathsf{Fin}].$$

Indeed, it suffices to show that $\mathsf{cat}\,\mathrm{CMon} = \mathsf{Fin}$, where *CMon* is the set operad of unital commutative monoids. Since $CMon(n) = \{*\}$ (one element), we have $\mathsf{cat}\,\mathrm{CMon}(m,n) = \{f : \underline{m} \to \underline{n}\} = \mathsf{Fin}(\underline{m},\underline{n})$ and we are done.

EXAMPLE 2 ($\mathsf{cat}\,uAss$). Let $\mathcal{P} = uAss$ be the operad of unital associative algebras. As in the previous case we can work in the set operad framework. The operad $\mathsf{cat}\,\mathrm{Mon}$ of unital monoids admits the following description. Its objects are the integers $\underline{n}$, $n \in \mathbb{N}$, and the morphisms, elements of $\mathsf{cat}\,\mathrm{Mon}(m,n)$, are the linear maps $f : \{1,\dots,m\} \to \{1,\dots,n\}$ equipped with a total ordering on each fiber $f^{-1}(i)$, $1 \le i \le n$, see Appendix B.5.3. Observe that for any composite $g \circ f$ the set $(g \circ f)^{-1}(i)$ inherits a total ordering. See [Pir02a] for more details. This category is denoted by ΔS in [FL91, Lod98] because any morphism can be written uniquely as a composite of a morphism in the simplicial category Δ and an isomorphism (i.e. a permutation).

It is an example of a "matched pair of categories". Hence we have

$$\mathrm{cat}\, uAss = \mathbb{K}[\Delta S].$$

Note that there is a shift of notation: in Δ the object with $n+1$ elements is usually denoted by $[n]$.

5.4.2 Group Associated to a Symmetric Operad

We consider a partition P of $\underline{n} := \{1, \ldots, n\}$ into k subsets. We order this partition by the minimum of each subset: $\mathrm{P} = (\mathrm{P}_1, \ldots, \mathrm{P}_k)$ with

$$\min(\mathrm{P}_1) < \min(\mathrm{P}_2) < \cdots < \min(\mathrm{P}_k).$$

Let $i_j = \#\mathrm{P}_j$. We denote by $\sqcup(i_1, \ldots, i_k)$ the set of ordered partitions whose jth part has cardinal i_j (see Sect. 8.2.1 for examples). Any ordered partition $\mathrm{P} \in \sqcup(i_1, \ldots, i_k)$ defines a $(i_1, \ldots, i_k)$-unshuffle σ_{P}.

Let $\mathscr{P}$ be a reduced operad, i.e. $\mathscr{P}(0) = 0$. We consider the series

$$\underline{a} := (a_0, a_1, \ldots, a_n, \ldots)$$

where $a_n \in \mathscr{P}(n+1)$. We denote by $G(\mathscr{P})$ the set of series for which a_0 is invertible for the multiplication in $\mathscr{P}(1)$. We define a binary operation $\underline{a}\ \underline{b}$ on this set as follows:

$$(\underline{a}\ \underline{b})_n := \sum_k \sum_{\substack{(i_1,\ldots,i_k)\\ i_1+\cdots+i_k=n}} \sum_{\mathrm{P}\in\sqcup(i_1,\ldots,i_k)} \gamma > (a_{k-1}; b_{i_1-1}, \ldots, b_{i_k-1}) \circ \sigma_{\mathrm{P}}.$$

Proposition 5.4.3. [LN12] *The binary operation* $(\underline{a}, \underline{b}) \mapsto \underline{a}\ \underline{b}$ *makes* $G(\mathscr{P})$ *into a group with unit* $1 = (\mathrm{id}, 0, 0, \ldots)$.

Proof. First, we remark that the symmetric group $\mathbb{S}_k$ is acting freely on the set

$$\bigcup_{\substack{(i_1,\ldots,i_k)\\ i_1+\cdots+i_k=n}} Sh^{-1}(i_1, \ldots, i_k).$$

Second, the quotient of this set by $\mathbb{S}_k$ is precisely

$$\bigcup_{\substack{(i_1,\ldots,i_k)\\ i_1+\cdots+i_k=n}} \sqcup(i_1, \ldots, i_k).$$

Hence the associativity property of the product follows readily from the associativity property of γ.

The existence of an inverse, that is for any $\underline{a}$ there exists $\underline{b}$ such that $\underline{a}\,\underline{b} = 1$, is achieved by induction. For instance, when $a_0 = \mathrm{id}$, we get

$$b_1 = -a_1, \qquad b_2 = -a_2 + a_1 \circ_1 a_1 + (a_1 \circ_1 b_1)^{[132]} + a_1 \circ_2 a_1. \qquad \square$$

Observe that for $\mathscr{P} = Com$, the group $G(\mathscr{P})$ is isomorphic, in characteristic zero, to the group of power series in one variable with no constant term and with first coefficient invertible, under the composition. More precisely the isomorphism is given by

$$(a_0, a_1, \ldots, a_n, \ldots) \mapsto a_0 x + \frac{a_1}{2!}x^2 + \frac{a_2}{3!}x^3 + \cdots + \frac{a_{n-1}}{n!}x^n + \cdots,$$

where $a_i \in \mathbb{K}$.

5.4.3 Pre-Lie Algebra Associated to a Symmetric Operad

Let $\mathscr{P}$ be an operad with $\mathscr{P}(0) = 0$ and consider the space $\bigoplus_{n\geq 1} \mathscr{P}(n)$, resp. $\prod_{n\geq 1} \mathscr{P}(n)$. We construct a bilinear operation $\{-,-\}$ as follows:

$$\{\mu, \nu\} := \sum_{i=1}^{i=m} \sum_{\mathrm{P}} (\mu \circ_i \nu)^{\sigma_{\mathrm{P}}}$$

for $\mu \in \mathscr{P}(m)$, $\nu \in \mathscr{P}(n)$ and the sum is extended over the ordered partitions $\mathrm{P} \in \sqcup\!\sqcup(1, \ldots, 1, \underbrace{n-i+1}_{i\text{th position}}, 1, \ldots, 1)$.

Proposition 5.4.4. *The binary operation $\{-,-\}$ makes $\bigoplus_n \mathscr{P}(n)$, resp. $\prod_n \mathscr{P}(n)$, into a pre-Lie algebra, and hence into a Lie algebra.*

Proof. From the properties of the partial operations it follows that the binary operation $\{-,-\}$ is pre-Lie (cf. definition Sect. 1.4). Indeed, computing explicitly the associator $\{\{\lambda, \mu\}, \nu\} - \{\lambda, \{\mu, \nu\}\}$ we get a sum over trees of type II (see Sect. 5.3.4) where the vertices at the upper level are decorated by either μ or ν. So this sum is symmetric in μ and ν and, as a consequence, the associator is symmetric in the last two variables as expected. $\square$

5.4.4 Hopf Algebra Associated to a Symmetric Operad

Since the space $\bigoplus_{n\geq 1} \mathscr{P}(n)$ is a pre-Lie algebra, a fortiori it is a Lie algebra. Taking the universal enveloping algebra of this Lie algebra gives a Hopf algebra (cf.

Sect. 1.1.10). This is in fact a combinatorial Hopf algebra in the sense of [LR10], which is cofree (by PBW theorem) and left-sided. So one can recover the pre-Lie algebra structure on the primitive part from this data. A direct construction of the Hopf algebra from the operad can also be performed, see for instance [Moe01, vdLM06].

5.5 Free Operad

By definition the *free operad* over the $\mathbb{S}$-module M is an operad $\mathscr{F}(M)$ equipped with an $\mathbb{S}$-module morphism $\eta(M) : M \to \mathscr{F}(M)$ which satisfies the following universal condition:

Any $\mathbb{S}$-module morphism $f : M \to \mathscr{P}$, where $\mathscr{P}$ is an operad, extends uniquely into an operad morphism $\tilde{f} : \mathscr{F}(M) \to \mathscr{P}$:

$$\begin{array}{ccc} M & \xrightarrow{\eta(M)} & \mathscr{F}(M) \\ & \searrow^{f} & \downarrow{\tilde{f}} \\ & & \mathscr{P}. \end{array}$$

In other words the functor $\mathscr{F} : \mathbb{S}\text{-Mod} \to \mathsf{Op}_{\mathbb{K}}$ is left adjoint to the forgetful functor $\mathsf{Op}_{\mathbb{K}} \to \mathbb{S}\text{-Mod}$. We will show that the free operad exists and we will construct it explicitly. Observe that the free operad over M is well-defined up to a unique isomorphism.

Another, less ad hoc, construction is given in Sect. 5.5.5.

5.5.1 The Tree Module and the Free Operad

We give an explicit construction of the free operad following [BJT97, Rez96]. We rely on the fact that the composition of $\mathbb{S}$-modules $- \circ -$ is linear on the left-hand side. The classical "tensor algebra" construction does not work here because $- \circ -$ is not linear on the right-hand side. For a more general construction which works when no linearity is assumed whatsoever, see Sect. 5.5.5 and [Val09]. In the following construction, one takes advantage of the left linearity to produce a particular colimit which gives the free operad.

Let M be an $\mathbb{S}$-module. By induction we define the functor $\mathscr{T}_n M : \mathsf{Vect} \to \mathsf{Vect}$ as follows:

$$\begin{aligned} \mathscr{T}_0 M &:= \mathrm{I}, \\ \mathscr{T}_1 M &:= \mathrm{I} \oplus M, \\ \mathscr{T}_2 M &:= \mathrm{I} \oplus \big(M \circ (\mathrm{I} \oplus M)\big), \\ &\cdots \\ \mathscr{T}_n M &:= \mathrm{I} \oplus (M \circ \mathscr{T}_{n-1} M), \\ &\cdots \end{aligned}$$

The transformation of functors $i_n : \mathcal{T}_{n-1}M \to \mathcal{T}_n M$ is defined inductively by $i_1 : \mathrm{I} \hookrightarrow \mathrm{I} \oplus M$ (inclusion in the first factor) for $n = 1$, and by $i_n = \mathrm{Id}_{\mathrm{I}} \oplus (\mathrm{Id}_M \circ i_{n-1})$ higher up. Observe that i_n is a split monomorphism. By definition the *tree module* $\mathcal{T}M$ over the $\mathbb{S}$-module M is:

$$\mathcal{T}M := \bigcup_n \mathcal{T}_n M = \mathrm{colim}_n \mathcal{T}_n M.$$

Observe that $\mathcal{T}_n M$ contains $M^{\circ n}$ but is strictly larger in general. In terms of trees (cf. Sect. 5.6) $\mathcal{T}_n M$ is the space of trees with at most n levels, whose vertices are labeled by I and M. We write i for any of the inclusion maps $\mathcal{T}_n M \rightarrowtail \mathcal{T}_m M$ and j_n or simply j for the inclusion of the second factor $M \circ \mathcal{T}_{n-1}M \rightarrowtail \mathcal{T}_n M$. This last map induces a transformation of functors $j : M \to \mathcal{T}M$.

Theorem 5.5.1 (The free operad construction). *There is an operad structure γ on $\mathcal{T}M$ such that $\mathcal{T}(M) := (\mathcal{T}M, \gamma, j)$ is the free operad on M, so $\mathcal{F}(M) \cong \mathcal{T}M$.*

Proof. We follow Appendix B of [BJT97] by Baues, Tonks and Jibladze word for word. In order not to confuse the composition of functors $\circ$ with the composition of transformations of functors, we denote this latter one by juxtaposition. The identity transformation of a functor F is denoted by Id_F or simply by Id.

The steps of the proof are as follows:

(1) we construct the map $\gamma : \mathcal{T}M \circ \mathcal{T}M \to \mathcal{T}M$,
(2) we prove that γ is associative and unital,
(3) we prove that $(\mathcal{T}M, \gamma, j)$ satisfies the universal property of a free operad.

(1) First, we construct the composition map $\gamma : \mathcal{T}M \circ \mathcal{T}M \to \mathcal{T}M$. For any integers n and m we construct a map

$$\gamma_{n,m} : \mathcal{T}_n M \circ \mathcal{T}_m M \to \mathcal{T}_{n+m} M$$

by induction on n as follows. For $n = 0$,

$$\gamma_{0,m} := \mathrm{Id} : \mathrm{I} \circ \mathcal{T}_m M = \mathcal{T}_m M \to \mathcal{T}_m M.$$

For higher n, $\gamma_{n,m}$ is the composite

$$\begin{aligned}\mathcal{T}_n M \circ \mathcal{T}_m M = (\mathrm{I} \oplus M \circ \mathcal{T}_{n-1}M) \circ \mathcal{T}_m M &\cong \mathcal{T}_m M \oplus (M \circ \mathcal{T}_{n-1}M) \circ \mathcal{T}_m M \\ &\cong \mathcal{T}_m M \oplus M \circ (\mathcal{T}_{n-1}M \circ \mathcal{T}_m M) \\ &\xrightarrow{(\mathrm{Id}, \mathrm{Id} \circ \gamma_{n-1,m})} \mathcal{T}_m M \oplus M \circ \mathcal{T}_{n+m-1} M \xrightarrow{i+j} \mathcal{T}_{n+m} M.\end{aligned}$$

Observe that, in this definition of the composite, we use the associativity isomorphism (cf. Sect. 5.1.8)

$$(M \circ \mathcal{T}_{n-1}M) \circ \mathcal{T}_m M \cong M \circ (\mathcal{T}_{n-1}M \circ \mathcal{T}_m M).$$

We prove that the map γ is compatible with the colimits on n and on m by induction. For $n = 0$ it is immediate since $\gamma_{0,m} = \mathrm{Id}$. From $n-1$ to n it is a consequence of the commutativity of the following diagrams:

$$\begin{array}{ccc}
\mathscr{T}_m M \oplus M \circ \mathscr{T}_{n-1} M \circ \mathscr{T}_m M & \xrightarrow{i+j(\mathrm{Id}\circ\gamma_{n-1,m})} & \mathscr{T}_{n+m} M \\
\downarrow{\scriptstyle i} & & \downarrow{\scriptstyle i} \\
\mathscr{T}_m M \oplus M \circ \mathscr{T}_n M \circ \mathscr{T}_m M & \xrightarrow{i+j(\mathrm{Id}\circ\gamma_{n,m})} & \mathscr{T}_{n+m+1} M
\end{array}$$

$$\begin{array}{ccc}
\mathscr{T}_m M \oplus M \circ \mathscr{T}_{n-1} M \circ \mathscr{T}_m M & \xrightarrow{i+j(\mathrm{Id}\circ\gamma_{n-1,m})} & \mathscr{T}_{n+m} M \\
\downarrow{\scriptstyle i} & & \downarrow{\scriptstyle i} \\
\mathscr{T}_{m+1} M \oplus M \circ \mathscr{T}_{n-1} M \circ \mathscr{T}_{m+1} M & \xrightarrow{i+j(\mathrm{Id}\circ\gamma_{n-1,m+1})} & \mathscr{T}_{n+m+1} M.
\end{array}$$

So we have proved that $\gamma_{n+1,m}(i \circ \mathrm{Id}) = i(\gamma_{n,m}) = \gamma_{n,m+1}(\mathrm{Id} \circ i)$. By passing to the colimit we get a well-defined map $\gamma : \mathscr{T} M \circ \mathscr{T} M \to \mathscr{T} M$.

(2) Let us show now that γ is associative. It is sufficient to prove that for any p, q, r we have the equality

$$\gamma_{p+q,r}(\gamma_{p,q} \circ \mathrm{Id}) = \gamma_{p,q+r}(\mathrm{Id} \circ \gamma_{q,r}) : \mathscr{T}_p M \circ \mathscr{T}_q M \circ \mathscr{T}_r M \to \mathscr{T}_{p+q+r} M.$$

We work by induction on p. For $p = 0$ it is immediate since $\gamma_{0,m} = \mathrm{Id}$. We leave it to the reader to write down the diagram which shows that associativity for $p-1$ implies associativity for p.

The map $i : \mathrm{I} = \mathscr{T}_0 M \to \mathscr{T} M$ is the unit. Indeed, it is sufficient to prove that the following diagram is commutative

$$\begin{array}{ccc}
\mathrm{I} \circ \mathscr{T}_{m+1} M & \xrightarrow{(i,\mathrm{Id})} & \mathscr{T}_n M \circ \mathscr{T}_{m+1} M \\
\downarrow{\scriptstyle i} & & \downarrow{\scriptstyle \gamma_{n,m+1}} \\
\mathscr{T}_{m+1} M \oplus M \circ \mathscr{T}_{n-1} M \circ \mathscr{T}_{m+1} M & \xrightarrow{i+j(\mathrm{Id}\circ\gamma_{n-1,m+1})} & \mathscr{T}_{n+m+1} M.
\end{array}$$

(3) Finally we prove that $(\mathscr{T} M, \gamma, j)$ is the free operad over M. Let $\mathscr{P}$ be an operad with composition $\gamma_{\mathscr{P}}$ and unit $\eta_{\mathscr{P}}$. It is sufficient to prove that there are morphisms of operads $\phi_{\mathscr{P}} : \mathscr{T}\mathscr{P} \to \mathscr{P}$ and $\psi_{\mathscr{P}} : \mathscr{P} \to \mathscr{T}\mathscr{P}$ natural in the variable $\mathscr{P}$ such that both composites

$$\mathscr{P} \xrightarrow{\psi_{\mathscr{P}}} \mathscr{T}\mathscr{P} \xrightarrow{\phi_{\mathscr{P}}} \mathscr{P}, \qquad \mathscr{T} M \xrightarrow{\mathscr{T}(\psi_M)} \mathscr{T}\mathscr{T} M \xrightarrow{\phi_{\mathscr{T} M}} \mathscr{T} M$$

are the identity ($\mathscr{T}$ left adjoint to the forgetful functor). The first one shows the existence of the extension of $f : M \to \mathscr{P}$ to $\mathscr{T} M$ and the second one shows its uniqueness.

We construct $\phi_{\mathscr{P}} : \mathscr{T}\mathscr{P} \to \mathscr{P}$ as follows. For $n = 0$, take $\phi_0 = \eta_{\mathscr{P}} : \mathscr{T}_0\mathscr{P} = \mathrm{I} \to \mathscr{P}$. For $n = 1$, take $\phi_1 = \eta_{\mathscr{P}} + \mathrm{Id}_{\mathscr{P}} : \mathscr{T}_1\mathscr{P} = \mathrm{I} \oplus \mathscr{P} \to \mathscr{P}$. By induction, take

$\phi_n = \eta_{\mathcal{P}} + \gamma_{\mathcal{P}}(\mathrm{Id}_{\mathcal{P}} \circ \phi_{n-1}) : \mathcal{T}_n\mathcal{P} = \mathrm{I} \oplus (\mathcal{P} \circ \mathcal{T}_{n-1}\mathcal{P}) \to \mathcal{P}$. Since $\phi_n \circ i = \phi_{n-1}$, we get at the colimit a transformation of functors $\phi_{\mathcal{P}} : \mathcal{T}\mathcal{P} \to \mathcal{P}$. The expected properties of $\phi_{\mathcal{P}}$ are straightforward to prove by induction. □

5.5.2 Examples

(a) Let $M = (0, W, 0, \ldots, 0, \ldots)$ where W is a vector space. The Schur functor is $M(V) = W \otimes V$. Since M is linear, that is $M(V \oplus V') = M(V) \oplus M(V')$, it follows that

$$\mathcal{T}_n M = \mathrm{I} \oplus M \oplus M \circ M \oplus \cdots \oplus M^{\circ n},$$

and therefore $\mathcal{T}M = (0, T(W), 0, \ldots)$. We recover that the free associative algebra is the tensor algebra, cf. Sect. 1.1.3.

(b) Let $M = (0, 0, \mathbb{K}[\mathbb{S}_2], 0, \ldots, 0, \ldots)$ where $\mathbb{K}[\mathbb{S}_2]$ is the regular representation. From the description of the free operad in terms of trees (see Sect. 5.6.1) it follows that

$$\mathcal{T}_n M = \bigoplus_{k \le n} (\mathcal{T}M)^{(k)} \cong \bigoplus_{k \le n} \mathbb{K}[PBT_k] \otimes \mathbb{K}[\mathbb{S}_k], \quad n \ge 1,$$

where PBT_k is the set of planar binary rooted trees with k leaves, cf. Appendix C.2. See Sect. 5.9.6 for the precise identification with trees.

(c) Let $M = (0, 0, M_2 \otimes \mathbb{K}[\mathbb{S}_2], 0, \ldots, 0, \ldots)$. The same argument as in the previous example shows that $(TM)(n) = (M_2)^{\otimes n-1} \otimes \mathbb{K}[PBT_n] \otimes \mathbb{K}[\mathbb{S}_n]$. It is helpful to think of its elements as binary operations decorating the vertices of a tree:

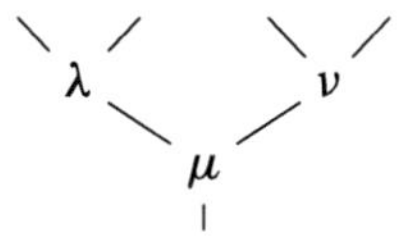

5.5.3 Weight-Grading of the Free Operad

We introduce a weight grading on the free operad: the weight is the number of generating operations needed in the construction of a given operation of the free operad.

Let M be an $\mathbb{S}$-module and let $\mathcal{T}(M)$ be the free operad on M. By definition the weight $w(\mu)$ of an operation μ of $\mathcal{T}(M)$ is defined as follows:

$$w(\mathrm{id}) = 0, \qquad w(\mu) = 1 \quad \text{when } \mu \in M(n),$$

and more generally

$$w(\nu; \nu_1, \ldots, \nu_n) = w(\nu) + w(\nu_1) + \cdots + w(\nu_n).$$

We denote by $\mathscr{T}(M)^{(r)}$ the $\mathbb{S}$-module of operations of weight r. So we get $\mathscr{T}(M)^{(0)} = \mathbb{K}\,\mathrm{id}$ (concentrated in arity 1) and $\mathscr{T}(M)^{(1)} = M$.

EXAMPLE 1. Suppose that $M = (0, W, 0, \ldots, 0, \ldots)$. Then $\mathscr{T}(M)$ is simply $(0, T(W), 0, \ldots, 0, \ldots)$ where $T(W)$ is the tensor algebra. The weight grading of the tensor operad corresponds to the weight grading of the tensor algebra, cf. Sect. 1.1.3.

EXAMPLE 2. Suppose that $M = (0, 0, E, 0, \ldots, 0, \ldots)$ where E is an $\mathbb{S}_2$-module. So the free operad $\mathscr{T}(M)$ is generated by binary operations. Then $\mathscr{T}(M)^{(r)}$ is concentrated in arity $r+1$ and is exactly $\mathscr{T}(M)(r+1)$. Indeed, an operation on $r+1$ variables in $\mathscr{T}(M)$ needs r binary operations to be constructed. This is the reason why, when dealing with binary operads, the weight is, in general, not mentioned.

5.5.4 Presentation of an Operad

Let M be an $\mathbb{S}$-module generated (as an $\mathbb{S}$-module) by elements μ_i. Let $\mathscr{I}$ be an ideal in the free operad $\mathscr{T}(M)$ and let r_j be generators of the ideal $\mathscr{I}$. Then a $(\mathscr{T}(M)/\mathscr{I})$-algebra is determined by the set of operations μ_i (with their symmetry) and the set of relations $r_j = 0$.

5.5.5 Another Construction of the Free Operad

In this subsection, taken out of [Val08], we give an outline of a construction of the free operad which has the advantage of working in any monoidal category satisfying some mild assumptions.

We denote by M_+ the augmented $\mathbb{S}$-module of M, that is

$$M_+ := \mathrm{I} \oplus M,$$

and we write $M^n := (M_+)^{\circ n} = (\underbrace{\cdots((M_+ \circ M_+) \circ M_+) \circ \cdots \circ M_+}_{n})$. The inclusions of I and M in M_+ are denoted respectively by $\eta : \mathrm{I} \to M_+$ and $\eta_M : M \to M_+$. The map

$$\eta_i : M^n \cong M^i \circ \mathrm{I} \circ M^{n-i} \xrightarrow{\mathrm{Id} \circ \eta \circ \mathrm{Id}} M^i \circ M_+ \circ M^{n-i} \cong M^{n+1}$$

is called the ith *degeneracy map*.

The colimit over the degeneracy maps, that is

$$\mathrm{I} = M^0 \longrightarrow M^1 \rightrightarrows M^2 \substack{\longrightarrow\\ \longrightarrow\\ \longrightarrow} M^3 \cdots,$$

is too large to be the free object, essentially because, in the composition $((xy)(zt))$, the order of parenthesizing (xy first or zt first?) should be irrelevant. So we are going to make a quotient of these spaces.

Since I is a unit for $\circ$ there are isomorphisms $\lambda : \mathrm{I} \circ M \cong M$ and $\rho : M \circ \mathrm{I} \cong M$. We consider the composite

$$\tau : M \xrightarrow{\lambda^{-1} \oplus \rho^{-1}} \mathrm{I} \circ M \oplus M \circ \mathrm{I} \xrightarrow{\eta \circ \eta_M - \eta_M \circ \eta} (M_+)^{\circ 2} = M^2.$$

For any $\mathbb{S}$-modules A and B there is a well-defined $\mathbb{S}$-module

$$R_{A;B} := \mathrm{Im}\left(A \circ \left(M \oplus M^2\right) \circ B \xrightarrow{\mathrm{Id}_A \circ (\tau + \mathrm{Id}_{M^2}) \circ \mathrm{Id}_B} A \circ M^2 \circ B\right).$$

We put

$$\widetilde{M^n} := M^n / \sum_{i=0}^{n-2} R_{M^i; M^{n-2-i}}.$$

It can be shown that, under this quotient, the degeneracy maps η_i, for $i = 1, \ldots, n$, are equal and define $\eta : \widetilde{M^n} \to \widetilde{M^{n+1}}$. By definition $\mathscr{F}(M)$ is the (sequential) colimit of

$$\widetilde{M} \xrightarrow{\eta} \widetilde{M^2} \xrightarrow{\eta} \widetilde{M^3} \xrightarrow{\eta} \cdots.$$

Theorem 5.5.2 [Val08] *In any monoidal category such that the monoidal product preserves sequential colimits and reflexive coequalizers, $\mathscr{F}(M)$ can be equipped with a monoid structure that is free over M.*

The advantage of this construction lies in its generalization to some types of bialgebras. In order to develop the same kind of arguments, the notion of an operad has to be replaced by the notion of properad, see Sect. 13.14.9. The aforementioned result produces the free properad, cf. [Val07b].

5.6 Combinatorial Definition of an Operad

In this section we give a fourth definition of an operad based on some combinatorial objects: the rooted trees. The main advantage of this presentation is to admit several important variations by changing the combinatorial objects and/or by decorating them. For instance, if we replace the rooted trees by ladders, then we get unital associative algebras in place of operads. If we take planar rooted trees, then we get nonsymmetric operads. If we take nonrooted trees, then we get cyclic operads, cf. Sect. 13.14 and [GK95a]. A far-reaching generalization has been given by Borisov and Manin in [BM08], see also [Lei04].

5.6.1 The Monad of Trees

A *reduced rooted tree* is a nonplanar rooted tree such that each vertex has one input or more, cf. Appendix C.3. Let X be a finite set. For any tree $t \in RT(X)$ (i.e. we are given a bijection from the set of leaves of t to X) we denote by $\mathrm{vert}(t)$ the set of vertices of t and by $\mathrm{in}(v)$ the set of inputs of the vertex $v \in \mathrm{vert}(t)$. Let M be an $\mathbb{S}$-module with $M(0) = 0$ or, equivalently, a functor $M : \mathsf{Bij} \to \mathsf{Vect}$, $X \mapsto M(X)$, such that $M(\emptyset) = 0$, cf. Sect. 5.1.13.

We define the *treewise tensor product* $M(t)$ as follows:

$$M(t) := \bigotimes_{v \in \mathrm{vert}(t)} M\big(\mathrm{in}(v)\big).$$

See Sect. 5.1.14 for the precise meaning of $\otimes_{v \in \mathrm{vert}(t)}$. Using this notation we define a functor

$$\mathbb{T} : \mathbb{S}\text{-Mod} \to \mathbb{S}\text{-Mod}$$

by

$$\mathbb{T}(M)(X) := \bigoplus_{t \in RT(X)} M(t).$$

It is helpful to think about an element of $\mathbb{T}(M)(X)$ as a sum of rooted trees where each vertex v is decorated by an element of $M(\mathrm{in}(v))$ and each leaf is decorated by an element of X.

First, we construct a transformation of functors

$$\iota : \mathrm{Id}_{\mathbb{S}\text{-Mod}} \to \mathbb{T}$$

as follows. For any $\mathbb{S}$-module M we have to say what is the $\mathbb{S}$-module morphism $M \to \mathbb{T}(M)$, i.e. for any finite set X a linear map $M(X) \to \mathbb{T}(M)(X)$. In the set of trees $RT(X)$, there is a particular one $t = cor$ which is the corolla. We have $M(cor) = M(X)$ by definition, since the corolla has only one vertex. Hence $M(X)$ is a direct summand of $\mathbb{T}(M)(X)$. The expected map is the corresponding inclusion.

Second, we construct a transformation of functors

$$\alpha : \mathbb{T} \circ \mathbb{T} \to \mathbb{T}$$

as follows. The *substitution* of trees consists in replacing the vertices of a tree by given trees (with matching inputs) like in Fig. 5.3.

In order to perform the substitution in the tree t we need, for any $v \in \mathrm{vert}(t)$ a tree t_v and a bijection $\mathrm{in}(v) \cong \mathrm{leaves}(t_v)$.

Lemma 5.6.1. *The substitution of trees defines a transformation of functors α : $\mathbb{T} \circ \mathbb{T} \to \mathbb{T}$ which is associative and unital. So $(\mathbb{T}, \alpha, \iota)$ is a monad.*

Proof. From the definition of $\mathbb{T}$ we get

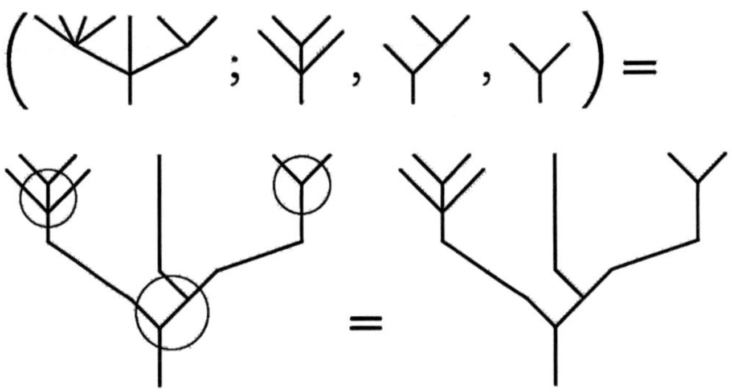

Fig. 5.3 Substitution

$$
\begin{aligned}
\mathbb{T}\big(\mathbb{T}(M)\big)(X) &= \bigoplus_{t\in RT(X)} \mathbb{T}(M)(t) \\
&= \bigoplus_{t\in RT(X)} \Big(\bigotimes_{v\in \mathrm{vert}(t)} \mathbb{T}(M)\big(\mathrm{in}(v)\big)\Big) \\
&= \bigoplus_{t\in RT(X)} \Big(\bigotimes_{v\in \mathrm{vert}(t)} \Big(\bigoplus_{s\in RT(\mathrm{in}(v))} M(s)\Big)\Big).
\end{aligned}
$$

The decoration of the vertex v in t is an element of $\mathbb{T}(M)(\mathrm{in}(v))$, that is a tree whose leaves are labeled by $\mathrm{in}(v)$. This is exactly the data which permits us to perform the substitution. So we get an element of $\mathbb{T}(M)(X)$. As a result we have defined an $\mathbb{S}$-module morphism $\alpha(M) : \mathbb{T}(\mathbb{T}(M)) \to \mathbb{T}(M)$. Obviously this morphism is functorial in M so we have constructed a transformation of functors $\alpha : \mathbb{T}\circ\mathbb{T} \to \mathbb{T}$. The substitution process is clearly associative, so α is associative.

Recall that the unit ι consists in identifying an element μ of $M(X)$ with the corolla with vertex decorated by μ. Substituting a vertex by a corolla does not change the tree. Substituting a corolla by a vertex gives the former tree. Hence α is also unital.

We have proved that $(\mathbb{T}, \alpha, \iota)$ is a monad. □

5.6.2 Combinatorial Definition

The *combinatorial definition of an operad* consists in defining it as an algebra over the monad $(\mathbb{T}, \alpha, \iota)$, cf. Appendix B.4. In other words an operad is an $\mathbb{S}$-module $\mathscr{P}$ together with an $\mathbb{S}$-module map $\mathbb{T}(\mathscr{P}) \to \mathscr{P}$ compatible with α in the obvious sense.

Proposition 5.6.2. [GJ94] *The combinatorial definition of an operad is equivalent to the partial definition of an operad, and therefore to all the other definitions.*

Proof. It suffices to verify that the partial operation is indeed a substitution and that the substitution of trees satisfies the two axioms (I) and (II) of partial operations defining an operad. □

Proposition 5.6.3. *For any $\mathbb{S}$-module M, $\mathbb{T}(M)$ is an operad which is the free operad $\mathscr{T}(M)$ over M.*

Proof. Since $\mathbb{T}(M)$ is an algebra over the monad $(\mathbb{T}, \alpha, \iota)$, by Proposition 5.6.2 it is an operad. Checking that it is free is analogous to the proof of Proposition 5.2.1. □

5.6.3 Comparison of the Two Constructions of the Free Operad

Recall that in Theorem 5.5.1 we constructed the free operad on M inductively as a colimit: $\mathscr{T}(M) = \mathrm{colim}_n \mathscr{T}_n M$, where

$$\mathscr{T}_0 M := \mathrm{I},$$
$$\mathscr{T}_1 M := \mathrm{I} \oplus M,$$
$$\mathscr{T}_n M := \mathrm{I} \oplus (M \circ \mathscr{T}_{n-1} M).$$

Since $\mathscr{T}(M)$ and $\mathbb{T}(M)$ are both the free operad on M we know that they are isomorphic. We make this isomorphism explicit as follows.

The map $\mathscr{T}_0 M = \mathrm{I} \to \mathbb{T}(M)$ is given by the operation $\mathrm{id} \in \mathbb{T}(M)(1)$. The map $M \to \mathbb{T}(M)$ is given by $\mu \mapsto$ corolla, where the number of leaves of the corolla is the arity of μ, and the single vertex is decorated by μ. By induction we suppose that $\mathscr{T}_{n-1} M \to \mathbb{T}(M)$ has been constructed. The map $M \circ \mathscr{T}_{n-1} M \to \mathbb{T}(M)$ is obviously given by the composition in $\mathbb{T}(M)$ of the images of M and of $\mathscr{T}_{n-1} M$. Notice that, under the above isomorphism, the $\mathbb{S}_n$-module $\mathscr{T}_n M$ corresponds to linear combination of trees with at most n levels. More details are given in Sect. 5.9.6.

5.7 Type of Algebras

We make explicit the relationship between the notion of algebraic operad and some types of algebras. We suppose that we are in characteristic zero.

5.7.1 Type of Algebras and Presentation of an Operad

Let P-alg be a category of algebras presented as follows. An object of P-alg is a vector space A equipped with some n-ary operations $\mu_i : A^{\otimes n} \to A$ (possibly for various n's), called the *generating operations* satisfying some relations $r_j = 0$. Let us suppose that the relations are multilinear, that is of the form

$$\sum_\phi \phi(a_1, \ldots, a_n) = 0 \quad \text{for all } a_1, \ldots, a_n \in A,$$

where ϕ is a composite of the generating operations μ_i. An element like $r = \sum \phi$ is called a *relator*. Let us denote by M the $\mathbb{S}$-module which is, in arity n, the $\mathbb{S}_n$-module spanned by the generating n-ary operations. We take into account the symmetries of these operations to determine the $\mathbb{S}_n$-module structure. A relator determines an operation in the free operad $\mathcal{T}M$. Let R be the sub-$\mathbb{S}$-module of $\mathcal{T}M$ spanned by all the relators, and let (R) be the operadic ideal of $\mathcal{T}M$ generated by R. Then we get the operad $\mathcal{T}M/(R)$.

Let us denote by $\mathcal{P}(V)$ the free $\mathtt{P}$-algebra over V. It defines a functor $\mathcal{P}$: $\mathsf{Vect} \to \mathsf{Vect}$.

Lemma 5.7.1. *The functor $\mathcal{P}$ is a Schur functor whose arity n component is the multilinear part of the free $\mathtt{P}$-algebra $\mathcal{P}(\mathbb{K}x_1 \oplus \cdots \oplus \mathbb{K}x_n)$. Moreover $\mathcal{P}$ is an algebraic operad.*

Proof. Let us first prove that the free algebra $\mathcal{P}(V)$ of the given type $\mathtt{P}$ is equipped with a monoid structure. Since $\mathcal{P}(V)$ is free over V, it comes with a natural map $V \to \mathcal{P}(V)$ that we denote by $\eta(V)$. Consider the map $\mathrm{Id}_{\mathcal{P}(V)} : \mathcal{P}(V) \to \mathcal{P}(V)$ as a well-defined map from the vector space $W = \mathcal{P}(V)$ to the algebra $\mathcal{P}(V)$ of type $\mathtt{P}$. By the universal property, there exists a lifting of $\mathrm{Id}_{\mathcal{P}(V)}$ denoted $\gamma(V)$: $\mathcal{P}(\mathcal{P}(V)) = \mathcal{P}(W) \to \mathcal{P}(V)$. It is clear that this morphism of algebras of type $\mathtt{P}$ is functorial in V.

Again, from the universal property of free algebras, we deduce that γ is associative. From the fact that $\gamma(V)$ is a lifting of $\mathrm{Id}_{\mathcal{P}(V)}$ we deduce that γ is unital. Hence $(\mathcal{P}, \gamma, \eta)$ is a monoid in the category of endofunctors of Vect.

Let us now show that $\mathcal{P}$ is a Schur functor. Since the relations are multilinear, the free $\mathtt{P}$-algebra over V is the direct sum of its homogeneous components. By the Schur Lemma (cf. Sect. A.2.3) the homogeneous component of degree n (in characteristic zero) is of the form $\mathcal{P}(n) \otimes_{\mathbb{S}_n} V^{\otimes n}$ for some $\mathbb{S}_n$-module $\mathcal{P}(n)$. Taking $V = V_n = \mathbb{K}x_1 \oplus \cdots \oplus \mathbb{K}x_n$ we verify that $\mathcal{P}(n)$ is the multilinear part of the free $\mathtt{P}$-algebra over V_n as an $\mathbb{S}_n$-module, cf. Lemma 5.1.1. So it follows that the free $\mathtt{P}$-algebra over V is of the form

$$\mathcal{P}(V) = \bigoplus_{n \geq 0} \mathcal{P}(n) \otimes_{\mathbb{S}_n} V^{\otimes n},$$

as expected. □

Lemma 5.7.2. *Let $\mathtt{P}$ be a type of algebras defined by the $\mathbb{S}$-module of generating operations M and the $\mathbb{S}$-module of relators $R \subset \mathcal{T}M$. Then the operad $\mathcal{P}$ constructed above coincides with the quotient operad $\mathcal{T}M/(R)$.*

Proof. By definition of the operad $\mathcal{T}M/(R)$ out of the type of algebras P, it follows that $(\mathcal{T}M/(R))(V)$ is the free P-algebra over V. By construction we also know that $\mathcal{P}(V)$ is a free P-algebra over V. Since the identification $(\mathcal{T}M/(R))(V) = \mathcal{P}(V)$ is functorial in V, we are done. □

Proposition 5.7.3. *In characteristic zero, a type of algebras whose relations are multilinear determines an operad. The category of algebras over this operad is equivalent to the category of algebras of the given type.*

Proof. From the preceding results we know that the type P determines the operad $\mathcal{P}$. From the identification of $\mathcal{P}$ with $\mathcal{T}M/(R)$ it follows that the two categories of algebras $\mathcal{P}$-alg and P-alg are equivalent. □

5.7.2 Examples

The associative algebras, the commutative algebras and the Lie algebras are examples of types of algebras with multilinear relations. The operad that they determine is denoted by *Ass*, *Com* and *Lie* respectively. Though they will be studied in detail later on, let us make some comments on the cases of commutative algebras and associative algebras.

Since the free commutative algebra over $\{x_1, \ldots, x_n\}$ is the polynomial algebra (modulo the constants), it follows that $Com(n) = \mathbb{K}$ for $n \geq 1$, and its generator μ_n is the n-ary operation determined by

$$\mu_n(x_1, \ldots, x_n) = x_1 \cdots x_n \in \mathbb{K}[x_1, \ldots, x_n].$$

The action of $\mathbb{S}_n$ is trivial, hence $Com(n)$ is the one-dimensional trivial representation. From the classical "composition" of polynomials, it follows that

$$\gamma(\mu_k; \mu_{i_1}, \ldots, \mu_{i_k}) = \mu_{i_1+\cdots+i_k}.$$

Since the free associative algebra over $\{x_1, \ldots, x_n\}$ is the noncommutative polynomial algebra (modulo the constants), it follows that $Ass(n) = \mathbb{K}[\mathbb{S}_n]$, and the n-ary operation μ_σ corresponding to $\sigma \in \mathbb{S}_n$ is determined by

$$\mu^\sigma(x_1, \ldots, x_n) = x_{\sigma^{-1}(1)} \cdots x_{\sigma^{-1}(n)} \in \mathbb{K}\langle x_1, \ldots, x_n\rangle.$$

The action of $\mathbb{S}_n$ is by multiplication, so $Ass(n)$ is the regular representation. From the classical "composition" of noncommutative polynomials, it follows that

$$\gamma(\sigma; \sigma_1, \ldots, \sigma_k) = \tilde{\sigma} \circ (\sigma_1, \ldots, \sigma_k),$$

where $\tilde{\sigma}$ is the block permutation associated to σ.

5.7.3 Kernels

Let $\mathscr{Q}$-alg $\to$ $\mathscr{P}$-alg be a functor between two types of algebras, which is supposed to commute with the forgetful functor to vector spaces:

$$\begin{array}{ccc} \mathscr{Q}\text{-alg} & \longrightarrow & \mathscr{P}\text{-alg} \\ & \searrow \quad \swarrow & \\ & \mathsf{Vect}_{\mathbb{K}}. & \end{array}$$

By looking at the free algebras, we check that it comes from a morphism of operads $\alpha : \mathscr{P} \to \mathscr{Q}$. We know that $\mathrm{I} \oplus \operatorname{Ker}\alpha$ is an operad, so there is a new type of algebras $(\mathrm{I} \oplus \operatorname{Ker}\alpha)$-alg. In many examples $\mathscr{P}$ and $\mathscr{Q}$ are presented by a small number of generators and relations, but no such small presentation is known for $\mathrm{I} \oplus \operatorname{Ker}\alpha$ in general. These are examples where the use of operads is a necessity.

5.7.4 A Universal Presentation

We know that a group can always be presented as follows: the generators are its elements and the relations are given by the table of multiplication. Similarly an operad can always be presented as follows. Choose a linear basis for $\mathscr{P}(n)$, $n \geq 1$, and take the composite products as relations.

5.7.5 Non-examples (?)

Here are two examples of types of algebras, which, a priori, do not fall directly into the operad theory because the relations are not multilinear. However minor changes will make them accessible.

A *Jordan algebra* A is a vector space equipped with a binary operation which satisfies the relation

$$\big(a^2 b\big)a = a^2(ba).$$

It seems to lie outside our framework since the relation fails to be multilinear. However it suffices to multilinearize it and we get the following type of algebras: one binary symmetric operation ab and one relation

$$(ab)(dc) + (ac)(db) + (bc)(da) = \big((ab)d\big)c + \big((ac)d\big)b + \big((bc)d\big)a.$$

More information is to be found in Sect. 13.10.

A *divided power algebra* is an augmented commutative algebra equipped with unary operations $\gamma_n(x)$ which bear the formal properties of the operations $x^n/n!$. A priori such a structure cannot be encoded by an operad, however taking invariants in place of coinvariants in the construction of the Schur functor permits us to solve the problem. See Sect. 13.1.12 for more details.

5.8 Cooperad

In our treatment of Koszul duality for associative algebras we put algebras and coalgebras on the same footing. In order to play the same game with operads we need to introduce the notion of cooperad.

We define the notion of cooperad as a comonoid in the monoidal category of $\mathbb{S}$-modules with the composite product. Cooperads are used to encode categories of coalgebras. We introduce the important notion of conilpotent cooperad. In the same way as for operads, we give an equivalent combinatorial definition of a conilpotent cooperad in terms of trees.

5.8.1 Algebraic Cooperad

Let $\mathscr{C}$ be an $\mathbb{S}$-module. A *cooperad* is a structure of comonoid on $\mathscr{C}$ in the monoidal category ($\mathbb{S}$-Mod, $\bar{\circ}$, I), where $(\mathscr{P} \bar{\circ} \mathscr{Q})(n) := \bigoplus_r (\mathscr{P}(r) \otimes \mathscr{Q}^{\otimes r})^{\mathbb{S}_r}(n)$, see Sect. 5.1.15. Explicitly it consists into two morphisms of $\mathbb{S}$-modules (equivalently transformations of Schur functors)

$$\Delta : \mathscr{C} \to \mathscr{C} \bar{\circ} \mathscr{C} \quad \text{(decomposition)} \quad \text{and} \quad \varepsilon : \mathscr{C} \to \mathrm{I} \quad \text{(counit)},$$

which satisfy the axioms of coassociativity:

$$\begin{array}{ccc}
\mathscr{C} & \xrightarrow{\Delta} & \mathscr{C} \bar{\circ} \mathscr{C} \\
\downarrow{\scriptstyle \Delta} & & \downarrow{\scriptstyle \mathrm{Id} \bar{\circ} \Delta} \\
 & & \mathscr{C} \bar{\circ} (\mathscr{C} \bar{\circ} \mathscr{C}) \\
 & & \swarrow{\scriptstyle \cong} \\
\mathscr{C} \bar{\circ} \mathscr{C} & \xrightarrow{\Delta \bar{\circ} \mathrm{Id}} & (\mathscr{C} \bar{\circ} \mathscr{C}) \bar{\circ} \mathscr{C}
\end{array}$$

and counitality:

$$\begin{array}{ccccc}
 & & \mathscr{C} & & \\
 & \swarrow{\scriptstyle \cong} & \downarrow{\scriptstyle \Delta} & \searrow{\scriptstyle \cong} & \\
\mathrm{I} \bar{\circ} \mathscr{C} & \xleftarrow[\varepsilon \bar{\circ} \mathrm{Id}]{} & \mathscr{C} \bar{\circ} \mathscr{C} & \xrightarrow[\mathrm{Id} \bar{\circ} \varepsilon]{} & \mathscr{C} \bar{\circ} \mathrm{I}.
\end{array}$$

We observe that the $\mathbb{S}$-module I associated to identity functor $\mathsf{Vect} \to \mathsf{Vect}$ is a cooperad.

From Corollary 5.1.4, it follows that Δ is made up of $\mathbb{S}_n$-module morphisms $\Delta(n)$:

$$\mathscr{C}(n) \to \mathscr{C} \bar{\circ} \mathscr{C}(n) = \bigoplus_{k \geq 0} \Big(\mathscr{C}(k) \otimes \Big(\bigoplus \mathrm{Ind}_{\mathbb{S}_{i_1} \times \cdots \times \mathbb{S}_{i_k}}^{\mathbb{S}_n} (\mathscr{C}(i_1) \otimes \cdots \otimes \mathscr{C}(i_k)) \Big) \Big)^{\mathbb{S}_k}$$

where the second sum is extended to all the k-tuples $(i_1, \dots, i_k)$ satisfying $i_1 + \cdots + i_k = n$. So a cooperad concentrated in arity 1 is a coassociative coalgebra.

A cooperad is said to be *coaugmented* if there is a cooperad morphism $\eta : \mathrm{I} \to \mathscr{C}$ such that $\varepsilon\eta = \mathrm{Id}_{\mathrm{I}}$. The image of $1 \in I(1) = \mathbb{K}$ is denoted $\mathrm{id} \in \mathscr{C}(1)$ and is called the *identity cooperation*. The cokernel of η is denoted by $\overline{\mathscr{C}}$ and $\mathscr{C} \cong \mathrm{I} \oplus \overline{\mathscr{C}}$.

Observe that, by the counitality assumption, the component of $\Delta(n)(\mu)$ in the two extreme summands ($k = 1$ and $k = n$ respectively) is $(\mathrm{id}; \mu)$ and $(\mu; \mathrm{id}^{\otimes n})$ respectively. We will sometimes adopt the following abuse of notation

$$\Delta(\mu) = \sum (\nu; \nu_1, \dots, \nu_k)$$

where $\nu \in \mathscr{C}(k)$, $\nu_j \in \mathscr{C}(i_j)$ for the image of the cooperation μ by the decomposition map, and

$$\Delta(\mu) = (\mathrm{id}; \mu) + \left(\mu; \mathrm{id}^{\otimes n}\right) + \bar{\Delta}(\mu).$$

The map $\bar{\Delta}$ is called the *reduced decomposition map*.

To define the exact dual of the notion of operad, one should instead consider the monoidal product

$$\mathscr{C} \hat{\circ} \mathscr{C}(n) = \prod_{k \geq 0} \left(\mathscr{C}(k) \otimes \left(\prod \mathrm{Ind}_{\mathbb{S}_{i_1} \times \cdots \times \mathbb{S}_{i_k}}^{\mathbb{S}_n} \left(\mathscr{C}(i_1) \otimes \cdots \otimes \mathscr{C}(i_k) \right) \right) \right)^{\mathbb{S}_k}$$

in the category of $\mathbb{S}$-modules, where the sums are replaced by products. In that case, a cooperad is defined as a comonoid $\Delta : \mathscr{C} \to \mathscr{C} \hat{\circ} \mathscr{C}$.

When $\mathscr{C}(0) = 0$, the right-hand side product is equal to a sum. The decomposition map $\Delta : \mathscr{C} \to \mathscr{C} \bar{\circ} \mathscr{C} \subset \mathscr{C} \hat{\circ} \mathscr{C}$ of some cooperads is made up of sums of elements. In this case, we are back to the previous definition. In this book, we will mainly encounter cooperads of this first type, so we work with this definition.

Working over a field of characteristic 0, we can identify invariants with coinvariants, see Appendix A.1, and work with $\circ$ instead of $\bar{\circ}$. But this more general definition plays a key role in characteristic p, for instance.

5.8.2 From Cooperads to Operads and Vice Versa

Let $\mathscr{C}$ be a cooperad and let $\mathscr{P}(n) = \mathscr{C}(n)^* = \mathrm{Hom}(\mathscr{C}(n), \mathbb{K})$. Since $\mathscr{C}(n)$ is a right $\mathbb{S}_n$-module, its dual $\mathscr{P}(n)$ is a left $\mathbb{S}_n$-module. We make it into a right $\mathbb{S}_n$-module by the classical formula $\mu^\sigma := \sigma^{-1} \cdot \mu$, for $\mu \in \mathscr{P}(n)$ and $\sigma \in \mathbb{S}_n$. The transpose of the counit ε gives a unit η. The decomposition map Δ gives a composition map γ by dualization followed by the natural map from invariants to coinvariants, cf. Sect. 5.1.15.

In the other way round, let $\mathscr{P}$ be an operad such that each $\mathscr{P}(n)$ is finite dimensional. This condition ensures that $\mathscr{P}(n)^* \otimes \mathscr{P}(m)^* \to (\mathscr{P}(n) \otimes \mathscr{P}(m))^*$ is an isomorphism, as in Sect. 1.2.2. So the linear dual $\mathscr{C} := \mathscr{P}^*$ of $\mathscr{P}$ gives rise to a cooperad $\mathscr{C} \to \mathscr{C} \hat{\circ} \mathscr{C}$. If we further suppose that $\mathscr{P}(0) = 0$ and that the preimage under the composition map of any element in $\mathscr{P}$ is finite, then the decomposition map of $\mathscr{C}$ lives in $\mathscr{C} \bar{\circ} \mathscr{C}$.

5.8.3 Coalgebra over a Cooperad

By definition a *coalgebra* over the cooperad $\mathscr{C}$, or $\mathscr{C}$*-coalgebra* for short, is a vector space C equipped with a map $\Delta_C : C \to \widehat{\mathscr{C}}(C)$, where $\widehat{\mathscr{C}}(C) = \prod_n (\mathscr{C}(n) \otimes C^{\otimes n})^{\mathbb{S}_n}$, such that the following diagrams commute:

$$
\begin{array}{ccc}
C & \xrightarrow{\Delta_C} & \widehat{\mathscr{C}}(C) \\
{\scriptstyle \Delta_C}\downarrow & & \downarrow{\scriptstyle \widehat{\mathscr{C}}(\Delta_C)} \\
\widehat{\mathscr{C}}(C) & \xrightarrow{\Delta(C)} & \widehat{\mathscr{C}}(\widehat{\mathscr{C}}(C))
\end{array}
\qquad
\begin{array}{ccc}
C & & \\
{\scriptstyle \Delta(C)}\downarrow & \searrow{\scriptstyle =} & \\
\widehat{\mathscr{C}}(C) & \xrightarrow{\eta(C)} & C.
\end{array}
$$

So for any n, we have a map $\Delta_n : C \to (\mathscr{C}(n) \otimes C^{\otimes n})^{\mathbb{S}_n}$.

Let $\mathscr{C}^*$ be the operad obtained by linear dualization. The map Δ_n gives rise to an $\mathbb{S}_n$-equivariant map

$$\mathscr{C}^*(n) \longrightarrow \operatorname{Hom}\bigl(C, C^{\otimes n}\bigr),$$

and it is immediate to check that C is a coalgebra over the operad $\mathscr{C}^*$ in the sense of Sect. 5.2.15.

5.8.4 Conilpotent Coalgebra, Primitive Part

Let C be a coalgebra over a coaugmented cooperad $\mathscr{C}$. We denote the image under its structure map by

$$\Delta_C(x) = (x_1, x_2, \ldots) \in \prod_{n \geq 1} \bigl(\mathscr{C}(n) \otimes C^{\otimes n}\bigr)^{\mathbb{S}_n}.$$

We define the *coradical filtration* of C as follows:

$$F_1 C := \operatorname{Prim} C := \{x \in C \mid x_1 = x, \text{ and } x_k = 0 \text{ for any } k > 1\}.$$

The space $\operatorname{Prim} C$ is called the *primitive part* of C, and its elements are said to be *primitive*. Then we define the filtration by:

$$F_r C := \{x \in C \mid x_k = 0 \text{ for any } k > r\}.$$

We say that the coalgebra C is *conilpotent*, if this filtration is exhaustive $C = \bigcup_{r \geq 1} F_r C$.

Proposition 5.8.1. *The coalgebra C is conilpotent if and only if the decomposition map $\Delta_C : C \to \widehat{\mathscr{C}}(C)$ factors through $\mathscr{C}(C)$.*

Proof. By direct inspection. □

Any coaugmented coassociative coalgebra C is equivalent to an As^*-coalgebra structure on $\overline{C}$. In this case, the above definitions of filtration and conilpotent coalgebra coincide with the ones given in Sect. 1.2.4.

5.8.5 Conilpotent Cooperad

Let $(\mathscr{C}, \Delta, \varepsilon, \eta)$ be a coaugmented cooperad. Under the isomorphism $\mathscr{C} \cong \mathrm{I} \oplus \overline{\mathscr{C}}$, we consider the map $\tilde{\Delta} : \mathscr{C} \to \mathscr{C} \,\bar{\circ}\, \mathscr{C}$ defined by

$$\begin{cases} \mathrm{I} \;\to \mathrm{I} \,\bar{\circ}\, \mathrm{I}, \\ \mathrm{id} \mapsto \tilde{\Delta}(\mathrm{id}) := \mathrm{id} \,\bar{\circ}\, \mathrm{id} \end{cases} \quad \text{and} \quad \begin{cases} \overline{\mathscr{C}} \to \overline{\mathscr{C}} \,\bar{\circ}\, \mathscr{C}, \\ \mu \;\mapsto \tilde{\Delta}(\mu) := \bar{\Delta}(\mu) + \big(\mu; \mathrm{id}^{\otimes n}\big). \end{cases}$$

We iterate the map $\tilde{\Delta}$ on the right-hand side:

$$\tilde{\Delta}^0 := \mathrm{Id}_{\mathscr{C}}, \qquad \tilde{\Delta}^1 := \tilde{\Delta}, \quad \text{and} \quad \tilde{\Delta}^n := (\mathrm{Id} \circ \tilde{\Delta})\tilde{\Delta}^{n-1} : \mathscr{C} \to \mathscr{C}^{\circ(n+1)}.$$

EXAMPLE. Representing the elements of the cooperad As^* by corollas, we get:

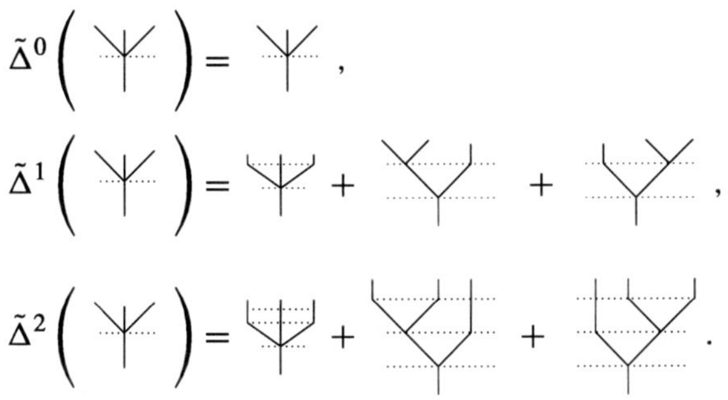

Higher up, no new term appears. We get the same three trees but with the top levels filled with trivial trees $|$. So in this example, the iteration of $\tilde{\Delta}$ on the element of $As^*(3)$ stabilizes at rank 2.

The composite

$$(\mathrm{Id} \,\bar{\circ}\, \eta)\tilde{\Delta}^{n-1} : \mathscr{C} \to \mathscr{C}^{\bar{\circ}n} \cong \mathscr{C}^{\bar{\circ}n} \circ \mathrm{I} \to \mathscr{C}^{\bar{\circ}(n+1)}$$

amounts to adding a level made up of identities $\mathrm{id} = |$ to the trees produced by $\tilde{\Delta}^{n-1}$. So the difference

$$\hat{\Delta}^n := \tilde{\Delta}^n - (\mathrm{Id} \,\bar{\circ}\, \eta)\tilde{\Delta}^{n-1}$$

contains the new leveled trees between the nth iteration of $\tilde{\Delta}$ and the $(n-1)$th iteration. Up to identification of the target spaces, we have $\tilde{\Delta}^n = \sum_{k=0}^{n} \hat{\Delta}^k$.

We define the *coradical filtration* of a coaugmented cooperad as follows.

$$F_0\mathscr{C} := \mathrm{I} \quad \text{and} \quad F_n\mathscr{C} := \ker \hat{\Delta}^n, \quad \text{for } n \geq 1.$$

Since for $n=1$, $\tilde{\Delta}^1 - (\mathrm{Id} \,\bar{\circ}\, \eta)\tilde{\Delta}^0 = \bar{\Delta}$, we get $F_1\mathscr{C} = \mathrm{I} \oplus \ker \bar{\Delta}$. We call the elements of $\overline{\mathscr{C}}$ which live in the kernel of $\bar{\Delta}$ the *primitive elements* of the cooperad $\mathscr{C}$. So we get a filtration of the cooperad $\mathscr{C}$:

$$F_0\mathscr{C} \subset F_1\mathscr{C} \subset F_2\mathscr{C} \subset \cdots \subset F_n\mathscr{C} \subset F_{n+1}\mathscr{C} \subset \cdots.$$

A coaugmented cooperad $\mathscr{C}$ is called *conilpotent* when the coradical filtration is exhaustive: $\mathrm{colim}_n\, F_n\mathscr{C} = \mathscr{C}$. This is equivalent to requiring that, for any element c of $\mathscr{C}$, the iteration of $\hat{\Delta}$ on c stabilizes at some point.

5.8.6 The Cofree Cooperad

By definition the cofree cooperad on M is the cooperad $\mathscr{F}^c(M)$, which is cofree in the category of conilpotent cooperads (cf. Sect. 1.2.5). Explicitly it means that for any $\mathbb{S}$-module morphism $\varphi : \mathscr{C} \to M$ sending id to 0, there exists a unique cooperad morphism $\tilde{\varphi} : \mathscr{C} \to \mathscr{F}^c(M)$ which renders the following diagram commutative

$$\begin{array}{ccc} \mathscr{C} & & \\ {\scriptstyle\tilde{\varphi}}\downarrow & \searrow^{\varphi} & \\ \mathscr{F}^c(M) & \longrightarrow & M. \end{array}$$

By dualizing what we have done for free operads in Sect. 5.5, we can prove the existence of the cofree cooperad and give an explicit construction by induction, which we denote by $\mathscr{T}^c(M)$.

The underlying $\mathbb{S}$-module of the cofree cooperad $\mathscr{T}^c(M)$ is the $\mathbb{S}$-module $\mathscr{T}M$ constructed in Sect. 5.5.1 and called the tree module. Let us recall that $\mathscr{T}M = \mathrm{colim}_n\, \mathscr{T}_n M$, where $\mathscr{T}_n M = \mathrm{I} \oplus (M \,\bar{\circ}\, \mathscr{T}_{n-1} M)$. The decomposition map

$$\Delta : \mathscr{T}M \to \mathscr{T}M \,\bar{\circ}\, \mathscr{T}M$$

is defined inductively on $\mathscr{T}_n M$ as follows. First we put

$$\Delta(\mathrm{id}) := \mathrm{id} \,\bar{\circ}\, \mathrm{id},$$

and for any $\mu \in M$ of arity n we put

$$\Delta(\mu) := \mathrm{id} \,\bar{\circ}\, \mu + \mu \,\bar{\circ}\, \mathrm{id}^{\otimes n} \in \mathrm{I} \,\bar{\circ}\, M \oplus M \,\bar{\circ}\, \mathrm{I} \subset \mathscr{T}_1 M \,\bar{\circ}\, \mathscr{T}_1 M.$$

This formula defines Δ on $\mathscr{T}_1 M$. Then we proceed by induction. We suppose that $\Delta : \mathscr{T}_{n-1}M \to \mathscr{T}_{n-1}M \mathbin{\bar{\circ}} \mathscr{T}_{n-1}M$ is defined and we construct Δ on $\mathscr{T}_n M$. For $\lambda = (\mu; \nu_1, \dots, \nu_k) \in M \mathbin{\bar{\circ}} \mathscr{T}_{n-1}M \subset \mathscr{T}_n M$, we put

$$\Delta(\lambda) = \Delta(\mu; \nu_1, \dots, \nu_k) := \mathrm{id} \mathbin{\bar{\circ}} (\mu; \nu_1, \dots, \nu_k) + \Delta^+(\mu; \nu_1, \dots, \nu_k),$$

where Δ^+ is the following composite:

$$\begin{aligned} M \mathbin{\bar{\circ}} \mathscr{T}_{n-1}M &\xrightarrow{\mathrm{Id}_M \mathbin{\bar{\circ}} \Delta} M \mathbin{\bar{\circ}} (\mathscr{T}_{n-1}M \mathbin{\bar{\circ}} \mathscr{T}_{n-1}M) \\ &\cong (M \mathbin{\bar{\circ}} \mathscr{T}_{n-1}M) \mathbin{\bar{\circ}} \mathscr{T}_{n-1}M \xrightarrow{j_n \mathbin{\bar{\circ}} i_n} \mathscr{T}_n M \mathbin{\bar{\circ}} \mathscr{T}_n M. \end{aligned}$$

Adopting the operadic version $\Delta(\nu_i) = \sum \nu_i^{(1)} \mathbin{\bar{\circ}} \nu_i^{(2)}$ of Sweedler's notation, where $\nu_i^{(2)}$ is in fact a tensor product of elements of M: $\nu_i^{(2)} = (\nu_{i,1}^{(2)}, \dots, \nu_{i,r}^{(2)})$, we get

$$\Delta(\mu; \nu_1, \dots, \nu_k) = \mathrm{id} \mathbin{\bar{\circ}} (\mu; \nu_1, \dots, \nu_k) + \sum \bigl(\mu; \nu_1^{(1)}, \dots, \nu_k^{(1)}\bigr) \mathbin{\bar{\circ}} \bigl(\nu_1^{(2)}, \dots, \nu_k^{(2)}\bigr).$$

So we have written $\Delta(\lambda)$ as $\sum \lambda^{(1)} \mathbin{\bar{\circ}} \lambda^{(2)}$. Observe that $\Delta(\lambda)$ contains $\mathrm{id} \mathbin{\bar{\circ}} \lambda$ but also $\lambda \mathbin{\bar{\circ}} (\mathrm{id}, \dots, \mathrm{id})$ as a summand. Indeed, by induction, $\Delta(\nu_i)$ contains a summand with $\nu_i^{(1)} = \nu_i$ and $\nu_i^{(2)} = (\mathrm{id}, \dots, \mathrm{id})$, hence in the sum we get a summand of the form

$$(\mu; \nu_1, \dots, \nu_k) \mathbin{\bar{\circ}} (\mathrm{id}, \dots, \mathrm{id}) = \lambda \mathbin{\bar{\circ}} (\mathrm{id}, \dots, \mathrm{id}).$$

Observe that the construction of Δ involves the associativity isomorphism, see Sect. 5.1.8. So, in the graded case there is a sign appearing in front of the sum. It is (-1) to the power

$$\bigl|\nu_1^{(2)}\bigr|\bigl|\nu_2^{(1)}\bigr| + \bigl(\bigl|\nu_1^{(2)}\bigr| + \bigl|\nu_2^{(2)}\bigr|\bigr)\bigl|\nu_3^{(1)}\bigr| + \cdots + \bigl(\bigl|\nu_1^{(2)}\bigr| + \cdots + \bigl|\nu_{k-1}^{(2)}\bigr|\bigr)\bigl|\nu_k^{(1)}\bigr|.$$

Here is an example for which we allow ourselves to leave out the commas to ease the notation. Let | be the tree representing the identity and let Y $\in M(2)$ be a binary operation. Since $\Delta($ Y $) = | \mathbin{\bar{\circ}}$ Y + Y $\mathbin{\bar{\circ}} (|, |)$, we get

where

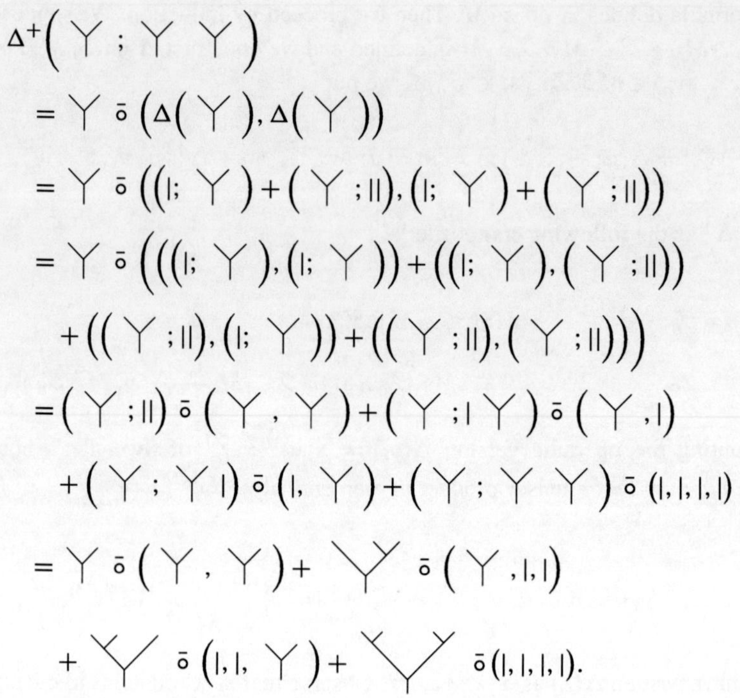

The map $\varepsilon : \mathscr{T}M \to M$ is defined by $\mathscr{T}_1 M = \mathrm{I} \oplus M \twoheadrightarrow M$. The coaugmentation map $\eta : M \to \mathscr{T}M$ is equal to the map j.

Proposition 5.8.2. *The above maps induce a coaugmented cooperad structure on* $\mathscr{T}^c(M) := (\mathscr{T}M, \Delta, \varepsilon, \eta)$.

Proof. We have constructed maps $\mathscr{T}_n M \to \mathscr{T}M \bar{\circ} \mathscr{T}M$. Since they commute with the map $i_n : \mathscr{T}_n M \to \mathscr{T}_{n+1} M$, they give rise to Δ on $\mathscr{T}M = \operatorname{colim}_n \mathscr{T}_n M$.

Coassociativity, counitality and coaugmentation can be proved by induction, by a check similar to the one done in Theorem 5.5.1. These properties can also be proved by using the explicit form given in Proposition 5.8.4. □

Theorem 5.8.3. *The conilpotent cooperad* $\mathscr{T}^c(M) := (\mathscr{T}M, \Delta, \varepsilon, \eta)$ *is cofree on* M *among conilpotent cooperads.*

The proof is postponed to the end of the next section.

5.8.7 Description of the Cofree Cooperad in Terms of Trees

Recall from Sect. 5.6.3 that the $\mathbb{S}$-module $\mathscr{T}M$ is isomorphic to the treewise tensor module $\mathbb{T}(M)$ made up of trees with vertices labeled by elements of M.

The decomposition Δ on such a tree t is constructed by "degrafting" as follows. A *cut* of the tree is *admissible* if the grafting of the pieces gives the original tree back. The degrafting $\Delta(t)$ of t is the sum of all the admissible cuttings $(r; s_1, \ldots, s_k)$, where r is the piece containing the root, and k is the number of leaves of r. Of course each vertex keeps its labeling.

EXAMPLE.

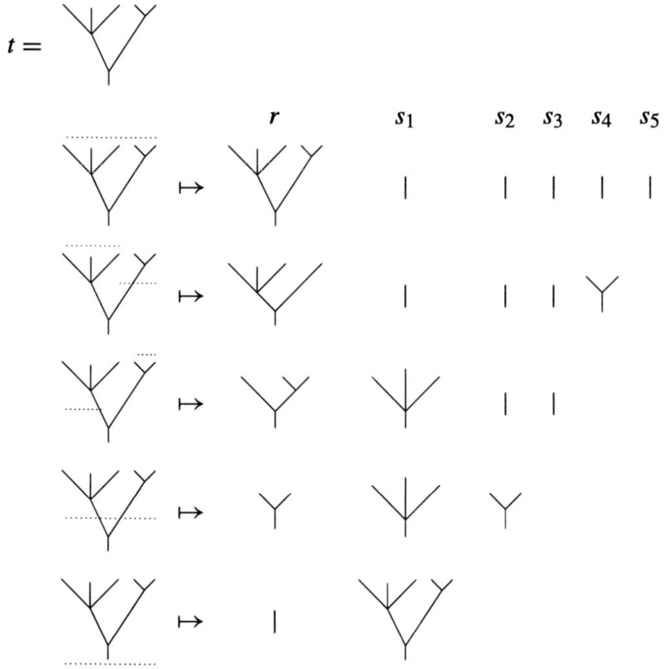

Proposition 5.8.4. *The cooperad $(\mathcal{T}^c M, \Delta, \varepsilon)$ is isomorphic to the treewise tensor module equipped with the decomposition map given by the admissible cuttings.*

Proof. Recall that $\mathcal{T}_n M$ is isomorphic to labeled trees with at most n levels. We prove the assertion by induction on n. For $n = 0$ and $n = 1$, the explicit form of Δ on $\mathcal{T}_0 M = \mathrm{I}$ and on $\mathcal{T}_0 M = \mathrm{I} \oplus M$ allows to conclude. Suppose that the result holds up to $n - 1$. Let t be a labeled tree in $\mathcal{T}_n M$. If not trivial, this tree can be written $t = (\mu; t_1, \ldots, t_k)$, where $t_1, \ldots, t_k$ are labeled sub-trees. By definition, $\Delta(t) := \mathrm{id}\,\bar{\circ}\,t + \Delta^+(\mu; t_1, \ldots, t_k)$. The first component gives the bottom cutting. The second component is given by the cuttings $\Delta(t_i)$ of the sub-trees t_i, where the bottom part is then grafted onto μ. Finally, we get all the admissible cuttings of the tree t. □

With this description of $\mathcal{T}^c(M)$, it is easy to see that M is the space of primitive elements. The coradical filtration is equal to $F_n \mathcal{T}^c(M) = \mathcal{T}_n M$, that is coincides with the defining filtration. So the coaugmented cooperad $\mathcal{T}^c(M)$ is conilpotent.

Proof of Theorem 5.8.3. Let $\mathscr{C}$ be a conilpotent cooperad and let $\varphi : \mathscr{C} \to M$ be an $\mathbb{S}$-module map, which sends id to 0. We claim that there is a unique morphism of cooperads

$$\tilde{\varphi} : \mathscr{C} \to \mathscr{T}^c(M)$$

which extends φ. We construct $\tilde{\varphi}_n : \mathscr{C} \to \mathscr{T}_n M$ by induction on n. For $n = 0$, we put $\tilde{\varphi}_0(\mathrm{id}) = \mathrm{id}$ and $\tilde{\varphi}_0$ is 0 in the other components. For $n = 1$, we put $\tilde{\varphi}_1(\mathrm{id}) = \mathrm{id}$ and $\tilde{\varphi} = \varphi : \mathscr{C} \to M \subset \mathscr{T}_1 M$. Let us suppose that $\tilde{\varphi}_{n-1}$ has been constructed. The image of id by $\tilde{\varphi}_n$ is $\mathrm{id} \in \mathrm{I} \subset \mathrm{I} \oplus M \bar{\circ} \mathscr{T}_{n-1} M = \mathscr{T}_n M$. The component in the other summand is equal to the composite

$$(\varphi \bar{\circ} \tilde{\varphi}_{n-1})\Delta : \mathscr{C} \to \mathscr{C} \bar{\circ} \mathscr{C} \to M \bar{\circ} \mathscr{T}_{n-1} M = \mathscr{T}_n M.$$

One can see that the map $\tilde{\varphi}_n$ is equal to the following composite

$$\tilde{\varphi}_n : \mathscr{C} \xrightarrow{\tilde{\Delta}^{n-1}} \mathscr{C}^{\bar{\circ} n} \xrightarrow{\tilde{\varphi}_1} (\mathrm{I} \oplus M)^{\bar{\circ} n} \twoheadrightarrow \mathscr{T}_n M,$$

where the last map is the projection of n-leveled trees into non-leveled trees.

Since the cooperad $\mathscr{C}$ is conilpotent, this process stabilizes, that is for any $c \in \mathscr{C}$, the image under the composite maps $\mathscr{C} \xrightarrow{\tilde{\varphi}_n} \mathscr{T}_n M \rightarrowtail \mathscr{T} M$ give the same image in the colimit $\mathscr{T} M$, for $n \geq N$. So the map $\tilde{\varphi}$ is well defined.

Since we want $\tilde{\varphi}$ to be a map of cooperads and to coincide with φ in the component M, we have no choice for $\tilde{\varphi}$. By the definition of $\tilde{\varphi}$ in terms of $\tilde{\Delta}$ and by the coassociativity of Δ, the map $\tilde{\varphi}$ is a morphism of cooperads. □

5.8.8 Combinatorial Definition of a Cooperad

In the same way as in Sect. 5.5, we give an equivalent definition of a conilpotent cooperad using a comonad of trees.

The adjunction

$$\mathscr{U} : \textsf{conil coOp} \rightleftarrows \mathbb{S}\textsf{-Mod} : \mathscr{T}^c$$

of the previous sections induces a comonad denoted by $\mathbb{T}^c$. Explicitly it is a comonoid in the category of endofunctors of $\mathbb{S}$-modules, see Appendix B.4.1. The underlying endofunctor is the same as in Sect. 5.6.1: $\mathbb{T}^c : M \mapsto \mathscr{T} M$. The coproduct and the counit maps

$$\Delta : \mathbb{T}^c \to \mathbb{T}^c \circ \mathbb{T}^c \quad \text{and} \quad \varepsilon : \mathbb{T}^c \to \mathrm{Id}_{\mathbb{S}\textsf{-Mod}}$$

are given as follows. For any $\mathbb{S}$-module M, $\mathbb{T}^c \circ \mathbb{T}^c(M) = \mathscr{T}(\mathscr{T} M)$ is made up of "trees of trees" with vertices labeled by M. Equivalently, it coincides with trees labeled by M equipped with a partition into subtrees. The map $\Delta(M)$ associates to a tree t labeled by M, the sum of all the partitioned trees coming from t. The map $\varepsilon(M) : \mathscr{T}M \to M$ is the projection onto corollas.

Similarly we consider the comonad $\overline{\mathbb{T}}^c$ made up of trees $\overline{\mathscr{T}}^c$ without the trivial tree.

Proposition 5.8.5. *Let $\overline{\mathscr{C}}$ be an $\mathbb{S}$-module. A coalgebra structure on $\overline{\mathscr{C}}$ over the comonad $\overline{\mathbb{T}}^c$ is equivalent to a conilpotent cooperad structure on $\mathscr{C} := \overline{\mathscr{C}} \oplus \mathrm{I}$.*

Proof. ($\Leftarrow$) Let $(\mathscr{C} = \overline{\mathscr{C}} \oplus \mathrm{I}, \Delta, \varepsilon, \eta)$ be a conilpotent cooperad. The map $\Delta_{\overline{\mathscr{C}}}$: $\overline{\mathscr{C}} \to \overline{\mathbb{T}}^c(\overline{\mathscr{C}}) = \overline{\mathscr{T}}\overline{\mathscr{C}}$ is given by the universal property of the conilpotent free cooperad applied to

$$\begin{array}{ccc} \overline{\mathscr{C}} & & \\ {\scriptstyle \Delta_{\overline{\mathscr{C}}}}\downarrow & \searrow^{\mathrm{Id}_{\overline{\mathscr{C}}}} & \\ \overline{\mathscr{T}}^c\overline{\mathscr{C}} & \twoheadrightarrow & \overline{\mathscr{C}}. \end{array}$$

($\Rightarrow$) In the other way round, let $\Delta_{\overline{\mathscr{C}}}: \overline{\mathscr{C}} \to \overline{\mathbb{T}}^c(\overline{\mathscr{C}}) = \overline{\mathscr{T}}\overline{\mathscr{C}}$ be a coalgebra over the comonad $\overline{\mathbb{T}}^c$. We view the trees of $\overline{\mathscr{T}_2\mathscr{C}}$ as 2-leveled trees by adding trivial trees | if necessary. By projecting onto this summand, we get a coassociative decomposition map $\Delta : \mathscr{C} \to \mathscr{C}\bar{\circ}\mathscr{C}$, where the image of id is defined by $\mathrm{id} \circ \mathrm{id}$. The unit and the coaugmentation maps come for free. In the end, it defines a conilpotent cooperad structure on $\mathscr{C}$. □

When $\mathscr{C}$ is a conilpotent cooperad, we denote by $\Delta_{\mathscr{C}} : \mathscr{C} \to \mathscr{T}^c(\overline{\mathscr{C}})$ the morphism of cooperads $\mathrm{Id}_{\mathrm{I}} \oplus \Delta_{\overline{\mathscr{C}}}$. In conclusion, we get the following result, which will play a crucial role in Sect. 10.3.

Proposition 5.8.6. *Let $(\mathscr{C}, \Delta, \varepsilon, \eta)$ be a conilpotent cooperad and let $\varphi : \mathscr{C} \to M$ be a morphism of $\mathbb{S}$-modules such that $\varphi(\mathrm{id}) = 0$. Its unique extension into a morphism of cooperads $\tilde{\varphi} : \mathscr{C} \to \mathscr{T}^c(M)$ is equal to the composite*

$$\tilde{\varphi} : \mathscr{C} \xrightarrow{\Delta_{\mathscr{C}}} \mathscr{T}^c(\overline{\mathscr{C}}) \xrightarrow{\mathscr{T}^c(\varphi)} \mathscr{T}^c(M),$$

where the map $\Delta_{\mathscr{C}}$ is given by the iterations $\tilde{\Delta}^n = \sum_{k=0}^{n} \hat{\Delta}^k$.

Proof. It is direct consequence of the above results. □

5.9 Nonsymmetric Operad

Replacing the category of $\mathbb{S}$-modules by the category of arity graded vector spaces gives the notion of nonsymmetric operad (called ns operad for short). To any ns operad one can associate an operad by tensoring with the regular representation in each arity. This section can be read independently of the rest of the chapter. We work over a field $\mathbb{K}$ though most of the notions and results of this section are valid over a commutative ring.

5.9.1 More on Arity Graded Modules

Let

$$M. = \{M_n\}_{n\geq 0}$$

be a graded vector space (or graded module). We denote by $\mathbb{N}$-Mod the category of graded vector spaces (or graded $\mathbb{K}$-modules if $\mathbb{K}$ is a commutative ring). The integer n is called the arity in this framework. The *Schur functor* $M : \mathsf{Vect} \to \mathsf{Vect}$ associated to $M.$ is, by definition,

$$M(V) := \bigoplus_{n\geq 0} M_n \otimes_{\mathbb{K}} V^{\otimes n}.$$

In literature the object $M.$ is sometimes called a *collection*. We refrain to call it a nonsymmetric $\mathbb{S}$-module. Recall that the sum, tensor product, composition and Hadamard product of arity graded spaces are given by

$$\begin{aligned}
(M. \oplus N.)_n &:= M_n \oplus N_n, \\
(M. \otimes N.)_n &:= \bigoplus_{i+j=n} M_i \otimes N_j, \\
(M. \circ N.)_n &:= \bigoplus_k M_k \otimes \Big(\bigoplus N_{i_1} \otimes \cdots \otimes N_{i_k}\Big), \\
(M. \underset{\mathrm{H}}{\otimes} N.)_n &:= M_n \otimes N_n,
\end{aligned}$$

where the second sum in line 3 is over all the k-tuples $(i_1, \ldots, i_k)$ satisfying $i_1 + \cdots + i_k = n$. Observe that the associativity property of the composition of graded modules involves the switching map, cf. Sect. 5.1.8. For any vector space V we have natural isomorphisms:

$$\begin{aligned}
(M \oplus N)(V) &= M(V) \oplus N(V), \\
(M \otimes N)(V) &= M(V) \otimes N(V), \\
(M \circ N)(V) &:= M\big(N(V)\big).
\end{aligned}$$

The Hilbert–Poincaré series of the arity graded module $M.$ is:

$$f^M(x) := \sum_{n\geq 0} \dim M_n x^n.$$

The generating series of a sum (resp. product, composition, Hadamard product) of arity graded modules is the sum (resp. product, composition, Hadamard product) of their respective generating series.

In the sequel we simply write M instead of $M_.$ whenever there is no ambiguity.

5.9.2 Monoidal Definition of a Nonsymmetric Operad

By definition a *nonsymmetric operad* (also called *non-Σ-operad* in the literature) $\mathcal{P} = (\mathcal{P}, \gamma, \eta)$ is an arity graded vector space $\mathcal{P} = \{\mathcal{P}_n\}_{n\geq 0}$ equipped with composition maps

$$\gamma_{i_1,\dots,i_k} : \mathcal{P}_k \otimes \mathcal{P}_{i_1} \otimes \cdots \otimes \mathcal{P}_{i_k} \longrightarrow \mathcal{P}_{i_1+\cdots+i_k}$$

and an element $\mathrm{id} \in \mathcal{P}_1$, such that the transformations of functors $\gamma : \mathcal{P} \circ \mathcal{P} \to \mathcal{P}$ and $\eta : \mathrm{I} \to \mathcal{P}$, deduced from this data, make $(\mathcal{P}, \gamma, \eta)$ into a monoid.

We often abbreviate "nonsymmetric operad" into "ns operad".

5.9.3 Classical Definition of a NS Operad

Obviously a nonsymmetric operad can be defined as an arity graded module $\mathcal{P}$ equipped with linear maps

$$\gamma_{i_1,\dots,i_k} : \mathcal{P}_k \otimes \mathcal{P}_{i_1} \otimes \cdots \otimes \mathcal{P}_{i_k} \longrightarrow \mathcal{P}_{i_1+\cdots+i_k}$$

and an element $\mathrm{id} \in \mathcal{P}_1$, such that the following diagram (in which the tensor signs are omitted) is commutative

$$\begin{array}{ccc}
\mathcal{P}_n \mathcal{P}_{r_1} \cdots \mathcal{P}_{r_n} & \longrightarrow & \\
\uparrow & & \\
\mathcal{P}_n \mathcal{P}_{i_1} \mathcal{P}_{j_{1,1}} \cdots \mathcal{P}_{j_{1,i_1}} \mathcal{P}_{i_2} \mathcal{P}_{j_{2,1}} \cdots\cdots \mathcal{P}_{i_n} \cdots \mathcal{P}_{j_{n,i_n}} & & \\
\downarrow \cong & & \\
\mathcal{P}_n \mathcal{P}_{i_1} \cdots \mathcal{P}_{i_n} \mathcal{P}_{j_{1,1}} \cdots \mathcal{P}_{j_{1,i_1}} \mathcal{P}_{j_{2,1}} \cdots\cdots \mathcal{P}_{j_{n,1}} \cdots \mathcal{P}_{j_{n,i_n}} & & \downarrow \\
\downarrow & & \\
\mathcal{P}_m \mathcal{P}_{j_{1,1}} \cdots \mathcal{P}_{j_{1,i_1}} \mathcal{P}_{j_{2,1}} \cdots\cdots \mathcal{P}_{j_{n,1}} \cdots \mathcal{P}_{j_{n,i_n}} & \longrightarrow & \mathcal{P}_\ell
\end{array}$$

where $r_k = j_{k,1} + \cdots + j_{k,i_k}$ for $k = 1$ to n, $m = i_1 + \cdots + i_n$ and $\ell = r_1 + \cdots + r_n$. Moreover the element id is such that the evaluation of $\gamma_n : \mathcal{P}_1 \otimes \mathcal{P}_n \to \mathcal{P}_n$ on (id, μ) is μ, and the evaluation of $\gamma_{1,\dots,1}$ on $(\mu; \mathrm{id}, \dots, \mathrm{id})$ is μ.

The equivalence between the monoidal definition and the classical definition is a straightforward check.

5.9.4 Partial Definition of a NS Operad

A nonsymmetric operad can be defined as an arity graded vector space $\mathscr{P}$ equipped with *partial compositions*:

$$\circ_i : \mathscr{P}_m \otimes \mathscr{P}_n \to \mathscr{P}_{m-1+n}, \qquad \text{for } 1 \leq i \leq m,$$

satisfying the relations

$$\begin{cases} \text{(I)} & (\lambda \circ_i \mu) \circ_{i-1+j} \nu = \lambda \circ_i (\mu \circ_j \nu), \quad \text{for } 1 \leq i \leq l,\ 1 \leq j \leq m, \\ \text{(II)} & (\lambda \circ_i \mu) \circ_{k-1+m} \nu = (\lambda \circ_k \nu) \circ_i \mu, \quad \text{for } 1 \leq i < k \leq l, \end{cases}$$

for any $\lambda \in \mathscr{P}_l$, $\mu \in \mathscr{P}_m$, $\nu \in \mathscr{P}_n$.

This definition (with different notations and grading) appears in Gerstenhaber's paper [Ger63] under the name "pre-Lie system".

The equivalence with the monoidal definition is given by constructing the map $\gamma_{i_1,\dots,i_n}$ as an iteration of the partial operations. In the other direction the partial operation $\circ_i$ is obtained by restriction:

$$\lambda \circ_i \mu = \gamma(\lambda; \text{id}, \dots, \text{id}, \mu, \text{id}, \dots, \text{id})$$

where μ is at the ith position.

5.9.5 Combinatorial Definition of a NS Operad

For any planar rooted tree t we denote by $\text{vert}(t)$ its set of vertices and by $|v|$ the number of inputs of the vertex $v \in \text{vert}(t)$, see Appendix C for details. Let M be an arity graded space with $M_0 = 0$. Recall that the integer n is called the *arity* of t and the integer $k = \#\text{vert}(t)$ is called the *weight* of t. For any tree t we define

$$M_t := \bigotimes_{v \in \text{vert}(t)} M_{|v|}.$$

We get a functor

$$\mathbb{PT} : \mathbb{N}\text{-Mod} \to \mathbb{N}\text{-Mod}$$

by $\mathbb{PT}(M)_n := \bigoplus_{t \in PT_n} M_t$. It is helpful to think about an element of $\mathbb{PT}(M)_n$ as a planar rooted tree where each vertex v is decorated by an element of $M_{|v|}$.

In the following example we have $\tau \in M_3$, $\lambda \in M_2$, $\mu \in M_1$, $\nu \in M_3$:

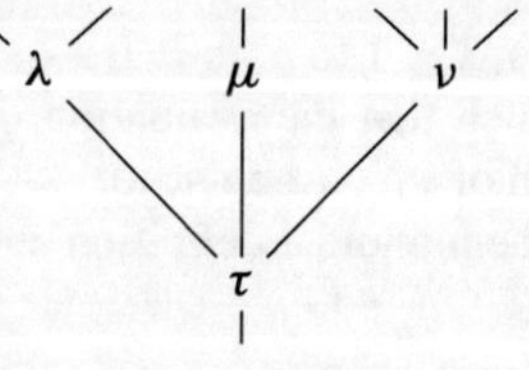

In particular the corolla enables us to define the transformation of functors $\eta : \mathrm{I}_{\mathbb{N}\text{-Mod}} \to \mathrm{P}\mathbb{T}$.

The *substitution* of trees consists in replacing the vertices by given trees (with matching inputs). The substitution of trees defines a transformation of functors $\alpha : \mathrm{P}\mathbb{T} \circ \mathrm{P}\mathbb{T} \to \mathrm{P}\mathbb{T}$ as follows. From the definition of $\mathrm{P}\mathbb{T}$ we get

$$\begin{aligned} \mathrm{P}\mathbb{T}(\mathrm{P}\mathbb{T}(M))_n &= \bigoplus_{t\in PT_n} \mathrm{P}\mathbb{T}(M)_t \\ &= \bigoplus_{t\in PT_n} \Big(\bigotimes_{v\in \mathrm{vert}(t)} \mathrm{P}\mathbb{T}(M)_{|v|} \Big) \\ &= \bigoplus_{t\in PT_n} \Big(\bigotimes_{v\in \mathrm{vert}(t)} \Big(\bigoplus_{s\in PT_{|v|}} M_s \Big)\Big). \end{aligned}$$

Under the substitution of trees we get an element of $\mathrm{P}\mathbb{T}(M)_n$, since at any vertex v of t we have an element of $\mathrm{P}\mathbb{T}(M)_{|v|} = \bigoplus_{s\in PT_{|v|}} M_s$, that is a tree s and its decoration. We substitute this data at each vertex of t to get a new decorated tree. Therefore we have defined an $\mathbb{S}$-module morphism $\alpha(M) : \mathrm{P}\mathbb{T}(\mathrm{P}\mathbb{T}(M)) \to \mathrm{P}\mathbb{T}(M)$.

The transformation of functors α is obviously associative and unital, so $(\mathrm{P}\mathbb{T}, \alpha, \eta)$ is a monad.

The *combinatorial definition of a ns operad* consists in defining it as a unital algebra over the monad $(\mathrm{P}\mathbb{T}, \alpha, \eta)$, cf. Appendix B.4. In other words a ns operad is an arity graded module $\mathscr{P}$ together with a map $\mathrm{P}\mathbb{T}(\mathscr{P}) \to \mathscr{P}$ which is compatible with α and η in the usual sense.

The combinatorial definition of a ns operad is equivalent to the partial definition of a ns operad, and therefore to all the other definitions.

5.9.6 Free NS Operad and Planar Trees

By definition the *free nonsymmetric operad* over the arity graded module M is the ns operad $\mathscr{T}(M)$ equipped with a graded module morphism $M \to \mathscr{T}(M)$ which satisfies the classical universal property. Explicitly it can be constructed inductively as in Sect. 5.5.1, i.e. $\mathscr{T}M = \bigcup_n \mathscr{T}_n M$, where $\mathscr{T}_0 M := I$ and $\mathscr{T}_n M := \mathrm{I} \oplus (M \circ \mathscr{T}_{n-1} M)$. It can also be constructed as a quotient as in Sect. 5.5.5, or, more explicitly, by using planar trees. In fact the graded module $\mathrm{P}\mathbb{T}(M)$ constructed above is endowed with a ns operad structure as follows. Let t and s be two decorated trees. The partial composition $t \circ_i s$ is the decorated tree obtained by grafting the tree s on the ith leaf of t and keeping the decorations.

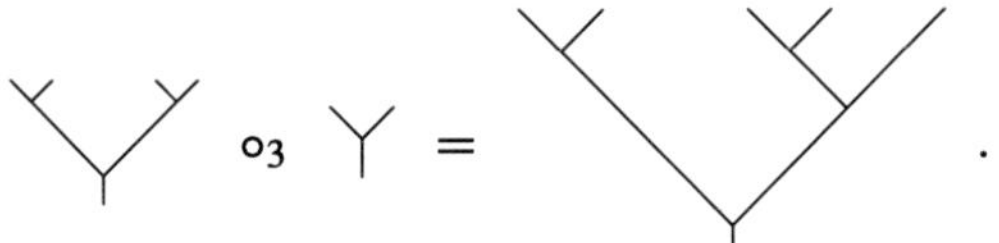

It is immediate to check that this partial composition makes $\mathrm{P}\mathbb{T}(M)$ into a ns operad, and that this ns operad is free over M. The map $\eta : M \to \mathrm{P}\mathbb{T}(M)$ consists in sending the operation $\mu \in M_n$ to the nth corolla decorated by μ. The isomorphism $\varphi : \mathscr{T}(M) \to \mathrm{PT}(M)$ is made explicit as follows. First, we have $\varphi(\mathrm{id}) = |$. Second, the generating operation $\mu \in M_k$ is sent to the kth corolla decorated by μ. Third, for $\omega_i \in \mathscr{T}_{n-1}M$, $i = 1, \ldots, k$, the generic element $(\mu; \omega_1, \ldots, \omega_k) \in M \circ \mathscr{T}_{n-1}M \subset \mathscr{T}_n M$ is mapped under φ to the tree $\varphi(\mu; \omega_1, \ldots, \omega_k)$ obtained by grafting the decorated trees $\varphi(\omega_i)$ to the leaves of the kth corolla (image of μ). It is immediate to check that we get an isomorphism.

5.9.7 Free NS Operad in the Graded Framework

In the construction of the free ns operad in terms of trees in the sign-graded framework, signs show up in the computation of composition. Here is an explicit example.

Let $\lambda, \mu, \nu \in M$ be three binary graded operations. In $\mathscr{T}M$ the tree

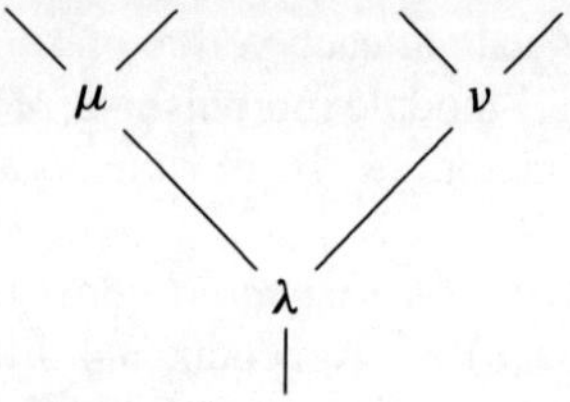

corresponds to the element $(\lambda; \mu, \nu) \in M \circ M$ which is to be interpreted as an element in $\mathscr{T}_2 M = \mathrm{I} \oplus M \circ (\mathrm{I} \oplus M)$. Viewed as an element of $\mathscr{T}_3 M$ via $i : \mathscr{T}_2 M \to \mathscr{T}_3 M$ it becomes

$$\big(\lambda; (\mu; \mathrm{id}, \mathrm{id}), (\nu; \mathrm{id}, \mathrm{id})\big) \in M \circ \big(\mathrm{I} \oplus M \circ (\mathrm{I} \oplus M)\big).$$

Now, let us identify the two composites from bottom to top that is

$(\lambda \circ_2 \nu) \circ_1 \mu =$

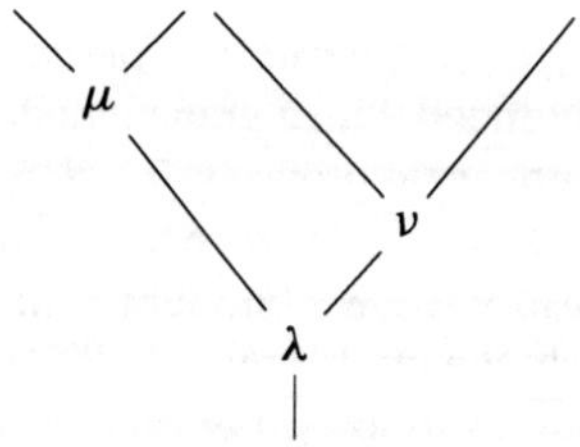

resp. $(\lambda \circ_1 \mu) \circ_3 \nu =$

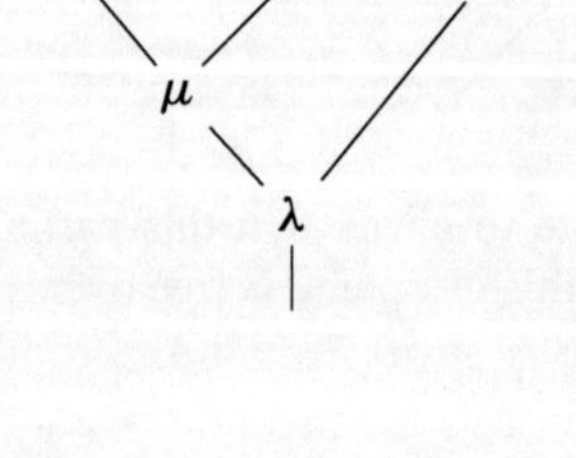

corresponding to the composite $((\lambda; \mathrm{id}, \nu); \mu, \mathrm{id}, \mathrm{id})$, resp. $((\lambda; \mu, \mathrm{id}); \mathrm{id}, \mathrm{id}, \nu)$. Under the associativity isomorphism, they are equal to

$$(-1)^{|\nu||\mu|}\big(\lambda; (\mathrm{id}; \mu), (\nu; \mathrm{id}, \mathrm{id})\big), \quad \text{resp.} \quad \big(\lambda; (\mu; \mathrm{id}, \mathrm{id}), (\mathrm{id}; \nu)\big), \quad \text{that is}$$
$$(-1)^{|\nu||\mu|}\big(\lambda; (\mu; \mathrm{id}, \mathrm{id}), (\nu; \mathrm{id}, \mathrm{id})\big), \quad \text{resp.} \quad \big(\lambda; (\mu; \mathrm{id}, \mathrm{id}); (\nu; \mathrm{id}, \mathrm{id})\big).$$

In conclusion we get

$$(\lambda \circ_2 \nu) \circ_1 \mu = (-1)^{|\nu||\mu|}(\lambda; \mu, \nu) \quad \text{and} \quad (\lambda \circ_1 \mu) \circ_3 \nu = (\lambda; \mu, \nu).$$

5.9.8 Algebra over a Nonsymmetric Operad

For any vector space A the graded module $\mathrm{End}(A)$, defined by $\mathrm{End}(A)_n := \mathrm{Hom}(A^{\otimes n}, A)$, is a ns operad for the composition of maps (cf. Sect. 5.2.11). By definition an *algebra over the ns operad* $\mathscr{P}$ is a morphism of ns operads $\mathscr{P} \to \mathrm{End}(A)$. Equivalently, a $\mathscr{P}$-algebra structure on A is a family of linear maps $\mathscr{P}_n \otimes A^{\otimes n} \to A$ compatible with the ns operad structure of $\mathscr{P}$.

5.9.9 Nonsymmetric Operad, Type of Algebras

Let us consider a type of algebras for which the generating operations have no symmetry, the relations are multilinear and, in these relations, the variables stay in the same order. Then this type of algebras can be faithfully encoded by a nonsymmetric operad.

The relationship between types of algebras and operads is slightly simpler in the nonsymmetric case, as shown by the following result.

Proposition 5.9.1. *A nonsymmetric operad $\mathscr{P}$ is completely determined by the free $\mathscr{P}$-algebra on one generator.*

Proof. For a nonsymmetric operad $\mathscr{P}$ the free algebra on one generator is

$$\mathscr{P}(\mathbb{K}) = \bigoplus_{n\geq 0} \mathscr{P}_n \otimes \mathbb{K}^{\otimes n} = \bigoplus_{n\geq 0} \mathscr{P}_n.$$

Hence $\mathscr{P}_n$ is the n-multilinear part of $\mathscr{P}(\mathbb{K})$. Using the ubiquity of the operations, see Sect. 5.2.13, it follows that the composition maps are completely determined by the $\mathscr{P}$-algebra structure of $\mathscr{P}(\mathbb{K})$. □

Remark that this statement is not true for symmetric operads. For instance *Ass* and *Com* have the same free algebra on one generator, namely the ideal (x) in the polynomial algebra $\mathbb{K}[x]$. It determines *As*, but not *Com*.

5.9.10 Hadamard Product of NS Operads

Let $\mathscr{P}$ and $\mathscr{Q}$ be two ns operads. The Hadamard tensor product $\mathscr{P} \underset{\mathrm{H}}{\otimes} \mathscr{Q}$ of the underlying graded modules has a natural operad structure:

$$\begin{aligned}
&(\mathscr{P} \underset{\mathrm{H}}{\otimes} \mathscr{Q})_k \otimes (\mathscr{P} \underset{\mathrm{H}}{\otimes} \mathscr{Q})_{i_1} \otimes \cdots \otimes (\mathscr{P} \underset{\mathrm{H}}{\otimes} \mathscr{Q})_{i_k} \\
&= \mathscr{P}_k \otimes \mathscr{Q}_k \otimes \mathscr{P}_{i_1} \otimes \mathscr{Q}_{i_1} \otimes \cdots \otimes \mathscr{P}_{i_k} \otimes \mathscr{Q}_{i_k} \\
&\cong \mathscr{P}_k \otimes \mathscr{P}_{i_1} \otimes \cdots \otimes \mathscr{P}_{i_k} \otimes \mathscr{Q}_k \otimes \mathscr{Q}_{i_1} \otimes \cdots \otimes \mathscr{Q}_{i_k} \\
&\longrightarrow \mathscr{P}_n \otimes \mathscr{Q}_n = (\mathscr{P} \underset{\mathrm{H}}{\otimes} \mathscr{Q})_n
\end{aligned}$$

for $n = i_1 + \cdots + i_k$. Observe that we use the switching map in the category **Vect** to put the factors $\mathscr{Q}_i$ in the correct position. Therefore, when **Vect** is replaced by another tensor category (cf. Appendix B.3) signs might be involved. The operad uAs is obviously a unit for this operation.

The ns operad $\mathscr{P} \underset{\mathrm{H}}{\otimes} \mathscr{Q}$ is called the *Hadamard product* of the ns operads $\mathscr{P}$ and $\mathscr{Q}$.

5.9.11 From NS Operads to Symmetric Operads and Vice Versa

Let $\mathscr{P}$ be a ns operad with $\mathscr{P}_n$ as the space of n-ary operations. The category of $\mathscr{P}$-algebras can be encoded by a symmetric operad. We still denote it by $\mathscr{P}$ and the space of n-ary operations by $\mathscr{P}(n)$. It is immediate that

$$\mathscr{P}(n) = \mathscr{P}_n \otimes \mathbb{K}[\mathbb{S}_n]$$

where the action of the symmetric group on $\mathscr{P}(n)$ is given by the regular representation $\mathbb{K}[\mathbb{S}_n]$. Indeed we have $(\mathscr{P}_n \otimes \mathbb{K}[\mathbb{S}_n]) \otimes_{\mathbb{K}[\mathbb{S}_n]} V^{\otimes n} = \mathscr{P}_n \otimes V^{\otimes n}$. The composition map $\gamma(i_1, \ldots, i_k)$ in the symmetric framework is given, up to a permutation of factors, by the tensor product of the composition map $\gamma_{i_1,\ldots,i_k}$ in the ns framework with the composition map of the symmetric operad Ass. Considered as a symmetric operad $\mathscr{P}$ is sometimes called a *regular operad*. Observe that the categories of algebras over a ns operad and over its associated operad are the same, so they encode the same type of algebras. We usually take the same notation for the ns operad and its associated symmetric operad, except in the case of associative algebras where we use As and Ass respectively in this book.

In conclusion we have constructed a functor

$$\mathsf{nsOp}_{\mathbb{K}} \longrightarrow \mathsf{Op}_{\mathbb{K}}.$$

This functor admits a right adjoint:

$$\mathsf{Op}_{\mathbb{K}} \longrightarrow \mathsf{nsOp}_{\mathbb{K}}, \qquad \mathscr{P} \mapsto \widetilde{\mathscr{P}}.$$

Explicitly we have $\widetilde{\mathscr{P}}_n = \mathscr{P}(n)$, in other words we forget the $\mathbb{S}_n$-module structure. We have $\widetilde{\mathscr{P}}(n) = \mathscr{P}(n) \otimes \mathbb{K}[\mathbb{S}_n]$ where the $\mathbb{S}_n$-module structure is given by the action on $\mathbb{K}[\mathbb{S}_n]$ (not the diagonal action). The composition maps

$$\widetilde{\gamma}_{i_1,\ldots,i_k} = \gamma(i_1,\ldots,i_k) : \mathscr{P}(k) \otimes \mathscr{P}(i_1) \otimes \cdots \otimes \mathscr{P}(i_k) \longrightarrow \mathscr{P}(i_1 + \cdots + i_k)$$

satisfy the axioms of a ns operad.

EXAMPLES. By direct inspection we see that $\widetilde{Com} = As$. In [ST09] Salvatore and Tauraso show that the operad $\widetilde{Lie}$ is a free ns operad. In [BL10] Bergeron and Livernet show that $\widetilde{preLie}$ is also free.

5.9.12 Nonsymmetric Operads as Colored Algebras

A *colored algebra* is a graded vector space $A = \{A_n\}_{n\geq 0}$ equipped with operations which are only defined under some conditions depending on the colors (elements of an index set). For instance let us suppose that we have operations (i.e. graded linear maps) $\circ_i : A_m \otimes A \to A$ defined only when $1 \leq i \leq m+1$ and a map $\mathbb{K} \to A_0$, $1_{\mathbb{K}} \mapsto 1$. Let us suppose that they satisfy the relations:

$$\begin{cases} \text{(I)} & (x \circ_i y) \circ_{i-1+j} z = x \circ_i (y \circ_j z), \quad i \leq j \leq i+m-1, \\ \text{(II)} & (x \circ_i y) \circ_{j+m-2} z = (x \circ_j z) \circ_i y, \quad i+m \leq j \leq l+m-1, \end{cases}$$

for any $x \in A_{l-1}$, $y \in A_{m-1}$, $z \in A$ and unital relations with respect to 1. It appeared in Gerstenhaber's paper [Ger63] as a *pre-Lie system* (our notation $\circ_i$ corresponds to his notation $\circ_{i-1}$). It also appears in [Ron11] by M. Ronco, where such a colored algebra is called a *grafting algebra*. Compared to that paper we have taken the opposite products and we have shifted the numbering of the operations by 1. Then it is obvious that under the change of notation $A_n = \mathscr{P}_{n+1}$, this is nothing but the notion of nonsymmetric operad. This point of view permits us to look at variations of colored algebras as variations of operads, cf. [Ron11], Sect. 13.14, and also to introduce the notion of colored operads, cf. [vdL03].

5.9.13 Category Associated to a NS Operad

As in Sect. 5.4.1 one can associate to any ns operad a category whose objects are indexed by the natural numbers. When the operad is set-theoretic, this construction can be done in the set-theoretic framework. The category associated to uAs can be identified to the linearized simplicial category $\mathbb{K}[\Delta]$ (cf. [Pir02a]). The category associated to As can be identified with the linearized presimplicial category $\mathbb{K}[\Delta^{pre}]$ (i.e. Δ without degeneracies).

5.9.14 Group Associated to a NS Operad

Let $\mathscr{P}$ be a ns operad such that $\mathscr{P}_0 = 0$ and $\mathscr{P}_1 = \mathbb{K}\,\mathrm{id}$. We consider the series

$$\underline{a} := (a_0, a_1, \ldots, a_n, \ldots)$$

where $a_n \in \mathscr{P}_{n+1}$ for any n and $a_0 = \mathrm{id} \in \mathscr{P}_1$. We denote by $G(\mathscr{P})$ this set of series. We define a binary operation $\underline{a}\,\underline{b}$ on this set as follows:

$$(\underline{a}\,\underline{b})_n := \sum_k \sum_{i_1+\cdots+i_k=n} \gamma(a_{k-1}; b_{i_1-1}, \ldots, b_{i_k-1}).$$

Proposition 5.9.2. *The binary operation $(\underline{a}, \underline{b}) \mapsto \underline{a}\,\underline{b}$ makes $G(\mathscr{P})$ into a group with unit* $1 = (\mathrm{id}, 0, 0, \ldots)$.

Proof. The associativity property follows readily from the associativity property of γ. The existence of an inverse, that is for any $\underline{a}$ there exists $\underline{b}$ such that $\underline{a}\,\underline{b} = 1$, is achieved by induction. For instance $b_1 = -a_1, b_2 = -a_2 - a_1 \circ (\mathrm{id}, b_1) - a_1 \circ (b_1, \mathrm{id})$, etc. □

Observe that for $\mathscr{P} = As$, the group $G(\mathscr{P})$ is nothing but the group of power series in one variable with constant coefficient equal to 1. This construction has been used in several instances, cf. [Fra08, Cha01a, vdL02, LN12].

5.9.15 Pre-Lie Algebra Associated to a NS Operad

Let $\mathscr{P}$ be a ns operad with $\mathscr{P}(0) = 0$ and consider the space $\mathscr{P}(\mathbb{K}) := \bigoplus_{n\geq 1} \mathscr{P}_n$, resp. $\widehat{\mathscr{P}}(\mathbb{K}) := \prod_{n\geq 1} \mathscr{P}_n$. We construct a bilinear operation $\{-,-\}$ as follows:

$$\{\mu, \nu\} := \sum_{i=1}^{i=m} (\mu \circ_i \nu)$$

for $\mu \in \mathscr{P}_m$, $\nu \in \mathscr{P}_n$. As in the case of symmetric operads, cf. Sect. 5.4.3, the relations satisfied by the partial operations imply that the binary operation $\{-,-\}$ makes $\mathscr{P}(\mathbb{K})$, resp. $\widehat{\mathscr{P}}(\mathbb{K})$, into a pre-Lie algebra, and hence a Lie algebra by anti-symmetrization.

We check easily that in the case of the ns operad As we get the pre-Lie algebra of polynomial vector fields on the affine line, cf. Sect. 1.4.3.

When $\mathscr{P} = \mathrm{End}(A)$, this Lie bracket on $C^\bullet_{Hoch}(A, A)$ was first constructed by Gerstenhaber in [Ger63] in his study of Hochschild cohomology of an associative algebra A with coefficients into itself, cf. Sect. 13.3.11.

5.9.16 Hopf Algebra Associated to a NS Operad

Let $\mathscr{P}$ be a ns operad with $\mathscr{P}_0 = 0$ and $\mathscr{P}_1 = \mathbb{K}\,\mathrm{id}$. We put $\widehat{\mathscr{P}} := \prod_{n\geq 2} \mathscr{P}_n$. On the cofree coalgebra $T^c(\widehat{\mathscr{P}})$ we define a product, compatible in the Hopf sense with the coproduct, as follows. Since $T^c(\widehat{\mathscr{P}})$ is cofree, by Sect. 1.2.5 it suffices to construct the map

$$T^c(\widehat{\mathscr{P}}) \otimes T^c(\widehat{\mathscr{P}}) \to \widehat{\mathscr{P}}.$$

On $T^c(\widehat{\mathscr{P}})^{\geq 2} \otimes T^c(\widehat{\mathscr{P}})^{\geq 1}$ it is trivial, on $\widehat{\mathscr{P}} \otimes T^c(\widehat{\mathscr{P}})$ it is given by

$$\mu \otimes (\mu_1, \ldots, \mu_k) \mapsto \sum \gamma(\mu; \mathrm{id}, \ldots, \mathrm{id}, \mu_1, \mathrm{id}, \ldots, \mathrm{id}, \mu_2, \mathrm{id}, \ldots, \mu_k, \mathrm{id}, \ldots) \in \widehat{\mathscr{P}},$$

whenever $\mu \in \mathscr{P}_k$ and where the sum is over all possibilities. The associativity of this product on $T^c(\widehat{\mathscr{P}})$ follows from the associativity property of γ. One way of proving this result without too many tedious computations is to use the notion of brace algebra, see Propositions 13.11.4 and 13.11.5. As a result we get a cofree Hopf algebra. It is an example of a combinatorial Hopf algebra which is cofree and left-sided in the sense of [LR10].

Similarly, starting from a conilpotent ns cooperad one can construct a combinatorial Hopf algebra which is free and left-sided, see [vdLM02].

5.9.17 Nonsymmetric Cooperad and Cutting

It is clear that all the cooperadic definitions and constructions can be performed in the nonsymmetric framework, that is over arity graded spaces instead of $\mathbb{S}$-modules as done in Sect. 5.8.7. Let us just give some details on the free nonsymmetric cooperad over an arity graded space of the form $M = (0, 0, M_2, M_3, \ldots)$. As a graded module $\mathscr{T}^c(M)$ is spanned by the planar rooted trees whose vertices are labeled by elements of M. In fact, if the vertex has k inputs (arity k), then its label is in M_k. The decomposition Δ on such a tree t is constructed by "degrafting" as follows. A *cut* of the tree is *admissible* if the grafting of the pieces gives the original tree back. The degrafting $\Delta(t)$ of t is the sum of all the admissible cuttings $(r; s_1, \ldots, s_k)$, where r is the piece containing the root, and k is the number of leaves of r. Of course each vertex keeps its labeling.

The explicit formula given in Sect. 5.8.6 is valid only when M is in even degree. Indeed, since the associativity isomorphism for composition is involved (cf. Sect. 5.1.8) signs appear in the formula in the general graded case. For instance if  is in degree 1, then the formula becomes:

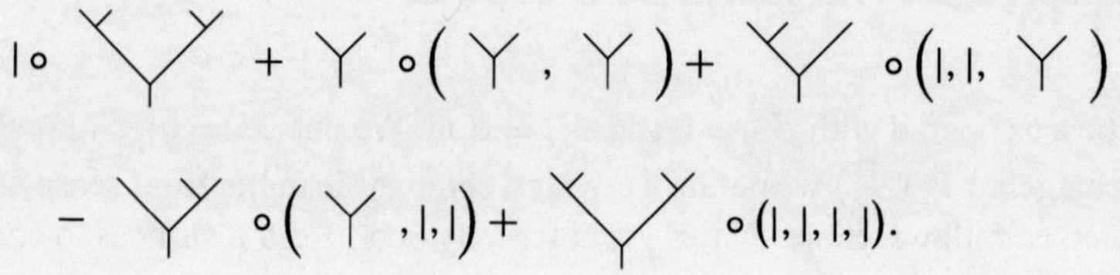

More generally, if t and s are elements of $\mathscr{T}^c(M)$, then the coproduct of $t \vee s = (\curlyvee\, ; t, s)$ is given by

$$\begin{aligned}
\Delta(t \vee s) &= \Delta(\curlyvee\, ; t, s) \\
&= (|; t \vee s) + \left(\curlyvee\, ; (t^{(1)}; t^{(2)}), (s^{(1)}; s^{(2)})\right) \\
&= (|; t \vee s) + (-1)^{|t^{(2)}|\,|s^{(1)}|}\left(\left(\curlyvee\, ; t^{(1)}, s^{(1)}\right); t^{(2)}, s^{(2)}\right) \\
&= (|; t \vee s) + (-1)^{|t^{(2)}|\,|s^{(1)}|}\left(t^{(1)} \vee s^{(1)}; t^{(2)}, s^{(2)}\right),
\end{aligned}$$

where $\Delta(t) = (t^{(1)}; t^{(2)})$ and $\Delta(s) = (s^{(1)}; s^{(2)})$. The sign comes from the exchange of $s^{(1)}$ and $t^{(2)}$.

For instance we get

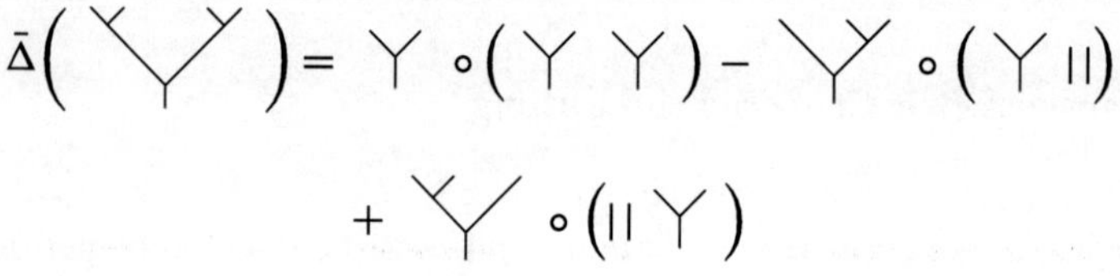

and

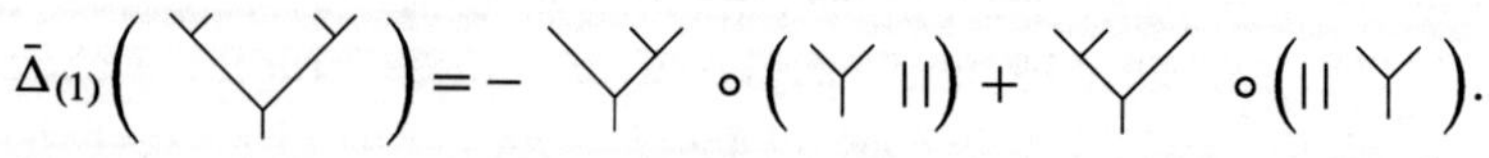

5.10 Résumé

We gave several equivalent definitions of an operad, which can be summarized as follows.

Definition 0. Given a *type of algebras* the algebraic operad is given by the functor "free algebra", which is a monad in Vect. If the relations are multilinear, then the endofunctor is a Schur functor.

Definition 1. The *monoidal definition.* An algebraic operad is a monoid $(\mathscr{P}, \gamma, \eta)$ in the monoidal category of $\mathbb{S}$-modules (resp. arity graded spaces). So $\gamma : \mathscr{P} \circ \mathscr{P} \to \mathscr{P}$ is associative and $\eta : \mathrm{I} \to \mathscr{P}$ is its unit. It is called a symmetric operad (resp. nonsymmetric operad).

Definition 2. The *classical definition.* A symmetric operad is a family of $\mathbb{S}_n$-modules $\mathscr{P}(n), n \geq 1$, and linear maps

$$\gamma(i_1, \ldots, i_k) : \mathscr{P}(k) \otimes \mathscr{P}(i_1) \otimes \cdots \otimes \mathscr{P}(i_k) \longrightarrow \mathscr{P}(i_1 + \cdots + i_k)$$

satisfying some axioms expressing equivariance under the action of the symmetric group and associativity of the composition. They ensure that the associated functor $V \mapsto \mathscr{P}(V) := \bigoplus_n \mathscr{P}(n) \otimes_{\mathbb{S}_n} V^{\otimes n}$ is a monoid.

Definition 3. The *partial definition.* A symmetric operad is a family of $\mathbb{S}_n$-modules $\mathscr{P}(n), n \geq 0$, and partial compositions

$$\circ_i : \mathscr{P}(m) \otimes \mathscr{P}(n) \to \mathscr{P}(m - 1 + n), \text{ for } 1 \leq i \leq m,$$

satisfying equivariance with respect to the symmetric groups and the axioms:

$$\begin{cases} \text{(I)} \quad (\lambda \circ_i \mu) \circ_{i-1+j} \nu = \lambda \circ_i (\mu \circ_j \nu), & \text{for } 1 \leq i \leq l,\ 1 \leq j \leq m, \\ \text{(II)} \quad (\lambda \circ_i \mu) \circ_{k-1+m} \nu = (\lambda \circ_k \nu) \circ_i \mu, & \text{for } 1 \leq i < k \leq l, \end{cases}$$

for any $\lambda \in \mathscr{P}(l)$, $\mu \in \mathscr{P}(m)$, $\nu \in \mathscr{P}(n)$. One assumes the existence of a unit element $\mathrm{id} \in \mathscr{P}(1)$.

Definition 4. The *combinatorial definition.* There exists a monad $\mathbb{T}$ over the category of $\mathbb{S}$-modules made out of rooted trees and substitution, such that a symmetric operad is an algebra (i.e. a representation) over $\mathbb{T}$. Nonsymmetric operads are obtained by replacing trees by planar trees.

Observe that Definitions 2, 3 and 4 can be thought of as various presentations of the monad $\mathbb{T}$. In Definition 2 the generators have two levels, in Definition 3 they involve only two variables, in Definition 4 every element is a generator.

Algebra over an Operad. A $\mathscr{P}$-algebra is a vector space A equipped with a linear map $\gamma_A : \mathscr{P}(A) \to A$ compatible with the operadic structure γ and η. It is equivalent to a morphism of operads

$$\mathscr{P} \to \mathrm{End}_A.$$

In order to get the analogous definitions for nonsymmetric operads, it suffices to replace the $\mathbb{S}$-modules by the arity graded modules (no action of the symmetric group anymore) where degree = arity. In definition 0 the relations should be such

Table 5.2 Various algebras over combinatorial monads

	Category	Product	Unit	Combinatorial objects
monoid	Set	$\times$	$\{*\}$	ladders
algebra	Vect	$\otimes$	$\mathbb{K}$	ladders
operad	$\mathbb{S}$-Mod	$\circ$	I	rooted trees
ns operad	$\mathbb{N}$-Mod	$\circ$	I	planar rooted trees

that the variables stay in the same order in the involved monomials. In Definition 4 the trees are supposed to be planar.

Monoids, unital associative algebras, symmetric operads, nonsymmetric operads, are all monoids in an ad hoc monoidal category. They can also be interpreted as algebras over a combinatorial monad (see Table 5.2).

5.11 Exercises

Exercise 5.11.1 (Identity operad). Let I be the $\mathbb{S}$-module corresponding to the identity functor from Vect to itself. What is $\mathrm{I}(n)$?

Exercise 5.11.2 (On $\mathrm{End}_{\mathbb{K}}$). Show that $\mathrm{End}_{\mathbb{K}} = uCom$ as an operad, and $\mathrm{End}(\mathbb{K}) = uAs$ as a ns operad.

Exercise 5.11.3 (A graded operad). Show that the category of algebras over the operad $\mathrm{End}_{s\mathbb{K}}$ can be described as follows. An $\mathrm{End}_{s\mathbb{K}}$-algebra is a graded vector space A with a bilinear map $A_n \otimes A_m \to A_{n+m+1}$, $x \otimes y \mapsto xy$ for any $n, m \geq 0$, such that

$$xy = -(-1)^{|x||y|}yx, \qquad (xy)z = (-1)^{|x|}x(yz).$$

Exercise 5.11.4 (Shifting degrees). Let M and N be two endofunctors of the category of graded vector spaces related by the formula $M(V) = N(sV)$ for any graded space V. Show that

$$s^{-1}N = (\mathrm{End}_{s^{-1}\mathbb{K}}) \underset{\mathrm{H}}{\otimes} M.$$

Exercise 5.11.5 (From Ass to Com). Show that the forgetful functor Com-alg $\to$ Ass-alg induces, on the space of n-ary operations, the augmentation map $\mathbb{K}[\mathbb{S}_n] \to \mathbb{K}$, $\sigma \mapsto 1$ for $\sigma \in \mathbb{S}_n$.

Exercise 5.11.6 (Explicit free operad). Show that a functor $F : \mathrm{Vect} \to \mathscr{P}$-alg gives a free $\mathscr{P}$-algebra $F(V)$ if and only if there exists $\phi_A : F(A) \to A$ for any $\mathscr{P}$-algebra A, and $\psi_V : V \to F(V)$ for any vector space V, such that ϕ_A is natural in A, ψ_V is natural in V, and both composites

$$F(V) \xrightarrow{F(\psi_V)} F\bigl(F(V)\bigr) \xrightarrow{\phi_{F(V)}} F(V) \text{ and } A \xrightarrow{\psi_A} F(A) \xrightarrow{\phi_A} A$$

give the identity.

Exercise 5.11.7 (Free operad). Show directly from the definition of a free operad that $\mathcal{T}M$ can be described in terms of planar binary trees when $M = (0, 0, M_2 \otimes \mathbb{K}[\mathbb{S}_2], 0, \ldots, 0, \ldots)$.

Exercise 5.11.8 (Plethysm). Let E, resp. F, be a representation of $\mathbb{S}_n$, resp. $\mathbb{S}_m$. Let $\widetilde{E}$, resp. $\widetilde{F}$, be the associated Schur functor. Show that $\widetilde{E} \circ \widetilde{F}$ is the Schur functor of a certain representation G of $\mathbb{S}_{m^n}$ and describe it explicitly. This representation is called the *plethysm* of E and F.

Exercise 5.11.9 (*Ass* explicit). Describe explicitly the map

$$\gamma(i_1, \ldots, i_k) : \mathbb{S}_k \times \mathbb{S}_{i_1} \times \cdots \times \mathbb{S}_{i_k} \longrightarrow \mathbb{S}_{i_1+\cdots+i_k}$$

which induces the map

$$\gamma(i_1, \ldots, i_k) : Ass(k) \otimes Ass(i_1) \otimes \cdots \otimes Ass(i_k) \longrightarrow Ass(i_1 + \cdots + i_k)$$

of the operad *Ass*.

Exercise 5.11.10 (Induction). Let E be an $\mathbb{S}_2$-module. Let $\mathbb{S}_2$ act on $E \otimes E$ via its action on the second variable only. Show that, as a vector space, $\mathrm{Ind}_{\mathbb{S}_2}^{\mathbb{S}_3}(E \otimes E) = 3E \otimes E$. Describe explicitly the action of $\mathbb{S}_3$ on $3E \otimes E$.

Exercise 5.11.11 (Arity 3). Describe explicitly the $\mathbb{S}_3$-representation $\mathcal{T}(\mathbb{K}\mu)(3)$ when μ is a binary operation, resp. a symmetric binary operation, resp. an antisymmetric binary operation.

HINT. You should obtain a space of dimension 12, resp. 3, resp. 3. The multiplicities of the isotypic components (trivial, hook, signature) are $(2, 4, 2)$, resp. $(1, 1, 0)$, resp. $(0, 1, 1)$.

Exercise 5.11.12 (Poisson algebra). A Poisson algebra is determined by a commutative product $(x, y) \mapsto x \cdot y$ and a Lie bracket $(x, y) \mapsto [x, y]$ related by the derivation property (Leibniz rule):

$$[x \cdot y, z] = x \cdot [y, z] + [x, z] \cdot y.$$

This gives a presentation of the operad *Pois* of Poisson algebras. Show that there is another presentation involving only one operation xy with no symmetry and only one relation (see Sect. 13.3.3 for the solution).

Exercise 5.11.13 (Invariants). Show that the map

$$V^{\otimes n} \to \left(\mathbb{K}[\mathbb{S}_n] \otimes V^{\otimes n}\right)^{\mathbb{S}_n}, \qquad v_1 \cdots v_n \mapsto \sum_{\sigma \in \mathbb{S}_n} \sigma \otimes (v_1 \cdots v_n)^{\sigma}$$

is an isomorphism. Deduce that the categories ΓAss-alg and Ass-alg are the same.

Exercise 5.11.14 (From operad to cooperad). Let $\mathscr{P}$ be an algebraic operad and let $\mathscr{P}^* := \{\mathscr{P}(n)^*\}_{n\geq 1}$ be the linear dual cooperad. Suppose we are given a linear basis for all the spaces $\mathscr{P}(n)$. The composition is completely determined by the constants $a^{\lambda}_{\mu\nu_1\cdots\nu_k}$ appearing in the formulas

$$\gamma(\mu; \nu_1, \ldots, \nu_k) = \sum a^{\lambda}_{\mu\nu_1\cdots\nu_k}\lambda$$

where $\lambda, \mu, \nu_1, \ldots, \nu_k$ are basis elements. Show that the decomposition map of the cooperad $\mathscr{P}^*$ is completely determined by the formulas:

$$\Delta(\lambda^*) = \sum a^{\lambda}_{\mu\nu_1\cdots\nu_k}\left(\mu^*; \nu_1^*, \ldots, \nu_k^*\right).$$

Exercise 5.11.15 (On $\widetilde{Ass}$ ★). Find a presentation for the $\widetilde{Ass}$-algebras (notation introduced in 5.9.11).

HINT. Use [Pir03].

Exercise 5.11.16 (Möbius basis ★). Let $\{M_\sigma\}_{\sigma\in\mathbb{S}_n}$ be the basis of $\mathbb{K}[\mathbb{S}_n]$ defined as

$$M_\sigma := \sum_{\sigma\leq\tau} \mu(\sigma, \tau)\tau.$$

Here $\leq$ stands for the weak Bruhat order on the symmetric group and $\mu(\sigma, \tau)$ is the Möbius function. Show that for any integer i satisfying $1 \leq i \leq n$ there are uniquely determined permutations $(\sigma, \tau)^i$ and $(\sigma, \tau)_i$ such that

$$M_\sigma \circ_i M_\tau = \sum_{(\sigma,\tau)^i\leq\omega\leq(\sigma,\tau)_i} M_\omega.$$

Cf. [AL07].

Exercise 5.11.17 (Category associated to $uMag$ ★). Let $uMag$ be the set-theoretic ns operad with one binary operation and a unit. Give a presentation of cat$uMag$ analogous to the classical presentation of the simplicial category $\mathbb{K}[\Delta] = \mathrm{cat}\,uAss$ (cf. [Pir02a]) in terms of faces and degeneracies.

HINT. Same but delete the relations $\sigma_j\sigma_i = \sigma_i\sigma_{j+1}, i \leq j$.

Exercise 5.11.18 (Regular $\mathbb{S}$-modules ★). Let M and N be two $\mathbb{S}$-modules such that $M(n) = M_n \otimes \mathbb{K}[\mathbb{S}_n]$ and $N(n) = N_n \otimes \mathbb{K}[\mathbb{S}_n]$, $n \geq 1$. Show that $M \circ N$ is such that $(M \circ N)(n) = (M \circ N)_n \otimes \mathbb{K}[\mathbb{S}_n]$. Compute $(M \circ N)_n$ out of the components M_i and N_j.

Exercise 5.11.19 (Right adjoint of Schur functor ★). Let $F : \mathsf{Vect} \to \mathsf{Vect}$ be an endofunctor of the category of vector spaces. Let

$$RF_n := \mathrm{Hom}_{End}\left(T^{(n)}, F\right)$$

where the endofunctor $T^{(n)}:\mathsf{Vect}\to\mathsf{Vect}$ is given by $T^{(n)}(V):=V^{\otimes n}$. Show that R is right adjoint to the Schur functor $\mathbb{S}\text{-}mod\to\mathrm{End}(\mathsf{Vect})$.

Exercise 5.11.20 (Non-morphism ★). Show that there is a morphism of $\mathbb{S}_n$-modules $F(n):Com(n)\to Ass(n)$, which identifies the trivial representation to its copy in the regular representation. Show that the resulting morphism of $\mathbb{S}$-modules $F:Com\to Ass$ is not a morphism of operads, i.e. $F(\mu\circ_1\mu)\neq F(\mu)\circ_1 F(\mu)$.

HINT. It follows from the fact that, in an associative algebra, the symmetrized product $a\cdot b:=ab+ba$ is not associative in general.

Exercise 5.11.21 (Composite with *Ass* ★). Let M be an $\mathbb{S}$-module and let Reg be the regular $\mathbb{S}$-module, that is $Reg(n):=\mathbb{K}[\mathbb{S}_n]$. This is the $\mathbb{S}$-module underlying the operad Ass (and several others). Show that the composite $\mathbb{S}$-module $M\circ Reg$ can be described as follows:

$$(M\circ Reg)(n)=\bigoplus_k M(k)\Bigg(\bigoplus_{i_1+\cdots+i_k=n}\mathbb{K}[\mathbb{S}_n]\Bigg)$$

where the action of $\mathbb{S}_k$ on the right sum is explicitly given by

$$\sigma\cdot(i_1,\ldots,i_k;\omega)=(i_{\sigma^{-1}(1)},\ldots,i_{\sigma^{-1}(k)};\sigma_\omega)$$

for $\sigma\in\mathbb{S}_k$, $\omega\in\mathbb{S}_n$, $i_1+\cdots+i_k=n$. The permutation $\sigma_\omega\in\mathbb{S}_n$ is the precomposition of ω by the action of σ on the "blocks" of size $i_1,\ldots,i_k$.

Exercise 5.11.22 (Trees and free operad ★). Let $\alpha,\beta,\gamma,\delta$ be binary operations in M. Show that the element corresponding to the tree

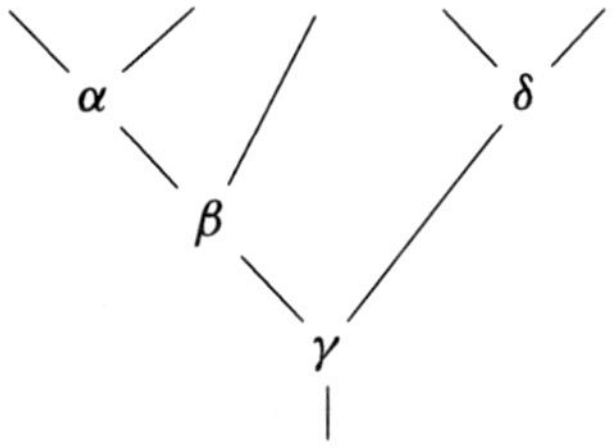

is $(\gamma;(\beta;\alpha,\mathrm{id}),(\delta;\mathrm{id},\mathrm{id}))\in\mathscr{T}_3M$.

Exercise 5.11.23 (Free operad and automorphisms of trees ★). Extend the results of Sect. 5.6 (combinatorial definition and free operad) to the case where the $\mathbb{S}$-module M has a nontrivial component $M(0)$.

HINT. In this case, the trees are not reduced and hence they have nontrivial automorphism groups, see Appendix C.4.

Exercise 5.11.24 (Hopf operad ★). Show that any set operad gives rise to a Hopf operad.

Exercise 5.11.25 (Modules over a Hopf operad ★). Show that the tensor product of left modules over a Hopf operad is still a left module.

Exercise 5.11.26 (Explicit enveloping algebra ★). Let $\alpha : \mathscr{P} \to \mathscr{Q}$ be a morphism of operads and let (A, γ_A) be a $\mathscr{P}$-algebra. Suppose that the operad $\mathscr{P}$ comes with a presentation $\mathscr{P} = \mathscr{P}(E, R)$. Show that the relative free $\mathscr{Q}$-algebra $\alpha_!(A) = \mathscr{Q} \circ_{\mathscr{P}} A$ introduced in Sect. 5.2.12 is isomorphic to the quotient of the free $\mathscr{Q}$-algebra over the space A by the relation which identifies the two $\mathscr{P}$-algebra structures. More precisely we have

$$\mathscr{Q}(A)/\big((\alpha(\mu); a_1, \ldots, a_k) - \gamma_A(\mu; a_1, \ldots, a_k); \mu \in E(k), a_1, \ldots, a_k \in A\big),$$

where the right-hand side stands for the ideal generated by the listed elements for any k.

Exercise 5.11.27 (Convolution operad ★). Show that any symmetric, resp. ns, operad $\mathscr{P}$ is isomorphic to the convolution operad $\mathrm{Hom}_{\mathbb{S}}(uAss^c, \mathscr{P})$, resp. $\mathrm{Hom}(uAs^c, \mathscr{P})$.

Chapter 6
Operadic Homological Algebra

If I could only understand the beautiful consequence following from the concise proposition $d^2 = 0$.

Henri Cartan on receiving the degree of Doctor Honoris Causa, Oxford University, 1980

The aim of this chapter is to develop homological algebra in the operadic context.

We introduce the notions of *differential graded $\mathbb{S}$-module*, *differential graded operad*, *differential graded $\mathscr{P}$-algebra*, *differential graded cooperad*, etc. Since the composite product $\circ$ defining operads is not linear on the right-hand side, these generalizations are not automatic. We define the *infinitesimal composite product*, as a linearization of the composite product, and the *infinitesimal composite of morphisms*, as a linearization of the composite of morphisms. This latter one plays a crucial role in the definition of the differential of the composite product of two dg $\mathbb{S}$-modules.

In the second part of this chapter, we transpose the results on twisting morphisms from the algebra setting to the operad setting. From a dg cooperad $\mathscr{C}$ and a dg operad $\mathscr{P}$, we construct a *convolution dg operad* $\mathrm{Hom}(\mathscr{C}, \mathscr{P})$, which induces a *dg (pre-)Lie convolution algebra* $\mathrm{Hom}_{\mathbb{S}}(\mathscr{C}, \mathscr{P})$. In this setting, we can consider the Maurer–Cartan equation, whose solutions are the *operadic twisting morphisms*.

As in the algebra case, the main homological constructions for dg operads come from this notion. First, with an operadic twisting morphism, one can twist the differential of the composite product $\mathscr{C} \circ \mathscr{P}$ to produce a *twisted composite product*. Then, the operadic twisting morphism bifunctor is represented by the *operadic bar* and *cobar constructions*. These constructions generalize the ones encountered in Chap. 2 at the algebra level. Once again, since the composite product is not bilinear, these constructions are more involved. We make this "adaptation" explicit.

As in the algebra case, an operadic Koszul morphism is defined as an operadic twisting morphism whose twisted composite product is acyclic. In the last part of the chapter, we state and prove the fundamental theorem of operadic twisting morphisms which says that a Koszul morphism corresponds to a quasi-isomorphism to

J.-L. Loday, B. Vallette, *Algebraic Operads*,
Grundlehren der mathematischen Wissenschaften 346,
DOI 10.1007/978-3-642-30362-3_6,

the bar construction and to a quasi-isomorphism from the cobar construction respectively.

The material of this chapter mainly comes from Ginzburg and Kapranov [GK94], Getzler and Jones [GJ94], and Fresse [Fre04].

In this chapter, we work over a ground field $\mathbb{K}$ of characteristic 0. Notice that all the constructions and some of the results hold true without this hypothesis.

6.1 Infinitesimal Composite

Whereas the tensor product $\otimes$ of $\mathbb{K}$-modules is linear on the left and on the right, the composite product $\circ$ of $\mathbb{S}$-modules is linear on the left but not on the right because the right hand side of the composite $M \circ N$ involves several components of N. In order to do homological algebra for $\mathbb{S}$-modules and operads, we will need a linearized version of the composite product and of the composite product of morphisms. First, we linearize the composite product of $\mathbb{S}$-modules to define the notion called infinitesimal composite product of $\mathbb{S}$-modules. We also need to introduce the infinitesimal composite of $\mathbb{S}$-module morphisms, for which the composite product remains unchanged but where the composite of two morphisms is linearized.

6.1.1 Infinitesimal Composite Product

In this section, we introduce the right linear part of the composite product as a particular case of a more general construction.

Let M, N_1 and N_2 be three $\mathbb{S}$-modules. The composite $M \circ (N_1 \oplus N_2)$ defines a functor $(\mathbb{S}\textsf{-Mod})^3 \to \mathbb{S}\textsf{-Mod}$. We consider the sub-functor which is linear in N_2. We denote the image of (M, N_1, N_2) under this sub-functor by $M \circ (N_1; N_2)$. This notation has to be taken as a whole, because $(N_1; N_2)$ has no meaning by itself. Explicitly, $M \circ (N_1; N_2)$ is the sub-$\mathbb{S}$-module of $\bigoplus_n M(n) \otimes_{\mathbb{S}_n} (N_1 \oplus N_2)^{\otimes n}$ where N_2 appears once and only once in each summand (see Sect. 5.1.4).

Graphically a typical element is of the form:

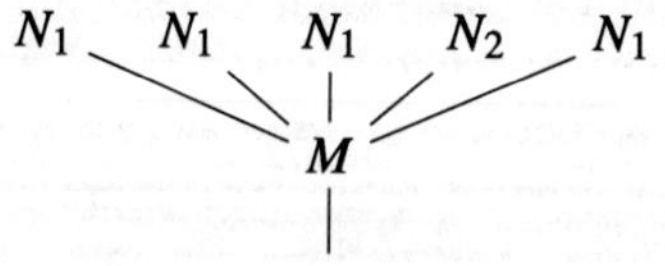

This functorial construction is linear in the variable M and in the variable N_2.

Proposition 6.1.1. *For any $\mathbb{S}$-modules M, M', N_1, N_2 and N'_2, we have*

$$(M \oplus M') \circ (N_1; N_2) = M \circ (N_1; N_2) \oplus M' \circ (N_1; N_2) \quad \textit{and}$$
$$M \circ \big(N_1; N_2 \oplus N'_2\big) = M \circ (N_1; N_2) \oplus M \circ \big(N_1; N'_2\big).$$

Proof. The first formula is obvious because $\circ$ is linear on the left hand side. The second one follows from the definition of $M \circ (N_1; N_2)$. □

Notice that $M \circ (N; N)$ is not isomorphic to $M \circ N$. There exists an epimorphism $M \circ (N; N) \rightarrowtail M \circ (N \oplus N) \twoheadrightarrow M \circ N$, where the second map is $\mathrm{Id}_M \circ (\mathrm{Id}_N + \mathrm{Id}_N)$.

We now apply this construction to the particular case $N_1 = \mathrm{I}$ (identity functor). Let M and N be two $\mathbb{S}$-modules. By definition, the *infinitesimal composite of M and N* is defined by the formula

$$M \circ_{(1)} N := M \circ (\mathrm{I}; N).$$

Elements of $M \circ_{(1)} N$ are of the form $(\mu; \mathrm{id}, \ldots, \mathrm{id}, \nu, \mathrm{id}, \ldots, \mathrm{id})$.

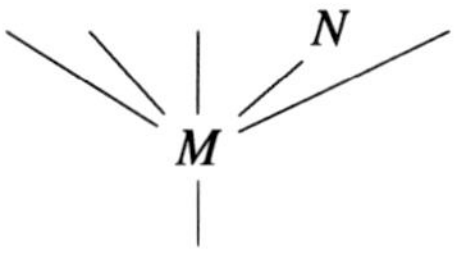

Notice that the infinitesimal composite of M with itself is naturally isomorphic, as $\mathbb{S}$-module, to the weight 2 part of the free operad on M, i.e. the treewise tensor module over M:

$$M \circ_{(1)} M \cong \mathscr{T}M^{(2)}.$$

The main property of the infinitesimal composite product is the following linearity property on both sides.

Corollary 6.1.2. *For any $\mathbb{S}$-modules M, M', N and N', we have*

$$\begin{aligned}
&(M \oplus M') \circ_{(1)} N = M \circ_{(1)} N \oplus M' \circ_{(1)} N,\\
&M \circ_{(1)} (N \oplus N') = M \circ_{(1)} N \oplus M \circ_{(1)} N',\\
&\mathrm{I} \circ_{(1)} N = N \quad \textit{and} \quad M \circ_{(1)} \mathrm{I} = M.
\end{aligned}$$

To any pair $f : M_1 \to M_2$ and $g : N_1 \to N_2$ of morphisms of $\mathbb{S}$-modules, we define a morphism $f \circ_{(1)} g : M_1 \circ_{(1)} N_1 \to M_2 \circ_{(1)} N_2$ by the formula

$$(f \circ_{(1)} g)(\mu; \mathrm{id}, \ldots, \nu, \ldots, \mathrm{id}) := \big(f(\mu); \mathrm{id}, \ldots, g(\nu), \ldots, \mathrm{id}\big).$$

6.1.2 Infinitesimal Composition Map of an Operad

When $(\mathscr{P}, \gamma, \eta)$ is an operad, we use the notion of infinitesimal composite product to define the infinitesimal part of the composition map. By definition, the *infinitesimal composition map* $\gamma_{(1)} : \mathscr{P} \circ_{(1)} \mathscr{P} \to \mathscr{P}$ of $\mathscr{P}$ is given by

$$\gamma_{(1)} : \mathscr{P} \circ_{(1)} \mathscr{P} = \mathscr{P} \circ (\mathrm{I}; \mathscr{P}) \rightarrowtail \mathscr{P} \circ (\mathrm{I} \oplus \mathscr{P}) \xrightarrow{\mathrm{Id}_{\mathscr{P}} \circ (\eta + \mathrm{Id}_{\mathscr{P}})} \mathscr{P} \circ \mathscr{P} \xrightarrow{\gamma} \mathscr{P}.$$

It is the restriction of the composition map of the operad $\mathscr{P}$ where we only compose two operations of $\mathscr{P}$.

6.1.3 Infinitesimal Composite of Morphisms

Instead of linearizing the underlying composite product of two $\mathbb{S}$-modules, we keep it unchanged but we linearize the composite of two morphisms.

Recall from Sect. 5.1.9 that for any pair $f : M_1 \to M_2$, $g : N_1 \to N_2$ of morphisms of $\mathbb{S}$-modules, their composite product $f \circ g : M_1 \circ N_1 \to M_2 \circ N_2$ is given explicitly by the formula

$$f \circ g(\mu; \nu_1, \ldots, \nu_k) := \big(f(\mu); g(\nu_1), \ldots, g(\nu_k)\big),$$

where $(\mu; \nu_1, \ldots, \nu_k)$ represents an element of

$$M_1(k) \otimes_{\mathbb{S}_k} \Big(\bigoplus \operatorname{Ind}_{\mathbb{S}_{i_1} \times \cdots \times \mathbb{S}_{i_k}}^{\mathbb{S}_n} \big(N_1(i_1) \otimes \cdots \otimes N_1(i_k)\big)\Big),$$

cf. Sect. 5.1.7.

We define the *infinitesimal composite of morphisms*

$$f \circ' g : M_1 \circ N_1 \to M_2 \circ (N_1; N_2)$$

by the formula

$$f \circ' g := \sum_i f \otimes \big(\mathrm{Id}_{N_1} \otimes \cdots \otimes \mathrm{Id}_{N_1} \otimes \underbrace{g}_{i\text{th position}} \otimes \mathrm{Id}_{N_1} \otimes \cdots \otimes \mathrm{Id}_{N_1}\big).$$

The notation $f \circ' g$, with the prime symbol, is called the *infinitesimal composite* of f and g. This notion should not be confused with $f \circ_{(1)} g : M_1 \circ_{(1)} N_1 \to M_2 \circ_{(1)} N_2$. For instance, the involved domains are different. The main property of this infinitesimal composite of morphisms is the following additivity property.

Proposition 6.1.3. *For any morphisms $f : M_1 \to M_2$ and $g, h : N_1 \to N_2$ of $\mathbb{S}$-modules, we have*

$$f \circ' (g + h) = f \circ' g + f \circ' h \quad \textit{in } \operatorname{Hom}\big(M_1 \circ N_1, M_2 \circ (N_1; N_2)\big).$$

When $N_1 = N_2 = N$, we still denote the composite

$$M_1 \circ N \to M_2 \circ (N; N) \to M_2 \circ N$$

by $f \circ' g$ since the context is obvious.

6.1.4 Infinitesimal Decomposition Map of a Cooperad

Dually, for any cooperad $(\mathscr{C}, \Delta, \eta)$, we consider the projection of the decomposition map to the infinitesimal part of the composite product $\mathscr{C} \circ \mathscr{C}$. This map is called the *infinitesimal decomposition map of* $\mathscr{C}$ and is defined by the following composite

$$\Delta_{(1)} := \mathscr{C} \xrightarrow{\Delta} \mathscr{C} \circ \mathscr{C} \xrightarrow{\mathrm{Id}_{\mathscr{C}} \circ' \mathrm{Id}_{\mathscr{C}}} \mathscr{C} \circ (\mathscr{C}; \mathscr{C}) \xrightarrow{\mathrm{Id}_{\mathscr{C}} \circ (\eta; \mathrm{Id}_{\mathscr{C}})} \mathscr{C} \circ (\mathrm{I}; \mathscr{C}) = \mathscr{C} \circ_{(1)} \mathscr{C}.$$

This map can be seen as a decomposition of an element of $\mathscr{C}$ into two parts.

6.2 Differential Graded $\mathbb{S}$-Module

In this section, we extend the notion of $\mathbb{S}$-module to the graded framework and to the differential graded framework. The notion of infinitesimal composite of morphisms plays a crucial role in the definition of the differential of the composite product.

6.2.1 Graded $\mathbb{S}$-Module

A *graded* $\mathbb{S}$*-module* M is an $\mathbb{S}$-module in the category of graded vector spaces. So the component of arity n is a graded $\mathbb{S}_n$-module $\{M_p(n)\}_{p\in\mathbb{Z}}$ for any n. Equivalently M can be considered as a family of $\mathbb{S}$-modules $\{M_p\}_{p\in\mathbb{Z}}$. By abuse of notation, the direct sum is also denoted by M:

$$M = M_\bullet := \cdots \oplus M_0 \oplus M_1 \oplus \cdots \oplus M_p \oplus \cdots.$$

Notice that a graded $\mathbb{S}$-module M is a family indexed by two labels: the degree p and the arity n. The degree of an element $\mu \in M(n)$ is denoted by $|\mu|$.

A *morphism* $f : M(n) \to N(n)$ *of degree* r *of graded* $\mathbb{S}_n$*-modules* is a family $\{f_p\}_{p\in\mathbb{Z}}$ of $\mathbb{S}_n$-equivariant maps $f_p : M_p(n) \to N_{p+r}(n)$. A *morphism* $f : M \to N$ *of degree* r *of graded* $\mathbb{S}$*-modules* is a family $\{f(n) : M(n) \to N(n)\}_{n\geq 0}$ of morphisms of degree r of graded $\mathbb{S}_n$-modules. The degree of f is the integer r and is denoted by $|f|$. We denote by $\mathrm{Hom}^r_{\mathbb{S}}(M, N)$ the set of morphisms of graded $\mathbb{S}$-modules of degree r.

The tensor product of $\mathbb{S}$-modules is extended to graded $\mathbb{S}$-modules by the formula:

$$(M \otimes N)_p(n) := \bigoplus_{i+j=n, q+r=p} \mathrm{Ind}^{\mathbb{S}_n}_{\mathbb{S}_i\times\mathbb{S}_j} \big(M_q(i) \otimes N_r(j)\big).$$

The composite product of two $\mathbb{S}$-modules is extended to graded $\mathbb{S}$-modules by the following formula:

$$(M \circ N)_p(n) := \bigoplus_{k\geq 0} M_q(k) \otimes_{\mathbb{S}_k} \Big(\bigoplus \mathrm{Ind}^{\mathbb{S}_n}_{\mathbb{S}_{i_1}\times\cdots\times\mathbb{S}_{i_k}} \big(N_{r_1}(i_1) \otimes \cdots \otimes N_{r_k}(i_k)\big)\Big),$$

where the first sum also runs over the $k+1$ tuples $(q, r_1, \ldots, r_k)$ such that $q+r_1+\cdots+r_k = p$ and where the second sum runs over the k-tuples $(i_1, \ldots, i_k)$ such that $i_1+\cdots+i_k = n$.

The $\mathbb{S}$-module $\mathrm{I} := (0, \mathbb{K}, 0, 0, \ldots)$ is considered as a graded $\mathbb{S}$-module concentrated in degree 0.

Proposition 6.2.1. *With these definitions, the category of graded* $\mathbb{S}$*-modules, with the product* $\circ$ *and the unit* I, *form a monoidal category (see Appendix* B.4*) denoted* gr $\mathbb{S}$-Mod.

We denote by $\mathbb{K}s$ the graded $\mathbb{S}$-module $(\mathbb{K}s, 0, 0, \ldots)$, where $\mathbb{K}s$ is concentrated in degree 1 and arity 0. By definition the *suspension* of a graded $\mathbb{S}$-module M is the shifted graded $\mathbb{S}$-module $sM := \mathbb{K}s \otimes M$, that is $sM_p(n) = M_{p-1}(n)$. We denote an elementary tensor $s \otimes m$ by sm.

Similarly let $\mathbb{K}s^{-1}$ be the graded $\mathbb{S}$-module $(\mathbb{K}s^{-1}, 0, 0, \ldots)$ concentrated in degree -1 and arity 0. By definition, the *desuspension* of a graded $\mathbb{S}$-module M is the shifted graded $\mathbb{S}$-module $s^{-1}M := \mathbb{K}s^{-1} \otimes M$, that is $s^{-1}M_p(n) = M_{p+1}(n)$. We denote an elementary tensor $s^{-1} \otimes m$ by $s^{-1}m$.

6.2.2 Differential Graded $\mathbb{S}$-Module

A *differential graded* $\mathbb{S}_n$-module $(M(n), d)$ is a graded $\mathbb{S}_n$-module $M(n)$ equipped with a *differential* d of $\mathbb{S}_n$-modules

$$\cdots \xleftarrow{d} M_0(n) \xleftarrow{d} M_1(n) \xleftarrow{d} M_2(n) \xleftarrow{d} \cdots \xleftarrow{d} M_p(n) \xleftarrow{d} \cdots$$

such that $d^2 = 0$. A *dg* $\mathbb{S}$*-module* is a family $\{M(n)\}_{n\geq 0}$ of differential graded $\mathbb{S}_n$-modules. The differential d has degree -1. The homology groups $H_\bullet(M)$ of a dg $\mathbb{S}$-module form a graded $\mathbb{S}$-module.

A *morphism* $f : (M, d_M) \to (N, d_N)$ *of differential graded* $\mathbb{S}$*-modules* is a morphism f of graded $\mathbb{S}$-modules of degree 0 which commutes with the differentials, that is $d_N \circ f = f \circ d_M$. The differential $\mathbb{S}$-modules with their morphisms form a category denoted by dg $\mathbb{S}$-Mod.

Let (M, d_M) and (N, d_N) be two dg $\mathbb{S}$-modules. Their composite product $M \circ N$ is a graded $\mathbb{S}$-module that we endow with a differential $d_{M\circ N}$ defined as follows

$$d_{M\circ N} := d_M \circ \mathrm{Id}_N + \mathrm{Id}_M \circ' d_N.$$

Observe that the second summand uses the infinitesimal composite of morphisms. Explicitly we get:

$$d_{M\circ N}(\mu; \nu_1, \ldots, \nu_k) = \big(d_M(\mu); \nu_1, \ldots, \nu_k\big) + \sum_{i=1}^{k} (-1)^{\varepsilon_i} \big(\mu; \nu_1, \ldots, d_N(\nu_i), \ldots, \nu_k\big),$$

where $\varepsilon_i = |\mu| + |\nu_1| + \cdots + |\nu_{i-1}|$. It is an easy exercise to check that $d_{M \circ N}{}^2 = 0$.

Proposition 6.2.2. *Under these definitions, the composite of dg $\mathbb{S}$-modules is a monoidal product. So the category* (dg $\mathbb{S}$-Mod, $\circ$, I) *is a monoidal category (see Appendix* B.4).

The notions of infinitesimal composite products and infinitesimal composite of morphisms extend to the differential graded framework. Their definitions are straightforward and require only the use of the Koszul sign rule. Hence, we leave it to the reader to fill in the details.

Proposition 6.2.3 (Operadic Künneth formula). *Let M, N be two dg $\mathbb{S}$-modules. Over a field $\mathbb{K}$ of characteristic* 0, *we have the following isomorphism of graded $\mathbb{S}$-modules*

$$H_\bullet(M \circ N) \cong H_\bullet(M) \circ H_\bullet(N).$$

Proof. Since we work over a characteristic zero field $\mathbb{K}$, the ring $\mathbb{K}[\mathbb{S}_n]$ is semi-simple by Maschke's theorem. Therefore every $\mathbb{K}[\mathbb{S}_n]$-module is projective, see, for instance, Chap. 4.2 of [Wei94]. We apply this result to the explicit formula of the composite product $\circ$. □

Corollary 6.2.4. *Let M be a dg $\mathbb{S}$-module. Over a field $\mathbb{K}$ of characteristic* 0, *we have the following equivalence*

$$H_\bullet(M) \cong \mathrm{I} \iff H_\bullet\big(M(V)\big) \cong V, \quad \forall V \in \mathsf{Mod}_{\mathbb{K}}.$$

Proof. By the preceding proposition, we have $H_\bullet(M(V)) \cong H_\bullet(M)(V)$, for any $\mathbb{K}$-module V. We conclude with the Schur Lemma, see Appendix A.2.3. □

6.3 Differential Graded Operad

We extend the notions of operad, algebra over an operad, cooperad and coalgebra over a cooperad to the differential graded context. This is a generalization of the dga algebra case of Sect. 1.5.

6.3.1 Differential Graded Operad

A *graded operad* $\mathscr{P}$ is a monoid in the monoidal category of graded $\mathbb{S}$-modules. The *composition map* $\gamma : \mathscr{P} \circ \mathscr{P} \to \mathscr{P}$ and the *unit map* $\eta : \mathrm{I} \to \mathscr{P}$ are supposed to be of degree 0. Let M be a graded $\mathbb{S}$-module concentrated in weight 1. The grading on

M induces a natural grading on the free operad $\mathcal{T}(M)$ on M, see Sect. 5.5.3. Since the composition product of the free operad preserves this grading, the free operad is a graded operad.

A *differential graded operad* $(\mathcal{P}, \gamma, \eta)$, *dg operad* for short, is a monoid in the monoidal category (dg $\mathbb{S}$-Mod, $\circ$, I), that is $(\mathcal{P}, \gamma, \eta)$ is a graded operad structure on a dg $\mathbb{S}$-module $(\mathcal{P}, d_{\mathcal{P}})$ such that $\gamma : \mathcal{P} \circ \mathcal{P} \to \mathcal{P}$ and $\eta : \mathrm{I} \to \mathcal{P}$ are morphisms of dg $\mathbb{S}$-modules of degree 0. The composite map γ is a morphism of dg $\mathbb{S}$-modules if and only if the following diagram commutes

$$\begin{array}{ccc} \mathcal{P} \circ \mathcal{P} & \xrightarrow{\gamma} & \mathcal{P} \\ \downarrow{\scriptstyle d_{\mathcal{P} \circ \mathcal{P}}} & & \downarrow{\scriptstyle d_{\mathcal{P}}} \\ \mathcal{P} \circ \mathcal{P} & \xrightarrow{\gamma} & \mathcal{P}. \end{array}$$

Explicitly, it means that $d_{\mathcal{P}}$ is a *derivation* on the graded operad $\mathcal{P}$:

$$d_{\mathcal{P}}(\gamma) = \gamma(d_{\mathcal{P} \circ \mathcal{P}}) = \gamma(d_{\mathcal{P}} \circ \mathrm{Id}_{\mathcal{P}}) + \gamma(\mathrm{Id}_{\mathcal{P}} \circ' d_{\mathcal{P}}),$$

where $\circ'$ is the infinitesimal composite introduced in Sect. 6.1.3. Observe that we have written the composition of morphisms as concatenation to avoid confusion with the two different meanings of $\circ$.

Applied to operations, it gives

$$\begin{aligned} d_{\mathcal{P}}\big(\gamma(\mu; \mu_1, \ldots, \mu_k)\big) := {} & \gamma\big(d_{\mathcal{P}}(\mu); \mu_1, \ldots, \mu_k\big) \\ & + \sum_{i=1}^{k} (-1)^{\varepsilon_i} \gamma\big(\mu; \mu_1, \ldots, d_{\mathcal{P}}(\mu_i), \ldots, \mu_k\big), \end{aligned}$$

where $\varepsilon_i = |\mu| + |\mu_1| + \cdots + |\mu_{i-1}|$. So a dg operad is a graded operad endowed with a square-zero derivation.

For example, let (A, d_A) be a chain complex. The differential of $\mathrm{End}_A(n)$ is given by

$$\partial_A(f) := [d_A, f] = d_A \circ f - (-1)^{|f|} f \circ d_{A^{\otimes n}} : A^{\otimes n} \to A,$$

where $d_{A^{\otimes n}}$ is, as usual, induced by d_A,

$$d_{A^{\otimes n}} := (d_A, \mathrm{id}, \ldots, \mathrm{id}) + \cdots + (\mathrm{id}, \ldots, \mathrm{id}, d_A, \mathrm{id}, \ldots, \mathrm{id}) + \cdots + (\mathrm{id}, \ldots, \mathrm{id}, d_A).$$

With this definition, $(\mathrm{End}_A, \partial_A)$ becomes a dg operad.

Proposition 6.3.1. *When the characteristic of the ground field* $\mathbb{K}$ *is* 0, *the underlying homology groups* $H(\mathcal{P})$ *of a dg operad* $\mathcal{P}$ *carry a natural operad structure.*

Proof. It is a direct application of the Künneth formula for the tensor product $\otimes$. □

An *augmented dg operad* is a dg operad $\mathcal{P}$ equipped with a morphism $\varepsilon : \mathcal{P} \to \mathrm{I}$ of dg operads (of degree 0), called the *augmentation morphism*. It sends $\mathrm{id} \in \mathcal{P}(1)$ to $\mathrm{id} \in \mathrm{I}(1)$.

The notion of infinitesimal composition product of Sect. 6.1 extends naturally to the differential graded framework.

6.3.2 Infinitesimal Tree Module

Like in Sect. 6.1, we need to linearize the tree module functor in order to do homological algebra with operads.

Let M and N be two $\mathbb{S}$-modules. The tree module on their direct sum $\mathscr{T}(M \oplus N)$ defines a functor from $(\mathbb{S}\text{-}\mathsf{Mod})^2$ to $\mathbb{S}$-Mod. We consider the sub-functor whose image is defined by the linear part of $\mathscr{T}(M \oplus N)$ in N and we denote it by $\overline{\mathscr{T}}(M; N)$. In fact, we will work with its augmented version

$$\mathscr{T}(M; N) := \mathrm{I} \oplus \overline{\mathscr{T}}(M; N) = \mathrm{I} \oplus N \oplus (M \circ_{(1)} N \oplus N \circ_{(1)} M) \oplus \cdots .$$

We call this construction the *infinitesimal tree module*. It is explicitly given by the sum of trees whose vertices are indexed by elements of M except exactly one which is indexed by an element of N.

Proposition 6.3.2. *For any graded $\mathbb{S}$-modules M, N and N', we have*

$$I \oplus \mathscr{T}\big(M; N \oplus N'\big) = \mathscr{T}(M; N) \oplus \mathscr{T}\big(M; N'\big).$$

Let E be an $\mathbb{S}$-module. The natural diagonal map, $\mathrm{diag} : E \to E \oplus E$, induces the following morphism

$$\Delta^E := \mathscr{T}(E) \xrightarrow{\mathscr{T}(\mathrm{diag})} \mathscr{T}(E \oplus E) \twoheadrightarrow \mathscr{T}(E; E).$$

This application singles out every vertex of a tree together with its indexing element of E.

6.3.3 Quasi-free Operad

We make explicit the notion of derivation on free operads.

Proposition 6.3.3. *Let E be a graded $\mathbb{S}$-module. Any derivation on the free operad $\mathscr{T}(E)$ is completely characterized by the image of the generators: $E \to \mathscr{T}(E)$. Explicitly, the unique derivation on $\mathscr{T}(E)$, which extends a morphism $\varphi : E \to \mathscr{T}(E)$, is given by the following composite*

$$d_\varphi = \mathscr{T}(E) \xrightarrow{\Delta^E} \mathscr{T}(E; E) \xrightarrow{\mathscr{T}(\mathrm{Id}_E; \varphi)} \mathscr{T}\big(E; \mathscr{T}(E)\big) \to \mathscr{T}\big(\mathscr{T}(E)\big) \to \mathscr{T}(E),$$

where the last map is the composition in the free operad, i.e. grafting of trees.

Proof. We prove by induction on n that the image under any derivation d of any element of $\mathscr{T}(E)^{(n)}$ is characterized by the restriction $d_{|E}: E \to \mathscr{T}(E)$. The property obviously holds for $n=1$. Suppose that it is true up to n. Any element of $\mathscr{T}(E)^{(n+1)}$ is the image under the composition map $\mathscr{T}(E) \circ \mathscr{T}(E) \to \mathscr{T}(E)$ of elements of $\mathscr{T}(E)^{(k)}$ with $k \leq n$. Hence the value of d on $\mathscr{T}(E)^{(n+1)}$ is well-determined.

Since the restriction of d_φ on E is equal to φ, it is enough to verify that d_φ is a derivation, which is left to the reader. Then one concludes by the uniqueness property. □

When representing an element of $\mathscr{T}(E)$ by a labeled tree, its image under d_φ is the sum of labeled trees, where φ has been applied once and only once to any vertex.

Such dg operads are called *quasi-free operads*. It means that the underlying graded operad, forgetting the differential, is free.

6.3.4 Minimal Model for Operads

Let $\mathscr{P}$ be a dg operad. A *model* for $\mathscr{P}$ is a dg operad $\mathscr{M}$ with a given epimorphism of dg operads $\mathscr{M} \twoheadrightarrow \mathscr{P}$ that induces an isomorphism in homology.

By definition, a *minimal operad* is a quasi-free operad $(\mathscr{T}(E), d)$

1. whose differential is *decomposable*, that is $d: E \to \mathscr{T}(E)^{(\geq 2)}$, and
2. such that the generating graded $\mathbb{S}$-module E admits a decomposition into

$$E = \bigoplus_{k \geq 1} E^{(k)}$$

satisfying

$$d\big(E^{(k+1)}\big) \subset \mathscr{T}\left(\bigoplus_{i=1}^{k} E^{(i)}\right).$$

A *minimal model* for the dg operad $\mathscr{P}$ is the data of a minimal operad $(\mathscr{T}(E), d)$ together with a quasi-isomorphism of dg operads

$$\big(T(E), d\big) \xrightarrow{\sim} \mathscr{P},$$

which is an epimorphism. Notice that this last condition is always satisfied when the differential of $\mathscr{P}$ is trivial.

Theorem 6.3.4 (Fundamental theorem of minimal models for operads [DCV11]). *When $\mathscr{P}$ admits a minimal model, it is unique up to a (non-unique) isomorphism.*

Proof. The proof follows the same steps as in the algebra case, see the proof of Theorem 1.5.2. Given two minimal models $\mathscr{M} = (\mathscr{T}(E), d) \xrightarrow{\sim} \mathscr{P}$ and $\mathscr{M}' = (\mathscr{T}(E'), d') \xrightarrow{\sim} \mathscr{P}$ of a common dg operad $\mathscr{P}$, one first proves the existence of a quasi-isomorphism $f : \mathscr{M} \xrightarrow{\sim} \mathscr{M}'$ of dg operads. Since the differentials d and d' are decomposable, this induces an isomorphism of graded modules $E \cong E'$. Therefore, the map f is an isomorphism of dg operads. □

As in the algebra case, the proof shows that the generating space E of minimal models is uniquely determined up to isomorphism. The automorphism groups of minimal models produce interesting groups to study. For instance, in the case of the Gerstenhaber operad, it is related to the Grothendieck–Teichmüller group, see [Kon99]. The Koszul duality theory for operads, developed in this book, induces minimal models with differential maps satisfying $d(E) \subset T(E)^{(2)}$. In this case, the differential is called *quadratic* and the minimal model is called a *quadratic model.*

The special case where $\mathscr{M}(0) = 0$ and $\mathscr{M}(1) = \mathbb{K}$ is treated in the literature by M. Markl in [Mar96b].

6.3.5 $\mathscr{P}$-Algebra in the Differential Graded Framework

In Sect. 5.2.3, we defined the notion of an algebra over an operad. We extend this definition to the dg framework here.

By definition, a *dg $\mathscr{P}$-algebra* is a chain complex $A = (A, d_A)$ endowed with a morphism of dg operads $f : \mathscr{P} \to \mathrm{End}_A$. When $(\mathscr{P}, d_{\mathscr{P}})$ is a dg operad, each $\mathscr{P}(n)$ is a chain complex whose differential is still denoted by $d_{\mathscr{P}}$.

Hence the differential maps are related by the formula

$$f\big(d_{\mathscr{P}}(\mu)\big) = \big[d_A, f(\mu)\big] := d_A \circ f(\mu) - (-1)^{|\mu|} f(\mu) \circ d_{A^{\otimes n}} : A^{\otimes n} \to A,$$

with $\mu \in \mathscr{P}(n)$.

The underlying homology $H(A, d_A)$ of a dg $\mathscr{P}$-algebra is called the *homotopy* of the dg $\mathscr{P}$-algebra A.

Proposition 6.3.5. *Let $\mathscr{P}$ be an operad. The homotopy $H(A)$ of a dg $\mathscr{P}$-algebra A carries a natural $\mathscr{P}$-algebra structure.*

Proof. There is a natural induced map $\mathscr{P} \to \mathrm{End}_{H(A)}$, which makes the homotopy $H(A)$ into a $\mathscr{P}$-algebra. □

When $\mathscr{P}$ is a dg operad and when the characteristic of the ground field $\mathbb{K}$ is 0, the homotopy $H(A)$ is an $H(\mathscr{P})$-algebra.

6.3.6 Derivation of a $\mathscr{P}$-Algebra

Since

$$\mathrm{Hom}_{\mathsf{dg\ S\text{-}Mod}}(\mathscr{P}, \mathrm{End}_A) \cong \mathrm{Hom}_{\mathsf{dg\ Mod}}\big(\mathscr{P}(A), A\big),$$

a structure of dg $\mathscr{P}$-algebra on A is equivalently given by a morphism of dg modules $\gamma_A : \mathscr{P}(A) \to A$, which satisfies the same commutative diagrams as in Sect. 5.2.3. Recall that the differential on $\mathscr{P}(A) = \mathscr{P} \circ A$ is equal to $d_{\mathscr{P}} \circ \mathrm{Id}_A + \mathrm{Id}_{\mathscr{P}} \circ' d_A$. The linear map $d_A : A \to A$ is called a *derivation of the $\mathscr{P}$-algebra A* if it makes the following diagram commutative:

$$\begin{array}{ccc} \mathscr{P} \circ A & \xrightarrow{\gamma_A} & A \\ {\scriptstyle d_{\mathscr{P}} \circ \mathrm{Id}_A + \mathrm{Id}_{\mathscr{P}} \circ' d_A} \downarrow & & \downarrow {\scriptstyle d_A} \\ \mathscr{P} \circ A & \xrightarrow{\gamma_A} & A. \end{array}$$

In other words, a dg $\mathscr{P}$-algebra is a $\mathscr{P}$-algebra with a square-zero derivation.

We denote the space of derivations on a $\mathscr{P}$-algebra A by $\mathrm{Der}(A)$. The composite dd' of two derivations does not give a derivation in general. But it is the case of the associated Lie bracket $[d, d'] := dd' - (-1)^{|d||d'|} d'd$, obtained by anti-symmetrization. In the particular case of the free $\mathscr{P}$-algebra over an n-dimensional vector space, we denote by

$$gl_n(\mathscr{P}) := \big(\mathrm{Der}\big(\mathscr{P}(\mathbb{K}x_1 \oplus \cdots \oplus \mathbb{K}x_n)\big), [\,,\,]\big)$$

the Lie algebra of derivations.

Proposition 6.3.6. *Any derivation d on a free $\mathscr{P}$-algebra $\mathscr{P}(V)$ is completely characterized by its restriction on the generators, $V \to \mathscr{P}(V)$:*

$$\mathrm{Der}\big(\mathscr{P}(V)\big) \cong \mathrm{Hom}\big(V, \mathscr{P}(V)\big).$$

Explicitly, given a map $\varphi : V \to \mathscr{P}(V)$, the unique derivation d_φ on the free $\mathscr{P}$-algebra $\mathscr{P}(V)$, which extends φ, is given by

$$d_\varphi = d_{\mathscr{P}} \circ \mathrm{Id}_V + (\gamma_{(1)} \circ \mathrm{Id}_V)(\mathrm{Id}_{\mathscr{P}} \circ' \varphi),$$

where the last term is equal to the following composite

$$\mathscr{P}(V) \xrightarrow{\mathrm{Id}_{\mathscr{P}} \circ' \varphi} \mathscr{P} \circ \big(V; \mathscr{P}(V)\big) \cong (\mathscr{P} \circ_{(1)} \mathscr{P})(V) \xrightarrow{\gamma_{(1)} \circ \mathrm{Id}_V} \mathscr{P}(V).$$

Proof. By the definition of a derivation, the following diagram commutes

$$\begin{array}{ccccc} \mathscr{P} \circ \mathrm{I}(V) & & \xrightarrow{\gamma \circ \mathrm{Id}_V} & & \mathscr{P}(V) \\ {\scriptstyle d_{\mathscr{P}} \circ \mathrm{Id}_V + \mathrm{Id}_{\mathscr{P}} \circ' d} \downarrow & & & & \downarrow {\scriptstyle d} \\ \mathscr{P}(V) \oplus \mathscr{P} \circ (V; \mathscr{P}(V)) & \longrightarrow & (\mathscr{P} \circ_{(1)} \mathscr{P})(V) & \xrightarrow{\gamma_{(1)} \circ \mathrm{Id}_V} & \mathscr{P}(V). \end{array}$$

Since the top map $\mathscr{P} \circ \mathrm{I}(V) \to \mathscr{P}(V)$ is an isomorphism, d is characterized by the image of the generators $\mathrm{I}(V) = V$. This diagram also shows that, if we denote by φ this restriction, then the total differential is equal to $d = d_{\mathscr{P}} \circ \mathrm{Id}_V + (\gamma_{(1)} \circ \mathrm{Id}_V)(\mathrm{Id}_{\mathscr{P}} \circ' \varphi)$. □

6.3.7 Differential Graded Cooperad

Dually, a *graded cooperad* $\mathscr{C}$ is a comonoid in the monoidal category (gr $\mathbb{S}$-Mod, $\circ$, I). It is defined by a coassociative *decomposition map* $\Delta : \mathscr{C} \to \mathscr{C} \circ \mathscr{C}$ and a *counit map* $\varepsilon : \mathscr{C} \to \mathrm{I}$, both of degree 0, on a graded $\mathbb{S}$-module $\mathscr{C}$. Consider a graded $\mathbb{S}$-module M concentrated in weight 1. This induces a natural grading on the cofree connected cooperad on M, see Sect. 5.8.6, which makes it into a graded cooperad.

A *differential graded cooperad* $(\mathscr{C}, \Delta, \varepsilon)$, *dg cooperad* for short, is a comonoid in the monoidal category (dg $\mathbb{S}$-Mod, $\circ$, I). The structure map Δ and ε commute with the respective differentials. It means, for instance, that the following diagram is commutative

$$\begin{array}{ccc} \mathscr{C} & \xrightarrow{\Delta} & \mathscr{C} \circ \mathscr{C} \\ \downarrow{\scriptstyle d_{\mathscr{C}}} & & \downarrow{\scriptstyle d_{\mathscr{C}\circ\mathscr{C}}} \\ \mathscr{C} & \xrightarrow{\Delta} & \mathscr{C} \circ \mathscr{C}. \end{array}$$

We define a *coderivation* $d_{\mathscr{C}}$ on a graded cooperad $\mathscr{C}$ to be a morphism of $\mathbb{S}$-modules $\mathscr{C} \to \mathscr{C}$ such that

$$\Delta(d_{\mathscr{C}}) = d_{\mathscr{C}\circ\mathscr{C}}(\Delta) = d_{\mathscr{C}} \circ \mathrm{Id}_{\mathscr{C}}(\Delta) + \mathrm{Id}_{\mathscr{C}} \circ' d_{\mathscr{C}}(\Delta).$$

Recall that for $(\mathscr{C}, \Delta)$ a cooperad, the image of an element $c \in \mathscr{C}$ under the decomposition map Δ is written

$$\Delta(c) = \sum (c; c_1, \ldots, c_k),$$

see Sect. 5.8.1. Under this notation, $d_{\mathscr{C}}$ is a coderivation if and only if

$$\Delta\big(d_{\mathscr{C}}(c)\big) = \sum \Bigg(\big(d_{\mathscr{C}}(c); c_1, \ldots, c_k\big) + \sum_{i=1}^{k} (-1)^{\nu_i} \big(c; c_1, \ldots, d_{\mathscr{C}}(c_i), \ldots, c_k\big) \Bigg),$$

where $\nu_i = |c| + |c_1| + \cdots + |c_{i-1}|$. The decomposition map $\Delta : \mathscr{C} \to \mathscr{C} \circ \mathscr{C}$ commutes with the differential $d_{\mathscr{C}}$ if and only if $d_{\mathscr{C}}$ is a coderivation of $\mathscr{C}$.

When a dg cooperad $\mathscr{C}$ is equipped with a *coaugmentation*, that is a morphism $\eta : \mathrm{I} \to \mathscr{C}$ of dg cooperads of degree 0, $\mathscr{C}$ is called a *coaugmented dg cooperad*. A coaugmented dg cooperad is conilpotent when its coradical filtration is exhaustive, as in Sect. 5.8.5.

The notion of infinitesimal decomposition coproduct of Sect. 6.1.4 extends naturally to the differential graded framework.

6.3.8 Quasi-cofree Cooperad

We make explicit coderivations on cofree cooperads as follows.

Let E be a graded $\mathbb{S}$-module. Recall from Sect. 5.8.8, that the $\mathbb{S}$-module $\mathcal{T}^c(\overline{\mathcal{T}}^c(E))$ is made up of trees of nontrivial trees, whose vertices are labeled by elements of E. The map

$$\Delta(E) : \mathcal{T}^c(E) \to \mathcal{T}^c\big(\overline{\mathcal{T}}^c(E)\big)$$

associates to a tree t labeled by E, the sum of all the partitioned trees coming from t.

Proposition 6.3.7. *Let E be a graded $\mathbb{S}$-module. Any coderivation on the cofree cooperad $\mathcal{T}^c(E)$ is completely characterized by its projection on the cogenerators* $\mathrm{proj}_E \circ d : \mathcal{T}^c(E) \to E$.

Explicitly, the unique coderivation on $\mathcal{T}^c(E)$ which extends a morphism $\varphi : \mathcal{T}^c(E) \to E$ is given by the following composite

$$d_\varphi = \mathcal{T}^c(E) \xrightarrow{\Delta(E)} \mathcal{T}^c\big(\overline{\mathcal{T}}^c(E)\big) \xrightarrow{\Delta_{\overline{\mathcal{T}}^c(E)}} \mathcal{T}^c\big(\overline{\mathcal{T}}^c(E); \overline{\mathcal{T}}^c(E)\big)$$
$$\xrightarrow{\mathcal{T}^c(\mathrm{proj}_E;\varphi)} \mathcal{T}^c(E; E) \to \mathcal{T}^c(E).$$

Proof. The proof of this proposition is dual to the proof of Proposition 6.3.3. □

With the tree representation of $\mathcal{T}^c(E)$, the coderivation d_φ consists in applying the map φ to any sub-tree of $\mathcal{T}^c(E)$. A square-zero coderivation is called a *codifferential*.

Any cooperad, whose underlying graded cooperad is cofree, is called a *quasi-cofree cooperad*.

6.3.9 $\mathscr{C}$-Coalgebra in the Differential Graded Framework

By definition, a *dg $\mathscr{C}$-coalgebra* is a chain complex $C = (C, d_C)$ equipped with a morphism $\Delta_C : C \to \widehat{\mathscr{C}}(C)$ of chain complexes which satisfies the same commutative diagrams as in Sect. 5.8.3. If the map Δ_C factors though $\mathscr{C} \to \mathscr{C}(C)$, then the dg $\mathscr{C}$-coalgebra is called *conilpotent*.

Any map d_C on C such that the following diagram commutes

$$\begin{array}{ccc} C & \xrightarrow{\Delta_C} & \widehat{\mathscr{C}}(C) = \mathscr{C}\hat{\circ}C \\ \downarrow{\scriptstyle d_C} & & \downarrow{\scriptstyle d_{\mathscr{C}} \circ \mathrm{Id}_C + \mathrm{Id}_{\mathscr{C}} \hat{\circ}' d_C} \\ C & \xrightarrow{\Delta_C} & \widehat{\mathscr{C}}(C) = \mathscr{C}\hat{\circ}C \end{array}$$

is called a *coderivation of the* $\mathscr{C}$*-coalgebra* C. A square-zero coderivation of a $\mathscr{C}$-coalgebra is called a *codifferential*. Hence a dg $\mathscr{C}$-coalgebra is a $\mathscr{C}$-coalgebra with a square-zero coderivation. We denote the set of coderivations on a $\mathscr{C}$-coalgebra C by $\mathrm{Coder}(C)$ and the set of codifferentials by $\mathrm{Codiff}(C)$.

Proposition 6.3.8. *Any coderivation d on a cofree $\mathscr{C}$-coalgebra $\mathscr{C}(V)$ is completely characterized by its projection onto the space of the cogenerators* $\mathrm{proj}_V \circ d : \mathscr{C}(V) \to V$.

$$\mathrm{Coder}\big(\mathscr{C}(V)\big) \cong \mathrm{Hom}\big(\mathscr{C}(V), V\big)$$

Explicitly, given a map $\varphi : \mathscr{C}(V) \to V$, the unique coderivation d_φ on the cofree $\mathscr{C}$-coalgebra $\mathscr{C}(V)$ which extends φ is given by

$$d_\varphi = d_{\mathscr{C}} \circ \mathrm{Id}_V + \big(\mathrm{Id}_{\mathscr{C}} \circ (\mathrm{Id}_V; \varphi)\big)(\Delta_{(1)} \circ \mathrm{Id}_V),$$

where the last term is equal to the following composite

$$\mathscr{C}(V) \xrightarrow{\Delta_{(1)} \circ \mathrm{Id}_V} (\mathscr{C} \circ_{(1)} \mathscr{C})(V) \cong \mathscr{C} \circ \big(V; \mathscr{C}(V)\big) \xrightarrow{\mathrm{Id}_{\mathscr{C}}(\mathrm{Id}_V;\varphi)} \mathscr{C} \circ (V; V) \to \mathscr{C}(V).$$

Proof. This proposition is the exact dual of Proposition 6.3.6. □

6.3.10 $\mathscr{P}$-Module in the Differential Graded Framework

In this section, we extend the notion of module over an operad of Sect. 5.2.2 to the dg framework. This notion will play an important role in the next section.

A *differential graded left (resp. right) module* over a dg operad $(\mathscr{P}, d_{\mathscr{P}})$ is a dg $\mathbb{S}$-module (M, d_M) equipped with a left action $\lambda : \mathscr{P} \circ M \to M$ (resp. right action $\rho : M \circ \mathscr{P} \to M$) which commutes with the respective differentials. In this case, the boundary map d_M is called a *derivation*. A dg left $\mathscr{P}$-module concentrated in arity 0 is nothing but a dg $\mathscr{P}$-algebra (see Sect. 6.3.5).

Let N be a dg $\mathbb{S}$-module. The free dg left (resp. right) $\mathscr{P}$-module on N is given by $(\mathscr{P} \circ N, d_{\mathscr{P}\circ N})$ (resp. $(N \circ \mathscr{P}, d_{N\circ\mathscr{P}})$). The following proposition extends Propositions 6.3.6 and 1.5.3. It shows that any derivation on a free $\mathscr{P}$-module is completely characterized by its restriction to the generators.

Proposition 6.3.9. *Let $(\mathscr{P}, d_{\mathscr{P}})$ be a dg operad and let N be a graded $\mathbb{S}$-module. There is a one-to-one correspondence between derivations on the dg free $\mathscr{P}$-modules $\mathscr{P} \circ N$ or $N \circ \mathscr{P}$ and their restriction to the space of generators N. More precisely,*

- ◇ *the unique derivation $d_\varphi : \mathscr{P} \circ N \to \mathscr{P} \circ N$ which extends $\varphi : N \to \mathscr{P} \circ N$ is given by*

$$d_\varphi = d_{\mathscr{P}} \circ \mathrm{Id}_N + (\gamma_{(1)} \circ \mathrm{Id}_N)(\mathrm{Id}_{\mathscr{P}} \circ' \varphi);$$

◇ *the unique derivation* $d_\varphi : N \circ \mathcal{P} \to N \circ \mathcal{P}$ *which extends* $\varphi : N \to N \circ \mathcal{P}$ *is given by*

$$d_\varphi = \mathrm{Id}_N \circ' d_{\mathcal{P}} + (\mathrm{Id}_N \circ \gamma)(\varphi \circ \mathrm{Id}_{\mathcal{P}}).$$

Proof. The proof is similar to the ones of Propositions 6.3.6 and 1.5.3. □

6.3.11 Weight Grading

In the next section, we will need an extra grading to prove the main theorem 6.6.1. We will require that the underlying $\mathbb{S}$-module of an operad, of a cooperad or of a module, has an extra grading, which we call *weight grading* to avoid confusion with the homological degree. This means that every dg $\mathbb{S}$-module is a direct sum of sub-dg $\mathbb{S}$-modules indexed by this weight. For instance, a *weight-graded dg operad*, *wgd operad* for short, is an operad structure on a weight-graded dg $\mathbb{S}$-module $\mathcal{P}$. Its product is supposed to preserve the weight grading. We denote by $\mathcal{P}_p^{(\omega)}$ the sub-module of degree p and weight ω of $\mathcal{P}$. Similarly, there is the notion of *weight-graded dg cooperad* $\mathcal{C}$, *wgd cooperad* for short, which is a cooperad structure on a weight graded dg $\mathbb{S}$-module $\mathcal{C}$. In this context, morphisms are supposed to respect the weight grading.

In this book, the weight grading and the homological degree are nonnegative gradings. A wdg operad is called *connected* if it decomposes as

$$\mathcal{P} := \mathbb{K}\mathrm{id} \oplus \mathcal{P}^{(1)} \oplus \cdots \oplus \mathcal{P}^{(\omega)} \oplus \cdots$$

with $\mathcal{P}^{(0)} = \mathbb{K}\mathrm{id}$ concentrated in degree 0. The same definition holds for wdg cooperad if it satisfies the same decomposition. A connected wdg cooperad is conilpotent, cf. Sect. 5.8.5.

6.4 Operadic Twisting Morphism

In this section, we define the notion of twisting morphism from a dg cooperad $\mathcal{C}$ to a dg operad $\mathcal{P}$. It is a solution of the Maurer–Cartan equation in the convolution dg (pre-)Lie algebra $\mathrm{Hom}_{\mathbb{S}}(\mathcal{C}, \mathcal{P})$. Then, we show how to twist the differential of the composite products $\mathcal{C} \circ \mathcal{P}$, $\mathcal{P} \circ \mathcal{C}$, and $\mathcal{P} \circ \mathcal{C} \circ \mathcal{P}$ with such a twisting morphism to obtain an operadic twisted complex.

6.4.1 Convolution Operad

Let $(\mathcal{C}, \Delta, \varepsilon)$ be a cooperad and $(\mathcal{P}, \gamma, \eta)$ be an operad. We consider the graded module

$$\mathrm{Hom}(\mathcal{C}, \mathcal{P}) := \{\mathrm{Hom}_{\mathbb{K}}(\mathcal{C}(n), \mathcal{P}(n))\}_{n \geq 0}.$$

It becomes a right $\mathbb{S}$-module under the action by conjugation:

$$(f^{\sigma})(x) := \left(f\left(x^{\sigma^{-1}}\right)\right)^{\sigma},$$

for $f \in \mathrm{Hom}(\mathscr{C}(n), \mathscr{P}(n))$, $\sigma \in \mathbb{S}_n$ and $x \in \mathscr{C}(n)$.

Proposition 6.4.1. *The* $\mathbb{S}$-*module* $\mathrm{Hom}(\mathscr{C}, \mathscr{P})$ *is an operad.*

Proof. The proof is a generalization of the arguments of Sect. 1.6.1. Let $f \in \mathrm{Hom}(\mathscr{C}(k), \mathscr{P}(k))$ and let $g_i \in \mathrm{Hom}(\mathscr{C}(i_l), \mathscr{P}(i_l))$ for $1 \le l \le k$. As usual $n = i_1 + \cdots + i_k$. We define the composition map $\gamma(f; g_1, \ldots, g_k)$ by the formula

$$\begin{aligned}
&\mathscr{C}(n) \xrightarrow{\Delta_{\mathscr{C}}} (\mathscr{C} \circ \mathscr{C})(n) \twoheadrightarrow \mathscr{C}(k) \otimes \mathscr{C}(i_1) \otimes \cdots \otimes \mathscr{C}(i_k) \otimes \mathbb{K}[\mathbb{S}_n] \\
&\xrightarrow{f \otimes g_1 \otimes \cdots \otimes g_k \otimes \mathrm{Id}} \mathscr{P}(k) \otimes \mathscr{P}(i_1) \otimes \cdots \otimes \mathscr{P}(i_k) \otimes \mathbb{K}[\mathbb{S}_n] \to (\mathscr{P} \circ \mathscr{P})(n) \\
&\xrightarrow{\gamma_{\mathscr{P}}} \mathscr{P}(n).
\end{aligned}$$

□

The operad $\mathrm{Hom}(\mathscr{C}, \mathscr{P})$ is called the *convolution operad* and is due to C. Berger and I. Moerdijk [BM03a]. This construction generalizes to operads the convolution algebra $\mathrm{Hom}(C, A)$ of Sect. 1.6.1 from a coalgebra C and an algebra A.

We suppose now that $(\mathscr{P}, d_{\mathscr{P}})$ is a dg operad and that $(\mathscr{C}, d_{\mathscr{C}})$ is a dg cooperad, For a homogeneous morphism $f : \mathscr{C} \to \mathscr{P}$ of dg $\mathbb{S}$-modules of degree $|f|$, we define its *derivative* $\partial(f)$ by the classical formula

$$\partial(f) = [d, f] := d_{\mathscr{P}} \circ f - (-1)^{|f|} f \circ d_{\mathscr{C}}.$$

Proposition 6.4.2. *The convolution operad* $(\mathrm{Hom}(\mathscr{C}, \mathscr{P}), \partial)$ *is a dg operad.*

Proof. The map ∂ squares to zero, $\partial^2 = 0$, and it is a derivation with respect to the operadic composition on $\mathrm{Hom}(\mathscr{C}, \mathscr{P})$. The proof is a straightforward generalization of Proposition 2.1.1. □

6.4.2 Operadic Convolution DG Lie Algebra

Recall from Sect. 5.4.3 the composite of functors

$$\mathsf{Op} \to \mathit{preLie}\text{-}\mathsf{alg} \to \mathit{Lie}\text{-}\mathsf{alg},$$

where the underlying objects are $\mathscr{P} \mapsto \prod_{n \ge 0} \mathscr{P}(n)$. It extends to the differential graded framework

$$\mathsf{dg\ Op} \to \mathsf{dg}\ \mathit{preLie}\text{-}\mathsf{alg} \to \mathsf{dg}\ \mathit{Lie}\text{-}\mathsf{alg}.$$

In the case of convolution operads, the pre-Lie product can be made explicit as follows.

Proposition 6.4.3. *The pre-Lie product $f \star g$ of two elements f and g in $\prod_{n\geq 0} \mathrm{Hom}(\mathscr{C}, \mathscr{P})(n)$ is equal to the following composite*

$$f \star g = \mathscr{C} \xrightarrow{\Delta_{(1)}} \mathscr{C} \circ_{(1)} \mathscr{C} \xrightarrow{f \circ_{(1)} g} \mathscr{P} \circ_{(1)} \mathscr{P} \xrightarrow{\gamma_{(1)}} \mathscr{P}.$$

Proof. The proof is a straightforward application of the definition of the composition map in the convolution operad, Proposition 6.4.1, with the definition of the pre-Lie product associated to an operad, Proposition 5.4.3. □

The space of invariant elements of $\mathrm{Hom}(\mathscr{C}(n), \mathscr{P}(n))$ under the conjugation action of the symmetric group $\mathbb{S}_n$ is equal to the subspace $\mathrm{Hom}_{\mathbb{S}_n}(\mathscr{C}(n), \mathscr{P}(n))$ of $\mathbb{S}_n$-equivariant morphisms from $\mathscr{C}(n)$ to $\mathscr{P}(n)$. We denote by

$$\mathrm{Hom}_{\mathbb{S}}(\mathscr{C}, \mathscr{P}) := \prod_{n\geq 0} \mathrm{Hom}_{\mathbb{S}_n}\big(\mathscr{C}(n), \mathscr{P}(n)\big)$$

the associated product of $\mathbb{S}$-equivariant maps.

Lemma 6.4.4. *The space* $\mathrm{Hom}_{\mathbb{S}}(\mathscr{C}, \mathscr{P})$ *is stable under the pre-Lie product associated to the convolution operad* $\mathrm{Hom}(\mathscr{C}, \mathscr{P})$.

Proof. The pre-Lie product was proved to be equal to the composite of $\mathbb{S}$-equivariant maps in the previous proposition. □

To sum up, we have proved the following result.

Proposition 6.4.5. *Let $\mathscr{P}$ be a dg operad and $\mathscr{C}$ be a dg cooperad. The product space of $\mathbb{S}$-equivariant maps*

$$\big(\mathrm{Hom}_{\mathbb{S}}(\mathscr{C}, \mathscr{P}), \star, \partial\big)$$

is a dg pre-Lie algebra. The associated Lie bracket induces a dg Lie algebra structure

$$\big(\mathrm{Hom}_{\mathbb{S}}(\mathscr{C}, \mathscr{P}), [\ ,\ ], \partial\big).$$

The latter dg Lie algebra associated to the convolution operad is called the *convolution dg Lie algebra.*

6.4.3 Maurer–Cartan Equation, Operadic Twisting Morphism

As usual, the Maurer–Cartan equation in the dg pre-Lie algebra $\mathrm{Hom}_{\mathbb{S}}(\mathscr{C}, \mathscr{P})$ reads

$$\partial(\alpha) + \alpha \star \alpha = 0.$$

A solution $\alpha : \mathscr{C} \to \mathscr{P}$ of degree -1 of this equation is called an *operadic twisting morphism*. We denote by $\mathrm{Tw}(\mathscr{C}, \mathscr{P})$ the space of twisting morphisms for $\mathscr{C}$ to $\mathscr{P}$. We will simply say twisting morphism if no confusion can arise. When $\mathscr{C}$ is a coaugmented dg cooperad, we require that the composition of an operadic twisting morphism with the coaugmentation map vanishes: $\alpha\eta = 0$. Respectively, when $\mathscr{P}$ is an augmented dg operad, we require that the composition of an operadic twisting morphism with the augmentation map vanishes: $\varepsilon\alpha = 0$.

Since 2 is invertible in the ground ring, we have $\alpha \star \alpha = \frac{1}{2}[\alpha, \alpha]$, this equation is equivalent to the classical Maurer–Cartan equation $\partial(\alpha) + \frac{1}{2}[\alpha, \alpha] = 0$ in the associated dg Lie algebra. When $\mathscr{C}$ and $\mathscr{P}$ are concentrated in arity 1, we recover the notion of twisting morphism between dga algebras and dga coalgebras of Sect. 2.1.

6.4.4 Twisted Structure on the Hom Space

Recall from Lemma 2.1.2 that, in any dg Lie algebra, every solution α of the Maurer–Cartan equation gives rise to a twisted differential $\partial_\alpha(f) := \partial(f) + [\alpha, f]$ on $\mathrm{Hom}_{\mathbb{S}}(\mathscr{C}, \mathscr{P})$. We denote this twisted chain complex by $\mathrm{Hom}^{\alpha}_{\mathbb{S}}(\mathscr{C}, \mathscr{P}) := (\mathrm{Hom}_{\mathbb{S}}(\mathscr{C}, \mathscr{P}), \partial_\alpha)$.

Proposition 6.4.6. *The triple* $(\mathrm{Hom}^{\alpha}_{\mathbb{S}}(\mathscr{C}, \mathscr{P}), [\ ,\ ], \partial_\alpha)$ *is a dg Lie algebra.*

Proof. It is a direct corollary of Lemma 2.1.2. □

We call this new dg Lie algebra the *twisted convolution Lie algebra*.

6.4.5 Twisted Composite Products

In this section, we study dg free $\mathscr{P}$-modules over $\mathscr{C}$.

In Proposition 6.3.9, we showed that the derivations on $\mathscr{P} \circ \mathscr{C}$ and on $\mathscr{C} \circ \mathscr{P}$ are characterized by their restrictions to $\mathscr{C}$. To any map $\alpha : \mathscr{C} \to \mathscr{P}$, of degree -1, we associate the following derivations.

On $\mathscr{C} \circ \mathscr{P}$: we consider the unique derivation which extends

$$\mathscr{C} \xrightarrow{\Delta_{(1)}} \mathscr{C} \circ_{(1)} \mathscr{C} \xrightarrow{\mathrm{Id}_{\mathscr{C}} \circ_{(1)} \alpha} \mathscr{C} \circ_{(1)} \mathscr{P} \to \mathscr{C} \circ \mathscr{P}.$$

It is explicitly given by $d^r_\alpha = (\mathrm{Id}_C \circ \gamma)([(\mathrm{Id}_{\mathscr{C}} \circ_{(1)} \alpha)\Delta_{(1)}] \circ \mathrm{Id}_{\mathscr{P}})$, that is

$$\begin{aligned} d^r_\alpha : \mathscr{C} \circ \mathscr{P} &\xrightarrow{\Delta_{(1)} \circ \mathrm{Id}_{\mathscr{P}}} (\mathscr{C} \circ_{(1)} \mathscr{C}) \circ \mathscr{P} \xrightarrow{(\mathrm{Id}_{\mathscr{C}} \circ_{(1)} \alpha) \circ \mathrm{Id}_{\mathscr{P}}} (\mathscr{C} \circ_{(1)} \mathscr{P}) \circ \mathscr{P} \\ &\cong \mathscr{C} \circ (\mathscr{P}; \mathscr{P} \circ \mathscr{P}) \xrightarrow{\mathrm{Id}_{\mathscr{C}} \circ (\mathrm{Id}_{\mathscr{P}}; \gamma)} \mathscr{C} \circ (\mathscr{P}; \mathscr{P}) \cong \mathscr{C} \circ \mathscr{P}. \end{aligned}$$

Using the tree representation, without the top leaves, it has the following form

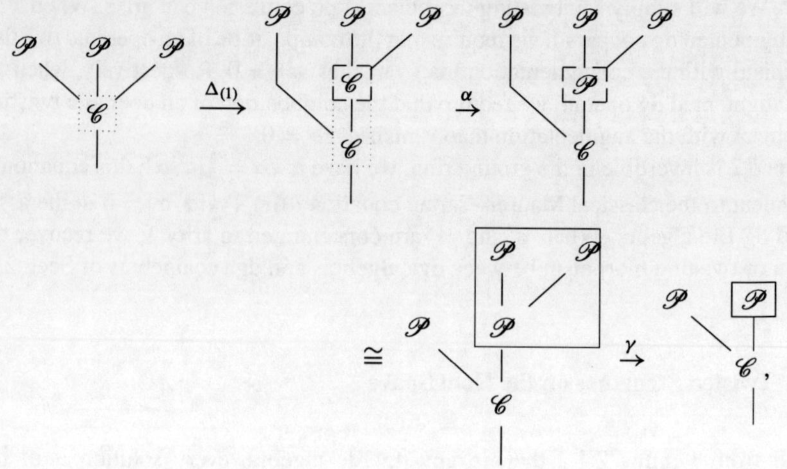

where the dotted box stands for $\Delta_{(1)}$, the dashed box stands for α and the final box stands for γ.

We consider the full derivation

$$d_\alpha = d_{\mathscr{C}} \circ \mathrm{Id}_{\mathscr{P}} + \mathrm{Id}_{\mathscr{C}} \circ' d_{\mathscr{P}} + d_\alpha^r,$$

where d^r is the twisting term.

On $\mathscr{P} \circ \mathscr{C}$: we consider the unique derivation which extends

$$\mathscr{C} \xrightarrow{\Delta} \mathscr{C} \circ \mathscr{C} \xrightarrow{\alpha \circ \mathrm{Id}_{\mathscr{C}}} \mathscr{P} \circ \mathscr{C}.$$

It is explicitly given by $d_\alpha^l = (\gamma_{(1)} \circ \mathrm{Id}_{\mathscr{C}})(\mathrm{Id}_{\mathscr{P}} \circ' [(\alpha \circ \mathrm{Id}_{\mathscr{C}})\Delta])$. We consider the full derivation

$$d_\alpha = d_{\mathscr{P}} \circ \mathrm{Id}_{\mathscr{C}} + \mathrm{Id}_{\mathscr{P}} \circ' d_{\mathscr{C}} + d_\alpha^l,$$

where d_α^l is the twisting term.

Lemma 6.4.7. *On $\mathscr{P} \circ \mathscr{C}$, the derivation d_α satisfies*

$${d_\alpha}^2 = d^l_{\partial(\alpha)+\alpha\star\alpha}$$

and on $\mathscr{C} \circ \mathscr{P}$, the derivation d_α satisfies

$${d_\alpha}^2 = d^r_{\partial(\alpha)+\alpha\star\alpha}.$$

In both cases, $\alpha \in \mathrm{Tw}(\mathscr{C}, \mathscr{P})$ if and only if ${d_\alpha}^2 = 0$.

Proof. The proof is based on the relations $[d^l_\alpha, d^l_\beta] = d^l_{[\alpha,\beta]}$ and $[d^r_\alpha, d^r_\beta] = d^r_{[\alpha,\beta]}$ and is similar to the proof of Lemma 2.1.4. □

When $\alpha : \mathscr{C} \to \mathscr{P}$ is a twisting morphism, the map d_α is a differential defining the chain complex

$$\mathscr{P} \circ_\alpha \mathscr{C} := \left(\mathscr{P} \circ \mathscr{C}, d_\alpha = d_{\mathscr{P}\circ\mathscr{C}} + d^l_\alpha\right),$$

which is called the *left twisted composite product.* In the other way round, the chain complex

$$\mathscr{C} \circ_\alpha \mathscr{P} := \left(\mathscr{C} \circ \mathscr{P}, d_\alpha = d_{\mathscr{C}\circ\mathscr{P}} + d^r_\alpha\right)$$

is called the *right twisted composite product.* Since the context is obvious, we denote the two differentials by the same symbol. It is a direct generalization of the twisted tensor product of Sect. 2.1.4. As in the case of dga (co)algebras, these two constructions are not isomorphic. In the operadic case, the underlying $\mathbb{S}$-modules $\mathscr{P} \circ \mathscr{C}$ and $\mathscr{C} \circ \mathscr{P}$ are not even isomorphic since the composite product is not symmetric.

These constructions are functorial as follows. We consider $\mathscr{C} \circ_\alpha \mathscr{P}$ and $\mathscr{C}' \circ_{\alpha'} \mathscr{P}'$ two twisted composite products. Let $g : \mathscr{P} \to \mathscr{P}'$ be a morphism of dg operads and let $f : \mathscr{C} \to \mathscr{C}'$ be a morphism of dg cooperads. These morphisms are *compatible* with the twisting morphisms if $\alpha' \circ f = g \circ \alpha$:

$$\begin{array}{ccc} \mathscr{C} & \xrightarrow{f} & \mathscr{C}' \\ \downarrow{\scriptstyle \alpha} & & \downarrow{\scriptstyle \alpha'} \\ \mathscr{P} & \xrightarrow{g} & \mathscr{P}'. \end{array}$$

We leave it to the reader to show that, in this case, $f \circ g : \mathscr{C} \circ_\alpha \mathscr{P} \to \mathscr{C}' \circ_{\alpha'} \mathscr{P}'$ induces a morphism of chain complexes.

When $\mathscr{C}$ and $\mathscr{P}$ are weight graded, we ask that the twisting morphisms α preserve the weight. In this case, if two among these three morphisms are quasi-isomorphisms, then so is the third one. This result comes essentially from Fresse [Fre04].

Lemma 6.4.8 (Comparison Lemma for twisted composite products). *Let $g : \mathscr{P} \to \mathscr{P}'$ be a morphism of wdg connected operads and $f : \mathscr{C} \to \mathscr{C}'$ be a morphism of wdg connected cooperads. Let $\alpha : \mathscr{C} \to \mathscr{P}$ and $\alpha' : \mathscr{C}' \to \mathscr{P}'$ be two twisting morphisms, such that f and g are compatible with α and α'.*

Right: *If two morphisms among f, g and $f \circ g : \mathscr{C} \circ_\alpha \mathscr{P} \to \mathscr{C}' \circ_{\alpha'} \mathscr{P}'$ are quasi-isomorphisms, then so is the third one.*

Left: *If two morphisms among f, g and $g \circ f : \mathscr{P} \circ_\alpha \mathscr{C} \to \mathscr{P}' \circ_{\alpha'} \mathscr{C}'$ are quasi-isomorphisms, then so is the third one.*

Proof. We prove this theorem later in Sect. 6.7. □

6.5 Operadic Bar and Cobar Construction

Are the functors $\mathrm{Tw}(\mathscr{C}, -)$ and $\mathrm{Tw}(-, \mathscr{P})$ representable ? In this section, we define the bar and the cobar constructions, which give a positive answer to this question. A direct corollary proves that they form a pair of adjoint functors. The constructions are the generalization to the operadic case of the constructions given in Sect. 2.2 in the algebra context. This section comes from [GJ94, GK94].

6.5.1 Bar Construction

In this section, we define a functor

$$\mathrm{B} : \{\text{aug. dg operads}\} \longrightarrow \{\text{conil. dg cooperads}\}$$

from the category of augmented dg operads to the category of conilpotent dg cooperads called the *bar construction*.

Let $\mathscr{P} := (\mathscr{P}, \gamma, \eta, \varepsilon)$ be an augmented operad. Consider its augmentation ideal, that is $\overline{\mathscr{P}} := \ker(\varepsilon : \mathscr{P} \to I)$. The $\mathbb{S}$-module $\mathscr{P}$ is naturally isomorphic to $\mathscr{P} = I \oplus \overline{\mathscr{P}}$. The bar construction $\mathrm{B}\mathscr{P}$ of $\mathscr{P}$ is a dg cooperad, whose underlying space is the cofree cooperad $\mathscr{T}^c(s\overline{\mathscr{P}})$ on the suspension of $\overline{\mathscr{P}}$.

We consider the map $\gamma_s : \mathbb{K}s \otimes \mathbb{K}s \to \mathbb{K}s$ of degree -1 defined by $\gamma_s(s \otimes s) := s$. The infinitesimal composition map, that is the restriction of the composition map of $\mathscr{P}$ on two operations, see Sect. 6.1.2, induces the following map

$$\begin{aligned} d_2 : \mathscr{T}^c(s\overline{\mathscr{P}}) \twoheadrightarrow \mathscr{T}^c(s\overline{\mathscr{P}})^{(2)} &\cong (\mathbb{K}s \otimes \overline{\mathscr{P}}) \circ_{(1)} (\mathbb{K}s \otimes \overline{\mathscr{P}}) \\ &\xrightarrow{\mathrm{Id} \otimes \tau \otimes \mathrm{Id}} (\mathbb{K}s \otimes \mathbb{K}s) \otimes (\overline{\mathscr{P}} \circ_{(1)} \overline{\mathscr{P}}) \\ &\xrightarrow{\gamma_s \otimes \gamma_{(1)}} \mathbb{K}s \otimes \overline{\mathscr{P}}. \end{aligned}$$

Since $\mathscr{T}^c(s\overline{\mathscr{P}})$ is a cofree cooperad, there exists a unique coderivation $\mathscr{T}^c \times (s\overline{\mathscr{P}}) \to \mathscr{T}^c(s\overline{\mathscr{P}})$ which extends d_2 by Proposition 6.3.7. By a slight abuse of notation, we still denote it by d_2.

Proposition 6.5.1. *The map d_2 is a differential*: ${d_2}^2 = 0$.

Proof. It is a direct consequence of the relation satisfied by the infinitesimal composition map $\gamma_{(1)}$ and the Koszul sign rules. □

The chain complex $\mathrm{B}\mathscr{P} := (\mathscr{T}^c(s\overline{\mathscr{P}}), d_2)$ is a conilpotent dg cooperad, called the *bar construction* of the augmented operad $\mathscr{P}$. It is a functor from the category of augmented operads to the category of coaugmented dg cooperads.

We extend this definition to dg operads as follows. Let $\mathscr{P} = (\mathscr{P}, d_{\mathscr{P}})$ be a dg operad. The differential $d_{\mathscr{P}}$ on $\mathscr{P}$ induces an internal differential d_1 on $\mathscr{T}^c(s\overline{\mathscr{P}})$, that is ${d_1}^2 = 0$. Since $(\mathscr{P}, d_{\mathscr{P}})$ is a dg operad, the infinitesimal composition map

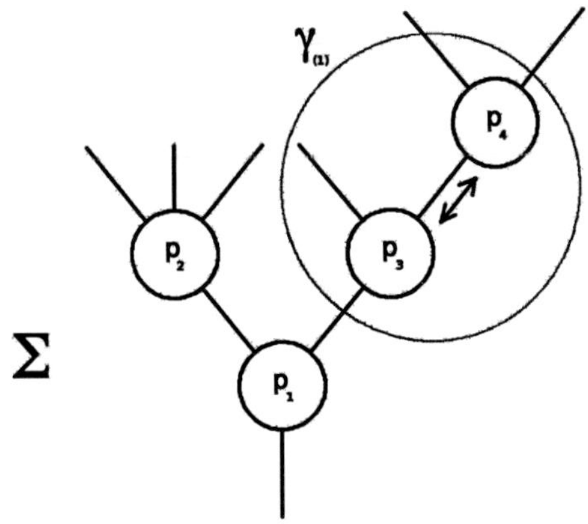

Fig. 6.1 The boundary map d_2 of the operadic bar construction

commutes with the differential $d_{\mathscr{P}}$. The differentials d_1 and d_2 anti-commute: $d_1 \circ d_2 + d_2 \circ d_1 = 0$. The total complex of this bicomplex is called the *bar construction*:

$$\mathrm{B}\mathscr{P} := \left(\mathscr{T}^c(s\overline{\mathscr{P}}), d = d_1 + d_2\right)$$

of the augmented dg operad $(\mathscr{P}, d_{\mathscr{P}})$.

Proposition 6.5.2. *Under the tree representation of the cofree cooperad, the boundary map d_2 is the sum of the contractions of the internal edges with the composition of the two elements indexing the two extremal vertices of each edge, see Fig.* 6.1.

Proof. Since the differential d_2 is the unique coderivation extending the (suspended) infinitesimal product of $\overline{\mathscr{P}}$, the proposition is a direct corollary of Proposition 6.3.7. □

Such a chain complex is an example of *graph complex* à la Kontsevich [Kon03], see Appendix C.4.2.

Proposition 6.5.3. *When the characteristic of the ground field* $\mathbb{K}$ *is* 0, *the operadic bar construction preserves quasi-isomorphisms.*

Proof. We use the same kind of filtration as in the proof of Proposition 2.2.3, but based on the number of vertices of trees this time. The characteristic 0 assumption ensures that we can apply the Künneth formula. □

6.5.2 Cobar Construction

Dually we construct a functor

$$\Omega : \{\text{coaug. dg cooperads}\} \longrightarrow \{\text{aug. dg operads}\}$$

from the category of coaugmented dg cooperads to the category of augmented dg operads called the *cobar construction.*

Let $\mathscr{C} := (\mathscr{C}, \Delta, \varepsilon, \eta)$ be a cooperad. Recall that the coaugmentation coideal of $\mathscr{C}$ is $\overline{\mathscr{C}} := \mathrm{coker}(\eta : \mathrm{I} \to \mathscr{C})$. In this case, $\mathscr{C}$ splits naturally as $\mathscr{C} \cong I \oplus \overline{\mathscr{C}}$. The cobar construction $\Omega\mathscr{C}$ of $\mathscr{C}$ is an augmented dg operad defined on the free operad $\mathscr{T}(s^{-1}\overline{\mathscr{C}})$ over the desuspension of $\overline{\mathscr{C}}$.

Consider $\mathbb{K}s^{-1}$ equipped with the diagonal map $\Delta_s : \mathbb{K}s^{-1} \to \mathbb{K}s^{-1} \otimes \mathbb{K}s^{-1}$ defined by the formula $\Delta_s(s^{-1}) := -s^{-1} \otimes s^{-1}$ of degree -1. The infinitesimal decomposition map $\Delta_{(1)}$ of $\mathscr{C}$, defined in Sect. 6.1.4 by the projection of Δ on $\mathscr{C} \circ_{(1)} \mathscr{C}$, induces a map d_2 on $s^{-1}\overline{\mathscr{C}} = \mathbb{K}s^{-1} \otimes \overline{\mathscr{C}}$ as follows

$$
\begin{aligned}
d_2 : \mathbb{K}s^{-1} \otimes \overline{\mathscr{C}} &\xrightarrow{\Delta_s \otimes \Delta_{(1)}} \left(\mathbb{K}s^{-1} \otimes \mathbb{K}s^{-1}\right) \otimes (\mathscr{C} \circ_{(1)} \mathscr{C}) \\
&\xrightarrow{\mathrm{Id} \otimes \tau \otimes \mathrm{Id}} \left(\mathbb{K}s^{-1} \otimes \overline{\mathscr{C}}\right) \circ_{(1)} \left(\mathbb{K}s^{-1} \otimes \overline{\mathscr{C}}\right) \\
&\cong \mathscr{T}\left(s^{-1}\overline{\mathscr{C}}\right)^{(2)} \rightarrowtail \mathscr{T}\left(s^{-1}\overline{\mathscr{C}}\right).
\end{aligned}
$$

Since we work with the free operad $\mathscr{T}(s^{-1}\overline{\mathscr{C}})$ over the desuspension $s^{-1}\overline{\mathscr{C}}$, the map $d_2 : s^{-1}\overline{\mathscr{C}} \to \mathscr{T}(s^{-1}\overline{\mathscr{C}})$ extends to a unique derivation on $\mathscr{T}(s^{-1}\overline{\mathscr{C}})$ by Proposition 6.3.3. We still denote it by $d_2 : \mathscr{T}(s^{-1}\overline{\mathscr{C}}) \to \mathscr{T}(s^{-1}\overline{\mathscr{C}})$.

Proposition 6.5.4. *The map d_2 is a differential*: $d_2{}^2 = 0$.

Proof. The infinitesimal decomposition map $\Delta_{(1)}$ shares dual relations with the infinitesimal composition map of an operad. These relations and Koszul sign rules give the result. □

Finally, $\Omega\mathscr{C} := (\mathscr{T}(s^{-1}\overline{\mathscr{C}}), d_2)$ is a dg operad, called the *cobar construction* of the cooperad $\mathscr{C}$. It is a functor from the category of coaugmented cooperads to the category of augmented dg operads.

One easily extends this functor to coaugmented dg cooperads $\mathscr{C} = (\mathscr{C}, d_{\mathscr{C}})$ by adding to d_2 the internal differential d_1 induced by the differential $d_{\mathscr{C}}$. Since the infinitesimal decomposition map $\Delta_{(1)}$ of $\mathscr{C}$ is a morphism of chain complexes, the differentials d_2 and d_1 anticommute and one has a well-defined bicomplex. The total complex of this bicomplex is called the *cobar construction* of the dg cooperad

$$
\Omega\mathscr{C} := \left(\mathscr{T}\left(s^{-1}\overline{\mathscr{C}}\right), d := d_1 + d_2\right).
$$

Proposition 6.5.5. *The elements of the free operad $\mathscr{T}(s^{-1}\overline{\mathscr{C}})$ can be represented by trees with vertices indexed by elements of $s^{-1}\overline{\mathscr{C}}$. The image of such a tree under the boundary map d_2, is the sum over the vertices v of the (desuspended) infinitesimal decomposition map of $\overline{\mathscr{C}}$ applied to the element indexing the vertex v, see Fig.* 6.2.

Proof. It is a direct corollary of the description of the free derivation on a free operad of Proposition 6.3.3. □

Once again, the cobar construction of a cooperad is a graph homology à la Kontsevich. The boundary map consists in expanding each vertex into two.

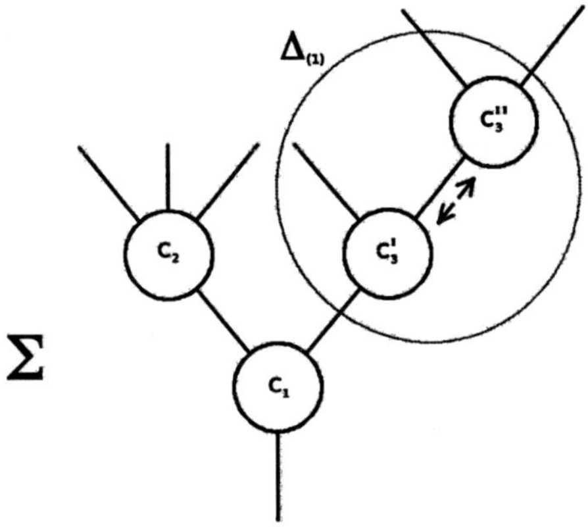

Fig. 6.2 The boundary map d_2 of the operadic cobar construction

A nonnegatively graded dg cooperad $\mathscr{C}$ is called 2-*connected* if $\mathscr{C}_0 = \mathrm{I}$ and $\mathscr{C}_1 = 0$.

Proposition 6.5.6. *The operadic cobar construction preserves quasi-isomorphisms between* 2*-connected dg cooperads.*

Proof. We use the same kind of filtration as in the proof of Proposition 2.2.5, but based on the number of vertices of trees this time. □

This result does not hold when the dg cooperads are not 2-connected as Proposition 2.4.3 shows.

6.5.3 Bar–Cobar Adjunction

We show that the operadic bar and cobar constructions form a pair of adjoint functors

$$\Omega : \{\text{aug. dg cooperads}\} \rightleftharpoons \{\text{coaug. dg operads}\} : \mathrm{B}.$$

More precisely, this adjunction is given by the set of twisting morphisms.

Theorem 6.5.7. *For every augmented dg operad* $\mathscr{P}$ *and every conilpotent dg cooperad* $\mathscr{C}$*, there exist natural isomorphisms*

$$\mathrm{Hom}_{\mathsf{dg\ Op}}(\Omega\mathscr{C}, \mathscr{P}) \cong \mathrm{Tw}(\mathscr{C}, \mathscr{P}) \cong \mathrm{Hom}_{\mathsf{dg\ coOp}}(\mathscr{C}, \mathrm{B}\mathscr{P}).$$

Proof. We make explicit the first isomorphism, the second one being dual. If we forget the differentials, then $\Omega\mathscr{C}$ is the free operad $\mathscr{T}(s^{-1}\overline{\mathscr{C}})$. Therefore any morphism of operads $f : \Omega\mathscr{C} \to \mathscr{P}$ is characterized by its restriction $\bar{f}$ on the generators $s^{-1}\overline{\mathscr{C}}$. We denote by $s^{-1}\bar{f} : \overline{\mathscr{C}} \to \mathscr{P}$ the induced map of degree -1. The map f is a morphism of dg operads. It means that it commutes with the differentials:

$$\begin{CD} s^{-1}\overline{\mathscr{C}} @>\bar{f}>> \mathscr{P} \\ @V{d_1+d_2}VV @VV{d_{\mathscr{P}}}V \\ s^{-1}\overline{\mathscr{C}} \oplus s^{-1}\overline{\mathscr{C}} \circ_{(1)} s^{-1}\overline{\mathscr{C}} @>{\bar{f}+\gamma_{(1)}(\bar{f}\circ_{(1)}\bar{f})}>> \mathscr{P}. \end{CD}$$

This is equivalent to $\partial(s^{-1}\bar{f}) + (s^{-1}\bar{f}) \star (s^{-1}\bar{f}) = 0$, which is the Maurer–Cartan equation for $s^{-1}\bar{f}$. □

Otherwise stated, the functor $\mathrm{Tw}(\mathscr{C}, -)$ is represented by $\Omega\mathscr{C}$ and the functor $\mathrm{Tw}(-, \mathscr{P})$ is represented by $\mathrm{B}\mathscr{P}$.

6.5.4 Universal Twisting Morphisms

We denote by $\upsilon : \mathscr{C} \to \mathrm{B}\Omega\mathscr{C}$ the *unit* of this adjunction, obtained with $\mathscr{P} = \Omega\mathscr{C}$ and $\mathrm{Id}_{\Omega\mathscr{C}}$ on the left-hand side. In this case, the corresponding twisting morphism is denoted by $\iota : \mathscr{C} \to \Omega\mathscr{C}$. It is universal among the set of twisting morphisms: any twisting morphism $\alpha : \mathscr{C} \to \mathscr{P}$ factorizes uniquely through ι.

Dually, we denote by $\varepsilon : \Omega\mathrm{B}\mathscr{P} \to \mathscr{P}$ the *counit*, obtained with $\mathscr{C} = \mathrm{B}\mathscr{P}$ and $\mathrm{Id}_{\mathrm{B}\mathscr{P}}$ on the right-hand side. The associated twisting morphism is denoted by $\pi : \mathrm{B}\mathscr{P} \to \mathscr{P}$. It satisfies the following universal property: any twisting morphism $\alpha : \mathscr{C} \to \mathscr{P}$ factors uniquely through π.

Proposition 6.5.8. *Any twisting morphism $\alpha : \mathscr{C} \to \mathscr{P}$ factorizes uniquely through the universal twisting morphisms π and ι as follows*

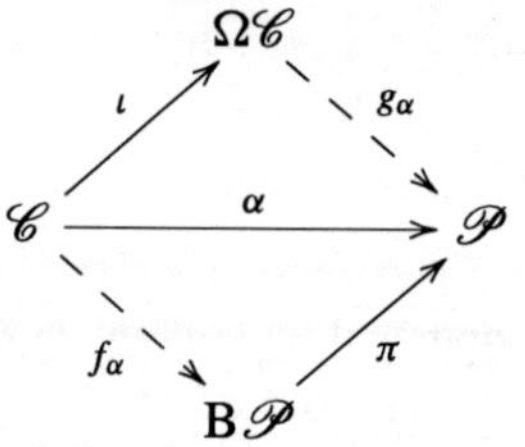

where g_α is a dg operad morphism and where f_α is a dg cooperads morphism.

6.5.5 Augmented Bar and Cobar Construction

Let $\mathscr{P}$ be an augmented dg operad. Recall that its bar construction $\mathrm{B}\mathscr{P}$ is a dg cooperad whose underlying space is $\mathscr{T}^c(s\overline{\mathscr{P}})$. The universal twisting morphism π is equal to the following composite

$$\pi : \mathscr{T}^c(s\overline{\mathscr{P}}) \twoheadrightarrow s\overline{\mathscr{P}} \xrightarrow{s^{-1}} \overline{\mathscr{P}} \rightarrowtail \mathscr{P}.$$

The associated twisted composite product $\mathrm{B}\mathscr{P} \circ_\pi \mathscr{P}$ is a dg $\mathbb{S}$-module, with the twisted differential d_π of Sect. 6.4.5, and is called the *augmented bar construction of* $\mathscr{P}$.

Dually, let $\mathscr{C}$ be a coaugmented dg cooperad. Its cobar construction $\Omega\mathscr{C} = (\mathscr{T}(s^{-1}\overline{\mathscr{C}}), d)$ is a dg operad and the universal twisting morphism ι is given explicitly by the composite

$$\mathscr{C} \twoheadrightarrow \overline{\mathscr{C}} \xrightarrow{s^{-1}} s^{-1}\overline{\mathscr{C}} \rightarrowtail \mathscr{T}\big(s^{-1}\overline{\mathscr{C}}\big).$$

The resulting dg $\mathbb{S}$-module $(\mathscr{C} \circ_\iota \Omega\mathscr{C}, d_\iota)$ is called the *coaugmented cobar construction of* $\mathscr{C}$.

Lemma 6.5.9. *The chain complexes*

$$\mathscr{P} \circ_\pi \mathrm{B}\mathscr{P}, \quad \mathrm{B}\mathscr{P} \circ_\pi \mathscr{P}, \quad \mathscr{C} \circ_\iota \Omega\mathscr{C}, \quad \Omega\mathscr{C} \circ_\iota \mathscr{C}$$

are acyclic.

Proof. Left composite product $\mathscr{P} \circ_\pi \mathrm{B}\mathscr{P}$. Since the composite product $\circ$ is linear on the left, we can use the same kind of contracting homotopy h as in the proof of Proposition 2.2.8. Let $(\mu; t_1, \ldots, t_k)$ represent an element of $\overline{\mathscr{P}} \circ \mathrm{B}\mathscr{P} = \overline{\mathscr{P}} \circ \mathscr{T}^c(s\overline{\mathscr{P}})$. We define its image under h by the formula $h(\mu; t_1, \ldots, t_k) := (\mathrm{I}; t')$, where t' is the tree obtained by grafting $t_1, \ldots, t_k$ above the vertex $s\mu$. The restriction of h to $\mathrm{I} \circ \mathrm{B}\mathscr{P}$ is null. Therefore, the map h is well defined from $\mathscr{P} \circ \mathrm{B}\mathscr{P}$ to itself and has degree 1. We leave it to the reader to verify that it is a homotopy from Id to 0: $hd_\pi + d_\pi h = \mathrm{Id}$.

Right composite product $\mathrm{B}\mathscr{P} \circ_\pi \mathscr{P}$. We need to refine the arguments since there is more than one element of $\mathscr{P}$ on the right-hand side of the product. For any $n \in \mathbb{N}$, we consider the subspace $(\mathrm{B}\mathscr{P} \circ_\pi \mathscr{P})^{(n)}$ made up of trees with exactly n nontrivial vertices, that is vertices indexed by elements of $\overline{\mathscr{P}}$. We define the following increasing filtration $F_i := \bigoplus_{n \leqslant i} (\mathrm{B}\mathscr{P} \circ_\pi \mathscr{P})^{(n)}$. Since it is stable under the differential d_π, it induces a spectral sequence $E^\bullet_{pq}$ such that

$$E^0_{pq} = F_p\big((\mathrm{B}\mathscr{P} \circ_\pi \mathscr{P})_{p+q}\big)/F_{p-1}\big((\mathrm{B}\mathscr{P} \circ_\pi \mathscr{P})_{p+q}\big) = (\mathrm{B}\mathscr{P} \circ_\pi \mathscr{P})^{(p)}_{p+q}.$$

The first differential d^0 is equal to $d^0 = d_1 \circ \mathrm{Id}_{\mathscr{P}} + \mathrm{Id}_{\mathrm{B}\mathscr{P}} \circ' d_{\mathscr{P}} + d'_\pi$, where d'_π is the part of d_π which does not change the number of $\overline{\mathscr{P}}$. That is, d'_π consists only in extracting an element of $\mathrm{B}\mathscr{P} = \mathscr{T}^c(s\overline{\mathscr{P}})$, desuspending it and composing it on the right with only units I in $\mathscr{P}$. For $p = 0$, we have $(E^0_{0q}, d^0) = (\mathrm{I}, 0)$. Therefore, $E^1_{00} = I$ and $E^1_{0q} = 0$ for $q \neq 0$.

When $p > 0$, we introduce again a contracting homotopy h. For any element of $\mathrm{B}\mathscr{P} \circ_\pi \mathscr{P}$, we choose a representative $(t; p_1, \ldots, p_k)$. For instance, with the species notation of Sect. 5.1.13, we choose p_1 to be the element with a leaf labeled by 1. We define its image under the map h as follows. If $p_1 = \mathrm{I}$, then $h(t; p_1, \ldots, p_k) := 0$. Otherwise, when $p_1 \in \overline{\mathscr{P}}$, $h(t; p_1, \ldots, p_k) := (-1)^{|t|}(t'; I, \ldots, I, p_2, \ldots, p_k)$, where t' is the element of $\mathrm{B}\mathscr{P}$ obtained by grafting the tree t with sp_1 above. This map has degree 1 and is a homotopy between Id and 0: $hd^0 + d^0h = \mathrm{Id}$, on $E^0_{p\bullet}$ for any $p > 0$. Therefore, the spectral sequence collapses, that is $E^1_{pq} = 0$ for $p > 0$ and any q.

Since the spectral sequence F_i is bounded below $F_{-1} = 0$ and exhaustive $\bigcup_{i\geq 0} F_i = \mathrm{B}\mathscr{P} \circ_\pi \mathscr{P}$, we can apply the classical theorem of convergence of spectral sequences (Theorem 1.5.1). It implies that $E^\bullet_{pq}$ converges to the homology of $\mathrm{B}\mathscr{P} \circ_\pi \mathscr{P}$. Hence, this homology is equal to I.

The proof of the other case is completely dual and left to the reader. □

This proof is extracted from E. Getzler and J.D.S. Jones [GJ94], see also B. Fresse [Fre04] and [Val07b].

6.6 Operadic Koszul Morphisms

We define operadic Koszul morphisms and state their main property, which relates them to the operadic bar and cobar constructions. As a corollary, we prove that the unit and the counit of the bar–cobar adjunction are quasi-isomorphisms. The proofs follow the same pattern as the ones of Chap. 2 for algebras.

6.6.1 Operadic Koszul Criterion

A twisting morphism $\alpha : \mathscr{C} \to \mathscr{P}$ is called a *Koszul morphism* when either its left twisted complex $\mathscr{P} \circ_\alpha \mathscr{C}$ or its right twisted complex $\mathscr{C} \circ_\alpha \mathscr{P}$ is acyclic. We denote the set of Koszul morphisms by $\mathrm{Kos}(\mathscr{C}, \mathscr{P})$. Under this terminology, the result of the previous section states that the universal twisting morphisms $\pi \in \mathrm{Kos}(\mathrm{B}\mathscr{P}, \mathscr{P})$ and $\iota \in \mathrm{Kos}(\mathscr{C}, \Omega\mathscr{C})$ are Koszul morphisms.

The main theorem about operadic twisting morphisms is the following criterion for Koszul morphisms.

Theorem 6.6.1 (Operadic twisting morphisms fundamental theorem). *Let $\mathscr{P}$ be a connected wgd operad and let $\mathscr{C}$ be a connected wgd cooperad. Let $\alpha : \mathscr{C} \to \mathscr{P}$ be an operadic twisting morphism. The following assertions are equivalent*:

1. *the right twisted composite product $\mathscr{C} \circ_\alpha \mathscr{P}$ is acyclic,*
2. *the left twisted composite product $\mathscr{P} \circ_\alpha \mathscr{C}$ is acyclic,*
3. *the morphism of dg cooperads $f_\alpha : \mathscr{C} \xrightarrow{\sim} \mathrm{B}\mathscr{P}$ is a quasi-isomorphism,*
4. *the morphism of dg operads $g_\alpha : \Omega\mathscr{C} \xrightarrow{\sim} \mathscr{P}$ is a quasi-isomorphism.*

Proof. The proof of this theorem is similar to the proof of Theorem 2.3.1. First notice that the bar and the cobar constructions are weight graded and connected. The universal twisting morphisms π and ι preserve the weight. Therefore, we can apply Lemma 6.4.8.

Let us first prove (1) $\Leftrightarrow$ (3) $\Leftrightarrow$ (4).

To prove (1) $\Leftrightarrow$ (3), we apply Lemma 6.7.1 to f_α, $\mathrm{Id}_{\mathscr{P}}$ and $f_\alpha \circ \mathrm{Id}_{\mathscr{P}} : \mathscr{C} \circ_\alpha \mathscr{P} \to \mathrm{B}\mathscr{P} \circ_\pi \mathscr{P}$. Since $\mathrm{B}\mathscr{P} \circ_\pi \mathscr{P}$ is always acyclic by Lemma 6.5.9, $f_\alpha \circ \mathrm{Id}_{\mathscr{P}}$ is a quasi-isomorphism if and only if $\mathscr{C} \circ_\alpha \mathscr{P}$ is acyclic, which happens if and only if f_α is a quasi-isomorphism.

To prove (1) $\Leftrightarrow$ (4), we apply Lemma 6.7.1 to $\mathrm{Id}_{\mathscr{C}}$, g_α and $\mathrm{Id}_{\mathscr{C}} \circ g_\alpha : \mathscr{C} \circ_\iota \Omega\mathscr{C} \to \mathscr{C} \circ_\alpha \mathscr{P}$. In this case, $\mathscr{C} \circ_\iota \Omega\mathscr{C}$ is acyclic by Lemma 6.5.9. Therefore, $\mathrm{Id}_{\mathscr{C}} \circ g_\alpha$ is a quasi-isomorphism if and only if the twisted composite product $\mathscr{C} \circ_\alpha \mathscr{P}$ is acyclic. Since $\mathrm{Id}_{\mathscr{C}}$ is a quasi-isomorphism, Lemma 6.5.9 shows that $\mathscr{C} \circ_\alpha \mathscr{P}$ is acyclic if and only if g_α is a quasi-isomorphism.

The proof of the equivalence (1) $\Leftrightarrow$ (2) $\Leftrightarrow$ (3) is similar and uses the two other cases of Lemma 6.5.9 and Lemma 6.7.2 this time. □

Corollary 6.6.2. *In the weight graded case, the right twisted composite product $\mathscr{C} \circ_\alpha \mathscr{P}$ is acyclic if and only if the left twisted composite product $\mathscr{P} \circ_\alpha \mathscr{C}$ is acyclic.*

This corollary shows that, under the weight grading assumption, the notion of Koszul morphism is equivalently defined with the left or with the right twisted composite product.

6.6.2 Bar–Cobar Resolution

Theorem 6.6.3. *The counit $\varepsilon : \Omega\mathrm{B}\mathscr{P} \xrightarrow{\sim} \mathscr{P}$ is a quasi-isomorphism of dg operads. Dually, the unit $\upsilon : \mathscr{C} \xrightarrow{\sim} \mathrm{B}\Omega\mathscr{C}$ is a quasi-isomorphism of dg cooperads.*

Proof. We prove this result in the weight graded case as a corollary of Theorem 6.6.1 but this result holds in full generality, see [Fre04]. As in the proof of Corollary 2.3.2, we apply the Koszul criterion (Theorem 6.6.1) to the following diagram

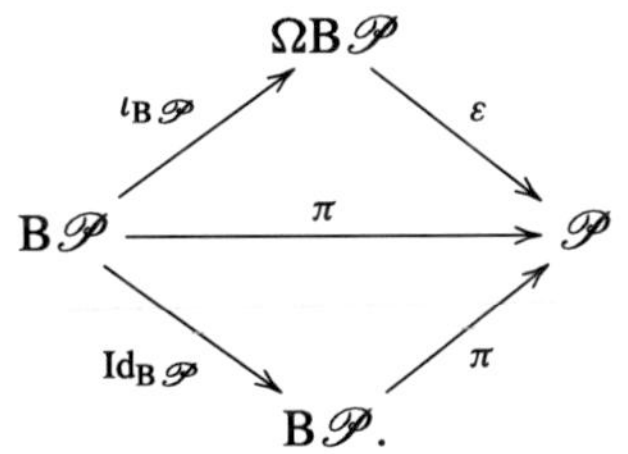

Since $\mathrm{Id}_{\mathrm{B}\mathscr{P}}$ is a quasi-isomorphism, or equivalently since π is a Koszul morphism, the morphism of dg operads ε is a quasi-isomorphism. The dual statement is proved with the same arguments applied to the Koszul morphism $\iota : \mathscr{C} \to \Omega\mathscr{C}$. □

Hence the counit of adjunction provides a canonical functorial resolution of dg operads. It is called the *bar–cobar resolution*. It is a quasi-free resolution which is not minimal in general.

Let us make it explicit. We denote by *proj* the projection $\mathcal{T}^c(s\overline{\mathcal{P}}) \twoheadrightarrow s\overline{\mathcal{P}}$ and by $\gamma_{\mathcal{P}}$ the morphism $\mathcal{T}(\overline{\mathcal{P}}) \to \overline{\mathcal{P}}$ coming from the combinatorial definition (Sect. 5.6) of the operad $\mathcal{P}$. It is explicitly given by the composition under γ of the operations of $\overline{\mathcal{P}}$ along the tree composition scheme.

Lemma 6.6.4. *The unit* $\varepsilon : \Omega \mathrm{B}\mathcal{P} \to \mathcal{P}$ *of the adjunction is equal to the composite*

$$\varepsilon : \mathcal{T}\big(s^{-1}\overline{\mathcal{T}}^c(s\overline{\mathcal{P}})\big) \xrightarrow{\mathcal{T}(s^{-1}proj)} \mathcal{T}\big(s^{-1}s\overline{\mathcal{P}}\big) \cong \mathcal{T}(\overline{\mathcal{P}}) \xrightarrow{\gamma_{\mathcal{P}}} \overline{\mathcal{P}}.$$

Proof. The proof follows directly from the definition of ε given in Sect. 6.5.4 and is left to the reader. □

6.7 Proof of the Operadic Comparison Lemmas

In this section, we prove the Comparison Lemmas at the level of operads. The results are the operadic generalizations of the Comparison Lemma for twisted tensor product (Lemma 2.5.1). Since the right twisted composite product is not isomorphic to the left twisted composite product, we give the following two versions of this theorem. It comes essentially from [Fre04], with slightly different hypotheses.

Lemma 6.7.1 (Comparison Lemma for right twisted composite product). *Let* $g : \mathcal{P} \to \mathcal{P}'$ *be a morphism of wdg connected operads and* $f : \mathcal{C} \to \mathcal{C}'$ *be a morphism of wdg connected cooperads. Let* $\alpha : \mathcal{C} \to \mathcal{P}$ *and* $\alpha' : \mathcal{C}' \to \mathcal{P}'$ *be two twisting morphisms, such that* f *and* g *are compatible with* α *and* α'.

If two morphisms among f, g *and* $f \circ g$ *are quasi-isomorphisms, then so is the third one.*

Proof. We denote by $\mathcal{M}$ (resp. $\mathcal{M}'$) the weight-graded chain complex $\mathcal{C} \circ_\alpha \mathcal{P}$ (resp. $\mathcal{C}' \circ_{\alpha'} \mathcal{P}'$). We define a filtration F_s on $\mathcal{M}^{(n)}$, where $n \in \mathbb{N}$ is the weight, by the formula

$$F_s\big(\mathcal{M}^{(n)}\big) := \bigoplus_{d+m\leq s} \big(\mathcal{C}_d^{(m)} \circ \mathcal{P}\big)^{(n)} = \bigoplus_{d+m\leq s} \mathcal{C}_d^{(m)} \circ \underbrace{\mathcal{P}}_{(n-m)},$$

where the total weight of the elements of $\mathcal{P}$ on the right-hand side of $\circ$ is equal to $(n-m)$. The differential d_α on $\mathcal{M} = \mathcal{C} \circ_\alpha \mathcal{P}$ is the sum of three terms $\mathrm{Id}_C \circ' d_{\mathcal{P}}$, $d_{\mathcal{C}} \circ \mathrm{Id}_{\mathcal{P}}$ and d_α^r. One has $\mathrm{Id}_C \circ' d_{\mathcal{P}} : F_s \to F_s$, $d_C \circ \mathrm{Id}_{\mathcal{P}} : F_s \to F_{s-1}$ and $d_\alpha^r : F_s \to F_{s-2}$. Therefore, F_s is a filtration of the chain complex $\mathcal{M}^{(n)}$ and we consider the associated spectral sequence $E_{st}^{\bullet}$. One has

$$E_{st}^0 = F_s\big(\mathcal{M}^{(n)}\big)_{s+t}/F_{s-1}\big(\mathcal{M}^{(n)}\big)_{s+t} = \bigoplus_{m=o}^{n} \mathcal{C}_{s-m}^{(m)} \circ \underbrace{\mathcal{P}}_{t+m;(n-m)}.$$

The notation means that the total homological degree of elements of $\mathscr{P}$ is equal to $t+m$ and their total weight is equal to $(n-m)$. The study of the differential d_α on the filtration F_s of $\mathscr{M}$ shows that $d_0 = \mathrm{Id}_C \circ' d_{\mathscr{P}}$ and that $d_1 = d_C \circ \mathrm{Id}_{\mathscr{P}}$. Since we are working over a field of characteristic 0, Maschke's theorem applies, which proves that any $\mathbb{K}[\mathbb{S}_*]$-module is flat. It follows that

$$E^2_{st} = \bigoplus_{m=o}^{n} H_{s-m}\big(\mathscr{C}_\bullet^{(m)}\big) \circ \underbrace{H}_{t+m}\big(\underbrace{\mathscr{P}_\bullet}_{(n-m)}\big).$$

Since $\mathscr{P}$ and $\mathscr{C}$ are weight graded and connected, the part $m=0$ is concentrated in $s=0$ and $t \geq 0$, where it is equal to $E^2_{0t} = H_t(\mathscr{P}_\bullet^{(n)})$. The part $m=n$ is concentrated in $t=-n$ and $s \geq n$, where it is equal to $E^2_{s-n} = H_{s-n}(\mathscr{C}_\bullet^{(n)})$. For any $0<m<n$, the nonvanishing part of $H_{s-m}(\mathscr{C}_\bullet^{(m)}) \otimes H_{t+m}(\mathscr{P}_\bullet^{(n-m)})$ is in $s \geq 1$ and $t \geq -n+1$.

The filtration F_s is exhaustive $\mathscr{M}^{(n)} = \bigcup_{s\geq 0} F_s(\mathscr{M}^{(n)})$ and bounded below $F_{-1}(\mathscr{M}^{(n)}) = \{0\}$, so the spectral sequence converges to the homology of $\mathscr{M}^{(n)}$ by the classical convergence Theorem 1.5.1.

$$E^\infty_{st}\big(\mathscr{M}^{(n)}\big) = H_{s+t}\big(\mathscr{M}^{(n)}\big).$$

We consider the same filtration on $\mathscr{M}'$ and we denote by Φ the morphism of chain complexes $\Phi := f \circ g$. We treat the three cases one after the other.

(1) *If f and g are quasi-isomorphisms, then $\Phi = f \circ g$ is a quasi-isomorphism.*

For every s, t and n, the maps $E^2_{st}(\mathscr{M}^{(n)}) \xrightarrow{H_\bullet(f)\otimes H_\bullet(g)} E^2_{st}(\mathscr{M}'^{(n)})$ are isomorphisms. By the convergence of the two spectral sequences, the maps

$$H_{s+t}\big(\mathscr{M}^{(n)}\big) = E^\infty_{st}\big(\mathscr{M}^{(n)}\big) \xrightarrow{H_\bullet(\Phi)} E^\infty_{st}\big(\mathscr{M}'^{(n)}\big) = H_{s+t}\big(\mathscr{M}'^{(n)}\big)$$

are again isomorphisms. So the map Φ is a quasi-isomorphism.

(2) *If $\Phi = f \circ g$ and g are quasi-isomorphisms, then f is a quasi-isomorphism.*

Let us work by induction on the weight n. When $n=0$, the map $f^{(0)} : \mathbb{K} \to \mathbb{K}$ is a quasi-isomorphism. Suppose now that the result is true up to rank $n-1$. We consider the mapping cone of $\Phi^{(n)}$: $cone(\Phi^{(n)}) := s^{-1}\mathscr{M}^{(n)} \oplus \mathscr{M}'^{(n)}$ and the associated filtration $F_s(cone(\Phi^{(n)})) := F_{s-1}(\mathscr{M}^{(n)}) \oplus F_s(\mathscr{M}'^{(n)})$, which satisfies $E^1_{\bullet t}(cone(\Phi^{(n)})) = cone(E^1_{\bullet t}(\Phi^{(n)}))$. The long exact sequence of the mapping cone reads

$$\cdots \to H_{s+1}\big(cone\big(E^1_{\bullet t}(\Phi^{(n)})\big)\big) \to H_s\big(E^1_{\bullet t}(\mathscr{M}^{(n)})\big)$$
$$\xrightarrow{H_s(E^1_{\bullet t}(\Phi^{(n)}))} H_s\big(E^1_{\bullet t}(\mathscr{M}'^{(n)})\big) \to H_s\big(cone\big(E^1_{\bullet t}(\Phi^{(n)})\big)\big) \to \cdots.$$

Therefore there is a long exact sequence (ξ_t)

$$(\xi_t) \qquad \cdots \to E^2_{s+1t}\big(cone(\Phi^{(n)})\big) \to E^2_{st}\big(\mathscr{M}^{(n)}\big) \xrightarrow{E^2_{st}(\Phi^{(n)})} E^2_{st}\big(\mathscr{M}'^{(n)}\big) \to E^2_{st}\big(cone(\Phi^{(n)})\big) \to \cdots$$

where $E^2_{st}(\Phi^{(n)})$ is given by $H_\bullet(f)\otimes H_\bullet(g)$.

When $t > -n$, we have seen that only $\mathscr{C}^{(m)}$ (and $\mathscr{C}'^{(m)}$) with $m < n$ are involved in E^2_{st}. In that case, since

$$E^2_{st}\big(\mathscr{M}^{(n)}\big) = \bigoplus_{m=o}^{n-1} H_{s-m}\big(C^{(m)}_\bullet\big) \circ \underbrace{H}_{t+m}\big(\underbrace{\mathscr{P}_\bullet}_{(n-m)}\big),$$

the induction hypothesis gives that

$$E^2_{st}\big(\mathscr{M}^{(n)}\big) \xrightarrow{H_\bullet(f)\otimes H_\bullet(g)} E^2_{st}\big(\mathscr{M}'^{(n)}\big)$$

is an isomorphism for every s and every $t > -n$. Using the long exact sequence (ξ_t) for $t > -n$, it gives $E^2_{st}(cone(\Phi^{(n)})) = 0$ for every s and every $t \neq -n$. The collapsing of the spectral sequence $E^\bullet_{st}(cone(\Phi^{(n)}))$ at rank 2 implies the equality $E^\infty_{st}(cone(\Phi^{(n)})) = E^2_{st}(cone(\Phi^{(n)}))$. The convergence of the spectral sequence $E^\bullet_{st}(cone(\Phi^{(n)}))$ shows that

$$E^2_{st}\big(cone(\Phi^{(n)})\big) = H_{s+t}\big(cone(\Phi^{(n)})\big) = 0$$

since $\Phi^{(n)}$ is a quasi-isomorphism. Since $E^2_{s-n}(cone(\Phi^{(n)})) = 0$, the long exact sequence (ξ_{-n}) gives the isomorphism

$$H_{s-n}\big(\mathscr{C}^{(n)}_\bullet\big) = E^2_{s-n}\big(\mathscr{M}^{(n)}\big) \xrightarrow{H_\bullet(f)} E^2_{s-n}\big(\mathscr{M}'^{(n)}\big) = H_{s-n}\big(\mathscr{C}'^{(n)}_\bullet\big),$$

for every s.

(3) *If $\Phi = f \circ g$ and f are quasi-isomorphisms, then g is a quasi-isomorphism.*

Once again, we work by induction on the weight n. For $n = 0$, the map $g^{(0)} : \mathbb{K} \to \mathbb{K}$ is an isomorphism. Suppose that the result if true up to rank $n-1$. When $s \geq 1$, we have seen that only $\mathscr{P}^{(n-m)}$ (and $\mathscr{P}'^{(n-m)}$) with $m > 0$ are involved in E^2_{st},

$$E^2_{st}\big(\mathscr{M}^{(n)}\big) = \bigoplus_{m=1}^{n} H_{s-m}\big(\mathscr{C}^{(m)}_\bullet\big) \circ \underbrace{H}_{t+m}\big(\underbrace{\mathscr{P}_\bullet}_{(n-m)}\big).$$

In this case, the induction hypothesis implies that $E^2_{st}(\mathscr{M}^{(n)}) \xrightarrow{H_\bullet(f)\circ H_\bullet(g)} E^2_{st}(\mathscr{M}'^{(n)})$ is an isomorphism for every $s \geq 1$ and every t. The long exact sequence (ξ_t) shows that $E^2_{st}(cone(\Phi^{(n)})) = 0$ for $s \geq 2$ and every t. The spectral sequence of the cone of $\Phi^{(n)}$ converges to its homology, which is null since $\Phi^{(n)}$ is a quasi-isomorphism. Therefore, $E^2_{1,t-1}(cone(\Phi^{(n)})) = E^2_{0,t}(cone(\Phi^{(n)})) = 0$ for every t. This implies

$E^2_{st}(cone(\Phi^{(n)})) = 0$ for every t and s. Finally, the beginning ($s = 0$) of the exact sequence (ξ_t) gives the isomorphism

$$H_t(\mathscr{P}^{(n)}_\bullet) = E^2_{0t}(\mathscr{M}^{(n)}) \xrightarrow{H_\bullet(g)} E^2_{0t}(\mathscr{M}'^{(n)}) = H_t(\mathscr{P}'^{(n)}_\bullet). \qquad \square$$

Lemma 6.7.2 (Comparison Lemma for left twisted composite product). *Let $g : \mathscr{P} \to \mathscr{P}'$ be a morphism of wdg connected operads and $f : \mathscr{C} \to \mathscr{C}'$ be a morphism of wdg connected cooperads. Let $\alpha : \mathscr{C} \to \mathscr{P}$ and $\alpha' : \mathscr{C}' \to \mathscr{P}'$ be two twisting morphisms, such that f and g are compatible with α and α'.*

If two morphisms among f, g and $g \circ f$ are quasi-isomorphisms, then so is the third one.

Proof. The proof of this case is similar to the previous one. We consider the following filtration this time

$$F_s\big((\mathscr{P} \circ \mathscr{C})^{(n)}\big) := \bigoplus_{d+m \le s} \Big(\mathscr{P}^{(n-m)} \circ \underbrace{\mathscr{C}}_{d;(m)}\Big)^{(n)},$$

where the total weight of the elements of $\mathscr{C}$ on the right-hand side of the composite product $\circ$ is equal to (m) and where the total degree of these elements is equal to d.

From now on, the same arguments apply and we leave it to the reader, as a good exercise to finish to proof. $\square$

6.8 Résumé

6.8.1 Infinitesimal Notions

▷ $M \circ (N_1; N_2)$:= sub-$\mathbb{S}$-module of $M \circ (N_1 \oplus N_2)$ linear in N_2.

$$M \circ_{(1)} N := M \circ (\mathrm{I}, N).$$

▷ *Infinitesimal composition map* of an operad: $\mathscr{P} \circ_{(1)} \mathscr{P} \xrightarrow{\gamma_{(1)}} \mathscr{P}$.

▷ *Infinitesimal composite of morphisms* $f : M_1 \to N_1$ and $g : M_2 \to N_2$

$$f \circ' g : M_1 \circ N_1 \to M_2 \circ (N_1; N_2).$$

EXAMPLE: $d_{M \circ N} = d_M \circ \mathrm{Id}_N + \mathrm{Id}_M \circ' d_N$.

6.8.2 Differential Operadic Notions

▷ *dg $\mathbb{S}$-modules,*

▷ *dg operads, quasi-free operads, dg $\mathscr{P}$-algebras,*

▷ *dg cooperads, quasi-free cooperads, dg $\mathscr{C}$-coalgebras.*

Table 6.1 Hierarchy of notions

$\alpha \in$	Defines:
$\mathrm{Hom}_{\mathbb{S}}(\mathscr{C}, \mathscr{P})_{-1}$	$d_\alpha : \mathscr{C} \circ \mathscr{P} \to \mathscr{C} \circ \mathscr{P}$
$\cup$	
$\mathrm{Tw}(\mathscr{C}, \mathscr{P})$	$(d_\alpha)^2 = 0$, chain complex $\mathscr{C} \circ_\alpha \mathscr{P}$
$\cup$	
$\mathrm{Kos}(\mathscr{C}, \mathscr{P})$	acyclicity of $\mathscr{C} \circ_\alpha \mathscr{P}$

6.8.3 Operadic Twisting Morphism

Convolution dg (pre-)Lie algebra: $\mathscr{C}$ dg cooperad and $\mathscr{P}$ dg operad:

$$\big(\mathrm{Hom}_{\mathbb{S}}(\mathscr{C}, \mathscr{P}), \star, \partial\big).$$

Operadic twisting morphism, $\mathrm{Tw}(\mathscr{C}, \mathscr{P})$*:* solution of degree -1 to the Maurer–Cartan equation

$$\partial(\alpha) + \alpha \star \alpha = \partial(\alpha) + \frac{1}{2}[\alpha, \alpha] = 0.$$

6.8.4 Twisted Composite Products

Any $\alpha \in \mathrm{Tw}(\mathscr{C}, \mathscr{P})$ induces

▷ a twisted differential $\partial_\alpha := \partial + [\alpha, -]$ in $\mathrm{Hom}_{\mathbb{S}}(\mathscr{C}, \mathscr{P})$,
▷ a differential $d_\alpha := d_{\mathscr{C} \circ \mathscr{P}} + d_\alpha^r$ on the right composite product $\mathscr{C} \circ \mathscr{P}$ defining the right twisted composite product $\mathscr{C} \circ_\alpha \mathscr{P}$.

Table 6.1 summarizes this hierarchy of notions.

6.8.5 Operadic Bar and Cobar Constructions and Koszul Criterion

$$\begin{array}{ccccc}
\mathrm{Hom}_{\mathsf{Op}}\big(\mathscr{T}(s^{-1}\overline{\mathscr{C}}), \mathscr{P}\big) & \cong & \mathrm{Hom}_{\mathbb{S}}(\overline{\mathscr{C}}, \overline{\mathscr{P}})_{-1} & \cong & \mathrm{Hom}_{\mathsf{coOp}}\big(\mathscr{C}, \mathscr{T}^c(s\overline{\mathscr{P}})\big) \\
\cup & & \cup & & \cup \\
\mathrm{Hom}_{\mathsf{dg\ Op}}(\Omega\mathscr{C}, \mathscr{P}) & \cong & \mathrm{Tw}(\mathscr{C}, \mathscr{P}) & \cong & \mathrm{Hom}_{\mathsf{dg\ coOp}}(\mathscr{C}, \mathrm{B}\mathscr{P}) \\
\cup & & \cup & & \cup \\
\text{q-Iso}_{\mathsf{dg\ Op}}(\Omega\mathscr{C}, \mathscr{P}) & \cong & \mathrm{Kos}(\mathscr{C}, \mathscr{P}) & \cong & \text{q-Iso}_{\mathsf{dg\ coOp}}(\mathscr{C}, \mathrm{B}\mathscr{P}).
\end{array}$$

Universal twisting morphisms: $\iota : \mathscr{C} \to \Omega\mathscr{C}$ and $\pi : \mathrm{B}\mathscr{P} \to \mathscr{P}$, which are Koszul.

Factorization of any operadic twisting morphism $\alpha : \mathscr{C} \to \mathscr{P}$:

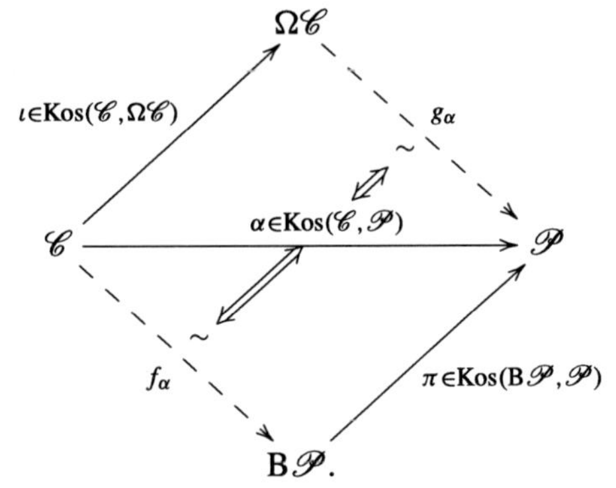

▷ $g_\alpha : \Omega\mathscr{C} \to \mathscr{P}$ morphism of dg operads,
▷ $f_\alpha : \mathscr{C} \to \mathrm{B}\mathscr{P}$ morphism of dg cooperads.

Operadic Twisting Morphism Fundamental Theorem. *The following assertions are equivalent*

▷ *a twisting morphism $\alpha : \mathscr{C} \to \mathscr{P}$ is Koszul,*
▷ *the morphism of dg operads $f_\alpha : \Omega\mathscr{C} \xrightarrow{\sim} \mathscr{P}$ is a quasi-isomorphism,*
▷ *the morphism of dg cooperads $g_\alpha : \mathscr{C} \xrightarrow{\sim} \mathrm{B}\mathscr{P}$ is a quasi-isomorphism.*

Corollary. *Bar–cobar resolutions*: $\varepsilon : \Omega\mathrm{B}\mathscr{P} \xrightarrow{\sim} \mathscr{P}$ *and* $\upsilon : \mathscr{C} \xrightarrow{\sim} \mathrm{B}\Omega\mathscr{C}$.

6.9 Exercises

Exercise 6.9.1 (Infinitesimal composite product). Let M, N and P be three $\mathbb{S}$-modules. Prove the following isomorphism

$$(M \circ_{(1)} N) \circ P \cong M(P, N \circ P).$$

Make the same isomorphism explicit when the $\mathbb{S}$-modules are differential graded $\mathbb{S}$-modules. (Be careful at the nontrivial signs appearing in this case.)

Exercise 6.9.2 (Cofree dg $\mathscr{C}$-comodule). The aim of this exercise is to prove the dual of Proposition 6.3.9. Let $\mathscr{C}$ be a dg cooperad and let N be an $\mathbb{S}$-module. Prove that any coderivation on the cofree dg $\mathscr{C}$-comodule $\mathscr{C} \circ N$ (resp. $N \circ \mathscr{C}$) is characterized by its projection onto generators: $\mathscr{C} \circ N \to N$ (resp. $N \circ \mathscr{C} \to N$). In the other way round, give the formulas for the unique coderivations which extend such maps.

Exercise 6.9.3 (Two-sided twisted composite product ★). Let $\mathscr{P}$ be a dg operad, $\mathscr{C}$ a dg cooperad and let $\alpha : \mathscr{C} \to \mathscr{P}$ be a twisting morphism. We consider the following twisted differential on $\mathscr{P} \circ \mathscr{C} \circ \mathscr{P}$, the free $\mathscr{P}$-bimodule on $\mathscr{C}$:

$$d_\alpha := d_{\mathscr{P}\circ\mathscr{C}\circ\mathscr{P}} + \mathrm{Id}_{\mathscr{P}} \circ' d^r_\alpha - d^l_\alpha \circ \mathrm{Id}_{\mathscr{P}},$$

where d^r_α and d^l_α were defined in Sect. 6.4.5.

◇ Prove that ${d_\alpha}^2 = 0$.

We denote this chain complex by

$$\mathscr{P} \circ_\alpha \mathscr{C} \circ_\alpha \mathscr{P} := (\mathscr{P} \circ \mathscr{C} \circ \mathscr{P}, d_\alpha).$$

◇ Show that there is an isomorphism of chain complexes

$$\big(\mathrm{Hom}^\alpha_{\mathbb{S}}(\mathscr{C}, \mathscr{P}), \partial_\alpha\big) \cong \big(\mathrm{Hom}_{\mathscr{P}\text{-biMod}}(\mathscr{P} \circ_\alpha \mathscr{C} \circ_\alpha \mathscr{P}, \mathscr{P}), \partial\big).$$

◇ Show that the following composite

$$\xi : \mathscr{P} \circ \mathscr{C} \circ \mathscr{P} \xrightarrow{\mathrm{Id}\circ\varepsilon\circ\mathrm{Id}} \mathscr{P} \circ \mathrm{I} \circ \mathscr{P} \cong \mathscr{P} \circ \mathscr{P} \xrightarrow{\gamma} \mathscr{P}$$

is a morphism of dg $\mathscr{P}$-bimodules.

◇ Under the same weight grading assumptions as in Theorem 6.6.1, prove that $\xi : \mathscr{P} \circ_\alpha \mathscr{C} \circ_\alpha \mathscr{P} \xrightarrow{\sim} \mathscr{P}$ is a quasi-isomorphism if and only if α is a Koszul morphism.

Exercise 6.9.4 (Unital convolution pre-Lie algebra ★). Let $(\mathscr{C}, \Delta, \varepsilon)$ be a cooperad and let $(\mathscr{P}, \gamma, \eta)$ be an operad.

Show that the composite $\eta \circ \varepsilon : \mathscr{C} \to \mathscr{P}$ is a unit for the pre-Lie product $\star$ defined in Sect. 6.4.2 on the convolution pre-Lie algebra $\mathrm{Hom}_{\mathbb{S}}(\mathscr{C}, \mathscr{P})$.

Exercise 6.9.5 (Operadic cobar construction and quasi-isomorphisms ★). Extend the results of Sect. 2.4 from conilpotent dg coalgebras to conilpotent dg cooperads, see [SV12].

Chapter 7
Koszul Duality of Operads

Les maths, c'est comme l'amour, ça ne s'apprend pas dans les livres mais en pratiquant.

Adrien Douady (private communication)

One of the aims of this chapter is to construct an explicit minimal model for a quadratic operad $\mathscr{P}$. The key point is the construction of the *Koszul dual cooperad* of $\mathscr{P}$, denoted by $\mathscr{P}^{¡}$. It permits to us to construct the *Koszul complex* $(\mathscr{P}^{¡} \circ \mathscr{P}, d_\kappa)$ out of a certain twisting morphism $\kappa : \mathscr{P}^{¡} \to \mathscr{P}$, and also to construct a differential graded operad $\Omega\mathscr{P}^{¡}$ by using the cobar construction. If the Koszul complex is acyclic, then $\Omega\mathscr{P}^{¡}$ is a minimal model of $\mathscr{P}$. So, under this assumption, we have a concrete algorithm to construct the minimal model of a quadratic operad. In this case we say that $\mathscr{P}$ is a Koszul operad. We will show in the following chapters that the dg operad $\Omega\mathscr{P}^{¡}$ plays an important role in several topics (homotopy transfer of structure for instance).

This result is in fact a corollary of the general theorem about operadic twisting morphisms and operadic Koszul morphisms proved in Chap. 6, when applied to κ.

The *Koszul dual operad* of $\mathscr{P}$, denoted by $\mathscr{P}^!$, is defined as being, up to suspension, the graded linear dual of the Koszul dual cooperad $\mathscr{P}^{¡}$. We make explicit the presentation of $\mathscr{P}^!$ out of a presentation of $\mathscr{P}$ in the binary case. Later on in the book we present several examples, among them the classical ones:

$$As^! = As, \qquad Com^! = Lie, \qquad Lie^! = Com, \qquad Pois^! = Pois, \qquad Leib^! = Zinb.$$

Koszul duality theory for binary quadratic operads is due to Victor Ginzburg and Mikhail Kapranov [GK94], see also Ezra Getzler and John D.S. Jones [GJ94]. The extension to quadratic operads (not necessarily binary) is due to Getzler [Get95] (see also Benoit Fresse [Fre04] and Martin Markl [Mar96a]).

Section 7.1 starts with operadic quadratic data and the construction of quadratic operads and quadratic cooperads out of them. In Sect. 7.2 we make precise the Koszul dual cooperad of a quadratic operad. In Sect. 7.3 we work out the bar and cobar constructions in this framework. Section 7.4 contains the main theorem

J.-L. Loday, B. Vallette, *Algebraic Operads*,
Grundlehren der mathematischen Wissenschaften 346,
DOI 10.1007/978-3-642-30362-3_7,

of the Koszul duality theory which asserts the equivalence between the acyclicity of the Koszul complexes $\mathscr{P}^{\text{i}} \circ_\kappa \mathscr{P}$, $\mathscr{P} \circ_\kappa \mathscr{P}^{\text{i}}$ and the fact that the natural maps $\mathscr{P}^{\text{i}} \rightarrowtail \mathrm{B}\mathscr{P}$, $\Omega\mathscr{P}^{\text{i}} \twoheadrightarrow \mathscr{P}$ are quasi-isomorphisms. In Sect. 7.5 we show that the generating series of a quadratic operad and its Koszul dual are related by a functional equation whenever the operad is Koszul. It is a criterion which is helpful for testing the Koszulity of an operad. In Sect. 7.6 we treat in detail the case of binary quadratic operads. In particular we show that if $\mathscr{P}$ is determined by the quadratic data (E, R) then its Koszul dual operad $\mathscr{P}^!$ is determined by $(E^* \otimes \mathrm{sgn}, R^\perp)$, where $R^\perp$ is explicitly described. In Sect. 7.7 we briefly treat the case of nonsymmetric quadratic operads along the same lines. Section 7.8 is devoted to quadratic-linear operads, that is quadratic operads whose space of relations is non-necessarily homogeneous.

In this chapter, we work over a ground field $\mathbb{K}$ of characteristic 0. Notice that all the constructions and some of the results hold true without this hypothesis.

7.1 Operadic Quadratic Data, Quadratic Operad and Cooperad

We start with an operadic quadratic data (E, R) to which we associate an operad and a cooperad

$$\begin{array}{ccc} & (E, R) & \\ \swarrow & & \searrow \\ \mathscr{P}(E, R) & & \mathscr{C}(E, R). \end{array}$$

7.1.1 Operadic Quadratic Data

By definition, an *operadic quadratic data* (E, R) is a graded $\mathbb{S}$-module E and a graded sub-$\mathbb{S}$-module $R \subseteq \mathscr{T}(E)^{(2)}$. Recall that the weight two part $\mathscr{T}(E)^{(2)}$ of the free operad is the graded sub-$\mathbb{S}$-module of the free operad $\mathscr{T}(E)$ which is spanned by the composites of two elements of E, see Sect. 5.5.3. The elements of E are called the *generating operations*. The elements of R are called the *relations* (or more appropriately the *relators*). A *morphism* of quadratic data $f : (E, R) \to (E', R')$ is a morphism of graded $\mathbb{S}$-modules $f : E \to E'$ such that $\mathscr{T}(f)(R) \subseteq R'$.

In many cases E is simply an $\mathbb{S}$-module, that is E is concentrated in degree 0.

In terms of "type of algebras", i.e. when the operad is presented by generators and relations, the quadraticity hypothesis says that the relators are made of elements involving only two compositions. For instance, if there are only binary generating operations, then the relations involve only three variables. The associative algebra case is a typical example (see Sects. 1.1.1 and 9.1). The interchange relation, cf. Sect. 13.10.4, and the Jordan algebra relation, cf. Sect. 13.10, are examples of a nonquadratic relation (they are cubic).

7.1.2 Quadratic Operad

The *quadratic operad*

$$\mathscr{P}(E,R) := \mathscr{T}(E)/(R)$$

associated to the quadratic data (E,R) is, by definition, the quotient of the free operad $\mathscr{T}(E)$ over E by the operadic ideal (R) generated by $R \subseteq \mathscr{T}(E)^{(2)}$, cf. Sect. 5.2.14. In other words, $\mathscr{P}(E,R)$ is the quotient operad of $\mathscr{T}(E)$ which is universal among the quotient operads $\mathscr{P}$ of $\mathscr{T}(E)$ such that the composite

$$R \rightarrowtail \mathscr{T}(E) \twoheadrightarrow \mathscr{P}$$

is 0. Since (R) is a homogeneous ideal with respect to the weight, it follows that $\mathscr{P}(E,R)$ is weight graded (cf. Sect. 5.5.3). We say that (E,R) is a *presentation* of the operad $\mathscr{P}(E,R)$. Explicitly, for $E = (0, E(1), E(2), E(3), \ldots)$ we obtain $\mathscr{T}(E) = \oplus_k \mathscr{T}(E)^{(k)}$, where

$$\begin{aligned}
\mathscr{T}(E)^{(0)} &= I = (0, \mathbb{K}\,\mathrm{id}, 0, 0, \ldots), \\
\mathscr{T}(E)^{(1)} &= E = \big(0, E(1), E(2), E(3), \ldots\big), \\
\mathscr{T}(E)^{(2)} &= \big(0, E(1)^{\otimes 2}, \ldots\big).
\end{aligned}$$

If, moreover, $E(1) = 0$, then we get

$$\mathscr{T}(E)^{(2)} = \big(0, 0, 0, 3\big(E(2) \oplus E(2)\big), \ldots\big).$$

Indeed, if μ and ν are two binary operations, then one can form three different ternary operations:

$$\mu\big(\nu(x,y),z\big), \qquad \mu\big(\nu(z,x),y\big), \qquad \mu\big(\nu(y,z),x\big).$$

The action of $\mathbb{S}_3$ is by permutation of the variables (x,y,z), see Sect. 7.6.2 for more details. The quotient $\mathscr{P}(E,R)$ of $\mathscr{T}(E)$ is such that

$$\begin{aligned}
\mathscr{P}(E,R)^{(0)} &= I, \\
\mathscr{P}(E,R)^{(1)} &= E, \\
\mathscr{P}(E,R)^{(2)} &= \big(0, E(1)^{\otimes 2}/R(1), \ldots\big).
\end{aligned}$$

If, moreover, $E(1) = 0$, then we get

$$\mathscr{P}(E,R)^{(2)} = \big(0, 0, 0, 3\big(E(2) \oplus E(2)\big)/R(3), \ldots\big).$$

7.1.3 Degree, Arity and Weight Gradings

Recall that an operation $\mu \in \mathscr{P}(n)$ is said to be of "arity" n. When $\mathscr{P}$ is a quadratic operad, μ has also a "degree" d (sometimes called homological degree) which comes from the fact that the space of generating operations E is graded. Moreover, since the ideal of relations is weight homogeneous, the space $\mathscr{P}(n)$ is also graded by the "weight" k which is the number of generating operations necessary to construct μ. So, any operation $\mu \in \mathscr{P}(n)$ has three different degrees called "arity", "degree" and "weight".

It should be noted that, in some papers (cf. for instance [GK95a, MSS02]), a quadratic operad stands for what we call here a *binary quadratic operad*, that is an operad which is generated by binary operations (E concentrated in arity 2), and which is quadratic (the relators belong to the weight 2 part). It turns out that for these operads the space $\mathscr{P}(n)$ is concentrated in weight $n-1$, so the weight is completely determined by the arity (and so is not mentioned in general). We prefer to dissociate the two hypotheses, because Koszul duality works for quadratic operads which are not necessarily binary, as shown in [Get95] (see also [Fre04]).

7.1.4 Quadratic Cooperad

The *quadratic cooperad* $\mathscr{C}(E,R)$ associated to the quadratic data (E,R) is, by definition, the sub-cooperad of the cofree cooperad $\mathscr{T}^c(E)$ which is *universal* among the sub-cooperads $\mathscr{C}$ of $\mathscr{T}^c(E)$ such that the composite

$$\mathscr{C} \rightarrowtail \mathscr{T}^c(E) \twoheadrightarrow \mathscr{T}^c(E)^{(2)}/R$$

is 0 (see Sect. 5.8.6 for the notion of cofree cooperad). More precisely it means that there exists a unique morphism of cooperads $\mathscr{C} \to \mathscr{C}(E,R)$ which makes the following diagram commutative:

$$\begin{array}{ccccc} \mathscr{C} & \rightarrowtail & \mathscr{T}^c(E) & \twoheadrightarrow & \mathscr{T}^c(E)^{(2)}/R \\ \downarrow & \nearrow & & & \\ \mathscr{C}(E,R). & & & & \end{array}$$

The cooperad $\mathscr{C}(E,R)$ is weight graded. In this framework, the elements of E are called the *generating cooperations* and the elements of R are called the *corelators*.

The cooperad $\mathscr{T}^c(E)$ has the same underlying $\mathbb{S}$-module as $\mathscr{T}(E)$, cf. Sect. 7.1.2. The weight graded sub-$\mathbb{S}$-module $\mathscr{C}(E,R)$ of $\mathscr{T}^c(E)$ is such that

$$\begin{aligned} \mathscr{C}(E,R)^{(0)} &= I, \\ \mathscr{C}(E,R)^{(1)} &= E, \\ \mathscr{C}(E,R)^{(2)} &= \big(0, R(1), R(2), R(3), \ldots\big). \end{aligned}$$

If, moreover, $E(1) = 0$, then we get

$$\mathscr{C}(E, R)^{(2)} = \big(0, 0, 0, R(3), \ldots\big).$$

The cooperad structure of $\mathscr{C}(E, R)$ is induced by the cooperad structure of $\mathscr{T}^c(E)$, cf. Sect. 5.8.6.

7.2 Koszul Dual (Co)Operad of a Quadratic Operad

We construct the Koszul dual cooperad and the Koszul dual operad of a quadratic operad.

7.2.1 Koszul Dual Cooperad

By definition, the *Koszul dual cooperad* of the quadratic operad $\mathscr{P} = \mathscr{P}(E, R)$ is the quadratic cooperad

$$\mathscr{P}^{¡} := \mathscr{C}\big(sE, s^2R\big)$$

introduced in Sect. 7.1.4. The sign $^{¡}$ is usually pronounced "anti-shriek". Here sE denotes the $\mathbb{S}$-module E whose degree is shifted by 1, cf. Sect. 6.2.1. Observe that, as an $\mathbb{S}$-module, $\mathscr{P}^{¡}$ can be identified with $\mathscr{C}(E, R)$. The decoration s indicates the change of grading. In diagrams the grading is usually clear and we do not mention s at the expense of modifying the signs in the explicitation of the formulas.

7.2.2 Operadic Suspension

Let $\mathscr{S} := \mathrm{End}_{s\mathbb{K}}$ be the operad of endomorphisms of the one-dimensional space put in degree one. As a representation of $\mathbb{S}_n$ the space $\mathscr{S}(n) = \mathrm{Hom}((s\mathbb{K})^{\otimes n}, s\mathbb{K})$ is the signature representation concentrated in degree $-n+1$ (since its generator sends s^n to s). See Exercise 5.11.3 for the description of algebras over this operad. Denote by $\mu \in \mathscr{S}(2)$ the generator of arity 2. The associativity relation $\mu \circ (\mu, \mathrm{id}) = \mu \circ (\mathrm{id}, \mu)$ implies the following equality when evaluated on (s, s, s):

$$\mu\big(\mu(s, s), s\big) = -\mu\big(s, \mu(s, s)\big).$$

Indeed, under the isomorphism

$$\mathscr{S}(2) \otimes \mathscr{S}(2) \otimes (s\mathbb{K})^{\otimes 3} \cong \mathscr{S}(2) \otimes s\mathbb{K} \otimes \mathscr{S}(2) \otimes (s\mathbb{K})^{\otimes 2}$$

the image of (μ, μ, s, s, s) is $-(\mu, s, \mu, s, s)$ because $|\mu| = -1$, $|s| = 1$.

We also adopt the notation $\mathscr{S}^{-1} := \mathrm{End}_{s^{-1}\mathbb{K}}$.

Let $\mathscr{P}$ be an operad. In general the suspension of the underlying $\mathbb{S}$-module is not an operad, however the Hadamard product $\mathscr{S} \underset{\mathrm{H}}{\otimes} \mathscr{P}$ is indeed an operad. We call it the *operadic suspension* of $\mathscr{P}$. It has the following property. For any graded vector space V one has an isomorphism

$$s\mathscr{P}(V) = \Big(\mathscr{S} \underset{\mathrm{H}}{\otimes} \mathscr{P}\Big)(sV).$$

Therefore, the space V is equipped with a $\mathscr{P}$-algebra structure if and only if the suspended space sV is equipped with a $\mathscr{S} \underset{\mathrm{H}}{\otimes} \mathscr{P}$-algebra structure. Similarly if $\mathscr{C}$ is a cooperad we define its suspension as being the cooperad $\mathscr{S}^c \underset{\mathrm{H}}{\otimes} \mathscr{C}$, where $\mathscr{S}^c$ is the cooperad $\mathrm{End}^c_{s\mathbb{K}}$. So, the space C is equipped with a $\mathscr{C}$-coalgebra structure if and only if the suspended space sC is equipped with a $\mathscr{S}^c \underset{\mathrm{H}}{\otimes} \mathscr{C}$-coalgebra structure.

7.2.3 Koszul Dual Operad

For any quadratic operad $\mathscr{P}$ we consider the operadic suspension of the cooperad $\mathscr{P}^{¡}$, that is the cooperad $\mathscr{S}^c \underset{\mathrm{H}}{\otimes} \mathscr{P}^{¡}$. Taking the linear dual of this cooperad, we get an operad. By definition the *Koszul dual operad* of the quadratic operad $\mathscr{P}$ is the quadratic operad

$$\mathscr{P}^! := \Big(\mathscr{S}^c \underset{\mathrm{H}}{\otimes} \mathscr{P}^{¡}\Big)^*.$$

We recall that the linear dualization mentioned here is the "arity-graded linearization", that is, we dualize each arity component individually. We observe that $\mathscr{P}^!$ is quadratic.

Proposition 7.2.1. *For any quadratic operad $\mathscr{P} = \mathscr{P}(E, R)$, generated by a reduced $\mathbb{S}$-module E of finite dimension in each arity, the Koszul dual operad $\mathscr{P}^!$ admits a quadratic presentation of the form*

$$\mathscr{P}^! = \mathscr{P}\Big(s^{-1}\mathscr{S}^{-1} \underset{\mathrm{H}}{\otimes} E^*, R^{\perp}\Big).$$

The proof of this statement is going to use the notions of Manin product and shuffle trees that will be introduced later (see Sects. 8.8.1 and 8.2 respectively). An ad hoc proof in the binary case will be given in Sect. 7.6.

Proof. The Koszul dual operad $\mathscr{P}^!$ is equal to $(\mathscr{S}^c \underset{\mathrm{H}}{\otimes} \mathscr{P}^{¡})^* = \mathscr{S}^{-1} \underset{\mathrm{H}}{\otimes} (\mathscr{P}^{¡})^*$. The main point is to use the following quadratic-linear presentation of $\mathscr{S}^{-1}$: $\mathscr{S}^{-1} = \mathscr{T}(F)/(U)$, where $F = \bigoplus_{n \geq 2} \mathbb{K}\nu_n$, with $\mathbb{K}\nu_n = \mathrm{Hom}((s^{-1}\mathbb{K})^n, s^{-1}\mathbb{K})$, that is

$|\nu_n| = 1-n$, and where U is made of all the relations of the form $\nu_n \circ_i \nu_m = \nu_{n+m-1}$, for $n, m \geq 2$ and $1 \leq i \leq n$.

Then we use the same ideas as in the proof of Proposition 8.8.1. For any shuffle tree $\mathbb{T}$ (not necessarily binary here), the maps

$$\mathscr{C}_{\mathbb{T}}^{\mathscr{S}} : \mathbb{T}(F) \to \mathscr{T}(F)(n) \twoheadrightarrow \mathscr{S}^{-1}(n)$$

are surjective. It follows that the Hadamard product is equal to the Manin white product:

$$\mathscr{P}^! = \mathscr{S}^{-1} \underset{\mathrm{H}}{\otimes} \left(\mathscr{P}^{¡}\right)^* = \mathscr{S}^{-1} \bigcirc \left(\mathscr{P}^{¡}\right)^*.$$

Since the generating space E has finite dimensional arity components, we can use the identification $(\mathscr{T}(sE)^{(2)})^* \cong \mathscr{T}(s^{-1}E^*)^{(2)}$. To this extent, we choose a tree basis of the free operad and the isomorphism reduces to $(E(n) \otimes E(m))^* \cong E(n)^* \otimes E(m)^*$. In general, one can use the basis provided by shuffle trees, see Sect. 8.2.5. Under this isomorphism, the submodule $s^2R \subset \mathscr{T}(sE)^{(2)}$ gives the orthogonal submodule $(s^2R)^{\perp} \subset \mathscr{T}(s^{-1}E^*)^{(2)}$. Since the Koszul dual cooperad $\mathscr{P}^{¡} = \mathscr{C}(sE, s^2R)$ is a quadratic cooperad, its linear dual is a quadratic operad with presentation $\mathscr{P}^! = \mathscr{P}(s^{-1}E^*, (s^2R)^{\perp})$.

By definition, Manin's white product $\mathscr{S}^{-1} \bigcirc (\mathscr{P}^{¡})^*$ is equal to the operad with presentation $\mathscr{T}(F \underset{\mathrm{H}}{\otimes} s^{-1}E^*)/(\Phi^{-1}(U \underset{\mathrm{H}}{\otimes} \mathscr{T}(s^{-1}E^*) + \mathscr{T}(F) \underset{\mathrm{H}}{\otimes} (s^2R)^{\perp}))$. The space of generators is equal to $s^{-1}\mathscr{S}^{-1} \underset{\mathrm{H}}{\otimes} E^*$. Since U is quadratic and linear and since R is quadratic, this latter operad is quadratic. The space of relations, denoted $R^{\perp}$, is obtained from $(s^2R)^{\perp}$ by proper (de)suspension of the operations indexing the vertices of the trees and by induced sign rule. □

If E is concentrated in degree 0, then so is $\mathscr{P}$ and sE is in degree $+1$. As a result $\mathscr{P}^!$ is concentrated in degree 0.

In Sect. 7.6 we describe an explicit way to construct a quadratic presentation of $\mathscr{P}^!$ out of a presentation of $\mathscr{P}$ in the binary case. In particular we will show that

$$Ass^! = Ass, \qquad Com^! = Lie, \qquad Lie^! = Com,$$

$$Pois^! = Pois, \qquad Leib^! = Zinb, \qquad Dend^! = Dias.$$

More examples will be worked out in Chap. 13.

Proposition 7.2.2. *For any quadratic operad $\mathscr{P}$, generated by a reduced $\mathbb{S}$-module E of finite dimension in each arity, we have*

$$\left(\mathscr{P}^!\right)^! = \mathscr{P}.$$

Proof. By direct inspection. The hypothesis about finite dimensionality ensures, as usual, that the natural map $E \to (E^*)^*$ is an isomorphism. □

7.3 Bar and Cobar Construction on an Operadic Quadratic Data

We make explicit the dg cooperad $\mathrm{B}\mathscr{P}$ and the dg operad $\Omega\mathscr{C}$ in the quadratic case. The Koszul dual objects are shown to be equal to the syzygy degree 0 homology group in both cases. From now on "quadratic data" means "operadic quadratic data".

7.3.1 Bar Construction on a Quadratic Operad

The bar construction $\mathrm{B}\mathscr{P} := \mathscr{T}^c(s\overline{\mathscr{P}})$ over the quadratic dg operad $\mathscr{P} = \mathscr{P}(E, R)$ is equipped with a weight grading and with a syzygy grading.

Since E has trivial internal differential, the differential on $\mathrm{B}\mathscr{P}$ reduces to d_2, which raises the syzygy degree by 1 and preserves the weight grading. So it forms a cochain complex with respect to the syzygy degree, which splits with respect to the weight grading. Hence the associated cohomology group is a direct sum over the weight grading.

The following diagram depicts this decomposition.

$$\begin{array}{ccccccccc}
0 & \leftarrow \cdots \leftarrow & & \cdots & & \leftarrow & \mathscr{T}(E)^{(3)} & & (3) \\
 & & 0 & \leftarrow & \mathscr{T}(E)^{(2)}/R & \leftarrow & \mathscr{T}(E)^{(2)} & & (2) \\
 & & & & 0 & \leftarrow & E & & (1) \\
 & & & & & & \mathrm{I} & & (0) \\
3 & & 2 & & 1 & & 0 & &
\end{array}$$

Proposition 7.3.1. *Let (E, R) be a quadratic data, $\mathscr{P} = \mathscr{P}(E, R)$ the associated quadratic operad and $\mathscr{P}^{¡} = \mathscr{C}(sE, s^2R)$ its Koszul dual cooperad. The natural cooperad inclusion $i : \mathscr{P}^{¡} \rightarrowtail \mathrm{B}\mathscr{P}$ induces an isomorphism of graded cooperads:*

$$i : \mathscr{P}^{¡} \xrightarrow{\cong} H^0(\mathrm{B}^\bullet\mathscr{P}).$$

Proof. The proof is mimicked on the proof of the algebra case, i.e. Proposition 3.3.1. □

7.3.2 Cobar Construction on a Quadratic Cooperad

Like the bar construction, the cobar construction $\Omega\mathscr{C} = T(s^{-1}\overline{\mathscr{C}})$ over the quadratic dg cooperad $\mathscr{C} = \mathscr{C}(V, R)$ has several degrees.

We introduce the same definitions as for the bar construction. We consider the *weight grading* $\Omega\mathscr{C}^{(n)}$ and the *syzygy degree* of $\Omega\mathscr{C}$ denoted by $\Omega_d\mathscr{C}$.

Since the internal differential of the cooperad $\mathscr{C}$ is trivial, the differential of the cobar construction $\Omega\mathscr{C}$ reduces to d_2, which lowers the syzygy degree by 1. Hence, $(\Omega_\bullet\mathscr{C}, d_2)$ becomes a chain complex. Its associated cohomology is a direct sum over the weight grading.

The following diagram depicts this decomposition.

$$\begin{array}{ccccccccc} 0 & \to & \cdots & \to & \cdots & \to & \mathscr{T}^c(E)^{(3)} & & (3) \\ & & 0 & \to & R & \to & \mathscr{T}^c(E)^{(2)} & & (2) \\ & & & & 0 & \to & E & & (1) \\ & & & & & & \mathrm{I} & & (0) \\ 3 & & 2 & & 1 & & 0 & & \end{array}$$

Proposition 7.3.2. *Let $\mathscr{C} = \mathscr{C}(E, R)$ be the quadratic cooperad associated to the quadratic data (E, R), and let $\mathscr{C}^{i} := \mathscr{P}(s^{-1}E, s^{-2}R)$ be its Koszul dual operad. The natural operad projection $p : \Omega\mathscr{C} \twoheadrightarrow \mathscr{C}^{i}$ induces an isomorphism of graded operads*:

$$p : H_0(\Omega_\bullet\mathscr{C}) \xrightarrow{\cong} \mathscr{C}^{i}.$$

7.4 Koszul Operads

We construct the Koszul complexes of a quadratic operad and we define the notion of "Koszul operad". Then we apply the theory of twisting morphisms to show that the Koszul dual cooperad permits us to construct the minimal model of an operad, when it is a Koszul operad.

7.4.1 The Natural Twisting Morphism κ

Since, for a given quadratic data (E, R), we have $\mathscr{P}(E, R)^{(1)} = E$ and $\mathscr{C}(E, R)^{(1)} = E$, we can define the morphism κ as the composite

$$\kappa : \mathscr{C}\big(sE, s^2R\big) \twoheadrightarrow sE \xrightarrow{s^{-1}} E \rightarrowtail \mathscr{P}(E, R).$$

Because of the degree shift the map κ is of degree -1.

Lemma 7.4.1. *We have $\kappa \star \kappa = 0$, and therefore κ is a twisting morphism.*

Proof. Since κ is 0 except in weight 1, the convolution product $\kappa \star \kappa$ is 0 except maybe on weight 2. Computing $\kappa \star \kappa$ explicitly in weight 2 we find that it is equal to the composite

$$\mathscr{C}\left(sE, s^2R\right)^{(2)} \xrightarrow{s^{-2}} R \to \mathscr{T}(E)^{(2)} \to \mathscr{T}(E)^{(2)}/R = \mathscr{P}(E,R)^{(2)},$$

which is 0 by definition.

Hence, the map κ is a twisting morphism in the framework of $\mathbb{S}$-modules by Sect. 6.4.3. □

7.4.2 The Koszul Complex of an Operad

By Sect. 6.4.5 there is, associated to the twisting morphism κ, a chain complex of $\mathbb{S}$-modules

$$\mathscr{P}^{¡} \circ_\kappa \mathscr{P} := \left(\mathscr{P}^{¡} \circ \mathscr{P}, d_\kappa\right)$$

that we call the *Koszul complex* of the operad $\mathscr{P}$. So, for any $n \geq 0$, we have a chain complex of $\mathbb{S}_n$-modules $((\mathscr{P}^{¡} \circ \mathscr{P})(n), d_\kappa)$, which we call the Koszul complex in arity n. We can also construct another Koszul complex, namely $\mathscr{P} \circ_\kappa \mathscr{P}^{¡}$, which is completely different, in general, from the first one. They both come from the complex $\mathscr{P} \circ_\kappa \mathscr{P}^{¡} \circ_\kappa \mathscr{P}$.

7.4.3 Koszul Operad

By definition, a quadratic operad $\mathscr{P}$ is called a *Koszul operad* if its associated Koszul complex $(\mathscr{P}^{¡} \circ \mathscr{P}, d_\kappa)$ is acyclic (here we mean that its homology is the identity functor).

We will show later that there exist many Koszul operads, namely the three graces *Ass*, *Com*, *Lie*, but also *Pois*, *Leib*, *Zinb*, *Dend* and many more.

7.4.4 The Minimal Model

The main advantage of the Koszul dual construction is to give the minimal model for $\mathscr{P}$ when $\mathscr{P}$ is Koszul. Recall that the advantage for a model to be minimal is that it is unique up to isomorphism, see Sect. 6.3.4.

Theorem 7.4.2 (Koszul criterion). *Let (E,R) be an operadic quadratic data, let $\mathscr{P} := \mathscr{P}(E,R)$ be the associated operad, let $\mathscr{P}^{¡} := \mathscr{C}(sE, s^2R)$ be the Koszul dual cooperad and let $\kappa : \mathscr{P}^{¡} \to \mathscr{P}$ be the associated twisting morphism. Then the following assertions are equivalent:*

1. *the right Koszul complex $\mathscr{P}^{¡} \circ_\kappa \mathscr{P}$ is acyclic,*
2. *the left Koszul complex $\mathscr{P} \circ_\kappa \mathscr{P}^{¡}$ is acyclic,*

3. *the inclusion* $i : \mathscr{P}^{¡} \rightarrowtail \mathrm{B}\mathscr{P}$ *is a quasi-isomorphism of dg cooperads,*
4. *the projection* $p : \Omega\mathscr{P}^{¡} \twoheadrightarrow \mathscr{P}$ *is a quasi-isomorphism of dg operads.*

Proof. First we remark that, by Theorem 6.6.3, the assertions (3) and (4) are equivalent. Let us prove the equivalence (1) $\Leftrightarrow$ (3). Consider the tensor map $i \circ \mathrm{Id} : \mathscr{P}^{¡} \circ \mathscr{P} \to \mathrm{B}\mathscr{P} \circ \mathscr{P}$. Since $\pi \circ i = \mathrm{Id} \circ \kappa$, the map $i \circ \mathrm{Id}$ is a morphism of chain complexes from $\mathscr{P}^{¡} \circ_\kappa \mathscr{P}$ to $\mathrm{B}\mathscr{P} \circ_\pi \mathscr{P}$. We have seen in Sect. 6.5.4 that the augmented bar construction is always acyclic. Therefore, the Koszul complex $\mathscr{P}^{¡} \circ_\kappa \mathscr{P}$ is acyclic if and only if $i \circ \mathrm{Id}$ is a quasi-isomorphism. The comparison Lemma 6.4.8 in the operadic framework implies that $\mathscr{P}$ is a Koszul operad if and only if i is a quasi-isomorphism. □

Corollary 7.4.3. *Let* $\mathscr{P}$ *be a quadratic operad. If* $\mathscr{P}$ *is Koszul, then* $\Omega\mathscr{P}^{¡}$ *is the minimal model of* $\mathscr{P}$.

Proof. By Theorem 7.4.2 the dg operad $\Omega\mathscr{P}^{¡}$ is a model of $\mathscr{P}$. By construction $\Omega\mathscr{P}^{¡}$ is a free operad, hence it is a quasi-free dg operad. Also by construction the differential of $\Omega\mathscr{P}^{¡}$ is quadratic (cf. Sect. 6.5.2), therefore we get a minimal model. □

EXAMPLE. If E is concentrated in arity one, then the operad $\mathscr{P}$ is completely determined by the associative algebra $\mathscr{P}(1)$. We recover the theory of Koszul duality for associative algebras as a particular case of the theory of Koszul duality for operads as done in Chap. 3.

Proposition 7.4.4. *Let* $\mathscr{P} = \mathscr{P}(E, R)$ *be a finitely generated quadratic operad. The operad* $\mathscr{P}$ *is Koszul if and only if its Koszul dual operad* $\mathscr{P}^!$ *is Koszul.*

Proof. Since $\mathscr{P}$ is a Koszul operad, the morphism of weight graded dg cooperads $i : \mathscr{P}^{¡} \xrightarrow{\sim} \mathrm{B}\mathscr{P}$ is a quasi-isomorphism by Theorem 7.4.2. Since E is finitely generated, the chain sub-complexes of fixed weight of $\mathrm{B}\mathscr{P}$ are finitely generated. Taking the weight graded linear dual, we get a quasi-isomorphism $\Omega\mathscr{P}^* \xrightarrow{\sim} (\mathscr{P}^{¡})^*$. Again by Theorem 7.4.2, it proves that $\mathscr{S} \underset{\mathrm{H}}{\otimes} \mathscr{P}^!$ is Koszul. One can prove that the bar constructions $\mathrm{B}(\mathcal{S} \underset{\mathrm{H}}{\otimes} \mathscr{P}^!)$ and $\mathrm{B}\mathscr{P}^!$, considered with their syzygy degree, are quasi-isomorphic. It proves that $\mathscr{P}^!$ is Koszul.

In the other way round, we use Proposition 7.2.2: $(\mathscr{P}^!)^! \cong \mathscr{P}$. □

7.4.5 Homotopy $\mathscr{P}$-Algebras

Let $\mathscr{P}$ be a Koszul operad. By definition a *homotopy* $\mathscr{P}$*-algebra* is an algebra over the dg operad $\Omega\mathscr{P}^{¡}$ of $\mathscr{P}$. Such an algebra is also called a $\mathscr{P}$*-algebra up to homotopy* or a $\mathscr{P}_\infty$*-algebra*, where $\mathscr{P}_\infty$ stands for $\Omega\mathscr{P}^{¡}$.

Hence, a homotopy $\mathscr{P}$-algebra structure on a dg module A is a morphism of dg operads $\Omega\mathscr{P}^{¡} \to \mathrm{End}_A$. Notice that any $\mathscr{P}$-algebra can be considered as a homotopy $\mathscr{P}$-algebra concentrated in degree 0. The $\mathscr{P}_\infty$-structure is given by the composition:

$$\mathscr{P}_\infty = \Omega\mathscr{P}^{¡} \xrightarrow{p} \mathscr{P} \to \mathrm{End}_A,$$

where ε is the augmentation map.

The dg operad $\mathscr{P}_\infty$ has a lot of properties which will be analyzed in Chap. 10.

7.4.6 Operadic Homology

Let A be an algebra over the Koszul operad $\mathscr{P}$. It is the quotient of a free $\mathscr{P}$-algebra $\mathscr{P}(V)$ for some space V. The boundary map $d_\kappa(V)$ on $\mathscr{P}^{¡} \circ \mathscr{P}(V) = \mathscr{P}^{¡}(\mathscr{P}(V))$ passes to the quotient to give a boundary map d and so a chain complex $C_\bullet^{\mathscr{P}}(A) := (\mathscr{P}^{¡}(A), d(A))$. By definition the *operadic homology* of A (with trivial coefficients) is the homology of the chain complex $C_\bullet^{\mathscr{P}}(A)$:

$$H_\bullet^{\mathscr{P}}(A) := H_\bullet\big(C_\bullet^{\mathscr{P}}(A)\big).$$

It will be studied in detail in Chap. 12. We only mention here a criterion for the operad $\mathscr{P}$ to be Koszul.

Proposition 7.4.5. *The acyclicity of the Koszul complex of a quadratic operad $\mathscr{P}$ is equivalent to the vanishing of the operadic homology of the free $\mathscr{P}$-algebra $\mathscr{P}(V)$ for any vector space V, more precisely*:

$$H_1^{\mathscr{P}}\big(\mathscr{P}(V)\big) = V, \quad \textit{and} \quad H_n^{\mathscr{P}}\big(\mathscr{P}(V)\big) = 0, \quad \textit{for } n \geq 2.$$

Proof. This is essentially a rephrasing of the definition of Koszulity. Indeed one has $C_\bullet^{\mathscr{P}}(\mathscr{P}(V)) = \mathscr{P}^{¡}(\mathscr{P}(V)) = \mathscr{P}^{¡} \circ \mathscr{P}(V)$ and $d(\mathscr{P}(V)) = d_\kappa(V)$. The vanishing of the homology for any V is equivalent to the vanishing of the (right) Koszul complex by Corollary 6.2.4. □

7.4.7 Koszul Theory in the Cooperad Setting

Though we are not going to work in this framework in the rest of the book, let us mention briefly the Koszul duality theory for cooperads.

By dualizing the preceding theory, we obtain the following results. An operadic quadratic data (E, R) gives rise to a cooperad $\mathscr{C} = \mathscr{C}(E, R)$, which has a Koszul dual operad $\mathscr{C}^{\text{¡}} := \mathscr{P}(s^{-1}E, s^{-2}R)$. There is a twisting morphism κ which is the composite:

$$\kappa : \mathscr{C}(E, R) \twoheadrightarrow E \xrightarrow{s^{-1}} s^{-1}E \rightarrowtail \mathscr{P}\big(s^{-1}E, s^{-2}R\big).$$

Its Koszul complex is $\mathscr{C} \circ_\kappa \mathscr{C}^{\text{¡}} := (\mathscr{C} \circ \mathscr{C}^{\text{¡}}, d_\kappa)$. The cooperad $\mathscr{C}$ is said to be Koszul if its Koszul complex is acyclic. Dually one can also define another Koszul complex: $\mathscr{C}^{\text{¡}} \circ_\kappa \mathscr{C}$.

The Koszul criterion in this case says that the following assertions are equivalent:

1. the Koszul complex $\mathscr{C} \circ_\kappa \mathscr{C}^{\text{¡}}$ is acyclic,
2. the Koszul complex $\mathscr{C}^{\text{¡}} \circ_\kappa \mathscr{C}$ is acyclic,
3. the inclusion $i : \mathscr{C} \rightarrowtail \mathrm{B}\mathscr{C}^{\text{¡}}$ is a quasi-isomorphism of dg operads,
4. the projection $p : \Omega\mathscr{C} \twoheadrightarrow \mathscr{C}^{\text{¡}}$ is a quasi-isomorphism of dg cooperads.

As a corollary, if $\mathscr{C}$ is a Koszul cooperad, then $\mathrm{B}\mathscr{C}^{\text{¡}}$ is a minimal model of $\mathscr{C}$. Observe that, for any quadratic operad $\mathscr{P}$, there is a natural identification

$$\big(\mathscr{P}^{\text{¡}}\big)^{\text{¡}} = \mathscr{P},$$

which does not require any finite dimensionality assumption.

7.5 Generating Series

The generating series of an $\mathbb{S}$-module is a formal power series which bears nice properties when the $\mathbb{S}$-module underlies a Koszul operad. It can be used to prove some relations between formal power series. It can also be used to prove that an operad is *not* a Koszul operad (but not the other way round).

7.5.1 Generating Series

Let $\mathscr{P} = \mathscr{P}(E, R)$ be a reduced quadratic operad (i.e. $E(0) = 0$), which is finitely generated, that is, $\dim E(n)$ is finite. From its presentation we deduce that $\mathscr{P}$ is weight graded, E being of weight 1. By convention the identity operation id is of weight 0. We denote by $\mathscr{P}^{(r)}(n)$ the subspace of n-ary operations of weight r. By definition the *generating series* (or *Hilbert–Poincaré series*) of the weight graded operad $\mathscr{P}$ is

$$f^{\mathscr{P}}(x, y) := \sum_{r \geq 0, n \geq 1} \frac{\dim \mathscr{P}^{(r)}(n)}{n!} y^r x^n.$$

Observe that $f^{\mathscr{P}}$ depends only on the weight graded $\mathbb{S}$-module $\mathscr{P}$. If $\mathscr{P}$ is binary, then $\mathscr{P}(n) = \mathscr{P}^{(n-1)}(n)$ and therefore there is no need for two variables. Putting $y = 1$ we recover the definition given in Sect. 5.1.10.

Theorem 7.5.1. *Let $\mathscr{P} = \mathscr{P}(E, R)$ be a connected quadratic operad. If $\mathscr{P}$ is Koszul, then the generating series of $\mathscr{P}$ and of its Koszul dual operad $\mathscr{P}^!$ are related by the following functional equation:*

$$f^{\mathscr{P}^!}\big(f^{\mathscr{P}}(x, y), -y\big) = x.$$

Proof. From the definition of $\mathscr{P}^!$ out of $\mathscr{P}^{¡}$ it follows that these two weight graded $\mathbb{S}$-modules have the same generating series: $f^{\mathscr{P}^!} = f^{\mathscr{P}^{¡}}$. The Koszul complex $\mathscr{K}$ of $\mathscr{P}$ splits has a sum of chain complexes, indexed by arity and weight:

$$\mathscr{K} = \bigoplus_{r,n} \mathscr{K}^{(r)}(n),$$

where we have

$$\mathscr{K}^{(r)}(n) : 0 \to \mathscr{P}^{¡(r)}(n) \to \cdots \to \big(\mathscr{P}^{¡(k)} \circ \mathscr{P}\big)^{(r)}(n) \to \cdots \to \mathscr{P}^{(r)}(n).$$

By the Koszulity of $\mathscr{P}$ these complexes are acyclic (in the unital sense). Hence the Euler–Poincaré characteristic is trivial. Putting all these identities together amounts to the formula:

$$f^{\mathscr{P}^!}\big(f^{\mathscr{P}}(x, y), -y\big) = x.$$

In fact it is a particular case of a more general formula of the same type valid for "properads", cf. Sect. 13.14.9, and proved in [Val07b]. □

7.5.2 Generating Series of Quadratic Algebras

Let us suppose that the operad $\mathscr{P}$ is in fact a quadratic algebra R, that is $\mathscr{P}(n) = 0$ when $n \neq 1$ and $\mathscr{P}(1) = R$. Let f^R be the generating series of the quadratic algebra R (cf. Sect. 3.5). It follows that $f^{\mathscr{P}}(x, y) = f^R(y)x$ and $f^{\mathscr{P}^!}(x, y) = f^{R^!}(y)x$. The functional equation reads

$$f^{R^!}(-y) f^R(y)x = x$$

and so we recover the functional equation for the generating series of Koszul algebras given in Sect. 3.5.

7.5.3 Example and Counterexample

In Proposition 13.11.7 we will show how to use Theorem 7.5.1 to compute the coefficients of the inverse of the generic formal power series.

The operad *PreLie* • *Nil* is an example of an operad which is not Koszul. This is proved by showing that its generating series and the generating series of its Koszul dual do not satisfy the functional equation, see [Val08].

7.6 Binary Quadratic Operads

We make explicit the Koszul dual operad of an operad presented by (finitely many) binary operations and quadratic relations.

7.6.1 Explicitation of a Binary Quadratic Operad

Let E be an $\mathbb{S}_2$-module (put in degree 0) and let $\underline{E}$ be the $\mathbb{S}$-module

$$\underline{E} := (0, 0, E, 0, \ldots, 0, \ldots).$$

The free operad on $\underline{E}$ is denoted by $\mathscr{T}(E)$ instead of $\mathscr{T}(\underline{E})$. It is clear from its construction, cf. Sect. 5.5, that $\mathscr{T}(E)(0) = 0$, $\mathscr{T}(E)(1) = \mathbb{K}$, $\mathscr{T}(E)(2) = E$. Since we are in the binary case we have $\mathscr{T}(E)^{(2)} = \mathscr{T}(E)(3)$.

Proposition 7.6.1. *The* $\mathbb{S}_3$*-module* $\mathscr{T}(E)(3)$ *is*

$$\mathscr{T}(E)(3) = E \otimes \mathrm{Ind}_{\mathbb{S}_2}^{\mathbb{S}_3} E \cong 3E \otimes E.$$

Proof. From the explicit construction of the free operad given in Sect. 5.5, we get

$$\mathscr{T}(E)(3) = \big(\mathrm{Id} \oplus \underline{E} \circ (\mathrm{Id} \oplus \underline{E})\big)(3) = E \otimes \mathrm{Ind}_{\mathbb{S}_2}^{\mathbb{S}_3} E.$$

Indeed, it suffices to look for the $V^{\otimes 3}$ component of $\mathscr{T}(E)(V)$. Since the quotient $\mathbb{S}_2 \backslash \mathbb{S}_3$ is a set with three elements, we get the second assertion (see Sect. 7.6.2 for more details on this isomorphism). □

7.6.2 The Space $\mathscr{T}(E)(3)$ Made Explicit

In the proof of Proposition 7.6.1 we use the "tensor type" construction of the free operad. If, instead, we use the "combinatorial" type as done in Sect. 5.6, then we can make explicit the isomorphism $\mathscr{T}(E)(3) \cong 3E \otimes E$ as follows. Let $\mu, \nu \in E$ be two binary operations. We denote by

$$\mu \circ_{\mathrm{I}} \nu, \quad \mu \circ_{\mathrm{II}} \nu, \quad \mu \circ_{\mathrm{III}} \nu \in \mathscr{T}(E)(3)$$

respectively the three operations on three variables defined by:

$$(\mu \circ_{\mathrm{I}} \nu)(x, y, z) := \mu\big(\nu(x, y), z\big),$$
$$(\mu \circ_{\mathrm{II}} \nu)(x, y, z) := \mu\big(\nu(z, x), y\big),$$
$$(\mu \circ_{\mathrm{III}} \nu)(x, y, z) := \mu\big(\nu(y, z), x\big).$$

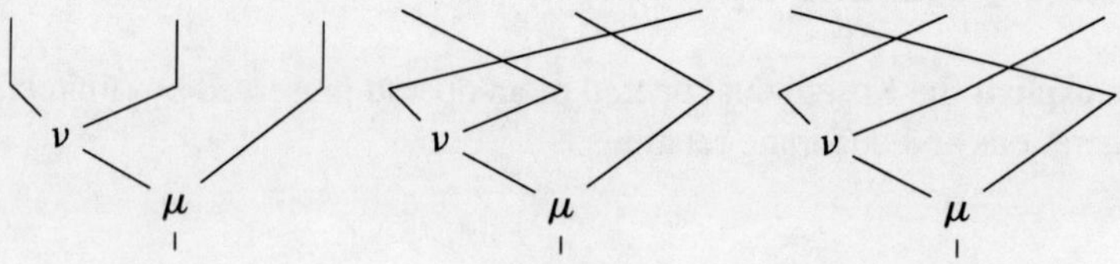

$(\mu \circ_{\mathrm{I}} \nu)$ $(\mu \circ_{\mathrm{II}} \nu)$ $(\mu \circ_{\mathrm{III}} \nu)$

These formulas determine the three copies of $E \otimes E$. If λ denotes the cyclic permutation on three variables defined by $\lambda(x, y, z) = (y, z, x)$, then we have

$$\mu \circ_{\mathrm{I}} \nu = (\mu \circ_1 \nu)^{\lambda^0}, \qquad \mu \circ_{\mathrm{II}} \nu = (\mu \circ_1 \nu)^{\lambda^1}, \qquad \mu \circ_{\mathrm{III}} \nu = (\mu \circ_1 \nu)^{\lambda^2}.$$

Let us now describe the action of $\mathbb{S}_3$ on $3E \otimes E = (E \otimes E)_{\mathrm{I}} \oplus (E \otimes E)_{\mathrm{II}} \oplus (E \otimes E)_{\mathrm{III}}$. It suffices to describe the action of the two cycles (12) and $\lambda = (123)$. If ω is a ternary operation, then $\omega^\sigma(x, y, z) = \omega(\sigma \cdot (x, y, z))$. Hence we get

$$\begin{aligned}(\mu \circ_{\mathrm{I}} \nu)^{(12)} &= \mu \circ_{\mathrm{I}} \nu^{(12)}\\ (\mu \circ_{\mathrm{II}} \nu)^{(12)} &= \mu \circ_{\mathrm{III}} \nu^{(12)}\\ (\mu \circ_{\mathrm{III}} \nu)^{(12)} &= \mu \circ_{\mathrm{II}} \nu^{(12)}\end{aligned}$$

and

$$(\mu \circ_{\mathrm{I,II\ or\ III}} \nu)^{(123)} = \mu \circ_{\mathrm{II,III\ or\ I}} \nu,$$

i.e. cyclic permutation of the indices.

7.6.3 Dualizing the Quadratic Data

For any finite dimensional right $\mathbb{S}_2$-module E its linear dual vector space $E^* = \mathrm{Hom}(E, \mathbb{K})$ is a left $\mathbb{S}_2$-module. We make it into a right $\mathbb{S}_2$-module as usual (it is purely formal here since $\mathbb{K}[\mathbb{S}_2]$ is commutative). Then we define

$$E^\vee := E^* \otimes \mathrm{sgn}_2 .$$

So, as a vector space, $E^\vee$ is isomorphic to E^* and the action of $\sigma \in \mathbb{S}_2$ on $f : E \to \mathbb{K}$ is given by

$$f^\sigma(e) = \mathrm{sgn}(\sigma) f\big(e^{\sigma^{-1}}\big).$$

We identify the $\mathbb{S}_3$-module $\mathcal{T}(E^\vee)(3)$ to the dual of $\mathcal{T}(E)(3)$ by means of the following scalar product:

$$\begin{aligned}&\langle -, - \rangle : \mathcal{T}\big(E^\vee\big)(3) \otimes \mathcal{T}(E)(3) \longrightarrow \mathbb{K},\\ &\big\langle \alpha^* \circ_u \beta^*, \mu \circ_v \nu \big\rangle := \alpha^*(\mu)\beta^*(\nu) \in \mathbb{K} \quad \text{if } u = v,\\ &\big\langle \alpha^* \circ_u \beta^*, \mu \circ_v \nu \big\rangle := 0 \quad \text{otherwise.}\end{aligned}$$

Here the indices u and v take values in the set {I, II, III}.
By definition the orthogonal space $R^{\perp} \subset \mathscr{T}(E^{\vee})(3)$ is

$$R^{\perp} := \{x \in \mathscr{T}(E^{\vee})(3) \mid \langle x, R\rangle = 0\}.$$

Theorem 7.6.2 (Koszul dual operad of a binary quadratic operad). *For any (finitely generated) binary quadratic operad $\mathscr{P} = \mathscr{P}(E, R)$ its Koszul dual operad is given by*

$$\mathscr{P}^{!} = \mathscr{P}(E^{\vee}, R^{\perp}).$$

We first prove a property of the scalar product introduced in Sect. 7.6.3.

Lemma 7.6.3. *The scalar product $\langle -, -\rangle$ is nondegenerate and $\mathbb{S}_3$-sign-invariant:*

$$\langle \Phi^{\sigma}, \omega^{\sigma}\rangle = \operatorname{sgn}(\sigma)\langle \Phi, \omega\rangle$$

for any $\sigma \in \mathbb{S}_3$.

Proof. The first assertion is obvious. To prove the second one, it suffices to check it for $\sigma = (12)$ and $\sigma = (123)$, $\Phi \in \mathscr{T}(E^{\vee})(3)$ and $\omega \in \mathscr{T}(E)(3)$.

Case $\sigma = (12)$. We use the following equality, which is a consequence of the definition of the action of $\mathbb{S}_2$ on $E^{\vee}$:

$$(\beta^*)^{(12)} = -(\beta^{(12)})^*, \quad \text{for any } \beta \in E.$$

We also use the action of $\mathbb{S}_3$ on $\mathscr{T}(E)(3)$ as described in Sect. 7.6.2. We compute:

$$\begin{aligned}\langle (\alpha^* \circ_u \beta^*)^{*(12)}, (\mu \circ_v \nu)^{(12)}\rangle &= \langle \alpha^* \circ_{u^{(12)}} \beta^{*(12)}, \mu \circ_{v^{(12)}} \nu^{(12)}\rangle \\ &= \langle \alpha^* \circ_u \beta^*, \mu \circ_v \nu\rangle.\end{aligned}$$

The last equality holds because the two elements are equal to 0 or 1 under the same conditions.

Case $\sigma = (123)$. The action of $\sigma = (123)$ on the index u of the element $\alpha \circ_u \beta$ is the cyclic permutation among I, II, III. The two elements $\langle (\alpha^* \circ_u \beta^*)^{\sigma}, (\mu \circ_v \nu)^{\sigma}\rangle = \langle \alpha^* \circ_{u^{\sigma}} \beta^*, \mu \circ_{v^{\sigma}} \nu\rangle$ and $\langle \alpha^* \circ_u \beta^*, \mu \circ_v \nu\rangle$ are equal because they are equal to 0 or 1 under the same conditions. Since the signature of σ is $+1$ we are done. □

Proof of Theorem 7.6.2. Let us recall from Sect. 7.2.3 that the Koszul dual operad $\mathscr{P}^{!}$ is

$$\mathscr{P}^{!} := \left(\mathscr{S}^c \underset{\mathrm{H}}{\otimes} \mathscr{P}^{¡}\right)^*.$$

We claim that

$$\mathcal{P}^{¡} = \mathcal{C}\big(sE, s^2R\big),$$
$$\mathcal{S}^c \underset{\mathrm{H}}{\otimes} \mathcal{P}^{¡} = \mathcal{C}\big(E \otimes \mathrm{sgn}_2, R'\big),$$
$$\big(\mathcal{S}^c \underset{\mathrm{H}}{\otimes} \mathcal{P}^{¡}\big)^* = \mathcal{P}\big(E^\vee, R^\perp\big).$$

Since $\mathcal{S}^c(n) = \mathrm{End}_{s\mathbb{K}}(n) = \mathrm{Hom}((s\mathbb{K})^{\otimes n}, s\mathbb{K})$, the generator in arity 2 is s^{-1}. Moreover, the action of $\mathbb{S}_2$ on this generator is by the signature representation. Therefore the space of cogenerators of the cooperad $\mathcal{S}^c \underset{\mathrm{H}}{\otimes} \mathcal{P}^{¡}$ is $E \otimes \mathrm{sgn}_2$, which is in degree 0. By hypothesis R is a sub-$\mathbb{S}_3$-space of $\mathcal{T}(E)(3) = 3E \otimes E$. Since $\mathrm{sgn}_2 \otimes \mathrm{sgn}_2$ is the trivial representation, there is a canonical isomorphism

$$\mathcal{T}(E)(3) = 3E \otimes E \xrightarrow{\approx} 3E \otimes \mathrm{sgn}_2 \otimes E \otimes \mathrm{sgn}_2 = \mathcal{T}(E \otimes \mathrm{sgn}_2)(3).$$

The space $R' \subset \mathcal{T}(E \otimes \mathrm{sgn}_2)(3)$ is the image of R under this isomorphism.

For a quadratic cooperad $\mathcal{C}(E, R)$, the associated linear dual operad $\mathcal{C}(E, R)^*$ is a quadratic operad $\mathcal{P}(E^*, S)$, where $S \subset \mathcal{T}(E^*)(3)$ is obtained as follows:

$$\begin{array}{ccccc} R^* & \twoheadleftarrow & \mathcal{T}(E)(3)^* & \hookleftarrow & \mathrm{Ker} \\ & & \downarrow \cong & & \downarrow \cong \\ & & \mathcal{T}(E^*)(3) & \hookleftarrow & S. \end{array}$$

The middle vertical isomorphism is obtained through the nondegenerate bilinear form

$$\langle -, - \rangle : \mathcal{T}\big(E^*\big)(3) \otimes \mathcal{T}(E)(3) \longrightarrow \mathbb{K},$$
$$\big\langle \alpha^* \circ_u \beta^*, \mu \circ_v \nu \big\rangle := \alpha^*(\mu)\beta^*(\nu) \in \mathbb{K} \quad \text{if } u = v,$$
$$\big\langle \alpha^* \circ_u \beta^*, \mu \circ_v \nu \big\rangle := 0 \quad \text{otherwise}.$$

Here the indices u and v take values in the set $\{\mathrm{I}, \mathrm{II}, \mathrm{III}\}$.

Applying this process to $\mathcal{C}(E \otimes \mathrm{sgn}_2, R')$ we get $\mathcal{P}^! = \mathcal{P}(E^\vee, R^\perp)$ as expected. □

7.6.4 Examples of Computations: *Com*, *Ass* and More

The operad $Com = \mathcal{P}(E, R)$ of commutative algebras is generated by a binary symmetric operation μ, i.e. $E = \mathbb{K}\mu$. Let us denote by u_{I}, respectively u_{II} and u_{III}, the elements $\mu \circ_{\mathrm{I}} \mu$, respectively $\mu \circ_{\mathrm{II}} \mu$ and $\mu \circ_{\mathrm{III}} \mu$, in $\mathcal{T}(\mathbb{K}\mu)(3)$. Its space of relations R admits the following $\mathbb{K}$-linear basis made of two elements $u_{\mathrm{I}} - u_{\mathrm{II}}$ and $u_{\mathrm{II}} - u_{\mathrm{III}}$.

The Koszul dual operad of *Com* is generated by the space $E^\vee = \mathbb{K}\mu^* \otimes \mathrm{sgn}_2 \cong \mathbb{K}\nu$, where $\nu^{(12)} = -\nu$. Again, we denote by v_{I}, respectively v_{II} and v_{III}, the elements $\nu \circ_{\mathrm{I}} \nu$, respectively $\nu \circ_{\mathrm{II}} \nu$ and $\nu \circ_{\mathrm{III}} \nu$ in $\mathcal{T}(\mathbb{K}\nu)(3)$. The orthogonal space of R under the scalar product $\langle -, - \rangle$ has dimension one with basis $v_{\mathrm{I}} + v_{\mathrm{II}} + v_{\mathrm{III}}$. De-

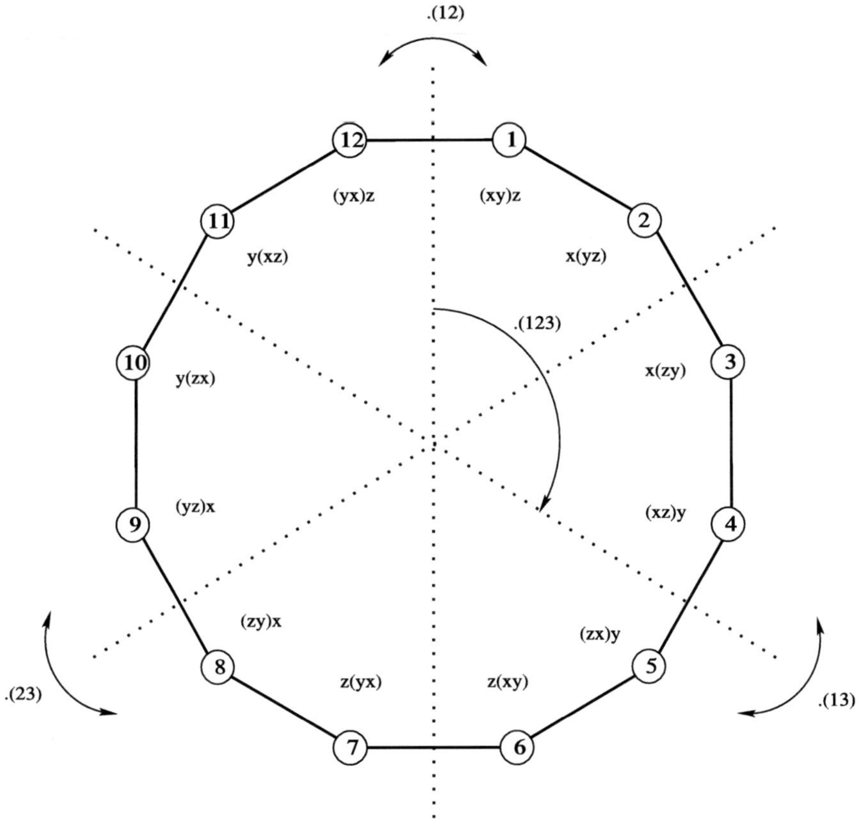

Fig. 7.1 The permutoassociahedron

noting the skew-symmetric generating operation ν by a bracket [,] the latter relation is nothing but the Jacobi relation: $[[a,b],c]+[[b,c],a]+[[c,a],b]=0$. Therefore we obtain $Com^! = Lie$.

More generally, when the generators of the binary quadratic operad $\mathscr{P}(E,R)$ form the regular representation $E=\mathbb{K}[\mathbb{S}_2]=\mathbb{K}\mu\oplus\mathbb{K}\mu'$, with $\mu' := \mu^{(12)}$, we adopt the following convention. Denote by $u_1,\ldots,u_{12}$ the corresponding 12 elements of $\mathscr{T}(E)(3)$:

1	$\mu \circ_{\mathrm{I}} \mu \leftrightarrow (xy)z$	**5**	$\mu \circ_{\mathrm{III}} \mu \leftrightarrow (zx)y$	**9**	$\mu \circ_{\mathrm{II}} \mu \leftrightarrow (yz)x$
2	$\mu' \circ_{\mathrm{II}} \mu \leftrightarrow x(yz)$	**6**	$\mu' \circ_{\mathrm{I}} \mu \leftrightarrow z(xy)$	**10**	$\mu' \circ_{\mathrm{III}} \mu \leftrightarrow y(zx)$
3	$\mu' \circ_{\mathrm{II}} \mu' \leftrightarrow x(zy)$	**7**	$\mu' \circ_{\mathrm{I}} \mu' \leftrightarrow z(yx)$	**11**	$\mu' \circ_{\mathrm{III}} \mu' \leftrightarrow y(xz)$
4	$\mu \circ_{\mathrm{III}} \mu' \leftrightarrow (xz)y$	**8**	$\mu \circ_{\mathrm{II}} \mu' \leftrightarrow (zy)x$	**12**	$\mu \circ_{\mathrm{I}} \mu' \leftrightarrow (yx)z$

This labeling corresponds to the labeling of the permutoassociahedron [Kap93]. Figure 7.1 represents it with the action of the symmetric group $\mathbb{S}_3$. Thanks to Lemma 7.6.3 one needs only to verify the orthogonality of a few elements (4 out of 36 in the next example).

Under these notations, the symmetric operad *Ass* which encodes associative algebras, has a quadratic presentation $\mathscr{P}(\mathbb{K}\mu \oplus \mathbb{K}\mu', R_{Ass})$, where the space of relations R_{Ass} has the following $\mathbb{K}$-linear basis $\{u_i - u_{i+1}, \text{ for } i = 1, 3, 5, 7, 9, 11\}$. Its Koszul dual operad is generated by $E^\vee = (\mathbb{K}\mu^* \oplus \mathbb{K}(\mu')^*) \otimes \mathrm{sgn}_2 \cong \mathbb{K}\nu \oplus \mathbb{K}\nu'$, where $\nu^{(12)} = -\nu'$. We consider the associated basis $v_1, \ldots, v_{12}$ of $\mathscr{T}(E^\vee)(3)$. In this basis, the orthogonal space $R_{Ass}^\perp$ is linearly spanned by $\{v_i + v_{i+1}, \text{ for } i = 1, 3, 5, 7, 9, 11\}$. The latter operad is isomorphic to *Ass* under the unique morphism of operads such that $\nu \mapsto \mu$ and $\nu' \mapsto -\mu'$. Therefore we obtain the isomorphism $Ass^! \cong Ass$.

The associative case is a particular case of the following one.

Proposition 7.6.4. *Let $\mathscr{P} = \mathscr{P}(\mathbb{K}[\mathbb{S}_2], R)$ be the operad generated by one binary operation (without symmetry) and relation*

$$(xy)z = \sum_{\sigma \in \mathbb{S}_3} a_\sigma \sigma \cdot \big(x(yz)\big), \quad a_\sigma \in \mathbb{K}.$$

Its Koszul dual operad is also presented by one operation and one relation. This relation is

$$\big(x(yz)\big) = \sum_{\sigma \in \mathbb{S}_3} \mathrm{sgn}(\sigma) a_\sigma \sigma^{-1} \cdot \big((xy)z\big).$$

Proof. We apply the same argument as in the associative case. □

Poisson algebras, Leibniz algebras and Zinbiel algebras give operads of this form.

7.6.5 Lie Structure on a Tensor Product of Algebras

Proposition 7.6.5. *Let $\mathscr{P} = \mathscr{P}(E, R)$ be a binary quadratic operad, supposed to be finitely generated, let $\mathscr{P}^!$ be its Koszul dual operad. Let A be a $\mathscr{P}$-algebra and let B be a $\mathscr{P}^!$-algebra. The binary operation on $A \otimes B$ defined by*

$$[a \otimes b, a' \otimes b'] := \sum_\mu \mu(a, a') \otimes \mu^\vee(b, b'),$$

where the sum is extended to a basis of E made of operations which are either symmetric ($\mu^{(12)} = \mu$) or antisymmetric ($\mu^{(12)} = -\mu$), is a Lie bracket. In other words $A \otimes B$ is naturally a Lie algebra. Equivalently, there is a morphism of operads

$$Lie \longrightarrow \mathscr{P} \underset{\mathrm{H}}{\otimes} \mathscr{P}^!.$$

Proof. We suppose that E is equipped with a basis and E^* is equipped with the dual basis. The dual of $\mu \in E$ is denoted $\mu^\vee \in E^\vee$. First, we know that when μ is symmetric (resp. antisymmetric), then $\mu^\vee$ is antisymmetric (resp. symmetric). Hence the bracket is antisymmetric.

The Jacobi identity is a consequence of Lemma 7.6.6 below. A more conceptual explanation of the existence of the functor $Lie \longrightarrow \mathcal{P} \otimes \mathcal{P}^!$ will be given in Sect. 8.8 in terms of Manin products. □

Lemma 7.6.6. *Let L be a finite dimensional vector space, with linear dual L^*. Let $R \subset L$ be a subvector space and let $R^\perp \subset L^*$ be its orthogonal. The element $u \in L^* \otimes L$, corresponding to id_L under the isomorphism $\mathrm{End}(L) \cong L^* \otimes L$, satisfies the following property:*

$$u \in R^\perp \otimes L + L^* \otimes R.$$

Proof. The image of an endomorphism $f : L \to L$ in $L^* \otimes L$ is denoted by $\tilde{f}$. So we have $u = \widetilde{\mathrm{id}_L}$. Let $\pi : L \twoheadrightarrow R$ be a section of the inclusion map $R \rightarrowtail L$. So we have $\pi|_R = \mathrm{id}_R$, $\mathrm{Im}\,\pi \subset R$ and $\mathrm{Ker}(\mathrm{id}_L - \pi) \subset R$. It follows that we have $\tilde{\pi} \in L^* \otimes R$ and $(\mathrm{id}_L - \pi)\tilde{} \in R^\perp \otimes L$. From the identity $\mathrm{id}_L = (\mathrm{id}_L - \pi) - \pi$ it follows that $u = \widetilde{\mathrm{id}_L} \in R^\perp \otimes L + L^* \otimes R$. □

REMARK. In practice, an efficient way to unravel a presentation of the dual operad $\mathcal{P}^!$, when $\mathcal{P}$ is given through a (small) presentation, consists in writing the Jacobi identity in $A \otimes B$ for generic algebras A and B.

7.6.6 Generating Series of a Binary Quadratic Operad

By definition the generating series of a binary quadratic operad is the generating series of its underlying $\mathbb{S}$-module (cf. Sect. 5.1.10):

$$f^{\mathcal{P}}(t) := \sum_{n \geq 1} \frac{\dim \mathcal{P}(n)}{n!} t^n.$$

Theorem 7.6.7. *Let $\mathcal{P}$ be a binary quadratic operad. If $\mathcal{P}$ is Koszul, then the generating series of $\mathcal{P}$ and of its Koszul dual operad $\mathcal{P}^!$ are related by the following functional equation:*

$$f^{\mathcal{P}^!}\big(-f^{\mathcal{P}}(t)\big) = -t.$$

Proof. Formerly it is a particular case of Theorem 7.5.1. Indeed, since $\mathcal{P}$ is binary we have $\mathcal{P}(n) = \mathcal{P}^{(n-1)}(n)$. Taking $x = t$ and $y = 1$, resp. $y = -1$, in the generating series with two variables we get $f^{\mathcal{P}}(t) = f^{\mathcal{P}}(t, 1)$, resp. $-f^{\mathcal{P}^!}(-t) = f^{\mathcal{P}}(t, -1)$. The functional equation reads:

$$-f^{\mathcal{P}^!}\big(-f^{\mathcal{P}}(t)\big) = t.$$

Of course one can show this formula directly by splitting the Koszul complex according to the arity and take the Euler–Poincaré characteristic. □

EXAMPLES. For the operad *Ass* of associative algebras we get

$$f^{Ass}(t) = \sum_{n\geq 1} t^n = \frac{t}{1-t}.$$

It is immediate to check that $f^{Ass}(-f^{Ass}(t)) = -t$.

For the operad *Com* of commutative algebras (resp. *Lie* of Lie algebras) we get

$$f^{Com}(t) = \sum_{n\geq 1} \frac{t^n}{n!} = \exp(t) - 1,$$

$$f^{Lie}(t) = \sum_{n\geq 1} \frac{t^n}{n} = -\log(1-t).$$

It is immediate to check that $f^{Com}(-f^{Lie}(t)) = -t$.

More examples are treated in Chap. 13.

Theorem 7.6.7 is helpful in proving that some operads are not Koszul. For instance the operad with one binary generating operation and with relation $(xy)z = 2x(yz)$ (2 supposed to be invertible in $\mathbb{K}$) is not Koszul. Another example is the operad $Nil \bullet preLie$, see Exercise 8.10.16 and [Val08].

7.7 Nonsymmetric Binary Quadratic Operad

A nonsymmetric binary quadratic operad $\mathscr{P} = \mathscr{P}(F, R)$ is completely determined by the space of generating operations $\mathscr{P}_2 = F$ and the space of relations $R \subset \mathscr{T}(F)^{(2)} = \mathscr{T}(F)(3) = F^{\otimes 2} \oplus F^{\otimes 2}$. Its Koszul dual operad is also a nonsymmetric binary quadratic operad, that we are going to make explicit.

7.7.1 Nonsymmetric Binary Quadratic Koszul Dual Operad

By convention, an element (μ, ν) in the first (resp. second) copy of $F^{\otimes 2}$ in the sum $F^{\otimes 2} \oplus F^{\otimes 2}$ corresponds to the composition $\mu \circ (\nu, \mathrm{id})$, resp. $\mu \circ (\mathrm{id}, \nu)$:

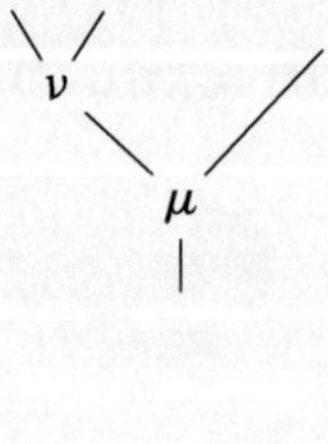

resp.

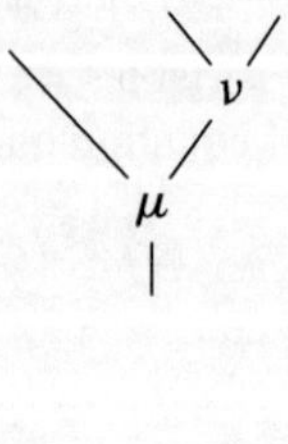

.

Consider $\mathscr{S} := \mathrm{End}_{s\mathbb{K}}$ (resp. $\mathscr{S}^c := (\mathrm{End}_{s\mathbb{K}})^c$) as a ns operad (resp. ns cooperad). An algebra over $\mathscr{S}$ is a graded vector space A with a bilinear map $A_n \otimes A_m \to A_{n+m+1}, x \otimes y \mapsto xy$ for any $n, m \geq 0$, such that $(xy)z = (-1)^{|x|}x(yz)$.

By definition the Koszul dual ns operad of a binary quadratic ns operad $\mathscr{P}$ is

$$\mathscr{P}^! := \left(\mathscr{S}^c \underset{\mathrm{H}}{\otimes} \mathscr{P}^{¡}\right)^*.$$

Theorem 7.7.1. *Let $\mathscr{P} = \mathscr{P}(F, R)$ be a (finitely generated) binary quadratic ns operad. Its Koszul dual ns operad is*

$$\mathscr{P}^! = \mathscr{P}\big(F^*, R^\perp\big),$$

where $R^\perp$ is the orthogonal space of R in $F^* \otimes F^* \bigoplus F^* \otimes F^*$ for the scalar product $\left[\begin{smallmatrix} 1 & 0 \\ 0 & -1 \end{smallmatrix}\right]$.

Proof. We claim that, in the ns framework,

$$\begin{aligned} \mathscr{P}^{¡} &= \mathscr{C}\big(sF, s^2R\big), \\ \mathscr{S}^c \underset{\mathrm{H}}{\otimes} \mathscr{P}^{¡} &= \mathscr{C}\big(F, R'\big), \\ \left(\mathscr{S}^c \underset{\mathrm{H}}{\otimes} \mathscr{P}^{¡}\right)^* &= \mathscr{P}\big(F^*, R^\perp\big). \end{aligned}$$

Since $\mathscr{S}^c_n = (\mathrm{End}_{s\mathbb{K}})_n = \mathrm{Hom}((s\mathbb{K})^{\otimes n}, s\mathbb{K})$, the generator in arity 2 is s^{-1}. Therefore the space of cogenerators of the cooperad $\mathscr{S}^c \underset{\mathrm{H}}{\otimes} \mathscr{P}^{¡}$ is F, which is in degree 0. By hypothesis R is a subspace of $\mathscr{T}(F)_3 = 2F \otimes F$ and there is a canonical isomorphism

$$\mathscr{T}(F)_3 = 2F \otimes F \xrightarrow{\approx} \mathscr{T}(F \otimes \mathrm{sgn}_2)_3.$$

The space $R' \subset \mathscr{T}(F)_3$ is the image of R under this isomorphism.

For a quadratic cooperad $\mathscr{C}(F, R)$, the associated linear dual operad $\mathscr{C}(F, R)^*$ is a quadratic operad $\mathscr{P}(F^*, S)$, where $S \subset \mathscr{T}(F^*)_3$ is obtained as follows:

$$\begin{array}{ccccc} R^* & \twoheadleftarrow & \mathscr{T}(F)_3^* & \leftarrowtail & \mathrm{Ker} \\ & & \downarrow \cong & & \downarrow \cong \\ & & \mathscr{T}(F^*)_3 & \leftarrowtail & S. \end{array}$$

The middle vertical isomorphism is obtained through the nondegenerate bilinear form $\langle -, - \rangle : \mathscr{T}(F^*)_3 \otimes \mathscr{T}(F)_3 \to \mathbb{K}$:

$$\begin{aligned} &\big\langle (\mu^*, \nu^*)_1, (\mu, \nu)_1 \big\rangle = 1, \\ &\big\langle (\mu^*, \nu^*)_2, (\mu, \nu)_2 \big\rangle = -1, \\ &\big\langle (\alpha^*, \beta^*)_i, (\mu, \nu)_j \big\rangle = 0 \quad \text{in all other cases.} \end{aligned}$$

Here the index 1 or 2 indicates the copy in which the element lies (first or second).

Applying this process to $\mathscr{C}(F, R')$ we get $\mathscr{P}^! = \mathscr{P}(F^*, R^\perp)$ as expected. □

7.7.2 The Example of the NS Operad *As*

The binary generating operation of *As* is xy. The relator is the associator $as(x, y, z) := (xy)z - x(yz)$. Let us still denote the dual operation by xy. The relator of the dual operad is of the form $a(xy)z - bx(yz)$. Since $\langle (xy)z, (xy)z\rangle = +1$ and $\langle x(yz), x(yz)\rangle = -1$, we get the condition $a = b$. So *As* is self-dual.

Proposition 7.7.2. *Let $\mathscr{P} = \mathscr{P}(F, R)$ be a (finitely generated) nonsymmetric binary quadratic operad and let $\mathscr{P}^!$ be its Koszul dual ns operad. Let A be a $\mathscr{P}$-algebra and let B be a $\mathscr{P}^!$-algebra. Then the binary operation on $A \otimes B$ defined by*

$$(a \otimes b)(a' \otimes b') := \sum_\mu \mu(a, a') \otimes \mu^*(b, b')$$

where the sum is extended to a basis of F, is associative. In other words $A \otimes B$ is naturally an associative algebra, that is, there is a morphism of nonsymmetric operads

$$As \longrightarrow \mathscr{P} \underset{\mathrm{H}}{\otimes} \mathscr{P}^!.$$

Proof. In the formula we use a basis of F and the dual basis for F^*. The associativity property is a consequence of Lemma 7.6.6. See Sect. 8.8 for a conceptual explanation. □

REMARK. Under the notation of Theorem 7.7.2 the algebra A (resp. B) can be considered as an algebra over the symmetric operad associated to $\mathscr{P}$ (resp. $\mathscr{P}^!$). Hence, by Theorem 7.6.5, $A \otimes B$ has a Lie algebra structure. It is immediate to verify that this is exactly the Lie algebra structure coming from the associative structure (see Sect. 1.1.9).

7.8 Koszul Duality for Inhomogeneous Quadratic Operads

So far we supposed that the operadic data (E, R) was weight homogeneous quadratic, that is $R \subset \mathscr{T}(E)^{(2)}$. In this section, we suppose that (E, R) is only inhomogeneous quadratic, that is

$$R \subset \mathscr{T}(E)^{(1)} \oplus \mathscr{T}(E)^{(2)}.$$

There exists an even more general case allowing also constant terms in the space of relations, cf. [HM10].

A Koszul duality theory still exists under this hypothesis. It generalizes to operads the results exposed in Sect. 3.6 for associative algebras. In this section, we state the results without proofs and refer the reader to [GCTV09] for details. As in the case of algebras, the main change from quadratic operad to quadratic-linear operad consists in the appearance of a differential in the Koszul dual cooperad. This generalization of Koszul duality theory was developed in order to treat the case of the operad BV encoding Batalin–Vilkovisky algebras, see Sect. 13.7.

7.8.1 Quadratic-Linear Operad

An operadic *quadratic-linear* data (E, R) is a graded $\mathbb{S}$-module E together with a degree homogeneous sub-$\mathbb{S}$-module $R \subset E \oplus \mathcal{T}(E)^{(2)}$. So, there may be linear terms in the space of relations. We still denote by $\mathcal{P} = \mathcal{P}(E, R) = \mathcal{T}(E)/(R)$ the associated quotient operad. We consider $q : \mathcal{T}(E) \twoheadrightarrow \mathcal{T}(E)^{(2)}$ the projection onto the quadratic part of the free operad. The image of R under q, denoted qR, is homogeneous quadratic, so (E, qR) is a quadratic data in the sense of Sect. 7.1.1. We denote by $q\mathcal{P}$ its associated quadratic operad: $q\mathcal{P} := \mathcal{P}(E, qR)$. We assume that R satisfies the relation

$$(ql_1) : R \cap E = \{0\}.$$

If it is not the case, removing some elements of E, one can choose another presentation of $\mathcal{P}$ which satisfies (ql_1). This condition amounts to the minimality of the space of generators of $\mathcal{P}$. Under this assumption, there exists a map $\varphi : qR \to E$ such that R is the graph of φ:

$$R = \bigl\{X - \varphi(X) \mid X \in qR\bigr\}.$$

The weight grading on $\mathcal{T}(E)$ induces a filtration which is compatible with the operadic ideal (R). Hence the quotient operad $\mathcal{P}$ is filtered: $F_n\mathcal{P} = \mathrm{Im}(\bigoplus_{k\leq n} \mathcal{T} \times (E)^{(k)})$. Since we assumed $R \cap E = \{0\}$, we get $F_1\mathcal{P} = \mathrm{I} \oplus E$. We denote by $\mathrm{gr}\,\mathcal{P}$ the graded operad associated to this filtration of $\mathcal{P}$, $\mathrm{gr}_n\,\mathcal{P} := F_n\mathcal{P}/F_{n-1}\mathcal{P}$. We denote by

$$p : q\mathcal{P} \twoheadrightarrow \mathrm{gr}\,\mathcal{P}$$

the resulting epimorphism. It is obviously an isomorphism in weights 0 and 1, but not necessarily in weight 2. A corollary of the present theory shows that p is an isomorphism provided that $q\mathcal{P}$ is Koszul, see Theorem 7.8.5.

7.8.2 Koszul Dual DG Cooperad in the Inhomogeneous Quadratic Framework

The map φ permits us to construct the composite map

$$\tilde{\varphi} : (q\mathcal{P})^{¡} = \mathcal{C}\bigl(sE, s^2qR\bigr) \twoheadrightarrow s^2qR \xrightarrow{s^{-1}\varphi} sE.$$

By Sect. 6.3.8, there exists a unique coderivation, $\widetilde{d_\varphi} : (q\mathscr{P})^{¡} \to \mathscr{T}^c(sE)$, which extends this composite.

Lemma 7.8.1.

(a) *The coderivation* $\widetilde{d_\varphi}$ *restricts to a coderivation* d_φ *on the sub-cooperad* $(q\mathscr{P})^{¡} = \mathscr{C}(sE, s^2qR) \subset \mathscr{T}^c(sE)$ *if* $\{R \circ_{(1)} E + E \circ_{(1)} R\} \cap \mathscr{T}(E)^{(2)} \subset qR$.

(b) *The coderivation* d_φ *squares to* 0 *if* R *satisfies the condition*

$$(ql_2) : \{R \circ_{(1)} E + E \circ_{(1)} R\} \cap \mathscr{T}(E)^{(2)} \subset R \cap \mathscr{T}(E)^{(2)}.$$

Since $R \cap \mathscr{T}(E)^{(2)} \subset qR$, condition (ql_2) implies the inclusion

$$\{R \circ_{(1)} E + E \circ_{(1)} R\} \cap \mathscr{T}(E)^{(2)} \subset qR.$$

Condition (ql_2) amounts to saying that one cannot create new quadratic relations in R by adding an element to the relations of the presentation.

Let (E, R) be a quadratic-linear data satisfying the conditions (ql_1) and (ql_2). By definition, the *Koszul dual dg cooperad* of $\mathscr{P} = \mathscr{P}(E, R)$ is the dg cooperad

$$\mathscr{P}^{¡} := \left((q\mathscr{P})^{¡}, d_\varphi\right) = \left(\mathscr{C}\left(sE, s^2qR\right), d_\varphi\right).$$

7.8.3 Koszulity in the Inhomogeneous Quadratic Operad Framework

An operadic quadratic-linear data (resp. a quadratic-linear operad) is said to be *Koszul* if it satisfies the conditions (ql_1) and (ql_2) and if the quadratic data (E, qR), or equivalently the quadratic operad $q\mathscr{P}$, is Koszul in the sense of Sect. 7.4.

Notice that for a homogeneous quadratic-linear data, Koszul in the classical sense is Koszul in this sense. In this case, the conditions (ql_1), (ql_2) are trivially satisfied and the inner coderivation d_φ vanishes.

7.8.4 Cobar Construction in the Inhomogeneous Quadratic Operadic Framework

Under the hypotheses (ql_1) and (ql_2), we have constructed a conilpotent dg cooperad $\mathscr{P}^{¡}$. Applying the cobar construction of Sect. 6.5.2, we get a dg operad $\Omega\mathscr{P}^{¡}$, whose differential is of the form $d_1 + d_2$. The internal derivation d_1 is the unique derivation which extends d_φ. The derivation d_2 is induced by the cooperad structure of $\mathscr{P}^{¡}$.

We consider the same map κ in this context

$$\kappa : \mathscr{P}^{¡} = \mathscr{C}\left(sE, s^2qR\right) \twoheadrightarrow sE \xrightarrow{s^{-1}} E \rightarrowtail \mathscr{P}.$$

Lemma 7.8.2. *The map κ is a twisting morphism in* $\mathrm{Hom}_{\mathbb{S}}(\mathscr{P}^{i}, \mathscr{P})$, *that is* $\partial(\kappa) + \kappa \star \kappa = 0$.

The twisting morphism κ induces a morphism of dg operads $g_\kappa : \Omega \mathscr{P}^{¡} \to \mathscr{P}$ by Theorem 6.5.7.

Theorem 7.8.3. *Let $\mathscr{P}$ be a quadratic-linear Koszul operad. The cobar construction of its Koszul dual dg cooperad is a resolution of $\mathscr{P}$: the morphism of dg cooperads $g_\kappa : \Omega \mathscr{P}^{i} \xrightarrow{\sim} \mathscr{P}$ is a quasi-isomorphism.*

Notice that, in the inhomogeneous case, this resolution is not minimal because of the internal differential d_1.

7.8.5 Operadic Poincaré–Birkhoff–Witt Theorem

Theorem 7.8.4 (Operadic Poincaré–Birkhoff–Witt theorem). *When a quadratic-linear operad $\mathscr{P}$ is Koszul, then the epimorphism $p : q\mathscr{P} \twoheadrightarrow \mathrm{gr}\, \mathscr{P}$ is an isomorphism of graded operads*

$$q\mathscr{P} \cong \mathrm{gr}\, \mathscr{P}.$$

Even if the Poincaré–Birkhoff–Witt theorem is a direct consequence of the proof of Proposition 7.8.3, it has the following two nontrivial consequences: Corollary 7.8.5 and Proposition 7.8.6.

Corollary 7.8.5. *Let $\mathscr{P}(E, R)$ be a quadratic-linear operad. If the quadratic operad $q\mathscr{P} = \mathscr{P}(E, qR)$ is Koszul, then conditions (ql_1) and (ql_2) are equivalent to conditions*

$$(ql_1{}') : (R) \cap E = \{0\} \quad \textit{and} \quad (ql_2{}') : R = (R) \cap \{E \oplus E^{\otimes 2}\}.$$

Conditions $(ql_1{}')$ and $(ql_2{}')$ amount to saying that the ideal generated by R does not create any new quadratic-linear relation. It is equivalent to the maximality of the space of relations in the presentation of the inhomogeneous quadratic operad. Such conditions can be hard to check in practice because one would have to compute the full ideal generated by R. But this proposition shows that if one finds a quadratic-linear presentation of an operad satisfying conditions (ql_1), (ql_2) and whose homogeneous quadratic data is Koszul, then the space of relations R is maximal.

REMARK. As in Sect. 3.6.5, this result can be interpreted as a Diamond Lemma for Gröbner bases, see Sect. 8.5.6.

$$(q\mathscr{P})^{¡} \text{ Koszul } \& \ (ql_2) \quad \Rightarrow \quad \mathscr{P}^{¡} \text{ Koszul } \& \ (ql_2{}').$$

7.8.6 Acyclicity of the Koszul Complex

As in the quadratic case, the Koszul complex associated to an inhomogeneous Koszul operad is acyclic.

Proposition 7.8.6. *If the quadratic-linear operad $\mathscr{P}(E, R)$ is Koszul, then the Koszul complexes $\mathscr{P}^{¡} \circ_\kappa \mathscr{P}$ and $\mathscr{P} \circ_\kappa \mathscr{P}^{¡}$ are acyclic.*

This result induces functorial quasi-free resolutions for $\mathscr{P}$-algebras, see Chap. 11, which are used to compute the (co)homology of $\mathscr{P}$-algebras, see Chap. 12.

7.8.7 Koszulity of the Operad BV

A Batalin–Vilkovisky algebra is a Gerstenhaber algebra $(A, \cdot, \langle -, - \rangle)$ with a square-zero degree 1 unary operator Δ satisfying the following quadratic-linear relation

$$\langle -, - \rangle := \Delta(- \cdot -) - \big(\Delta(-) \cdot -\big) - \big(- \cdot \Delta(-)\big).$$

We refer the reader to Sect. 13.7 for more details. Therefore the operad BV encoding BV-algebras admits the following quadratic-linear presentation: $BV \cong \mathscr{T}(E)/(R)$. The space of generators is equal to

$$E = \mathbb{K}m \oplus \mathbb{K}c \oplus \mathbb{K}\Delta,$$

where $\mathbb{K}m$ is a trivial representation of $\mathbb{S}_2$ of degree 0, $\mathbb{K}c$ a trivial representation of $\mathbb{S}_2$ of degree 1, and $\mathbb{K}\Delta$ is a one-dimensional graded vector space ($\mathbb{S}_1$-module) concentrated in degree 1. The space of relations R is the $\mathbb{K}[\mathbb{S}_3]$-module generated by the relators

$$\begin{cases} m \circ_1 m - m \circ_2 m, \\ c \circ_1 c + (c \circ_1 c)^{(123)} + (c \circ_1 c)^{(321)}, \\ \Delta^2, \\ c \circ_1 m - m \circ_2 c - (m \circ_1 c)^{(23)}, \\ c - \Delta \circ_1 m + m \circ_1 \Delta + m \circ_2 \Delta, \\ \Delta \circ_1 c + c \circ_1 \Delta + c \circ_2 \Delta. \end{cases}$$

Since there is no pure linear relation, this presentation satisfies condition (ql_1). The last relator is obtained from the previous one as follows. Denoting $\rho := c - \Delta \circ_1 m + m \circ_1 \Delta + m \circ_2 \Delta$ the last relator is equal to the composite of $\rho \otimes (\Delta \otimes \mathrm{id} + \mathrm{id} \otimes \Delta) + \Delta \otimes \rho$ in the operad BV. Therefore, it is not necessary in the definition of the operad BV, *but* we have to consider it to satisfy condition (ql_2).

To obtain the quadratic analog qBV, one has just to change the inhomogeneous relation $c - \Delta \circ_1 m + m \circ_1 \Delta + m \circ_2 \Delta$ by $\Delta \circ_1 m - m \circ_1 \Delta - m \circ_2 \Delta$. Hence, in the operad qBV, the operator Δ is a derivation with respect to both the product m and the bracket c. We view these relations as rewriting rules

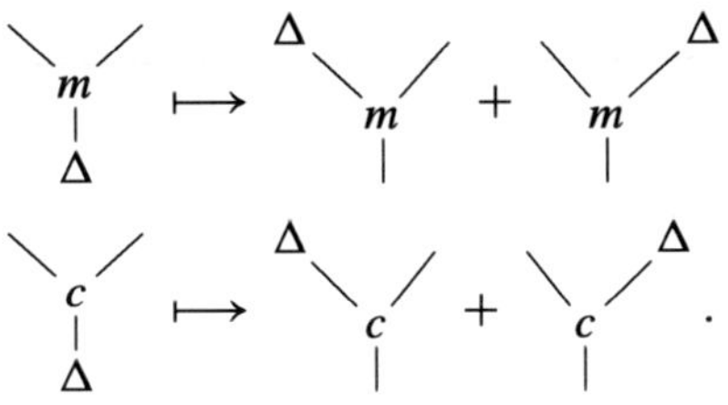

Theorem 7.8.7. [GCTV09] *The operad BV is a Koszul operad.*

Proof. The aforementioned presentation of the operad BV satisfies conditions (ql_1) and (ql_2). The quadratic analog qBV satisfies the conditions of the Distributive Law method (Sect. 8.6), so it is a homogeneous quadratic Koszul operad. □

Since the quadratic operad qBV is made of the operad *Gerst*, encoding Gerstenhaber algebras, and the algebra of dual number $D := T(\Delta)/(\Delta^2)$ by means of a distributive law, it is isomorphic to

$$qBV \cong Gerst \circ D \cong Com \circ \mathscr{S}^{-1}Lie \circ D,$$

where $\mathscr{S}^{-1}Lie \cong Lie \underset{\mathrm{H}}{\otimes} \mathrm{End}_{s^{-1}\mathbb{K}}$.

So the operadic PBW theorem (Sect. 7.8.5) gives the form of the underlying module of the operad BV.

Proposition 7.8.8. [Get94] *There is an isomorphism of* $\mathbb{S}$*-modules*

$$BV \cong Com \circ \mathscr{S}^{-1}Lie \circ D.$$

As a corollary of the distributive law method (Sect. 8.6), we get also

$$qBV^{¡} \cong D^{¡} \circ Gerst^{¡} \cong T^c(\delta) \circ \left(\mathscr{S}^{-2}\right)^c Com^c \circ \left(\mathscr{S}^{-1}\right)^c Lie^c,$$

where $\delta := s\Delta$ has degree 2 and where $(\mathscr{S}^{-i})^c\mathscr{C} = \mathrm{End}^c_{s^{-i}\mathbb{K}} \underset{\mathrm{H}}{\otimes} \mathscr{C}$. Under this isomorphism, we denote the elements of $qBV^{¡}$ simply by $\delta^d \otimes L_1 \odot \cdots \odot L_t$, where each $L_i \in Lie^c(n_i)$ and where $\odot$ stands for the *symmetric tensor product* $x \odot y = (-1)^{|x||y|} y \odot x$.

Proposition 7.8.9. [GCTV09] *The square-zero coderivation d_φ of the cooperad $qBV^{¡}$ is explicitly given by*

$$d_\varphi\left(\delta^d \otimes L_1 \odot \cdots \odot L_t\right) = \sum_{i=1}^{t} (-1)^{\varepsilon_i} \delta^{d-1} \otimes L_1 \odot \cdots \odot L_i' \odot L_i'' \odot \cdots \odot L_t,$$

where $L_i' \odot L_i''$ is Sweedler-type notation for the image of L_i under the binary part

$$\left(\mathscr{S}^{-1}\right)^c Lie^c \to \left(\mathscr{S}^{-1}\right)^c Lie^c(2) \otimes \left(\left(\mathscr{S}^{-1}\right)^c Lie^c \otimes \left(\mathscr{S}^{-1}\right)^c Lie^c\right)$$

of the decomposition map of the cooperad $(\mathcal{S}^{-1})^c Lie^c$. *The sign is given by* $\varepsilon_i = n_1 + \cdots + n_{i-1} + i - 1$. *The image of* d_φ *is equal to* 0 *when the exponent* $d = 0$ *and when* $L_i \in (\mathcal{S}^{-1})^c Lie^c(1) = \mathbb{K}\,\mathrm{id}$ *for all* i.

Proof. Let us denote by l the generator of $(\mathcal{S}^{-1})^c Lie^c(2)$. Inside $qBV^{\text{!`}}$, the element $\delta \otimes \mathrm{I} \otimes l$ is the homotopy for the derivation relation between $\Delta$ and $m$. So the coderivation $d_\varphi$ of $BV^{\text{!`}}$ is the unique coderivation which sends $\delta \otimes \mathrm{id} \otimes l$ to $\mathrm{id} \otimes \mu \otimes \mathrm{id}$, where μ is the generator of $(\mathcal{S}^{-2})^c Com^c(2)$. □

Notice that, up to the term δ^d, the transpose of d_φ is equal to the Chevalley–Eilenberg boundary map defining the homology of the free Lie algebra, see Sect. 13.2.7:

$$\begin{aligned} &{}^t d_\varphi\big((\delta^*)^d \otimes L_1^* \odot \cdots \odot L_t^*\big) \\ &\quad = \sum_{1 \le i < j \le t} \pm (\delta^*)^{d+1} \otimes \big[L_i^*, L_j^*\big] \odot L_1^* \odot \cdots \odot \widehat{L_i^*} \odot \cdots \odot \widehat{L_j^*} \odot \cdots \odot L_t^*. \end{aligned}$$

Finally, the Koszul dg cooperad of BV is

$$BV^{\text{!`}} = \big(qBV^{\text{!`}}, d_\varphi\big)$$

and Theorem 7.8.3 provides a quasi-free, but not minimal resolution,

$$BV_\infty := \Omega BV^{\text{!`}} \xrightarrow{\sim} BV$$

for the operad BV. This defines the notion of homotopy BV-algebra. We refer the reader to Sect. 13.7.7 for more details and applications.

7.9 Résumé

7.9.1 Quadratic Data and Koszul Dual Constructions for Operads

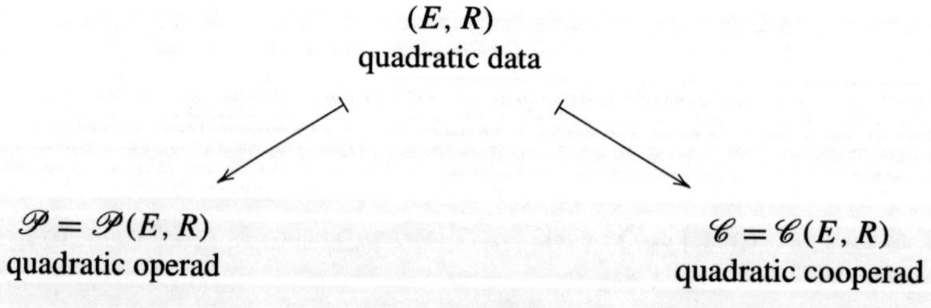

The quadratic operad: $\mathcal{P} = \mathcal{P}(E, R) = \mathcal{T}(E)/(R)$.
The quadratic cooperad: $\mathcal{C} = \mathcal{C}(E, R) \subset \mathcal{T}^c(E)$.
Koszul dual cooperad of an operad: $\mathcal{P}(E, R)^{\text{!`}} := \mathcal{C}(sE, s^2 R)$.

Koszul dual operad of a cooperad: $\mathscr{C}(E,R)^{\text{i}} := \mathscr{P}(s^{-1}E, s^{-2}R)$,

$$\left(\mathscr{P}^{\text{i}}\right)^{\text{i}} \cong \mathscr{P}.$$

Koszul dual operad of an operad: when E is finite dimensional,

$$\mathscr{P}^! := \left(\mathscr{S}^c \underset{\mathrm{H}}{\otimes} \mathscr{P}^{\text{i}}\right)^* \cong \mathscr{P}\left(s^{-1}\mathscr{S}^{-1} \underset{\mathrm{H}}{\otimes} E^*, R^{\perp}\right).$$

7.9.2 Koszul Duality Theory

Twisting morphism κ: $\mathscr{P}^{\text{i}} = \mathscr{C}(sE, s^2R) \twoheadrightarrow sE \xrightarrow{s^{-1}} E \rightarrowtail \mathscr{P}(E,R) = \mathscr{P}$.
Koszul complexes: $\mathscr{P} \circ_\kappa \mathscr{P}^{\text{i}}$ and $\mathscr{P}^{\text{i}} \circ_\kappa \mathscr{P}$,

$$\mathscr{P}^{\text{i}} \rightarrowtail \mathrm{B}\mathscr{P} \quad \text{and} \quad \Omega\mathscr{P}^{\text{i}} \twoheadrightarrow \mathscr{P},$$

with the *syzygy degree*: $H^0(\mathrm{B}^\bullet\mathscr{P}) \cong \mathscr{P}^{\text{i}}$ and $H_0(\Omega_\bullet\mathscr{P}^{\text{i}}) \cong \mathscr{P}$.

The quadratic data (E,R) is *Koszul* when one of the following equivalent assertions is satisfied:

1. the right Koszul complex $\mathscr{P}^{\text{i}} \circ_\kappa \mathscr{P}$ is acyclic,
2. the left Koszul complex $\mathscr{P} \circ_\kappa \mathscr{P}^{\text{i}}$ is acyclic,
3. the inclusion $i : \mathscr{P}^{\text{i}} \rightarrowtail \mathrm{B}\mathscr{P}$ is a quasi-isomorphism,
4. the projection $p : \Omega\mathscr{P}^{\text{i}} \twoheadrightarrow \mathscr{P}$ is a quasi-isomorphism,
5. $H^n(\mathrm{B}^\bullet\mathscr{P}) = 0$ for $n \geq 1$,
6. $H_n(\Omega_\bullet\mathscr{P}^{\text{i}}) = 0$ for $n \geq 1$,
7. $H^\bullet(\mathrm{B}^\bullet\mathscr{P})$ is a sub-cooperad of $\mathscr{T}^c(sE)$.

7.9.3 Generating Series or Hilbert–Poincaré Series

$$f^{\mathscr{P}}(t) := \sum_{n\geq 0} \frac{\dim \mathscr{P}(n)}{n!} t^n,$$

$$\mathscr{P} \text{ binary, quadratic and Koszul} \quad \Longrightarrow \quad f^{\mathscr{P}^!}\left(-f^{\mathscr{P}}(-t)\right) = t.$$

7.9.4 Binary Quadratic Operads

Quadratic data: (E,R) where $E \in \mathscr{T}(E)^{(2)}$.
Koszul dual operad of $\mathscr{P} = \mathscr{P}(E,R)$:

$$\mathscr{P}^! = \mathscr{P}\left(E^\vee, R^\perp\right)$$

where $R^\perp$ is the orthogonal of R for some bilinear form.

7.9.5 Inhomogeneous Koszul Duality Theory

Quadratic-linear data: (E, R), with $R \subset \mathcal{T}(E)^{(1)} \oplus \mathcal{T}(E)^{(2)}$.
Quadratic analogs: $qR := \mathrm{proj}_{\mathcal{T}(E)^{(2)}}(R)$ and $q\mathcal{P} := \mathcal{P}(E, qR)$.

$$(ql_1): R \cap E = \{0\} \quad \Rightarrow \quad R = \mathrm{Graph}(\varphi : qR \to E)$$

$$(ql_2): \{R \circ_{(1)} E + E \circ_{(1)} R\} \cap \mathcal{T}(E)^{(2)} \subset R \cap \mathcal{T}(E)^{(2)}$$

$$\Rightarrow \quad d_\varphi \text{ codifferential on } q\mathcal{P}^{\text{!`}}.$$

Koszul dual dg cooperad: $\mathcal{P}^{\text{!`}} := (q\mathcal{P}^{\text{!`}}, d_\varphi)$.

$\mathcal{P}(E, R)$ *Koszul operad* when (ql_1), (ql_2) and $q\mathcal{P}$ homogeneous quadratic Koszul operad. In this case:

- quasi-free resolution: $\Omega\mathcal{P}^{\text{!`}} \xrightarrow{\sim} \mathcal{P}$,
- Poincaré–Birkhoff–Witt theorem: $q\mathcal{P} \cong \mathrm{gr}\,\mathcal{P}$,
- Koszul complex: $\mathcal{P} \circ_\kappa \mathcal{P}^{\text{!`}}$ acyclic.

EXAMPLE: The operad coding the Batalin–Vilkovisky algebras.

7.10 Exercises

Exercise 7.10.1 (On the bilinear form $\langle -, - \rangle$). Let E be an $\mathbb{S}$-module concentrated in arity 2. Show that the space $\mathcal{T}(E)(3)$ is a quotient of the sum of two copies of $E \otimes E$, one corresponding to the tree  (i.e. $- \circ_1 -$) and the other one corresponding to the tree  (i.e. $- \circ_2 -$). Show that the quotient map is equivariant for the action of the symmetric group $\mathbb{S}_3$.

Under this description of $\mathcal{T}(E)(3)$ show that the bilinear form $\langle -, - \rangle$ used in the construction of the orthogonal space $R^\perp$ is sign-invariant under the action of $\mathbb{S}_3$ and satisfies

$$\langle \alpha^* \circ_i \beta^*, \mu \circ_j \nu \rangle = \delta_{ij}\alpha^*(\mu)\beta^*(\nu)$$

for any $\alpha^*, \beta^* \in E^\vee$; $\mu, \nu \in E$; $i, j = 1$ or 2; δ_{ij} being the Kronecker symbol.

Exercise 7.10.2 (Other choice of basis). Rewrite Sect. 7.6 by choosing the following basis for $\mathcal{T}(E)(3)$:

$$\mu \circ_1 \nu, \quad (\mu \circ_1 \nu)^{(23)}, \quad \mu \circ_2 \nu.$$

Exercise 7.10.3 (An example with graded operation). Show that the operad with one generating operation xy and one relation

$$(xy)z = -x(yz)$$

is not Koszul by looking at the generating series.

Show that if the product is supposed to be of degree one, then the operad is Koszul.

Exercise 7.10.4 (Koszul dual ns operads). Describe the Koszul dual operad of the following ns operads:

1. One binary generating operation, no relation (cf. Sect. 13.8.2).
2. Two generating operations $\prec$ and $\succ$, three relations (cf. Sect. 13.13.2):

$$(x \prec y) \prec z = x \prec (y \prec z),$$
$$(x \succ y) \prec z = x \succ (y \prec z),$$
$$(x \succ y) \succ z = x \succ (y \succ z).$$

3. Two generating operations $\prec$ and $\succ$, three relations (cf. Sect. 13.6):

$$(x \prec y) \prec z = x \prec (y \prec z + y \succ z),$$
$$(x \succ y) \prec z = x \succ (y \prec z),$$
$$(x \prec y + x \succ y) \succ z = x \succ (y \succ z).$$

4. n generating operations $\circ_i$ for $i = 1, \ldots, n$, and many relations:

$$(x \circ_i y) \circ_j z = x \circ_i (y \circ_j z), \quad \text{for any } i, j.$$

Exercise 7.10.5 (Koszul dual symmetric operads). Describe the Koszul dual operad of the following symmetric operads:

1. One binary generating operation $[x, y]$, one relation (cf. Sect. 13.5.1):

$$\big[[x, y], z\big] = \big[[x, z], y\big] + \big[x, [y, z]\big].$$

2. One binary generating operation $x \prec y$, one relation (cf. Sect. 13.5.2):

$$(x \prec y) \prec z = x \prec (y \prec z + z \prec y).$$

3. One binary generating operation $\{x, y\}$, one relation (cf. Sect. 13.4.1):

$$\big\{\{x, y\}, z\big\} - \big\{x, \{y, z\}\big\} = \big\{\{x, z\}, y\big\} - \big\{x, \{z, y\}\big\}.$$

4. One binary generating operation xy, two relations (cf. Sect. 13.4.6):

$$(xy)z = x(yz) = x(zy).$$

Exercise 7.10.6 (From Koszul ns operads to Koszul operads). Show that the functor $\mathsf{ns\,Op} \longrightarrow \mathsf{Op}$ of Sect. 5.9.11 commutes with the Koszul dual functor $\mathscr{P} \mapsto \mathscr{P}^!$ and that it preserves the Koszul property.

Exercise 7.10.7 (Every augmented operad is inhomogeneous Koszul). Extend Exercise 3.8.10 to reduced augmented operads.

Exercise 7.10.8 (Inhomogeneous operad ★). Let $\mathscr{P}$ be an inhomogeneous quadratic operad. Show that, if $\mathscr{P}$ is Koszul, then $f_\kappa : \mathscr{P}^{\text{!`}} \to \mathrm{B}\mathscr{P}$ is a quasi-isomorphism of dga coalgebras.

Chapter 8
Methods to Prove Koszulity of an Operad

Nous voulons, tant ce feu nous brûle le cerveau, Plonger au fond du gouffre, Enfer ou Ciel, qu'importe? Au fond de l'Inconnu pour trouver du nouveau!

Charles Baudelaire

This chapter extends to the operadic level the various methods, obtained in Chap. 4, to prove that algebras are Koszul. In Chap. 7, we have already given one method, based on the vanishing of the homology of the free $\mathscr{P}$-algebra.

We begin by generalizing the rewriting method for associative algebras given in Sect. 4.1 to nonsymmetric operads. To extend it much further to symmetric operads, we need to introduce the notion of shuffle operad, which sits in between the notion of operad and nonsymmetric operad. It consists of the same kind of compositions as in an operad but without the symmetric groups action. For instance, the free symmetric operad is isomorphic to the free shuffle operad as $\mathbb{K}$-modules, thereby providing a $\mathbb{K}$-linear basis in terms of shuffle trees for the first one.

With this notion of shuffle operad at hand, we adapt the rewriting method, the reduction by filtration method, the Diamond Lemma, the PBW bases and the Gröbner bases of Chap. 4 from associative algebras to operads.

We give then yet another method. Starting from two operads $\mathscr{A}$ and $\mathscr{B}$, one can sometimes cook up a third one on the underlying $\mathbb{S}$-module $\mathscr{A} \circ \mathscr{B}$, by means of a distributive law $\mathscr{B} \circ \mathscr{A} \to \mathscr{A} \circ \mathscr{B}$. One can interpret this data as a rewriting rule, which pulls the elements of $\mathscr{B}$ above those of $\mathscr{A}$. This interpretation allows us to show the same kind of results as the ones obtained by the aforementioned methods. For instance, we give a Diamond Lemma for distributive laws which proves that an operad obtained from two Koszul operads by means of a distributive law is again Koszul. Notice that, retrospectively, such a method applies to associative algebras as well.

Instead of Backelin's lattice criterion used in the algebra case, we introduce a partition poset method in the operad case. The idea is to associate a family of partition type posets to a set operad. The main theorem asserts that the linear operad

J.-L. Loday, B. Vallette, *Algebraic Operads*,
Grundlehren der mathematischen Wissenschaften 346,
DOI 10.1007/978-3-642-30362-3_8,

generated by the set operad is Koszul if and only if the homology groups of the operadic partition posets are concentrated in top dimension (Cohen–Macaulay posets). On the one hand, the many combinatorial criteria to show that a poset is Cohen–Macaulay provide ways to prove that an operad is Koszul. On the other hand, it gives a method to compute explicitly the homology groups of some partition type posets as $\mathbb{S}$-modules since the top homology groups are isomorphic to the Koszul dual cooperad.

In a last section, we extend the definition and the properties of Manin products to operads.

The material of this chapter mainly comes from Hoffbeck [Hof10c], Dotsenko and Khoroshkin [DK10], Markl [Mar96a], [Val07b], Ginzburg and Kapranov [GK94, GK95b], and [Val08].

In the first five sections of this chapter, we work with *reduced* $\mathbb{S}$-modules M: $M(0)=0$, respectively $M(0)=\emptyset$ in the set theoretic case.

8.1 Rewriting Method for Binary Quadratic NS Operads

In this section, we explain how the rewriting method for associative algebras of Sect. 4.1 extends to binary quadratic ns operads. It provides a short algorithmic method, based on the rewriting rules given by the relations, to prove that an operad is Koszul. The general theory for operads (not necessarily binary) requires new definitions that will be given in the following sections.

Let $\mathscr{P}(E, R)$ be a binary quadratic ns operad.

Step 1. We consider an ordered basis $\{\mu_1, \mu_2, \dots, \mu_k\}$ for the generating space E of binary operations. The ordering $\mu_1 < \mu_2 < \cdots < \mu_k$ will play a key role in the sequel.

Step 2. The ternary operations, which span the weight 2 part of the free ns operad, are of the form $\mu_i \circ_a \mu_j$, where $a = 1, 2$. We put a total order on this set as follows:

$$\begin{cases} \mu_i \circ_2 \mu_j < \mu_i \circ_1 \mu_j, & \text{for any } i, j, \\ \mu_i \circ_a \mu_j < \mu_k \circ_a \mu_l, & \text{whenever } i < k, \quad a = 1 \text{ or } 2, \text{ and for any } j, l, \\ \mu_i \circ_a \mu_j < \mu_i \circ_a \mu_l, & \text{whenever } j < l, \quad a = 1 \text{ or } 2. \end{cases}$$

The operad $\mathscr{P}$ is determined by the space of relations R, which is spanned by a set of relators written in this basis as

$$r = \lambda \mu_i \circ_a \mu_j - \sum \lambda_{k,b,l}^{i,a,j} \mu_k \circ_b \mu_l, \quad \lambda, \lambda_{k,b,l}^{i,a,j} \in \mathbb{K} \quad \text{and} \quad \lambda \neq 0,$$

so that the sum runs over the indices satisfying $\mu_i \circ_a \mu_j > \mu_k \circ_b \mu_l$. The operation $\mu_i \circ_a \mu_j$ is called the *leading term* of the relator (r). One can always suppose that λ is equal to 1, that the leading terms of the set of relators are all distinct and that there is no leading term of any other relator in the sum in the right-hand side. This is called a *normalized form* of the presentation.

Step 3. Observe that such a relator gives rise to a *rewriting rule* in the operad $\mathscr{P}$:

$$\mu_i \circ_a \mu_j \mapsto \sum \lambda_{k,b,l}^{i,a,j} \mu_k \circ_b \mu_l.$$

Given three generating binary operations μ_i, μ_j, μ_k, one can compose them in 5 different ways: they correspond to the 5 planar binary trees with 3 vertices. Such a monomial, i.e. decorated planar tree, is called *critical* if the two sub-trees with 2 vertices are leading terms.

Step 4. There are at least two ways of rewriting a critical monomial ad libitum, that is, until no rewriting rule is applicable any more. If all these ways lead to the same element, then the critical monomial is said to be *confluent*.

Conclusion. If each critical monomial is confluent, then the ns operad $\mathscr{P}$ is Koszul.

This assertion is a consequence of the following result.

Theorem 8.1.1 (Rewriting method for ns operads). *Let $\mathscr{P}(E, R)$ be a reduced quadratic ns operad. If its generating space E admits an ordered basis for which there exists a suitable order on planar trees such that every critical monomial is confluent, then the ns operad $\mathscr{P}$ is Koszul.*

In this case, the ns operad $\mathscr{P}$ admits a $\mathbb{K}$-linear basis made up of some planar trees called a PBW-basis, see Sect. 8.5. The proof is analogous to the proof of Theorem 4.1.1; it follows from Sect. 8.5.4.

EXAMPLE. Consider the ns operad *As* encoding associative algebras, see Chap. 9. It is generated by one operation of arity 2:  . The rewriting rule expressing associativity reads:

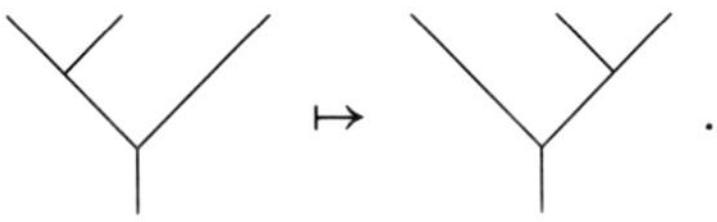

There is only one critical monomial (the left comb), which gives the following confluent graph (Fig. 8.1).

Therefore, the nonsymmetric operad *As* is Koszul. It admits a PBW basis made up of the right combs:

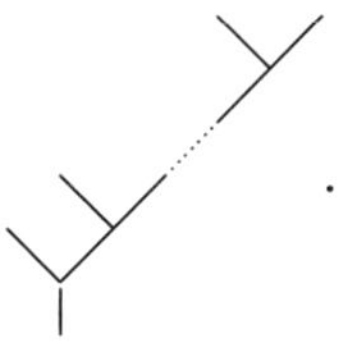

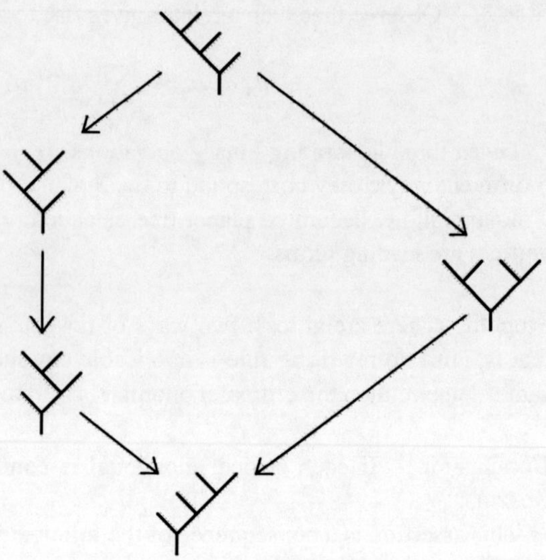

Fig. 8.1 The diamond for the nonsymmetric operad *As*

REMARK. Notice first that one recovers the associativity pentagon of monoidal categories, see Appendix B.3. Moreover, the Diamond Lemma applied to the ns operad *As* is exactly Mac Lane's coherence theorem for (nonunital) monoidal categories. The first one states that any graph built out of the left combs, under the associativity relation, is confluent. The second one states that any graph built with the associativity relation is commutative. With this remark in mind, the reading of [ML95, Sect. VII-2] enjoys another savor.

COUNTER-EXAMPLE. We consider the same example but with the modified associativity relation

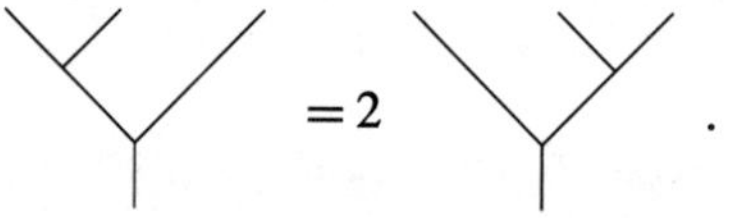

In this case, the above graph is not confluent because $2^3 \neq 2^2$. Whatever the suitable order is, the graph will never be confluent since it can be proved that this ns operad is not Koszul, see Exercise 8.10.9.

8.2 Shuffle Operad

The forgetful functor $\mathsf{Op} \to \mathsf{ns\,Op}$, from symmetric operads to nonsymmetric operads, forgets the action of the symmetric groups, see Sect. 5.9.11. It factors through the category of shuffle operads $\mathsf{Op}_{\sqcup\!\sqcup}$. Shuffle operads have the advantage of being based on arity-graded vector spaces like nonsymmetric operads, while retaining the whole structure of a symmetric operad.

The notion of shuffle operad is due to Dotsenko–Khoroshkin [DK10] after ideas of Hoffbeck [Hof10c].

8.2.1 Shuffle Composite Product

For any subset X of $\underline{n} := \{1, \dots, n\}$ we denote by $\min(X)$ the smallest element of X. Any partition P of $\underline{n}$ into k subsets can be written uniquely

$$\mathrm{P} = (\mathrm{P}_1, \dots, \mathrm{P}_k)$$

under the requirement

$$\min(\mathrm{P}_1) < \min(\mathrm{P}_2) < \cdots < \min(\mathrm{P}_k).$$

Writing the elements of P_i in order and concatenating them for all i, it defines the preimages of $\{1, \dots, n\}$ under a permutation σ_{P} of $\mathbb{S}_n$.

$$\{1, 3, 4\} < \{2, 7\} < \{5, 6, 8\} \quad \mapsto \quad [1\ 4\ 2\ 3\ 6\ 7\ 5\ 8].$$

The associated permutation σ_{P} is a $(i_1, \dots, i_k)$-unshuffle, cf. Sect. 1.3.2, where $i_j := |\mathrm{P}_j|$,

$$\begin{aligned}
&\sigma^{-1}(1) < \cdots < \sigma^{-1}(i_1),\\
&\sigma^{-1}(i_1 + 1) < \cdots < \sigma^{-1}(i_1 + i_2),\\
&\vdots\\
&\sigma^{-1}(i_1 + \cdots + i_{k-1} + 1) < \cdots < \sigma^{-1}(n),
\end{aligned}$$

satisfying the extra property

$$\sigma^{-1}(1) < \sigma^{-1}(i_1 + 1) < \cdots < \sigma^{-1}(i_1 + \cdots + i_{k-1} + 1).$$

Such unshuffles are called *pointed* $(i_1, \dots, i_k)$*-unshuffles*, or simply *pointed unshuffles* when the underlying type is understood. We denote the associated set by $\sqcup\!\sqcup(i_1, \dots, i_k)$. For instance, the set $\sqcup\!\sqcup(2, 1)$ has two elements, namely [1 2 3] and [1 3 2]; the set $\sqcup\!\sqcup(1, 2)$ has only one element, namely [1 2 3].

For any arity-graded spaces M and N, we define the *shuffle composite product* $M \circ_{\sqcup\!\sqcup} N$ as follows:

$$(M \circ_{\sqcup\!\sqcup} N)_n := \bigoplus_{\substack{k \geq 1, (i_1, \dots, i_k)\\ i_1 + \cdots + i_k = n}} M_k \otimes N_{i_1} \otimes \cdots \otimes N_{i_k} \otimes \mathbb{K}\big[\sqcup\!\sqcup(i_1, \dots, i_k)\big].$$

Proposition 8.2.1. *The shuffle composition product makes the category of arity-graded spaces* $(\mathbb{N}\text{-}\mathsf{Mod}, \circ_{\sqcup\!\sqcup}, \mathrm{I})$ *into a monoidal category.*

Proof. The associativity of $\circ_{\sqcup}$ is proved by direct inspection, see [DK10, Sect. 2] for more details. □

8.2.2 Shuffle Trees

A *shuffle tree* is a reduced planar rooted tree equipped with a labeling of the leaves by integers $\{1, 2, \ldots, n\}$ satisfying some condition stated below. First, we label each edge of the tree as follows. The leaves are already labeled. Any other edge is the output of some vertex v. We label this edge by $\min(v)$ which is the minimum of the labels of the inputs of v. Second, the condition for a labeled tree to be called a shuffle tree is that, for each vertex, the labels of the inputs, read from left to right, are increasing.

EXAMPLE. See Fig. 8.2.

The relationship between shuffle trees and pointed unshuffles is the following: Any shuffle tree with two levels corresponds to a partition written in order and vice versa.

EXAMPLE. See Fig. 8.3.

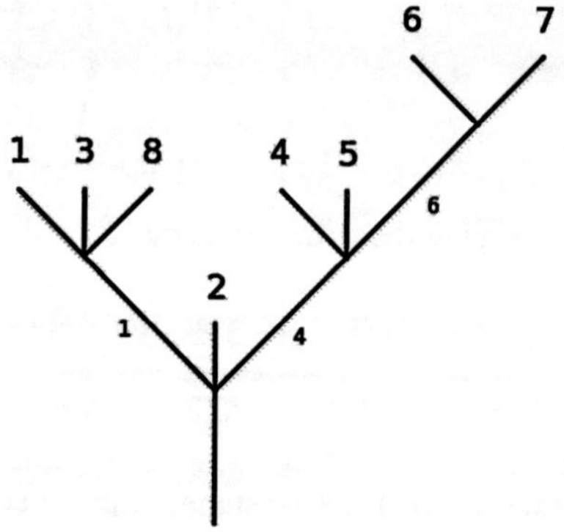

Fig. 8.2 Example of a shuffle tree

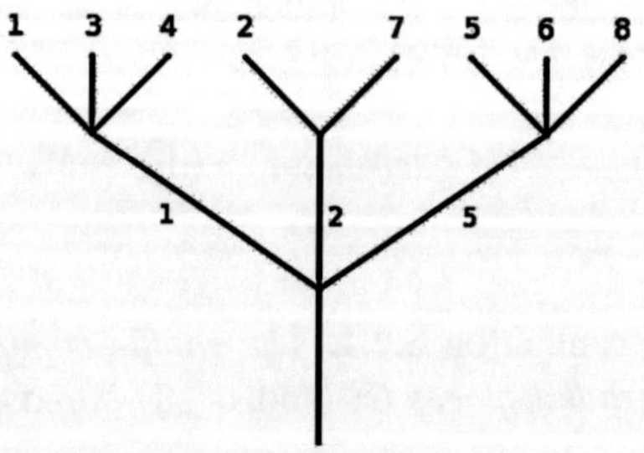

Fig. 8.3 Example of a 2-leveled shuffle tree

8.2.3 Monoidal Definition of Shuffle Operad

By definition a *shuffle operad* is a monoid $(\mathscr{P}, \gamma_{\sqcup\!\sqcup}, \eta)$ in the monoidal category $(\mathbb{N}\text{-}\mathsf{Mod}, \circ_{\sqcup\!\sqcup}, \mathrm{I})$. Explicitly, it is an arity-graded vector space $\mathscr{P}$ equipped with an associative composition map $\gamma_{\sqcup\!\sqcup} : \mathscr{P} \circ_{\sqcup\!\sqcup} \mathscr{P} \to \mathscr{P}$ and a unit map $\eta : \mathrm{I} \to \mathscr{P}$.

Equivalently a shuffle operad can be defined by maps

$$\gamma_\sigma : \mathscr{P}_k \otimes \mathscr{P}_{i_1} \otimes \cdots \otimes \mathscr{P}_{i_k} \to \mathscr{P}_n$$

for any pointed unshuffle $\sigma \in \sqcup\!\sqcup(i_1, \ldots, i_k)$. Assembling these maps, we get $\gamma : \mathscr{P} \circ_{\sqcup\!\sqcup} \mathscr{P} \to \mathscr{P}$. Associativity of γ can be written explicitly in terms of the individual maps γ_σ as in Proposition 5.3.1.

8.2.4 Partial Definition of Shuffle Operad

Let m and i, $1 \le i \le m$, be positive integers. Any monotonic injection

$$\{i+1, i+2, \ldots, i+n-1\} \to \{i+1, i+2, \ldots, m+n-1\}$$

is completely determined by a $(n-1, m-i)$-unshuffle ω that we let act on the set $\{i+1, \ldots, m+n-1\}$. This data is equivalent to a partition of type

$$\begin{aligned} P = \big(&\{1\}, \ldots, \{i-1\}, \big\{i, \omega^{-1}(i+1), \ldots, \omega^{-1}(i+n-1)\big\}, \\ &\big\{\omega^{-1}(i+n)\big\}, \ldots, \big\{\omega^{-1}(m+n-1)\big\}\big), \end{aligned} \tag{8.1}$$

where all the subsets but one are made up of one element.

Such a partition is equivalent to a shuffle tree with two vertices:

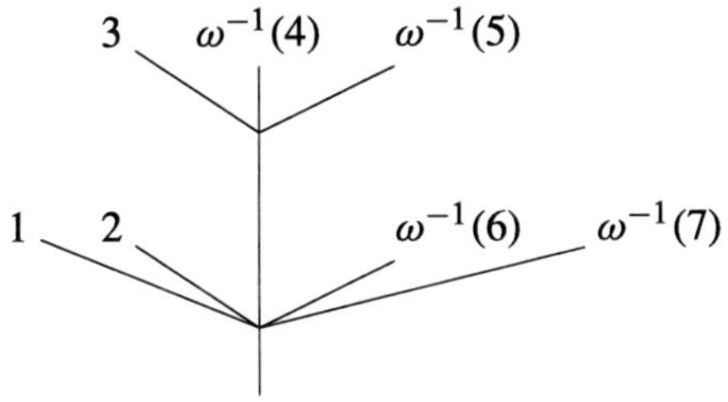

The associated pointed unshuffle σ_P is of type $(1, \ldots, 1, n, 1, \ldots, 1)$ and it determines a map

$$\gamma_{\sigma_P} : \mathscr{P}(m) \otimes \mathscr{P}(1) \otimes \cdots \otimes \mathscr{P}(1) \otimes \mathscr{P}(n) \otimes \mathscr{P}(1) \otimes \cdots \otimes \mathscr{P}(1) \to \mathscr{P}(m+n-1).$$

By evaluating γ_{σ_P} on the elements $\mathrm{id} \in \mathscr{P}(1)$ we get a map:

$$\circ_{i,\omega} : \mathscr{P}(m) \otimes \mathscr{P}(n) \to \mathscr{P}(m+n-1).$$

They are called the *partial shuffle products* in the shuffle operad framework.

The partial operations $\circ_{i,\omega}$ generate, under composition, all the shuffle compositions. They satisfy some relations. For instance if the permutation shuffles

$$\sigma \in Sh(m,n), \quad \lambda \in Sh(m+n,r), \quad \delta \in Sh(n,r), \quad \gamma \in Sh(m,n+r)$$

satisfy the relation

$$(\sigma \times 1_r)\lambda = (1_m \times \delta)\gamma \quad \text{in } \mathbb{S}_{m+n+r},$$

then the partial operations satisfy the relation

$$(x \circ_{1,\sigma} y) \circ_{1,\lambda} z = x \circ_{1,\gamma} (y \circ_{1,\delta} z)$$

for any $x \in \mathscr{P}_{1+m}$, $y \in \mathscr{P}_{1+n}$, $z \in \mathscr{P}_{1+r}$.

We leave it to the reader to find the complete set of relations, which presents a shuffle operad out of the partial shuffle products.

8.2.5 Combinatorial Definition of Shuffle Operad

The *combinatorial definition of a shuffle operad* is the same as the combinatorial definition of a ns operad, cf. Sect. 5.9.5, except that we have to replace the planar rooted trees by the shuffle trees. The only subtle point is the substitution of shuffle trees, which is obtained as follows.

Let t be a shuffle tree and v be a vertex of t whose inputs are labeled by $(i_1, \ldots, i_k)$. So we have $i_1 < i_2 < \cdots < i_k$. Let s be a shuffle tree with k leaves and let $(j_1, \ldots, j_k)$ be the sequence of labels of the leaves. So $(j_1, \ldots, j_k)$ is a permutation of $\underline{k}$. Then, the substitution of s at v gives a new planar rooted tree, cf. Sect. 5.9.5, whose labeling is obtained as follows: each label j_l is changed into i_{j_l} for $l = 1, \ldots, k$ and the other labels are unchanged.

EXAMPLE. See Fig. 8.4.

Proposition 8.2.2. *An algebra over the monad of shuffle trees is a shuffle operad.*

Proof. The proof is left to the reader as a good exercise. □

As a result we get a description of the free shuffle operad over a reduced arity graded module M as follows. The underlying reduced arity-graded module $\mathscr{T}_{\sqcup}M$ is spanned by the shuffle trees with vertices indexed by elements of M, respecting the number of inputs. Its shuffle composition $\gamma_{\sqcup}$ is defined by the grafting of shuffle trees, as the example of Fig. 8.5 shows.

Theorem 8.2.3. *The shuffle operad $(\mathscr{T}_{\sqcup}M, \gamma_{\sqcup})$ is free over M among the shuffle operads.*

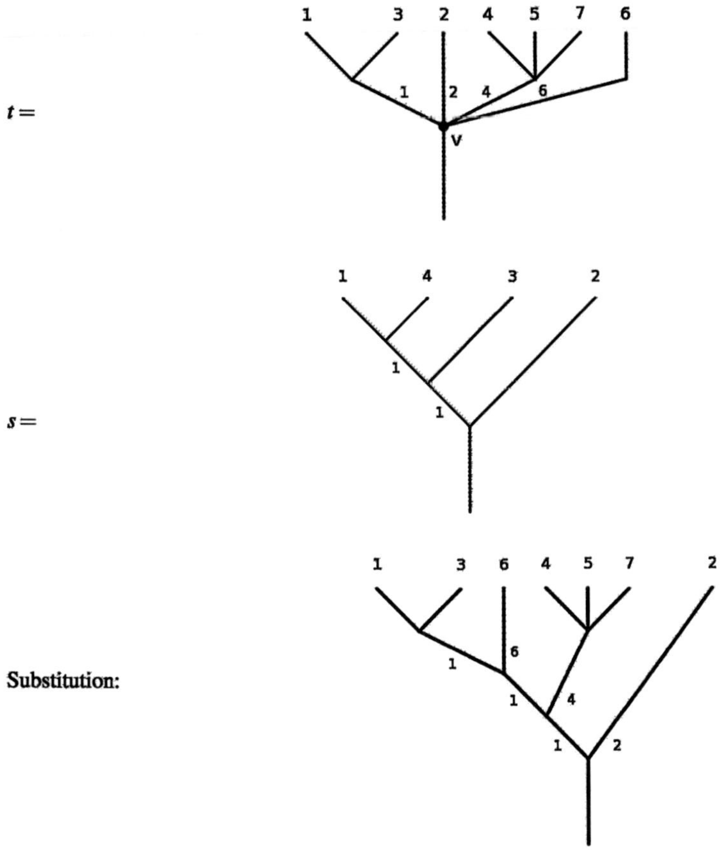

Fig. 8.4 Example of substitution of shuffle trees

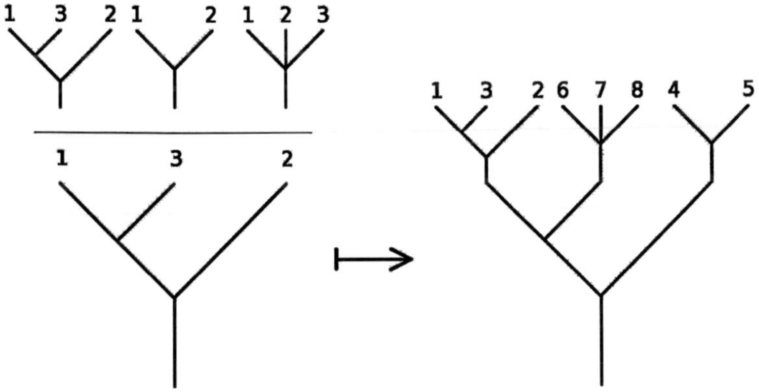

Fig. 8.5 Example of shuffle composition in the free shuffle operad

Proof. The shuffle tree space is equal to the same colimit as in Sect. 5.5.1 or equivalently as in Sect. 5.5.5 but applied to the shuffle composite product $\circ_{\sqcup\sqcup}$ instead of the composite product $\circ$. Thus, the proof of the present case follows from the same arguments. □

8.2.6 Group and Pre-Lie Algebra Associated to a Shuffle Operad

In Sect. 5.4.2, resp. Sect. 5.4.3, we associate to any symmetric operad a group, resp. a pre-Lie algebra. These constructions consist in summing over operations which are determined by two-level shuffle trees. Hence they make sense for shuffle operads as well.

8.2.7 From Symmetric Operads to Shuffles Operads

For any $\mathbb{S}$-module $\mathscr{P} = \{\mathscr{P}(n)\}_{n\geq 1}$, we denote by $\mathscr{P}^f$ the underlying arity-graded module:

$$\left(\mathscr{P}^f\right)_n := \mathscr{P}(n).$$

This is the forgetful functor from $\mathbb{S}$-Mod to $\mathbb{N}$-Mod. The crucial property of the shuffle composite product states that for any $\mathbb{S}$-modules $\mathscr{P}$ and $\mathscr{Q}$ there is an isomorphism

$$(\mathscr{P} \circ \mathscr{Q})^f \cong \mathscr{P}^f \circ_{\sqcup\sqcup} \mathscr{Q}^f.$$

Recall that the set of unshuffles $Sh^{-1}_{i_1,\ldots,i_k}$ provides representatives for the quotient of $\mathbb{S}_n$ under the left action by $\mathbb{S}_{i_1} \times \cdots \times \mathbb{S}_{i_k}$. The set of pointed unshuffles $\sqcup\sqcup(i_1, \ldots, i_k)$ provides representatives for the quotient of $Sh^{-1}_{i_1,\ldots,i_k}$ by the action of $\mathbb{S}_k$.

This equality enables us to define the composite

$$\gamma_{\sqcup\sqcup} : \mathscr{P}^f \circ_{\sqcup\sqcup} \mathscr{P}^f \cong (\mathscr{P} \circ \mathscr{P})^f \xrightarrow{\gamma} \mathscr{P}^f,$$

when $\mathscr{P} = (\mathscr{P}, \gamma, \eta)$ is an operad. Then, it is straightforward to see that $(\mathscr{P}^f, \gamma_{\sqcup\sqcup}, \eta)$ is a shuffle operad.

Proposition 8.2.4. *The forgetful functor*

$$(\mathbb{S}\text{-Mod}, \circ) \longrightarrow (\mathbb{N}\text{-Mod}, \circ_{\sqcup\sqcup})$$

is a strong monoidal functor, see Appendix B.3.3. *Therefore it induces the following functor*

$$\mathsf{Op} \longrightarrow \mathsf{Op}_{\sqcup\sqcup}, \qquad (\mathscr{P}, \gamma, \eta) \mapsto \left(\mathscr{P}^f, \gamma_{\sqcup\sqcup}, \eta\right).$$

Proof. It is straightforward to check the axioms of strong monoidal functors with the previous discussion. □

8.2.8 From Shuffle Operads to NS Operads

For any partition $n = i_1 + \cdots + i_k$, the identity permutation is an element of $\sqcup\!\sqcup(i_1, \ldots, i_k)$. Hence, for any arity graded modules M and N we have $M \circ N \subset M \circ_{\sqcup\!\sqcup} N$ and

$$\bigoplus M_k \otimes N_{i_1} \otimes \cdots \otimes N_{i_k} \subset \bigoplus M_k \otimes N_{i_1} \otimes \cdots \otimes N_{i_k} \otimes \mathbb{K}\big[\sqcup\!\sqcup(i_1, \ldots, i_k)\big].$$

Proposition 8.2.5. *The functor*

$$(\mathbb{N}\text{-Mod}, \circ_{\sqcup\!\sqcup}) \longrightarrow (\mathbb{N}\text{-Mod}, \circ)$$

is a monoidal functor, see Appendix B.3.3. *Therefore it induces the following functor*

$$\mathsf{Op}_{\sqcup\!\sqcup} \longrightarrow \mathsf{nsOp}.$$

Proof. It is a straightforward to check the axioms of monoidal functors with the previous discussion. □

Finally, we have two functors

$$\mathsf{Op} \longrightarrow \mathsf{Op}_{\sqcup\!\sqcup} \longrightarrow \mathsf{nsOp},$$

whose composite is the forgetful functor $\mathscr{P} \to \tilde{\mathscr{P}}$ mentioned in Sect. 5.9.11.

8.2.9 From Shuffle Operads to Permutads

Some families of shuffle trees are closed under substitution. The first example, mentioned above, is the set of planar trees which is in bijection with the shuffle trees with numbering $\{1, 2, 3, \ldots\}$. The associated monad defines the nonsymmetric operads.

Another example of a family of shuffle trees closed under substitution is made up of the "shuffle left combs", that is shuffle trees whose underlying planar tree is a left comb, see Fig. 8.6.

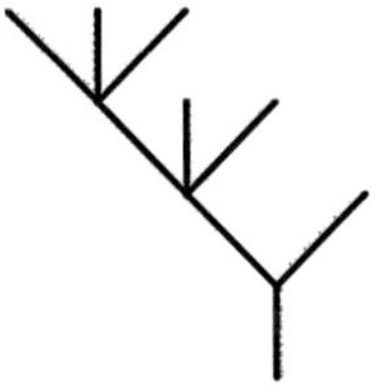

Fig. 8.6 Example of a left comb

An algebra over this monad is a permutad, see Sect. 13.14.7. It can be shown to coincide with the notion of shuffle algebra introduced by M. Ronco in [Ron11].

8.2.10 Koszul Duality Theory for Shuffle Operads

In the preceding chapters, we have developed the Koszul duality theory for symmetric and nonsymmetric operads following a certain pattern (twisting morphisms, bar and cobar constructions, twisting composite products, Koszul morphisms, comparison lemma). The same scheme applies to shuffle operads mutatis mutandis. In this way, one can develop the Koszul duality theory for shuffle operads as well. We leave the details to the reader as a very good exercise.

8.3 Rewriting Method for Operads

With the help of shuffle trees, we settle the rewriting method of Sects. 4.1 and 8.1 for (symmetric) operads. It works in the same way as in the algebra case except that one has to use a suitable order on shuffle trees in this case.

Let $\mathscr{P}(E, R)$ be a quadratic operad (not necessarily binary), for instance, the operad *Lie* encoding Lie algebras

$$Lie = \mathscr{P}\big(\mathbb{K}_{\mathrm{sgn}} c, c \circ (c \otimes \mathrm{id}) + c \circ (c \otimes \mathrm{id})^{(123)} + c \circ (c \otimes \mathrm{id})^{(321)}\big),$$

where $\mathbb{K}_{\mathrm{sgn}}$ stands for the signature representation of $\mathbb{S}_2$.

Step 1. We choose an ordered $\mathbb{K}$-linear basis $\{e_i\}_{i=1,\dots,m}$ for the $\mathbb{S}$-module of generators E.

Step 2. We consider, for instance, the induced path-lexicographic ordered basis on the shuffle trees $\mathscr{T}_{\sqcup}^{(2)}$ with 2 vertices, see Fig. 8.7. (For the complete definition of the path-lexicographic order, we refer the reader to Sect. 8.4.) Notice that one can use any other suitable order on shuffle trees, see loc. cit.

We consider the induced $\mathbb{K}$-linear basis of the $\mathbb{S}$-module R. Any element is of the form

$$r = \lambda t(e_i, e_j) - \sum_{t'(k,l) < t(i,j)} \lambda_{t'(k,l)}^{t(i,j)} t'(e_k, e_l); \quad \lambda \neq 0,$$

where the notation $t(i, j)$ (resp. $t(e_i, e_j)$) represents a shuffle tree with 2 vertices labeled by i and j (resp. by e_i and e_j). The monomial tree $t(e_i, e_j)$ is called the *leading term* of r. As usual, we can always change this basis for one with *normalized form*.

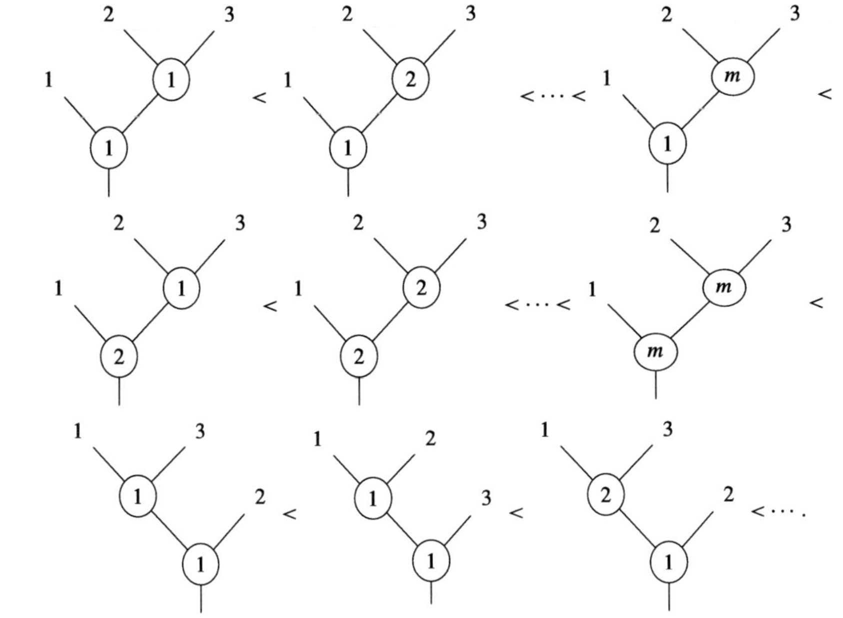

Fig. 8.7 Path-lexicographic order on 2-vertices binary shuffle trees

In the example of the operad *Lie*, the space of relations R admits the following normalized basis:

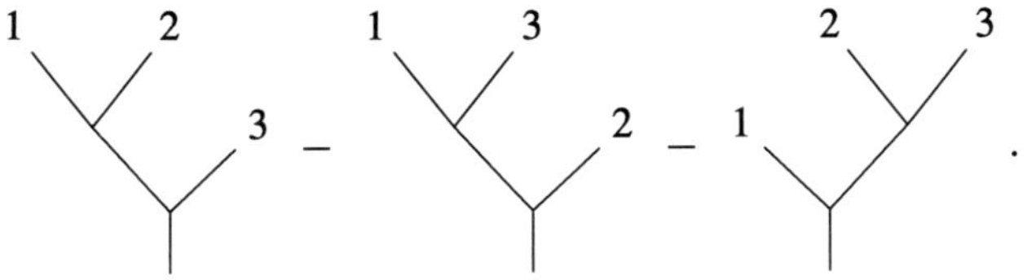

Since there is only one label $e_1 = c$ for the vertices, we do not mention it here.

Step 3. These choices provide rewriting rules of the form

$$t(e_i, e_j) \mapsto \sum_{t'(k,l) < t(i,j)} \lambda_{t'(k,l)}^{t(i,j)} t'(e_k, e_l),$$

$$\text{leading term} \mapsto \text{sum of lower and non-leading terms},$$

for any relator r in the normalized basis of R. A tree monomial $t(e_i, e_j, e_k)$ with 3 vertices is called *critical* if its two shuffle subtrees with 2 vertices are leading terms.

In the case of the operad *Lie*, there is one rewriting rule

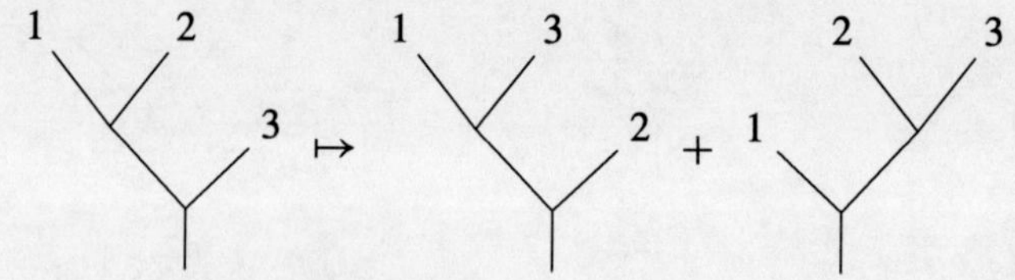

and only one critical tree monomial

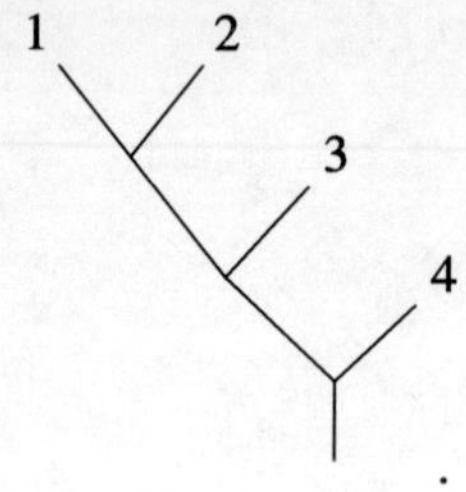

Step 4. Any critical tree monomial gives rise to a graph made up of the successive applications of the rewriting rules aforementioned. Any critical monomial is called *confluent* if the associated graph has only one terminal vertex (confluent graph).

In the guiding example of the operad *Lie*, the graph associated to the only critical monomial is confluent, see Fig. 8.8.

REMARK. Notice that the graph given here is a compact version of the full rewriting graph of the operad *Lie*: when it is possible to apply the rewriting rule to two different trees of a sum, we have only drawn one arrow, applying the two rewriting rules at once. There is yet another way to draw this rewriting diagram, which gives the Zamolodchikov tetrahedron equation. It leads to the categorical notion of *Lie 2-algebra*, see Baez–Crans [BC04, Sect. 4].

Conclusion. If each critical monomial is confluent, then the operad $\mathscr{P}$ is Koszul.

It is a consequence of the following result.

Theorem 8.3.1 (Rewriting method for operads). *Let $\mathscr{P}(E, R)$ be a quadratic operad. If its generating space E admits a $\mathbb{K}$-linear ordered basis, for which there exists a suitable order on shuffle trees, such that every critical monomial is confluent, then the operad $\mathscr{P}$ is Koszul.*

In this case, the operad $\mathscr{P}$ admits an induced shuffle tree basis sharing nice properties, called a PBW basis, see Sect. 8.5.

Therefore, the operad *Lie* is Koszul and admits a PBW basis, see Sect. 13.2.4 for more details.

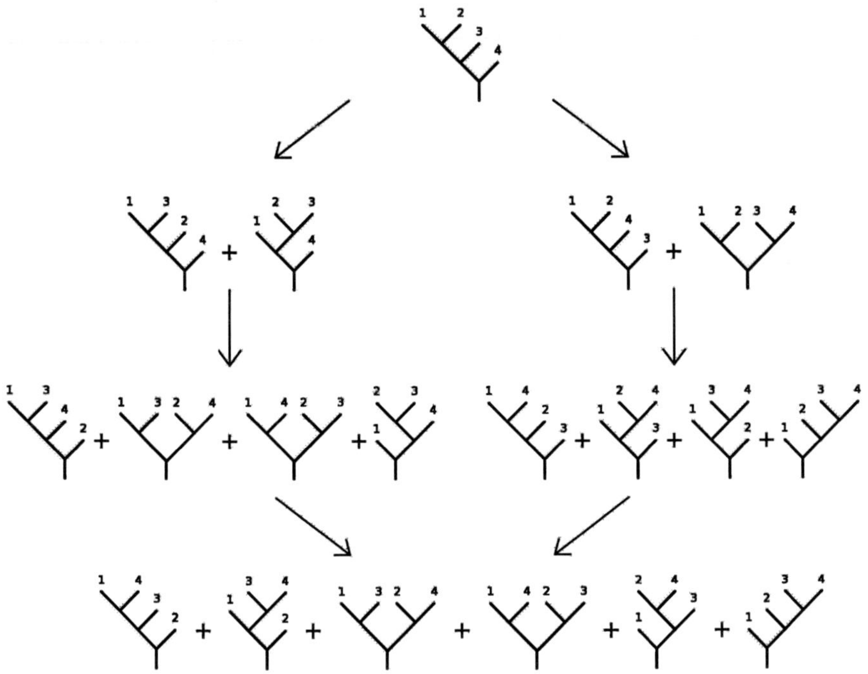

Fig. 8.8 The diamond for the operad *Lie*

8.4 Reduction by Filtration

In this section, we extend to operads the "reduction by filtration" method for algebras as described in Sect. 4.2. The only real new points lie in the use of the notion of shuffle operad and in suitable orders for the free (shuffle) operad. The proofs follow the same pattern as in Chap. 4 for associative algebras, so we skip most of them.

8.4.1 Suitable Order on Shuffle Trees

We consider the set of shuffle trees with vertices labeled by $\{1, \ldots, m\}$, where m can be infinite. We denote it simply by $\mathscr{T}_{\sqcup\sqcup}$. Notice that it labels a basis of the shuffle operad $\mathscr{T}_{\sqcup\sqcup}(E)$, on a reduced arity-graded module such that $\dim E_n = m$, for any $n \geq 1$. We choose a bijection between this set and the set of nonnegative integers $\mathbb{N}$, with its total order. (We require the identity tree | being sent to 0.) It endows the set of labeled shuffle trees with a total order denoted by $\mathscr{T}_{\sqcup\sqcup}{}^{p}$. We consider the partial shuffle products

$$\circ_{i,\omega} : \mathscr{T}_{\sqcup\sqcup}{}^{p} \times \mathscr{T}_{\sqcup\sqcup}{}^{q} \to \mathscr{T}_{\sqcup\sqcup}{}^{\chi(\sigma;p,q)}.$$

We ask that all these maps are strictly increasing, with respect to the lexicographic order on the left-hand side. In this case, we say that the order on labeled shuffle trees is a *suitable order*.

EXAMPLE (Path-lexicographic order). For simplicity, we restrict ourselves to the set of labeled shuffle trees whose vertices are at least trivalent, and such that m is finite. To any tree of arity n, we associate a sequence of $n+1$ words as follows. The n first words are obtained by reading the tree from the root to each leaf and by recording the labels indexing the vertices. The last word is given by the ordered labeling of the leaves, or equivalently by the image of the inverse of the associated pointed unshuffle. For example, one associates to the following tree

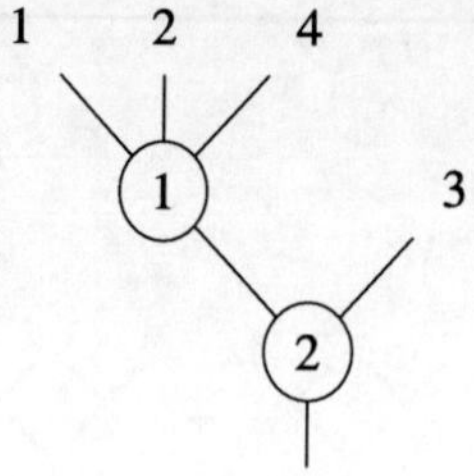

the sequence

$$(21, 21, 2, 21; 1324).$$

We leave it to the reader to verify that such a sequence characterizes the reduced labeled shuffle tree. We consider the following total order on this type of sequences.

1. We order them according to the number of elements of the sequence, that is the arity.
2. We consider the lexicographic order, with reversed order for the last word.

The example $m=2$ is given in Fig. 8.9. This ordering endows the set of reduced labeled shuffle trees with a suitable order, see [Hof10c, Proposition 3.5] and [DK10, Proposition 5]. These two references provide other examples of suitable orders on reduced trees. We refer the reader to Exercise 8.10.4 for an example of a suitable order on labeled shuffle trees.

8.4.2 Associated Graded Shuffle Operad

Let $\mathscr{P}(E,R)=\mathscr{T}(E)/(R)$ be a homogeneous quadratic operad, i.e. the $\mathbb{S}$-module of relations satisfies $R\subset \mathscr{T}(E)^{(2)}$. Suppose that the generating space E comes equipped with an extra datum: a decomposition into $\mathbb{S}$-modules $E\cong E_1\oplus\cdots\oplus E_m$. Notice that E and the E_i can be degree graded $\mathbb{S}$-modules.

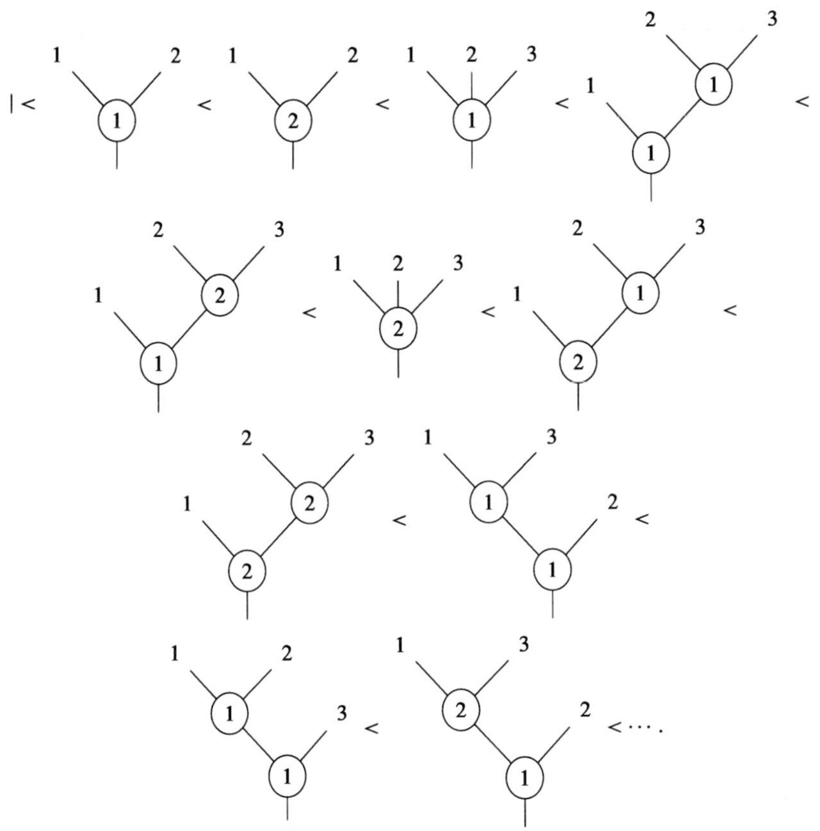

Fig. 8.9 Path-lexicographic order in the case $m = 2$

Since the forgetful functor $(\mathbb{S}\text{-}\mathsf{Mod}, \circ) \to (\mathbb{N}\text{-}\mathsf{Mod}, \circ_{\sqcup})$, $\mathscr{M} \mapsto \mathscr{M}^f$, is a strong monoidal functor by Proposition 8.2.4, we automatically get the isomorphisms of arity-graded modules

$$\big(\mathscr{T}(E)\big)^f \cong \mathscr{T}_{\sqcup}\big(E^f\big) \quad \text{and} \quad \big(\mathscr{P}(E,R)\big)^f \cong \mathscr{P}_{\sqcup}\big(E^f, R^f\big).$$

In other words, the underlying space of the free operad is equal to the free shuffle operad and the underlying space of the quadratic operad $\mathscr{P}$ is equal to the quadratic shuffle operad, which we denote by $\mathscr{P}_{\sqcup}$.

We consider a suitable order on the set of labeled shuffle trees. In this case, we say that the operad $\mathscr{P}$ is equipped with an *extra ordered grading*. Notice that if the generating space E is concentrated in some arities, it is enough to consider a suitable order only on the induced trees. For example, if $E(1) = 0$, then the shuffle trees are reduced and the aforementioned path-lexicographic order applies.

The total order on shuffle trees induces the following filtration of the free shuffle operad $F_p\mathcal{T}_{\sqcup\sqcup}(E) := \bigoplus_{q=0}^{p} \mathcal{T}_{\sqcup\sqcup}(E)^q$. This image under the canonical projection $\mathcal{T}_{\sqcup\sqcup}(E) \twoheadrightarrow \mathcal{P}_{\sqcup\sqcup} = \mathcal{P}_{\sqcup\sqcup}(E, R)$ defines a filtration on the quadratic shuffle operad denoted by $\mathrm{F}_p\mathcal{P}_{\sqcup\sqcup}$. The associated χ-graded arity-graded module is denoted by $\mathrm{gr}_\chi\, \mathcal{P}_{\sqcup\sqcup}$ or simply by $\mathrm{gr}\,\mathcal{P}_{\sqcup\sqcup}$.

Proposition 8.4.1. *Any quadratic operad $\mathcal{P}(E, R)$ equipped with an extra order grading induces a shuffle operad structure on the χ-graded arity-graded module* $\mathrm{gr}\,\mathcal{P}_{\sqcup\sqcup}$.

Proof. By the definition of a suitable order, any partial shuffle product of the shuffle operad $\mathcal{P}_{\sqcup\sqcup}$ induces a well-defined partial shuffle product on the graded module $\mathrm{gr}\,\mathcal{P}_{\sqcup\sqcup}$:

$$\bar{\circ}_{i,\omega} : \mathrm{gr}_p\, \mathcal{P}_{\sqcup\sqcup} \otimes \mathrm{gr}_q\, \mathcal{P}_{\sqcup\sqcup} \to \mathrm{gr}_{\chi(\sigma;p,q)}\, \mathcal{P}_{\sqcup\sqcup},$$

for any i and ω. These data endow the graded module $\mathrm{gr}\,\mathcal{P}_{\sqcup\sqcup}$ with a shuffle operad structure. □

Notice that, since the extra grading on $\mathcal{T}_{\sqcup\sqcup}(E)$ refines the weight grading, the shuffle operad $\mathrm{gr}\,\mathcal{P}_{\sqcup\sqcup}$ is also weight graded.

8.4.3 The Koszul Property

Theorem 8.4.2. *Let $\mathcal{P} = \mathcal{P}(E, R)$ be a quadratic operad equipped with an extra ordered grading. If the shuffle operad $\mathrm{gr}\,\mathcal{P}_{\sqcup\sqcup}$ is Koszul, then the operad $\mathcal{P}$ is also Koszul.*

Proof. The proof is essentially the same as the proof of Theorem 4.2.1, once the following point is understood. The isomorphism

$$\mathrm{B}\mathcal{P} \cong \mathrm{B}_{\sqcup\sqcup}\mathcal{P}_{\sqcup\sqcup}$$

from the bar construction of the operad $\mathcal{P}$ to the shuffle bar construction of the associated shuffle operad $\mathcal{P}_{\sqcup\sqcup}$, as differential graded arity-graded module, follows from Proposition 8.2.4 again. So if the homology of $\mathrm{B}_{\sqcup\sqcup}\mathcal{P}_{\sqcup\sqcup}$ is concentrated in syzygy degree 0, then so is the homology of $\mathrm{B}\mathcal{P}$. To conclude, it is enough to apply the methods of Theorem 4.2.1 to the shuffle operad $\mathcal{P}_{\sqcup\sqcup}$ to prove that $\mathrm{gr}\,\mathcal{P}_{\sqcup\sqcup}$ Koszul implies $\mathcal{P}_{\sqcup\sqcup}$ Koszul. □

From now on, we concentrate on trying to prove that the shuffle operad $\mathrm{gr}\,\mathcal{P}_{\sqcup\sqcup}$ is Koszul in order to show that the operad $\mathcal{P}$ is Koszul.

8.4.4 Quadratic Analog

We consider the kernel R_{lead} of the restriction to R of the projection $\mathscr{T}_{\sqcup\!\sqcup}(E) \twoheadrightarrow \operatorname{gr} \mathscr{P}_{\sqcup\!\sqcup}$. Any relator in R can be written $r = T_1 + \cdots + T_p$, where any T_i is a tree monomial in $\mathscr{T}_{\sqcup\!\sqcup}(E)^{(2)}$, and such that $T_i < T_{i+1}$ and $T_p \neq 0$. This latter term T_p is called the *leading term* or r. The space R_{lead} is linearly spanned by the leading terms of the elements of R.

We consider the following quadratic shuffle algebra

$$\mathring{\mathscr{P}}_{\sqcup\!\sqcup} := \mathscr{T}_{\sqcup\!\sqcup}(E)/(R_{\text{lead}}),$$

which is the best candidate for being a quadratic presentation of the shuffle operad $\operatorname{gr} \mathscr{P}_{\sqcup\!\sqcup}$.

Proposition 8.4.3. *Let $\mathscr{P} = \mathscr{P}(E, R)$ be a quadratic operad equipped with an extra ordered grading. There is a commutative diagram of epimorphisms of χ-graded, thus weight graded, shuffle operads*

$$\begin{array}{ccc} \mathscr{T}_{\sqcup\!\sqcup}(E) & & \\ \downarrow & \searrow & \\ \psi : \mathring{\mathscr{P}}_{\sqcup\!\sqcup} = \mathscr{T}_{\sqcup\!\sqcup}(E)/(R_{\text{lead}}) & \twoheadrightarrow & \operatorname{gr} \mathscr{P}_{\sqcup\!\sqcup}. \end{array}$$

If the shuffle operad $\mathring{\mathscr{P}}_{\sqcup\!\sqcup}$ is Koszul and if the canonical projection $\mathring{\mathscr{P}}_{\sqcup\!\sqcup} \cong \operatorname{gr} \mathscr{P}_{\sqcup\!\sqcup}$ is an isomorphism, then the operad $\mathscr{P}$ is Koszul.

8.4.5 Diamond Lemma

When the quadratic shuffle operad $\mathring{\mathscr{P}}_{\sqcup\!\sqcup}$ is Koszul, it is enough to check that the canonical projection $\psi : \mathring{\mathscr{P}}_{\sqcup\!\sqcup} \twoheadrightarrow \operatorname{gr} \mathscr{P}_{\sqcup\!\sqcup}$ is injective in weight 3, to show that it is an isomorphism.

Theorem 8.4.4 (Diamond Lemma for quadratic operads). *Let $\mathscr{P} = \mathscr{P}(E, R)$ be a quadratic operad equipped with an extra ordered grading. Suppose that the quadratic shuffle operad $\mathring{\mathscr{P}}_{\sqcup\!\sqcup} = \mathscr{T}_{\sqcup\!\sqcup}(E)/(R_{\text{lead}})$ is Koszul. If the canonical projection $\mathring{\mathscr{P}}_{\sqcup\!\sqcup} \twoheadrightarrow \operatorname{gr} \mathscr{P}_{\sqcup\!\sqcup}$ is injective in weight 3, then it is an isomorphism. In this case, the operad $\mathscr{P}$ is Koszul and its underlying arity-graded module is isomorphic to $\mathscr{P} \cong \mathring{\mathscr{P}}_{\sqcup\!\sqcup}$.*

8.4.6 The Inhomogeneous Case

For an inhomogeneous quadratic operad $\mathscr{P} = \mathscr{P}(E, R)$, that is $R \subset E \oplus \mathscr{T}(E)^{(2)}$, we require that the presentation satisfies the conditions (ql_1) and (ql_2) of Sect. 7.8.

We suppose that the associated homogeneous quadratic operad $q\mathscr{P} := \mathscr{P}(E, qR)$ admits an extra ordered grading. In this case, there exists a commutative diagram

$$\begin{array}{ccccc}
q\mathring{\mathscr{P}}_{\sqcup\!\sqcup} & & & & \\
\downarrow{\scriptstyle \mathsf{Op}_{\sqcup\!\sqcup}} & & & & \\
\mathrm{gr}_\chi\, q\mathscr{P}_{\sqcup\!\sqcup} & \xrightarrow[\mathbb{N}\text{-Mod}]{\cong} & q\mathscr{P} & & \\
\downarrow{\scriptstyle \mathsf{Op}_{\sqcup\!\sqcup}} & & \downarrow{\scriptstyle \mathsf{Op}} & & \\
\mathrm{gr}_\chi\, \mathscr{P}_{\sqcup\!\sqcup} & \xrightarrow[\mathbb{N}\text{-Mod}]{\cong} & \mathrm{gr}\,\mathscr{P} & \xrightarrow[\mathbb{S}\text{-Mod}]{\cong} & \mathscr{P},
\end{array}$$

where the type of the morphisms is indicated on the arrows. The morphisms of the first column preserve the χ-grading and the morphisms of the second column preserve the weight grading.

Theorem 8.4.5 (Diamond Lemma for inhomogeneous quadratic operads). *Let $\mathscr{P} = \mathscr{P}(E, R)$ be a quadratic-linear operad with a presentation satisfying conditions (ql_1) and (ql_2). We suppose that $\mathscr{T}_{\sqcup\!\sqcup}(E)$ comes equipped with an extra ordered grading.*

If the quadratic operad $q\mathring{\mathscr{P}}_{\sqcup\!\sqcup}$ is Koszul and if the canonical projection $q\mathring{\mathscr{P}}_{\sqcup\!\sqcup} \twoheadrightarrow \mathrm{gr}_\chi\, q\mathscr{P}_{\sqcup\!\sqcup}$ is injective in weight 3, then the operad $\mathscr{P}$ is Koszul and all the maps of the above diagram are isomorphisms, in particular:

$$q\mathring{\mathscr{P}}_{\sqcup\!\sqcup} \cong \mathrm{gr}_\chi\, q\mathscr{P}_{\sqcup\!\sqcup} \cong q\mathscr{P} \cong \mathrm{gr}\,\mathscr{P} \cong \mathscr{P}.$$

8.4.7 Reduction by Filtration Method for Nonsymmetric Operads

The same theory holds for nonsymmetric operads. In this simpler case, there is no need to use the notions of shuffle operad and shuffle trees. One works with the set of planar trees PT from the very beginning and one remains in the context of nonsymmetric operads. The only point is to consider a suitable order on planar trees, that is a total order such that any partial composite product

$$\circ_i : PT_k \times PT_l \to PT_n$$

is an increasing map. The adaptation of the path-lexicographic order gives an example of such a suitable order. Notice that reduced planar trees are particular examples of shuffle trees and that partial composite products are particular examples of partial shuffle products. We leave the details to the reader as a good exercise.

8.5 PBW Bases and Gröbner Bases for Operads

In this section, we study the particular case of the preceding section when the generating space E is equipped with an extra grading $E \cong E_1 \oplus \cdots \oplus E_m$ such that each sub-space E_i is one-dimensional. This gives rise to the notion of Poincaré–Birkhoff–Witt basis for (shuffle) operads. Quadratic operads which admit such a basis share nice properties. For instance, they are Koszul operads.

We introduce the equivalent notion of (quadratic) Gröbner basis, which is to the (quadratic) ideal (R) what PBW basis is to the quotient operad $\mathscr{T}(E)/(R)$.

For operads, the notion of PBW basis is due to Hoffbeck [Hof10c] and the notion of Gröbner basis is due to Dotsenko and Khoroshkin [DK10].

8.5.1 Ordered Bases

Let $\mathscr{P} = \mathscr{P}(E, R)$ be a quadratic operad with a decomposition of the generating space $E \cong E_1 \oplus \cdots \oplus E_m$ into one-dimensional vector spaces. This data is equivalent to an ordered basis $\{e_1, \ldots, e_m\}$ of the $\mathbb{K}$-module E. Together with a suitable order on shuffle trees $\mathscr{T}_{\sqcup\sqcup}$, it induces a totally ordered basis of $\mathscr{T}_{\sqcup\sqcup}(E)$, made up of tree monomials. In this case, we say that $\mathscr{T}_{\sqcup\sqcup}(E)$ is equipped with a *suitable ordered basis*.

In this basis, the space of relations is equal to

$$R = \left\{ \lambda t(e_i, e_j) - \sum_{t'(k,l) < t(i,j)} \lambda_{t'(k,l)}^{t(i,j)} t'(e_k, e_l); \lambda \neq 0 \right\},$$

where the notation $t(i, j)$ (resp. $t(e_i, e_j)$) stands for a shuffle tree with two vertices indexed by i and j (resp. by e_i and e_j). We denote by $\overline{T}^{(2)}$ the subset of $\mathscr{T}_{\sqcup\sqcup}$ made up of labeled shuffle trees which appear as leading terms of some relators. We denote by $T^{(2)}$ the complement of $\overline{T}^{(2)}$ in $\mathscr{T}_{\sqcup\sqcup}^{(2)}$. Notice that the space of relations admits a normalized basis of the form

$$R = \left\langle t(e_i, e_j) - \sum_{t'(k,l) \in T^{(2)} < t(i,j)} \lambda_{t'(k,l)}^{t(i,j)} t'(e_k, e_l) \right\rangle.$$

Proposition 8.5.1. *Let $\mathscr{P}$ be a quadratic operad $\mathscr{P}(E, R)$, with $\mathscr{T}_{\sqcup\sqcup}(E)$ equipped with a suitable ordered basis. The associated quadratic shuffle operad $\mathring{\mathscr{P}}_{\sqcup\sqcup}$ is equal to the quadratic shuffle operad $\mathscr{P}_{\sqcup\sqcup}(E, R_{\mathrm{lead}})$, with $R_{\mathrm{lead}} \cong \langle t(e_i, e_j); t(i, j) \in \overline{T}^{(2)} \rangle$.*

8.5.2 Quadratic Monomial Shuffle Operads

A *quadratic monomial shuffle operad* is a quadratic shuffle operad $\mathring{\mathscr{P}}_{\sqcup\sqcup} = \mathscr{P}_{\sqcup\sqcup}(E, R)$ with a (non-necessarily ordered) basis $\{e_i\}_{1 \leq i \leq m}$ of E such that the space of relations R is linearly spanned by a set of trees $\{t(e_i, e_j)\}_{t(i,j)\in\overline{T}^{(2)}}$, where $\overline{T}^{(2)} \subset \mathcal{T}_{\sqcup\!\sqcup}^{(2)}$. Hence the complement $T^{(2)}$ of $\overline{T}^{(2)}$ in $\mathcal{T}_{\sqcup\!\sqcup}^{(2)}$ labels a basis of the quotient $\mathring{\mathcal{P}}_{\sqcup\!\sqcup}^{(2)} = \mathcal{T}_{\sqcup\!\sqcup}(E)^{(2)}/R$.

We set $T^{(0)} := \{|\}$ and $T^{(1)} := \mathcal{T}_{\sqcup\!\sqcup}^{(1)}$. For any $n \geq 2$, we define the subset $T^{(n)} \subset \mathcal{T}_{\sqcup\!\sqcup}^{(n)}$ as the set of labeled shuffle trees with n vertices such that for any internal edge, the associated two-vertices subtree is in $T^{(2)}$. Finally, we consider $T = \bigsqcup_{n\in\mathbb{N}} T^{(n)} \subset \mathcal{T}_{\sqcup\!\sqcup}$. We define the subset $\overline{T} \subset \mathcal{T}_{\sqcup\!\sqcup}$ in the same way, that is a tree lives in $\overline{T}^{(n)} \subset \mathcal{T}_{\sqcup\!\sqcup}^{(n)}$ if for any internal edge, the associated two-vertices subtree is in $\overline{T}^{(2)}$.

Proposition 8.5.2. *Any quadratic monomial shuffle operad $\mathring{\mathcal{P}}_{\sqcup\!\sqcup} = \mathcal{P}_{\sqcup\!\sqcup}(E, R)$ admits a basis labeled by the subset $T \subset \mathcal{T}_{\sqcup\!\sqcup}$. Its Koszul dual shuffle cooperad $(\mathring{\mathcal{P}}_{\sqcup\!\sqcup})^{¡}$ admits a basis labeled by $\overline{T} \subset \mathcal{T}_{\sqcup\!\sqcup}$.*

Theorem 8.5.3. *Any quadratic monomial shuffle operad is Koszul.*

Proof. By the preceding proposition, the underlying space of the Koszul complex is equal to $(\mathring{\mathcal{P}}_{\sqcup\!\sqcup})^{¡} \circ_{\sqcup\!\sqcup} \mathring{\mathcal{P}}_{\sqcup\!\sqcup}$, which admits a basis of the following form. We consider the set of labeled shuffle trees with a horizontal partition into two parts; any two-vertices subtrees in the upper part belong to $T^{(2)}$ and any two-vertices subtrees in the lower part belong to $\overline{T}^{(2)}$. Finally, we conclude with the same argument as in the proof of Theorem 4.3.4. The differential map amounts to pulling up one top vertex from the lower part to the upper part. When it is 0, it produces a cycle element, which is easily shown to be a boundary element. Hence, this chain complex is acyclic. □

This result is a key point in the PBW basis theory because it simplifies the statements of Sect. 8.4. When the decomposition $E \cong E_1 \oplus \cdots \oplus E_m$ is made up of one-dimensional sub-spaces, the quadratic analog $\mathring{\mathcal{P}}_{\sqcup\!\sqcup}$ is always a Koszul shuffle operad.

8.5.3 PBW Basis

The image of the monomial basis $\{t(e_{\bar{i}})\}_{t\in T}$ of the shuffle operad $\mathring{\mathcal{P}}_{\sqcup\!\sqcup}$, given in Proposition 8.5.2, under the successive morphisms of graded (and also arity-graded) modules $\mathring{\mathcal{P}}_{\sqcup\!\sqcup} \twoheadrightarrow \operatorname{gr}\mathcal{P}_{\sqcup\!\sqcup} \cong \mathcal{P}_{\sqcup\!\sqcup} \cong \mathcal{P}$ provides a family of elements $\{\bar{t}(e_{\bar{i}})\}_{t\in T}$, which span the operad $\mathcal{P}$. When these elements are linearly independent, they form a basis of the operad $\mathcal{P}$, called a *Poincaré–Birkhoff–Witt basis*, or *PBW basis* for short. This is equivalent to having an isomorphism $\psi : \mathring{\mathcal{P}}_{\sqcup\!\sqcup} \cong \operatorname{gr}\mathcal{P}_{\sqcup\!\sqcup}$. We say that an operad $\mathcal{P} = \mathcal{P}(E, R)$ admits a PBW basis if the free shuffle operad $\mathcal{T}_{\sqcup\!\sqcup}(E)$ admits a suitable ordered basis such that the associated elements $\{\bar{t}(e_{\bar{i}})\}_{t\in T}$ form a basis of the operad $\mathcal{P}$.

EXAMPLE. The operad *Com* is generated by a one-dimensional space concentrated in arity 2: $E = \mathbb{K}\bullet$. We consider the suitable order obtained by restriction of the path-lexicographic order Sect. 8.4.1 on binary shuffle trees. On trees with 2 vertices, it is equal to

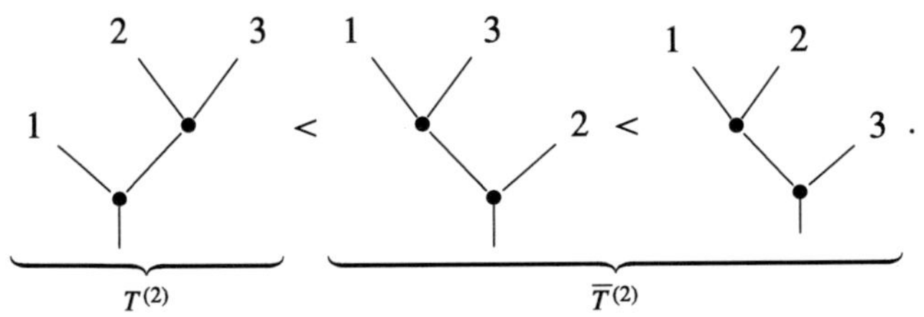

The space of relations is the linear span of the three terms made up of the difference of such two trees. Hence the set $T^{(2)}$ is made up of the first tree. It generates the set T made up of the left combs:

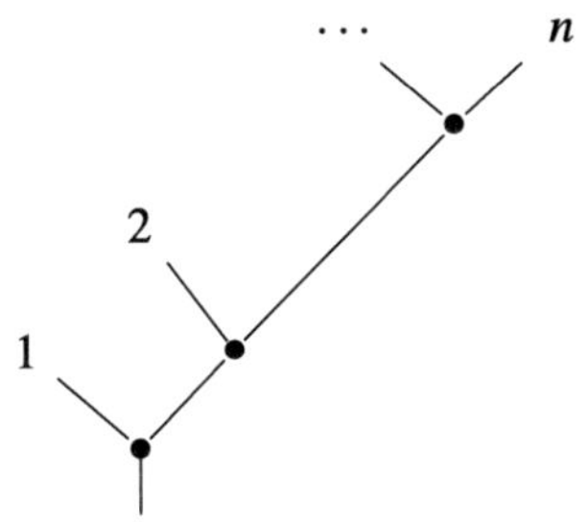

Since $\dim_{\mathbb{K}} Com(n) = 1$, this forms a PBW basis of the operad *Com*.

Theorem 8.5.4. *Any quadratic operad endowed with a PBW basis is Koszul.*

As for algebras, the existence of a PBW basis gives a purely algebraic condition to prove that an algebra is Koszul, without having to compute any homology group.

8.5.4 Diamond Lemma for PBW Bases

The following Diamond Lemma gives an easy way to prove that one has a PBW basis.

Theorem 8.5.5 (Diamond Lemma for PBW bases of operads). *Let $\mathcal{P} = \mathcal{P}(E, R)$ be a quadratic operad, with $\mathcal{T}_{\sqcup\sqcup}(E)$ equipped with a suitable ordered basis $\{t(e_{\bar{i}})\}_{t \in T}$. If the associated elements $\{\bar{t}\}_{t \in T^{(3)}}$ are linearly independent in $\mathcal{P}^{(3)}$, then the elements $\{\bar{t}(e_{\bar{i}})\}_{t \in T}$ form a PBW basis of $\mathcal{P}$. In that case, the operad $\mathcal{P}$ is Koszul.*

To check the assumption of this theorem, one uses the rewriting method of Sect. 8.3: one shows that every critical monomial is confluent. Notice that this theorem and Theorem 4.3.7 give a proof of the rewriting method Theorem 8.3.1.

8.5.5 Partial Shuffle Product of Elements of a PBW Basis

The composite $\mathring{\mathscr{P}}_{\sqcup} \twoheadrightarrow \operatorname{gr}\mathscr{P}_{\sqcup} \cong \mathscr{P}_{\sqcup} \cong \mathscr{P}$ is an epimorphism of arity-graded modules, but not of (shuffle) operads. Therefore, the partial shuffle product of two elements of the generating family $\{\bar{t}(e_{\bar{i}})\}_{t\in T}$ is not always equal to an element of this family, but sometimes to a sum of lower terms.

Proposition 8.5.6. *The elements $\{\bar{t}(e_{\bar{i}})\}_{t\in T}$ satisfy the following properties.*

(1) *Let σ be a partial pointed shuffle and let $t, s \in T$ be a pair of trees with matching arity. If the partial shuffle product of trees is not in T, $t \circ_\sigma s \notin T$, then the partial shuffle product $\bar{t} \circ_\sigma \bar{s}$ in the operad $\mathscr{P}$ is equal to a linear combination of strictly lower terms labeled by T:*

$$\bar{t} \circ_\sigma \bar{s} = \sum_{u\in T, u< t\circ_\sigma s} \lambda_u^{t,s} \bar{u},$$

with $\lambda_u^{t,s} \in \mathbb{K}$.

(2) *Any shuffle tree lies in T, if and only if any shuffle subtree $s \subset t$ lies in T.*

In the above example of the operad *Com*, only the partial shuffle composite $i = k$ and $\sigma = \mathrm{id}$ produces an element of the basis. The other composites are equal to the left comb in *Com*, which is a strictly lower but in the basis.

We leave it to the reader to prove that a tree basis T satisfying conditions (1) and (2) is a PBW basis. This equivalent definition is the definition originally given by Hoffbeck in [Hof10c].

8.5.6 Gröbner Bases for Operads

Following [DK10], we introduce the notion of Gröbner basis for an operadic ideal I of the free (shuffle) operad. In the quadratic case, when $I = (R)$, it is equivalent to a PBW basis for the quotient operad $\mathscr{P} = \mathscr{T}(E)/(R)$.

Any element $t \in \mathscr{T}_{\sqcup}(E)$ of the free shuffle operad is a linear combination of tree monomials. When $\mathscr{T}_{\sqcup}(E)$ is equipped with a suitable ordered basis, we denote by t_{lead} the leading term of t. For any subset $M \subset \mathscr{T}_{\sqcup}(E)$, we consider the set made up of the leading terms of any element of M and we denote it by $\mathrm{Lead}(M)$. Under this notation, the space of relations R_{lead} of Sect. 8.5.1 is equal to the linear span of $\mathrm{Lead}(R)$: $R_{\mathrm{lead}} = \langle \mathrm{Lead}(R)\rangle$.

A *Gröbner basis of an ideal* I *in* $\mathcal{T}_{\sqcup}(E)$ is a set $G \subset I$ which generates the ideal I, i.e. $(G) = I$, such that the leading terms of G and the leading terms of the elements of I generate the same ideal: $(\mathrm{Lead}(G)) = (\mathrm{Lead}(I))$.

Proposition 8.5.7. *Let $\mathcal{P} = \mathcal{P}(E, R)$ be a quadratic operad such that $\mathcal{T}_{\sqcup}(E)$ is equipped with a suitable ordered basis. The elements $\{\bar{t}(e_{\bar{i}})\}_{t\in T}$ form a PBW basis of $\mathcal{P}$ if and only if the elements*

$$\left\{t(e_i, e_j) - \sum_{t'(k,l)\in T^{(2)}<t(i,j)} \lambda_{t'(k,l)}^{t(i,j)}\, t'(e_k, e_l)\right\}_{t(i,j)\in \overline{T}^{(2)}},$$

spanning R, form a Gröbner basis of the ideal (R) in $\mathcal{T}_{\sqcup}(E)$.

In the quadratic case, the two notions of PBW basis and Gröbner basis are equivalent dual notions. The terminology "PBW basis" refers to the basis of the quotient operad while the terminology "Gröbner basis" refers to the ideal (R). We refer to [DK10] for more details on Gröbner bases for operads.

8.5.7 PBW Bases for Inhomogeneous Quadratic Operads

Following Sect. 8.4.6, we say that *an inhomogeneous quadratic operad $\mathcal{P}$ admits a PBW basis* if there exists a presentation $\mathcal{P} = \mathcal{P}(E, R)$, satisfying conditions (ql_1) and (ql_2), such that the associated quadratic algebra $q\mathcal{P} = \mathcal{P}(E, qR)$ admits a PBW basis. In this case, the image $\{\bar{t}(e_{\bar{i}})\}_{t\in T} \subset \mathcal{P}$ of the tree basis elements $\{t(e_{\bar{i}})\}_{\bar{i}\in T}$ of the quadratic monomial shuffle operad $q\mathring{\mathcal{P}}_{\sqcup}$ gives a basis of the inhomogeneous quadratic operad $\mathcal{P}$. Such a result is once again proved using the following version of the Diamond Lemma.

Theorem 8.5.8 (Diamond Lemma for PBW bases of inhomogeneous operads). *Let $\mathcal{P} = \mathcal{P}(E, R)$ be an inhomogeneous quadratic operad with a quadratic-linear presentation satisfying conditions (ql_1) and (ql_2) and such that $\mathcal{T}_{\sqcup}(E)$ is equipped with a suitable ordered basis.*

If the images of the tree elements $\{t(e_{\bar{i}})\}_{\bar{i}\in T^{(3)}}$ in $q\mathcal{P}$ are linearly independent, then the images $\{\bar{t}(e_{\bar{i}})\}_{t\in T}$ of the elements $\{t(e_{\bar{i}})\}_{t\in T}$ form a basis of $\mathcal{P}$, and the operad $\mathcal{P}$ is Koszul.

In the inhomogeneous case too, the notion of PBW basis is equivalent and dual to that of Gröbner basis.

Proposition 8.5.9. *Let $\mathcal{P}$ be an inhomogeneous quadratic operad with a quadratic-linear presentation $\mathcal{P} = \mathcal{P}(E, R)$ satisfying conditions (ql_1) and (ql_2) and such that $\mathcal{T}_{\sqcup}(E)$ is equipped with a suitable ordered basis. Let $\varphi : qR \to E$ be* *the linear map whose graph gives R. The elements* $\{\bar{t}(e_{\bar{i}})\}_{t\in T} \subset \mathscr{P}$ *form a PBW basis of* $\mathscr{P}$ *if and only if the elements*

$$\left\{(\mathrm{Id}-\varphi)\left(t(e_i, e_j) - \sum_{t'(k,l)\in T^{(2)} < t(i,j)} \lambda_{t'(k,l)}^{t(i,j)}\, t'(e_k, e_l)\right)\right\}_{t(i,j)\in \overline{T}^{(2)}}$$

spanning R, *form a Gröbner basis of the ideal* (R) *in* $\mathscr{T}_{\sqcup}(E)$.

8.5.8 PBW/Gröbner Bases for Nonsymmetric Operads

Once again, the same results hold for nonsymmetric operads. One has just to replace the set of shuffle trees by the one of planar trees.

8.6 Distributive Laws

In this section, we show how to build a new operad out of two operads by means of an extra datum, called *distributive law*. When the two first operads are quadratic, we describe how to get a distributive law from these presentations. The main result states that, in this case, the resulting operad is Koszul if the two first ones are Koszul.

The notion of distributive law goes back to Jon Beck [Bec69]. Its application to Koszul operads was first introduced by Martin Markl in [Mar96a, FM97] and then refined in [Val07b, Dot07].

8.6.1 Definition of a Distributive Law

Let $\mathscr{A} = (\mathscr{A}, \gamma_{\mathscr{A}}, \iota_{\mathscr{A}})$ and $\mathscr{B} = (\mathscr{B}, \gamma_{\mathscr{B}}, \iota_{\mathscr{B}})$ be two operads. In order to put an operad structure on the composite $\mathscr{A} \circ \mathscr{B}$, one needs a morphism of $\mathbb{S}$-modules

$$\Lambda : \mathscr{B} \circ \mathscr{A} \to \mathscr{A} \circ \mathscr{B}.$$

It is called a *distributive law* if the following diagrams are commutative:

$$\text{(I)}\qquad \begin{array}{ccccc} \mathscr{B}\circ\mathscr{A}\circ\mathscr{A} & \xrightarrow{\Lambda\circ \mathrm{Id}_{\mathscr{A}}} & \mathscr{A}\circ\mathscr{B}\circ\mathscr{A} & \xrightarrow{\mathrm{Id}_{\mathscr{A}}\circ\Lambda} & \mathscr{A}\circ\mathscr{A}\circ\mathscr{B} \\ \downarrow{\scriptstyle \mathrm{Id}_{\mathscr{B}}\circ\gamma_{\mathscr{A}}} & & & & \downarrow{\scriptstyle \gamma_{\mathscr{A}}\circ \mathrm{Id}_{\mathscr{B}}} \\ \mathscr{B}\circ\mathscr{A} & & \xrightarrow{\quad\Lambda\quad} & & \mathscr{A}\circ\mathscr{B}, \end{array}$$

$$
\text{(II)}\qquad
\begin{array}{ccccc}
\mathscr{B}\circ\mathscr{B}\circ\mathscr{A} & \xrightarrow{\mathrm{Id}_{\mathscr{B}}\circ\Lambda} & \mathscr{B}\circ\mathscr{A}\circ\mathscr{B} & \xrightarrow{\Lambda\circ\mathrm{Id}_{\mathscr{B}}} & \mathscr{A}\circ\mathscr{B}\circ\mathscr{B} \\
\downarrow{\scriptstyle \gamma_{\mathscr{B}}\circ\mathrm{Id}_{\mathscr{A}}} & & & & \downarrow{\scriptstyle \mathrm{Id}_{\mathscr{A}}\circ\gamma_{\mathscr{B}}} \\
\mathscr{B}\circ\mathscr{A} & & \xrightarrow{\qquad\Lambda\qquad} & & \mathscr{A}\circ\mathscr{B},
\end{array}
$$

$$
\text{(i)}\quad
\begin{array}{ccc}
 & \mathscr{B} & \\
\swarrow{\scriptstyle \mathrm{Id}_{\mathscr{B}}\circ\iota_{\mathscr{A}}} & & \searrow{\scriptstyle \iota_{\mathscr{A}}\circ\mathrm{Id}_{\mathscr{B}}} \\
\mathscr{B}\circ\mathscr{A} & \xrightarrow{\ \Lambda\ } & \mathscr{A}\circ\mathscr{B}
\end{array}
\qquad
\text{(ii)}\quad
\begin{array}{ccc}
 & \mathscr{A} & \\
\swarrow{\scriptstyle \iota_{\mathscr{B}}\circ\mathrm{Id}_{\mathscr{A}}} & & \searrow{\scriptstyle \mathrm{Id}_{\mathscr{A}}\circ\iota_{\mathscr{B}}} \\
\mathscr{B}\circ\mathscr{A} & \xrightarrow{\ \Lambda\ } & \mathscr{A}\circ\mathscr{B}.
\end{array}
$$

The associativity isomorphisms and the identifications like $\mathscr{B}\circ \mathrm{I}\cong\mathscr{B}\cong \mathrm{I}\circ\mathscr{B}$ are implicit in these diagrams.

Proposition 8.6.1. *If $\Lambda : \mathscr{B}\circ\mathscr{A}\to\mathscr{A}\circ\mathscr{B}$ is a distributive law for the operad structures of $\mathscr{A}$ and $\mathscr{B}$, then $\mathscr{A}\circ\mathscr{B}$ is an operad for the composition*

$$\gamma_\Lambda := (\gamma_{\mathscr{A}}\circ\gamma_{\mathscr{B}})(\mathrm{Id}_{\mathscr{A}}\circ\Lambda\circ\mathrm{Id}_{\mathscr{B}}) : (\mathscr{A}\circ\mathscr{B})\circ(\mathscr{A}\circ\mathscr{B})\to\mathscr{A}\circ\mathscr{B},$$

and for the unit

$$\iota_\Lambda := \iota_{\mathscr{A}}\circ\iota_{\mathscr{B}} : \mathrm{I}\to\mathscr{A}\circ\mathscr{B}.$$

Proof. In order to simplify the notation we write $\mathscr{A}\mathscr{B}$ in place of $\mathscr{A}\circ\mathscr{B}$. The associativity condition for γ_Λ is the commutativity of the outer square diagram. It follows from the commutativity of the inner diagrams, which are either straightforward, or follows from the associativity of $\gamma_{\mathscr{A}}$ and $\gamma_{\mathscr{B}}$, or from the hypothesis.

$$
\begin{array}{ccccccc}
\mathscr{A}\mathscr{B}\mathscr{A}\mathscr{B}\mathscr{A}\mathscr{B} & \to & \mathscr{A}\mathscr{B}\mathscr{A}\mathscr{A}\mathscr{B}\mathscr{B} & \to & \mathscr{A}\mathscr{B}\mathscr{A}\mathscr{B}\mathscr{B} & \to & \mathscr{A}\mathscr{B}\mathscr{A}\mathscr{B} \\
\downarrow & & \downarrow & & \downarrow & & \downarrow \\
\mathscr{A}\mathscr{A}\mathscr{B}\mathscr{B}\mathscr{A}\mathscr{B} & \to & \mathscr{A}\mathscr{A}\mathscr{B}\mathscr{A}\mathscr{B}\mathscr{B} & \text{(I)} & \downarrow & & \downarrow \\
\downarrow & \text{(II)} & \mathscr{A}\mathscr{A}\mathscr{A}\mathscr{B}\mathscr{B}\mathscr{B} & \to & \mathscr{A}\mathscr{A}\mathscr{B}\mathscr{B}\mathscr{B} & \to & \mathscr{A}\mathscr{A}\mathscr{B}\mathscr{B} \\
\downarrow & & \downarrow & & & & \downarrow \\
\mathscr{A}\mathscr{A}\mathscr{B}\mathscr{A}\mathscr{B} & \to & \mathscr{A}\mathscr{A}\mathscr{A}\mathscr{B}\mathscr{B} & & \text{Ass. of } \gamma_{\mathscr{A}},\gamma_{\mathscr{B}} & & \downarrow \\
\downarrow & & \downarrow & & & & \downarrow \\
\mathscr{A}\mathscr{B}\mathscr{A}\mathscr{B} & \to & \mathscr{A}\mathscr{A}\mathscr{B}\mathscr{B} & & \longrightarrow & & \mathscr{A}\mathscr{B}.
\end{array}
$$

In this diagram the maps are composite products of Λ, $\gamma_{\mathscr{A}}$, $\gamma_{\mathscr{B}}$ and the identities. The exact combination is clear from the source and the target. For instance the leftmost arrow of the first row is $\mathrm{Id}_{\mathscr{A}}\circ\mathrm{Id}_{\mathscr{B}}\circ\mathrm{Id}_{\mathscr{A}}\circ\Lambda\circ\mathrm{Id}_{\mathscr{B}}$.

The left unit property is proved in the same way by the following commutative diagram.

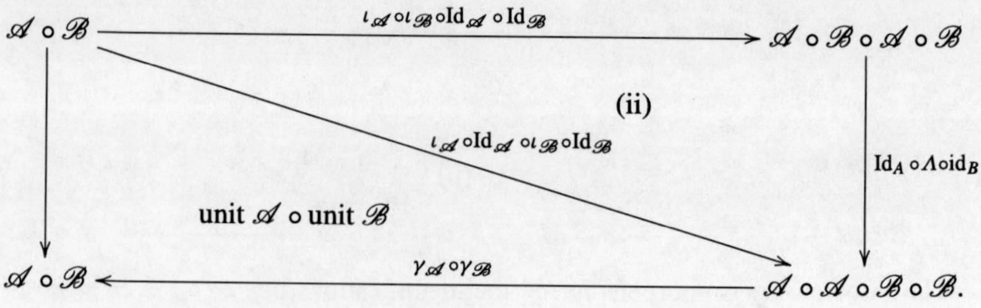

The proof of the right unit property uses (i). □

This notion of distributive law is the exact application to Schur functors of the general one for monads due to Beck [Bec69]. For a converse result, we refer the reader to Exercise 8.10.10.

8.6.2 Distributive Laws for Quadratic Data

Let $\mathscr{A} := \mathscr{P}(V, R) = \mathscr{T}(V)/(R)$ and $\mathscr{B} := \mathscr{P}(W, S) = \mathscr{T}(W)/(S)$ be two quadratic operads. Their coproduct $\mathscr{A} \vee \mathscr{B}$ in the category of operads is again a quadratic operad with presentation $\mathscr{A} \vee \mathscr{B} = \mathscr{P}(V \oplus W, R \oplus S)$. We suppose now that there exists a compatibility relation between the generating operations V of $\mathscr{A}$ and the generating operations W of $\mathscr{B}$ of the following form.

Under the notation $\circ_{(1)}$, introduced in Sect. 6.1.1, we have a natural isomorphism of $\mathbb{S}$-modules:

$$\mathscr{T}(V \oplus W)^{(2)} = \mathscr{T}(V)^{(2)} \oplus V \circ_{(1)} W \oplus W \circ_{(1)} V \oplus \mathscr{T}(W)^{(2)}.$$

Let $\lambda : W \circ_{(1)} V \to V \circ_{(1)} W$ be a morphism of $\mathbb{S}$-modules that we call a *rewriting rule*. The graph of λ gives the weight 2 space of relation

$$D_\lambda := \left(T - \lambda(T), T \in W \circ_{(1)} V \right) \subset \mathscr{T}(V \oplus W)^{(2)}.$$

Finally, we consider the operad $\mathscr{A} \vee_\lambda \mathscr{B}$ which is, by definition, the quotient of the coproduct $\mathscr{A} \vee \mathscr{B}$ by the ideal generated by D_λ. This operad admits the following quadratic presentation:

$$\mathscr{A} \vee_\lambda \mathscr{B} = \mathscr{P}(V \oplus W, R \oplus D_\lambda \oplus S).$$

So, in this operad, the map λ has to been seen as a rewriting rule, which allows us to move the operations of V under the operations of W, see the figures of Sect. 8.6.3 below. It remains to see whether this *local* rewriting rule induces or not a *global* distributive law $\mathscr{B} \circ \mathscr{A} \to \mathscr{A} \circ \mathscr{B}$.

The composite $\mathscr{T}(V) \circ \mathscr{T}(W) \to \mathscr{T}(V \oplus W) \twoheadrightarrow \mathscr{T}(V \oplus W)/(R \oplus D_\lambda \oplus S)$ induces the following epimorphism of $\mathbb{S}$-modules

$$p : \mathscr{A} \circ \mathscr{B} \twoheadrightarrow \mathscr{A} \vee_\lambda \mathscr{B}.$$

Similarly, there exists a morphism of $\mathbb{S}$-modules $\mathscr{B} \circ \mathscr{A} \to \mathscr{A} \vee_\lambda \mathscr{B}$.

Proposition 8.6.2. *Let $\mathscr{A} = \mathscr{P}(V, R)$ and $\mathscr{B} = \mathscr{P}(W, S)$ be two quadratic operads. For any morphism of $\mathbb{S}$-modules $\lambda : W \circ_{(1)} V \to V \circ_{(1)} W$, such that $p : \mathscr{A} \circ \mathscr{B} \twoheadrightarrow \mathscr{A} \vee_\lambda \mathscr{B}$ is an isomorphism, the composite*

$$\Lambda : \mathscr{B} \circ \mathscr{A} \to \mathscr{A} \vee_\lambda \mathscr{B} \xrightarrow{p^{-1}} \mathscr{A} \circ \mathscr{B}$$

is a distributive law.

In that case, the map $p : (\mathscr{A} \circ \mathscr{B}, \gamma_\Lambda, \iota_\Lambda) \to \mathscr{A} \vee_\lambda \mathscr{B}$ is an isomorphism of operads.

Proof. Conditions (i) and (ii) of Sect. 8.6.1 are trivially satisfied. The inclusion $i : \mathscr{A} \rightarrowtail \mathscr{A} \vee \mathscr{B} \twoheadrightarrow \mathscr{A} \vee_\lambda \mathscr{B}$ satisfies properties like the following commutative diagram:

$$\begin{array}{ccccc}
\mathscr{B} \circ \mathscr{A} \circ \mathscr{A} & \xrightarrow{p \circ \mathrm{Id}_{\mathscr{A}}} & (\mathscr{A} \vee_\lambda \mathscr{B}) \circ \mathscr{A} & \xrightarrow{\mathrm{Id} \circ i} & (\mathscr{A} \vee_\lambda \mathscr{B}) \circ (\mathscr{A} \vee_\lambda \mathscr{B}) \\
\downarrow{\scriptstyle \mathrm{Id}_{\mathscr{B}} \circ \gamma_{\mathscr{A}}} & & & & \downarrow{\scriptstyle \gamma} \\
\mathscr{B} \circ \mathscr{A} & & \xrightarrow{\quad p \quad} & & (\mathscr{A} \vee_\lambda \mathscr{B}),
\end{array}$$

where Id and γ refer to $\mathscr{A} \vee_\lambda \mathscr{B}$.

Therefore, "Condition (I) of a distributive law" is a consequence of the commutativity of the diagram:

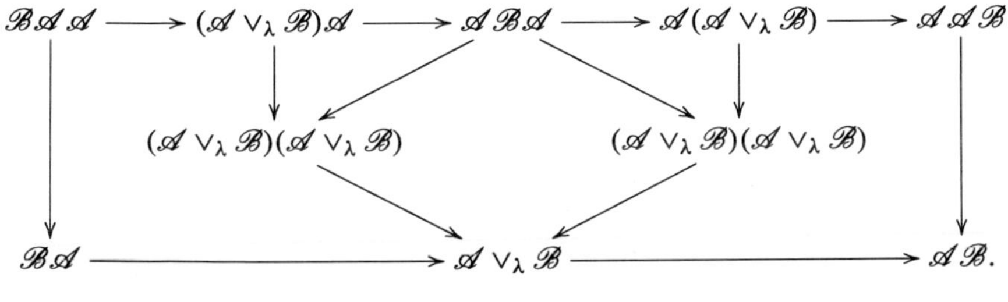

The commutativity of the middle square comes from the associativity of the composite of $\mathscr{A} \vee_\lambda \mathscr{B}$: $\mathscr{A}\mathscr{B}\mathscr{A} \rightarrowtail (\mathscr{A} \vee_\lambda \mathscr{B})^{\circ 3} \to \mathscr{A} \vee_\lambda \mathscr{B}$. □

Notice the similarity with the rewriting method: the elements of $\mathscr{A} \circ \mathscr{B}$ play the same role as the chosen monomials which can form a PBW basis. They always linearly span the final operad but it remains to prove that they actually form a basis of it. The rest of this section is written in the same way as the previous sections.

For a more general study of distributive laws, see P.-L. Curien [Cur12].

8.6.3 The Example of the Operad *Pois*

Let us consider the example of the operad *Pois* encoding Poisson algebras, see Sect. 13.3 for more details. A Poisson algebra is a vector space (or dg module) endowed with an associative and commutative product $\bullet$ and with a Lie bracket $[\,,\,]$, which satisfy the Leibniz relation $[x \bullet y, z] = [x, z] \bullet y + x \bullet [y, z]$. The graphical representation of this relation is

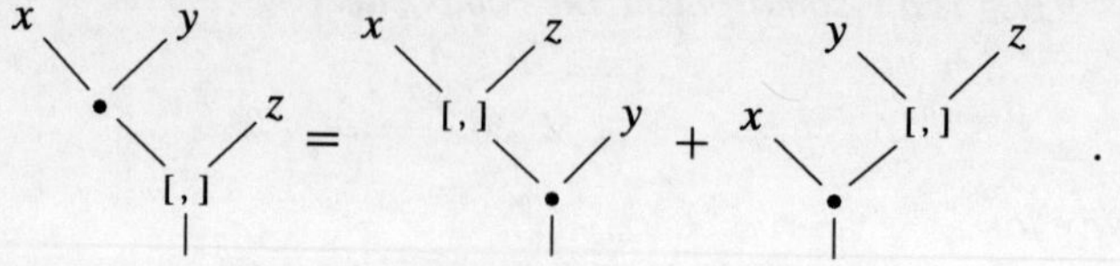

It is an example of the preceding definition. Here the operad $\mathcal{A} = Com$ is the operad of commutative algebras with $V = \mathbb{K}\bullet$, the operad $\mathcal{B} = Lie$ is the operad of Lie algebras with $W = \mathbb{K}[\,,\,]$ and the rewriting rule λ is equal to

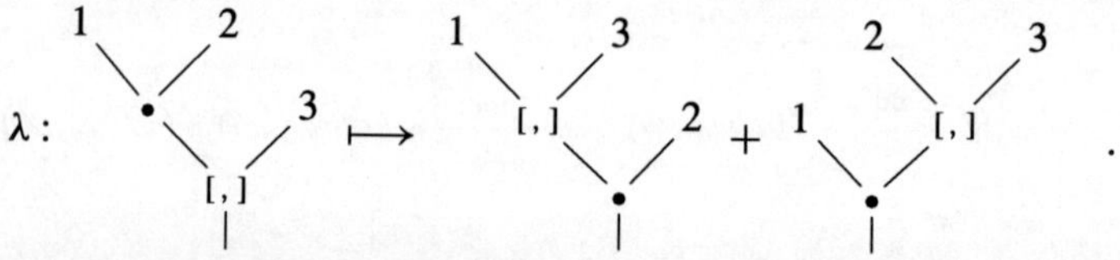

Hence the $\mathbb{S}$-module $Com \circ Lie$ linearly spans the operad *Pois*. Actually, the rewriting rule λ induces a distributive law and the $\mathbb{S}$-module $Com \circ Lie$ is isomorphic to *Pois*. Since the proof is quite involved, we postpone it to Sect. 8.6.7 after Theorem 8.6.5.

8.6.4 Koszul Duality of Operads with Trivial Distributive Law

Let $\mathcal{A} = \mathcal{P}(V, R)$ and $\mathcal{B} = \mathcal{P}(W, S)$ be two quadratic operads and consider the trivial rewriting rule $\lambda \equiv 0$. On the level of $\mathbb{S}$-modules $\mathcal{A} \vee_0 \mathcal{B} \cong \mathcal{A} \circ \mathcal{B}$, so it induces a global distributive law, which is the trivial one $\Lambda \equiv 0$. This yields the isomorphism of operads $\mathcal{A} \vee_0 \mathcal{B} \cong (\mathcal{A} \circ \mathcal{B}, \gamma_0)$.

Proposition 8.6.3. *Let $\mathcal{A} = \mathcal{P}(V, R)$ and $\mathcal{B} = \mathcal{P}(W, S)$ be two quadratic operads. The underlying module of the Koszul dual cooperad of the operad defined by the trivial rewriting rule is equal to*

$$(\mathcal{A} \vee_0 \mathcal{B})^{i} \cong \mathcal{B}^{i} \circ \mathcal{A}^{i}.$$

If moreover, the operads $\mathcal{A}$, $\mathcal{B}$ are Koszul, then the operad $\mathcal{A} \vee_0 \mathcal{B}$ is Koszul.

Proof. The isomorphism $(\mathscr{A} \vee_0 \mathscr{B})^{¡} \cong \mathscr{B}^{¡} \circ \mathscr{A}^{¡}$ is a direct consequence of Proposition 7.3.1. The Koszul complex of the operad $\mathscr{A} \vee_0 \mathscr{B}$ is isomorphic to $(\mathscr{B}^{¡} \circ \mathscr{A}^{¡}) \circ (\mathscr{A} \circ \mathscr{B})$. We filter this chain complex by the weight of the elements of $\mathscr{B}^{¡}$. Hence the first term of the associated spectral sequence is equal to

$$\left(E^0, d^0\right) \cong \mathscr{B}^{¡} \circ \left(\mathscr{A}^{¡} \circ_{\kappa_{\mathscr{A}}} \mathscr{A}\right) \circ \mathscr{B}.$$

So its homology is equal to $(E^1, d^1) \cong \mathscr{B}^{¡} \circ_{\kappa_{\mathscr{B}}} \mathscr{B}$. It finally gives $E^2 \cong \mathrm{I}$ and we conclude by the convergence theorem of spectral sequences (Theorem 1.5.1). □

8.6.5 Distributive Law Implies Koszul

The Koszul property is stable under the construction of operads via distributive laws.

Theorem 8.6.4. *Let $\mathscr{A} = \mathscr{P}(V, R)$ and $\mathscr{B} = \mathscr{P}(W, S)$ be two quadratic operads endowed with a rewriting rule $\lambda : W \circ_{(1)} V \to V \circ_{(1)} W$ which induces a distributive law.*

The operads $\mathscr{A}$ and $\mathscr{B}$ are Koszul if and only if the operad $\mathscr{A} \vee_\lambda \mathscr{B}$ is Koszul.

Proof. (⇒) We follow the same ideas as in the proof of Theorem 4.2.1. We consider the bar construction $\mathrm{B}^{-\bullet}(\mathscr{A} \vee_\lambda \mathscr{B})$ with the opposite of the syzygy degree as homological degree and we introduce the following filtration. Since λ defines a distributive law, the operad $\mathscr{A} \vee_\lambda \mathscr{B}$ is isomorphic to $(\mathscr{A} \circ \mathscr{B}, \gamma_\Lambda)$. The bar construction $\mathrm{B}(\mathscr{A} \circ \mathscr{B}, \gamma_\Lambda)$ is made up of linear combinations of trees with vertices labeled by 2-leveled trees made up of elements of $\mathscr{A}$ in the bottom part and elements of $\mathscr{B}$ in the top part. We say that one internal edge carries an inversion if it links a nontrivial element of $\mathscr{B}$ in the lower vertex with a nontrivial element of $\mathscr{A}$ in the upper vertex. For such a tree, we define its number of inversions by the sum, over the internal edges, of inversions. The filtration F_p of $\mathrm{B}(\mathscr{A} \vee_\lambda \mathscr{B})$ is equal to the sub-$\mathbb{S}$-module generated by trees with number of inversions less than or equal to p. This filtration is stable under the boundary map. The first term of the associated spectral sequence (E^0, d^0) is isomorphic to the bar construction $\mathrm{B}(\mathscr{A} \circ \mathscr{B}, \gamma_0) \cong \mathrm{B}(\mathscr{A} \vee_0 \mathscr{B})$ of the operad defined by the trivial rewriting rule. Since this latter operad is Koszul by Proposition 8.6.3, the homology of its bar construction is concentrated in syzygy degree 0. Therefore, the second page E^1 is concentrated on the diagonal $E^1_{p\,-p}$. So it collapses at rank 1 and the limit E^∞ is also concentrated on the diagonal $E^\infty_{p\,-p}$. The filtration $\mathrm{F_p}$ being exhaustive and bounded below, the classical convergence theorem of spectral sequences (Theorem 1.5.1) ensures that the homology of $\mathrm{B}(\mathscr{A} \vee_\lambda \mathscr{B})$ is concentrated in syzygy degree 0. Hence the operad $\mathscr{A} \vee_\lambda \mathscr{B}$ is Koszul.

(⇐) The bar constructions $\mathrm{B}\mathscr{A}$ and $\mathrm{B}\mathscr{B}$ are sub-chain complexes of the bar construction $\mathrm{B}(\mathscr{A} \vee_\lambda \mathscr{B})$. If the homology of the latter one is concentrated in syzygy degree 0 ($\mathscr{A} \vee_\lambda \mathscr{B}$ Koszul operad), it is also true for $\mathrm{B}\mathscr{A}$ and $\mathrm{B}\mathscr{B}$ ($\mathscr{A}$, $\mathscr{B}$ Koszul operads). □

8.6.6 The Diamond Lemma for Distributive Laws

We prove an analog of the Diamond Lemma for distributive laws. We denote by $(\mathscr{A}\circ\mathscr{B})^{(\omega)}$ the sub-$\mathbb{S}$-module of $\mathscr{A}\circ\mathscr{B}$ made up of elements of total weight ω. The map $p:\mathscr{A}\circ\mathscr{B}\twoheadrightarrow\mathscr{A}\vee_\lambda\mathscr{B}$ is always injective on the components of weight 0, 1 and 2. As in the case of the Diamond Lemma for PBW bases (Theorem 8.5.5), if we have an isomorphism in weight (3), then we have an isomorphism in any weight.

Theorem 8.6.5 (Diamond Lemma for distributive laws). *Let $\mathscr{A}=\mathscr{P}(V,R)$ and $\mathscr{B}=\mathscr{P}(W,S)$ be two Koszul operads endowed with a rewriting rule $\lambda: W\circ_{(1)}V\to V\circ_{(1)}W$ such that the restriction of $p:\mathscr{A}\circ\mathscr{B}\twoheadrightarrow\mathscr{A}\vee_\lambda\mathscr{B}$ on $(\mathscr{A}\circ\mathscr{B})^{(3)}$ is injective. In this case, the morphism p is an isomorphism, the map λ defines a distributive law and the induced operad $(\mathscr{A}\circ\mathscr{B},\gamma_\Lambda)$ is Koszul.*

Proof. We use the same filtration as in the proof of Theorem 8.6.4, together with the ideas of the proof of Theorem 4.2.4.

Step 1. Since the map $p:\mathscr{A}\circ\mathscr{B}\twoheadrightarrow\mathscr{A}\vee_\lambda\mathscr{B}$ is an isomorphism in weight 1, 2 and 3, the underlying $\mathbb{S}$-modules of the components of syzygy degree 0, 1 and 2 of the bar constructions $\mathrm{B}^\bullet(\mathscr{A}\vee_\lambda\mathscr{B})$ and $\mathrm{B}^\bullet(\mathscr{A}\vee_0\mathscr{B})$ are isomorphic. Therefore the three first lines of the first term of the spectral sequence are equal to $\bigoplus_{p\in\mathbb{N}}E^0_{p\,p+q}\cong\mathrm{B}^{-\bullet}(\mathscr{A}\vee_0\mathscr{B})$, for $p+q=-\bullet=0,1,2$. It implies the isomorphisms of $\mathbb{S}$-modules $\bigoplus_{p\in\mathbb{N}}E^1_{p\,-p}\cong\mathscr{B}^{\text{!`}}\circ\mathscr{A}^{\text{!`}}$ and $\bigoplus_{p\in\mathbb{N}}E^1_{p\,-p}\cong 0$. Finally, the convergence of the spectral sequence, Theorem 1.5.1, gives the isomorphisms

$$H^0\big(\mathrm{B}^\bullet(\mathscr{A}\vee_\lambda\mathscr{B})\big)\cong(\mathscr{A}\vee_\lambda\mathscr{B})^{\text{!`}}\cong\bigoplus_{p\in\mathbb{N}}E^\infty_{p\,-p}\cong\bigoplus_{p\in\mathbb{N}}E^1_{p\,-p}\cong\mathscr{B}^{\text{!`}}\circ\mathscr{A}^{\text{!`}}.$$

Step 2. Let us first construct the induced decomposition coproduct on $\mathscr{B}^{\text{!`}}\circ\mathscr{A}^{\text{!`}}$ from the one of the cooperad $(\mathscr{A}\vee_0\mathscr{B})^{\text{!`}}$. It is given by the following composite

$$\begin{aligned}&\mathscr{B}^{\text{!`}}\circ\mathscr{A}^{\text{!`}}\xrightarrow{\Delta\circ\mathrm{Id}}\mathscr{B}^{\text{!`}}\circ\mathscr{B}^{\text{!`}}\circ\mathscr{A}^{\text{!`}}\twoheadrightarrow\mathscr{B}^{\text{!`}}\circ\big(\mathrm{I}\circ\mathscr{A}^{\text{!`}}\oplus\overline{\mathscr{B}}^{\text{!`}}\circ\mathscr{A}^{\text{!`}}\big)\xrightarrow{\mathrm{Id}\circ(\mathrm{Id}\circ\Delta+\mathrm{Id})}\\&\mathscr{B}^{\text{!`}}\circ\big(\mathrm{I}\circ\mathscr{A}^{\text{!`}}\circ\mathscr{A}^{\text{!`}}\oplus\overline{\mathscr{B}}^{\text{!`}}\circ\mathscr{A}^{\text{!`}}\big)\cong\mathscr{B}^{\text{!`}}\circ\big(\mathscr{A}^{\text{!`}}\circ\mathrm{I}\circ\mathscr{A}^{\text{!`}}\oplus\mathrm{I}\circ\overline{\mathscr{B}}^{\text{!`}}\circ\mathscr{A}^{\text{!`}}\big)\\&\to\big(\mathscr{B}^{\text{!`}}\circ\mathscr{A}^{\text{!`}}\big)\circ\big(\mathscr{B}^{\text{!`}}\circ\mathscr{A}^{\text{!`}}\big).\end{aligned}$$

This is proved using the following commutative diagram of cooperads

$$\begin{array}{ccc}(\mathscr{A}\vee_0\mathscr{B})^{\text{!`}} & \rightarrowtail & \mathscr{T}^c(sV\oplus sW)\\ \downarrow\cong & & \uparrow\\ \mathscr{B}^{\text{!`}}\circ\mathscr{A}^{\text{!`}} & \rightarrowtail & \mathscr{T}^c(sW)\circ\mathscr{T}^c(sV).\end{array}$$

We consider the same kind of number of inversions, changing this time B by Ω, $\mathscr{A}$ by $\mathscr{B}^{\text{!`}}$ and $\mathscr{B}$ by $\mathscr{A}^{\text{!`}}$. We consider the decreasing filtration induced by the number of inversions on the components of fixed total weight of the cobar construction $\Omega((\mathscr{A} \vee_\lambda \mathscr{B})^{¡}) \cong \Omega(\mathscr{B}^{¡} \circ \mathscr{A}^{¡})$, so that the induced spectral sequence converges. One concludes by the same arguments as before. The first term of the spectral sequence is equal to $(E^0, d^0) \cong \Omega((\mathscr{A} \vee_0 \mathscr{B})^{¡})$, since d^0 corresponds to the aforementioned cooperad structure on $\mathscr{B}^{¡} \circ \mathscr{A}^{¡}$. Since the operad $\mathscr{A} \vee_0 \mathscr{B}$ is Koszul, the spectral sequence collapses at rank 1 and

$$\begin{aligned}\mathscr{A} \vee_\lambda \mathscr{B} \cong H_0\big(\Omega\big((\mathscr{A} \vee_\lambda \mathscr{B})^{¡}\big)\big) &\cong \bigoplus_{p \in \mathbb{N}} E^\infty_{p-p} \\ &\cong \bigoplus_{p \in \mathbb{N}} E^1_{p-p} \cong H_0\big(\Omega(\mathscr{A} \vee_0 \mathscr{B})\big) \cong \mathscr{A} \circ \mathscr{B}.\end{aligned}$$

□

This methods gives an easy way to prove that the operad $\mathscr{A} \vee_\lambda \mathscr{B}$ is isomorphic to $(\mathscr{A} \circ \mathscr{B}, \gamma_\Lambda, \iota_\Lambda)$.

8.6.7 Example: The Poisson Operad

In the case of the operad *Pois* introduced in Sect. 8.6.3, we leave it to the reader to check that $Com \circ Lie(4) \cong Pois(4)$. Hence the Diamond Lemma implies that the operad *Pois* is isomorphic to the operad defined on $Com \circ Lie$ by the distributive law.

8.6.8 Counter-Example

The following operad, introduced in [Mer04], gives an example of an operad defined by a rewriting rule λ but which does not induce a distributive law Λ.

We consider *noncommutative Poisson algebras*, whose definition is the same as Poisson algebras except that we do not require the associative product $*$ to be commutative. So this related operad is equal to $NCPois \cong Ass \vee_\lambda Lie$.

There are two ways to rewrite the critical monomial $[x * y, z * t]$ depending on which side we apply the relation first:

$$\begin{array}{ccc} & [x * y, z * t] & \\ \swarrow{=} & & \searrow{=} \\ [x, z * t] * y + x * [y, z * t] & & [x * y, z] * t + z * [x * y, t] \\ \downarrow & & \downarrow \\ \begin{array}{l}[x, z] * t * y + z * [x, t] * y \\ +x * [y, z] * t + x * z * [y, t]\end{array} & = & \begin{array}{l}[x, z] * y * t + x * [y, z] * t \\ +z * x * [y, t] + z * [x, t] * y.\end{array}\end{array}$$

This yields the relation

$$[x,z]*t*y+x*z*[y,t]=[x,z]*y*t+z*x*[y,t],$$

which produces a nontrivial element in the kernel of the map $(Ass \circ Lie)^{(3)} \to Ass \vee_\lambda Lie \cong NCPois$. Therefore the map λ does not induce a distributive law in this case and the $\mathbb{S}$-module $NCPois \cong Ass \vee_\lambda Lie$ is not isomorphic to $Ass \circ Lie$.

We refer the reader to Exercise 8.10.12 for another counterexample.

8.6.9 Distributive Laws and the Koszul Duals

When an operad is given by a distributive law, it is also the case for its Koszul dual operad.

Proposition 8.6.6. *Let $\mathscr{A} = \mathscr{P}(V,R)$ and $\mathscr{B} = \mathscr{P}(W,S)$ be two quadratic operads endowed with a rewriting rule $\lambda : W \circ_{(1)} V \to V \circ_{(1)} W$ which induces a distributive law.*

The underlying $\mathbb{S}$-module of the Koszul dual cooperad of $\mathscr{A} \vee_\lambda \mathscr{B}$ is isomorphic to

$$(\mathscr{A} \vee_\lambda \mathscr{B})^{¡} \cong \mathscr{B}^{¡} \circ \mathscr{A}^{¡}.$$

Proof. It is a direct corollary of the preceding proof and Proposition 8.6.3:

$$\bigoplus_{p\in\mathbb{N}} E^{\infty}_{p,-p} \cong (\mathscr{A} \vee_\lambda \mathscr{B})^{¡} \cong \bigoplus_{p\in\mathbb{N}} E^{0}_{p,-p} \cong (\mathscr{A} \vee_0 \mathscr{B})^{¡} \cong \mathscr{B}^{¡} \circ \mathscr{A}^{¡}.$$ □

Proposition 8.6.7. *Let $\mathscr{A} := \mathscr{P}(V,R)$ and $\mathscr{B} := \mathscr{P}(W,S)$ be two finitely generated binary quadratic operads and let $\lambda : W \circ_{(1)} V \to V \circ_{(1)} W$ be a rewriting rule. The Koszul dual operad of $\mathscr{A} \vee_\lambda \mathscr{B}$ has the following presentation:*

$$(\mathscr{A} \vee_\lambda \mathscr{B})^! \cong \mathscr{B}^! \vee_{{}^t\lambda} \mathscr{A}^!,$$

where ${}^t\lambda$ is the transpose of λ.

Proof. We apply the general formula of Theorem 7.6.2 for the Koszul dual operad of a finitely generated binary quadratic operad: $(\mathscr{A} \vee_\lambda \mathscr{B})^! \cong \mathscr{P}(V^\vee \oplus W^\vee, R^\perp \oplus D_\lambda^\perp \oplus S^\perp)$. The space $D_\lambda^\perp$ is isomorphic to the graph $D_{{}^t\lambda}$ of the transpose map ${}^t\lambda$, which concludes the proof. □

Theorem 8.6.8. *Under the hypotheses of the previous proposition, if, moreover, the rewriting rule λ induces a distributive law, then the transpose rewriting rule ${}^t\lambda$ induces a distributive law on the Koszul dual operad*

$$\mathscr{B}^! \circ \mathscr{A}^! \cong \mathscr{B}^! \vee_{{}^t\lambda} \mathscr{A}^!.$$

Proof. By the definition of the Koszul dual operad of Sect. 7.2.3 (linear dual of the Koszul dual cooperad up to suspension and signature representations), it is a direct corollary of the two previous propositions. □

8.7 Partition Poset Method

In this section, we construct a family of posets associated to a set operad. Since they are a generalization of the poset of partitions of a set, they are called *operadic partition posets*. The main theorem asserts that the induced linear operad is Koszul if and only if every poset of the family is Cohen–Macaulay, i.e. its homology is concentrated in top dimension. When it is the case, these top homology groups are isomorphic to the Koszul dual cooperad. Notice that the proof relies on the properties of the simplicial bar construction of operads.

On the one hand, the various combinatorial ways to prove that a poset is Cohen–Macaulay give simple ways of proving that an operad is Koszul. On the other hand, this result provides a means to compute the homology of partition type posets.

The construction of the operadic partition posets was first described by M. Méndez and J. Yang in [MY91]. The case of the operad *Com* and the properties of the simplicial bar construction of operads were studied by B. Fresse in [Fre04]. The Koszul–Cohen–Macaulay criterion was proved in [Val07a].

8.7.1 Partition Poset

We denote by $\underline{n}$ the set $\{1, \ldots, n\}$. Recall that a *partition* of the set $\underline{n}$ is a non-ordered collection of subsets $I_1, \ldots, I_k$, called *blocks*, which are nonempty, pairwise disjoint and whose union is equal to $\underline{n}$.

For any integer n, there is a partial order $\leqslant$ on the set of partitions of $\underline{n}$ defined by the *refinement of partitions*. Let π and ρ be two partitions of $\underline{n}$. We have $\pi \leqslant \rho$ when π is *finer* than ρ, that is when the blocks of π are contained in the blocks of ρ, for instance, $\{\{1\}, \{3\}, \{2, 4\}\} \leqslant \{\{1, 3\}, \{2, 4\}\}$. This partially ordered set is called the *partition poset* (or partition lattice) and denoted by $\Pi(n)$. See Fig. 8.10 for the example of $\Pi(3)$.

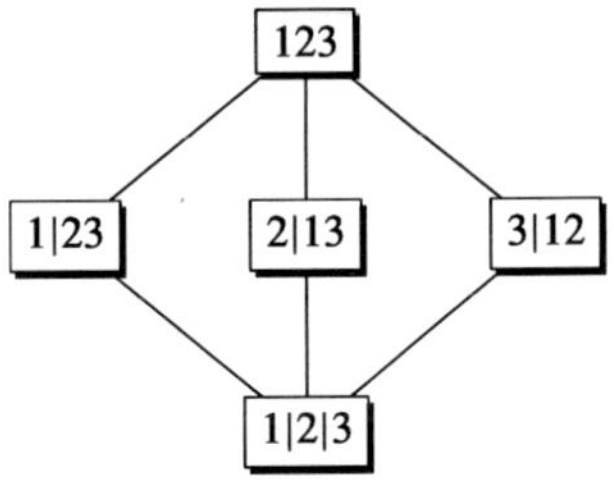

Fig. 8.10 Hasse diagram of $\Pi(3)$

The single block partition $\{\{1,\dots,n\}\}$ is the maximal partition of $\underline{n}$, and the collection $\{\{1\},\dots,\{n\}\}$ is the minimal partition.

The set of partitions of $\underline{n}$ is equipped with a left action of the symmetric group $\mathbb{S}_n$. Let $\sigma : \underline{n} \to \underline{n}$ be a permutation, the image of the partition $\{\{i_1^1,\dots,i_{j_1}^1\},\dots,\{i_1^k,\dots,i_{j_k}^k\}\}$ under σ is the partition

$$\{\{\sigma(i_1^1),\dots,\sigma(i_{j_1}^1)\},\dots,\{\sigma(i_1^k),\dots,\sigma(i_{j_k}^k)\}\}.$$

8.7.2 Operadic Partitions

We introduce the notion of partitions enriched with operadic elements.

Let P be a set operad. Recall from Sects. 5.1.13, 5.3.5 and 5.6.1 that it can be either considered as a functor Bij → Set or as an $\mathbb{S}$-set, together with an associative and unital composition product. In one way, we use $\mathrm{P}(n) := \mathrm{P}(\underline{n})$ and, in the other way round, we use

$$\mathrm{P}(X) := \mathrm{P}(n) \times_{\mathbb{S}_n} \mathsf{Bij}(\underline{n}, X) = \left(\bigsqcup_{\substack{f:\text{bijection}\\ \underline{n}\to X}} \mathrm{P}(n) \right)_{\mathbb{S}_n},$$

where the action of $\sigma \in \mathbb{S}_n$ on $(f;\mu)$ for $\mu \in M(n)$ is given by $(f\sigma;\mu^\sigma)$.

An element of $\mathsf{Bij}(\underline{n}, X)$ can be seen as an ordered sequence of elements of X, each element appearing only once. Therefore, an element in $\mathrm{P}(X)$ can be thought of as a sequence of elements of X indexed by an operation of $\mathrm{P}(n)$ with respect to the symmetry of this operation.

A P-*partition* of $\underline{n}$ is a set of elements called *blocks* $\{B_1,\dots,B_k\}$ such that each B_j belongs to $\mathrm{P}(I_j)$, for $\{I_1,\dots,I_k\}$ a partition of $\underline{n}$. A P-partition of $\underline{n}$ corresponds to a classical partition of $\underline{n}$ enriched by the operations of P.

EXAMPLES.

1. The operad *Com* for commutative algebras comes from a set operad which has only one element in arity $n \geq 1$, with trivial action of the symmetric group. Therefore, $Com(X)$ has only one element, which corresponds to the set X itself. As a consequence, a *Com*-partition of $\underline{n}$ is a classical partition of $\underline{n}$.
2. The set operad *Perm* is defined by the set $Perm(n) := \{e_1^n,\dots,e_n^n\}$ of n elements in arity n, where the action of the symmetric group is $e_k^n \cdot \sigma := e_{\sigma^{-1}(k)}^n$, see Sect. 13.4.6. The operadic composition is given by

$$e_k^n \circ_i e_l^m := \begin{cases} e_{k+m-1}^{n+m-1} & \text{for } i < k, \\ e_{k+l-1}^{n+m-1} & \text{for } i = k, \\ e_k^{n+m-1} & \text{for } i > k. \end{cases}$$

We leave it to the reader as a good exercise to check that *Perm* is an operad. As a species, *Perm*(X) can represented by the set of *pointed sets* associated to $X = \{x_1, \ldots, x_n\}$:

$$\big\{\{\bar{x}_1, \ldots, x_n\}, \{x_1, \bar{x}_2, \ldots, x_n\}, \ldots, \{x_1, \ldots, \bar{x}_n\}\big\}.$$

The element e_k^n "singles out the kth element of X". Finally a *Perm*-partition is a *pointed partition* like $1\bar{3}4|\bar{2}6|57\bar{8}$.

The natural action of the symmetric group on $\underline{n}$ induces a right action of $\mathbb{S}_n$ on the set of P-partitions of $\underline{n}$. For instance,

$$\{1\bar{3}4|\bar{2}6|57\bar{8}\}^{(123)} = \bar{2}34|\bar{1}6|57\bar{8}.$$

8.7.3 Operadic Partition Poset

We now define a partial order on the set of operadic partitions as follows. We consider the following natural map on P-partitions. Let $\{B_1, \ldots, B_t\}$ be a P-partition of a set X associated to a partition $\{I_1, \ldots, I_t\}$. Each element B_j in $\mathrm{P}(I_j)$ can be represented as the class of an element $[\nu_j \times (x_1^j, \ldots, x_{i_j}^j)]$, where $\nu_j \in \mathrm{P}(i_j)$ and $I_j = \{x_1^j, \ldots, x_{i_j}^j\}$.

Lemma 8.7.1. *The map $\tilde{\gamma}$ given by the formula*

$$\tilde{\gamma} : \mathrm{P}(t) \times \big(\mathrm{P}(I_1) \times \cdots \times \mathrm{P}(I_t)\big) \to \mathrm{P}(I),$$
$$\nu \times (B_1, \ldots, B_t) \mapsto \big[\gamma(\nu; \nu_1, \ldots, \nu_t) \times \quad \big(x_1^1, \ldots, x_{i_t}^t\big)\big],$$

is well defined and equivariant under the action of $\mathbb{S}_t$.

Proof. It is a direct consequence of the equivariance, under the action of the symmetric groups, in the definition of the composition of a set operad, see Sect. 5.3.5. □

Let $\pi = \{B_1, \ldots, B_r\}$ and $\rho = \{C_1, \ldots, C_s\}$ be two P-partitions of $\underline{n}$ associated to two partitions $\{I_1, \ldots, I_r\}$ and $\{J_1, \ldots, J_s\}$ of $\underline{n}$. The P-partition π is a *refinement* of ρ if, for any $k \in \{1, \ldots, s\}$, there exist $\{p_1, \ldots, p_t\} \subset \{1, \ldots, r\}$ such that $\{I_{p_1}, \ldots, I_{p_t}\}$ is a partition of J_k and if there exists an element ν in $\mathrm{P}(t)$ such that $C_k = \tilde{\gamma}(\nu \times (B_{p_1}, \ldots, B_{p_t}))$. We denote this relation by $\pi \leqslant \rho$.

Proposition 8.7.2. *When* P *is a reduced set operad,* $\mathrm{P}(0) = \emptyset$, *such that* $\mathrm{P}(1) = \{\mathrm{id}\}$, *the relation $\leqslant$ defines a partial order on the set of* P*-partitions.*

Proof. The symmetry of $\leqslant$ comes from the unit of the operad, the reflexivity comes from $\mathrm{P}(1) = \{\mathrm{id}\}$ and the transitivity comes from the associativity of the operad. □

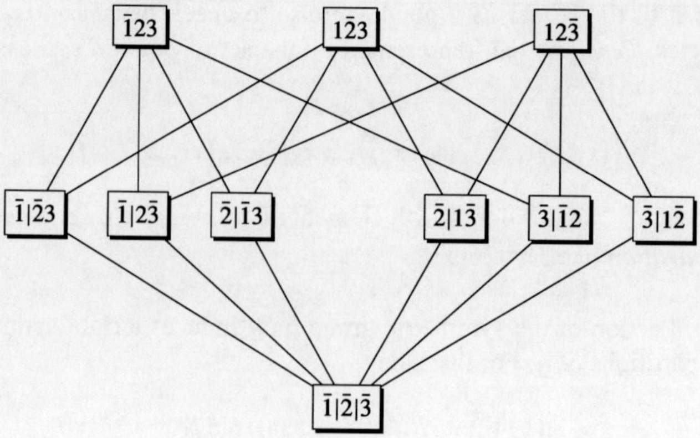

Fig. 8.11 Hasse diagram of $\Pi_{Perm}(3)$

We call this poset the P-*partition poset* associated to the operad P and we denote it by $\Pi_{\mathrm{P}}(n)$, for any integer $n \geq 1$.

EXAMPLES.

1. In the case of the operad *Com*, the *Com*-partition poset is exactly the classical partition poset.
2. In the case of the operad *Perm*, a pointed partition π is less than a pointed partition ρ if the underlying partition of π refines that of ρ and if the pointed elements of ρ belong to the pointed elements of π. See Fig. 8.11 for the example of $\Pi_{Perm}(3)$.

Since the set P(1) is reduced to the identity, the poset $\Pi_{\mathrm{P}}(n)$ has only one minimal element corresponding to the partition $\{\{1\}, \ldots, \{n\}\}$, where $\{i\}$ represents the unique element of $\mathrm{P}(\{i\})$. Following the classical notations, we denote this element by $\hat{0}$. The set of maximal elements is $\mathrm{P}(\underline{n}) \cong \mathrm{P}(n)$. Hence, the number of maximal elements of $\Pi_{\mathrm{P}}(n)$ is equal to the number of elements of $\mathrm{P}(n)$.

8.7.4 Graded Posets

We denote by $\mathrm{Min}(\Pi)$ and $\mathrm{Max}(\Pi)$ the sets of minimal and maximal elements of Π. When $\mathrm{Min}(\Pi)$ and $\mathrm{Max}(\Pi)$ have only one element, the poset is said to be *bounded*. In this case, we denote the unique element of $\mathrm{Min}(\Pi)$ by $\hat{0}$ and the unique element of $\mathrm{Max}(\Pi)$ by $\hat{1}$. For a pair $x \leqslant y$ in Π, we consider the *closed interval* $\{z \in \Pi \mid x \leqslant z \leqslant y\}$, denoted by $[x, y]$, and the *open interval* $\{z \in \Pi \mid x < z < y\}$, denoted by (x, y). For any $\alpha \in \mathrm{Min}(\Pi)$ and any $\omega \in \mathrm{Max}(\Pi)$, the closed interval $[\alpha, \omega]$ is a bounded poset. If Π is a bounded poset, the *proper part* $\overline{\Pi}$ of Π is the open interval $(\hat{0}, \hat{1})$.

For two elements $x < y$, we say that y *covers* x if there is no z such that $x < z < y$. The *covering relation* is denoted by $x \prec y$. A *chain* $\lambda_0 < \lambda_1 < \cdots < \lambda_l$ is a totally ordered sequence of elements of Π. Its *length* is equal to l. A *maximal chain* between x and y, is a chain $x = \lambda_0 \prec \lambda_1 \prec \cdots \prec \lambda_l = y$ which cannot be lengthened. A *maximal chain of* Π is a maximal chain between a minimal element of Π and a maximal element of Π. A poset is *pure* if, for any $x \leqslant y$, all maximal chains between x and y have the same length. If a poset is both bounded and pure, it is called a *graded* poset. For example, the partition poset of Sect. 8.7.1 is graded.

For more details on posets, we refer the reader to Chap. 3 of [Sta97a].

If the operad P is quadratic and generated by a homogeneous $\mathbb{S}$-set E concentrated in arity k, that is $E_n = \emptyset$ for $n \neq k$, then we have $\mathrm{P}(n) = \emptyset$ for $n \neq i(k-1)+1$ with $i \in \mathbb{N}$. Therefore, the P-partitions have restricted block size. The possible lengths for the blocks are $i(k-1)+1$ with $i \in \mathbb{N}$.

Proposition 8.7.3. *Let* P *be a set-theoretic quadratic operad generated by a homogeneous* $\mathbb{S}$*-set* E *concentrated in arity* k*, with* $k \geq 2$*, then all the maximal chains of* $\Pi_{\mathcal{P}}$ *have the same length.*

For any $\omega \in \mathrm{Max}(\Pi_{\mathrm{P}}(n)) = \mathrm{P}(\underline{n})$*, the subposets* $[\hat{0}, \omega]$ *are graded posets.*

Proof. If the operad P is generated by operations of arity k with $k \geq 2$, the set P(1) is reduced to the identity operations and the P-partition poset is well defined. Since P is generated only by operations of arity k, every block of size $i(k-1)+1$ can be refined if and only if $i > 1$.

The length of maximal chains between $\hat{0}$ and ω is equal to $i+1$ if $n = i(k-1)+1$. Hence, each closed interval of the form $[\hat{0}, \omega]$, for $\omega \in \mathrm{Max}(\Pi_{\mathrm{P}}(n)) = \mathrm{P}(\underline{n})$ is bounded and pure. It is also graded by definition. □

8.7.5 Order Complex

We consider the set of chains $\lambda_0 < \lambda_1 < \cdots < \lambda_l$ of a poset $(\Pi, \leqslant)$ such that $\lambda_0 \in \mathrm{Min}(\Pi)$ and $\lambda_l \in \mathrm{Max}(\Pi)$. This set is denoted by $\Delta_\bullet(\Pi)$, or simply by $\Delta(\Pi)$. More precisely, a chain $\lambda_0 < \lambda_1 < \cdots < \lambda_l$ of length l belongs to $\Delta_l(\Pi)$.

The set $\Delta(\Pi)$ is equipped with the following face maps. For $0 < i < l$, the face map $d_i : \Delta_l(\Pi) \to \Delta_{l-1}(\Pi)$ is given by the omission of λ_i

$$d_i(\lambda_0 < \lambda_1 < \cdots < \lambda_l) := \lambda_0 < \lambda_1 < \cdots < \lambda_{i-1} < \lambda_{i+1} < \cdots < \lambda_l.$$

On the free module $\mathbb{K}[\Delta(\Pi)]$, we consider the induced linear maps still denoted d_i and we define the maps $d_0 = d_l = 0$ by convention. The module $\mathbb{K}[\Delta(\Pi)]$ (resp. the set $\Delta(\Pi)$) is a presimplicial module (resp. a presimplicial set), that is $d_i \circ d_j = d_{j-1} \circ d_i$, for $i < j$. These relations ensure that $d := \sum_{0 \leq i \leq l} (-1)^i d_i$ satisfies $d^2 = 0$. The chain complex $(\mathbb{K}[\Delta_\bullet(\Pi)], d)$ is called the *order complex* of Π. By definition, the *homology of a poset* $(\Pi, \leqslant)$ is the homology of its order complex, denoted by $H(\Pi)$.

The *reduced* homology of a poset is defined as follows. We denote by $\tilde{\Delta}_l(\Pi)$ the set of chains $\lambda_0 < \lambda_1 < \cdots < \lambda_l$, with no restriction on λ_0 and λ_l. The face maps d_i are defined by the omission of λ_i, for $0 \le i \le l$. By convention, this complex is augmented by $\tilde{\Delta}_{-1}(\Pi) = \{\emptyset\}$, that is $\mathbb{K}[\tilde{\Delta}_{-1}(\Pi)] = \mathbb{K}$. The associated homology groups are denoted by $\widetilde{H}(\Pi)$.

The relation between the two definitions is given by the following formula

$$\Delta_l(\Pi) = \bigsqcup_{(\alpha,\omega)\in \mathrm{Min}(\Pi)\times \mathrm{Max}(\Pi)} \tilde{\Delta}_{l-2}\big((\alpha,\omega)\big),$$

which induces a canonical isomorphism of presimplicial complexes. Therefore, we have

$$H_l(\Pi) = \bigoplus_{(\alpha,\omega)\in \mathrm{Min}(\Pi)\times \mathrm{Max}(\Pi)} \widetilde{H}_{l-2}\big((\alpha,\omega)\big).$$

If Π is bounded, its homology is equal to the reduced homology of its proper part, up to a degree shift.

The action of a group G on a poset $(\Pi, \leqslant)$ is *compatible* with the partial order $\leqslant$ if for every $g \in G$ and for every pair $x \leqslant y$, we still have $g \cdot x \leqslant g \cdot y$. In this case, the modules $\mathbb{K}[\Delta_l(\Pi)]$ are G-modules. Since, the chain map commutes with the action of G, the homology groups $H(\Pi)$ are also G-modules.

For example, the action of the symmetric group on the operadic partition posets is compatible with the partial order. So the module $\mathbb{K}[\Delta_\bullet(\Pi_{\mathrm{P}}(n))]$ is an $\mathbb{S}_n$-presimplicial module and $H_\bullet(\Pi_{\mathrm{P}}(n))$ is an $\mathbb{S}_n$-module.

8.7.6 Cohen–Macaulay Poset

Let Π be a graded poset. It is said to be *Cohen–Macaulay* over $\mathbb{K}$ if the homology of each interval is concentrated in top dimension, i.e. for every $x \leqslant y$, if m is the length of maximal chains between x and y, we have

$$H_l\big([x,y]\big) = \widetilde{H}_{l-2}\big((x,y)\big) = 0, \quad \text{for } l \neq m.$$

There are many sufficient combinatorial conditions for a poset to be Cohen–Macaulay, e.g. modular, distributive, shellable. For a comprehensive survey on these notions, we refer the reader to the article by A. Björner, A.M. Garsia and R.P. Stanley [BGS82]. We recall the one used in the sequel.

A pure poset is *semi-modular* if for every triple x, y and t such that x and y cover t, there exists z covering both x and y. A *totally semi-modular* poset is a pure poset such that each interval is semi-modular.

Proposition 8.7.4. [Bac76, Far79] *A totally semi-modular poset is Cohen–Macaulay.*

For instance, the partition posets are totally semi-modular (it is actually much more). We also leave it to the reader to prove that the pointed partition posets, coming from the operad *Perm*, are semi-modular.

8.7.7 Koszul–Cohen–Macaulay Criterion

The main theorem of this section requires the following assumption. For any element $(\nu_1, \ldots, \nu_t)$ of $\mathrm{P}(i_1) \times \cdots \times \mathrm{P}(i_t)$, we denote by $\gamma_{\nu_1,\ldots,\nu_t}$ the following map defined by the composition of the set operad P:

$$\begin{aligned}\gamma_{\nu_1,\ldots,\nu_t} : \mathrm{P}(t) &\to \mathrm{P}(i_1 + \cdots + i_t),\\ \nu &\mapsto \gamma(\nu; \nu_1, \ldots, \nu_t).\end{aligned}$$

A set operad P is called a *basic-set operad* if the maps $\gamma_{\nu_1,\ldots,\nu_t}$ are injective, for any $(\nu_1, \ldots, \nu_t)$ in $\mathrm{P}(i_1) \times \cdots \times \mathrm{P}(i_t)$. The operads *Com* and *Perm* are examples of basic-set operads.

Theorem 8.7.5 (Koszul–Cohen–Macaulay criterion). *Let* P *be a quadratic basic-set operad generated by a homogeneous* $\mathbb{S}$*-set concentrated in arity* k*, with* $k \geq 2$.

1. *The linear operad* $\mathcal{P}(n) = \mathbb{K}[\mathrm{P}(n)]$ *is a Koszul operad if and only if, for every* $n \geq 1$ *and every* $\omega \in \mathrm{Max}(\Pi_{\mathrm{P}}(n))$*, the interval* $[\hat{0}, \omega]$ *is Cohen–Macaulay.*
2. *The top homology groups are isomorphic to the Koszul dual cooperad*

$$H_{\mathrm{top}}\big(\Pi_{\mathrm{P}}(n)\big) \cong \mathcal{P}^{¡}(n).$$

We postpone the proof until after the next section about the simplicial bar construction of operads, on which it relies.

EXAMPLE. The aforementioned theorem provides a proof that the operads *Com* and *Perm* are Koszul. On the level of poset homology, it shows that the homology groups of the partition posets and of the pointed partition posets are concentrated in top dimension. The last point of Theorem 8.7.5 shows that they are respectively isomorphic to

$$H_{n-1}\big(\Pi_{Com}(n)\big) \cong Lie(n)^* \otimes \mathrm{sgn}_{\mathbb{S}_n} \quad \text{and} \quad H_{n-1}\big(\Pi_{Perm}(n)\big) \cong preLie(n)^* \otimes \mathrm{sgn}_{\mathbb{S}_n}$$

as $\mathbb{S}_n$-modules. The first result has already been proved with more classical methods. (We refer the reader to the prolog of [Fre04] for complete reference.) The second result was proved in [Val07a] using this operadic method. We refer the reader to Sect. 13.4 for more details about the operad *preLie* encoding pre-Lie algebras. For instance, this operad admits a basis labeled by rooted trees, which, in turn, induces a basis for the homology groups of the pointed partition poset.

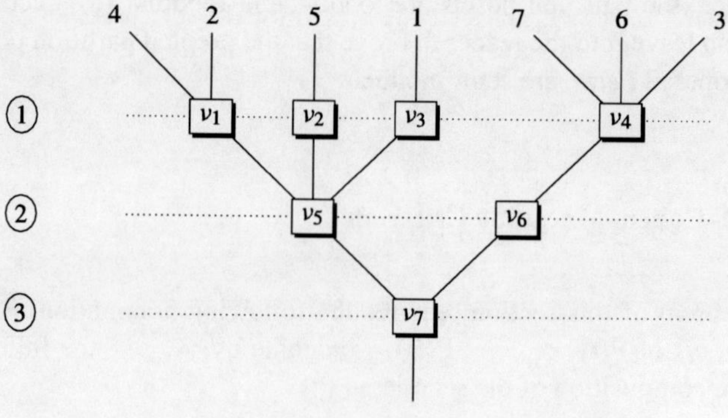

Fig. 8.12 An example of a 3-leveled tree with vertices indexed by elements of P

8.7.8 Simplicial and Normalized Bar Construction of Operads

In order to prove the Koszul–Cohen–Macaulay criterion, we introduce the simplicial and normalized bar construction of an operad.

Let $(\mathscr{P}, \gamma, \iota)$ be an operad. Its *simplicial bar construction* is the simplicial $\mathbb{S}$-module defined by $(\mathrm{C}\mathscr{P})_l := \mathscr{P}^{\circ l}$, equipped with the face and degeneracy maps

$$d_i = \mathrm{Id}^{\circ(i-1)} \circ \gamma \circ \mathrm{Id}^{\circ(l-i-1)} : \mathscr{P}^{\circ l} \to \mathscr{P}^{\circ(l-1)},$$

for $1 \le l \le l-1$, $d_0 = d_l = 0$, and by the face maps

$$s_j = \mathrm{Id}^{\circ j} \circ \iota \circ \mathrm{Id}^{\circ(l-j)} : \mathscr{P}^{\circ l} \to \mathscr{P}^{\circ(l+1)},$$

for $0 \le j \le l$.

Such a definition CP also holds for a set operad P, except for the face maps d_0 and d_l. The set $(\mathrm{CP})_l(n)$, which provides a basis for $(\mathrm{C}\mathscr{P})_l(n)$, is made up of l-leveled trees where the vertices are indexed by operations of P and where the leaves are labeled by $1, \dots, n$. See Fig. 8.12 for an example.

We consider the *normalized bar construction* $\mathrm{N}\mathscr{P}$, given as usual by the quotient of the simplicial bar construction under the images of the degeneracy maps. Recall that the two chains complexes associated to the simplicial and to the normalized bar construction respectively are quasi-isomorphic.

Let $(\mathrm{NP})_l(n)$ be the subset of $(\mathrm{CP})_l(n)$ made up of l-leveled trees with at least one nontrivial operation on each level. When P is an augmented set operad, it means that there is at least one nontrivial operation on each level. This set is stable under the face maps and is a presimplicial set.

Lemma 8.7.6. *The presimplicial* $\mathbb{S}$*-set* NP *provides a basis for the normalized bar construction* $\mathrm{N}\mathscr{P}$, *where* $\mathscr{P}(n) = \mathbb{K}[\mathrm{P}(n)]$ *is the linear operad associated to* P.

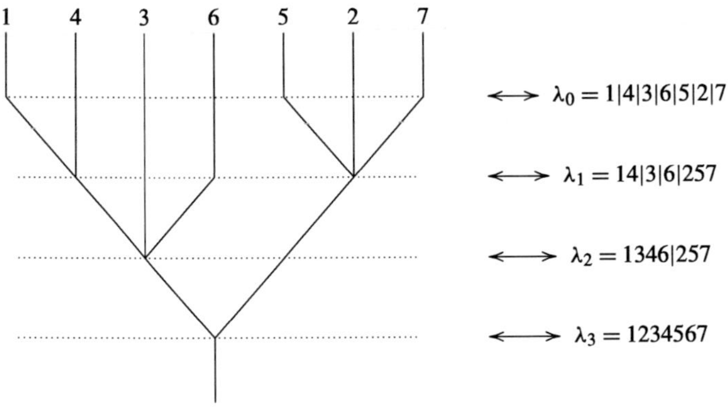

Fig. 8.13 An example of the image of Ψ in the case of the operad *Com*

8.7.9 Normalized Bar Construction and Order Complex

Theorem 8.7.7. *Let* P *be an augmented basic-set operad. For any* $n \geq 1$, *the presimplicial* $\mathbb{S}_n$*-set* $\Delta(\Pi_{\mathrm{P}}(n))$ (*resp. the presimplicial* $\mathbb{S}_n$*-module* $\mathbb{K}[\Delta(\Pi_{\mathrm{P}}(n))]$) *is in bijection with* (*resp. is isomorphic to*) *the presimplicial* $\mathbb{S}_n$*-set* $\mathrm{N}(\mathrm{P})(n)$ (*resp. the normalized bar construction* $\mathrm{N}(\mathscr{P})(n)$).

Proof. We define a bijection Ψ between $\mathrm{N}_l(\mathrm{P})(\underline{n})$ and $\Delta_l(\Pi_{\mathrm{P}}(n))$ as follows. Let $\mathbb{T}$ be a non-planar tree with l levels and n leaves whose vertices are indexed by elements of P. To such a tree, we associate a maximal chain of P-partitions of $\underline{n}$: we cut the tree $\mathbb{T}$ along the ith level and look upward. We get t indexed and labeled subtrees. By composing the operations indexing the vertices along the scheme given by the subtree, each of them induces an element of $\mathrm{P}(I_j)$, where $\{I_1, \ldots, I_t\}$ is a partition of $\underline{n}$. For every $0 \leq i \leq l$, the union of these blocks forms a P-partition λ_i of $\underline{n}$. Figure 8.13 shows an example in the case of the operad *Com*.

When $\mathbb{T}$ is a tree of $\mathrm{N}(\mathrm{P})(\underline{n})$, that is with at least one nontrivial operation on each level, λ_i is a strict refinement of λ_{i+1}. Since $\lambda_0 = \hat{0}$ and $\lambda_l \in \mathrm{P}(\underline{n})$, the chain $\lambda_0 < \lambda_1 < \cdots < \lambda_l$ is maximal. The image of the tree $\mathbb{T}$ under Ψ is this maximal chain $\lambda_0 < \lambda_1 < \cdots < \lambda_l$.

The surjectivity of the map Ψ comes from the definition of the partial order between the P-partitions. Since P is a basic-set operad, the injectivity of the maps $\gamma_{\nu_1,\ldots,\nu_t}$ induces the injectivity of Ψ. Therefore, Ψ is a bijection.

Composing the ith and the $(i+1)$th levels of the tree $\mathbb{T}$ corresponds, via Ψ, to removing the ith partition of the chain $\lambda_0 < \lambda_1 < \cdots < \lambda_l$. Therefore, the map Ψ commutes with the face maps. Moreover Ψ preserves the action of the symmetric group $\mathbb{S}_n$. □

8.7.10 Bar Construction and Normalized Bar Construction

In [Fre04], Fresse defined the following morphism $\mathscr{L} : \mathrm{B}\mathscr{P} \to \mathrm{N}\mathscr{P}$ of dg $\mathbb{S}$-modules between the bar construction and the normalized bar construction of an operad. Recall that an element of $\mathrm{B}\mathscr{P}$ is a tree $\mathbb{T}$ with l vertices labeled by elements of $s\overline{\mathscr{P}}$. Its image under the map $\mathscr{L}$ is given by the sum over all l-leveled trees with one and only one nontrivial vertex on each level and which give $\mathbb{T}$ after forgetting the levels (together with proper desuspension and sign convention). We consider here the bar construction as a chain complex with the homological degree given by the number of vertices of the underlying tree.

Proposition 8.7.8. *For any reduced operad $\mathscr{P}$, such that $\mathscr{P}(1) = \mathbb{K}\,\mathrm{id}$, the levelization morphism $\mathscr{L} : \mathrm{B}\mathscr{P} \to \mathrm{N}\mathscr{P}$ is a quasi-isomorphism of dg $\mathbb{S}$-modules.*

Proof. The idea of the proof is to apply an adequate version of the Comparison Lemma 6.7.1 to the quasi-isomorphism $\mathscr{L} \circ \mathrm{Id} : \mathrm{B}\mathscr{P} \circ \mathscr{P} \to \mathrm{N}\mathscr{P} \circ \mathscr{P}$. The normalized bar construction is not a cooperad but a right $\mathscr{P}$-comodule. We refer the reader to Sect. 4.7 of [Fre04] for the details. □

8.7.11 Proof of the Koszul–Cohen–Macaulay Criterion

Lemma 8.7.9. *Let* P *be a quadratic basic-set operad generated by a homogeneous $\mathbb{S}$-set E concentrated in arity k with $k \geq 2$.*

1. *The operad $\mathscr{P}$ is Koszul if and only if $H_l(\Pi_{\mathrm{P}}(m(k-1)+1)) = 0$ for $l \neq m$.*
2. *We have $H_m(\Pi_{\mathrm{P}}(m(k-1)+1)) \cong (\mathscr{P}^{\text{!`}})^{(m)}$.*

Proof. Since the operad is quadratic, it is weight graded and both the bar and the simplicial bar constructions split with respect to this extra grading denoted by (m). In both cases, the part of weight (m) is equal to the part of arity $m(k-1)+1$. The levelization morphism $\mathscr{L}$ of Sect. 8.7.10 preserves this grading. Therefore $\mathscr{L}^{(m)} : (\mathrm{B}\mathscr{P}(m(k-1)+1))^{(m)} \to \mathrm{N}\mathscr{P}(m(k-1)+1)^{(m)}$ is a quasi-isomorphism, for every $m \geq 0$, by Proposition 8.7.8. Recall from the Koszul criterion (Theorem 7.4.2) that the operad $\mathscr{P}$ is Koszul if and only if $H_l((\mathrm{B}_\bullet\mathscr{P}(m(k-1)+1))^{(m)}) = 0$ for $l \neq m$. By the preceding quasi-isomorphism, it is equivalent to asking that $H_l((\mathrm{N}_\bullet\mathscr{P}(m(k-1)+1))^{(m)}) = 0$ for $l \neq m$. And by Theorem 8.7.7, it is equivalent to $H_l(\Pi_{\mathrm{P}}(m(k-1)+1)) = 0$ for $l \neq m$.

The levelization quasi-isomorphism $\mathscr{L}^{(m)}$ and the isomorphism of Theorem 8.7.7 show that $H_m(\Pi_{\mathrm{P}}(m(k-1)+1)) \cong H_m((\mathrm{B}_\bullet\mathscr{P}(m(k-1)+1))^{(m)})$. Recall from Proposition 7.3.1 that the top homology group of $(\mathrm{B}_\bullet\mathscr{P}(m(k-1)+1))^{(m)}$ is equal to the Koszul dual cooperad $H^0((\mathrm{B}^\bullet\mathscr{P})^{(m)}) = (\mathscr{P}^{\text{!`}})^{(m)}$. □

This lemma implies the equivalence between Koszul set operads and Cohen–Macaulay posets.

Proof of Theorem 8.7.5. ($\Rightarrow$) Proposition 8.7.3 shows that Π_{P} is pure. Hence each interval $[\hat{0}, \omega]$ is graded.

If the operad $\mathscr{P}$ is Koszul, then Lemma 8.7.9 implies that the homology of each poset $\Pi_{\mathrm{P}}(m(k-1)+1)$ is concentrated in top dimension m. Since

$$H_l\big(\Pi_{\mathrm{P}}\big(m(k-1)+1\big)\big) = \bigoplus_{\omega \in \mathrm{Max}(\Pi_{\mathrm{P}}(m(k-1)+1)} \widetilde{H}_{l-2}\big((\hat{0}, \omega)\big),$$

we have $H_l([\hat{0}, \omega]) = \widetilde{H}_{l-2}((\hat{0}, \omega)) = 0$ for every $\omega \in \mathrm{Max}(\Pi_{\mathrm{P}}(m(k-1)+1))$ and every $l \neq m$. Let $x \leqslant y$ be two elements of $\Pi_{\mathrm{P}}(m(t-1)+1)$. We denote the P-partition x by $\{B_1, \ldots, B_r\}$ and the P-partition y by $\{C_1, \ldots, C_s\}$. Each $C_t \in \mathrm{P}(I_t)$ is refined by some B_p. For $1 \leq t \leq s$, we consider the subposet $[x_t, y_t]$ of $\Pi_{\mathrm{P}}(I_t)$, where $y_t = C_t$ and y_t the corresponding set of B_p. There exists one $\omega_t \in \mathrm{Max}(\Pi_{\mathrm{P}}(|x_t|))$ such that the poset $[x_t, y_t]$ is isomorphic to $[\hat{0}, \omega]$, which is a subposet of $\Pi_{\mathrm{P}}(|x_t|)$. (The notation $|x_t|$ stands for the number of B_p in x_t.) This decomposition gives, with Künneth Theorem, the following formula

$$\widetilde{H}_{l-1}\big((x, y)\big) \cong \bigoplus_{l_1+\cdots+l_s=l} \bigotimes_{t=1}^{s} \widetilde{H}_{l_t-1}\big((x_t, y_t)\big) \cong \bigoplus_{l_1+\cdots+l_s=l} \bigotimes_{t=1}^{s} \widetilde{H}_{l_t-1}\big((\hat{0}, \omega)\big).$$

(We can apply Künneth formula since we are working with chain complexes of free modules over an hereditary ring $\mathbb{K}$. The extra Tor terms in Künneth formula come from homology groups of lower dimension which are null.) If we define m_t by $|x_t| = m_t(k-1)+1$, the homology groups $\widetilde{H}_{l_t-1}((\hat{0}, \omega))$ vanish for $l_t \neq m_t - 1$. Therefore, if l is different from $\sum_{t=1}^{s}(m_t - 1)$, we have $\widetilde{H}_{l-1}((x, y)) = 0$. Since the length of maximal chains between x and y is equal to $m = \sum_{t=1}^{s}(m_t - 1) + 1$, see Proposition 8.7.3, the homology of the interval $[x, y]$ is concentrated in top dimension.

($\Leftarrow$) Conversely, if the poset Π_{P} is Cohen–Macaulay over the ring $\mathbb{K}$, we have $\widetilde{H}_{l-2}((\hat{0}, \omega)) = 0$, for every $m \geqslant 1$, $l \neq m$, and every $\omega \in \mathrm{Max}(\Pi_{\mathrm{P}}(m(k-1)+1))$. Therefore, we get

$$H_l\big(\Pi_{\mathrm{P}}\big(m(k-1)+1\big)\big) = \bigoplus_{\omega \in \mathrm{Max}(\Pi_{\mathrm{P}}(m(k-1)+1))} \widetilde{H}_{l-2}\big((\hat{0}, \omega)\big) = 0,$$

for $l \neq m$. Finally we conclude by (1) of Lemma 8.7.9. □

Recall that one can generalize the Koszul duality of operads over a Dedekind ring, see [Fre04] for more details. The proof of the Koszul–Cohen–Macaulay criterion also holds in that case. So it provides a method for proving that an operad is Koszul over Dedekind rings, not only fields.

8.7.12 Applications

- ◇ The triptych made up of the operad *Com*, the partition poset and the operad *Lie* plays a fundamental role in Goodwillie calculus in homotopy theory; we refer the reader to G. Arone and M. Mahowald [AM99] and to M. Ching [Chi05].
- ◇ M. Mendez in his thesis [Mén89] and F. Chapoton and M. Livernet in [CL07] explained (independently) how to associate incidence Hopf algebras to operadic partition posets. In the particular case of the operad *NAP* of [Liv06], these two last authors recover Connes–Kreimer Hopf algebra involved in renormalization theory [CK98].
- ◇ In [DK07], V. Dotsenko and A. Khoroshkin introduced the operad encoding a pair of compatible Lie brackets, which is an algebraic structure related to integrable Hamiltonian equations, like the KdV-equations. The Koszul dual operad is a basic-set operad, which was shown to be Koszul by H. Strohmayer in [Str08] using the present poset method. Notice that the associated posets are not totally semimodular this time, but only CL-shellable, that is yet another sufficient condition for being Cohen–Macaulay. Furthermore, he applied this result in the context of bi-Hamiltonian geometry in [Str10].

8.8 Manin Products

In this section, we extend the definition of Manin black and white products for quadratic algebras (Sect. 4.5) to operads. The conceptual approach followed here allows us to define the Manin products of pairs of operads given by any presentation, not necessarily quadratic. We explain how to compute some black products of operads. Finally, we study the behavior of Manin products of operads under Koszul duality; for instance, we state in the operadic context, the adjunction property between the black product and the white product.

Manin products for operads were first defined in the binary quadratic case by V. Ginzburg and M.M. Kapranov in [GK94, GK95b]. The more conceptual and general definition given here comes from [Val08].

8.8.1 White Product for Operads

Let V and W be two $\mathbb{S}$-modules. Let us denote by $i_V : V \rightarrowtail \mathscr{T}(V)$ the canonical inclusion of V into the free operad on V. There is a natural map $i_V \underset{\mathrm{H}}{\otimes} i_W : V \underset{\mathrm{H}}{\otimes} W \to \mathscr{T}(V) \underset{\mathrm{H}}{\otimes} \mathscr{T}(W)$. Recall from Sect. 5.3.2 that the Hadamard product of two operads is again an operad. So the Hadamard product $\mathscr{T}(V) \underset{\mathrm{H}}{\otimes} \mathscr{T}(W)$ is an operad. By the universal property of the free operad, there exists a (unique) morphism of operads Φ completing the following commutative diagram

$$\begin{array}{ccc} V \underset{\mathrm{H}}{\otimes} W & \xrightarrow{i_{V \underset{\mathrm{H}}{\otimes} W}} & \mathcal{T}(V \underset{\mathrm{H}}{\otimes} W) \\ & \searrow^{i_V \underset{\mathrm{H}}{\otimes} i_W} & \downarrow \exists ! \Phi \\ & & \mathcal{T}(V) \underset{\mathrm{H}}{\otimes} \mathcal{T}(W). \end{array}$$

Let $\mathcal{P} = \mathcal{T}(V)/(R)$ and $\mathcal{Q} = \mathcal{T}(W)/(S)$ be two operads defined by generators and relations (not necessarily quadratic). We denote by $\pi_{\mathcal{P}} : \mathcal{T}(V) \twoheadrightarrow \mathcal{P}$ and by $\pi_{\mathcal{Q}} : \mathcal{T}(W) \twoheadrightarrow \mathcal{Q}$ the respective projections. The composite $(\pi_{\mathcal{P}} \underset{\mathrm{H}}{\otimes} \pi_{\mathcal{Q}}) \circ \Phi$ is a morphism of operads. Hence its kernel is an operadic ideal and it admits the following factorization

$$\begin{array}{ccccc} \mathcal{T}(V \underset{\mathrm{H}}{\otimes} W) & \xrightarrow{\Phi} & \mathcal{T}(V) \underset{\mathrm{H}}{\otimes} \mathcal{T}(W) & \xrightarrow{\pi_{\mathcal{P}} \underset{\mathrm{H}}{\otimes} \pi_{\mathcal{Q}}} & \mathcal{P} \underset{\mathrm{H}}{\otimes} \mathcal{Q} \\ & \searrow & & \nearrow_{\overline{\Phi}} & \\ & & \mathcal{T}(V \underset{\mathrm{H}}{\otimes} W)/\operatorname{Ker}((\pi_{\mathcal{P}} \underset{\mathrm{H}}{\otimes} \pi_{\mathcal{Q}}) \circ \Phi). & & \end{array}$$

A direct inspection shows that the kernel of $(\pi_{\mathcal{P}} \underset{\mathrm{H}}{\otimes} \pi_{\mathcal{Q}}) \circ \Phi$ is the ideal generated by $\Phi^{-1}(R \underset{\mathrm{H}}{\otimes} \mathcal{T}(W) + \mathcal{T}(V) \underset{\mathrm{H}}{\otimes} S)$, that is

$$\operatorname{Ker}\left(\left(\pi_{\mathcal{P}} \underset{\mathrm{H}}{\otimes} \pi_{\mathcal{Q}}\right) \circ \Phi\right) = \left(\Phi^{-1}\left(R \underset{\mathrm{H}}{\otimes} \mathcal{T}(W) + \mathcal{T}(V) \underset{\mathrm{H}}{\otimes} S\right)\right).$$

We define the *Manin white product of* $\mathcal{P}$ *and* $\mathcal{Q}$ as being the quotient operad

$$\mathcal{P} \bigcirc \mathcal{Q} := \mathcal{T}\left(V \underset{\mathrm{H}}{\otimes} W\right)/\left(\Phi^{-1}\left(R \underset{\mathrm{H}}{\otimes} \mathcal{T}(W) + \mathcal{T}(V) \underset{\mathrm{H}}{\otimes} S\right)\right).$$

By definition, the white product comes equipped with a natural monomorphism of operads $\overline{\Phi} : \mathcal{P} \bigcirc \mathcal{Q} \rightarrowtail \mathcal{P} \underset{\mathrm{H}}{\otimes} \mathcal{Q}$.

The map Φ has the following form. The image of a tree, with vertices labeled by elements of $V \underset{\mathrm{H}}{\otimes} W$, under the map Φ is the tensor product of two copies of the same tree, with vertices labeled by the corresponding elements of V, resp. W.

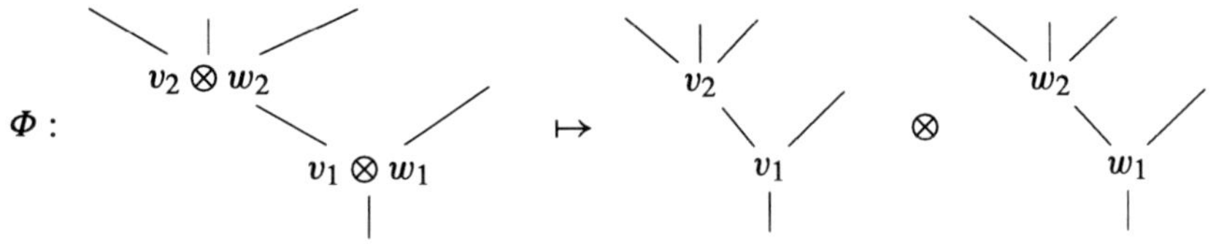

This description allows us to show that, in some cases, the white product of two operads is equal to their Hadamard product.

We suppose now that the generating space V is concentrated in arity 2, i.e. the operad $\mathscr{P}$ is binary. To any shuffle binary tree $\mathbb{T}$ of arity n, we associate the $\mathbb{K}$-module $\mathbb{T}(V)$, made up of copies of $\mathbb{T}$ with vertices labeled by elements of V. We consider the following map

$$\mathscr{C}_{\mathbb{T}}^{\mathscr{P}} : \mathbb{T}(V) \to \mathscr{T}(V)(n) \twoheadrightarrow \mathscr{P}(n),$$

which consists in composing in $\mathscr{P}$ the operations of V along the composition scheme given by the tree $\mathbb{T}$. (Recall that the set of binary shuffle trees provides a basis of the free operad over an $\mathbb{S}$-module concentrated in arity 2, see Sect. 8.2.5.)

Proposition 8.8.1. *Let $\mathscr{P} = \mathscr{P}(V, R)$ be a binary operad. If the maps $\mathscr{C}_{\mathbb{T}}$ are surjective for any shuffle binary tree $\mathbb{T}$, then, for any binary operad $\mathscr{Q} = \mathscr{P}(W, S)$, the white product with $\mathscr{P}$ is isomorphic to the Hadamard product:*

$$\mathscr{P} \bigcirc \mathscr{Q} \cong \mathscr{P} \underset{\mathrm{H}}{\otimes} \mathscr{Q}.$$

Proof. From the definition of the white product, we only have to show that the composite $(\pi_{\mathscr{P}} \underset{\mathrm{H}}{\otimes} \pi_{\mathscr{Q}}) \circ \Phi$ is surjective in this case. Let $\mu \otimes \nu$ be an elementary tensor of $\mathscr{P}(n) \underset{\mathrm{H}}{\otimes} \mathscr{Q}(n)$. There exists a shuffle binary tree $\mathbb{T}$ of arity n such that ν lives in the image of $\mathscr{C}_{\mathbb{T}}^{\mathscr{Q}}$. By the assumption, the element μ lives in the image of $\mathscr{C}_{\mathbb{T}}^{\mathscr{P}}$. Therefore, the element $\mu \otimes \nu$ is the image of the tree $\mathbb{T}$ with vertices labeled by elements of $V \underset{\mathrm{H}}{\otimes} W$ under the map $(\pi_{\mathscr{P}} \underset{\mathrm{H}}{\otimes} \pi_{\mathscr{Q}}) \circ \Phi$. □

The operads *Com* and *Perm* satisfy the assumption of this proposition. Therefore, the operad *Com* is the unit object with respect to the white product $\mathscr{P} \bigcirc Com = \mathscr{P} \underset{\mathrm{H}}{\otimes} Com = \mathscr{P}$ in the category of binary operads.

The category of binary quadratic operads is stable under Manin white product. Since it is associative and symmetric, the category of binary quadratic operads, endowed with the white product and the operad *Com* as a unit, forms a symmetric monoidal category.

Let $A = A(V, R)$ and $B = A(W, S)$ be two quadratic algebras, which we consider as operads concentrated in arity 1. We leave it to the reader to check that the white product of A and B as operads is equal to their white product as algebras, defined in Sect. 4.5.1.

8.8.2 Black Product for Cooperads

Dualizing the previous arguments and working in the opposite category, we get the notion of *black product for cooperads* as follows.

Let $(\mathscr{C}, \Delta_{\mathscr{C}})$ and $(\mathscr{D}, \Delta_{\mathscr{D}})$ be two cooperads. Their Hadamard product $\mathscr{C} \underset{\mathrm{H}}{\otimes} \mathscr{D}$ is again a cooperad. The coproduct is given by the composite

$$\mathscr{C} \underset{\mathrm{H}}{\otimes} \mathscr{D} \xrightarrow{\Delta_{\mathscr{C}} \underset{\mathrm{H}}{\otimes} \Delta_{\mathscr{D}}} (\mathscr{C} \circ \mathscr{C}) \underset{\mathrm{H}}{\otimes} (\mathscr{D} \circ \mathscr{D}) \twoheadrightarrow (\mathscr{C} \underset{\mathrm{H}}{\otimes} \mathscr{D}) \circ (\mathscr{D} \underset{\mathrm{H}}{\otimes} \mathscr{C}),$$

where the second map is a treewise projection. We denote by $p_V : \mathscr{T}^c(V) \twoheadrightarrow V$ the canonical projection from the conilpotent cofree cooperad to V. Since $\mathscr{T}^c(V) \underset{\mathrm{H}}{\otimes} \mathscr{T}^c(W)$ is a conilpotent cooperad, by the universal property of the conilpotent cofree cooperad, there exists a (unique) morphism of cooperads $\Psi : \mathscr{T}^c(V) \underset{\mathrm{H}}{\otimes} \mathscr{T}^c(W) \to \mathscr{T}^c(V \underset{\mathrm{H}}{\otimes} W)$ which factors the map $p_V \underset{\mathrm{H}}{\otimes} p_W$.

$$\begin{array}{ccc} V \underset{\mathrm{H}}{\otimes} W & \xleftarrow{\ p_{V \underset{\mathrm{H}}{\otimes} W}\ } & \mathscr{T}^c(V \underset{\mathrm{H}}{\otimes} W) \\ & \nwarrow^{p_V \underset{\mathrm{H}}{\otimes} p_W} & \uparrow {\scriptstyle \exists ! \Psi} \\ & & \mathscr{T}^c(V) \underset{\mathrm{H}}{\otimes} \mathscr{T}^c(W). \end{array}$$

Let $\mathscr{C} = \mathscr{C}(V, R)$ and $\mathscr{D} = \mathscr{C}(W, S)$ be two cooperads defined by cogenerators and corelators, not necessarily quadratic. Let us denote by $\iota_{\mathscr{C}} : \mathscr{C} \rightarrowtail \mathscr{T}^c(V)$ and by $\iota_{\mathscr{D}} : \mathscr{D} \rightarrowtail \mathscr{T}^c(W)$ the canonical inclusions. The composite of morphisms of cooperads $\Psi \circ (\iota_{\mathscr{C}} \underset{\mathrm{H}}{\otimes} \iota_{\mathscr{D}})$ factors through its image

$$\begin{array}{ccccc} \mathscr{T}^c(V \underset{\mathrm{H}}{\otimes} W) & \xleftarrow{\ \Psi\ } & \mathscr{T}^c(V) \underset{\mathrm{H}}{\otimes} \mathscr{T}^c(W) & \xleftarrow{\ \iota_{\mathscr{C}} \underset{\mathrm{H}}{\otimes} \iota_{\mathscr{D}}\ } & \mathscr{C} \underset{\mathrm{H}}{\otimes} \mathscr{D} \\ & \nwarrow & & \swarrow_{\bar{\Psi}} & \\ & & \mathscr{C} \bullet \mathscr{D} & & \end{array}$$

which we define to be *Manin black product of the cooperads* $\mathscr{C}$ *and* $\mathscr{D}$. It is the cooperad cogenerated by $V \underset{\mathrm{H}}{\otimes} W$ with corelations $\Psi(R \underset{\mathrm{H}}{\otimes} S)$:

$$\mathscr{C} \bullet \mathscr{D} := \mathscr{C}\Bigl(V \underset{\mathrm{H}}{\otimes} W, \Psi\Bigl(R \underset{\mathrm{H}}{\otimes} S\Bigr)\Bigr).$$

8.8.3 Black Products for Operads

Since it is easier to work with operads than cooperads, we will consider the linear dual of the previous definition. This provides a notion of black product for operads.

From now on, we will work in the category of finitely generated binary quadratic operads (concentrated in degree 0).

Recall from Sect. 7.6.3 the notation $V^\vee := V^* \otimes \mathrm{sgn}_2$, where the $\mathbb{S}$-module V is concentrated in arity 2. In that section, we introduced a basis of $\mathcal{T}(V)(3)$ and a scalar product $\langle -, - \rangle$, which induces an isomorphism $\theta_V : \mathcal{T}(V)(3) \xrightarrow{\cong} \mathcal{T}(V^\vee)(3)^*$. We define the morphism $\widetilde{\Psi}$ by the following composite

$$\begin{array}{ccc}
\mathcal{T}(V)(3) \underset{\mathrm{H}}{\otimes} \mathcal{T}(W)(3) & \xrightarrow{\ \widetilde{\Psi}\ } & \mathcal{T}(V \underset{\mathrm{H}}{\otimes} W \underset{\mathrm{H}}{\otimes} \mathrm{sgn}_{\mathbb{S}_2})(3) \\
\Big\downarrow {\scriptstyle \theta_V \underset{\mathrm{H}}{\otimes} \theta_W} & & \Big\uparrow {\scriptstyle \theta^{-1}_{V \underset{\mathrm{H}}{\otimes} W}} \\
\mathcal{T}(V^\vee)(3)^* \underset{\mathrm{H}}{\otimes} \mathcal{T}(W^\vee)(3)^* & & \mathcal{T}((V \underset{\mathrm{H}}{\otimes} W \underset{\mathrm{H}}{\otimes} \mathrm{sgn}_{\mathbb{S}_2})^\vee)(3)^* \\
\Big\downarrow {\scriptstyle \simeq} & & \Big\uparrow {\scriptstyle \simeq} \\
(\mathcal{T}(V^\vee)(3) \underset{\mathrm{H}}{\otimes} \mathcal{T}(W^\vee)(3))^* & \xrightarrow{\ {}^t\Phi\ } & \mathcal{T}(V^\vee \underset{\mathrm{H}}{\otimes} W^\vee)(3)^*,
\end{array}$$

where $\simeq$ stands for the natural isomorphism for the linear dual of a tensor product, since the modules are finite dimensional, and where Φ is the map defined in Sect. 8.8.1 applied to $V^\vee$ and $W^\vee$. The morphism $\widetilde{\Psi}$ defined here is a twisted version of Ψ.

The black product of two finitely generated binary quadratic operads $\mathcal{P} = \mathcal{P}(V, R)$ and $\mathcal{Q} = \mathcal{P}(W, S)$ is equal to

$$\mathcal{P} \bullet \mathcal{Q} := \mathcal{T}\Big(V \underset{\mathrm{H}}{\otimes} W \underset{\mathrm{H}}{\otimes} \mathrm{sgn}_{\mathbb{S}_2}\Big) / \Big(\widetilde{\Psi}\Big(R \underset{\mathrm{H}}{\otimes} S\Big)\Big).$$

8.8.4 Examples of Computations of Black Products

In order to compute black products for operads (and white products by Koszul duality), we use the basis and notations given in Sect. 7.6.4.

Proposition 8.8.2. *The following isomorphism holds*

$$\mathit{PreLie} \bullet \mathit{Com} \cong \mathit{Zinb}.$$

Proof. We denote by μ the generating operation of the operad *PreLie*. In a pre-Lie algebra the binary operation is such that its associator is right symmetric, that is $\mu(\mu(a, b), c) - \mu(a, \mu(b, c)) = \mu(\mu(a, c), b) - \mu(a, \mu(c, b))$. This relation corresponds to $v_i - v_{i+1} + v_{i+2} - v_{i+3}$ for $i = 1, 5, 9$ with our conventions. We denote by ν the commutative generating operation of *Com* and by w_1, w_5, w_9 the associated generators of $\mathcal{T}(\mathbb{K}\nu)(3)$. The associativity relation of ν reads in this basis:

$w_1 - w_5 = 0$ and $w_5 - w_9 = 0$. We have

(1) $\widetilde{\Psi}((v_1 - v_2 + v_3 - v_4) \otimes (w_1 - w_5)) = \widetilde{\Psi}(v_1 \otimes w_1 + v_4 \otimes w_5)$,
(2) $\widetilde{\Psi}((v_1 - v_2 + v_3 - v_4) \otimes (w_5 - w_9)) = \widetilde{\Psi}((v_2 - v_3) \otimes w_9 - v_4 \otimes w_5)$,
(3) $\widetilde{\Psi}((v_5 - v_6 + v_7 - v_8) \otimes (w_1 - w_5)) = \widetilde{\Psi}((v_7 - v_6) \otimes w_1 - v_5 \otimes w_5)$,
(4) $\widetilde{\Psi}((v_5 - v_6 + v_7 - v_8) \otimes (w_5 - w_9)) = \widetilde{\Psi}(v_5 \otimes w_5 + v_8 \otimes w_9)$,
(5) $\widetilde{\Psi}((v_9 - v_{10} + v_{11} - v_{12}) \otimes (w_1 - w_5)) = \widetilde{\Psi}(-v_{12} \otimes w_1 + (v_{10} - v_{11}) \otimes w_5)$,
(6) $\widetilde{\Psi}((v_9 - v_{10} + v_{11} - v_{12}) \otimes (w_5 - w_9)) = \widetilde{\Psi}((v_{11} - v_{10}) \otimes w_5 - v_9 \otimes w_9)$.

Using for instance Fig. 7.1, one can see that the action of (132) sends (1) to (4), (3) to (6) and (5) to (2). The image of (1) under (13) is (3). Therefore, we only need to make (1) and (2) explicit. If we identify $(\mathbb{K}\mu \oplus \mathbb{K}\mu') \otimes \mathbb{K}\nu \otimes \mathbb{K}\,\mathrm{sgn}_{\mathbb{S}_2}$ with $\mathbb{K}\gamma \oplus \mathbb{K}\gamma'$ via the isomorphism of $\mathbb{S}_2$-modules

$$\mu \otimes \nu \otimes 1 \mapsto \gamma,$$
$$\mu' \otimes \nu \otimes 1 \mapsto -\gamma',$$

the morphism $\widetilde{\Psi}$ becomes

$$\widetilde{\Psi}\big((\mu \circ_{\mathrm{I}} \mu) \otimes (\nu \circ_{\mathrm{I}} \nu)\big) = \widetilde{\Psi}(v_1 \otimes w_1) = \gamma \circ_{\mathrm{I}} \gamma = z_1 \quad \text{and}$$
$$\widetilde{\Psi}\big((\mu' \circ_{\mathrm{II}} \mu) \otimes (\nu \circ_{\mathrm{II}} \nu)\big) = \widetilde{\Psi}(v_2 \otimes w_1) = -\gamma' \circ_{\mathrm{I}} \gamma = -z_2.$$

The images of the other elements are obtained from these two by the action of $\mathbb{S}_3$. For instance, we have $\widetilde{\Psi}(v_3 \otimes w_1) = -z_3$, $\widetilde{\Psi}(v_4 \otimes w_1) = z_4$ and $\widetilde{\Psi}(v_5 \otimes w_5) = z_5$.

So, we get

$$\widetilde{\Psi}(v_1 \otimes w_1 + v_4 \otimes w_5) = \gamma \circ_{\mathrm{I}} \gamma - \gamma \circ_{\mathrm{III}} \gamma',$$
$$\widetilde{\Psi}\big((v_2 - v_3) \otimes w_9 - v_4 \otimes w_5\big) = -\gamma' \circ_{\mathrm{II}} \gamma - \gamma' \circ_{\mathrm{II}} \gamma' + \gamma \circ_{\mathrm{III}} \gamma'.$$

Finally, if we represent the operation $\gamma(x, y)$ by $x \prec y$, then we have

$$(x \prec y) \prec z = (x \prec z) \prec y,$$
$$(x \prec z) \prec y = x \prec (z \prec y) + x \prec (y \prec z),$$

where we recognize the axioms of a Zinbiel algebra, see Sect. 13.5.2. □

8.8.5 Manin Products and Koszul Duality

Theorem 8.8.3. *For any pair of finitely generated binary quadratic operads $\mathcal{P} = \mathcal{P}(V, R)$ and $\mathcal{Q} = \mathcal{P}(W, S)$, their black and white products are sent to one another under Koszul duality*

$$(\mathcal{P} \bullet \mathcal{Q})^! = \mathcal{P}^! \circ \mathcal{Q}^!.$$

Proof. Let us denote by $\langle -,-\rangle_E$ the scalar product on $\mathcal{T}(E^\vee)(3) \otimes \mathcal{T}(E)(3)$. The orthogonal space of $\widetilde{\Psi}(R \underset{\mathrm{H}}{\otimes} S)$ for $\langle -,-\rangle_{V \underset{\mathrm{H}}{\otimes} W \underset{\mathrm{H}}{\otimes} \mathrm{sgn}}$ is $\Phi^{-1}(R^\perp \underset{\mathrm{H}}{\otimes} \mathcal{T}(W^\vee) + \mathcal{T}(V^\vee) \underset{\mathrm{H}}{\otimes} S^\perp)$. By definition of the transpose of Φ, we have

$$\begin{aligned}\Big\langle X, \widetilde{\Psi}\Big(r \underset{\mathrm{H}}{\otimes} s\Big)\Big\rangle_{V \underset{\mathrm{H}}{\otimes} W \underset{\mathrm{H}}{\otimes} \mathrm{sgn}} &= \Big\langle \Phi(X), r \underset{\mathrm{H}}{\otimes} s\Big\rangle_{(\mathcal{T}(V) \underset{\mathrm{H}}{\otimes} \mathcal{T}(V^\vee)) \times (\mathcal{T}(W) \underset{\mathrm{H}}{\otimes} \mathcal{T}(W^\vee))} \\ &= \big(\langle -, r\rangle_V . \langle -, s\rangle_W\big) \circ \Phi(X),\end{aligned}$$

for every $(r,s) \in R \times S$ and every $X \in \mathcal{T}(V^\vee \underset{\mathrm{H}}{\otimes} W^\vee)$.

Therefore, we have

$$\begin{aligned}&\widetilde{\Psi}\Big(R \underset{\mathrm{H}}{\otimes} S\Big)^\perp \\ &= \Big\{X \in \mathcal{T}\Big(V^\vee \underset{\mathrm{H}}{\otimes} W^\vee\Big)(3) \mid \forall (r,s) \in R \times S \big(\langle -, r\rangle_V . \langle -, s\rangle_W\big) \circ \Phi(X) = 0\Big\} \\ &= \Big\{X \in \mathcal{T}\Big(V^\vee \underset{\mathrm{H}}{\otimes} W^\vee\Big)(3) \mid \Phi(X) \in R^\perp \underset{\mathrm{H}}{\otimes} \mathcal{T}(W^\vee) + \mathcal{T}(V^\vee) \underset{\mathrm{H}}{\otimes} S^\perp\Big\} \\ &= \Phi^{-1}\Big(R^\perp \underset{\mathrm{H}}{\otimes} \mathcal{T}(W^\vee) + \mathcal{T}(V^\vee) \underset{\mathrm{H}}{\otimes} S^\perp\Big).\end{aligned}$$ □

Since the operad *Com* is the unit object for the white product, the operad *Lie* is the unit object for the black product, that is

$$\mathit{Lie} \bullet \mathcal{P} = \mathcal{P},$$

for any finitely generated binary quadratic operad $\mathcal{P}$. The black product is also an associative product. Therefore the category of finitely generated binary quadratic operads, equipped with the black product and the operad *Lie* as unit, is a symmetric monoidal category.

Corollary 8.8.4. *The following isomorphism holds*

$$\mathit{Perm} \bigcirc \mathit{Lie} \cong \mathit{Leib}.$$

Proof. It is a direct corollary of Proposition 8.8.2 and Theorem 8.8.3. □

Contrarily to associative algebras, Manin black or white product of two Koszul operads is not necessarily a Koszul operad. A counterexample is given in Exercise 8.10.16.

8.8.6 Remark

In Sect. 7.2.3, we defined the Koszul dual operad as $\mathcal{P}^! := (\mathcal{S}^c \underset{\mathrm{H}}{\otimes} \mathcal{P}^{¡})^*$. Let Com_{-1} be the operad which is spanned by sgn_n (signature representation) of degree $n-1$

in arity n. Then we have $\mathscr{P}^! = (\mathscr{P}^{¡})^* \bigcirc Com_{-1}$. This formula explains the presentation of $\mathscr{P}^!$ in terms of the presentation of $\mathscr{P}$ given in Proposition 7.2.1.

8.8.7 Adjunction and Internal (Co)Homomorphism

Theorem 8.8.5. *There is a natural bijection*

$$\mathrm{Hom}_{\mathsf{Quad\text{-}Op}}(\mathscr{P} \bullet \mathscr{Q}, \mathscr{R}) \cong \mathrm{Hom}_{\mathsf{Quad\text{-}}Op}\big(\mathscr{P}, \mathscr{Q}^! \bigcirc \mathscr{R}\big),$$

where Quad-Op *stands for the category of finitely generated binary quadratic operads.*

In other words, the functors

$$- \bullet \mathscr{Q} : \mathsf{Quad\text{-}Op} \rightleftharpoons \mathsf{Quad\ Op} : \mathscr{Q}^! \bigcirc -$$

form a pair of adjoint functors, for any finitely generated binary quadratic operad $\mathscr{Q}$.

Proof. Let $\mathscr{P}$, $\mathscr{Q}$ and $\mathscr{R}$ be three operads presented by $\mathscr{P} = \mathscr{P}(V, R)$, $\mathscr{Q} = \mathscr{P}(W, S)$ and $\mathscr{R} = \mathscr{P}(X, T)$. There is a bijection between maps $f : V \underset{\mathrm{H}}{\otimes} W \underset{\mathrm{H}}{\otimes} \mathrm{sgn}_{\mathbb{S}_2} \to X$ and maps $g : V \to W^\vee \underset{\mathrm{H}}{\otimes} X$. Using the same arguments as in the proof of Theorem 8.8.3, we can see that

$$\begin{aligned}
&\Big\langle \Big(\Phi^{-1}\Big(S^\perp \underset{\mathrm{H}}{\otimes} \mathscr{T}(X) + \mathscr{T}\big(W^\vee\big) \underset{\mathrm{H}}{\otimes} T\Big)\Big)^\perp, \mathscr{T}(g)(R)\Big\rangle_{W^\vee \underset{\mathrm{H}}{\otimes} X} \\
&\quad = \Big\langle \widetilde{\Psi}\Big(S \underset{\mathrm{H}}{\otimes} T^\perp\Big), \mathscr{T}(g)(R)\Big\rangle_{W^\vee \underset{\mathrm{H}}{\otimes} X} \\
&\quad = \Big\langle T^\perp, \mathscr{T}(f)\Big(\widetilde{\Psi}\Big(R \underset{\mathrm{H}}{\otimes} S\Big)\Big)\Big\rangle_X.
\end{aligned}$$

Therefore $\mathscr{T}(f)(\widetilde{\Psi}(R \underset{\mathrm{H}}{\otimes} S)) \subset T$ is equivalent to $\mathscr{T}(g)(R) \subset \Phi^{-1}(S^\perp \underset{\mathrm{H}}{\otimes} \mathscr{T}(X) + \mathscr{T}(W^\vee) \underset{\mathrm{H}}{\otimes} T)$, which concludes the proof. □

In other words, $\mathsf{Hom}(B, C) := B^! \bigcirc C$ is the internal "Hom" functor in the monoidal category of finitely generated binary quadratic operads with the black product. Dually, $\mathsf{CoHom}(A, B) := A \bullet B^!$ is the internal "coHom" (or inner) functor in the monoidal category of finitely generated binary quadratic operads with the white product.

For another point of view on this type of adjunction and coHom objects in the general operadic setting, we refer the reader to the paper [BM08] by D. Borisov and Yu.I. Manin.

Let $\mathscr{P} = \mathscr{P}(V, R)$ be any finitely generated binary quadratic operad. We apply Proposition 8.8.5 to the three operads *Lie*, $\mathscr{P}$ and $\mathscr{P}$. Since *Lie* is the unit object for the black product, we have a natural bijection $\mathrm{Hom}_{\mathsf{Quad\ Op}}(\mathscr{P}, \mathscr{P}) \cong \mathrm{Hom}_{\mathsf{Quad\ Op}}(Lie, \mathscr{P}^! \bigcirc \mathscr{P})$. The image of the identity of $\mathscr{P}$ under this bijection provides a canonical morphism of operads $Lie \to \mathscr{P}^! \bigcirc \mathscr{P}$. The composite with $\overline{\Phi} : \mathscr{P}^! \bigcirc \mathscr{P} \to \mathscr{P} \underset{\mathrm{H}}{\otimes} \mathscr{P}^!$ gives a morphism of operads from $Lie \to \mathscr{P} \underset{\mathrm{H}}{\otimes} \mathscr{P}^!$. We leave it to the reader to verify that this morphism of operads is equal to the one given in Theorem 7.6.5, thereby providing a more conceptual proof of this property.

8.8.8 Hopf Operads

Proposition 8.8.6. *Let $\mathscr{P} = \mathscr{P}(V, R)$ be a finitely generated binary quadratic operad. The black product $\mathscr{P}^! \bullet \mathscr{P}$ is a Hopf operad.*

Proof. The proof is similar to the proof of Proposition 4.5.4. □

8.8.9 Manin Products for Nonsymmetric Operads

We can perform the same constructions in the category of arity-graded vector spaces. Some constructions of this type have been devised in [EFG05]. This defines *Manin black and white products for nonsymmetric operads*, which we denote by $\blacksquare$ and by $\square$. (The details are left to the reader.) Notice that the black or white product of two ns operads does not give in general the "same" result as the black or white product of the associated symmetric operads. One can also introduce the notion of Manin products for shuffle operads. In this case, the forgetful functor from operads to shuffle operads commutes with Manin white product, see Exercise 8.10.17.

The black and white products for ns operads are associative products. For any finitely generated binary quadratic ns operad $\mathscr{P}$, the ns operad As of associative algebras is the unit object for both products, that is $As \,\square\, \mathscr{P} = As \blacksquare \mathscr{P} = \mathscr{P}$. Therefore, the category of finitely generated binary quadratic ns operads carries two symmetric monoidal category structures provided by the black and white product and the operad As.

The following result is proved with the same argument as in the symmetric operad framework.

Theorem 8.8.7. *There is a natural bijection*

$$\mathrm{Hom}_{\mathsf{Quad\text{-}nsOp}}(\mathscr{P} \blacksquare \mathscr{Q}, \mathscr{R}) \cong \mathrm{Hom}_{\mathsf{Quad\text{-}nsOp}}\big(\mathscr{P}, \mathscr{Q}^! \,\square\, \mathscr{R}\big),$$

where Quad-nsOp *stands for the category of finitely generated binary quadratic ns operads.*

As a direct corollary, there exists a morphism of ns operads $As \to \mathscr{P}^! \,\square\, \mathscr{P}$ for any finitely generated binary quadratic ns operad $\mathscr{P}$. This yields a canonical morphism of ns operads $As \to \mathscr{P}^! \underset{\mathrm{H}}{\otimes} \mathscr{P}$, which is the one given in Theorem 7.7.2. This result will be crucial in the study of operations on the cohomology of algebras over a ns operad, see Sect. 13.3.11.

8.9 Résumé

8.9.1 Shuffle Operads

Shuffle monoidal product $\circ_{\sqcup\!\sqcup}$ on $\mathbb{N}$-Mod: Monoidal functors

$$(\mathbb{S}\text{-Mod}, \circ) \longrightarrow (\mathbb{N}\text{-Mod}, \circ_{\sqcup\!\sqcup}) \longrightarrow (\mathbb{N}\text{-Mod}, \circ),$$

the first one being strong. Induced functors

$$\mathsf{Op} \longrightarrow \mathsf{Op}_{\sqcup\!\sqcup} \longrightarrow \mathsf{ns\ Op}.$$

Shuffle trees $\mathscr{T}_{\sqcup\!\sqcup}$: $\mathbb{K}$-linear basis of the free shuffle operad and free operad.
Partial shuffle product:

$$\begin{aligned} \circ_\sigma : P_k \otimes P_l &\to P_n, \\ \mu \otimes \nu &\mapsto \mu \circ_\sigma \nu. \end{aligned}$$

8.9.2 Rewriting Method

Let $\mathscr{P}(E, R)$ be a quadratic operad such that $E = \oplus_{i=1}^m \mathbb{K}e_i$ is a vector space equipped with a finite ordered basis. We consider a suitable order on shuffle trees with 2 vertices $\mathscr{T}_{\sqcup\!\sqcup}^{(2)}$ indexed by the $\{1, \dots, m\}$.

Typical relation:

$$t(e_i, e_j) = \sum_{t'(k,l) < t(i,j)} \lambda_{t'(k,l)}^{t(i,j)} t'(e_k, e_l).$$

The element $t(e_i, e_j)$ is called a *leading term*. A tree monomial with 3 vertices $t(e_i, e_j, e_k)$ is called *critical* if both 2-vertices subtrees $t'(e_i, e_j)$ and $t''(e_j, e_k)$ are leading terms.

Theorem. *Confluence for all the critical tree monomials $\Rightarrow$ Koszulity of the operad.*

8.9.3 Reduction by Filtration and Diamond Lemma

Let $\mathscr{P} = \mathscr{P}(E, R)$ be a quadratic operad. Any grading on $E \cong E_1 \oplus \cdots \oplus E_m$ together with a suitable order on shuffle trees induce a filtration on the shuffle operad $\mathscr{P}_{\sqcup}$ and

$$\psi : \mathring{\mathscr{P}}_{\sqcup} = \mathscr{T}_{\sqcup}(E)/(R_{\text{lead}}) \twoheadrightarrow \operatorname{gr} \mathscr{P}_{\sqcup},$$

with $R_{\text{lead}} = \langle \text{leading term}(r), r \in R \rangle$.

DIAMOND LEMMA FOR QUADRATIC OPERADS.

$$\mathring{\mathscr{P}}_{\sqcup} \textit{ Koszul and } (\mathring{\mathscr{P}}_{\sqcup})^{(3)} \rightarrowtail (\operatorname{gr} \mathscr{P}_{\sqcup})^{(3)} \implies \mathscr{P} \textit{ Koszul and } \mathscr{P} \cong \operatorname{gr} \mathscr{P}_{\sqcup} \cong \mathring{\mathscr{P}}_{\sqcup}.$$

INHOMOGENEOUS CASE.

$$\begin{array}{c} q\mathring{\mathscr{P}}_{\sqcup} \textit{ Koszul and} \\ (q\mathring{\mathscr{P}}_{\sqcup})^{(3)} \rightarrowtail (\operatorname{gr}_{\chi} q\mathscr{P}_{\sqcup})^{(3)} \end{array} \implies \begin{array}{c} \mathscr{P} \textit{ Koszul and} \\ q\mathring{\mathscr{P}}_{\sqcup} \cong \operatorname{gr}_{\chi} q\mathscr{P}_{\sqcup} \cong q\mathscr{P} \cong \operatorname{gr} \mathscr{P} \cong \mathscr{P}. \end{array}$$

8.9.4 PBW Basis, Gröbner Basis and Diamond Lemma

Particular case:

$$\forall i \in I = \{1, \ldots, m\}, \dim(E_i) = 1 \quad \Leftrightarrow \quad \{e_i\}_{i \in I} \ \mathbb{K}\text{-linear basis of } E,$$

$$\mathring{\mathscr{P}}_{\sqcup} \text{ monomial shuffle operad} \quad \Rightarrow \quad \mathring{\mathscr{P}}_{\sqcup} \text{ Koszul and basis } \{t(e_{\bar{i}})\}_{t \in T}.$$

PBW basis of $\mathscr{P}(E, R)$:

$$\text{basis } \{\bar{t}(e_{\bar{i}})\}_{t \in T} := \text{image of } \{t(e_{\bar{i}})\}_{t \in T} \text{ under } \mathring{\mathscr{P}}_{\sqcup} \twoheadrightarrow \operatorname{gr} \mathscr{P}_{\sqcup} \cong \mathscr{P}_{\sqcup} \cong \mathscr{P}.$$

MAIN PROPERTIES OF PBW BASES.

$$\mathscr{P}(E, R) \text{ PBW basis} \quad \Rightarrow \quad \mathscr{P}(E, R) \textit{ Koszul algebra.}$$

DIAMOND LEMMA.

$$\{\bar{t}(e_{\bar{i}})\}_{t \in T^{(3)}} \textit{ linearly independent} \implies \{\bar{t}(e_{\bar{i}})\}_{t \in T} \textit{ PBW basis.}$$

GRÖBNER BASIS.

$$\textit{Gröbner basis of } (R) \subset \mathscr{T}(E) \iff \textit{PBW basis of } \mathscr{T}(E)/(R).$$

PBW bases for inhomogeneous quadratic algebras:

$$q\mathscr{P} = \mathscr{P}(E, qR) \text{ PBW basis} \quad \Rightarrow \quad \mathscr{P}(E, R) \text{ PBW basis.}$$

PBW/Gröbner bases for nonsymmetric operads: We consider planar trees instead of shuffle trees.

8.9.5 Distributive Laws

A *distributive law* $\Lambda : \mathscr{B} \circ \mathscr{A} \to \mathscr{A} \circ \mathscr{B}$ induces an operad structure on $\mathscr{A} \circ \mathscr{B}$ by

$$\gamma_\Lambda := (\gamma_{\mathscr{A}} \circ \gamma_{\mathscr{B}})(\mathrm{Id}_{\mathscr{A}} \circ \Lambda \circ \mathrm{Id}_{\mathscr{B}}).$$

LOCAL TO GLOBAL. For $\mathscr{A} := \mathscr{P}(V, R)$ and $\mathscr{B} := \mathscr{P}(W, S)$, any (local) rewriting rule $\lambda : W \circ_{(1)} V \to V \circ_{(1)} W$ induces (global) distributive laws if $p : \mathscr{A} \circ \mathscr{B} \twoheadrightarrow \mathscr{A} \vee_\lambda \mathscr{B}$ is an isomorphism.

DIAMOND LEMMA FOR DISTRIBUTIVE LAW.

$$\begin{array}{ccc} \mathscr{A}, \mathscr{B} \textit{ Koszul and} & & \mathscr{A} \vee_\lambda \mathscr{B} \textit{ Koszul and} \\ (\mathscr{A} \circ \mathscr{B})^{(3)} \rightarrowtail (\mathscr{A} \vee_\lambda \mathscr{B})^{(3)} & \Longrightarrow & \mathscr{A} \vee_\lambda \mathscr{B} \cong (\mathscr{A} \circ \mathscr{B}, \gamma_\Lambda). \end{array}$$

8.9.6 Partition Poset Method

Set operad $\mathrm{P} \longrightarrow$ family of *operadic partition posets* $\{\Pi_{\mathrm{P}}(n)\}_n$.

Theorem.

◇ $\mathscr{P} := \mathbb{K}[\mathrm{P}]$ *Koszul iff* $\Pi_{\mathrm{P}}(n)$ *Cohen–Macaulay, for all* n,
◇ $H_{\mathrm{top}}(\Pi_{\mathrm{P}}) \cong \mathscr{P}^{¡}$.

8.9.7 Manin Black and White Products for Operads

$$\mathscr{P}(V, R) \bigcirc \mathscr{P}(W, S) := \mathscr{P}\Big(V \underset{\mathrm{H}}{\otimes} W, \Phi^{-1}\Big(R \underset{\mathrm{H}}{\otimes} \mathscr{T}(W) + \mathscr{T}(V) \underset{\mathrm{H}}{\otimes} S\Big)\Big),$$

$$\mathscr{C}(V, R) \bullet \mathscr{C}(W, S) := \mathscr{C}\Big(V \underset{\mathrm{H}}{\otimes} W, \Psi\Big(R \underset{\mathrm{H}}{\otimes} S\Big)\Big),$$

$$\mathscr{P}(V, R) \bullet \mathscr{P}(W, S) = \mathscr{P}\Big(V \underset{\mathrm{H}}{\otimes} W \underset{\mathrm{H}}{\otimes} \mathrm{sgn}_{\mathbb{S}_2}, \widetilde{\Psi}\Big(R \underset{\mathrm{H}}{\otimes} S\Big)\Big).$$

For finitely generated quadratic operads:

$$(\mathscr{P} \bullet \mathscr{Q})^! = \mathscr{P}^! \bigcirc \mathscr{Q}^!.$$

Among finitely generated binary quadratic operads:
Unit for the white product: *Com*. Unit for the black product: *Lie*.

Theorem. $\mathrm{Hom}_{\mathsf{Quad\text{-}Op}}(\mathscr{P} \bullet \mathscr{Q}, \mathscr{R}) \cong \mathrm{Hom}_{\mathsf{Quad\text{-}Op}}(\mathscr{P}, \mathscr{Q}^! \bigcirc \mathscr{R})$.

$$\textit{Lie} \to \mathscr{P}^! \bigcirc \mathscr{P}; \qquad \mathscr{P}^! \bullet \mathscr{P} \quad \textit{is a Hopf operad.}$$

Manin black and white product for nonsymmetric operads: ■ and □.

8.10 Exercises

Exercise 8.10.1 (Dendriform operad). Using the method described in Sect. 8.1, show that the dendriform operad, see Sect. 13.6.1, is a Koszul operad.

Exercise 8.10.2 (Unshuffles and pointed unshuffles). For a given partition $n = i_1 + \cdots + i_k$, show that the set of unshuffles

$$\{\sigma \in Sh^{-1}_{i_{\tau(1)},\ldots,i_{\tau(k)}};\ \tau \in \mathbb{S}_k\}$$

admits a left action of $\mathbb{S}_k$ and that the set of pointed unshuffles $\sqcup(i_1,\ldots,i_k)$ gives one representative in every orbit.

Exercise 8.10.3 (From shuffle operads to symmetric operads). We denote by $M \mapsto M^{tr}$ the functor from arity-graded modules to $\mathbb{S}$-modules, which associates to M, the same underlying module, with the trivial action of the symmetric groups.

There is a natural isomorphism of arity-graded modules $M^{tr} \circ N^{tr} \cong (M \circ_{\sqcup} N)^{tr}$ given by

$$\begin{aligned}
&\bigoplus M^{tr}(k) \otimes_{\mathbb{S}_k} \big(N^{tr}(i_1) \otimes \cdots \otimes N^{tr}(i_k) \otimes_{\mathbb{S}_{i_1}\times\cdots\times\mathbb{S}_{i_k}} \mathbb{K}[\mathbb{S}_n]\big)\\
&\xrightarrow{\cong} \bigoplus M^{tr}(k) \otimes_{\mathbb{S}_k} \big(N^{tr}(i_1) \otimes \cdots \otimes N^{tr}(i_k) \otimes \mathbb{K}[\mathbb{S}_{i_1} \times \cdots \times \mathbb{S}_{i_k} \backslash \mathbb{S}_n]\big)\\
&\xrightarrow{\cong} \bigoplus M^{tr}(k)\big(N^{tr}(i_1) \otimes \cdots \otimes N^{tr}(i_k) \otimes \mathbb{K}[\sqcup_{i_1,\ldots,i_k}]\big).
\end{aligned}$$

Show that this isomorphism does not commute with the action of the symmetric groups. (Therefore, it does not induce a monoidal functor and does not send a shuffle operad to a symmetric operad.)

Exercise 8.10.4 (Suitable order on trees ★). We consider the following generalization of the path-lexicographic ordering of Sect. 8.4.1. It applies to the set of shuffle trees with bivalent vertices and whose vertices are labeled by $\{1,\ldots,m\}$, m being finite. To any such tree of arity n, we associate a sequence of $n+1$ words as follows. The n first words are obtained by reading the tree from the root to each leaf and by recording the labels indexing the vertices. If one encounters an arity 4 vertex labeled by 2, for instance, then the letter will be 2_4. The last word is given by the ordered labeling of the leaves, or equivalently the image of the inverse of the associate pointed unshuffle.

Prove that a such a sequence characterizes the labeled shuffle tree.

We say that $i_j < k_l$ when $(i, j) < (k, l)$ with the lexicographic order. We consider the following total order on this type of sequences.

1. We order them according to the total number of "letters" composing the words of the sequence.
2. We use the length of the last word (arity).
3. We consider the lexicographic order.

For $m = 2$, it gives

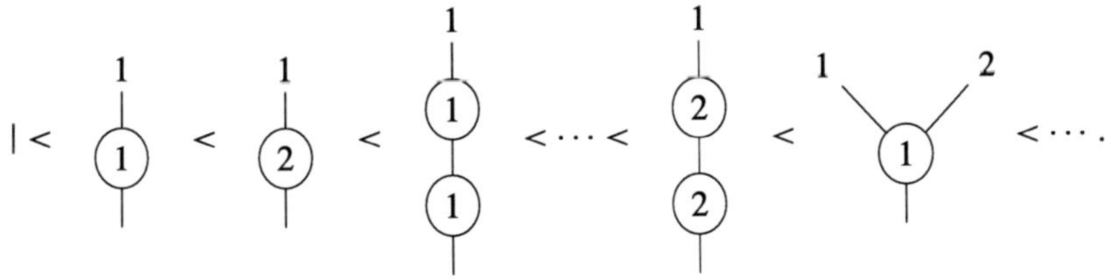

Show that this provides a suitable order.

Exercise 8.10.5 (Koszul dual operad with extra ordered grading). Let $\mathscr{P} = \mathscr{P}(E, R)$ be a finitely generated binary quadratic operad endowed with an extra ordered grading. Give a presentation of the Koszul dual operad $\mathscr{P}^!$ of $\mathscr{P}$ following the methods of Sect. 4.2.6.

Refine this result as in Sect. 4.3.9: when $\mathscr{P} = \mathscr{P}(E, R)$ is a finitely generated binary quadratic operad endowed with a PBW basis, give a presentation of the Koszul dual operad $\mathscr{P}^!$ of $\mathscr{P}$.

Exercise 8.10.6 (Reduction by filtration method for nonsymmetric operads). Write the entire Sect. 8.4 for nonsymmetric operads as proposed in Sect. 8.4.7.

Exercise 8.10.7 (Computations of PBW bases). Describe a PBW basis for the quadratic operad *Pois* of Poisson algebras (Sect. 13.3), for the quadratic operad *Perm* of permutative algebras and for the quadratic operad *preLie* of pre-Lie algebras (Sect. 13.4).

Exercise 8.10.8 (Computations of PBW bases in the inhomogeneous framework ★). Describe a PBW basis for the inhomogeneous operad BV encoding Batalin–Vilkovisky algebras, see Sects. 7.8.7 or 13.7 for the definition.

Exercise 8.10.9 (Parametrized operad *As*). We consider the nonsymmetric operad $\hbar As$ generated a binary product  , which satisfies the associativity relation

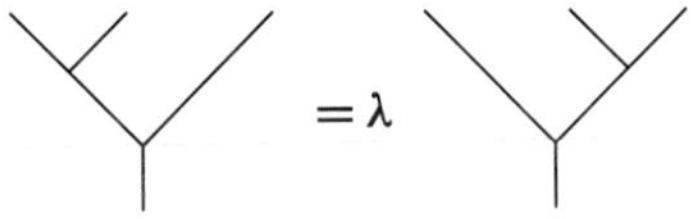

with parameter λ.

Show that the ns operad λAs is Koszul if and only if $\lambda = 0, 1$ or ∞, where the latter case means

 $= 0.$

Exercise 8.10.10 (From operad structure to distributive law). Let $\mathscr{A} = (\mathscr{A}, \gamma_{\mathscr{A}}, \iota_{\mathscr{A}})$ and $\mathscr{B} = (\mathscr{B}, \gamma_{\mathscr{B}}, \iota_{\mathscr{B}})$ be two operads. Let $\mathscr{A} \circ \mathscr{B}$ be endowed with an operad structure $(\gamma, \iota_{\mathscr{A}\circ\mathscr{B}})$ such that $\iota_{\mathscr{A}\circ\mathscr{B}} = \iota_{\mathscr{A}} \circ \iota_{\mathscr{A}}$.

Suppose that

$$\mathscr{A} \xrightarrow{\mathrm{Id}_{\mathscr{A}} \circ \iota_{\mathscr{B}}} \mathscr{A} \circ \mathscr{B} \xleftarrow{\iota_{\mathscr{A}} \circ \mathrm{Id}_{\mathscr{B}}} \mathscr{B}$$

are morphisms of operads and that the following diagram commutes

$$\begin{array}{ccc} & \mathscr{A} \circ \mathscr{B} & \\ {}^{\mathrm{Id}_{\mathscr{A}} \circ \iota_{\mathscr{B}} \circ \iota_{\mathscr{A}} \circ \mathrm{Id}_{\mathscr{B}}}\swarrow & & \searrow^{=} \\ \mathscr{A} \circ \mathscr{B} \circ \mathscr{A} \circ \mathscr{B} & \xrightarrow{\quad\gamma\quad} & \mathscr{A} \circ \mathscr{B}. \end{array}$$

Prove that

$$\Lambda := \mathscr{B} \circ \mathscr{A} \xrightarrow{\iota_{\mathscr{A}} \circ \mathrm{Id}_{\mathscr{B}} \circ \mathrm{Id}_{\mathscr{A}} \circ \iota_{\mathscr{B}}} \mathscr{A} \circ \mathscr{B} \circ \mathscr{A} \circ \mathscr{B} \xrightarrow{\gamma} \mathscr{A} \circ \mathscr{B}$$

is a distributive law (cf. [Bec69]).

Give another proof of Proposition 8.6.2 using this result. (Define the operadic composition on $\mathscr{A} \circ \mathscr{B}$ by transporting the operadic composition of $\mathscr{A} \vee_\lambda \mathscr{B}$ under the isomorphism p.)

Exercise 8.10.11 (Distributive law for the operad *Pois*). Prove the isomorphism $Com \circ Lie(4) \cong Pois(4)$.

Exercise 8.10.12 (Counter-Example, [Vladimir Dotsenko]). Let $\mathscr{A} = Com$ and $\mathscr{B} = Nil$, where a *Nil*-algebra is a vector space equipped with an antisymmetric nilpotent operation $[x, y]$, i.e. satisfying the quadratic relation $[[x, y], z] = 0$ for any x, y, z. We denote by $x \cdot y$ the commutative binary operation of *Com*. We consider the rewriting law λ given by the Leibniz relation:

$$[x \cdot y, z] = x \cdot [y, z] + [x, z] \cdot y$$

like in the Poisson case.

Show that this rewriting rule does not induce a distributive law.

HINT. Compute $[[x \cdot y, z], t]$ in two different ways.

Exercise 8.10.13 (Multi-pointed partition posets ★). A *multi-pointed partition* $\{B_1, \ldots, B_k\}$ of $\underline{n}$ is a partition of $\underline{n}$ on which at least one element of each block B_i is pointed, like $1\bar{3}\bar{4}|\bar{2}6|\bar{5}7\bar{8}$ for instance. Let π and ρ be two multi-pointed partitions. We say the ρ is larger than π, $\pi \leqslant \rho$, when the underlying partition of π is a refinement of the underlying partition of ρ and when, for each block of π, its pointed elements are either all pointed or all unpointed in ρ. For instance

$$1\bar{3}\bar{4}|\bar{2}6|\bar{5}7\bar{8} \leqslant 12\bar{3}\bar{4}6|\bar{5}7\bar{8}.$$

Prove that it is a graded poset and compute its top homology groups.

HINT. Introduce the operad *ComTrias* encoding *commutative trialgebras*, see [Val07a]. It is an algebraic structure defined by two binary operations $*$ and $\bullet$ such that the product $*$ is permutative (encoded by the operad *Perm*)

$$(x * y) * z = x * (y * z) = x * (z * y),$$

the product $\bullet$ is associative and commutative

$$\begin{cases} x \bullet y = y \bullet x, \\ (x \bullet y) \bullet z = x \bullet (y \bullet z), \end{cases}$$

and such that the two operations $*$ and $\bullet$ satisfy the following compatibility relations

$$\begin{cases} x * (y \bullet z) = x * (y * z), \\ (x \bullet y) * z = x \bullet (y * z). \end{cases}$$

Equivalently it is a triassociative algebra, cf. [LR04], satisfying some commutativity property.

Exercise 8.10.14 (Nijenhuis operad ★). Consider the following quadratic operad

$$Nij := \mathscr{P}\big(\mathbb{K}[\mathbb{S}_2]m \oplus \mathbb{K}c,\ R_{preLie} \oplus R_{Jacobi} \oplus R_{comp}\big),$$

where $\mathbb{K}c$ stands for the trivial representation of $\mathbb{S}_2$ and where $|m| = 0$ and $|c| = 1$. The space of relations are given by

$$\begin{cases} R_{preLie}\colon\ m \circ_1 m - m \circ_2 m - (m \circ_1 m)^{(23)} - (m \circ_2 m)^{(23)}, \\ R_{Jacobi}\colon\ c \circ_1 c + (c \circ_1 c)^{(123)} + (c \circ_1 c)^{(321)}, \\ R_{comp}\colon\ m \circ_1 c + m \circ_2 c + (m \circ_1 c)^{(12)} - c \circ_2 m - (c \circ_2 m)^{(12)}. \end{cases}$$

In plain words, the binary product m is a pre-Lie product and the binary product c is a "degree 1 Lie bracket".

Show that this operad is Koszul. For all the methods proposed in this chapter, try to see whether they can be applied. Notice that this operad provides the first example which requires the use of PBW bases.

REMARK. The introduction and the study of this operad are prompted by Nijenhuis geometry. For more details on the application of its Koszul resolution in this direction, we refer the reader to [Mer05, Str09].

Exercise 8.10.15 (Computations of Manin products ★). Prove that $Perm \bigcirc Ass \cong Dias$. Then conclude that $preLie \bullet Ass \cong Dend$. Prove this last isomorphism by a direct computation. (We refer to Sect. 13.6 for the definitions of the operads *Dias* and *Dend*.)

Exercise 8.10.16 (Counter-example ★). Consider the operad

$$Nil := \mathscr{P}\big(\mathrm{sgn}_{\mathbb{S}_2}, \mathscr{T}(\mathrm{sgn}_{\mathbb{S}_2})(3)\big)$$

of skew-symmetric nilpotent algebras. It is generated by the signature representation in arity 2, with all possible relations.

Compute the black product $Nil \bullet preLie$. (We refer to Exercise 8.10.14 and to Sect. 13.4 for the definition of the operad *preLie*.)

Show that the operad $Nil \bullet preLie$ is not Koszul, though the two operads *Nil* and *preLie* are.

Exercise 8.10.17 (Manin products for shuffle operads ★). Following the same method as in Sect. 8.8, define the notion of Manin white product $\bigcirc_{\sqcup\!\sqcup}$ for shuffle operads with presentation.

Prove the existence of a canonical isomorphism of shuffle operads of the form

$$(\mathscr{P} \bigcirc \mathscr{Q})^f \cong \mathscr{P}^f \bigcirc_{\sqcup\!\sqcup} \mathscr{Q}^f,$$

for any pair of (symmetric) operads $\mathscr{P} = \mathscr{P}(V, R)$ and $\mathscr{Q} = \mathscr{P}(W, S)$.

Chapter 9
The Operads As and A_∞

For me it all begins with Poincaré.
Jim Stasheff in "The pre-history of operads"
Contemp Math 202 (1997), 9–14

In this chapter, we first treat in detail the operad encoding the category of associative algebras along the lines of the preceding chapters. This is a particularly important example, because associative algebras are everywhere in mathematics, and because it will serve as a toy-model in the theory of operads.

In Sect. 9.1, we describe the nonsymmetric operad As and then the symmetric operad Ass, where the action of the symmetric group is taken into account. They both encode the category of associative algebras. Then we compute the Koszul dual cooperad $As^{¡}$. We show that As is a Koszul operad by analyzing in detail the Koszul complex. We show that the operadic homology of associative algebras is precisely Hochschild homology.

In Sect. 9.2, we proceed with the computation of the minimal model of As, that is $\Omega As^{¡}$. We show that this operad is exactly the operad A_∞, constructed by Jim Stasheff, encoding the category of "homotopy associative algebras", also known as A_∞-algebras. We describe this differential graded operad in terms of the Stasheff polytope (associahedron).

In Sect. 9.3, we study the bar–cobar construction on As, denoted $\Omega \mathrm{B}\, As$. We show that this ns operad can be understood in terms of a cubical decomposition of the Stasheff polytope (Boardman–Vogt W-construction). We compare $\Omega \mathrm{B}\, As$ and A_∞.

In Sect. 9.4, we deal with the *Homotopy Transfer Theorem*. In its simplest form it says that, starting with a dga algebra, any homotopy retract acquires a structure of A_∞-algebra. More generally A_∞-algebras are invariant under homotopy equivalence. These results are due to the work of V.K.A.M. Gugenheim, Hans Munkholm, Jim Stasheff, Tornike Kadeishvili, Alain Prouté, Serguei Merkulov, Martin Markl, Maxim Kontsevich and Yan Soibelman. In the next chapter this theorem is extended to any Koszul operad.

We make this chapter as self-contained as possible, so there is some redundancy with other parts of the book. We refer to Stasheff's paper [Sta10] for historical refer-

J.-L. Loday, B. Vallette, *Algebraic Operads*,
Grundlehren der mathematischen Wissenschaften 346,
DOI 10.1007/978-3-642-30362-3_9,

ences on the subject, linking twisting morphisms, also known as twisting cochains, to A_∞-algebra structures, and giving tribute to N. Berikashvili, K.T. Chen and T. Kadeishvili.

9.1 Associative Algebras and the Operad *Ass*

We study the nonsymmetric operad *As* encoding the category of (nonunital) associative algebras. We show that its Koszul dual is itself: $As^! = As$ and that it is a Koszul nonsymmetric operad. We also study the associated symmetric operad, denoted by *Ass*.

9.1.1 Associative Algebra

By definition an *associative algebra* over $\mathbb{K}$ is a vector space A equipped with a binary operation

$$\mu : A \otimes A \to A, \qquad \mu(x, y) = xy$$

satisfying the *associativity relation*

$$(xy)z = x(yz)$$

for any $x, y, z \in A$. This relation may also be written as

$$\mu \circ (\mu, \mathrm{id}) = \mu \circ (\mathrm{id}, \mu),$$

and, in terms of partial compositions, as

$$\mu \circ_1 \mu = \mu \circ_2 \mu.$$

There is an obvious notion of morphism between associative algebras and we denote by *As*-alg the category of associative algebras.

Here we work in the monoidal category Vect of vector spaces over $\mathbb{K}$, but, because of the form of the relation, we could as well work in the monoidal category of sets, resp. topological sets, resp. simplicial sets. Then we would obtain the notion of monoid, resp. topological monoid, resp. simplicial monoid.

It is sometimes helpful to assume the existence of a unit, cf. Sect. 1.1.1, but here we work with nonunital associative algebras.

9.1.2 The Nonsymmetric Operad *As*

Since, in the definition, of an associative algebra, the generating operation μ does not satisfy any symmetry property, and since, in the associativity relation, the variables stay in the same order, the category of associative algebras can be encoded by a nonsymmetric operad, that we denote by As.

Let us denote by μ_n the n-ary operation defined as

$$\mu_n(x_1, \dots, x_n) := x_1 \dots x_n.$$

The space of n-ary operations As_n is one-dimensional spanned by μ_n, because the free associative algebra over V is $\overline{T}(V) = \bigoplus_n V^{\otimes n}$. Therefore $As_n = \mathbb{K}\mu_n$. Since $\dim As_n = 1$, the generating series of the ns operad As is

$$f^{As}(t) = \sum_{n \geq 1} t^n = \frac{t}{1-t}.$$

CLASSICAL DEFINITION OF As. Under the classical definition of a nonsymmetric operad the composition

$$\gamma : As_k \otimes As_{i_1} \otimes \cdots \otimes As_{i_k} \to As_{i_1+\cdots+i_k}$$

is simply given by the identification

$$\mathbb{K} \otimes \mathbb{K} \otimes \cdots \otimes \mathbb{K} \cong \mathbb{K}, \qquad \mu_k \otimes \mu_{i_1} \otimes \cdots \otimes \mu_{i_k} \mapsto \mu_{i_1+\cdots+i_k}.$$

It simply follows from the composition of noncommutative polynomials.

PARTIAL DEFINITION OF As. The partial composition is given by

$$\mu_m \circ_i \mu_n = \mu_{m-1+n}$$

for any i because

$$x_1 \cdots x_{i-1}(x_i \cdots x_{m-1})x_m \cdots x_{m-1+n} = x_1 \cdots x_{m-1+n}.$$

QUADRATIC PRESENTATION. The free ns operad on a binary operation $\mu = \mu_2$ is spanned by the planar binary trees (each internal vertex being labeled by μ): $\mathscr{T}(\mathbb{K}\mu)_n = \mathbb{K}[PBT_n]$ (cf. Sect. 5.9.5). The space $\mathscr{T}(\mathbb{K}\mu)_3 = \mathbb{K}[PBT_3]$ is 2-dimensional and spanned by the trees  and  corresponding to the operation $\mu \circ (\mu, \mathrm{id})$ and $\mu \circ (\mathrm{id}, \mu)$ respectively. The relator is the *associator*

$$as := -\mu \circ (\mu, \mathrm{id}) + \mu \circ (\mathrm{id}, \mu) \in \mathscr{T}(\mathbb{K}\mu)^{(2)},$$

i.e. $-$  $+$ . There is an identification

$$As = \mathscr{P}(\mathbb{K}\mu, \mathbb{K}\,as) = \mathscr{T}(\mathbb{K}\mu)/(as)$$

where (as) is the operadic ideal generated by the associator. In the quotient any tree with n leaves gives rise to the same element, that we have denoted by μ_n.

9.1.3 The Operad *Ass*

We denote by *Ass* the symmetric operad encoding the category of associative algebras. The categories of *As*-algebras and *Ass*-algebras are the same, that is the category of associative algebras, since the action of $\mathbb{S}_n$ on $Ass(n)$ is free (see below).

The free associative algebra over the vector space V is known to be the reduced tensor module $\overline{T}(V) = \bigoplus_{n\geq 1} V^{\otimes n}$ equipped with the concatenation product. It is called the *reduced tensor algebra*, cf. Sect. 1.1.3. So we have $Ass(V) = \overline{T}(V)$. If V is generated by the elements $x_1, \ldots, x_n$, then $\overline{T}(V)$ is the algebra of noncommutative polynomials in $x_1, \ldots, x_n$ modulo the constants: $\mathbb{K}\langle x_1, \ldots, x_n\rangle/\mathbb{K}1$. The composition γ on *Ass*, i.e. the map $\gamma(V) : \overline{T}(\overline{T}(V)) \to \overline{T}(V)$, is given by substitution of polynomials: if $P(X_1, \ldots, X_k)$ is a polynomial in the variables X_i and if each X_i is a polynomial in the variables x_j, then $P(X_1(x_1, \ldots, x_n), \ldots, X_k(x_1, \ldots, x_n))$ is a polynomial in the variables x_j called the composite. This composition is obviously associative.

From the polynomial description of the free associative algebra it follows that the space of n-ary operations is $Ass(n) \cong \mathbb{K}[\mathbb{S}_n]$ equipped with the right action given by multiplication in $\mathbb{S}_n$. The n-ary operation $\mu^\sigma \in Ass(n)$ corresponding to the permutation $\sigma \in \mathbb{S}_n$ is

$$\mu^\sigma(x_1, \ldots, x_n) := \mu_n(x_{\sigma^{-1}(1)}, \ldots, x_{\sigma^{-1}(n)}) = x_{\sigma^{-1}(1)} \cdots x_{\sigma^{-1}(n)}.$$

Hence $Ass(n)$ is the regular representation of $\mathbb{S}_n$:

$$Ass(n) \otimes_{\mathbb{S}_n} V^{\otimes n} = \mathbb{K}[\mathbb{S}_n] \otimes_{\mathbb{S}_n} V^{\otimes n} = V^{\otimes n}.$$

The composition in the operad *Ass* is given by the composition of polynomials. It is induced by the maps

$$\gamma(i_1, \ldots, i_k) : \mathbb{S}_k \times \mathbb{S}_{i_1} \times \cdots \times \mathbb{S}_{i_k} \to \mathbb{S}_{i_1+\cdots+i_k}$$

given by concatenation of the permutations and block permutation by the elements of $\mathbb{S}_k$. Here is an example with $k = 2$, $i_1 = 2$, $i_2 = 3$:

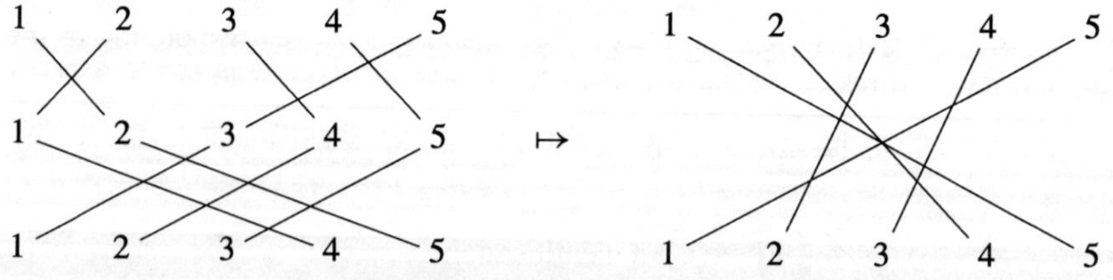

$$([21]; [21], [231]) \mapsto [54321].$$

The partial composition $\circ_i$ is easily deduced from the composition map γ in the polynomial framework. It simply consists in substituting the ith variable for a polynomial.

As a symmetric operad, *Ass* is presented as $Ass = \mathscr{T}(E_{Ass}, R_{Ass})$, where $E_{Ass} \cong \mathbb{K}[\mathbb{S}_2]$. We have denoted by μ the operation corresponding to $\mathrm{id}_{\mathbb{S}_2}$ and so the other linear generator is $\mu^{(12)}$. Under our convention, these two operations correspond to xy and yx respectively. The space of relations R_{Ass} is the sub-$\mathbb{S}_3$-module of $\mathscr{T}(E_{Ass})^{(2)}$ generated by $\mu \circ_1 \mu - \mu \circ_2 \mu$. It is clear that $\mathscr{T}(E_{Ass})^{(2)}$ is $2 \times 6 = 12$-dimensional spanned by the elements $(\mu \circ_1 \mu)^\sigma$, $(\mu \circ_2 \mu)^\sigma$, for $\sigma \in \mathbb{S}_3$ and that R_{Ass} is 6-dimensional spanned by the elements $(\mu \circ_1 \mu - \mu \circ_2 \mu)^\sigma$, for $\sigma \in \mathbb{S}_3$.

The characteristic of *Ass* in the algebra of symmetric functions is

$$F^{Ass} = 1 + p_1 + \cdots + p_1^n + \cdots = \frac{1}{1 - p_1}$$

where p_1 is the classical power symmetric function (cf. for instance [Mac95]).

9.1.4 Other Presentations of *Ass*

There are other presentations of the symmetric operad *Ass* which might be useful in certain problems. We will give only two of them. The second one has the advantage of showing that the Poisson operad is the limit, in a certain sense, of a family of operads all isomorphic to *Ass* (so *Pois* is a "tropical" version of *Ass*), see Sect. 13.3.4.

First, we take a generating operation μ_n of arity n for any $n \geq 2$. We take the following relations:

$$\mu_n \circ (\mu_{i_1}, \ldots, \mu_{i_n}) = \mu_{i_1 + \cdots + i_n}$$

for any tuples $(i_1, \ldots, i_n)$. Then obviously the associated operad is *As*. Considering the analogy with the presentation of groups, it is like presenting a group by taking its elements as generators and taking the table of multiplication as relations.

Here is another presentation in the symmetric framework.

Proposition 9.1.1 (Livernet–Loday, unpublished). *If* 2 *is invertible in* $\mathbb{K}$, *then the operad Ass admits the following presentation*:

- *a symmetric operation* $x \cdot y$ *and an antisymmetric operation* $[x, y]$ *as generating operations*,
- *the following relations*:

$$[x \cdot y, z] = x \cdot [y, z] + [x, z] \cdot y,$$
$$(x \cdot y) \cdot z - x \cdot (y \cdot z) = \big[y, [x, z]\big].$$

Proof. The equivalence between the two presentations is simply given by

$$x \cdot y = xy + yx,$$
$$[x, y] = xy - yx,$$

and, of course, $2xy = x \cdot y + [x, y]$. Observe that the Jacobi relation need not be put as an axiom in this presentation since it is a consequence of the second axiom and the commutativity property of the operation $x \cdot y$. □

For more information about the operation $x \cdot y$ see Sect. 13.10.

9.1.5 The Cooperad As^{i} and Koszul Duality

We compute the Koszul dual ns cooperad As^{i} of the ns operad As. By definition As is the quotient of the free ns operad generated by one binary operation μ, that is $\mathscr{T}(\mathbb{K}\mu)$, quotiented by the operadic ideal generated by the relator

$$as := -\mu \circ (\mu, \mathrm{id}) + \mu \circ (\mathrm{id}, \mu) \in \mathscr{T}(\mathbb{K}\mu)_3.$$

Let us denote by $\mu^c := s\mu$. So μ^c is a cooperation of arity 2 and degree 1. The cofree ns cooperad over $M = (0, 0, \mathbb{K}\mu^c, 0, \ldots)$, that is $\mathscr{T}^c(\mathbb{K}\mu^c)$, can be identified, as a graded vector space, with the vector space spanned by the planar binary trees, cf. Appendix C.1.1. The isomorphism

$$\psi : \mathbb{K}[PBT_n] \cong \mathscr{T}^c\big(\mathbb{K}\mu^c\big)_n$$

is given by

$\psi(|) := \mathrm{id}$, $\psi($  $) := \mu^c$, $\psi(r \vee s) := \big(\mu^c; \psi(r), \psi(s)\big)$.

For instance we have $\psi($ 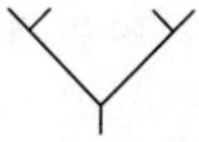 $) = (\mu^c; \mu^c, \mu^c)$.

Since the generator μ^c has homological degree 1, there are signs involved in the explicitation of ψ, see Sect. 5.9.7. Let us consider the map $\varphi : \mathbb{S}_{n-1} \to PBT_n$ constructed in Appendix C.1.3 which follows from the identification of the symmetric group $\mathbb{S}_{n-1}$ with the set of planar binary trees with levels $\widetilde{PBT}_n$. We denote by $\tilde{t}$ the permutation in the pre-image of $t \in PBT_n$ which corresponds to the planar binary tree with *upward* levels. It means that, among the leveled trees representing t, we choose the tree whose levels of the vertices, which are at the same level in t, go upward when moving from left to right. For instance, if $t =$ 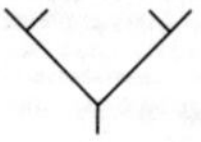, then $\tilde{t} =$ 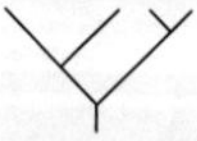 and the permutation is [231].

Observe that the element $\tilde{t}$ is easy to interpret in terms of the construction described in Sect. 5.5.5. We define

$\mu_1^c := |$, $\mu_2^c :=$ , and $\mu_n^c := -\sum_{t \in PBT_n} \mathrm{sgn}(\tilde{t})t$ for $n \geq 3$.

In low dimension we get

$\mu_3^c = -$ … , $\mu_4^c = -$ …

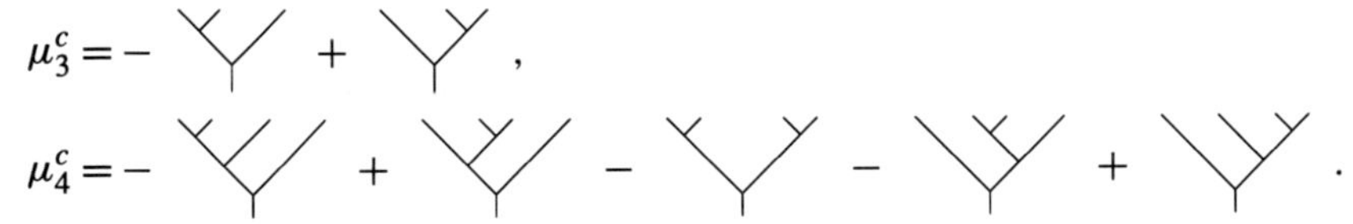

In this example, the involved permutations are [123], [213], [231], [312], [321]. The degree of μ_n^c is $n-1$.

Lemma 9.1.2. *Let* Δ *be the decomposition map of the cofree ns cooperad* $\mathscr{T}^c(\mathbb{K}\mu^c)$. *We have*

$$\Delta(\mu_n^c) = \sum_{i_1+\cdots+i_k=n} (-1)^{\sum(i_j+1)(k-j)} \left(\mu_k^c; \mu_{i_1}^c, \ldots, \mu_{i_k}^c\right).$$

Proof. This is a tedious but straightforward computation. The necessity of the sign $\operatorname{sgn}(\tilde{t})$ in the definition of μ_n^c comes from the formula for Δ in the graded framework (cf. Sect. 5.8.6), which is

$$\Delta(r \vee s) = (r \vee s) + (-1)^{|t^{(2)}||s^{(1)}|}\left(t^{(1)} \vee s^{(1)}; t^{(2)}, s^{(2)}\right).$$

Here are examples of this computation in low dimension under the convention $\Delta(t) = (|; t) + \bar{\Delta}(t) + (t; | \cdots |)$:

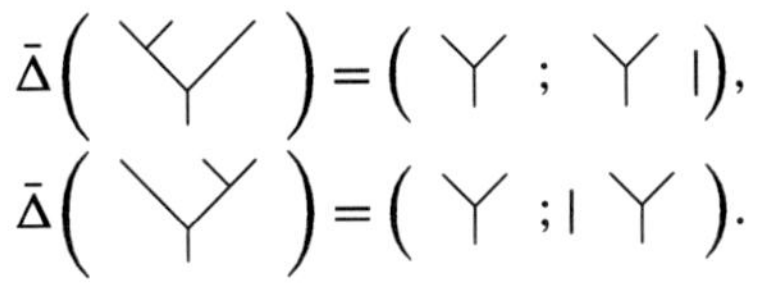

As a consequence we get $\bar{\Delta}(\mu_3^c) = -(\mu_2^c; \mu_2^c, \mu_1^c) + (\mu_2^c; \mu_1^c, \mu_2^c)$.
Then

As a consequence we get

$$\begin{aligned}\bar\Delta(\mu_4^c) = {} & (\mu_2^c;\mu_3^c,\mu_1^c) + (\mu_3^c;\mu_2^c,\mu_1^c,\mu_1^c) - (\mu_3^c;\mu_1^c,\mu_2^c,\mu_1^c)\\ & + (\mu_2^c;\mu_1^c,\mu_3^c) + (\mu_3^c;\mu_1^c,\mu_1^c,\mu_2^c) - (\mu_2^c;\mu_2^c,\mu_2^c).\end{aligned}$$ □

Proposition 9.1.3. *The ns cooperad $As^{i} \subset \mathcal{T}^c(\mathbb{K}\mu^c)$ is such that*

$$(As^{i})_n = \mathbb{K}\mu_n^c.$$

Proof. By Lemma 9.1.2 the elements μ_n^c, $n \geq 1$, span a sub-cooperad of $\mathcal{T}^c(\mathbb{K}\mu^c)$. Since $\mu_3^c = as$, this sub-cooperad is universal among the sub-cooperads whose projection to the quotient space $\mathcal{T}^c(\mathbb{K}\mu^c)^{(2)}/\mathbb{K}\,as$ vanishes. Therefore this cooperad is As^{i}. □

Proposition 9.1.4. *The quadratic ns operad As is self-dual for Koszul duality, that is*

$$As^! = As.$$

Proof. Recall from Sect. 7.7.1 that $As^! = (\mathcal{S}^c \underset{H}{\otimes} As^{i})^*$. Denoting by  the generator and the cogenerator (depending on the context) the relation (or the corelation) can be written:

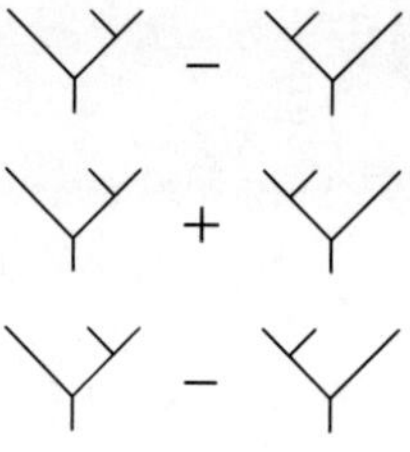

$As^!$ — by definition,

$\mathcal{S}^c \underset{H}{\otimes} As^{i}$ + by sign rule,

$\left(\mathcal{S}^c \underset{H}{\otimes} As^{i}\right)^*$ — by linear duality.

Hence $As^! = As$.

Of course we could as well apply directly Theorem 7.7.1 as follows. Since, in the presentation of As, the space of weight two operations in the free operad is of dimension 2 and the dimension of the space of relators R is of dimension 1, the orthogonal space $R^\perp$ is of dimension 1. Since R is orthogonal to itself for the quadratic form $\begin{bmatrix} 1 & 0 \\ 0 & -1 \end{bmatrix}$, it follows that $R^\perp = R$ and therefore $As^! = As$. □

We now study the Koszulity of the ns operad As.

Theorem 9.1.5. *The ns operad As is a Koszul ns operad.*

Proof. We give here a proof based on the analysis of the Koszul complex. Let us describe the Koszul complex of the ns operad As following p. 70. We consider the arity n sub-chain complex of $As^{i} \circ As$, that is $(As^{i} \circ As)_n$. It is a finite complex of length n which reads:

$$As^{\mathrm{i}}{}_n \otimes As_1 \otimes \cdots \otimes As_1 \to \cdots \to \bigoplus As^{\mathrm{i}}{}_k \otimes As_{i_1} \otimes \cdots \otimes As_{i_k} \to \cdots \to As^{\mathrm{i}}{}_1 \otimes As_n$$

where $i_1 + \cdots + i_k = n$ and $i_j \geq 1$. Observe that each component $As^{\mathrm{i}}{}_k \otimes As_{i_1} \otimes \cdots \otimes As_{i_k}$ is one-dimensional.

In order to describe the boundary map we need to compute $(\mathrm{Id} \circ_{(1)} \kappa)(\Delta_{(1)}(\mu_k^c))$. It is the coproduct of μ_k^c, but keeping on the right side only the terms which involve copies of id and one copy of μ^c (identified to μ under κ). So we get $\sum_j \pm \mu_{k-1}^c \circ (\mathrm{id}, \ldots, \mathrm{id}, \mu, \mathrm{id}, \ldots, \mathrm{id})$. Then we have to apply the associativity isomorphism to

$$\sum_j \pm\big(\mu_{k-1}^c \circ (\mathrm{id}, \ldots, \mathrm{id}, \mu, \mathrm{id}, \ldots, \mathrm{id})\big) \circ (\mu_{i_1}, \ldots, \mu_{i_k})$$

to get

$$\begin{aligned}&\sum_j \pm \mu_{k-1}^c \circ \big(\mathrm{id} \circ \mu_{i_1}, \ldots, \mathrm{id} \circ \mu_{i_{j-1}}, \mu \circ (\mu_{i_j}, \mu_{i_{j+1}}), \mathrm{id} \circ \mu_{i_{j+2}}, \ldots, \mathrm{id} \circ \mu_{i_k}\big)\\ &\quad = \sum_j \pm \mu_{k-1}^c \circ (\mu_{i_1}, \ldots, \mu_{i_{j-1}}, \mu_{i_j + i_{j+1}}, \mu_{i_{j+2}}, \ldots, \mu_{i_k}).\end{aligned}$$

There is no ambiguity to denote the generator of $As^{\mathrm{i}}{}_k \otimes As_{i_1} \otimes \cdots \otimes As_{i_k}$ by $[i_1, \ldots, i_k]$, and we get

$$d\big([i_1, \ldots, i_k]\big) = \sum_j \pm [i_1, \ldots, i_j + i_{j+1}, \ldots, i_k].$$

The boundary map of this Koszul complex can be identified with the boundary map of the augmented chain complex (shifted by one) of the cellular simplex Δ^{n-2}:

$$C_{n-2}\big(\Delta^{n-2}\big) \to \cdots \to C_{k-2}\big(\Delta^{n-2}\big) \to \cdots \to C_0\big(\Delta^{n-2}\big) \to \mathbb{K}.$$

Compared to the classical way of indexing the vertices of the simplex Δ^{n-2} by integers $0, \ldots, n-1$, the vertex number i corresponds to the chain $[i+1, n-i-1]$ in $(As^{\mathrm{i}} \circ As)_n$. Here is the simplex Δ^2:

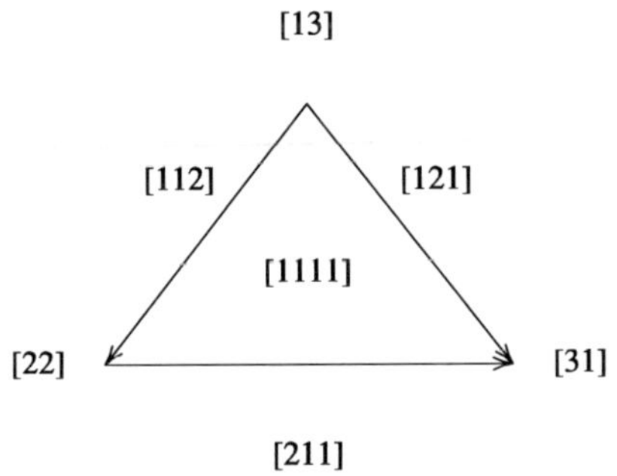

Since the simplex is contractible, its associated augmented chain complex is acyclic for any $n \geq 2$. For $n = 1$ the complex reduces to the space $As^{\mathrm{i}}{}_1 \otimes As_1 = \mathbb{K}$. □

REMARK. We could have also use the poset method (cf. Sect. 8.7), to prove the acyclicity of the Koszul complex. Or (see below), we could compute the Hochschild homology of the free algebra $\overline{T}(V)$ by providing an explicit homotopy.

Modulo all the apparatus, the shortest proof is the rewriting system method (cf. Sect. 8.1) since the only critical monomial is $((xy)z)t$ and the confluent property is immediate to verify:

- on one hand $((xy)z)t \mapsto (x(yz))t \mapsto x((yz)t) \mapsto x(y(zt))$ (left side of the pentagon),
- one the other hand $((xy)z)t \mapsto (xy)(zt) \mapsto x(y(zt))$ (right side of the pentagon).

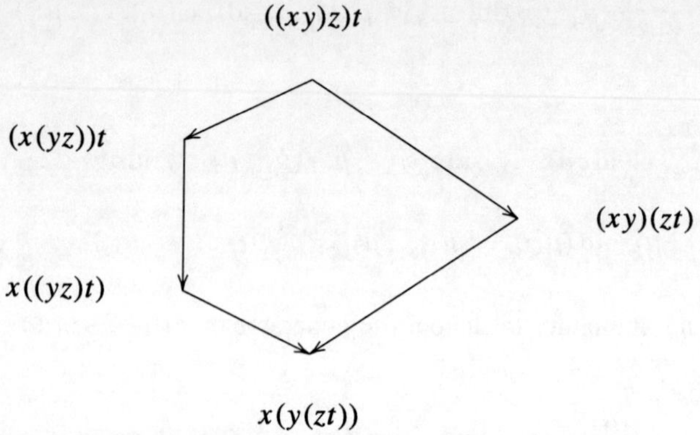

9.1.6 Hochschild Homology of Associative Algebras

Since we know that $As^! = As$, we can describe explicitly the chain complex $C_\bullet^{As}(A)$, which gives the operadic homology of the associative algebra A, cf. Sect. 12.1.2. Let us introduce the boundary map

$$b' : A^{\otimes n} \to A^{\otimes n-1}$$

by the formula

$$b'(a_1, \dots, a_n) = \sum_{i=1}^{n-1} (-1)^{i-1} (a_1, \dots, a_i a_{i+1}, \dots, a_n).$$

It is well known (and easy to check from the associativity of the product) that $(b')^2 = 0$. The resulting chain complex $(A^{\otimes \bullet}, b')$ is the *nonunital Hochschild complex*, up to a shift of degree.

Proposition 9.1.6. *The operadic chain complex of the associative algebra A is the nonunital Hochschild complex of A:*

$$C_{\bullet-1}^{As}(A) = \left(A^{\otimes \bullet}, b'\right).$$

Therefore the operadic homology of an associative algebra is the Hochschild homology up to a shift of degree.

Proof. By definition (Sect. 12.1.2) the complex $C_\bullet^{As}(A)$ is given by

$$\cdots \to As^{\mathrm{i}}{}_n \otimes A^{\otimes n} \to \cdots \to As^{\mathrm{i}}{}_1 \otimes A.$$

Since $As^{\mathrm{i}}{}_n$ is one-dimensional, we get $A^{\otimes n}$ in degree $n-1$. The boundary map is obtained as follows (cf. Sect. 6.6): we consider all the possibilities of "splitting" μ_n^c using one copy of μ^c on the right-hand side (infinitesimal decomposition map), that is

$$\Delta_{(1)}(\mu_n^c) = \sum \pm \mu_{n-1}^c \otimes (\mathrm{id}, \ldots, \mathrm{id}, \mu^c, \mathrm{id}, \ldots, \mathrm{id})$$

and then we apply the element $(\mathrm{id}, \ldots, \mathrm{id}, \mu^c, \mathrm{id}, \ldots, \mathrm{id})$ to $(a_1, \ldots, a_n)$ after replacing the cooperation μ^c by the operation μ under κ. This gives precisely the boundary map b' since $\mu(a_i, a_{i+1}) = a_i a_{i+1}$. □

9.1.7 Homology and Cohomology with Coefficients

In the literature, homology and cohomology with coefficients appear most of the time for unital associative algebras. The comparison between unital and nonunital cases is not completely straightforward, see for instance [LQ84, Lod98] Sect. 1.4. We describe briefly the unital case.

In order to construct a homology or cohomology with coefficients one needs a notion of "representation" (the coefficients). In the associative case it is the notion of bimodule because of the following fact. For any (unital) algebra A and any abelian extension

$$0 \to M \to A' \to A \to 0$$

the space M is a (unital) bimodule over A. Recall that here A' is a (unital) algebra and the product of two elements of M is 0 (abelian hypothesis).

Let A be a unital associative algebra and M a unital A-bimodule. The Hochschild complex $C_\bullet(A, M)$ is defined as $C_n(A, M) := M \otimes A^{\otimes n}$ and the boundary map $b : C_n(A, M) \to C_{n-1}(A, M)$ is given by the formula $b = \sum_{i=0}^{i=n} (-1)^i d_i$, where

$$\begin{aligned} d_0(m, a_1, \ldots, a_n) &:= (ma_1, a_2, \ldots, a_n), \\ d_i(m, a_1, \ldots, a_n) &:= (m, a_1, \ldots a_i a_{i+1}, \ldots, a_n), \quad 1 \le i \le n-1, \\ d_n(m, a_1, \ldots, a_n) &:= (a_n m, a_1, \ldots, a_{n-1}). \end{aligned}$$

The homology groups of $C_\bullet(A, M)$ are called the *Hochschild homology groups* of the unital algebra A with coefficients in the bimodule M.

The Hochschild complex of cochains $C^\bullet(A, M)$ is defined as $C^n(A, M) :=$-$\mathrm{Hom}(A^{\otimes n}, M)$ and the boundary map $b : C^n(A, M) \to C^{n+1}(A, M)$ is given by the formula $b = \sum_{i=0}^{i=n}(-1)^i d_i$, where

$$\begin{aligned} d_0(f)(a_1, \dots, a_{n+1}) &:= a_1 f(a_2, \dots, a_{n+1}), \\ d_i(f)(a_1, \dots, a_{n+1}) &:= f(a_1, \dots a_i a_{i+1}, \dots, a_{n+1}), \quad 1 \le i \le n, \\ d_n(f)(a_1, \dots, a_{n+1}) &:= f(a_1, \dots, a_n)a_{n+1}. \end{aligned}$$

The homology groups of $C^\bullet(A, M)$ are called the *Hochschild cohomology groups* of the unital algebra A with coefficients in the bimodule M.

These groups appear as obstruction groups in many questions and there is an extensive literature about them (for a first approach see for instance [Lod98]). For instance there is a classification theorem for abelian extensions of A by M:

$$H^2(A, M) \cong \mathscr{E}xt(A, M).$$

Similarly there is a classification theorem for crossed modules of A by M:

$$H^3(A, M) \cong \mathscr{X}Mod(A, M).$$

Historically these complexes were constructed by hand by Hochschild. Of course they can be viewed as coming from the operad theory. The advantage is to produce similar complexes and theorems for Koszul operads without having to do ad hoc constructions and proofs in each case. This is the theme of Chap. 12.

9.1.8 Other Homology Theories for Associative Algebras

The operad As has more structure: it is a cyclic operad (cf. Sect. 13.14.6). As such there exists a homology theory for associative algebras which takes into account this extra structure, it is called *cyclic homology*. It was first devised by Alain Connes [Con85] and further studied in [LQ84] and [Tsy83] (see also the monograph [Lod98]), mainly for unital associative algebras. Cyclic homology is encoded by the cyclic category, denoted by ΔC. This notation accounts for the structure of this category which is made up of the simplicial category Δ and the cyclic groups. There is a similar category, where the cyclic groups are replaced by the symmetric groups: ΔS, cf. [FL91, Lod98]. It turns out that ΔS is precisely the category associated to the operad uAs encoding the category of unital associative algebras by the method in Sect. 5.4.1, as shown by Pirashvili in [Pir02a], cf. Example 2.

The operad As can also be considered as a permutad (cf. Sect. 13.14.7). We refer to [LR12] for more on this theme.

9.2 Associative Algebras Up to Homotopy

In the 1960s, Jim Stasheff introduced in [Sta63] the notion of A_∞-algebra, also called "associative algebra up to strong homotopy". The idea is that associativity of the binary operation m_2 is satisfied only "up to higher homotopy". It has a meaning algebraically whenever one works with a chain complex (A, d). It means that the associator of m_2 is not zero, but there exists a ternary operation m_3 (the homotopy) such that

$$m_2 \circ (m_2, \mathrm{id}) - m_2 \circ (\mathrm{id}, m_2) = d \circ m_3 + m_3 \circ \big((d, \mathrm{id}, \mathrm{id}) + (\mathrm{id}, d, \mathrm{id}) + (\mathrm{id}, \mathrm{id}, d)\big).$$

But then, mixing m_2 and m_3 leads to introduce a 4-ary operation m_4 satisfying some relation, and so on, and so on. The whole algebraic structure, discovered by Jim Stasheff, is encoded into the notion of A_∞-algebra, that we recall below. The relevant operad, denoted A_∞, can be described in terms of the Stasheff polytope.

On the other hand the operad theory gives an algorithm to construct explicitly the minimal model of As, which is the dgns operad $As_\infty := \Omega As^{¡}$. It gives rise to the notion of "homotopy associative algebra". It turns out that, not surprisingly, $A_\infty = As_\infty$.

9.2.1 A_∞-Algebra [Sta63]

We have seen in Sect. 2.2.1 that any associative algebra A gives rise to a dga coalgebra $(\overline{T}^c(sA), d)$, which is its bar construction. It is natural to look for a converse statement. Given a graded vector space A together with a codifferential m on the cofree coalgebra $\overline{T}^c(sA)$, what kind of structure do we have on A? The answer is given by the notion of A_∞-algebra. Here are the details.

By definition an A_∞-*algebra* is a graded vector space $A = \{A_k\}_{k\in\mathbb{Z}}$ equipped with a codifferential map $m : \overline{T}^c(sA) \to \overline{T}^c(sA)$ (so $|m| = -1$, $m \circ m = 0$) and m is a coderivation, cf. Sect. 1.2.7. Observe that, since $\overline{T}^c(sA)$ is cofree, the coderivation m is completely determined by its projection $proj \circ m : \overline{T}^c(sA) \to sA$, that is by a family of maps $A^{\otimes n} \to A, n \geq 1$.

To any A_∞-algebra (A, m), we associate an n-ary operation m_n on A as the following composite:

$$\begin{array}{ccccccc} (sa_1)\cdots(sa_n) & \quad & (sA)^{\otimes n} & \xrightarrow{m_|} & sA & \quad & sa \\ \Big\uparrow & & \cong\Big\uparrow & & \Big\downarrow\cong & & \Big\downarrow \\ a_1\cdots a_n & & A^{\otimes n} & \xrightarrow{m_n} & A & & a \end{array}$$

where $m_|$ is the restriction of m to $(sA)^{\otimes n}$ composed with the projection onto sA. The map m_n is of degree $n-2$ since the degrees of the involved maps in the composition are $n-1$ and -1 respectively.

Lemma 9.2.1. *An* A_∞-algebra *is equivalent to a graded vector space* $A = \{A_k\}_{k\in\mathbb{Z}}$ *equipped with an n-ary operation*

$$m_n : A^{\otimes n} \to A \quad \textit{of degree } n-2 \textit{ for all } n \geq 1,$$

which satisfy the following relations:

$$(rel_n) \qquad \sum_{p+q+r=n} (-1)^{p+qr} m_k \circ_{p+1} m_q = 0, \quad n \geq 1,$$

where $k = p + 1 + r$.

Proof. First, recall that

$$m_k \circ_{p+1} m_q = m_k \circ \left(\mathrm{id}^{\otimes p} \otimes m_q \otimes \mathrm{id}^{\otimes r}\right).$$

Let (A, m) be an A_∞-algebra as defined above. Since m is a coderivation, and since $\overline{T}^c(sA)$ is cofree over sA, by Proposition 1.2.2, m is completely determined by its composite

$$\overline{T}^c(sA) \xrightarrow{m} \overline{T}^c(sA) \xrightarrow{proj} sA.$$

The condition $m^2 = 0$ is equivalent to the vanishing of the composite

$$\overline{T}^c(sA) \xrightarrow{m} \overline{T}^c(sA) \xrightarrow{m} \overline{T}^c(sA) \twoheadrightarrow sA.$$

For each $n \geq 1$, its restriction to the n-tensors gives the relation $(rel)_n$. The signs come from the fact that we establish the formula on $A^{\otimes n}$ instead of $(sA)^{\otimes n}$. □

We sometimes write m_n^A in place of m_n if it is necessary to keep track of the underlying chain complex. Let us make the relation (rel_n) explicit for $n = 1, 2, 3$:

$$\begin{aligned}
&m_1 \circ m_1 = 0,\\
&m_1 \circ m_2 - m_2 \circ (m_1, \mathrm{id}) - m_2 \circ (\mathrm{id}, m_1) = 0,\\
&m_1 \circ m_3 + m_2 \circ (m_2, \mathrm{id}) - m_2 \circ (\mathrm{id}, m_2)\\
&\quad + m_3 \circ (m_1, \mathrm{id}, \mathrm{id}) + m_3 \circ (\mathrm{id}, m_1, \mathrm{id}) + m_3 \circ (\mathrm{id}, \mathrm{id}, m_1) = 0.
\end{aligned}$$

9.2.2 Homotopy and Operadic Homology of an A_∞-Algebra

Let A be an A_∞-algebra, e.g. a dga algebra. The relation $(rel)_1$ implies that (A, m_1) is a chain complex. We prefer to denote the differential m_1 by $-d$ and consider an A_∞-algebra as being a chain complex (A, d) equipped with higher operations, see below. The homology of the underlying chain complex (A, d) is called the *homotopy of the* A_∞*-algebra* A. The homology of the chain complex $(\overline{T}^c(sA), m^A)$ is called the *operadic homology* of the A_∞-algebra A. If A is a dga algebra, then the operadic homology is the *Hochschild homology* (of the nonunital dga algebra).

Proposition 9.2.2. *The homotopy of an* A_∞*-algebra is a graded associative algebra.*

Proof. This is an immediate consequence of the relations (rel_2) and (rel_3) recalled in the introduction of this section and in Sect. 9.2.4. □

A finer statement can be found in Corollary 9.4.5.

9.2.3 Examples

We already mentioned that a (nonunital) dga algebra (A, d) is an example of A_∞-algebra. Indeed it suffices to take $m_1 = -d$, $m_2 = \mu$ and $m_n = 0$ for $n \geq 3$. Observe that if A is a dga algebra, then $(T^c(s\bar{A}), m)$ is precisely the cobar construction on A.

When $A = C^\bullet_{\text{sing}}(X)$ is the singular cochain complex of a topological space X, it is endowed with an associative cup product. This associative algebra structure transfers to an A_∞-algebra structure on the singular cohomology $H^\bullet_{\text{sing}}(X)$. These operations were originally defined by Massey in [Mas58].

The singular chains on a based loop space ΩX of the connected topological space X is an A_∞-algebra, cf. [Sta63].

9.2.4 The Operad A_∞

Let A be an A_∞-algebra. For $n = 1$ the relation (rel_n) reads as follows:

$$m_1 \circ m_1 = 0.$$

So $d := m_1$ is a differential on A. The derivative (cf. p. 25) of the map m_n is

$$\partial(m_n) := dm_n - (-1)^{n-2} m_n\big((d, \text{id}, \ldots, \text{id}) + \cdots + (\text{id}, \ldots, \text{id}, d)\big).$$

Using this notation the relations (rel_n) become:

$$\begin{aligned}
&(rel'_2) : \partial(m_2) = 0,\\
&(rel'_3) : \partial(m_3) = -m_2 \circ (m_2, \text{id}) + m_2 \circ (\text{id}, m_2),\\
&(rel'_4) : \partial(m_4) = -m_2 \circ (m_3, \text{id}) + m_3 \circ (m_2, \text{id}, \text{id}) - m_3 \circ (\text{id}, m_2, \text{id})\\
&\qquad\qquad + m_3 \circ (\text{id}, \text{id}, m_2) - m_2 \circ (\text{id}, m_3).
\end{aligned}$$

More generally, for any $n \geq 2$, the relation (rel_n) can be written as:

$$(rel'_n) : \quad \partial(m_n) = - \sum_{\substack{n=p+q+r\\ k=p+1+r\\ k>1, q>1}} (-1)^{p+qr} m_k \circ \big(\text{id}^{\otimes p} \otimes m_q \otimes \text{id}^{\otimes r}\big).$$

Therefore an A_∞-algebra can be seen as a chain complex (A, d) equipped with linear maps: $m_n : (A, d)^{\otimes n} \to (A, d)$, for $n \geq 2$, satisfying some relations. In other words, an A_∞-algebra is an algebra over some dgns operad, denoted A_∞. The generating operations of this operad correspond to the operations $m_n \in \mathrm{End}(A)$, $n \geq 2$, of degree $n-2$. So, the nonsymmetric operad A_∞ is free over these operations: $A_\infty = \mathscr{T}(\oplus_{n\geq 2}\mathbb{K}\mu_n)$. The differential structure is precisely given by the relations (rel'_n), $n \geq 2$.

Here we work in the homological framework (degree of d is -1), but one can of course define A_∞-algebras in the cohomological framework, see for instance [Kel01].

9.2.5 The Associahedron (Stasheff Polytope)

Let us recall from Appendix C that the associahedron $\mathscr{K}^n$ is a cell complex of dimension n homeomorphic to a ball, whose cells are in bijection with the planar trees with $n+2$ leaves. The chain complex associated to $\mathscr{K}^n$ is denoted by $C_\bullet(\mathscr{K}^n)$. A tree $t \in PT_{n+2,n-k+1}$ has $n+2$ leaves and $(n-k+1)$ internal vertices. It gives a chain in $C_k(\mathscr{K}^n)$. Since $\mathscr{K}^n$ is contractible, the homology of $C_\bullet(\mathscr{K}^n)$ is trivial except in dimension 0, where it is $\mathbb{K}$. We will identify the dg vector space of n-ary operations of the operad A_∞ (and of the operad As_∞) to $C_\bullet(\mathscr{K}^{n-2})$.

Proposition 9.2.3. *The dg operad A_∞ encoding the category of A_∞-algebras is a dgns operad whose dg module of n-ary operations is the chain complex of the associahedron*:

$$(A_\infty)_n = C_\bullet\big(\mathscr{K}^{n-2}\big).$$

Proof. Since, as an operad, A_∞ is free with one generator in each arity $n \geq 2$, there is an isomorphism $(A_\infty)_n \cong \mathbb{K}[PT_n]$, where PT_n is the set of planar trees with n leaves. The generating operation m_n corresponds to the n-corolla under this isomorphism. On the other hand, the cells of the associahedron $\mathscr{K}^{n-2}$ are in bijection with PT_n, whence the identification of vector spaces $\mathbb{K}[PT_n] = C_\bullet(\mathscr{K}^{n-2})$. The boundary map of $(A_\infty)_n$ is given by formula (rel'_n). Once translated in terms of cells of the associahedron it gives precisely the boundary of the big cell of $\mathscr{K}^{n-2}$ since the facettes are labeled by the planar trees with two internal vertices. □

EXAMPLES. (rel'_3) and (rel'_4):

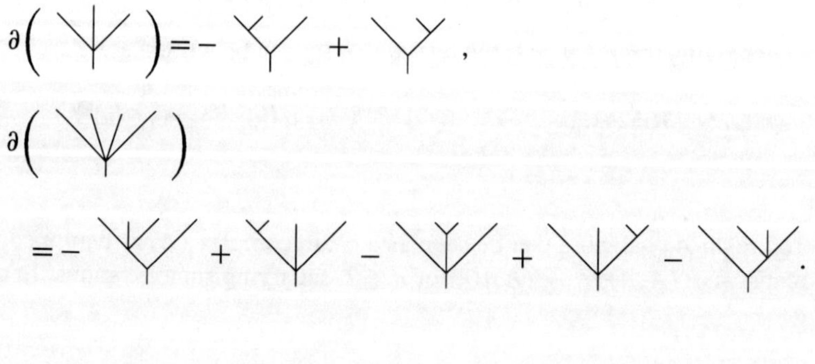

Fig. 9.1 Interval

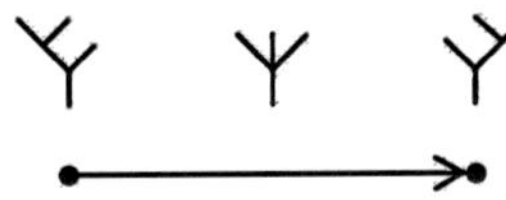

Fig. 9.2 Pentagon

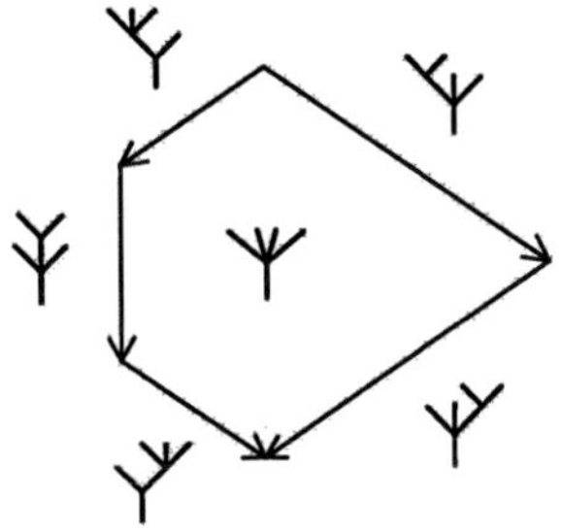

These formulas are the algebraic translation of the cell boundaries shown in Figs. 9.1 and 9.2.

Proposition 9.2.4. *The operad $As_\infty := \Omega As^{¡}$ is a dgns operad whose space of n-ary operations is the chain complex of the associahedron*:

$$\left(\Omega As^{¡}\right)_n = (As_\infty)_n = C_\bullet\left(\mathscr{K}^{n-2}\right).$$

Proof. We work in the nonsymmetric operad context. As a ns operad $As_\infty = \mathscr{T}(s^{-1}\overline{As^{¡}})$ is free on $\overline{As^{¡}}$, that is free on the generating operations $\tilde{\mu}_n^c := s^{-1}\mu_n^c, n \geq 2$ of degree $n-2$. Hence, as in the previous case, there is an isomorphism of vector spaces $(As_\infty)_n \cong \mathbb{K}[PT_n] = C_\bullet(\mathscr{K}^{n-2})$. By definition of the cobar construction, the boundary map on $\mathscr{T}(s\overline{As^{¡}})_n$ is induced by the cooperad structure of $As^{¡}$, more precisely by the infinitesimal decomposition map $\Delta_{(1)}$, cf. Sect. 6.1.4. Let us show that this boundary map is precisely the boundary map of the chain complex $C_\bullet(\mathscr{K}^{n-2})$.

As a linear generator of $\mathscr{T}(s^{-1}\overline{As^{¡}})_n$ the element $\tilde{\mu}_n^c$ corresponds to the big cell (n-corolla) $t(n)$. The degree of $\tilde{\mu}_n^c$ is $n-2$. It is mapped to $\mu_n \in (A_\infty)_n$. The image of μ_n^c under the map $\Delta_{(1)}$ is

$$\Delta_{(1)}\left(\mu_n^c\right) = \sum_{p+1+r=k} (-1)^{(q+1)r}\mu_k^c \otimes \big(\underbrace{\mathrm{id}, \ldots, \mathrm{id}}_{p}, \mu_q^c, \underbrace{\mathrm{id}, \ldots, \mathrm{id}}_{r}\big),$$

for $n = p+q+r$ by Lemma 9.1.2. Under the isomorphism $\mathscr{T}(s\overline{As^{¡}})_n \cong C_\bullet(\mathscr{K}^{n-2})$ this is precisely the boundary map of the associahedron. The evaluation of the boundary on the other cells follows from this case. □

Corollary 9.2.5. *The categories of A_∞-algebras and As_∞-algebras are the same*:

$$A_\infty = As_\infty := \Omega As^{¡}.$$

Proof. From the description of the operad A_∞ given in Proposition 9.2.3 and the description of the operad As_∞ given in Proposition 9.2.4 it is clear that we have a bijection

$$(A_\infty)_n = \mathbb{K}[PT_n] = (As_\infty)_n$$

which is compatible with the operad structure. Moreover both differentials coincide with the differential map in the chain complex of the associahedron, therefore the two dgns operads are identical. □

9.2.6 Infinity-Morphism

Let A and B be two A_∞-algebras. An *∞-morphism* $f : A \rightsquigarrow B$ (sometimes called *A_∞-morphism* in the literature) is by definition a morphism of dga coalgebras

$$f : \left(\overline{T}^c(sA), m^A\right) \to \left(\overline{T}^c(sB), m^B\right).$$

We adopt the notation $\rightsquigarrow$ to avoid confusion between morphism and ∞-morphism between two A_∞-algebras. Since $\overline{T}^c(sB)$ is cofree, an ∞-morphism is equivalent to a map $\overline{T}^c(sA) \to sB$, that is a family of maps of degree $n-1$

$$f_n : A^{\otimes n} \to B, \quad n \geq 1,$$

such that f_1 is a chain map, i.e. $d_B \circ f_1 = f_1 \circ d_A$ and such that

$$\sum_{\substack{p+1+r=k\\ p+q+r=n}} (-1)^{p+qr} f_k \circ \big(\underbrace{\mathrm{Id}_A, \ldots, \mathrm{Id}_A}_{p}, m_q^A, \underbrace{\mathrm{Id}_A, \ldots, \mathrm{Id}_A}_{r}\big)$$
$$- \sum_{\substack{k\geq 2\\ i_1+\cdots+i_k=n}} (-1)^\varepsilon m_k^B \circ (f_{i_1}, \ldots, f_{i_k}) = \partial(f_n),$$

in $\mathrm{Hom}(A^{\otimes n}, B)$, for $n \geq 2$. The sign ε is given by the formula

$$\varepsilon = (k-1)(i_1-1) + (k-2)(i_2-1) + \cdots + 2(i_{k-2}-1) + (i_{k-1}-1).$$

Under the tree notation, this relation becomes

$\partial(f_n)$

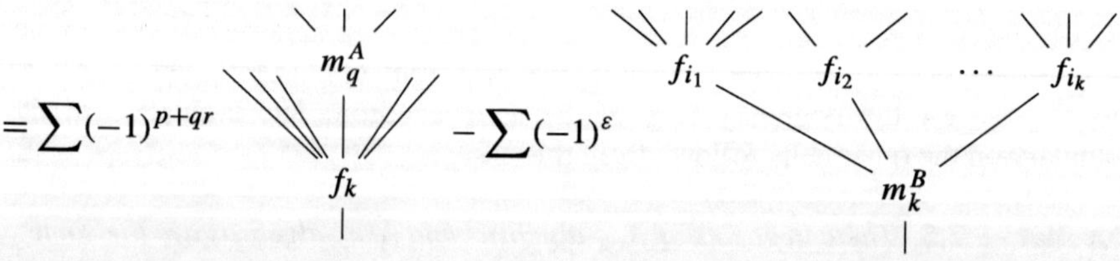

From the definition, it follows that the composite of two ∞-morphisms is again an ∞-morphism. The category of A_∞-algebras equipped with the ∞-morphisms is denoted ∞-A_∞-alg. It is a good exercise to write down the explicit formulas for $(g \circ f)_n$ in terms of the f_i and g_i for $A \overset{f}{\rightsquigarrow} B \overset{g}{\rightsquigarrow} C$.

An ∞-morphism f is called an *∞-quasi-isomorphism* when f_1 is a quasi-isomorphism. One can prove that ∞-quasi-isomorphisms are homotopy invertible. We refer to Chap. 10 for details on ∞-morphisms.

9.2.7 A_∞-Coalgebra

By definition a *conilpotent A_∞-coalgebra* in the monoidal category of sign-graded vector spaces is a graded vector space C equipped with a degree -1 square zero derivation on the tensor algebra:

$$\Delta : \overline{T}\big(s^{-1}C\big) \to \overline{T}\big(s^{-1}C\big).$$

Since the tensor algebra is free, the differential map Δ is completely determined by a family of maps

$$\Delta^n : C \to C^{\otimes n}$$

of degree $-n+2$ for all $n \geq 1$. These maps satisfy the following relations

$$\sum_{p+q+r=n} (-1)^{p+qr}\big(\mathrm{id}^{\otimes p} \otimes \Delta^q \otimes \mathrm{id}^{\otimes r}\big)\Delta^k = 0,$$

where $k = p + 1 + q$.

In the case where $\Delta^n = 0$ for $n > 2$, C is a noncounital dga coalgebra and $(T(s^{-1}C), \Delta)$ is its bar construction.

9.3 The Bar–Cobar Construction on *As*

In this section, we study the bar–cobar construction on the ns operad *As*, that is the dgns operad $\Omega \mathrm{B} As$. We give a geometric interpretation of the operad monomorphism $As_\infty = \Omega As^{\text{i}} \rightarrowtail \Omega \mathrm{B} As$.

9.3.1 The NS Operad $\Omega \mathrm{B} As$

The operad $\Omega \mathrm{B} As$ is a dgns operad which is free as a ns operad. We will describe the chain complex $(\Omega \mathrm{B} As)_n$ in terms of the cubical realization $\mathscr{K}^{n-2}_{cub}$ of the associahedron (Boardman–Vogt W-construction, see also [BM03a]), as described in Appendix C.2.2.

Proposition 9.3.1. *The chain complex $(\Omega \mathrm{B} As)_n$ is precisely the chain complex of the cubical realization of the associahedron*:

$$(\Omega \mathrm{B} As)_n = C_\bullet\big(\mathscr{K}^{n-2}_{cub}\big).$$

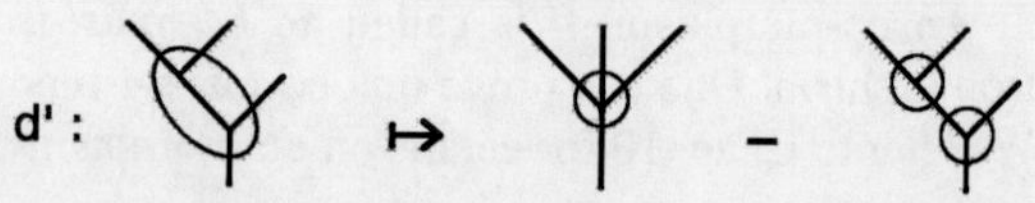

Fig. 9.3 The boundary map of $C_\bullet(\mathscr{K}^1_{cub})$

Proof. In Appendix C.2.2, it is shown that the chain complex $C_\bullet(\mathscr{K}^{n-2}_{cub})$ is spanned by the circled trees with n leaves:

$$C_\bullet(\mathscr{K}^{n-2}_{cub}) = \mathbb{K}[CPT_n].$$

We construct a map $(\Omega \mathrm{B} As)_n \to \mathbb{K}[CPT_n]$ as follows. Since $\Omega \mathrm{B} As = \mathscr{T}(s^{-1} \times \overline{\mathrm{B} As}) \cong \mathscr{T}(s^{-1}\overline{\mathscr{T}^c}(sAs))$, a linear generator is a planar tree whose vertices are labeled by linear generators of $\overline{\mathrm{B} As}$. Since $\overline{\mathrm{B} As}$ is spanned by planar trees, we get trees of trees, that is circled trees. The trees inside the circles are elements of $\overline{\mathrm{B} As}$.

We need now to show that the differential structure of $\Omega \mathrm{B} As$ corresponds exactly to the boundary map of the chain complex $C_\bullet(\mathscr{K}^{n-2}_{cub})$. By definition (cf. Sect. 6.5), the boundary map in $\Omega \mathrm{B} As$ has two components: d_1 coming from the boundary map of $\mathrm{B} As$ and d_2 coming from the cooperad structure of $\mathrm{B} As$. Since $\mathrm{B} As$ is dual to $\Omega As^{¡}$, it follows that $(\mathrm{B} As)_n$ can be identified with the chain complex $C_\bullet(\mathscr{K}^{n-2})$.

On the geometric side, the boundary map of $C_\bullet(\mathscr{K}^{n-2}_{cub})$ is made up of two components d'_1 and d'_2. Let us focus on the top-cells of $C_\bullet(\mathscr{K}^{n-2}_{cub})$, which are encoded by planar binary trees with one circle. Its boundary is made up of two kinds of elements: those which have only one circle (and one edge less), and those which have two circles (and the same number of edges). The elements of the first type account for d_1 and the elements of the second type account for d_2. Under the identification of linear generators we verify that $d_1 = d'_1$ and $d_2 = d'_2$. An example is given in Fig. 9.3. □

Proposition 9.3.2. *There is a sequence of dgns operad morphisms*

$$As_\infty = \Omega As^{¡} \xrightarrow{\sim} \Omega \mathrm{B} As \xrightarrow{\sim} As$$

which are quasi-isomorphisms. In arity n they are given by the quasi-isomorphisms

$$C_\bullet(\mathscr{K}^{n-2}) \xrightarrow{\sim} C_\bullet(\mathscr{K}^{n-2}_{cub}) \xrightarrow{\sim} \mathbb{K}\mu_n$$

where the first one is given by the cellular homeomorphism $\mathscr{K}^{n-2} \to \mathscr{K}^{n-2}_{cub}$ and the second one is the augmentation map.

Proof. The augmentation map $\Omega As^{¡} \to As$ has a dual which is $As^{¡} \to \mathrm{B} As$. Taking the cobar construction gives $\Omega As^{¡} \to \Omega \mathrm{B} As$. Under the identification with the chain complex of the two cellular decompositions of the associahedron, we get, in arity $n+2$, a chain map $C_\bullet(\mathscr{K}^n) \to C_\bullet(\mathscr{K}^n_{cub})$. It is given by the identification of $\mathscr{K}^n$ with itself, i.e. the big cell of $\mathscr{K}^n$ is sent to the sum of the top-cells $\mathscr{K}^n_{cub}$.

The augmentation map $\Omega As^{\text{i}} \to As$ is obtained by taking the homology. Since the associahedron is a convex polytope, it is homeomorphic to a ball and so its homology is the homology of a point. In degree 0, each vertex (i.e. each planar binary tree) is sent to μ_n, the generator of As_n. □

9.4 Homotopy Transfer Theorem for the Operad *As*

The notion of associative algebra is not stable under homotopy equivalence in general. Indeed, if V is a chain complex homotopy equivalent to a dga algebra A, the product induced on V is not necessarily associative. However the algebra structure on A can be transferred to an A_∞-algebra structure on V. We will see that, more generally, an A_∞-algebra structure on A can be transferred to an A_∞-algebra structure on V. So the category of A_∞-algebras is stable under homotopy equivalence. An interesting particular case is when V is the homotopy of A, that is $V = H_\bullet(A, d)$. Then, we obtain some new operations on $H(A)$ which generalize the higher Massey products.

9.4.1 Transferring the Algebra Structure

Let (A, d_A) be a dga algebra. We suppose that (V, d_V) is a *homotopy retract* of the chain complex (A, d_A):

$$h \circlearrowright (A, d_A) \underset{i}{\overset{p}{\rightleftarrows}} (V, d_V),$$

$$\mathrm{Id}_A - ip = d_A h + h d_A,$$

the map $i : V \to A$ being a quasi-isomorphism. From this hypothesis it follows immediately that the homology of (V, d_V) is the same as the homology of (A, d_A) and so $H_\bullet(V, d_V)$ is a graded associative algebra. It is natural to ask oneself what kind of algebraic structure there exists on V which implies that its homology is a graded associative algebra. One can define a binary operation $m_2 : V \otimes V \to V$ by the formula

$$m_2(u, v) := p\mu\big(i(u), i(v)\big)$$

where μ stands for the product in A:

$m_2 =$ 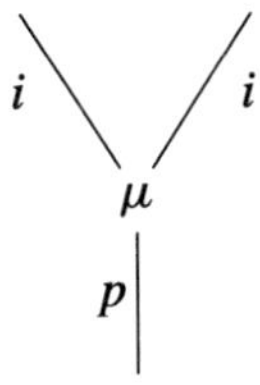

It is straightforward to see that there is no reason for m_2 to be associative. However the obstruction to associativity is measured as follows.

Lemma 9.4.1. *The ternary operation m_3 on V defined by the formula*

$$m_3(u,v,w) := -p\mu\big(i(u), h\mu\big(i(v), i(w)\big)\big) + p\mu\big(h\mu\big(i(u), i(v)\big), i(w)\big),$$

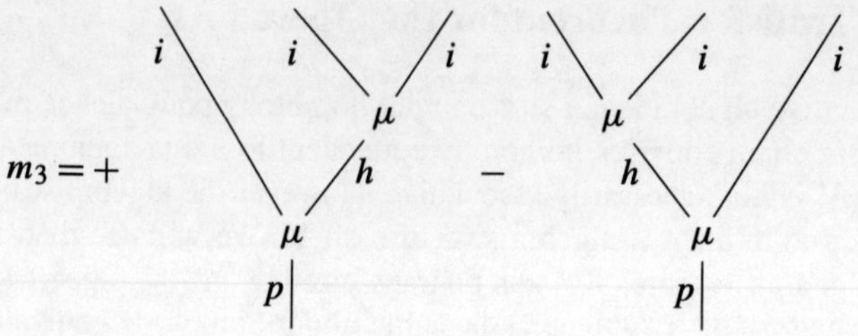

satisfies the relation

$$\partial(m_3) = -m_2 \circ (m_2, \mathrm{id}) + m_2 \circ (\mathrm{id}, m_2)$$

(as already mentioned $\partial(m_3) := d_V m_3 + m_3 d_{V^{\otimes 3}}$).

Proof. In order to ease the computation, we write the proof in the case of a deformation retract. So we think of V as a subspace of A (whence suppressing the notation i), so p becomes an idempotent of A satisfying $dh = \mathrm{id} - p - hd$. Let us compute $\partial(p\mu(h\mu(i,i),i))$, that is $\partial(p\mu(h\mu, \mathrm{id}))$ under our convention:

$$\begin{aligned}
&\partial\big(p\mu(h\mu, \mathrm{id})\big) \\
&\quad = dp\mu(h\mu, \mathrm{id}) + \big[\big(p\mu\big(h\mu(d, \mathrm{id}), \mathrm{id}\big)\big) + \big(p\mu\big(h\mu(\mathrm{id}, d)\big), \mathrm{id}\big) + \big(p\mu(h\mu, d)\big)\big] \\
&\quad = p\mu(dh\mu, \mathrm{id}) - p\mu(h\mu, d) + [\cdots] \\
&\quad = p\mu(\mu, \mathrm{id}) - p\mu(p\mu, \mathrm{id}) \\
&\qquad - \big(p\mu\big(h\mu(d, \mathrm{id}), \mathrm{id}\big)\big) - \big(p\mu\big(h\mu(\mathrm{id}, d)\big), \mathrm{id}\big) - p\mu(h\mu, d) + [\cdots] \\
&\quad = p\mu(\mu, \mathrm{id}) - m_2(m_2, \mathrm{id}).
\end{aligned}$$

Similarly one gets

$$\partial\big(p\mu(\mathrm{id}, h\mu)\big) = p\big(\mu(\mathrm{id}, \mu)\big) - m_2(\mathrm{id}, m_2).$$

Since μ is associative the terms $p\mu(\mu, \mathrm{id})$ and $p\mu(\mathrm{id}, \mu)$ are equal, and we get the expected formula. □

9.4.2 Geometric Interpretation of m_3

Let us consider the cubical decomposition of the interval, whose cells are labeled by the "circled trees", cf. Appendix C.2.2.

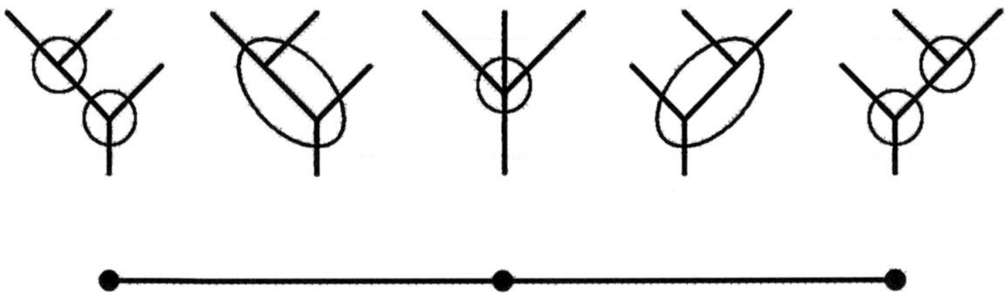

The two summands of m_3 correspond to the two 1-cells of this cubical decomposition and the formula $\partial(m_3) = m_2 \circ (\mathrm{id}, m_2) - m_2 \circ (m_2, \mathrm{id})$ is simply the computation of this boundary.

9.4.3 Higher Structure on the Homotopy Retract

Lemma 9.4.1 suggests that (V, d) inherits an A_∞-structure from the associative structure of (A, d). The geometric interpretation shows us the route to construct m_n explicitly: use the cubical decomposition of the Stasheff polytope:

$$m_n := \sum_{t \in PBT_{n+2}} \pm m_t,$$

where, for any planar binary tree (pb tree) t, the n-ary operation m_t is obtained by putting i on the leaves, μ on the vertices, h on the internal edges and p on the root, as in the case

$$m_3 = -m_{\curlyvee\!\curlyvee} + m_{\curlyvee\!\curlyvee}.$$

Theorem 9.4.2 (T. Kadeishvili [Kad80]). *Let*

$$h \circlearrowright (A, d_A) \underset{i}{\overset{p}{\rightleftarrows}} (V, d_V),$$

$$\mathrm{Id}_A - ip = d_A h + h d_A,$$

be a retract. If (A, d_A) is a dga algebra, then (V, d_V) inherits an A_∞-algebra structure $\{m_n\}_{n \geq 2}$, which extends functorially the binary operation of A.

Proof. The proof is analogous to the proof of Lemma 9.4.1. It is done by induction on the size of the trees. □

This statement is a particular case of Theorem 9.4.7 which gives an even more precise result.

Corollary 9.4.3. *Let*

$$h \circlearrowright (A, d_A) \underset{i}{\overset{p}{\rightleftarrows}} (V, d_V),$$

$$i = \textit{quasi-isomorphism}, \qquad \mathrm{Id}_A - ip = d_A h + h d_A,$$

be a homotopy retract (e.g. deformation retract). The homotopy class of the dga algebra (A, d) (considered as an A_∞-algebra) is equal to the homotopy class of the A_∞-algebra (V, d).

Lemma 9.4.4. *Let $\mathbb{K}$ be a field. Under a choice of sections any chain complex admits its homology as a deformation retract.*

Proof. Since we are working over a field, we can choose sections in the chain complex (A, d) so that $A_n \cong B_n \oplus H_n \oplus B_{n-1}$ where H_n is the homology and B_n the space of boundaries in degree n. The boundary map is 0 on $B_n \oplus H_n$ and identifies B_{n-1} with its copy in A_{n-1}. The homotopy h is 0 on $H_n \oplus B_{n-1}$ and identifies B_n with its copy in A_{n+1}. These choices make $(H_\bullet(A), 0)$ a deformation retract of (A, d):

	Id	$-ip$	dh	hd
B_n	Id	0	Id	0
H_n	Id	$-$Id	0	0
B_{n-1}	Id	0	0	Id

□

Corollary 9.4.5 (T. Kadeishvili [Kad80]). *For any dga algebra (A, d), there is an A_∞-algebra structure on $H_\bullet(A, d)$, with 0 differential, such that these two A_∞-algebras are homotopy equivalent.*

Proof. We apply Theorem 9.4.2 to the deformation retract constructed in Lemma 9.4.4. The homotopy equivalence of these two A_∞-algebras follows from the existence of an ∞-quasi-isomorphism. □

9.4.4 MacLane Invariant of a Crossed Module

Though it was constructed decades before m_3, the MacLane invariant of a crossed module can be interpreted as a nonlinear variation of m_3. Let us recall the framework. A *crossed module* is a group homomorphism $\mu : M \to N$ together with an action of N on M, denoted ${}^n m$, such that the following relations hold

$$\text{a)} \quad \mu({}^n m) = n\mu(m)n^{-1},$$

$$\text{b)} \quad {}^{\mu(m)}m' = mm'm^{-1}.$$

Let $Q := \operatorname{Coker} \mu$ and $L := \operatorname{Ker} \mu$. From the axioms it is easily seen that L is abelian and equipped with a Q-module structure. In [ML95, Chap. IV, Sect. 8], MacLane constructed an element $\alpha \in H^3(Q, L)$ as follows. Let P be the image of μ. We choose set-theoretic sections i and j such that $i(1) = 1$, $j(1) = 1$:

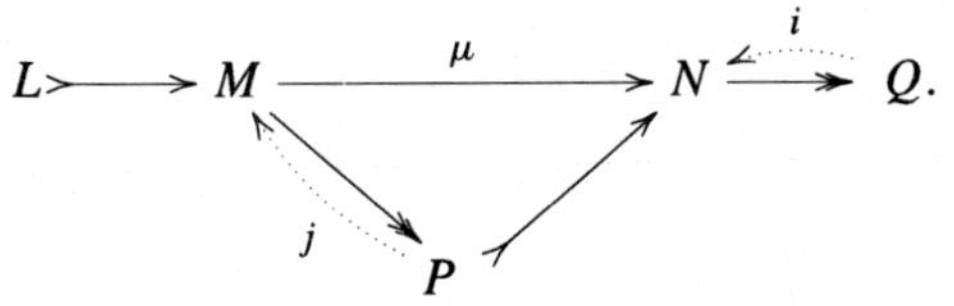

They permit us to construct set bijections: $N \cong P \times Q$ and $M \cong L \times P$. Hence, viewing the crossed module as a nonabelian chain complex (so μ plays the role of the differential) its homology $L \xrightarrow{0} Q$ can be seen as a (nonlinear) deformation retract:

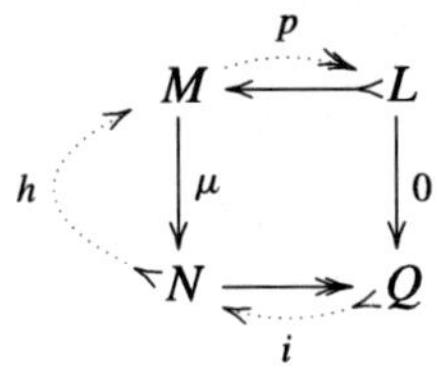

where p and h are the following composites:

$$p : M \cong L \times P \twoheadrightarrow L, \qquad h : N \cong P \times Q \twoheadrightarrow P \xrightarrow{j} M.$$

For $u, v \in Q$ we define

$$\varphi(u, v) := h\big(i(u)i(v)\big) = j\big(i(u)i(v)i(uv)^{-1}\big) \in M,$$

so that $i(u)i(v) = \mu(\varphi(u, v))i(uv)$. We compute as in the linear case:

$$\big(i(u)i(v)\big)i(w) = \mu\big(\varphi(u, v)\big)\mu\big(\varphi(uv, w)\big)i(uvw),$$
$$i(u)\big(i(v)i(w)\big) = \mu\big({}^u\varphi(v, w)\big)\mu\big(\varphi(u, vw)\big)i(uvw).$$

Comparing these two equalities, it follows that there exists a unique element $m_3(u, v, w) \in L = \operatorname{Ker} \mu$ such that

$${}^u\varphi(v, w)\varphi(u, vw) = m_3(u, v, w)\varphi(u, v)\varphi(uv, w) \in M.$$

MacLane showed that this element is a 3-cocycle and that its cohomology class in $H^3(Q, L)$ does not depend on the choice of the sections i and j. Moreover any morphism of crossed modules inducing an isomorphism on the kernel and the cokernel gives rise to the same invariant.

The topological interpretation is the following, cf. [Lod82]. The crossed module $\mu : M \to N$ defines a simplicial group whose classifying space is a topological space with only π_1, equal to Q, and π_2, equal to L. So its homotopy type is completely determined by the Postnikov invariant which is an element of $H^3(BQ, L) = H^3(Q, L)$. It is precisely the MacLane invariant. So, as in the linear case, the triple (Q, L, m_3) completely determines the homotopy type of the crossed module.

9.4.5 Massey Product

We know that the homology of a dga algebra (A, d) is a graded associative algebra $H_\bullet(A)$. Corollary 9.4.5 tells us that we have more structure: for any $n \geq 3$ there is an n-ary operation m_n which is nontrivial in general. They are called *Massey products* because they generalize the classical Massey products constructed in algebraic topology (cf. [Mas58, Kra66, May69]). Let X be a connected topological space, and let $C^\bullet_{\mathrm{sing}}(X)$ be the singular cochain complex with homology $H^\bullet_{\mathrm{sing}}(X)$. The product structure is given by the cup product of cochains. Then the *triple Massey product* $\langle x, y, z\rangle$ is classically defined for homology classes x, y, z such that $x \cup y = 0 = y \cup z$ as follows. Let us still denote by x, y, z the cycles representing the homology classes. Because of the hypothesis there exist chains a and b such that $(-1)^{|x|} x \cup y = da$, $(-1)^{|y|} y \cup z = db$. Then the chain

$$\langle x, y, z\rangle := (-1)^{|x|} x \cup b + (-1)^{|a|} a \cup z$$

is a cycle. Its homology class is well-defined in the quotient $H/(xH + Hz)$ and called the *triple Massey product* of x, y, z.

The prototypical example of a nonzero triple Massey product is given by the Borromean rings. We consider the complement (in the 3-sphere) of the Borromean rings:

Each "ring" (i.e. solid torus) determines a 1-cocycle: take a loop from the base-point with linking number one with the circle. Since any two of the three circles are disjoint, the cup product of the associated cohomology classes is 0. The nontriviality of the triple Massey product of these three cocycles detects the entanglement of the three circles (cf. [Mas58, Sta97b]).

Lemma 9.4.6. *For any cohomology classes x, y, z such that $x \cup y = 0 = y \cup z$ the triple Massey product is given by the operation m_3:*

$$\langle x, y, z\rangle = -(-1)^{|x|+|y|} m_3(x, y, z).$$

Proof. Let (A, d) be a dga algebra. We denote by $\cup$ its product (formerly denoted by μ) as well as the product induced on homology (formerly denoted by m_2). We use the homotopy equivalence data described in the proof of Lemma 9.4.5. If x, y, z are cycles and a, b chains such that $(-1)^{|x|} x \cup y = d(a)$, $(-1)^{|y|} y \cup z = d(b)$, then we can choose a and b such that $hd(a) = a$, $hd(b) = b$. Therefore one has

$$\begin{aligned}
-m_3(x, y, z) &= p\mu\big(i(x), h\mu\big(i(y), i(z)\big)\big) - p\mu\big(h\mu\big(i(x), i(y)\big), i(z)\big)\\
&= x \cup h(y \cup z) - h(x \cup y) \cup z\\
&= (-1)^{|y|} x \cup hd(b) - (-1)^{|x|} hd(a) \cup z\\
&= (-1)^{|y|} x \cup b - (-1)^{|x|} a \cup z\\
&= (-1)^{|x|+|y|} \langle x, y, z\rangle.
\end{aligned}$$

□

9.4.6 A Quadruple Massey Product

Let x, y, z, t be cycles in the dga algebra (A, d) such that there exist chains a, b, c satisfying

$$(-1)^{|x|} xy = d(a), \qquad (-1)^{|y|} yz = d(b), \qquad (-1)^{|z|} zt = d(c),$$

and such that there exist chains α, β, β', γ satisfying:

$$\begin{aligned}
(-1)^{|a|} az &= d(\alpha), \qquad & (-1)^{|x|} xb &= d(\beta),\\
(-1)^{|b|} bt &= d\big(\beta'\big), \qquad & (-1)^{|y|} yc &= d(\gamma).
\end{aligned}$$

One can check that the element

$$\langle x, y, z, t\rangle := (-1)^{|x|} \alpha t + (-1)^{|x|} \beta t + (-1)^{|x|} ac + (-1)^{|x|} x\beta' + (-1)^{|x|} x\gamma$$

is a cycle and defines a homology class in $H_\bullet(A, d)$. The five elements of this sum correspond to the five 0-cells of the cellular decomposition of the pentagon, cf. Fig. 9.2. In fact one can check that

$$\langle x, y, z, t\rangle = \pm m_4(x, y, z, t).$$

9.4.7 Homotopy Invariance of A_∞-Algebras

The homotopy transfer theorem for dga algebras can be generalized into a homotopy transfer theorem for A_∞-algebras.

Theorem 9.4.7. [Kad82] *Let*

$$h \circlearrowright (A, d_A) \underset{i}{\overset{p}{\rightleftarrows}} (V, d_V),$$

$$i = \textit{quasi-isomorphism}, \qquad \mathrm{Id}_A - ip = d_A h + h d_A,$$

be a homotopy retract. If (A, d_A) is an A_∞-algebra, then (V, d_V) inherits a A_∞-algebra structure $\{m_n\}_{n\geq 2}$ such that the quasi-isomorphism i extends to an ∞-quasi-isomorphism.

Proof. In the proof of Lemma 9.4.1 we used the fact that the binary product $\mu = \mu_2$ on A is associative. Suppose now that it is only associative up to homotopy, that is, there exists a ternary operation μ_3 on A such that

$$\partial(\mu_3) = -\mu_2 \circ (\mathrm{id}, \mu_2) + \mu_2 \circ (\mu_2, \mathrm{id}).$$

Then one needs to modify the ternary operation m_3 on V by adding the extra term $p\mu_3 i$:

$$m_3 := -p\mu_2\big(i, h\mu_2(i, i)\big) + p\mu_2\big(h\mu_2(i, i), i\big) + p\mu_3(i, i, i).$$

After this modification we get the formula

$$\partial(m_3) = -m_2 \circ (m_2, \mathrm{id}) + m_2 \circ (\mathrm{id}, m_2)$$

as in Lemma 9.4.1. Observe that the term which has been added corresponds to the corolla of the figure in Sect. 9.4.2.

Similarly the higher order operations m_n are defined by using not only the binary trees, but all the planar trees, with vertices indexed by the operations μ_n given by the A_∞-algebra structure of A. The proof is done by induction on the size of the trees, see for instance [KS00] by Kontsevich and Soibelman or [Mer99] by Merkulov. It uses the fact that the homotopy retract determines a morphism of dg cooperads $\mathrm{B}\,\mathrm{End}_A \to \mathrm{B}\,\mathrm{End}_V$. The A_∞-structure of A is encoded by a morphism of dg cooperads $As^{¡} \to \mathrm{B}\,\mathrm{End}_A$. By composition we get the expected A_∞-structure on V. □

This result is a particular case of a more general statement valid for any Koszul operad. Its complete proof is given in Theorem 10.3.1.

9.4.8 Variations on the Homotopy Transfer Theorem

There are various proofs and several generalizations of the Homotopy Transfer Theorem. The proof given here follows the method of Kontsevich and Soibelman [KS00], see also [Mer99] by Sergei Merkulov. Another method, closer to the original proof of Kadeishvili, consists in applying the Perturbation Lemma, see [HK91].

In [Mar06], Martin Markl showed that p can also be extended to an ∞-quasi-isomorphism, and h to an ∞-homotopy.

Table 9.1 The binary operation m_2 on (α, β)

$\alpha\backslash\beta$	a*	b*	c*	d*	e*	u^*	v^*	w^*	x^*	y^*	Z^*
a*	a*					u^*	v^*				Z^*
b*		b*						w^*			
c*			c*						x^*		
d*				d*						y^*	
e*					e*						
u^*		u^*						Z^*		Z^*	
v^*			v^*						$-Z^*$		
w^*				w^*						Z^*	
x^*					x^*						
y^*					y^*						
Z^*					Z^*						

9.5 An Example of an A_∞-Algebra with Nonvanishing m_3

Let us consider the cochain complex of the Stasheff polytope $\mathscr{K}^2$ (pentagon). We denote by $\mathtt{a}^*, \dots, u^*, \dots, Z^*$ the cochains which are linear dual of the cells $\mathtt{a}, \dots, u, \dots, Z$ of $\mathscr{K}^2$:

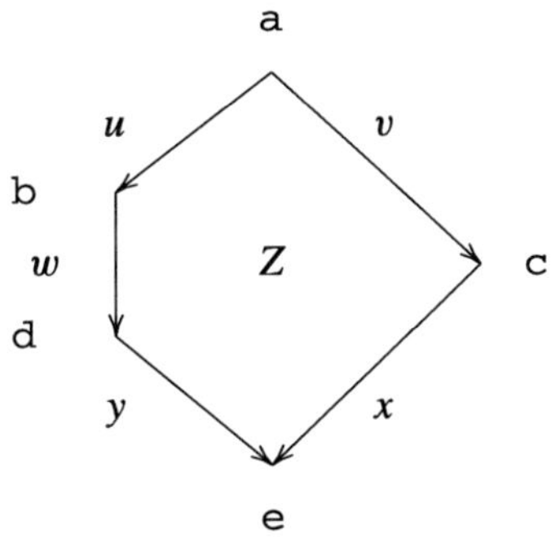

We make it into an A_∞-algebra as follows. First, we put $m_n = 0$ for any $n \geq 4$. Second, m_3 is zero except on the triple of 1-cochains (u^*, w^*, y^*) where

$$m_3\big(u^*, w^*, y^*\big) = Z^*.$$

Third, the binary operation m_2 on (α, β) is given by Table 9.1.

In Table 9.1, empty space means 0. This defines a cohomologically graded A_∞-algebra: $|\partial| = +1$ and $|m_3| = +1$. We leave it to the reader to verify the relations, that is

$$\partial(m_3) = m_2 \circ (m_2, \mathrm{id}) - m_2 \circ (\mathrm{id}, m_2)$$

and, since $\partial(m_4) = 0$,

$$m_2(m_3, \mathrm{id}) - m_3(m_2, \mathrm{id}, \mathrm{id}) + m_3(\mathrm{id}, m_2, \mathrm{id}) - m_3(\mathrm{id}, \mathrm{id}, m_2) + m_2(\mathrm{id}, m_3) = 0.$$

For the first relation, it suffices to evaluate the two sides on (u^*, w^*, y^*). For the second relation there are six cases to consider (one per face of the cube).

The topological interpretation of these formulas is better seen on the dual statement, that is $C_\bullet(\mathscr{K}^2)$ is an A_∞-coalgebra. The coproducts Δ^2 and Δ^3 applied to the 2-cell Z are shown in the following pictures:

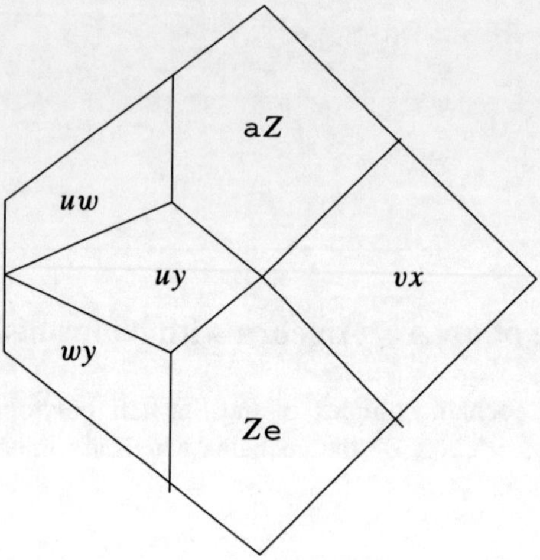

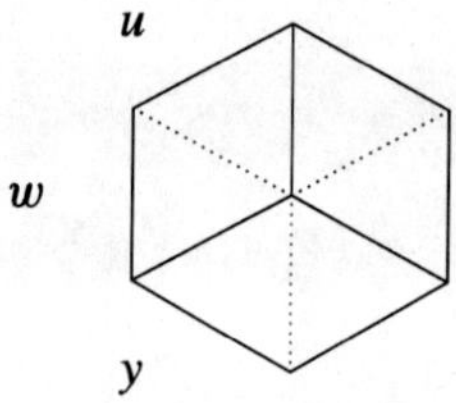

The identity

$$(\Delta^3, \mathrm{id})\Delta^2 - (\Delta^2, \mathrm{id}, \mathrm{id})\Delta^3 + (\mathrm{id}, \Delta^2, \mathrm{id})\Delta^3 - (\mathrm{id}, \mathrm{id}, \Delta^2)\Delta^3 + (\mathrm{id}, \Delta^3)\Delta^2 = 0$$

which is equivalent to

$$(\Delta^2, \mathrm{id}, \mathrm{id})\Delta^3 + (\mathrm{id}, \mathrm{id}, \Delta^2)\Delta^3 = (\Delta^3, \mathrm{id})\Delta^2 + (\mathrm{id}, \Delta^2, \mathrm{id})\Delta^3 + (\mathrm{id}, \Delta^3)\Delta^2$$

amounts to the identification of the following unions of cells:

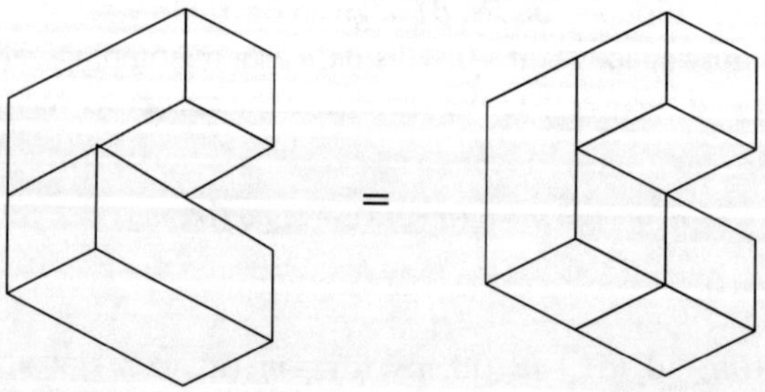

More generally, the cochain complex $C^\bullet(\mathscr{K}^n)$ can be shown to be an A_∞-algebra for any n, cf. [Lod12].

9.6 Résumé

9.6.1 The Operad *Ass*

Symmetric operad Ass encoding associative algebras: $Ass(n) = \mathbb{K}[\mathbb{S}_n]$.
Nonsymmetric operad As encoding associative algebras: $As_n = \mathbb{K}\mu_n$.
Composition rule: $\mu_m \circ_i \mu_n = \mu_{m-1+n}$.

9.6.2 Homotopy Associative Algebras

A an A_∞-algebra: $(T^c(sA), m^A) =$ dg coalgebra, equivalently, the chain complex (A, d) is equipped with an n-ary operation $m_n : A^{\otimes n} \to A, n \geq 2$, satisfying the relations:

$$\partial(m_n) = \sum (-1)^{p+qr} m_k \circ \left(\mathrm{id}^{\otimes p} \otimes m_q \otimes \mathrm{id}^{\otimes r}\right).$$

Dgns operad A_∞: $A_\infty = \Omega As^{\text{i}}$, $(A_\infty)_n = C_\bullet(\mathscr{K}^{n-2})$.

9.6.3 Cobar–Bar Construction

$$\left(\Omega \mathrm{B} As^{\text{i}}\right)_n = C_\bullet\left(\mathscr{K}_{cub}^{n-2}\right).$$

9.6.4 Homotopy Transfer Theorem

If (V, d) is a homotopy retract (e.g. deformation retract) of (A, d), then any A_∞-algebra structure on (A, d) can be transferred through explicit formulas to an A_∞-algebra structure on (V, d), so that they represent the same homotopy class. Whence the slogan

"A_∞-algebras are stable under homotopy equivalence".

EXAMPLE. If (A, d) is a dga algebra, then $H_\bullet(A)$ is an A_∞-algebra with trivial differential. This structure induces the Massey products.

9.7 Exercises

Exercise 9.7.1 (Explicit homotopy). Construct the homotopy from Id to 0 for the Koszul complex $(As^{¡} \circ As, d_\kappa)$.

Exercise 9.7.2 (Acyclicity again). Consider the small chain complex of length one

$$C_\bullet : \qquad \cdots \to 0 \to C_1 = \mathbb{K} \xrightarrow{\text{id}} C_0 = \mathbb{K}.$$

Show that the arity n part of the Koszul complex of As is isomorphic to $(C_\bullet)^{\otimes n}$. Deduce another proof of the acyclicity of the Koszul complex of As. Compare with Exercise 3.8.7.

Exercise 9.7.3 (An explicit computation). Let As^* be the linear dual cooperad of As. Compute explicitly $\Delta(\mu_4^c)$, $\bar{\Delta}(\mu_4^c)$ and $\Delta_{(1)}(\mu_4^c)$.

Exercise 9.7.4 (Totally diassociative algebras [Zha12]). A *totally diassociative algebra* is defined by two binary operations $x * y$ and $x \cdot y$ such that any composition is associative:

$$\begin{aligned}(x * y) * z &= x * (y * z),\\ (x \cdot y) * z &= x \cdot (y * z),\\ (x * y) \cdot z &= x * (y \cdot z),\\ (x \cdot y) \cdot z &= x \cdot (y \cdot z).\end{aligned}$$

We denote by $As^{(2)}$ the ns operad encoding these algebras. Show that $As_n^{(2)}$ is of dimension 2^{n-1}. Show that $As^{(2)}$ is self-dual for Koszul duality and is Koszul (apply the rewriting method). Describe explicitly the chain complex $(As_\infty^{(2)})_n$ in terms of the associahedron. What is an $As_\infty^{(2)}$-algebra?

Exercise 9.7.5 (Higher Massey products ★). In [Kra66], David Kraines defines higher Massey products for families of cochains $\underline{a} = \{a(i,j)\}_{1 \le i \le j \le k}$, in a cochain complex, satisfying

$$d\big(a(i,j)\big) = \sum_{r=i}^{j-1} (-1)^{|a(i,r)|} a(i,r) a(r+1,j)$$

as

$$c(\underline{a}) := \sum_{r=1}^{k-1} (-1)^{|a(1,r)|} a(i,r) a(r+1,k).$$

Interpret this construction and its properties in terms of A_∞-algebras. Compare with [May69, BM03b].

Exercise 9.7.6 (Classifying space of a crossed module ★). Let $G.$ be a simplicial group whose homotopy groups are trivial except $\pi_0 G = Q$ and $\pi_n G = L$ for some fixed n. Show that one can construct an analog $m_n : G^n \to L$ of MacLane invariant, which is a cocycle and gives a well-defined element in $H^n(Q, L)$ (cohomology of the discrete group Q with values in the Q-module L). Show that it is the Postnikov invariant of the classifying space $B|G.|$.

Chapter 10
Homotopy Operadic Algebras

The theory of algebras up to homotopy defined by operad action is a subject whose time has come.

J.-P. May 1995, review of Hinich–Schechtman "Homotopy Lie algebra"

When a chain complex is equipped with some compatible algebraic structure, its homology inherits this algebraic structure. The purpose of this chapter is to show that there is some hidden algebraic structure behind the scene. More precisely if the chain complex contains a smaller chain complex, which is a deformation retract, then there is a finer algebraic structure on this small complex. Moreover, the small complex with this new algebraic structure is homotopy equivalent to the starting data.

The operadic framework enables us to state explicitly this transfer of structure result as follows. Let $\mathscr{P}$ be a quadratic operad. Let A be a chain complex equipped with a $\mathscr{P}$-algebra structure. Let V be a deformation retract of A. Then the $\mathscr{P}$-algebra structure of A can be transferred into a $\mathscr{P}_\infty$-algebra structure on V, where $\mathscr{P}_\infty = \Omega\mathscr{P}^{¡}$. If $\mathscr{P}$ is Koszul, then the two objects are homotopy equivalent. Over a field, the homology can be made into a deformation retract, whence the hidden algebraic structure on the homology.

In fact this result is a particular case of a more general result, called the Homotopy Transfer Theorem, which will be stated in full in this chapter. This HTT has a long history and, in a sense, it goes back to the discovery of spectral sequences by Jean Leray and Jean-Louis Koszul in the 1940s.

This chapter is organized as follows. In Sect. 10.1, we define the notion of *homotopy $\mathscr{P}$-algebra* as an algebra over the Koszul resolution $\mathscr{P}_\infty := \Omega\mathscr{P}^{¡}$. Using the operadic bar–cobar adjunction, we give three equivalent definitions. The definition in terms of twisting morphisms is treated in details, as well as the definition in terms of square-zero coderivation on the cofree $\mathscr{P}^{¡}$-coalgebra.

Using this last definition, we define the notion of *infinity morphism*, also denoted ∞-*morphism*, between homotopy $\mathscr{P}$-algebras. An ∞-morphism is not only a map

J.-L. Loday, B. Vallette, *Algebraic Operads*,
Grundlehren der mathematischen Wissenschaften 346,
DOI 10.1007/978-3-642-30362-3_10, © Springer-Verlag Berlin Heidelberg 2012

but is made up of a collection of maps parametrized by $\mathcal{P}^{¡}$. This notion is well suited to the homotopy theory of $\mathcal{P}_\infty$-algebras.

The aforementioned Homotopy Transfer Theorem is the subject of Sect. 10.3. It states precisely that any $\mathcal{P}_\infty$-algebra structure can be transferred through a homotopy retract to produce a homotopy equivalent $\mathcal{P}_\infty$-algebra structure. It is proved by explicit formulas.

In Sect. 10.4, we study the properties of ∞-morphisms. When the first component of an ∞-morphism is invertible (respectively is a quasi-isomorphism), then it is called an *∞-isomorphism* (respectively an *∞-quasi-isomorphism*). We prove that the class of ∞-isomorphisms is the class of invertible ∞-morphisms. Any $\mathcal{P}_\infty$-algebra is shown to be decomposable into the product of a *minimal $\mathcal{P}_\infty$-algebra* and an *acyclic trivial $\mathcal{P}_\infty$-algebra*. Using this result, we prove that being ∞-quasi-isomorphic is an equivalence relation, called the *homotopy equivalence*.

In Sect. 10.5, we study the same kind of generalization "up to homotopy" but for operads this time. We introduce the notions of *homotopy operad* and *infinity morphism*, or *∞-morphism*, of homotopy operads. One key ingredient in the HTT is actually an explicit ∞-morphism between endomorphism operads. The functor from operads to Lie algebras is extended to a functor between homotopy operads to homotopy Lie algebras. This allows us to show that the relations between associative algebras, operads, pre-Lie algebra, and Lie algebras extend to the homotopy setting. Finally, we study homotopy representations of operads.

Throughout this chapter, we apply the various results to A_∞-algebras, already treated independently in the previous chapter, and to L_∞-algebras. In this chapter, the generic operad $\mathcal{P}$ is a Koszul operad.

The general study of homotopy algebras using the Koszul resolution $\mathcal{P}_\infty = \Omega\mathcal{P}^{¡}$ of $\mathcal{P}$ goes back to Ginzburg and Kapranov in [GK94] and to Getzler and Jones in [GJ94]. Many particular cases have been treated in the literature; we refer the reader to the survey given in Part *I* of the book [MSS02] of Markl, Shnider and Stasheff.

In this chapter, we work over a ground field $\mathbb{K}$ of characteristic 0. Notice that all the constructions and some of the results hold true without this hypothesis.

10.1 Homotopy Algebras: Definitions

In this section, we introduce the notion of homotopy $\mathcal{P}$-algebra, i.e. $\mathcal{P}_\infty$-algebra, for a Koszul operad $\mathcal{P}$. We give four equivalent definitions. We treat in detail the examples of homotopy associative algebras, or A_∞-algebras, and homotopy Lie algebras, or L_∞-algebras.

10.1.1 $\mathcal{P}_\infty$-Algebras

A *homotopy $\mathcal{P}$-algebra* is an algebra over the Koszul resolution $\Omega\mathcal{P}^{¡}$ of $\mathcal{P}$. It is sometimes called a $\mathcal{P}$-algebra up to homotopy or strong homotopy $\mathcal{P}$-algebra in the literature. We also call it a *$\mathcal{P}_\infty$-algebra*, where $\mathcal{P}_\infty$ stands for the dg operad

$\Omega\mathcal{P}^{\text{!`}}$. Hence, a homotopy $\mathcal{P}$-algebra structure on a dg module $A$ is a morphism of dg operads $\mathcal{P}_\infty = \Omega\mathcal{P}^{\text{!`}} \to \operatorname{End}_A$. The set of homotopy $\mathcal{P}$-algebra structures on A is equal to $\operatorname{Hom}_{\mathsf{dg\ Op}}(\Omega\mathcal{P}^{\text{!`}}, \operatorname{End}_A)$.

Notice that a $\mathcal{P}$-algebra is a particular example of homotopy $\mathcal{P}$-algebra. It occurs when the structure morphism factors through $\mathcal{P}$:

$$\Omega\mathcal{P}^{\text{!`}} \xrightarrow{\sim} \mathcal{P} \to \operatorname{End}_A.$$

10.1.2 Interpretation in Terms of Twisting Morphism

Let us now make this notion explicit. We saw in Proposition 6.5.7 that a morphism of dg operads from the quasi-free operad $\Omega\mathcal{P}^{\text{!`}}$ to End_A is equivalent to a twisting morphism in the convolution algebra

$$\mathfrak{g} = \mathfrak{g}_{\mathcal{P},A} := \operatorname{Hom}_{\mathbb{S}}\big(\mathcal{P}^{\text{!`}}, \operatorname{End}_A\big).$$

Explicitly, we recall from Sect. 6.5.3 the following correspondence

$$\begin{array}{ccc} \operatorname{Hom}_{\mathsf{Op}}\big(\Omega\mathcal{P}^{\text{!`}}, \operatorname{End}_A\big) & \cong & \operatorname{Hom}_{\mathbb{S}}\big(\overline{\mathcal{P}}^{\text{!`}}, \operatorname{End}_A\big)_{-1} \\ \cup & & \cup \\ \operatorname{Hom}_{\mathsf{dg\ Op}}\big(\Omega\mathcal{P}^{\text{!`}}, \operatorname{End}_A\big) & \cong & \operatorname{Tw}\big(\mathcal{P}^{\text{!`}}, \operatorname{End}_A\big). \end{array}$$

As a direct consequence, we get the following description of homotopy algebra structures.

Proposition 10.1.1. *A homotopy $\mathcal{P}$-algebra structure on the dg module A is equivalent to a twisting morphism in* $\operatorname{Tw}(\mathcal{P}^{\text{!`}}, \operatorname{End}_A)$.

Let us make explicit the notion of twisting morphism $\operatorname{Tw}(\mathcal{P}^{\text{!`}}, \operatorname{End}_A)$. Suppose first that the operad $\mathcal{P}$ is homogeneous quadratic. The internal differential of $\mathcal{P}^{\text{!`}}$ is trivial. For any element $\varphi \in \operatorname{Hom}_{\mathbb{S}}(\mathcal{P}^{\text{!`}}, \operatorname{End}_A)$ and for any element $\mu^c \in \mathcal{P}^{\text{!`}}$, the Maurer–Cartan equation becomes $\partial_A(\varphi(\mu^c)) + (\varphi \star \varphi)(\mu^c) = 0$, where ∂_A stands for the differential of End_A induced by the differential of A. Using Sweedler type notation of Sect. 6.1.4, we denote by $\sum(\mu^c_{(1)} \circ_i \mu^c_{(2)})^\sigma$ the image of μ^c under the infinitesimal decomposition map $\Delta_{(1)} : \mathcal{P}^{\text{!`}} \to \mathcal{P}^{\text{!`}} \circ_{(1)} \mathcal{P}^{\text{!`}}$ of the cooperad $\mathcal{P}^{\text{!`}}$. If we denote by m the image of μ^c under φ, we get the following equation in End_A:

$$\sum \pm (m_{(1)} \circ_i m_{(2)})^\sigma = \partial_A(m).$$

This formula describes the general relations satisfied by the operations of a homotopy $\mathcal{P}$-algebra.

Proposition 10.1.2. *The convolution pre-Lie algebra $\mathfrak{g}$ is endowed with a weight grading such that $\mathfrak{g} \cong \prod_{n\geq 0} \mathfrak{g}^{(n)}$.*

Proof. The Koszul dual dg cooperad $\mathscr{P}^{¡}$ is weight graded, $\mathscr{P}^{¡} = \bigoplus_{n\geq 0} \mathscr{P}^{¡\,(n)}$. Therefore the convolution pre-Lie algebra $\mathfrak{g}$ is graded by

$$\mathfrak{g}^{(n)} := \mathrm{Hom}_{\mathbb{S}}\big(\mathscr{P}^{¡\,(n)}, \mathrm{End}_A\big)$$

and the direct sum on $\mathscr{P}$ gives the product $\mathfrak{g} \cong \prod_{n\geq 0} \mathfrak{g}^{(n)}$. □

Hence, any twisting morphism φ in $\mathfrak{g}$ decomposes into a series $\varphi = \varphi_1 + \cdots + \varphi_n + \cdots$ with $\varphi_n \in \mathfrak{g}^{(n)}_{-1}$, since $I = \mathscr{P}^{¡\,(0)}$ and $\varphi_0 = 0$. Under this notation, the Maurer–Cartan equation is equivalent to

$$-\sum_{\substack{k+l=n\\ k,l<n}} \varphi_k \star \varphi_l = \partial(\varphi_n),$$

for any $n \geq 1$.

Proposition 10.1.3. *The differential of the convolution dg pre-Lie algebra* $\mathfrak{g}$ *splits into two terms* $\partial = \partial_0 + \partial_1$, *where* $\partial_0 = \partial_A$ *preserves the weight grading and where* ∂_1 *raises it by* 1.

Proof. The term ∂_1 is equal to $\partial_1(\varphi) := -(-1)^{|\varphi|}\varphi(d_{\mathscr{P}^{¡}})$. Since $d_{\mathscr{P}^{¡}}$ lowers the weight grading of the Koszul dual dg cooperad $\mathscr{P}^{¡}$ by one, ∂_1 raises the weight grading of the convolution pre-Lie algebra by one. □

Under the weight grading decomposition, the Maurer–Cartan equation reads

$$-\partial_1(\varphi_{n-1}) - \sum_{\substack{k+l=n\\ k,l<n}} \varphi_k \star \varphi_l = \partial_A(\varphi_n),$$

so the left-hand side relation holds up to the homotopy φ_n in $\mathfrak{g}^{(n)}$.

10.1.3 The Example of $\mathscr{P}$-Algebras

The $\mathscr{P}$-algebras are characterized among the $\mathscr{P}_\infty$-algebras by the following particular solutions to the Maurer–Cartan equation.

Proposition 10.1.4. *A* $\mathscr{P}_\infty$*-algebra is a* $\mathscr{P}$*-algebra if and only if its twisting morphism is concentrated in weight* 1.

Proof. Let $\mathscr{P} = \mathscr{P}(E, R)$ be a quadratic operad. A $\mathscr{P}$-algebra A is a $\mathscr{P}_\infty$-algebra whose structure map factors through $\mathscr{P}$. The map $\Omega\mathscr{P}^{¡} \twoheadrightarrow \mathscr{P}$ sends the elements of $\mathscr{P}^{¡\,(n)}$ to 0 for $n \geq 2$. So the nontrivial part under this morphism is the image of

$\mathscr{P}^{¡(1)} \mapsto \mathscr{P}^{(1)} = E$, that is the generating operations of $\mathscr{P}$. In this case, the only nontrivial components of the Maurer–Cartan equation are for $\mu^c \in \mathscr{P}^{¡(1)} \cong E$ and for $\mu^c \in \mathscr{P}^{¡(2)} \cong R$. The first one is equivalent, for the internal differential of A, to be a derivation with respect to the operations of E, and the second one is equivalent for these operations to satisfy the relations of R. □

The following proposition gives a first result on the algebraic structure of the homotopy $H(A)$ of a $\mathscr{P}_\infty$-algebra A.

Proposition 10.1.5. *The homotopy of a $\mathscr{P}_\infty$-algebra A, that is the homology $H(A)$ of the underlying chain complex, has a natural $\mathscr{P}$-algebra structure.*

Proof. Let A be a $\mathscr{P}_\infty$-algebra with structure map $\varphi \in \mathrm{Tw}(\mathscr{P}^{¡}, \mathrm{End}_A)$. The image under φ of any element in $\mathscr{P}^{¡(1)}$ gives operations in End_A for which d is a derivation. Therefore these operations are stable on homology. Since the differential on $H(A)$ is null, they define a $\mathscr{P}$-algebra structure on $H(A)$. □

Considering only the $\mathscr{P}$-algebra structure on $H(A)$, we are losing a lot of data. We will see in Sect. 10.3 that we can transfer a full structure of $\mathscr{P}_\infty$-algebra on $H(A)$, which faithfully contains the homotopy type of A.

10.1.4 $\mathscr{P}_{(n)}$-Algebras

The preceding section motivates the following definition. A *$\mathscr{P}_{(n)}$-algebra* A is a homotopy $\mathscr{P}$-algebra such that the structure map $\varphi : \mathscr{P}^{¡} \to \mathrm{End}_A$ vanishes on $\mathscr{P}^{¡(k)}$ for $k > n$. It is equivalent to a truncated solution of the Maurer–Cartan equation in the convolution algebra $\mathfrak{g}$. Under this definition a $\mathscr{P}$-algebra is a $\mathscr{P}_{(1)}$-algebra.

10.1.5 Example: Homotopy Associative Algebras, Alias A_∞-Algebras

We pursue the study of homotopy associative algebras, started in Sec. 9.2, but in terms of twisting morphism this time.

Consider the nonsymmetric Koszul operad As, see Sect. 9.1.5. We proved in Sect. 9.2 that an algebra over $\Omega As^{¡}$, i.e. a *homotopy associative algebra* or *A_∞-algebra*, is a chain complex (A, d_A) equipped with maps $m_n : A^{\otimes n} \to A$ of degree $n-2$, for any $n \geq 2$, which satisfy

$$\sum_{\substack{p+q+r=n\\ k=p+r+1>1,\, q>1}} (-1)^{p+qr+1} m_k \circ_{p+1} m_q = \partial_A(m_n) = d_A \circ m_n - (-1)^{n-2} m_n \circ d_{A^{\otimes n}}.$$

Notice that an associative algebra is an A_∞-algebra such that the higher homotopies m_n vanish for $n \geq 3$.

The Koszul dual nonsymmetric cooperad of As is one-dimensional in each arity $As^{¡}_n = \mathbb{K}\mu^c_n$, where the degree of μ^c_n is $n-1$. The image under the infinitesimal decomposition map of μ^c_n is

$$\Delta_{(1)}\big(\mu^c_n\big) = \sum_{\substack{p+q+r=n\\ k=p+r+1>1,\, q>1}} (-1)^{r(q+1)}\big(\mu^c_k; \underbrace{\mathrm{id}, \ldots, \mathrm{id}}_{p}, \mu^c_q, \underbrace{\mathrm{id}, \ldots, \mathrm{id}}_{r}\big).$$

Since the operad As is a nonsymmetric operad, the convolution algebra is given by $\mathrm{Hom}(As^{¡}, \mathrm{End}_A)$, without the action of the symmetric groups. It is isomorphic to the following dg module

$$\prod_{n\geq 1} \mathrm{Hom}\big(As^{¡}, \mathrm{End}_A\big)(n) \cong \prod_{n\geq 1} \mathrm{Hom}\big((sA)^{\otimes n}, sA\big) \cong \prod_{n\geq 1} s^{-n+1}\,\mathrm{Hom}\big(A^{\otimes n}, A\big).$$

The right-hand side is the direct product of the components of the nonsymmetric operad End_{sA}. Therefore, it is endowed with the pre-Lie operation of Sect. 5.9.15. For an element $f \in \mathrm{Hom}(A^{\otimes n}, A)$ and an element $g \in \mathrm{Hom}(A^{\otimes m}, A)$, it is explicitly given by

$$f \star g := \sum_{i=1}^{n} (-1)^{(i-1)(m+1)+(n+1)|g|} f \circ_i g.$$

This particular dg pre-Lie algebra was constructed by Murray Gerstenhaber in [Ger63].

Proposition 10.1.6. *The convolution dg pre-Lie algebra* $\mathfrak{g}_{As,A} = \mathrm{Hom}(As^{¡}, \mathrm{End}_A)$ *is isomorphic to the dg pre-Lie algebra* $(\prod_{n\geq 1} s^{-n+1}\,\mathrm{Hom}(A^{\otimes n}, A), \star)$, *described above.*

Proof. We denote by $\tilde{f} \in \mathrm{Hom}(As^{¡}, \mathrm{End}_A)(n)$ and by $\tilde{g} \in \mathrm{Hom}(As^{¡}, \mathrm{End}_A)(m)$ the maps which send μ^c_n to f and μ^c_m to g. Then the only nonvanishing component of the pre-Lie product $\tilde{f} \star \tilde{g}$ in the convolution algebra $\mathrm{Hom}(As^{¡}, \mathrm{End}_A)$ is equal to the composite

$$\mu^c_{n+m-1} \mapsto \sum_{i=1}^{n} (-1)^{(n-i)(m+1)}\big(\mu^c_n; \mathrm{id}, \ldots, \mathrm{id}, \underbrace{\mu^c_m}_{i\text{th place}}, \mathrm{id}, \ldots, \mathrm{id}\big)$$

$$\mapsto \sum_{i=1}^{n} (-1)^{(n-i)(m+1)+(n-1)(|g|-m+1)} f \circ_i g = f \star g.$$ □

Under this explicit description of the convolution pre-Lie algebra $\mathrm{Hom}(As^{¡}, \mathrm{End}_A)$, one can see that a twisting morphism in $\mathrm{Tw}(As^{¡}, \mathrm{End}_A)$ is exactly an A_∞-algebra structure on the dg module A.

10.1.6 Example: Homotopy Lie Algebras, Alias L_∞-Algebras

Applying Definition 10.1.1 to the operad $\mathcal{P} = Lie$, a *homotopy Lie algebra* is an algebra over the Koszul resolution $\Omega Lie^{¡} \xrightarrow{\sim} Lie$ of the operad *Lie*. It is also called an *L_∞-algebra*, or *strong homotopy Lie algebra* in the literature.

Recall that an n-multilinear map f is called *skew-symmetric* if it satisfies the condition $f = \operatorname{sgn}(\sigma) f^\sigma$ for any $\sigma \in \mathbb{S}_n$.

Proposition 10.1.7. *An L_∞-algebra structure on a dg module (A, d_A) is a family of skew-symmetric maps $\ell_n : A^{\otimes n} \to A$ of degree $|\ell_n| = n-2$, for all $n \geq 2$, which satisfy the relations*

$$\sum_{\substack{p+q=n+1\\ p,q>1}} \sum_{\sigma \in Sh_{p,q}^{-1}} \operatorname{sgn}(\sigma)(-1)^{(p-1)q}(\ell_p \circ_1 \ell_q)^\sigma = \partial_A(\ell_n),$$

where ∂_A is the differential in End_A *induced by d_A.*

Proof. Recall that there is a morphism of operads $Lie \to Ass$ defined by $[\ ,\ ] \mapsto \mu - \mu^{(12)}$. Its image under the bar construction functor induces a morphism of dg cooperads $\mathrm{B}Lie \to \mathrm{B}Ass$. By Proposition 7.3.1, the morphism between the syzygy degree 0 cohomology groups of the bar constructions gives a morphism between the Koszul dual cooperads $Lie^{¡} \to Ass^{¡}$. If we denote the elements of these two cooperads by $Lie^{¡}(n) \cong \mathbb{K}\ell_n^c \otimes \operatorname{sgn}_{\mathbb{S}_n}$ and by $Ass^{¡}(n) \cong \mathbb{K}\mu_n^c \otimes \mathbb{K}[\mathbb{S}_n]$ with $|\ell_n^c| = |\mu_n^c| = n-1$, this map is explicitly given by $\ell_n^c \mapsto \sum_{\sigma \in \mathbb{S}_n} \operatorname{sgn}(\sigma)(\mu_n^c)^\sigma$. Hence, the formula for the infinitesimal decomposition map of the cooperad $Ass^{¡}$ induces

$$\Delta_{(1)}(\ell_n^c) = \sum_{\substack{p+q=n+1\\ p,q>1}} \sum_{\sigma \in Sh_{p,q}^{-1}} \operatorname{sgn}(\sigma)(-1)^{(p+1)(q+1)}\left(\ell_p^c \circ_1 \ell_q^c\right)^\sigma,$$

since the (p,q)-unshuffles split the surjection $\mathbb{S}_{p+q} \twoheadrightarrow (\mathbb{S}_p \times \mathbb{S}_q)\backslash \mathbb{S}_{p+q}$, cf. Sect. 1.3.2. Let us denote by ℓ_n the image under the structure morphism $\Phi : \Omega Lie^{¡} \to \operatorname{End}_A$ of the generators $-s^{-1}\ell_n^c$. The $\mathbb{S}_n$-module $Lie^{¡}(n)$ being the one-dimensional signature representation, the map ℓ_n is skew-symmetric. The commutation of the structure morphism Φ with the differentials reads

$$\sum_{\substack{p+q=n+1\\ p,q>1}} \sum_{\sigma \in Sh_{p,q}^{-1}} \operatorname{sgn}(\sigma)(-1)^{(p-1)q}(\ell_p \circ_1 \ell_q)^\sigma = \partial_A(\ell_n).$$

□

As in the case of A_∞-algebras, we can denote the differential of A by $\ell_1 := -d_A$, and include it in the relations defining an L_∞-algebra as follows

$$\sum_{p+q=n+1} \sum_{\sigma \in Sh_{p,q}^{-1}} \operatorname{sgn}(\sigma)(-1)^{(p-1)q}(\ell_p \circ_1 \ell_q)^\sigma = 0.$$

This way of writing the definition of a homotopy Lie algebra is more compact but less explicit about the role of the boundary map $\ell_1 = -d_A$.

In the next proposition, we extend to homotopy algebras the anti-symmetrization construction of Sect. 1.1.9, which produces a Lie bracket from an associative product.

Proposition 10.1.8. [LS93] *Let $(A, d_A, \{m_n\}_{n\geq 2})$ be an A_∞-algebra structure on a dg module A. The anti-symmetrized maps $\ell_n : A^{\otimes n} \to A$, given by*

$$\ell_n := \sum_{\sigma\in\mathbb{S}_n} \operatorname{sgn}(\sigma) {m_n}^\sigma,$$

endow the dg module A with an L_∞-algebra structure.

Proof. It is a direct corollary of the proof of the previous proposition. The map of cooperads $Lie^{¡} \to Ass^{¡}$ induces a morphism of dg operads $\Omega Lie^{¡} \to \Omega Ass^{¡}$, given by $\ell_n^c \mapsto \sum_{\sigma\in\mathbb{S}_n} \operatorname{sgn}(\sigma)(\mu_n^c)^\sigma$. Hence, the pullback of a morphism $\Omega Ass^{¡} \to \operatorname{End}_A$, defining an A_∞-algebra structure on A, produces a morphism of dg operads $\Omega Lie^{¡} \to \operatorname{End}_A$, which is the expected L_∞-algebra structure on A. □

10.1.7 The Convolution Algebra Encoding L_∞-Algebras

The underlying module of the convolution pre-Lie algebra $\operatorname{Hom}_{\mathbb{S}}(Lie^{¡}, \operatorname{End}_A)$ is isomorphic to

$$\prod_{n\geq 1} \operatorname{Hom}_{\mathbb{S}}\big(Lie^{¡}, \operatorname{End}_A\big)(n) \cong \prod_{n\geq 1} \operatorname{Hom}\big(S^n(sA), sA\big) \cong \prod_{n\geq 1} s^{-n+1} \operatorname{Hom}\big(\Lambda^n A, A\big),$$

where $\Lambda^n A$ is the coinvariant space of $A^{\otimes n}$ with respect to the signature representation. Explicitly, it is the quotient of $A^{\otimes n}$ by the relations

$$a_1 \otimes \cdots \otimes a_n - \operatorname{sgn}(\sigma)\varepsilon a_{\sigma(1)} \otimes \cdots \otimes a_{\sigma(n)}$$

for $a_1, \ldots, a_n \in A$ and for $\sigma \in \mathbb{S}_n$ with ε the Koszul sign given by the permutation of the graded elements $a_1, \ldots, a_n$.

We endow the right-hand side with the following binary product

$$f \star g := \sum_{\substack{p+q=n+1\\ p,q>1}} \sum_{\sigma\in Sh_{p,q}^{-1}} \operatorname{sgn}(\sigma)(-1)^{(p-1)|g|}(f \circ_1 g)^\sigma,$$

for $f \in \operatorname{Hom}(\Lambda^p A, A)$ and $g \in \operatorname{Hom}(\Lambda^q A, A)$.

This product is called the *Nijenhuis–Richardson product* from [NR66, NR67].

Proposition 10.1.9. *For any dg module A, the Nijenhuis–Richardson product endows the space $\prod_{n\geq 1} s^{-n+1}\operatorname{Hom}(\Lambda^n A, A)$ with a dg pre-Lie algebra structure, which is isomorphic to the convolution dg pre-Lie algebra* $\operatorname{Hom}_{\mathbb{S}}(\mathit{Lie}^{\text{i}}, \operatorname{End}_A)$.

Proof. The proof is similar to the proof of Proposition 10.1.6 with the explicit form of the infinitesimal decomposition map of the cooperad Lie^{i} given above. We first check that the two products are sent to one another under this isomorphism. As a consequence, the Nijenhuis–Richardson product is a pre-Lie product. □

Under this explicit description of the convolution pre-Lie algebra $\operatorname{Hom}_{\mathbb{S}}(\mathit{Lie}^{\text{i}}, \operatorname{End}_A)$, we leave it to the reader to verify that a twisting element is exactly an L_∞-algebra structure on the dg module A.

For other examples of homotopy algebras, we refer to Chap. 13, where examples of algebras over operads are treated in detail.

10.1.8 Equivalent Definition in Terms of Square-Zero Coderivation

In this section, we give a third equivalent definition of the notion of $\mathscr{P}_\infty$-algebra. A structure of $\mathscr{P}_\infty$-algebra can be faithfully encoded as a square-zero coderivation as follows.

By Proposition 6.3.8, we have the following isomorphisms

$$\operatorname{Hom}_{\mathbb{S}}\big(\mathscr{P}^{\text{i}}, \operatorname{End}_A\big) \cong \operatorname{Hom}_{\mathsf{Mod}_{\mathbb{K}}}\big(\mathscr{P}^{\text{i}}(A), A\big) \cong \operatorname{Coder}\big(\mathscr{P}^{\text{i}}(A)\big),$$

where $\operatorname{Coder}(\mathscr{P}^{\text{i}}(A))$ stands for the module of coderivations on the cofree $\mathscr{P}^{\text{i}}$-coalgebra $\mathscr{P}^{\text{i}}(A)$. Let us denote by $\varphi \mapsto d^r_\varphi$ the induced isomorphism from left to right.

Proposition 10.1.10. *The map* $\operatorname{Hom}_{\mathbb{S}}(\mathscr{P}^{\text{i}}, \operatorname{End}_A) \cong \operatorname{Coder}(\mathscr{P}^{\text{i}}(A))$ *is an isomorphism of Lie algebras*:

$$\big[d^r_\alpha, d^r_\beta\big] = d^r_{[\alpha,\beta]}.$$

Proof. Let $\bar{\varphi}$ be the image of φ under the first isomorphism $\operatorname{Hom}_{\mathbb{S}}(\mathscr{P}^{\text{i}}, \operatorname{End}_A) \cong \operatorname{Hom}_{\mathsf{Mod}_{\mathbb{K}}}(\mathscr{P}^{\text{i}}(A), A)$. If we denote by proj_A the canonical projection $\mathscr{P}^{\text{i}}(A) \twoheadrightarrow A$, then Proposition 6.3.8 gives $\operatorname{proj}_A(d^r_\varphi) = \bar{\varphi}$. A direct computation shows that $\operatorname{proj}_A([d^r_\alpha, d^r_\beta]) = \overline{[\alpha,\beta]}$, which concludes the proof. □

We consider the sum

$$d_\varphi := d_{\mathscr{P}^{\text{i}}(A)} + d^r_\varphi$$

of d^r_φ with the internal differential on $\mathscr{P}^{\text{i}}(A)$.

Proposition 10.1.11. *A structure of $\mathscr{P}_\infty$-algebra on a dg module A is equivalent to a square-zero coderivation on the cofree $\mathscr{P}^{¡}$-coalgebra $\mathscr{P}^{¡}(A)$.*

Explicitly, an element $\varphi \in \mathrm{Hom}_{\mathbb{S}}(\mathscr{P}^{¡}, \mathrm{End}_A)$, such that $\varphi(\mathrm{id}) = 0$, satisfies the Maurer–Cartan equation $\partial(\varphi) + \varphi \star \varphi = 0$ if and only if ${d_\varphi}^2 = 0$.

Proof. Any $\varphi \in \mathrm{Tw}(\mathscr{P}^{¡}, \mathrm{End}_A)$ induces a coderivation d_φ^r of degree -1 on $\mathscr{P}^{¡}(A)$. Since φ is a twisting morphism, it vanishes on the counit of $\mathscr{P}^{¡}$. As a consequence d_φ^r vanishes on $A \subset \mathscr{P}^{¡}(A)$. Under the same notation as in Proposition 10.1.10, we have

$$\begin{aligned}\mathrm{proj}_A\big({d_\varphi}^2\big) &= \mathrm{proj}_A\big(d_{\mathscr{P}^{¡}(A)} \circ d_\varphi^r + d_\varphi^r \circ d_{\mathscr{P}^{¡}(A)} + (d_\varphi^r)^2\big)\\ &= d_A \circ \bar{\varphi} + \bar{\varphi} \circ d_{\mathscr{P}^{¡}(A)} + \overline{\varphi \star \varphi}\end{aligned}$$

in $\mathrm{Hom}_{\mathsf{Mod}_{\mathbb{K}}}(\mathscr{P}^{¡}(A), A)$. We conclude with the relation

$$\overline{\partial(\varphi) + \varphi \star \varphi} = \partial(\bar{\varphi}) + \overline{\varphi \star \varphi} = d_A \circ \bar{\varphi} + \bar{\varphi} \circ d_{\mathscr{P}^{¡}(A)} + \overline{\varphi \star \varphi}. \qquad \square$$

In this case, $(\mathscr{P}^{¡}(A), d_\varphi)$ becomes a quasi-cofree $\mathscr{P}^{¡}$-coalgebra. This proposition shows that a homotopy $\mathscr{P}$-algebra structure on a dg module A is equivalent to a dg $\mathscr{P}^{¡}$-coalgebra structure on $\mathscr{P}^{¡}(A)$, where the structure maps are encoded into the coderivation. We call *codifferential* any degree -1 square-zero coderivation on a $\mathscr{P}^{¡}$-coalgebra. So the set of $\mathscr{P}_\infty$-algebra structures is equal to the set of codifferentials $\mathrm{Codiff}(\mathscr{P}^{¡}(A))$.

For example, we get the definitions of A_∞-algebras and of L_∞-algebras in terms a square-zero coderivations.

Proposition 10.1.12. *An A_∞-algebra structure on a dg module A is equivalent to a codifferential on the noncounital cofree associative coalgebra $\overline{T}^c(sA)$.*

Similarly, an L_∞-algebra structure on A is equivalent to a codifferential on the noncounital cofree cocommutative coalgebra $\overline{S}^c(sA)$.

Proof. Since the Koszul dual nonsymmetric cooperad $As^{¡}$ is isomorphic to $As^* \underset{\mathrm{H}}{\otimes} \mathrm{End}^c_{s^{-1}\mathbb{K}}$ by Sect. 7.2.3, the quasi-cofree $As^{¡}$-coalgebra $As^{¡}(A)$ is isomorphic to the desuspension of the noncounital cofree associative coalgebra $s^{-1}\overline{T}^c(sA)$.

In the same way, since the Koszul dual cooperad $Lie^{¡}$ is isomorphic to $Com^* \underset{\mathrm{H}}{\otimes} \mathrm{End}^c_{s^{-1}\mathbb{K}}$ by Sect. 7.2.3, the quasi-cofree $Lie^{¡}$-coalgebra $Lie^{¡}(A)$ is isomorphic to the desuspension of the noncounital cofree cocommutative coalgebra $s^{-1}\overline{S}^c(sA)$. $\square$

10.1.9 Rosetta Stone

Using the bar–cobar adjunction of Sect. 6.5.3, a $\mathscr{P}_\infty$-algebra structure on a dg module A is equivalently defined by a morphism of dg cooperads $\mathscr{P}^{¡} \to \mathrm{B}\,\mathrm{End}_A$.

Notice that the endomorphism operad End_A is unital but non-necessarily augmented. So by the bar construction of End_A, we mean $\mathrm{B}\,\mathrm{End}_A := \mathscr{T}^c(s\,\mathrm{End}_A)$, endowed with the same differential map as in Sect. 6.5.1. With this definition, the bar–cobar adjunction still holds.

The four equivalent definitions of homotopy $\mathscr{P}$-algebras are summed up in the following theorem.

Theorem 10.1.13 (Rosetta Stone). *The set of $\mathscr{P}_\infty$-algebra structures on a dg module A is equivalently given by*

$$\begin{aligned}\mathrm{Hom}_{\mathsf{dg\ Op}}\big(\Omega\mathscr{P}^{¡}, \mathrm{End}_A\big) &\cong \mathrm{Tw}\big(\mathscr{P}^{¡}, \mathrm{End}_A\big)\\ &\cong \mathrm{Hom}_{\mathsf{dg\ Coop}}\big(\mathscr{P}^{¡}, \mathrm{B}\,\mathrm{End}_A\big) \cong \mathrm{Codiff}\big(\mathscr{P}^{¡}(A)\big).\end{aligned}$$

10.2 Homotopy Algebras: Morphisms

In this section, we make the notion of morphism of $\mathscr{P}_\infty$-algebras explicit. Then we introduce and study the more general notion of infinity-morphism, denoted ∞-morphism, of $\mathscr{P}_\infty$-algebras, which will prove to be more relevant to the homotopy theory of $\mathscr{P}_\infty$-algebras. The data of an ∞-morphism does not consist in only one map but in a family of maps parametrized by the elements of the Koszul dual cooperad $\mathscr{P}^{¡}$. More precisely, these maps live in $\mathrm{End}^A_B := \{\mathrm{Hom}(A^{\otimes n}, B)\}_{n\in\mathbb{N}}$, the space of multilinear maps between two $\mathscr{P}_\infty$-algebras.

The examples of ∞-morphisms of A_∞-algebras and of L_∞-algebras are given.

10.2.1 Morphisms of $\mathscr{P}_\infty$-Algebras

A *morphism $f : A \to B$ between $\mathscr{P}_\infty$-algebras* is a morphism of algebras over the operad $\mathscr{P}_\infty$ as in Sect. 5.2.3.

In terms of twisting morphisms, they are described as follows. Let A and B be two $\mathscr{P}_\infty$-algebras, whose associated twisting morphisms are denoted by $\varphi : \mathscr{P}^{¡} \to \mathrm{End}_A$ and $\psi : \mathscr{P}^{¡} \to \mathrm{End}_B$ respectively. We denote by End^A_B the $\mathbb{S}$-module defined by

$$\mathrm{End}^A_B := \big\{\mathrm{Hom}\big(A^{\otimes n}, B\big)\big\}_{n\in\mathbb{N}}.$$

In other words, a morphism of $\mathscr{P}_\infty$-algebras is map $f : A \to B$ such that the following diagram commutes

$$\begin{array}{ccc}\mathscr{P}^{¡} & \xrightarrow{\ \varphi\ } & \mathrm{End}_A\\ \big\downarrow{\scriptstyle\psi} & & \big\downarrow{\scriptstyle f_*}\\ \mathrm{End}_B & \xrightarrow{\ f^*\ } & \mathrm{End}^A_B,\end{array}$$

where f_* is the pushforward by f

$$g \in \operatorname{Hom}(A^{\otimes n}, A) \mapsto f_*(g) := fg \in \operatorname{Hom}(A^{\otimes n}, B),$$

and where f^* is the pullback by f

$$g \in \operatorname{Hom}(B^{\otimes n}, B) \mapsto f^*(g) := g(f, \ldots, f) \in \operatorname{Hom}(A^{\otimes n}, B).$$

In this case, the two homotopy $\mathscr{P}$-algebra structures strictly commute under f.

10.2.2 Infinity-Morphisms of $\mathscr{P}_\infty$-Algebras

We use the third equivalent definition of homotopy algebras to define the notion of ∞-*morphism* of homotopy algebras, which is an enhancement of the previous one.

By Proposition 10.1.11, a homotopy $\mathscr{P}$-algebra structure on A (resp. on B) is equivalent to a dg $\mathscr{P}^{¡}$-coalgebra structure on $\mathscr{P}^{¡}(A)$ (resp. on $\mathscr{P}^{¡}(B)$), with codifferential denoted by d_φ (resp. d_ψ).

By definition, an *∞-morphism of $\mathscr{P}_\infty$-algebras* is a morphism

$$F : (\mathscr{P}^{¡}(A), d_\varphi) \to (\mathscr{P}^{¡}(B), d_\psi)$$

of dg $\mathscr{P}^{¡}$-coalgebras. The composite of two ∞-morphisms is defined as the composite of the associated morphisms of dg $\mathscr{P}^{¡}$-coalgebras:

$$F \circ G := \mathscr{P}^{¡}(A) \to \mathscr{P}^{¡}(B) \to \mathscr{P}^{¡}(C).$$

Therefore $\mathscr{P}_\infty$-algebras with their ∞-morphisms form a category, which is denoted by ∞-$\mathscr{P}_\infty$-alg. An ∞-morphism between $\mathscr{P}_\infty$-algebras is denoted by

$$A \rightsquigarrow B$$

to avoid confusion with the above notion of morphism.

Proposition 10.2.1. *Let $\mathscr{C}$ be a cooperad. Any morphism $F : \mathscr{C}(V) \to \mathscr{C}(W)$ of cofree $\mathscr{C}$-coalgebras is completely characterized by its projection $\bar{f}$ onto the cogenerators $\bar{f} := \mathrm{proj}_W \circ F : \mathscr{C}(V) \to W$.*

Explicitly, the unique morphism $F : \mathscr{C}(V) \to \mathscr{C}(W)$ of $\mathscr{C}$-coalgebras which extends a map $\bar{f} : \mathscr{C}(V) \to W$ is given by the following composite

$$F = \mathscr{C}(V) = \mathscr{C} \circ V \xrightarrow{\Delta \circ \mathrm{Id}_V} \mathscr{C} \circ \mathscr{C} \circ V \xrightarrow{\mathrm{Id}_{\mathscr{C}} \circ \bar{f}} \mathscr{C} \circ W = \mathscr{C}(W).$$

Proof. The proof uses the same ideas as in Proposition 6.3.8. So it is left to the reader as an exercise. □

This proposition shows that an ∞-morphism of $\mathscr{P}_\infty$-algebras is equivalently given by a map $\bar{f} : \mathscr{P}^{¡}(A) \to B$, whose induced morphism $F : \mathscr{P}^{¡}(A) \to \mathscr{P}^{¡}(B)$ of $\mathscr{P}^{¡}$-coalgebras commutes with the differentials. Any such map $\bar{f}$ is equivalent to a map $f : \mathscr{P}^{¡} \to \mathrm{End}^A_B$. So an ∞-morphism is made out of a family of maps, from $A^{\otimes n} \to B$, parametrized by $\mathscr{P}^{¡}(n)$, for any n.

The next section makes the relation satisfied by an ∞-morphism explicit in terms of this associated map f.

10.2.3 Infinity-Morphisms in Terms of Twisting Morphisms

The module $\mathrm{Hom}_{\mathbb{S}}(\mathscr{P}^{¡}, \mathrm{End}_A)$ with its pre-Lie convolution product $\star$ form the pre-Lie algebra $\mathfrak{g}_A := (\mathrm{Hom}_{\mathbb{S}}(\mathscr{P}^{¡}, \mathrm{End}_A), \star)$.

The module $\mathrm{Hom}_{\mathbb{S}}(\mathscr{P}^{¡}, \mathrm{End}_B)$ is an associative algebra with the associative product $\circledcirc$ defined by

$$\psi \circledcirc \xi := \mathscr{P}^{¡} \xrightarrow{\Delta} \mathscr{P}^{¡} \circ \mathscr{P}^{¡} \xrightarrow{\psi \circ \xi} \mathrm{End}_B \circ \mathrm{End}_B \xrightarrow{\gamma_{\mathrm{End}_B}} \mathrm{End}_B .$$

(In general, this is not a *graded* associative algebra.) We denote this associative algebra by

$$\mathfrak{g}_B := \big(\mathrm{Hom}_{\mathbb{S}}\big(\mathscr{P}^{¡}, \mathrm{End}_B\big), \circledcirc\big).$$

Observe that the convolution product $\star$ is defined by the *infinitesimal* decomposition map $\Delta_{(1)}$, while the product $\circledcirc$ is defined by the *total* decomposition map Δ.

The composite of maps endows the $\mathbb{S}$-module End^A_B with a left module structure over the operad End_B:

$$\lambda : \mathrm{End}_B \circ \mathrm{End}^A_B \to \mathrm{End}^A_B,$$

and an infinitesimal right module structure over the operad End_A,

$$\rho : \mathrm{End}^A_B \circ_{(1)} \mathrm{End}_A \to \mathrm{End}^A_B .$$

They induce the following two actions on $\mathfrak{g}^A_B := \mathrm{Hom}_{\mathbb{S}}(\mathscr{P}^{¡}, \mathrm{End}^A_B)$:

◇ for $\varphi \in \mathfrak{g}_A$ and $f \in \mathrm{End}^A_B$, we define $f * \varphi : \mathscr{P}^{¡} \to \mathrm{End}^A_B$ by

$$f * \varphi := \mathscr{P}^{¡} \xrightarrow{\Delta_{(1)}} \mathscr{P}^{¡} \circ_{(1)} \mathscr{P}^{¡} \xrightarrow{f \circ_{(1)} \varphi} \mathrm{End}^A_B \circ_{(1)} \mathrm{End}_A \xrightarrow{\rho} \mathrm{End}^A_B;$$

◇ for $\psi \in \mathfrak{g}_B$ and $f \in \mathrm{End}^A_B$, we define $\psi \circledast f : \mathscr{P}^{¡} \to \mathrm{End}^A_B$ by

$$\psi \circledast f := \mathscr{P}^{¡} \xrightarrow{\Delta} \mathscr{P}^{¡} \circ \mathscr{P}^{¡} \xrightarrow{\psi \circ f} \mathrm{End}_B \circ \mathrm{End}^A_B \xrightarrow{\lambda} \mathrm{End}^A_B .$$

Proposition 10.2.2. *The module $(\mathfrak{g}_B^A, *)$ is a right module over the pre-Lie algebra $(\mathfrak{g}_A, \star)$, see Sect. 1.4.4. The module $(\mathfrak{g}_B^A, \circledast)$ is a left module over the associative algebra $(\mathfrak{g}_B, \circledcirc)$.*

Proof. The right action $*$ coincides with the pre-Lie subalgebra structure on $\mathfrak{g}_A \oplus \mathfrak{g}_B^A$ of the pre-Lie algebra $(\mathfrak{g}_{A\oplus B}, \star)$. In the same way, the left action $\circledast$ coincides with the associative subalgebra structure on $\mathfrak{g}_B \oplus \mathfrak{g}_B^A$ of the associative algebra $(\mathfrak{g}_{A\oplus B}, \circledcirc)$. □

Theorem 10.2.3. *Let $\varphi \in \mathrm{Tw}(\mathscr{P}^{¡}, \mathrm{End}_A)$ and $\psi \in \mathrm{Tw}(\mathscr{P}^{¡}, \mathrm{End}_B)$ be two $\mathscr{P}_\infty$-algebras. An ∞-morphism $F : \mathscr{P}^{¡}(A) \to \mathscr{P}^{¡}(B)$ of $\mathscr{P}_\infty$-algebras is equivalent to a morphism of dg $\mathbb{S}$-modules $f : \mathscr{P}^{¡} \to \mathrm{End}_B^A$ such that*

$$f * \varphi - \psi \circledast f = \partial(f)$$

in $\mathrm{Hom}_{\mathbb{S}}(\mathscr{P}^{¡}, \mathrm{End}_B^A)$:

$$\begin{array}{ccc}
\mathscr{P}^{¡} & \xrightarrow{\ \Delta\ } & \mathscr{P}^{¡} \circ \mathscr{P}^{¡} \\
 & \searrow^{\partial(f)} & \downarrow{\scriptstyle \psi\circ f} \\
\downarrow{\scriptstyle \Delta_{(1)}} & & \mathrm{End}_B \circ \mathrm{End}_B^A \\
 & & \downarrow{\scriptstyle \lambda} \\
\mathscr{P}^{¡} \circ_{(1)} \mathscr{P}^{¡} \xrightarrow{f\circ_{(1)}\varphi} \mathrm{End}_B^A \circ_{(1)} \mathrm{End}_A & \xrightarrow{\ \rho\ } & \mathrm{End}_B^A .
\end{array}$$

Proof. The morphism $F : \mathscr{P}^{¡}(A) \to \mathscr{P}^{¡}(B)$ of dg $\mathscr{P}^{¡}$-coalgebras commutes with the differentials d_φ and d_ψ if and only if $\mathrm{proj}_B(d_\psi \circ F - F \circ d_\varphi) = 0$. Using the explicit form of F given by the previous lemma, this relation is equivalent to the following commutative diagram

$$\begin{array}{ccccc}
\mathscr{P}^{¡}(A) & \xrightarrow{\Delta\circ\mathrm{Id}} & \mathscr{P}^{¡} \circ \mathscr{P}^{¡}(A) & \xrightarrow{\mathrm{Id}\circ\bar{f}} & \mathscr{P}^{¡}(B) \\
\downarrow{\scriptstyle d_\varphi} & & & & \downarrow{\scriptstyle d_B+\bar{\psi}} \\
\mathscr{P}^{¡}(A) & & \xrightarrow{\qquad\quad \bar{f} \qquad\quad} & & B .
\end{array}$$

We conclude with the explicit form of d_φ given in Proposition 6.3.8. □

Proposition 10.2.4. *Let $\varphi \in \mathrm{Tw}(\mathscr{P}^{¡}, \mathrm{End}_A)$, $\psi \in \mathrm{Tw}(\mathscr{P}^{¡}, \mathrm{End}_B)$, and $\zeta \in \mathrm{Tw}(\mathscr{P}^{¡}, \mathrm{End}_C)$ be three $\mathscr{P}_\infty$-algebras. Let $f \in \mathrm{Hom}_{\mathbb{S}}(\mathscr{P}^{¡}, \mathrm{End}_B^A)$ and $g \in \mathrm{Hom}_{\mathbb{S}}(\mathscr{P}^{¡}, \mathrm{End}_C^B)$ be two ∞-morphisms.*

Under the isomorphism between codifferentials on cofree $\mathscr{P}^{¡}$-coalgebras and twisting morphisms from $\mathscr{P}^{¡}$, the composite of the two ∞-morphisms f and g is *equal to*

$$g \circledcirc f := \mathscr{P}^{i} \xrightarrow{\Delta} \mathscr{P}^{i} \circ \mathscr{P}^{i} \xrightarrow{g \circ f} \mathrm{End}^{B}_{C} \circ \mathrm{End}^{A}_{B} \to \mathrm{End}^{A}_{C},$$

where the last map is the natural composition of morphisms.

Proof. By the adjunction $\mathrm{Hom}_{\mathbb{S}}(\mathscr{P}^{¡}, \mathrm{End}^{A}_{B}) \cong \mathrm{Hom}(\mathscr{P}^{¡}(A), B)$, f is equivalent to a map $\bar{f} : \mathscr{P}^{¡}(A) \to B$. This latter one is equivalent to a morphism of dg $\mathscr{P}^{¡}$-coalgebras $F : \mathscr{P}^{¡}(A) \to \mathscr{P}^{¡}(B)$ by Proposition 10.2.1. Respectively, $g : \mathscr{P}^{¡} \to \mathrm{End}^{B}_{C}$ is equivalent to a morphism of dg $\mathscr{P}^{¡}$-coalgebras $G : \mathscr{P}^{¡}(B) \to \mathscr{P}^{¡}(C)$. By the formula given in Proposition 10.2.1, the projection of the composite $G \circ F$ onto the space of cogenerators C is equal to

$$\mathscr{P}^{¡}(A) \cong \mathscr{P}^{¡} \circ A \xrightarrow{\Delta \circ \mathrm{Id}_A} \mathscr{P}^{¡} \circ \mathscr{P}^{¡} \circ A \xrightarrow{\mathrm{Id} \circ \bar{f}} \mathscr{P}^{¡}(B) \xrightarrow{\bar{g}} C.$$

We conclude by using the adjunction $\mathrm{Hom}_{\mathbb{S}}(\mathscr{P}^{¡}, \mathrm{End}^{A}_{C}) \cong \mathrm{Hom}(\mathscr{P}^{¡}(A), C)$ once again. □

Since the cooperad $\mathscr{P}^{¡}$ is weight graded, any map $f \in \mathrm{Hom}_{\mathbb{S}}(\mathscr{P}^{¡}, \mathrm{End}^{A}_{B})$ decomposes according to this weight, $f_{(n)} : \mathscr{P}^{¡\,(n)} \to \mathrm{End}^{A}_{B}$. Since Δ preserves this weight, the square in the diagram of Theorem 10.2.3 applied to $\mathscr{P}^{¡\,(n)}$ involves only the maps $f_{(k)}$ up to $k = n-1$. Therefore, the term $f_{(n)}$ is a homotopy for the relation $f * \varphi - \psi \circledast f = \partial(f_{(n)})$ in $\mathrm{Hom}_{\mathbb{S}}(\mathscr{P}^{¡\,(n)}, \mathrm{End}^{A}_{B})$.

The first component $f_{(0)} : \mathrm{I} \to \mathrm{Hom}(A, B)$ of an ∞-morphism is equivalent to a chain map $f_{(0)}(\mathrm{id}) : A \to B$ between the underlying chain complexes. In order to lighten the notation, we still denote this latter map by $f_{(0)}$.

10.2.4 Infinity-Isomorphism and Infinity-Quasi-isomorphism

An ∞-morphism f is called an ∞-*isomorphism* (resp. ∞-*quasi-isomorphism*) if its first component $f_{(0)} : A \to B$ is an isomorphism (resp. a quasi-isomorphism) of chain complexes. We will show later in Sect. 10.4.1 that ∞-isomorphisms are the isomorphisms of the category ∞-$\mathscr{P}_\infty$-alg.

10.2.5 Infinity-Morphisms and $\mathscr{P}$-Algebras

Proposition 10.2.5. *A morphism of $\mathscr{P}_\infty$-algebras is an ∞-morphism with only one nonvanishing component, namely the first one $f_{(0)} : A \to B$.*

Proof. Let $f : \mathscr{P}^{¡}(A) \to B$ be a morphism of dg modules such that $f_{(n)} = 0$ for $n \geq 1$. Since $\mathscr{P}^{(0)} = \mathrm{I}$, the first component $f_{(0)}$ of f is morphism of dg modules from A to B. In this particular case, the relation $\rho((f \circ_{(1)} \varphi)(\Delta_r)) - \lambda((\psi \circ f)(\Delta_l)) = \partial(f)$ applied to $\mathscr{P}^{¡\,(n)}$ for $n \geq 1$ is equivalent to $f_*(\varphi) = \psi(f^*)$. □

The category of $\mathscr{P}_\infty$-algebras with their morphisms forms a non-full subcategory of the category of $\mathscr{P}_\infty$-algebras with the ∞-morphisms.

One can also consider the category of $\mathscr{P}$-algebras with ∞-morphisms. It forms a full subcategory of ∞-$\mathscr{P}_\infty$-alg, which is denoted by ∞-$\mathscr{P}$-alg. Altogether these four categories assemble to form the following commutative diagram

$$\begin{array}{ccc} \mathscr{P}\text{-alg} & \xrightarrow{\text{not full}} & \infty\text{-}\mathscr{P}\text{-alg} \\ \downarrow{\scriptstyle \text{f.f.}} & & \downarrow{\scriptstyle \text{f.f.}} \\ \mathscr{P}_\infty\text{-alg} & \xrightarrow{\text{not full}} & \infty\text{-}\mathscr{P}_\infty\text{-alg}, \end{array}$$

where the vertical functors are full and faithful.

10.2.6 Infinity-Morphisms of A_∞-Algebras and L_∞-Algebras

Proposition 10.2.6. *An ∞-morphism $f : A \rightsquigarrow B$ of A_∞-algebras is a family of maps $\{f_n : A^{\otimes n} \to B\}_{n\geq 1}$ of degree $n-1$ which satisfy: $d_B \circ f_1 = f_1 \circ d_A$, that is f_1 is a chain map, and for $n \geq 2$,*

$$\sum_{\substack{p+1+r=k\\ p+q+r=n}} (-1)^{p+qr} f_k \circ \big(\underbrace{\mathrm{Id}_A, \ldots, \mathrm{Id}_A}_{p}, m_q^A, \underbrace{\mathrm{Id}_A, \ldots, \mathrm{Id}_A}_{r}\big)$$
$$- \sum_{\substack{k\geq 2\\ i_1+\cdots+i_k=n}} (-1)^{\varepsilon} m_k^B \circ (f_{i_1}, \ldots, f_{i_k}) = \partial(f_n),$$

in $\mathrm{Hom}(A^{\otimes n}, B)$.

Under the tree representation, this relation becomes

$\partial(f_n)$

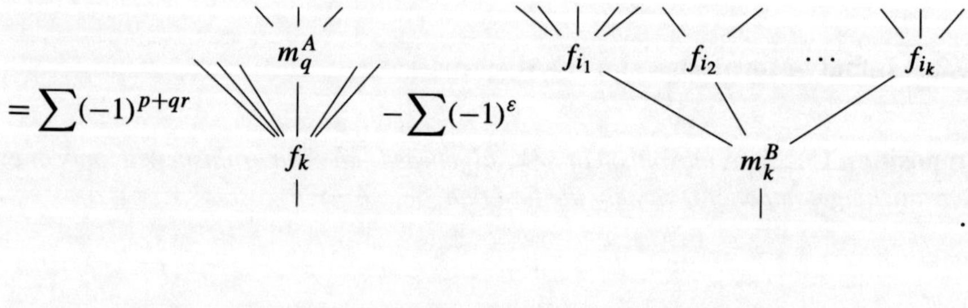

.

Proof. Let $\varphi \in \mathrm{Tw}(As^{¡}, \mathrm{End}_A)$ and $\psi \in \mathrm{Tw}(As^{¡}, \mathrm{End}_B)$ be two A_∞-algebra structures. Recall that $As^{¡}_n = \mathbb{K}\mu^c_n$ with $|\mu^c_n| = n-1$. We denote by $m^A_n \in \mathrm{Hom}(A^{\otimes n}, A)$ the image of μ^c_n under φ and by $m^B_n \in \mathrm{Hom}(B^{\otimes n}, B)$ the image of μ^c_n under ψ.

An ∞-morphism $f : As^{¡} \to \mathrm{End}^A_B$ between A and B is a family of maps $\{f_n : A^{\otimes n} \to B\}_{n\geq 1}$ of degree $n-1$. For $n \geq 2$, the formula of the infinitesimal decomposition map $\Delta_{(1)}$ of the cooperad $As^{¡}$ shows that the image of μ^c_n under $f * \varphi$ in End^A_B is equal to

$$(f * \varphi)(\mu^c_n) = \sum_{\substack{p+1+r=k\\ p+q+r=n}} (-1)^{p+qr} f_k \circ \big(\underbrace{\mathrm{Id}_A, \ldots, \mathrm{Id}_A}_{p}, m^A_q, \underbrace{\mathrm{Id}_A, \ldots, \mathrm{Id}_A}_{r}\big).$$

On the other hand, the formula of the decomposition map Δ of the cooperad $As^{¡}$, given in Lemma 9.1.2, shows that the image of μ^c_n under $\psi \circledast f$ in End^A_B is equal to

$$(\psi \circledast f)(\mu^c_n) = \sum_{\substack{k\geq 2\\ i_1+\cdots+i_k=n}} (-1)^{\varepsilon} m^B_k \circ (f_{i_1}, \ldots, f_{i_k}),$$

where $\varepsilon = (k-1)(i_1-1)+(k-2)(i_2-1)+\cdots+2(i_{k-2}-1)+(i_{k-1}-1)$. Therefore we find the same formula as in Sect. 9.2.6. □

The case of L_∞-algebras is similar.

Proposition 10.2.7. *An ∞-morphism $f : A \rightsquigarrow B$ of L_∞-algebras, is a family of maps $\{f_n : \Lambda^n A \to B\}_{n\geq 1}$ of degree $n-1$ which satisfy: $d_A \circ f_1 = f_1 \circ d_A$, that is f_1 is a chain map, and for $n \geq 2$,*

$$\sum_{\substack{p+q=n+1\\ p,q>1}} \sum_{\sigma\in Sh^{-1}_{p,q}} \mathrm{sgn}(\sigma)(-1)^{(p-1)|q|}\big(f_p \circ_1 \ell^A_q\big)^\sigma$$
$$- \sum_{\substack{k\geq 2\\ i_1+\cdots+i_k=n}} \sum_{\sigma\in Sh^{-1}_{(i_1,\ldots,i_k)}} \mathrm{sgn}(\sigma)(-1)^{\varepsilon}\ell^B_k \circ (f_{i_1}, \ldots, f_{i_k})^\sigma = \partial(f_n),$$

in $\mathrm{Hom}(\Lambda^n A, B)$.

Proof. The proof relies on the explicit morphism of cooperads $Lie^{¡} \to Ass^{¡}$ given in the proof of Proposition 10.1.7. The results for A_∞-algebras transfer to L_∞ under this morphism. □

So far, we can see why L_∞-algebras are very close to A_∞-algebras: the Koszul dual cooperad $Lie^{¡}$ of Lie is the antisymmetrized version of $Ass^{¡}$.

10.3 Homotopy Transfer Theorem

In this section, we prove that a homotopy $\mathscr{P}$-algebra structure on a dg module induces a homotopy $\mathscr{P}$-algebra structure on any homotopy equivalent dg module, with explicit formulas. This structure is called "the" transferred $\mathscr{P}_\infty$-algebra structure. We make the examples of A_∞-algebras and L_∞-algebras explicit.

When A is a $\mathscr{P}_\infty$-algebra, we have seen in Proposition 10.1.5 that its homotopy $H(A)$ carries a natural $\mathscr{P}$-algebra structure. When working over a field, the homotopy $H(A)$ can be made into a deformation retract of A. It enables us to transfer the $\mathscr{P}_\infty$-algebra structure from A to $H(A)$. These higher operations, called the operadic Massey products, extend the $\mathscr{P}$-algebra structure of $H(A)$. They contain the full homotopy data of A, since this $\mathscr{P}_\infty$-algebra $H(A)$ is homotopy equivalent to A.

A meaningful example is given by applying the Homotopy Transfer Theorem to $\mathscr{P} = D$, the algebra of dual numbers on one generator. In this case, a D-algebra A is a bicomplex and the transferred structured on $H(A)$ corresponds to the associated spectral sequence.

Recall that the particular case $\mathscr{P} = As$ has been treated independently in Sect. 9.4. It serves as a paradigm for the general theory developed here.

The Homotopy Transfer Theorem for A_∞-algebras and L_∞-algebras has a long history in mathematics, often related to the Perturbation Lemma. We refer the reader to the survey of Jim Stasheff [Sta10] and references therein. Its extension to the general operadic setting has been studied in the PhD thesis of Charles Rezk [Rez96]. One can find in the paper [Bat98] of Michael Batanin the case of algebras over non-symmetric simplicial operads. A version of HTT was recently proved for algebras over the bar–cobar construction $\Omega \mathrm{B}\mathscr{P}$ by Joseph Chuang and Andrey Lazarev in [CL10] and by Sergei Merkulov in [Mer10a]. Using a generalization of the Perturbation Lemma, it was proved for $\mathscr{P}_\infty$-algebras by Alexander Berglund in [Ber09]. The existence part of the theorem can also be proved by model category arguments, see Clemens Berger and Ieke Moerdijk [BM03a] and Benoit Fresse [Fre09b].

10.3.1 The Homotopy Transfer Problem

Let (V, d_V) be a homotopy retract of (W, d_W):

$$h \circlearrowright (W, d_W) \underset{i}{\overset{p}{\rightleftarrows}} (V, d_V),$$

$$\mathrm{Id}_W - ip = d_W h + h d_W,$$

where the chain map i is a quasi-isomorphism.

The transfer problem is the following one: given a structure of $\mathscr{P}_\infty$-algebra on W, does there exist a $\mathscr{P}_\infty$-algebra structure on V such that i extends to an ∞-quasi-isomorphism of $\mathscr{P}_\infty$-algebras? We will show that this is always possible and we say that the $\mathscr{P}_\infty$-algebra structure of W has been transferred to V.

Theorem 10.3.1 (Homotopy Transfer Theorem). *Let $\mathscr{P}$ be a Koszul operad and let (V, d_V) be a homotopy retract of (W, d_W). Any $\mathscr{P}_\infty$-algebra structure on W can be transferred into a $\mathscr{P}_\infty$-algebra structure on V such that i extends to an ∞-quasi-isomorphism.*

Proof. To prove this theorem, we use the third definition of a $\mathscr{P}_\infty$-algebra given in the Rosetta Stone (Theorem 10.1.13):

$$\mathrm{Hom}_{\mathsf{dg\ Op}}\big(\Omega\mathscr{P}^{\text{!`}}, \mathrm{End}_A\big) \cong \mathrm{Tw}\big(\mathscr{P}^{\text{!`}}, \mathrm{End}_A\big) \cong \mathrm{Hom}_{\mathsf{dg\ Coop}}\big(\mathscr{P}^{\text{!`}}, \mathrm{B}\,\mathrm{End}_A\big).$$

The plan of the proof is the following one. First, we show in Proposition 10.3.2 that the homotopy retract data between V and W induces a morphism of dg cooperads $\mathrm{B}\,\mathrm{End}_W \to \mathrm{B}\,\mathrm{End}_V$. Since a $\mathscr{P}_\infty$-algebra structure on W is equivalently given by a morphism of dg cooperads $\mathscr{P}^{\text{!`}} \to \mathrm{B}\,\mathrm{End}_W$, the composite

$$\mathscr{P}^{\text{!`}} \to \mathrm{B}\,\mathrm{End}_W \to \mathrm{B}\,\mathrm{End}_V$$

defines a $\mathscr{P}_\infty$-algebra structure on V.

We give an explicit formula for this transferred structure in Theorem 10.3.3. The extension of i into an ∞-quasi-isomorphism is provided through an explicit formula in Theorem 10.3.6. □

10.3.2 The Morphism of DG Cooperads $\mathrm{B}\,\mathrm{End}_W \to \mathrm{B}\,\mathrm{End}_V$

Let (V, d_V) be a homotopy retract of (W, d_W). We consider the map defined by $\mu \in \mathrm{End}_W(n) \mapsto p\mu i^{\otimes n} \in \mathrm{End}_V(n)$. Since i and p are morphisms of dg modules, this map is a morphism of dg $\mathbb{S}$-modules. But it does not commute with the operadic composition maps in general. For $\mu_1 \in \mathrm{End}_W(k)$ and $\mu_2 \in \mathrm{End}_W(l)$, with $k + l - 1 = n$, and for $1 \le j \le k$, we have

$$\big(p\mu_1 i^{\otimes k}\big) \circ_j \big(p\mu_2 i^{\otimes l}\big) = p\big(\mu_1 \circ_j (ip\mu_2)\big)i^{\otimes n},$$

which is not equal to $p(\mu_1 \circ_j \mu_2)i^{\otimes n}$ because ip is not equal to Id_W. Since ip is homotopic to Id_W, we will show that this morphism commutes with the operadic composition maps only up to homotopy. In the previous example, we have to consider the homotopy $\mu_1 \circ_j (h\mu_2)$ to get

$$p\big(\mu_1 \circ_j \big(\partial(h)\mu_2\big)\big)i^{\otimes n} = p(\mu_1 \circ_j \mu_2)i^{\otimes n} - p\big(\mu_1 \circ_j (ip\mu_2)\big)i^{\otimes n}.$$

Therefore the idea for defining the morphism of dg cooperads $\Psi : \mathrm{B}\,\mathrm{End}_W \to \mathrm{B}\,\mathrm{End}_V$ is to label the internal edges by the homotopy h as follows. A basis of $\mathscr{T}^c(s\,\mathrm{End}_W)$ is given by trees labeled by elements of $s\,\mathrm{End}_W$. Let $t := t(s\mu_1, \ldots, s\mu_k)$ be such a tree, where the vertices $1, \ldots, k$ are read for bottom to top and from left to right. The image of $\mathscr{T}$ under Ψ is defined by the suspension of the following composite: we label every leaf of the tree $t(\mu_1, \ldots, \mu_k)$ with $i : V \to W$, every internal edge by h and the root by p.

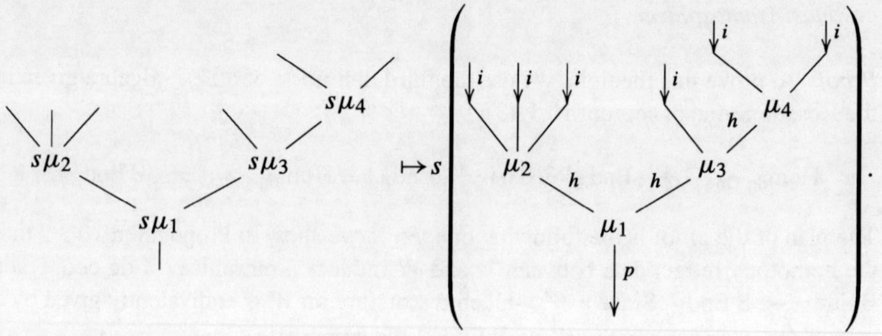

This composite scheme defines a map in $s\,\mathrm{End}_V$. Since $\mathcal{T}^c(s\,\mathrm{End}_V)$ is a conilpotent cofree cooperad, this map $\mathcal{T}^c(s\,\mathrm{End}_W) \to s\,\mathrm{End}_V$ extends to a unique morphism of cooperads $\Psi : \mathcal{T}^c(s\,\mathrm{End}_W) \to \mathcal{T}^c(s\,\mathrm{End}_V)$. Since the degree of h is $+1$, the degree of Ψ is 0. The next result states that this morphism of cooperads commutes with the differentials.

Proposition 10.3.2. [vdL03] *Let (V, d_V) be a homotopy retract of (W, d_W). The map $\Psi : \mathrm{B}\,\mathrm{End}_W \to \mathrm{B}\,\mathrm{End}_V$, defined above, is a morphism of dg cooperads.*

Proof. In this proof, by a slight abuse of notation, we denote the above defined map $\mathcal{T}^c(s\,\mathrm{End}_W) \to s\,\mathrm{End}_V$ by Ψ. By Proposition 10.5.3, we have to check that

$$(\xi): \quad \partial(\Psi) - \Psi\,\mathcal{T}^c(\mathrm{Id}; \gamma_{End_W})\Delta' + \gamma_{End_V}\,\mathcal{T}^c(\Psi)\bar{\Delta} = 0,$$

where the map

$$\gamma_{End_V} : \mathcal{T}^c(s\mathrm{End}_V) \twoheadrightarrow \mathcal{T}^c(s\mathrm{End}_V)^{(2)} \to s\mathrm{End}_V$$

(respectively γ_{End_W}) is given by the partial compositions of the operad End_V (respectively End_W), see Sect. 10.5.1. So it vanishes on $\mathcal{T}^c(s\,\mathrm{End}_V)^{(\geq 3)}$ (respectively on $\mathcal{T}^c(s\,\mathrm{End}_W)^{(\geq 3)}$). We apply Equation (ξ) to a tree $t = t(s\mu_1, \ldots, s\mu_k)$.

1. The first term $\partial(\Psi)(t)$ is equal to the sum over the internal edges e of t of trees $st(\mu_1, \ldots, \mu_k)$, where every internal edge is labeled by h except e, which is labeled by $\partial(h) = d_W h + h d_W$.
2. In the second term, one singles out a subtree with two vertices out of t, composes it in End_W and then one applies Ψ to the resulting tree. Therefore it is equal to the sum over the internal edges e of t of trees $st(\mu_1, \ldots, \mu_k)$, where every internal edge is labeled by h except e, which is labeled by Id_W.
3. The third term consists in splitting the tree t into two parts, applying Ψ to the two induced subtrees and then composing the two resulting images in End_V. Hence it is equal to the sum over all internal edges e of t of trees $st(\mu_1, \ldots, \mu_k)$, where every internal edge is labeled by h except e, which is labeled by ip.

Finally, Equation (ξ) applied to the tree t is equal to the sum, over all internal edges e of t, of trees $st(\mu_1, \ldots, \mu_k)$, where every internal edge is labeled by h except e, which is labeled by

$$d_W h + h d_W - \mathrm{Id}_W + ip = 0.$$

It concludes the proof. □

10.3.3 Transferred Structure

Let $\varphi \in \mathrm{Tw}(\mathscr{P}^{\text{i}}, \mathrm{End}_W)$ be a $\mathscr{P}_\infty$-algebra structure on W. We define a transferred structure of $\mathscr{P}_\infty$-algebra $\psi \in \mathrm{Tw}(\mathscr{P}^{\text{i}}, \mathrm{End}_V)$ on V as follows.

By Sect. 6.5.4, the twisting morphism φ is equivalent to a morphism of dg cooperads $f_\varphi : \mathscr{P}^{\text{i}} \to \mathrm{B}\,\mathrm{End}_W$. We compose it with the morphism of dg cooperads $\Psi : \mathrm{B}\,\mathrm{End}_W \to \mathrm{B}\,\mathrm{End}_V$. The resulting composite Ψf_φ is a morphism of dg cooperads, which gives the expected twisting morphism $\psi \in \mathrm{Tw}(\mathscr{P}^{\text{i}}, \mathrm{End}_V)$ by Sect. 6.5.4 again.

$$\begin{array}{ccc} \mathscr{P}^{\text{i}} & \xrightarrow{f_\varphi} & \mathrm{B}\,\mathrm{End}_W \\ & \searrow^{f_\psi} & \downarrow \Psi \\ & & \mathrm{B}\,\mathrm{End}_V\,. \end{array}$$

The associated twisting morphism $\psi : \mathscr{P}^{\text{i}} \to \mathrm{End}_V$ is equal to the projection of Ψf_φ on End_V. By a slight abuse of notation, we still denote it by

$$\psi = \Psi f_\varphi : \mathscr{P}^{\text{i}} \to \mathrm{End}_V.$$

Theorem 10.3.3 (Explicit formula). *Let $\mathscr{P}$ be a Koszul operad, let $\varphi \in \mathrm{Tw}(\mathscr{P}^{i}, \mathrm{End}_W)$ be a $\mathscr{P}_\infty$-algebra structure on W, and let (V, d_V) be a homotopy retract of (W, d_W).*

The transferred $\mathscr{P}_\infty$-algebra structure $\psi \in \mathrm{Tw}(\mathscr{P}^{\text{i}}, \mathrm{End}_V)$, defined above, on the dg module V is equal to the composite

$$\mathscr{P}^{i} \xrightarrow{\Delta_{\mathscr{P}^{i}}} \mathscr{T}^c\big(\overline{\mathscr{P}}^{i}\big) \xrightarrow{\mathscr{T}^c(s\varphi)} \mathscr{T}^c(s\mathrm{End}_W) \xrightarrow{\Psi} \mathrm{End}_V,$$

where $\Delta_{\overline{\mathscr{P}}^{i}}$ is the structure map corresponding to the combinatorial definition of the cooperad $\mathscr{P}^{i}$, see Sect. 5.8.8.

Proof. By Proposition 5.8.6, the unique morphism of dg cooperads $f_\varphi : \mathscr{P}^{\text{i}} \to \mathrm{B}\,\mathrm{End}_W = \mathscr{T}^c(s\,\mathrm{End}_W)$, which extends $s\varphi : \mathscr{P}^{\text{i}} \to s\,\mathrm{End}_W$, is equal to

$$\mathscr{P}^{\text{i}} \xrightarrow{\Delta_{\mathscr{P}^{\text{i}}}} \mathscr{T}^c\big(\overline{\mathscr{P}}^{\text{i}}\big) \xrightarrow{\mathscr{T}^c(s\varphi)} \mathscr{T}^c(s\mathrm{End}_W).$$ □

So the transferred structure given here is the composite of three distinct maps. The first map depends only on the cooperad $\mathscr{P}^{¡}$, that is on the type of algebraic structure we want to transfer. The second map depends only the starting $\mathscr{P}_\infty$-algebra structure. And the third map depends only on the homotopy retract data.

10.3.4 Examples: A_∞ and L_∞-Algebras Transferred

In the case of A_∞-algebras, we recover the formulas given in Sect. 9.4.

Theorem 10.3.4. *Let $\{m_n : W^{\otimes n} \to W\}_{n\geq 2}$ be an A_∞-algebra structure on W. The transferred A_∞-algebra structure $\{m'_n : V^{\otimes n} \to V\}_{n\geq 2}$ on a homotopy retract V is equal to*

$$m'_n = \sum_{PT_n} \pm$$

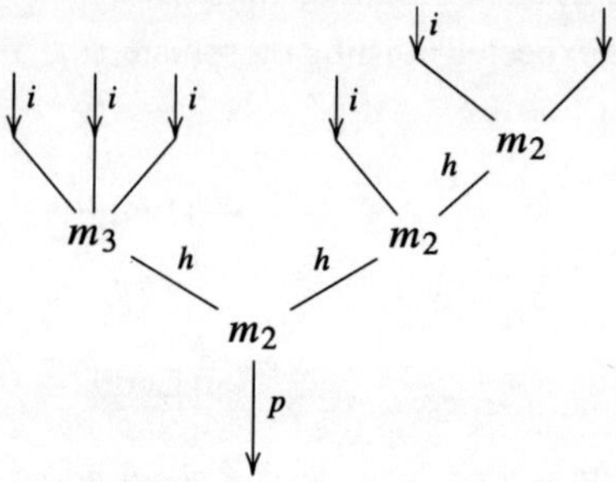

,

where the sum runs over the set PT_n of planar rooted trees with n leaves.

Proof. The combinatorial definition of the (nonsymmetric) cooperad $As^{¡}$ is given by

$$\Delta_{As^{¡}} : \mu_n^c \mapsto \sum_{t\in \mathscr{PT}_n} \pm t(\mu^c) \in \mathscr{T}^c(\overline{As}^{¡}),$$

where the sum runs over planar rooted trees t with n leaves and whose vertices with k inputs are labeled by μ_k^c. We conclude with Theorem 10.3.3. □

Theorem 10.3.5. *Let $\{\ell_n : W^{\otimes n} \to W\}_{n\geq 2}$ be an L_∞-algebra structure on W. The transferred L_∞-algebra structure $\{l_n : V^{\otimes n} \to V\}_{n\geq 2}$ on a homotopy retract V is equal to*

$$l_n = \sum_{t\in RT_n} \pm pt(\ell, h) i^{\otimes n},$$

where the sum runs over rooted trees t with n leaves and where the notation $t(\ell, h)$ stands for the n-multilinear operation on V defined by the composition scheme t with vertices labeled by the ℓ_k and internal edges labeled by h.

Proof. By Sect. 7.2.3, the Koszul dual cooperad $\mathit{Lie}^{¡}$ is isomorphic to $\mathrm{End}^c_{s^{-1}\mathbb{K}} \underset{\mathrm{H}}{\otimes} \mathit{Com}^*$. Therefore, the decomposition map of the combinatorial definition of the cooperad $\mathit{Lie}^{¡}$ is given, up to signs, by the one of Com^*, which is made up of nonplanar rooted trees. □

10.3.5 Infinity-Quasi-isomorphism

We define a map $\widetilde{\Psi} : \mathscr{T}^c(s\mathrm{End}_W) \to \mathrm{End}^V_W$ by the same formula as Ψ except for the root, which is labeled by the homotopy h and not by p this time. We consider the map $i_\infty : \mathscr{P}^{¡} \to \mathrm{End}^V_W$ defined by the following composite:

$$i_\infty : \overline{\mathscr{P}^{¡}} \xrightarrow{\Delta_{\mathscr{P}^{¡}}} \mathscr{T}\big(\overline{\mathscr{P}^{¡}}\big) \xrightarrow{\mathscr{T}(s\varphi)} \mathscr{T}(s\mathrm{End}_W) \xrightarrow{\widetilde{\Psi}} \mathrm{End}^V_W$$

and by $\mathrm{id} \in \mathrm{I} \mapsto i \in \mathrm{Hom}(V, W) \subset \mathrm{End}^V_W$.

Theorem 10.3.6. *Let $\mathscr{P}$ be a Koszul operad, let (W, φ) be a $\mathscr{P}_\infty$-algebra, and let (V, d_V) be a homotopy retract of (W, d_W).*

The map $i_\infty : \mathscr{P}^{¡} \to \mathrm{End}^V_W$ is an ∞-quasi-isomorphism between the $\mathscr{P}_\infty$-algebra (V, ψ), with the transferred structure, and the $\mathscr{P}_\infty$-algebra (W, φ).

Proof. Using Theorem 10.2.3, we have to prove that $i_\infty * \psi - \varphi \circledast i_\infty = \partial(i_\infty)$.

The first term $i_\infty * \psi$ is equal to

$$\big(\big(\widetilde{\Psi}\mathscr{T}^c(s\varphi)\Delta_{\mathscr{P}^{¡}}\big) \circ_{(1)} \big(\Psi\mathscr{T}^c(s\varphi)\Delta_{\mathscr{P}^{¡}}\big)\big)\Delta_{(1)} + i_*\Psi\mathscr{T}^c(s\varphi)\Delta_{\mathscr{P}^{¡}}.$$

It is equal to the composite $\widehat{\Psi}\mathscr{T}^c(s\varphi)\Delta_{\mathscr{P}^{¡}}$, where $\widehat{\Psi}$ is defined as $\widetilde{\Psi}$, except that either one internal edge or the root is labeled by ip instead of h. To prove this, we use the formula of $\Delta_{\mathscr{P}^{¡}}$ given in Sect. 5.8 in terms of the iterations of $\tilde{\Delta}$.

The second term $-\varphi \circledast i_\infty$ is equal to $-(\varphi \circ (\widetilde{\Psi}\mathscr{T}^c(s\varphi)\Delta_{\mathscr{P}^{¡}}))\Delta$. It is equal to $-\check{\Psi}\mathscr{T}^c(s\varphi)\Delta_{\mathscr{P}^{¡}}$, where $\check{\Psi}$ is defined as Ψ, except for the root, which is labeled by the identity of W.

The right-hand side $\partial(i_\infty)$ is equal to $d_{\mathrm{End}^V_W} i_\infty - i_\infty d_{\mathscr{P}^{¡}}$. The latter term $i_\infty d_{\mathscr{P}^{¡}}$ is equal to $\widetilde{\Psi}\mathscr{T}^c(s\varphi)\Delta_{\mathscr{P}^{¡}} d_{\mathscr{P}^{¡}}$. Since $d_{\mathscr{P}^{¡}}$ is a coderivation of the cooperad $\mathscr{P}^{¡}$, we get $i_\infty d_{\mathscr{P}^{¡}} = \widetilde{\Psi}\mathscr{T}^c(s\varphi; s\varphi d_{\mathscr{P}^{¡}})\Delta_{\mathscr{P}^{¡}}$, where the notation $\mathscr{T}^c(f; g)$ was introduced in Sect. 6.3.2. The other term $d_{\mathrm{End}^V_W} i_\infty$ is equal to

$$-\widetilde{\Psi}\mathscr{T}^c(s\varphi; sd_{\mathrm{End}_W}\varphi)\Delta_{\mathscr{P}^{¡}} - \mathring{\Psi}\mathscr{T}^c(s\varphi)\Delta_{\mathscr{P}^{¡}} + i_*\Psi\mathscr{T}^c(s\varphi)\Delta_{\mathscr{P}^{¡}} - \check{\Psi}\mathscr{T}^c(s\varphi)\Delta_{\mathscr{P}^{¡}},$$

where $\mathring{\Psi}$ is defined as $\widetilde{\Psi}$, except that one internal edge is labeled by $[d_W, h]$ instead of h. Since φ is a twisting morphism, $\varphi \in \mathrm{Tw}(\mathscr{P}^{¡}, \mathrm{End}_W)$, it satisfies the Maurer–Cartan equation $-d_{\mathrm{End}_W}\varphi - \varphi d_{\mathscr{P}^{¡}} = (\varphi \circ_{(1)} \varphi)\Delta_{(1)}$. Therefore

$$-\widetilde{\Psi}\mathscr{T}^c(s\varphi; sd_{\mathrm{End}_W}\varphi) - \widetilde{\Psi}\mathscr{T}^c(s\varphi; s\varphi d_{\mathscr{P}^{¡}}) = \overline{\Psi}\mathscr{T}^c(s\varphi)\Delta_{\mathscr{P}^{¡}},$$

where $\overline{\Psi}$ is defined as $\widetilde{\Psi}$ except that one internal edge is labeled by the identity of W instead of h.

We conclude by using $[d_W, h] = \mathrm{Id}_W - ip$. □

This theorem provides a homotopy control of the transferred structure: the starting $\mathcal{P}_\infty$-algebra and the transferred one are related by an explicit ∞-quasi-isomorphism. Therefore the two $\mathcal{P}_\infty$-algebras are homotopy equivalent, see Sect. 10.4.4.

Theorem 10.3.7. *Let $\mathcal{P}$ be a Koszul operad and let $i : V \xrightarrow{\sim} W$ be a quasi-isomorphism. Any $\mathcal{P}_\infty$-algebra structure on W can be transferred into a $\mathcal{P}_\infty$-algebra structure on V such that i extends to an ∞-quasi-isomorphism.*

Proof. Since we work over a field, any quasi-isomorphism $i : V \xrightarrow{\sim} W$ extends to a homotopy retract

$$h \circlearrowright (W, d_W) \underset{i}{\overset{p}{\rightleftarrows}} (V, d_V).$$

One shows this fact by refining the arguments of Lemma 9.4.4. So this result is equivalent to Theorem 10.3.1. □

10.3.6 Operadic Massey Products

In this section, we suppose the characteristic of the ground field $\mathbb{K}$ to be 0. Let (A, d) be a chain complex. Recall from Lemma 9.4.4, that, under a choice of sections, the homology $(H(A), 0)$ is a deformation retract of (A, d)

$$h \circlearrowright (A, d_A) \underset{i}{\overset{p}{\rightleftarrows}} \big(H(A), 0\big).$$

Lemma 10.3.8. *By construction, these maps also satisfy the following* side conditions:

$$h^2 = 0, \qquad ph = 0, \qquad hi = 0.$$

Proof. It is a straightforward consequence of the proof of Lemma 9.4.4. □

As a consequence, when A carries a $\mathcal{P}_\infty$-algebra structure, its homotopy $H(A)$ is endowed with a $\mathcal{P}_\infty$-algebra structure, such that the map i extends to an ∞-quasi-isomorphism, by Theorem 10.3.1. In this case, we can prove the same result for the map p as follows.

To the homotopy h relating Id_A and ip, we associate the degree one map $h^n : A^{\otimes n} \to A^{\otimes n}$ defined by

$$h^n := \frac{1}{n!} \sum_{\sigma \in \mathbb{S}_n} h^\sigma,$$

where

$$\begin{aligned} h^\sigma := &\ \mathrm{id} \otimes \cdots \otimes \mathrm{id} \otimes \underbrace{h}_{\sigma(1)} \otimes \mathrm{id} \otimes \cdots \otimes \mathrm{id} \\ &+ \mathrm{id} \otimes \cdots \otimes \mathrm{id} \otimes \underbrace{ip}_{\sigma(1)} \otimes \mathrm{id} \otimes \cdots \otimes \mathrm{id} \otimes \underbrace{h}_{\sigma(2)} \otimes \mathrm{id} \otimes \cdots \otimes \mathrm{id} + \cdots \\ &+ i \circ p \otimes \cdots \otimes ip \otimes \underbrace{h}_{\sigma(n)} \otimes ip \otimes \cdots \otimes ip. \end{aligned}$$

The map h^n is a symmetric homotopy relating $\mathrm{Id}_A^{\otimes n}$ and $(ip)^{\otimes n}$, that is

$$\partial(h^n) = \mathrm{Id}_A^{\otimes n} - (ip)^{\otimes n} \quad \text{and} \quad h^n \sigma = \sigma h^n, \quad \forall \sigma \in \mathbb{S}_n.$$

We denote by H the sum $\mathrm{H} := \sum_{n \geq 1} h^n : \overline{T}(A) \to \overline{T}(A)$.

We define the map Δ^{lev} as follows. To any element $\mu^c \in \mathscr{P}^{\mathrm{i}}$, its image under $\Delta_{\mathscr{P}^{\mathrm{i}}}$ is a sum of trees. To any of these trees, we associate the sum of all the leveled trees obtained by putting one and only one nontrivial vertex per level. (Notice that this operation might permute vertices and therefore it might yield signs.) The image of μ^c under Δ^{lev} is the sum of all these leveled trees.

Proposition 10.3.9. *Let $\mathbb{K}$ be a field of characteristic 0. Let $\mathscr{P}$ be a Koszul operad and let (A, φ) be a $\mathscr{P}_\infty$-algebra. The chain map $p : A \to H(A)$ extends to an ∞-quasi-isomorphism p_∞ given by the formula*

$$p_\infty := p_* \mathscr{T}^c_{\mathrm{lev}}(\varphi, H) \Delta^{\mathrm{lev}}$$

on $\overline{\mathscr{P}^i}$ and by $\mathrm{id} \in \mathscr{P}^i \mapsto p$. The map $\mathscr{T}^c_{\mathrm{lev}}(\varphi, H)$ first labels the vertices of a leveled tree by φ and the levels (including the leaves) by H and then composes the associated maps in $\mathrm{End}^A_{H(A)}$.

Proof. The map p_∞ given by this formula is well defined thanks to the conilpotency of the Koszul dual cooperad $\mathscr{P}^{\mathrm{i}}$. Let us denote by ψ the transferred $\mathscr{P}_\infty$-algebra structure on $H(A)$. By Theorem 10.2.3, we have to prove that $p_\infty * \varphi - \psi \circledast p_\infty = \partial(p_\infty)$. The arguments are similar to the arguments used in the proofs of Theorem 10.3.6 and Theorem 10.4.1 but use the side conditions of Lemma 10.3.8. The computations are left to the reader as a good exercise. □

Theorem 10.3.10 (Higher structures). *Let $\mathbb{K}$ be a field of characteristic 0. Let $\mathscr{P}$ be a Koszul operad and let A be a $\mathscr{P}_\infty$-algebra.*

◇ *There is a $\mathscr{P}_\infty$-algebra structure on the homology $H(A)$ of the underlying chain complex of A, which extends its $\mathscr{P}$-algebra structure.*
◇ *The embedding $i : H(A) \rightarrowtail A$ and the projection $p : A \twoheadrightarrow H(A)$, associated to the choice of sections for the homology, extend to ∞-quasi-isomorphisms of $\mathscr{P}_\infty$-algebras.*
◇ *The $\mathscr{P}_\infty$-algebra structure on the homotopy $H(A)$ is independent of the choice of sections of $H(A)$ into A in the following sense: any two such transferred structures are related by an ∞-isomorphism, whose first map is the identity on $H(A)$.*

Proof. The explicit form of the transferred $\mathscr{P}_\infty$-algebra structure on $H(A)$, given in Theorem 10.3.3, proves that it extends the $\mathscr{P}$-algebra structure given in Proposition 10.1.5.

The embedding $H(A) \rightarrowtail A$ extends to an ∞-quasi-isomorphism by Theorem 10.3.6. The projection $A \twoheadrightarrow H(A)$ extends to an ∞-quasi-isomorphism by Proposition 10.3.9.

Let (i, p) and (i', p') be the maps associated to two decompositions of the chain complex A. They induce two $\mathscr{P}_\infty$-algebra structures on $H(A)$ such that i, i', p and p' extend to ∞-quasi-isomorphisms by Theorem 10.3.6 and Proposition 10.3.9. The composite $p'_\infty i_\infty$ defines an ∞-quasi-isomorphism, from $H(A)$ with the first transferred structure to $H(A)$ with the second transferred structure, such that the first component is equal to $p'i = \mathrm{Id}_{H(A)}$. □

These higher $\mathscr{P}_\infty$-operations on the homotopy of a $\mathscr{P}_\infty$-algebra are called the *operadic Massey products*.

EXAMPLES. The case of the operad *As* has already been treated in Sect. 9.4.5. The terminology "Massey products" comes from the example given by the singular cochains $C^\bullet_{\mathrm{sing}}(X)$ of a topological space X endowed with its associative cup product [Mas58]. The case of the operad *Lie* was treated by Retakh in [Ret93].

Though the differential on $H(A)$ is equal to 0, the $\mathscr{P}_\infty$-algebra structure on $H(A)$ is *not* trivial in general. In this case, the relations satisfied by the $\mathscr{P}_\infty$-algebra operations on $H(A)$ do not involve any differential. Hence the operations of weight 1 satisfy the relations of a $\mathscr{P}$-algebra. But the higher operations exist and contain the homotopy data of A.

10.3.7 An Example: HTT for the Dual Numbers Algebra

The homotopy transfer theorem (HTT) can be applied to reduced operads which are concentrated in arity 1, that is to unital associative algebras. Recall that for such an operad, an algebra over it is simply a left module. The algebra of dual numbers $D := \mathbb{K}[\varepsilon]/(\varepsilon^2 = 0)$ is obviously Koszul and its Koszul dual coalgebra is the free coalgebra on one cogenerator $D^{¡} := \mathbb{K}[s\varepsilon]$. Observe that the element $(s\varepsilon)^n$ is in degree n and that the coproduct is given by

$$\Delta\big((s\varepsilon)^n\big) = \sum_{\substack{i+j=n \\ i\geq 0, j\geq 0}} (s\varepsilon)^i (s\varepsilon)^j .$$

From Sect. 2.2.2 we can compute $D_\infty := \Omega D^{¡}$. It follows that a D_∞-module is a chain complex (A, d) equipped with linear maps

$$t_n : A \to A, \quad \text{for } n \geq 1, \quad |t_n| = n - 1,$$

such that for any $n \geq 1$ the following identities hold

$$\partial(t_n) = -\sum_{\substack{i+j=n \\ i\geq 1, j\geq 1}} (-1)^i t_i t_j .$$

Observe that, denoting $t_0 := d$, this identity becomes

$$\sum_{\substack{i+j=n \\ i\geq 0, j\geq 0}} (-1)^i t_i t_j = 0, \quad \text{for any } n \geq 0.$$

Such a structure $(A, t_0, \ldots, t_n, \ldots)$ is called a *chain multicomplex*. As expected a D-module is a particular case of chain multicomplex for which $t_n = 0$ for $n \geq 2$.

The HTT can be written for any homotopy retract whose big chain complex is a chain multicomplex $(A, \{t_n\}_{n\geq 0})$ and it gives a chain multicomplex structure on the small chain complex $(V, \{t'_n\}_{n\geq 0})$. The explicit formulas are as follows:

$$t'_n := p\Big(\sum \pm t_{j_1} h t_{j_2} h \cdots h t_{j_k}\Big) i,$$

for any $n \geq 1$, where the sum runs over all the families $(j_1, \ldots, j_k)$ such that $j_1 + \cdots + j_k = n$.

SPECTRAL SEQUENCE. Let us look at a particular case. Any first quadrant bicomplex $(C_{\bullet,\bullet}, d^v, d^h)$ gives rise to a chain complex (A, d), which is a left module over D by declaring that $A_n := \bigoplus_p C_{p,n}, d := d^v$ and the action of ε is induced by d^h. More precisely, since $d^h d^v + d^h d^v = 0$, the restriction of ε to A_n is $(-1)^n d^h$.

It is well known that any first quadrant bicomplex gives rise to a spectral sequence $\{(E^n, d^n)\}_{n\geq 1}$ where $E^1 = H_\bullet(C, d^v)$ and $E^n = H_\bullet(E^{n-1}, d^{n-1})$. We claim that, after choosing sections which make $(E^1, 0)$ into a deformation retract of (C, d^v), cf. Lemma 9.4.4, the chain multicomplex structure of E^1 gives the spectral sequence. More precisely the map d^n is induced by t'_n.

The advantage of this point of view on bicomplexes, versus spectral sequences, is that the HTT can be applied to bicomplexes equipped with a deformation retract whose boundary map is not necessarily trivial.

For instance the cyclic bicomplex of a unital associative algebra, which involves the boundary maps b, b' and the cyclic operator, cf. [LQ84, Lod98], admits a deformation retract made up of the columns involving only b. Applying the HTT to it gives a chain multicomplex for which

$$t'_0 = b, \qquad t'_1 = 0, \qquad t'_2 = B, \qquad t'_n = 0, \quad \text{for } n \geq 3.$$

So, we get automatically Connes' boundary map B and we recover the fact that, in cyclic homology theory, the (b, B)-bicomplex is quasi-isomorphic to the cyclic bicomplex.

The details for this section can be found in [LV12], where direct explicit proofs are given. Similar results based simplicial technics can be found in [Mey78] and based on the perturbation lemma in [Lap01].

10.4 Inverse of ∞-Isomorphisms and ∞-Quasi-isomorphisms

In this section, we first prove that the ∞-isomorphisms are the invertible ∞-morphisms in the category ∞-$\mathcal{P}_\infty$-alg. We give the formula for the inverse of an ∞-morphism. Then we show that any $\mathcal{P}_\infty$-algebra is ∞-isomorphic to the product of a $\mathcal{P}_\infty$-algebra whose internal differential is null, with a $\mathcal{P}_\infty$-algebra whose structure operations are null and whose underlying chain complex is acyclic. Applying these two results, we prove that any ∞-quasi-isomorphism admits an ∞-quasi-isomorphism in the opposite direction. So, being ∞-quasi-isomorphic defines an equivalence relation among $\mathcal{P}_\infty$-algebras, which is called the homotopy equivalence.

10.4.1 Inverse of Infinity-Isomorphisms

Here we find the formula for the inverse of an ∞-isomorphism. We use the maps $\hat{\Delta}^k : \mathcal{P}^{¡} \to (\mathcal{P}^{¡})^{\circ(k+1)}$ introduced in Sect. 5.8.5.

Theorem 10.4.1. *Let $\mathcal{P}$ be a Koszul operad and let A and B be two $\mathcal{P}_\infty$-algebras. Any ∞-isomorphism $f : A \rightsquigarrow B$ admits a unique inverse in the category ∞-$\mathcal{P}_\infty$-alg. When f is expressed in terms of $f : \mathcal{P}^{¡} \to \mathrm{End}^A_B$, its inverse is given by the formula $(f^{-1})_{(0)} := (f_{(0)})^{-1} : B \to A$ and by*

$$f^{-1} := \sum_{k=0}^{\infty} (-1)^{k+1} \left(f_{(0)}^{-1}\right)_* \left(\left(f_{(0)}^{-1}\right)^* (f)\right)^{\circ(k+1)} \hat{\Delta}^k,$$

on $\overline{\mathcal{P}^{¡}}$, where the right-hand side is equal to the composite

$$\mathcal{P}^{¡} \xrightarrow{\hat{\Delta}^k} \left(\mathcal{P}^{¡}\right)^{\circ(k+1)} \xrightarrow{((f_{(0)}^{-1})^*(f))^{\circ(k+1)}} (\mathrm{End}_B)^{\circ(k+1)} \to \mathrm{End}_B \xrightarrow{(f_{(0)}^{-1})_*} \mathrm{End}^B_A.$$

For example, when $\hat{\Delta}$ produces an element of the form

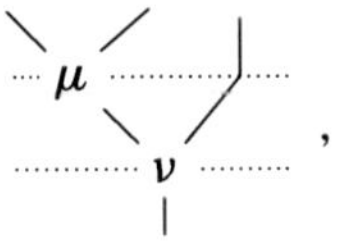

the associated composite in End_A^B is

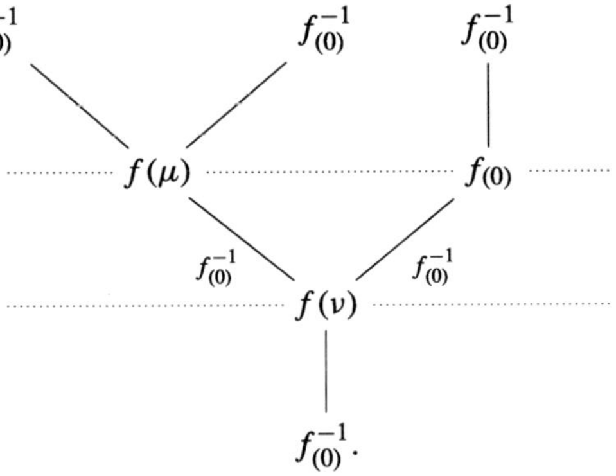

Proof. Let us denote by $g : \mathscr{P}^{\text{i}} \to \mathrm{End}_A^B$ the above defined map. It is well defined thanks to the conilpotency (Sect. 5.8.5) of the Koszul dual cooperad $\mathscr{P}^{\text{i}}$.

We first show that g is an ∞-morphism. Let us denote by $\varphi \in \mathrm{Tw}(\mathscr{P}^{\text{i}}, \mathrm{End}_A)$ and by $\psi \in \mathrm{Tw}(\mathscr{P}^{\text{i}}, \mathrm{End}_B)$ the respective $\mathscr{P}_\infty$-algebra structures. By Theorem 10.2.3, we prove now that $g * \psi - \varphi \circledast g = \partial(g)$ in $\mathrm{Hom}_{\mathbb{S}}(\mathscr{P}^{\text{i}}, \mathrm{End}_A^B)$. By Proposition 5.8.6, the map g is equal to the composite

$$\mathscr{P}^{\text{i}} \xrightarrow{\Delta_{\mathscr{P}^{\text{i}}}} \mathscr{T}^c\big(\overline{\mathscr{P}^{\text{i}}}\big) \xrightarrow{\mathscr{T}^c(f)} \mathscr{T}^c\big(\mathrm{End}_B^A\big) \xrightarrow{\Theta} \mathrm{End}_A^B,$$

where the map Θ amounts to labeling the leaves, the internal edges and the root of the trees by f_0^{-1} and to composing all the maps along the tree scheme. It also multiplies the elements by $(-1)^k$, where k is the minimal number of levels on which the tree can be put.

The derivative $\partial(g)$ is equal to

$$\partial(g) = d_{\mathrm{End}_A^B} \Theta \mathscr{T}^c(f) \Delta_{\mathscr{P}^{\text{i}}} - \Theta \mathscr{T}^c(f) \Delta_{\mathscr{P}^{\text{i}}} d_{\mathscr{P}^{\text{i}}}.$$

Since $d_{\mathscr{P}^{\text{i}}}$ is a coderivation for the cooperad $\mathscr{P}^{\text{i}}$, we get

$$\partial(g) = \Theta \mathscr{T}^c(f; d_{\mathrm{End}_B^A} f - f d_{\mathscr{P}^{\text{i}}}) \Delta_{\mathscr{P}^{\text{i}}}.$$

By Theorem 10.2.3, since the map f is an ∞-morphism, it satisfies $\partial(f) = f * \varphi - \psi \circledast f$. So we get

$$\partial(g) = \Theta\mathcal{T}^c(f; f * \varphi)\Delta_{\mathcal{P}^{\mathrm{i}}} - \Theta\mathcal{T}^c(f; \psi \circledast f)\Delta_{\mathcal{P}^{\mathrm{i}}}.$$

In $f * \varphi$, there are two kinds of terms involving either $f_{(0)}$ or $f_{(>1)}$. Therefore the term $\Theta\mathcal{T}^c(f; f * \varphi)\Delta_{\mathcal{P}^{\mathrm{i}}}$ splits into two sums on the trees produced by $\Delta_{\mathcal{P}^{\mathrm{i}}}$. Because of the sign based on the number of levels, almost all the terms cancel. Only remains the trees with φ labeling the vertex above the root. Hence, we get

$$\Theta\mathcal{T}^c(f; f * \varphi)\Delta_{\mathcal{P}^{\mathrm{i}}} = -\varphi \circledast g.$$

Using the same kind of arguments, one proves that $\Theta\mathcal{T}^c(f; \psi \circledast f)\Delta_{\mathcal{P}^{\mathrm{i}}}$ is made up of trees with ψ labeling any vertex at the top of the tree, that is

$$\Theta\mathcal{T}^c(f; \psi \circledast f)\Delta_{\mathcal{P}^{\mathrm{i}}} = g * \psi.$$

By Proposition 10.2.4, it is enough to prove that $f \circledcirc g = \mathrm{Id}_B$ and that $g \circledcirc f = \mathrm{Id}_A$. Since $g_{(0)} := (f_{(0)})^{-1}$, these two relations are satisfied on $\mathrm{I} = \mathcal{P}^{\mathrm{i}\,(0)}$. Higher up, since $\Delta(\mu) = \tilde{\Delta}(\mu) + (\mathrm{id}; \mu)$, for any $\mu \in \mathcal{P}^{\mathrm{i}}$, we have

$$\begin{aligned}(f \circledcirc g)(\mu) = &\sum_{k=0}^{\infty} (-1)^k \big((f_{(0)}^{-1})^*(f)\big)^{\circ(k+1)} \hat{\Delta}^k(\mu)\\ &+ \sum_{k=0}^{\infty} (-1)^{k+1} \big((f_{(0)}^{-1})^*(f)\big)^{\circ(k+1)} \hat{\Delta}^k(\mu) = 0.\end{aligned}$$

In the same way, since $\Delta(\mu) = \bar{\Delta}(\mu) + (\mathrm{id}; \mu) + (\mu; \mathrm{id}, \dots, \mathrm{id})$, for any $\mu \in \mathcal{P}^{\mathrm{i}}$, we have

$$\begin{aligned}(g \circledcirc f)(\mu) = &\sum_{k=1}^{\infty} (-1)^k \big((f_{(0)}^{-1})_*(f)\big)^{\circ(k+1)} \hat{\Delta}^k(\mu)\\ &+ \big(f_{(0)}^{-1}\big)_*(f)(\mu)\\ &+ \sum_{k=0}^{\infty} (-1)^{k+1} \big((f_{(0)}^{-1})_*(f)\big)^{\circ(k+1)} \hat{\Delta}^k(\mu) = 0.\end{aligned}$$ □

Remark. The formula which gives the inverse of an ∞-isomorphism is related to the inverse under composition of power series as follows. Let us consider the nonsymmetric cooperad As^* and the $\mathbb{K}$-modules $A = B = \mathbb{K}$. There is a bijection between the power series $f(x) = a_0 x + a_1 x^2 + \cdots$ with coefficients in $\mathbb{K}$ and the elements of $\mathrm{Hom}(As^*, \mathrm{End}_{\mathbb{K}})$, given by $\tilde{f} := \mu_n^c \mapsto a_{n-1} 1_n$, where μ_n^c is the generating element of $As^*(n)$ and where 1_n is the generating element of $\mathrm{End}_{\mathbb{K}}(n)$. This map is an isomorphism of associative algebras: $\widetilde{g \circ f} = \tilde{g} \circledcirc \tilde{f}$. So a power series is invertible for the composition if and only if a_0 is invertible. This condition is equivalent to $\tilde{f}_{(0)}$ invertible in $\mathrm{Hom}(\mathbb{K}, \mathbb{K})$. When $a_1 = 1$, the formula given in Theorem 10.4.1

induces the formula for the inverse of the power series f. For yet another approach to this formula, see Sect. 13.11.7.

10.4.2 Decomposition: Minimal ⊕ Acyclic Trivial

By definition, a $\mathscr{P}_\infty$-algebra (A, d_A, φ) is called

◇ *minimal* when $d_A = 0$;
◇ *acyclic* when the underlying chain complex (A, d_A) is acyclic;
◇ *trivial* when the structure map is trivial: $\varphi = 0$.

Lemma 10.4.2. *Let $(H, 0, \varphi)$ be a minimal $\mathscr{P}_\infty$-algebra and let $(K, d_K, 0)$ be an acyclic trivial $\mathscr{P}_\infty$-algebra. Their product in the category ∞-$\mathscr{P}_\infty$-*alg *exists and its underlying chain complex is the direct sum $H \oplus K$.*

Proof. We consider the following $\mathscr{P}_\infty$-structure on $H \oplus K$:

$$\mathscr{P}^{\text{i}} \xrightarrow{\varphi} \mathrm{End}_H \rightarrowtail \mathrm{End}_{H\oplus K}.$$

It satisfies the Maurer–Cartan equation in $\mathrm{Hom}_{\mathbb{S}}(\mathscr{P}^{\text{i}}, \mathrm{End}_{H\oplus K})$, since φ satisfies the Maurer–Cartan equation in $\mathrm{Hom}_{\mathbb{S}}(\mathscr{P}^{\text{i}}, \mathrm{End}_H)$. To any $\mathscr{P}_\infty$-algebra B with two ∞-morphisms, $f \in \mathrm{Hom}_{\mathbb{S}}(\mathscr{P}^{\text{i}}, \mathrm{End}_H^B)$ from B to H and $g \in \mathrm{Hom}_{\mathbb{S}}(\mathscr{P}^{\text{i}}, \mathrm{End}_K^B)$ from B to K respectively, we associate the following morphism

$$\mathscr{P}^{\text{i}} \xrightarrow{f+g} \mathrm{End}_H^B \oplus \mathrm{End}_K^B \cong \mathrm{End}_{H\oplus K}^B.$$

We leave it to the reader to check that this composite is an ∞-morphism, which satisfies the universal property of products. □

Theorem 10.4.3 (Minimal model for $\mathscr{P}_\infty$-algebras). *Let $\mathbb{K}$ be a field of characteristic 0 and let $\mathscr{P}$ be a Koszul operad. In the category of $\mathscr{P}_\infty$-algebras with ∞-morphisms, any $\mathscr{P}_\infty$-algebra is ∞-isomorphic to the product of a minimal $\mathscr{P}_\infty$-algebra, given by the transferred structure on its homotopy, with an acyclic trivial $\mathscr{P}_\infty$-algebra.*

Proof. Let (A, d_A, φ) be a $\mathscr{P}_\infty$-algebra. As in Lemma 9.4.4, we decompose the chain complex A with respect to its homology and boundary: $A_n \cong B_n \oplus H_n \oplus B_{n-1}$. We denote by $K_n := B_n \oplus B_{n-1}$ the acyclic sub-chain complex of A, so that A is the direct sum of the two dg modules $A \cong H(A) \oplus K$. By Theorems 10.3.1 and 10.3.10, the homotopy $H(A)$ is endowed with a minimal $\mathscr{P}_\infty$-algebra structure and we consider the trivial $\mathscr{P}_\infty$-algebra structure on K.

We define an ∞-morphism f from A to K as follows. Let q denote the projection from A to K and let of $f_{(0)}$ be equal to q. Higher up, f is given by the composite $f = (qh)_* \varphi$

$$f : \overline{\mathscr{P}^{¡}} \xrightarrow{\varphi} \mathrm{End}_A \xrightarrow{(qh)_*} \mathrm{End}^A_K .$$

Since the $\mathscr{P}_\infty$-algebra structure on K is trivial, we only have to check the equality $f * \varphi = \partial(f)$, by Theorem 10.2.3. This equation reads on $\overline{\mathscr{P}^{¡}}$:

$$(qh)_* (\varphi \circ_{(1)} \varphi)\Delta_{(1)} + q_* \varphi = d_{\mathrm{End}^A_K} (q\, h)_* \varphi - (qh)_* \varphi\, d_{\mathscr{P}^{¡}} .$$

Since μ is a twisting morphism, we have

$$(qh)_* (\varphi \circ_{(1)} \varphi)(\Delta_{(1)}) = -(qh)_* d_{\mathrm{End}_A} \varphi - (qh)_* \varphi\, d_{\mathscr{P}^{¡}} .$$

We conclude by using the equality $q(h \circ d_A + d_A \circ h) = q$.

The ∞-morphism p_∞ from A to $H(A)$ of Proposition 10.3.9 together with the ∞-morphism f from A to K induce an ∞-morphism from A to $H(A) \oplus K$, since this latter space is the product of $H(A)$ and K by Lemma 10.4.2. The first component of this ∞-morphism is equal to $p + q : A \cong H(A) \oplus K$, which is an isomorphism. □

10.4.3 Inverse of Infinity-Quasi-isomorphisms

Theorem 10.4.4. *Let $\mathscr{P}$ be a Koszul operad and let A and B be two $\mathscr{P}_\infty$-algebras. If there exists an ∞-quasi-isomorphism $A \overset{\sim}{\rightsquigarrow} B$, then there exists an ∞-quasi-isomorphism in the opposite direction $B \overset{\sim}{\rightsquigarrow} A$, which is the inverse of $H(A) \xrightarrow{\cong} H(B)$ on the level on holomogy.*

Proof. Let $f : A \overset{\sim}{\rightsquigarrow} B$ denote an ∞-quasi-isomorphism. By Theorem 10.3.6 and Proposition 10.3.9, the following composite g of ∞-quasi-isomorphisms

$$H(A) \overset{i^A_\infty}{\rightsquigarrow} A \overset{f}{\rightsquigarrow} B \overset{p^B_\infty}{\rightsquigarrow} H(B)$$

is an ∞-isomorphism. It admits an inverse ∞-isomorphism $g^{-1} : H(B) \rightsquigarrow H(A)$ by Theorem 10.4.1. The ∞-quasi-isomorphism $B \overset{\sim}{\rightsquigarrow} A$ is given by the following composite of ∞-quasi-isomorphisms

$$B \overset{p^B_\infty}{\rightsquigarrow} H(B) \overset{g^{-1}}{\rightsquigarrow} H(A) \overset{i^A_\infty}{\rightsquigarrow} A .$$ □

10.4.4 Homotopy Equivalence

We define the following relation among $\mathscr{P}_\infty$-algebras: *a $\mathscr{P}_\infty$-algebra A is homotopy equivalent to a $\mathscr{P}_\infty$-algebra B* if there exists an ∞-quasi-isomorphism from

A to B. The previous section shows that it is an equivalence relation. We denote it by $A \sim B$.

Under this terminology, the Higher Structure Theorem 10.3.10 implies that in the homotopy class of any $\mathscr{P}_\infty$-algebra, there is a minimal $\mathscr{P}_\infty$-algebra.

10.5 Homotopy Operads

In this section, we relax the notion of operad, up to homotopy, thereby defining homotopy operads. As for associative algebras and homotopy associative algebras, the relations satisfied by the partial compositions of an (nonunital) operad are relaxed up to a full hierarchy of higher homotopies. We introduce the notion of ∞-*morphism* for homotopy operads. We have already used this notion, without saying it, in Proposition 10.3.2, where the morphism Ψ is an ∞-morphism of operads.

We describe a functor from homotopy operads to homotopy Lie algebras.

Finally, we show that a homotopy representation of an operad, that is a homotopy morphism from $\mathscr{P}$ to End_A, is equivalent to a $\Omega \mathrm{B}\mathscr{P}$-algebra structure on A.

The notions of homotopy operad and ∞-morphism come from the work of Pepijn Van der Laan [vdL02, vdL03].

10.5.1 Definition

A *homotopy operad* is a graded $\mathbb{S}$-module $\mathscr{P}$ with a square-zero coderivation d of degree -1 on the cofree conilpotent cooperad $\mathscr{T}^c(s\mathscr{P})$. By extension, we call the dg cooperad $(\mathscr{T}^c(s\mathscr{P}), d)$ the *bar construction* of the homotopy operad $\mathscr{P}$ and we denote it by $\mathrm{B}\mathscr{P}$. Hence any nonunital operad $\mathscr{P}$ is a homotopy operad and the associated bar construction coincides with the classical bar construction of Sect. 6.5.1.

For any graded $\mathbb{S}$-module M, recall from Proposition 6.3.7 that any coderivation d_γ on the cofree cooperad $\mathscr{T}^c(M)$ is completely characterized by its projection onto the space of cogenerators $\gamma = \mathrm{proj} \circ d_\gamma : \overline{\mathscr{T}}^c(M) \to M$.

Let α and β be maps in $\mathrm{Hom}_{\mathbb{S}}(\overline{\mathscr{T}}^c(M), M)$. Their *convolution product* $\alpha \star \beta$ is defined by the composite

$$\alpha \star \beta := \overline{\mathscr{T}}^c(M) \xrightarrow{\Delta'} \overline{\mathscr{T}}^c\big(M; \overline{\mathscr{T}}^c(M)\big) \xrightarrow{\overline{\mathscr{T}}^c(\mathrm{Id}_M;\beta)} \overline{\mathscr{T}}^c(M; M) \to \overline{\mathscr{T}}^c(M) \xrightarrow{\alpha} M,$$

where the first map Δ' singles out every nontrivial subtree of a tree whose vertices are indexed by M, see Sect. 6.3.8.

Lemma 10.5.1. *For any map γ of degree -1 in* $\mathrm{Hom}_{\mathbb{S}}(\overline{\mathscr{T}}^c(M), M)$, *the associated coderivation d_γ on the cofree cooperad $\mathscr{T}^c(M)$ satisfies*

$$(d_\gamma)^2 = d_{\gamma \star \gamma}.$$

Proof. Since γ has degree -1, the composite $d_\gamma \circ d_\gamma$ is equal to $\frac{1}{2}[d_\gamma, d_\gamma]$; so it is a coderivation of $\overline{\mathcal{T}}^c(M)$. By Proposition 6.3.7, it is completely characterized by its projection onto M: $\mathrm{proj}((d_\gamma)^2) = \gamma \star \gamma$. □

Any degree -1 square-zero coderivation d on the cofree cooperad $\mathcal{T}^c(s\mathcal{P})$ is equal to the sum $d = d_1 + d_\gamma$, where d_1 is the coderivation which extends an internal differential $d_{\mathcal{P}}$ on $\mathcal{P}$ and where d_γ is the unique coderivation which extends the restriction $\gamma := \mathrm{proj}(d) : \mathcal{T}^c(s\mathcal{P})^{(\geq 2)} \to s\mathcal{P}$.

Proposition 10.5.2. *Let $(\mathcal{P}, d_{\mathcal{P}})$ be a dg $\mathbb{S}$-module. A structure of homotopy operad on $\mathcal{P}$ is equivalently defined by a map $\gamma : \mathcal{T}^c(s\mathcal{P})^{(\geq 2)} \to s\mathcal{P}$ of degree -1 such that*

$$\partial(\gamma) + \gamma \star \gamma = 0$$

in $\mathrm{Hom}_{\mathbb{S}}(\overline{\mathcal{T}}^c(s\mathcal{P}), s\mathcal{P})$.

Proof. Any coderivation $d : \mathcal{T}^c(s\mathcal{P}) \to \mathcal{T}^c(s\mathcal{P})$ defining a structure of homotopy operad satisfies $d^2 = 0$, which is equivalent to $d_1 d_\gamma + d_\gamma d_1 + d_\gamma d_\gamma = 0$. By projecting onto the space of cogenerators, this relation is equivalent to $d_{s\mathcal{P}} \gamma + \gamma d_1 + \gamma \star \gamma = 0$ in $\mathrm{Hom}_{\mathbb{S}}(\overline{\mathcal{T}}^c(s\mathcal{P}), s\mathcal{P})$. □

Hence a structure of homotopy operad on a dg $\mathbb{S}$-module $\mathcal{P}$ is a family of maps $\{\gamma_n : \mathcal{T}^c(s\mathcal{P})(n) \to s\mathcal{P}\}_{n\geq 2}$, which "compose" any tree with n vertices labeled by elements of $s\mathcal{P}$. The map $\gamma \star \gamma$ composes first any nontrivial subtree of a tree with γ and then composes the remaining tree with γ once again.

When the $\mathbb{S}$-module $\mathcal{P}$ is concentrated in arity 1, a homotopy operad structure on $\mathcal{P}$ is nothing but a homotopy associative algebra on $\mathcal{P}(1)$. If the structure map γ vanishes on $\mathcal{T}^c(s\mathcal{P})^{(\geq 3)}$, then the only remaining product

$$\gamma_2 : \mathcal{T}^c(s\mathcal{P})^{(2)} \cong s\mathcal{P} \circ_{(1)} s\mathcal{P} \to s\mathcal{P}$$

satisfies the same relations as the partial compositions (Sect. 5.3.4) of an operad. In this case $\mathcal{P}$ is a nonunital operad.

One can translate this definition in terms of operations $\{\mathcal{T}^{(n)}(\mathcal{P}) \to \mathcal{P}\}_{n\geq 2}$, without suspending the $\mathbb{S}$-module $\mathcal{P}$. This would involve extra signs as usual.

10.5.2 Infinity-Morphisms of Homotopy Operads

Let $(\mathcal{P}, \gamma)$ and $(\mathcal{Q}, \nu)$ be two homotopy operads. By definition, an ∞-*morphism of homotopy operads* between $\mathcal{P}$ and $\mathcal{Q}$ is a morphism

$$F : \mathrm{B}\mathcal{P} := \big(\mathcal{T}^c(s\mathcal{P}), d\big) \to \mathrm{B}\mathcal{Q} := \big(\mathcal{T}^c(s\mathcal{Q}), d'\big)$$

of dg cooperads. We denote it by $\mathscr{P} \rightsquigarrow \mathscr{Q}$. Homotopy operads with their ∞-morphisms form a category, which is denoted by $\infty\text{-}\mathsf{Op}_\infty$.

For any $\mathbb{S}$-module M, we consider the morphism of $\mathbb{S}$-modules

$$\Delta' : \mathscr{T}^c(M) \to \mathscr{T}^c\big(M; \mathscr{T}^c(M)^{(\geq 2)}\big),$$

which singles out one subtree with at least two vertices. We also consider the morphism of $\mathbb{S}$-modules

$$\bar{\Delta} : \mathscr{T}^c(M) \to \mathscr{T}^c\big(\overline{\mathscr{T}}^c(M)\big)^{(\geq 2)}$$

which is defined by the projection of $\Delta(M) : \mathscr{T}^c(M) \to \mathscr{T}^c(\overline{\mathscr{T}}^c(M))$, see Sect. 6.3.8, onto $\mathscr{T}^c(\overline{\mathscr{T}}^c(M))^{(\geq 2)}$. In words, it splits a tree into all partitions of subtrees with at least two nontrivial subtrees.

Proposition 10.5.3. *Let $(\mathscr{P}, \gamma)$ and $(\mathscr{Q}, \nu)$ be two homotopy operads. An ∞-morphism of homotopy operads between $\mathscr{P}$ and $\mathscr{Q}$ is equivalently given by a morphism $f : \overline{\mathscr{T}}^c(s\mathscr{P}) \to s\mathscr{Q}$ of graded $\mathbb{S}$-modules, which satisfies*

$$f\,\mathscr{T}^c(\mathrm{Id}; \gamma)\Delta' - \nu\mathscr{T}^c(f)\bar{\Delta} = \partial(f)$$

in $\mathrm{Hom}_{\mathbb{S}}(\overline{\mathscr{T}}^c(s\mathscr{P}), s\mathscr{Q})$:

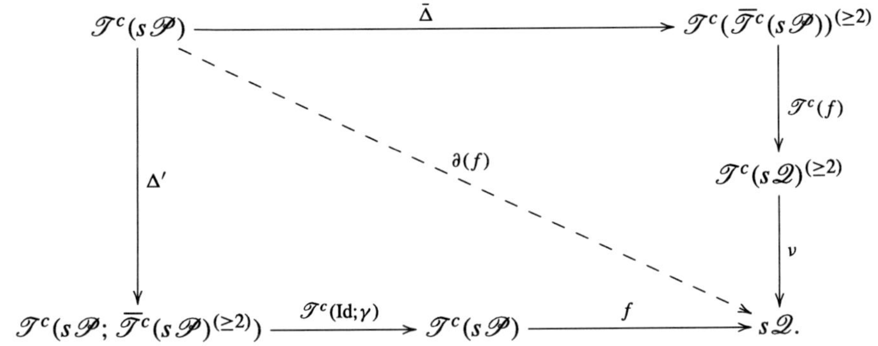

Proof. The universal property of cofree conilpotent cooperads states that every morphism $F : \mathscr{T}^c(s\mathscr{P}) \to \mathscr{T}^c(s\mathscr{Q})$ of cooperads is completely characterized by its projection onto the space of the cogenerators $f : \overline{\mathscr{T}}^c(s\mathscr{P}) \to s\mathscr{Q}$. Explicitly, the unique morphism of cooperads F which extends a map $f : \overline{\mathscr{T}}^c(s\mathscr{P}) \to s\mathscr{Q}$ is equal to the composite

$$F : \mathscr{T}^c(s\mathscr{P}) \xrightarrow{\Delta(s\mathscr{P})} \mathscr{T}^c\big(\overline{\mathscr{T}}^c(s\mathscr{P})\big) \xrightarrow{\mathscr{T}^c(f)} \mathscr{T}^c(s\mathscr{Q}).$$

The map f defines an ∞-morphism of homotopy operads if and only if the map F commutes with the differentials $d_1 + d_\gamma$ on $\mathscr{T}^c(s\mathscr{P})$ and $d_1' + d_\nu$ on $\mathscr{T}^c(s\mathscr{Q})$ respectively. Since F is a morphism of cooperads and since $d_1 + d_\gamma$ and $d_1' + d_\nu$

are coderivations, the relation $(d_1' + d_\nu)F = F(d_1 + d_\gamma)$ holds if and only if $\mathrm{proj}((d_1' + d_\nu)F - F(d_1 + d_\gamma)) = 0$. By the aforementioned universal property of cofree cooperads and by Proposition 6.3.7, we have

$$\mathrm{proj}\big((d_1' + d_\nu)F - F(d_1 + d_\gamma)\big) = \partial(f) + \nu \mathcal{T}^c(f)\bar{\Delta} - f\mathcal{T}^c(\mathrm{Id};\gamma)\Delta'. \quad \square$$

Therefore an ∞-morphism $\mathcal{P} \rightsquigarrow \mathcal{Q}$ of homotopy operads is a family of maps, which associate to any tree t labeled by elements of $s\mathcal{P}$ an element of $s\mathcal{Q}$. Since an operad is a particular case of homotopy operad, one can consider ∞-morphism of operads. Proposition 10.3.2 gives such an example.

Proposition 10.5.4. *A morphism of operads is an ∞-morphism with only one nonvanishing component, namely the first one*:

$$s\mathcal{P} \cong \overline{\mathcal{T}}^c(s\mathcal{P})^{(1)} \to s\mathcal{Q}.$$

Proof. By straightforward application of the definitions. $\square$

As for homotopy algebras, one can define four categories by considering either operads or homotopy operads for the objects and morphisms or infinity morphisms for the maps.

$$\begin{array}{ccc} \mathsf{Op} & \xrightarrow{\text{not full}} & \infty\text{-}\mathsf{Op} \\ \downarrow{\scriptstyle \text{f.f.}} & & \downarrow{\scriptstyle \text{f.f.}} \\ \mathsf{Op}_\infty & \xrightarrow{\text{not full}} & \infty\text{-}\mathsf{Op}_\infty \end{array}$$

(A morphism of homotopy operads is an ∞-morphism with nonvanishing components except for the first one.)

10.5.3 From Homotopy Operads to Homotopy Lie Algebras

We define a functor from homotopy operads to L_∞-algebras, which extends the functor from operads to *Lie*-algebras constructed in Proposition 5.4.3

$$\begin{array}{ccc} \mathsf{Op} & \longrightarrow & \mathit{Lie}\text{-}\mathsf{alg} \\ \downarrow & & \downarrow \\ \infty\text{-}\mathsf{Op}_\infty & \dashrightarrow & \infty\text{-}L_\infty\text{-}\mathsf{alg}. \end{array}$$

Let $\mathcal{P}$ be a dg $\mathbb{S}$-module. We consider either the direct sum of its components $\bigoplus_n \mathcal{P}(n)$ or the direct product $\prod_n \mathcal{P}(n)$. By a slight abuse of notation, we still denote it by $\mathcal{P}$ in this section.

By Proposition 10.1.12, an L_∞-algebra structure on $\mathscr{P}$ is equivalent to a degree -1 square-zero coderivation on the cofree cocommutative coalgebra $\overline{S}^c(s\mathscr{P})$. Recall that its underlying space is the space of invariant elements of the cofree coalgebra $\overline{T}^c(s\mathscr{P})$ under the permutation action. We denote its elements with the symmetric tensor notation:

$$\overline{S}^c(s\mathscr{P})^{(n)} := s\mu_1 \odot \cdots \odot s\mu_n \in (s\mathscr{P})^{\odot n}.$$

Let t be a tree with n vertices and let $\mu_1, \ldots, \mu_n$ be n elements of $\mathscr{P}$. We denote by $t(s\mu_1, \ldots, s\mu_n)$ the sum of all the possible ways of labeling the vertices of $\mathscr{T}$ with $s\mu_1, \ldots, s\mu_n$ according to the arity. We consider the following morphism

$$\Theta : \overline{S}^c(s\mathscr{P}) \to \overline{\mathscr{T}}^c(s\mathscr{P}); \qquad \Theta_n(s\mu_1 \odot \cdots \odot s\mu_n) := \sum_t t(s\mu_1, \ldots, s\mu_n),$$

where t runs over the set of n-vertices trees.

Proposition 10.5.5. *Let $(\mathscr{P}, \gamma)$ be a homotopy operad. The maps $\ell_n : S^c(s\mathscr{P})^{(n)} \to s\mathscr{P}$ of degree -1, defined by the composite $\ell_n := \gamma \circ \Theta_n$, endow the dg modules $\bigoplus_n \mathscr{P}(n)$, respectively $\prod_n \mathscr{P}(n)$, with an L_∞-algebra structure.*

Proof. We consider the "partial" coproduct δ' on the cofree cocommutative coalgebra $\overline{S}^c(s\mathscr{P})$ defined by

$$\begin{aligned} \delta' : \overline{S}^c(s\mathscr{P}) &\to \overline{S}^c\big(s\mathscr{P}; \overline{S}^c(s\mathscr{P})\big), \\ s\mu_1 \odot \cdots \odot s\mu_n &\mapsto \sum_{p=1}^{n-1} \sum_{\sigma \in Sh_{p,q}} \pm(s\mu_{\sigma(1)} \odot \cdots \odot s\mu_{\sigma(p)}) \\ &\qquad \odot s\mu_{\sigma(p+1)} \odot \cdots \odot s\mu_{\sigma(n)}, \end{aligned}$$

where the sign comes from the permutation of the graded elements, as usual. The unique coderivation on $\overline{S}^c(s\mathscr{P})$, which extends $d_{s\mathscr{P}} + \ell$ is equal to the following composite

$$d_1 + d_\ell := \overline{S}^c(s\mathscr{P}) \xrightarrow{\delta'} \overline{S}^c\big(s\mathscr{P}, \overline{S}^c(s\mathscr{P})\big) \xrightarrow{\overline{S}^c(\mathrm{Id}, d_{s\mathscr{P}}+\ell)} \overline{S}^c(s\mathscr{P}, s\mathscr{P}) \to \overline{S}^c(s\mathscr{P}).$$

Under the isomorphism $s\mathscr{P} \cong \mathscr{P}$, the map ℓ defines an L_∞-algebra structure on $\mathscr{P}$ if and only if this coderivation squares to zero.

The following commutative diagram

$$\begin{array}{ccccccc} \overline{S}^c(s\mathscr{P}) & \xrightarrow{\delta'} & \overline{S}^c(s\mathscr{P}, S^c(s\mathscr{P})^{(\geq 2)}) & \xrightarrow{\overline{S}^c(\mathrm{Id},\ell)} & \overline{S}^c(s\mathscr{P}, s\mathscr{P}) & \longrightarrow & \overline{S}^c(s\mathscr{P}) \\ \downarrow{\scriptstyle \Theta} & & & & & & \downarrow{\scriptstyle \Theta} \\ \overline{\mathscr{T}}^c(s\mathscr{P}) & \xrightarrow{\Delta'} & \overline{\mathscr{T}}^c(s\mathscr{P}, \mathscr{T}^c(s\mathscr{P})^{(\geq 2)}) & \xrightarrow{\overline{\mathscr{T}}^c(\mathrm{Id},\gamma)} & \overline{\mathscr{T}}^c(s\mathscr{P}; s\mathscr{P}) & \longrightarrow & \overline{\mathscr{T}}^c(s\mathscr{P}) \end{array}$$

proves that Θ commutes with the coderivations d_ℓ and d_γ. Hence Θ commutes with the full coderivations $d_1 + d_\ell$ and $d_1 + d_\gamma$. Since $d_1 + d_\ell$ is a coderivation, it squares to zero if and only if the projection of $(d_1 + d_l)^2$ onto the space of cogenerators $s\mathscr{P}$ vanishes. This projection is equal to the projection of $(d_1 + d_\gamma)^2\Theta$ on $s\mathscr{P}$, which is equal to zero, by the definition of a homotopy operad. □

This proposition includes and generalizes Proposition 10.1.8. If $\mathscr{P}$ is concentrated in arity 1, then it is an A_∞-algebra. In this case, the class of trees considered are only ladders. We recover the formula of Proposition 10.1.8, which associates an L_∞-algebra to an A_∞-algebra.

Proposition 10.5.6. *Let $(\mathscr{P}, \gamma)$ and $(\mathscr{Q}, \nu)$ be two homotopy operads and let $F : \mathrm{B}\mathscr{P} \to \mathrm{B}\mathscr{Q}$ be an ∞-morphism. The unique morphism of cocommutative coalgebras, which extends*

$$\overline{S}^c(s\mathscr{P}) \xrightarrow{\Theta} \overline{\mathscr{T}}^c(s\mathscr{P}) \xrightarrow{F} \overline{\mathscr{T}}^c(s\mathscr{Q}) \twoheadrightarrow s\mathscr{Q},$$

commutes with the differentials. In other words, it defines an ∞-morphism of L_∞-algebras. So there is a well-defined functor ∞-$\mathsf{Op}_\infty \to \infty$-$L_\infty$-$\mathsf{alg}$.

Proof. Let us denote by $\widetilde{F} : \overline{S}^c(s\mathscr{P}) \to \overline{S}^c(s\mathscr{Q})$ this unique morphism of cocommutative coalgebras. We first prove that the map Θ commutes with the morphisms F and $\widetilde{F}$ respectively. Let us introduce the structure map $\delta : \overline{S}^c(s\mathscr{P}) \to \overline{S}^c(\overline{S}^c(s\mathscr{P}))$, defined by

$$\begin{aligned}&\delta(s\mu_1 \odot \cdots \odot s\mu_n)\\ &\quad := \sum \pm (s\mu_{\sigma(1)} \odot \cdots \odot s\mu_{\sigma(i_1)}) \odot \cdots \odot (s\mu_{\sigma(i_1+\cdots+i_{n-1}+1)} \odot \cdots \odot s\mu_{\sigma(n)}),\end{aligned}$$

where the sum runs over $k \geq 1$, $i_1 + \cdots + i_k = n$ and $\sigma \in Sh_{i_1,\ldots,i_k}$. If we denote by f the projection of F onto the space of cogenerators, then the unique morphism of cocommutative coalgebras $\widetilde{F}$ extending $f\Theta$ is equal to $\widetilde{F} = \overline{S}^c(f\Theta)\delta$. Since the morphism F is equal to the composite $F = \overline{\mathscr{T}}^c(f)\Delta$, the following diagram is commutative

$$\begin{array}{ccccc}
 & & \widetilde{F} & & \\
\overline{S}^c(s\mathscr{P}) & \xrightarrow{\delta} & \overline{S}^c(\overline{S}^c(s\mathscr{P})) & \xrightarrow{\overline{S}^c(f)} & \overline{S}^c(s\mathscr{Q}) \\
\downarrow{\scriptstyle\Theta} & & \downarrow{\scriptstyle\Theta(\Theta)} & & \downarrow{\scriptstyle\Theta} \\
\overline{\mathscr{T}}^c(s\mathscr{P}) & \xrightarrow{\Delta} & \overline{\mathscr{T}}^c(\overline{\mathscr{T}}^c(s\mathscr{P})) & \xrightarrow{\overline{\mathscr{T}}^c(f)} & \overline{\mathscr{T}}^c(s\mathscr{Q}). \\
 & & F & &
\end{array}$$

Let us denote by $\tilde{d}_\gamma$, and by $\tilde{d}_\nu$ respectively, the induced square-zero coderivations on $\overline{S}^c(s\mathscr{P})$, and on $\overline{S}^c(s\mathscr{Q})$ respectively. Since $\widetilde{F}$ is a morphism of cocommutative coalgebras, to prove that it commutes with the coderivations $\tilde{d}_1 + \tilde{d}_\gamma$ and

$\tilde{d}_1' + \tilde{d}_\nu$ respectively, it is enough to prove by projecting onto the space $s\mathcal{Q}$ of cogenerators. To this end, we consider the following commutative diagram.

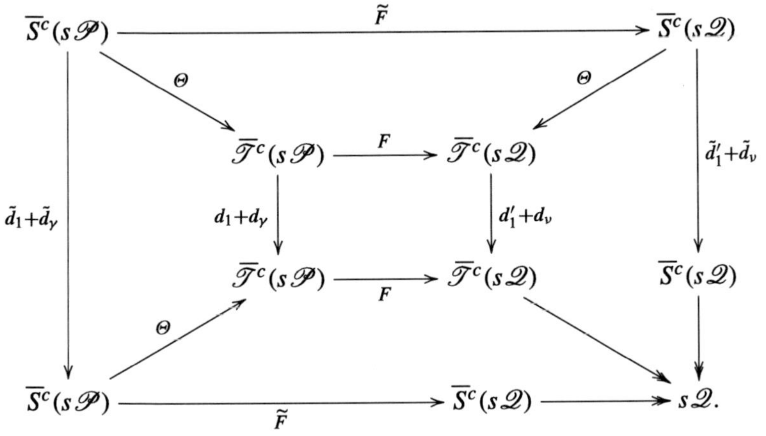

Since the internal diagram is commutative, the external one also commutes, which concludes the proof. □

10.5.4 From Homotopy Operads to Homotopy pre-Lie Algebras

The two aforementioned Propositions 10.5.5 and 10.5.6 extend from *Lie*-algebras and L_∞-algebras to *preLie*-algebras and $preLie_\infty$-algebras respectively, see Sect. 13.4.

Finally, we have the following commutative diagram of categories, which sums up the relations between the various algebraic structures encountered so far.

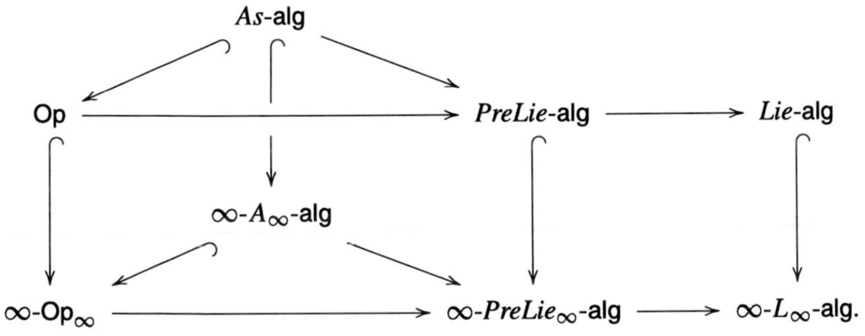

10.5.5 Homotopy Algebra Structures vs ∞-Morphisms of Operads

Since a $\mathcal{P}$-algebra structure on a dg module A is given by a morphism of dg operads $\mathcal{P} \to \mathrm{End}_A$, it is natural to ask what ∞-morphisms $\mathcal{P} \rightsquigarrow \mathrm{End}_A$ between $\mathcal{P}$ and

End_A do model, cf. [Lad76, vdL03]. The next proposition shows that a homotopy representation of an operad $\mathscr{P}$ is a $\Omega\mathrm{B}\mathscr{P}$-algebra.

Proposition 10.5.7. *For any operad $\mathscr{P}$ and any dg module A, there is a natural bijection between ∞-morphisms from $\mathscr{P}$ to End_A and $\Omega\mathrm{B}\mathscr{P}$-algebra structures on A:*

$$\mathrm{Hom}_{\infty\text{-}\mathsf{Op}}(\mathscr{P}, \mathrm{End}_A) \cong \mathrm{Hom}_{\mathsf{dg\ Op}}(\Omega\mathrm{B}\mathscr{P}, \mathrm{End}_A).$$

Proof. By definition, the first set is equal to $\mathrm{Hom}_{\mathsf{dg\ Coop}}(\mathrm{B}\mathscr{P}, \mathrm{B}\,\mathrm{End}_A)$. The natural bijection with $\mathrm{Hom}_{\mathsf{dg\ Op}}(\Omega\mathrm{B}\mathscr{P}, \mathrm{End}_A)$ given by the bar–cobar adjunction of Theorem 6.5.7 concludes the proof. □

By pulling back along the morphism of dg operads $\Omega\mathscr{P}^{¡} \to \Omega\mathrm{B}\mathscr{P}$, any $\Omega\mathrm{B}\mathscr{P}$-algebra A determines a $\mathscr{P}_\infty$-algebra. So an ∞-morphism of operads from $\mathscr{P}$ to End_A induces a $\mathscr{P}_\infty$-algebra structure on A.

Any $\mathscr{P}_\infty$-algebra structure on A is a morphism of dg operads $\Omega\mathscr{P}^{¡} \to \mathrm{End}_A$, which is a particular ∞-morphism of operads from $\Omega\mathscr{P}^{¡}$ to End_A by Proposition 10.5.4.

$$\begin{aligned}\mathrm{B} : \mathrm{Hom}_{\mathsf{dg\ Op}}\big(\Omega\mathscr{P}^{¡}, \mathrm{End}_A\big) &\to \mathrm{Hom}_{\mathsf{dg\ coOp}}\big(\mathrm{B}\Omega\mathscr{P}^{¡}, \mathrm{B}\,\mathrm{End}_A\big)\\ &\cong \mathrm{Hom}_{\infty\text{-}\mathsf{Op}}\big(\Omega\mathscr{P}^{¡}, \mathrm{End}_A\big).\end{aligned}$$

Again, what do ∞-morphisms $\Omega\mathscr{P}^{¡} \rightsquigarrow \mathrm{End}_A$ model? Any ∞-morphism of operads $\mathrm{B}\Omega\mathscr{P}^{¡} \to \mathrm{B}\,\mathrm{End}_A$ induces a $\mathscr{P}_\infty$-algebra structure on A by pulling back along the unit of adjunction $\upsilon_{\mathscr{P}^{¡}} : \mathscr{P}^{¡} \to \mathrm{B}\Omega\mathscr{P}^{¡}$ and by using the bar–cobar adjunction. Let us denote this map by

$$\upsilon^* : \mathrm{Hom}_{\infty\text{-}\mathsf{Op}}\big(\Omega\mathscr{P}^{¡}, \mathrm{End}_A\big) \to \mathrm{Hom}_{\mathsf{dg\ Op}}\big(\Omega\mathscr{P}^{¡}, \mathrm{End}_A\big).$$

So the set of $\mathscr{P}_\infty$-algebra structures on A is a "retract" of the set of ∞-morphisms from $\mathscr{P}_\infty = \Omega\mathscr{P}^{¡}$ to End_A.

We sum up the hierarchy of homotopy notions in the following table.

$$\begin{array}{ccc}
\{\mathscr{P}\text{-algebra structures on } A\} & = & \mathrm{Hom}_{\mathsf{Op}}(\mathscr{P}, \mathrm{End}_A)\\
\Big\downarrow & & \Big\downarrow\\
 & & \mathrm{Hom}_{\infty\text{-}\mathsf{Op}}(\mathscr{P}, \mathrm{End}_A)\\
 & & \Big\downarrow \cong\\
\{\Omega\mathrm{B}\mathscr{P}\text{-algebra structures on } A\} & = & \mathrm{Hom}_{\mathsf{dg\ Op}}(\Omega\mathrm{B}\mathscr{P}, \mathrm{End}_A)\\
\Big\downarrow & & \Big\downarrow\\
\{\mathscr{P}_\infty\text{-algebra structures on } A\} & = & \mathrm{Hom}_{\mathsf{dg\ Op}}(\Omega\mathscr{P}^{¡}, \mathrm{End}_A)\\
 & & \upsilon^*\Big\uparrow\Big\downarrow\\
 & & \mathrm{Hom}_{\infty\text{-}\mathsf{Op}}(\Omega\mathscr{P}^{¡}, \mathrm{End}_A)
\end{array}$$

10.6 Résumé

10.6.1 Homotopy $\mathscr{P}$-Algebras

Homotopy $\mathscr{P}$-algebra: algebra over $\mathscr{P}_\infty := \Omega \mathscr{P}^{¡}$.

ROSETTA STONE. The set of $\mathscr{P}_\infty$-algebra structures on A is equal to

$$\begin{aligned}\mathrm{Hom}_{\mathsf{dgOp}}\big(\Omega\mathscr{P}^{¡}, \mathrm{End}_A\big) &\cong \mathrm{Tw}\big(\mathscr{P}^{¡}, \mathrm{End}_A\big)\\ &\cong \mathrm{Hom}_{\mathsf{dgCoop}}\big(\mathscr{P}^{¡}, \mathrm{B}\,\mathrm{End}_A\big) \cong \mathrm{Codiff}\big(\mathscr{P}^{¡}(A)\big).\end{aligned}$$

10.6.2 Infinity-Morphisms

Let (A, φ) and (B, ψ) be two $\mathscr{P}_\infty$-algebras.

Infinity-morphism or ∞-morphism $A \rightsquigarrow B$: morphism of dg $\mathscr{P}^{¡}$-coalgebras

$$F : \big(\mathscr{P}^{¡}(A), d_\varphi\big) \to \big(\mathscr{P}^{¡}(B), d_\psi\big) \quad \Longleftrightarrow \quad f : \mathscr{P}^{¡} \to \mathrm{End}^A_B,$$
$$\text{such that } f * \varphi - \psi \circledcirc f = \partial(f) \text{ in } \mathrm{Hom}_{\mathbb{S}}\big(\mathscr{P}^{¡}, \mathrm{End}^A_B\big).$$

Category of $\mathscr{P}_\infty$-algebras with ∞-morphisms denoted ∞-$\mathscr{P}_\infty$-alg.

$$\begin{array}{ccc}\mathscr{P}\text{-}\mathsf{alg} & \xrightarrow{\text{not full}} & \infty\text{-}\mathscr{P}\text{-}\mathsf{alg}\\ \Big\downarrow{\scriptstyle \text{f.f.}} & & \Big\downarrow{\scriptstyle \text{f.f.}}\\ \mathscr{P}_\infty\text{-}\mathsf{alg} & \xrightarrow{\text{not full}} & \infty\text{-}\mathscr{P}_\infty\text{-}\mathsf{alg}\end{array}$$

∞-isomorphism: when $f_{(0)} : A \to B$ is an isomorphism.

Theorem. *∞-isomorphisms are the isomorphisms of the category* ∞-$\mathscr{P}_\infty$-alg.

∞-quasi-isomorphism: when $f_{(0)} : A \to B$ is a quasi-isomorphism.

10.6.3 Homotopy Transfer Theorem

Homotopy data: let (V, d_V) be a homotopy retract of (W, d_W)

$$h \circlearrowright (W, d_W) \underset{i}{\overset{p}{\rightleftarrows}} (V, d_V).$$

Proposition. *Any homotopy retract gives rise to a morphism* $\Psi : \mathrm{B}\,\mathrm{End}_W \to \mathrm{B}\,\mathrm{End}_V$ *of dg cooperads.*

Algebraic data: let φ be a $\mathscr{P}_\infty$-algebra structure on W.

HOMOTOPY TRANSFER THEOREM. There exists a $\mathscr{P}_\infty$-algebra structure on V such that i extends to an ∞-quasi-isomorphism.

$$\begin{array}{ccc} \mathscr{P}^{\text{i}} & \xrightarrow{f_\varphi} & \mathrm{B}\,\mathrm{End}_W \\ & \searrow_{f_\psi} & \downarrow \Psi \\ & & \mathrm{B}\,\mathrm{End}_V . \end{array}$$

EXPLICIT TRANSFERRED STRUCTURE.

$$\mathscr{P}^{\text{i}} \xrightarrow{\Delta_{\mathscr{P}^{\text{i}}}} \mathscr{T}^c(\overline{\mathscr{P}}^{\text{i}}) \xrightarrow{\mathscr{T}^c(s\varphi)} \mathscr{T}^c(s\mathrm{End}_W) \xrightarrow{\Psi} \mathrm{End}_V .$$

OPERADIC MASSEY PRODUCTS. They are the higher operations in the particular case: $W = A$, a $\mathscr{P}_\infty$-algebra, and $V = H(A)$.

CHAIN MULTICOMPLEX. Particular case where $\mathscr{P} = D := \mathbb{K}[\varepsilon]/(\varepsilon^2)$.

$$\begin{array}{rcl} D\textit{-algebra on } A & \longleftrightarrow & \textit{bicomplex}, \\ \textit{transferred } D_\infty\textit{-algebra on } H(A) & \longleftrightarrow & \textit{spectral sequence}. \end{array}$$

10.6.4 Homotopy Theory of Homotopy Algebras

DECOMPOSITION: MINIMAL $\oplus$ ACYCLIC TRIVIAL. Any $\mathscr{P}_\infty$-algebra A is ∞-isomorphic to a product

$$A \overset{\cong}{\rightsquigarrow} M \oplus K$$

in ∞-$\mathscr{P}_\infty$-alg, where M is minimal, i.e. $d_M = 0$, and where K is acyclic trivial, i.e. acyclic underlying chain complex and trivial $\mathscr{P}_\infty$-algebra structure.

HOMOTOPY EQUIVALENCE. If there exists an ∞-quasi-isomorphism $A \overset{\sim}{\rightsquigarrow} B$, then there exists an ∞-quasi-isomorphism $B \overset{\sim}{\rightsquigarrow} A$.

10.6.5 Homotopy Operads

Homotopy operad: degree -1 square-zero coderivation on $\mathscr{T}^c(s\mathscr{P})$

$$\Longleftrightarrow \quad \gamma : \mathscr{T}^c(\mathscr{P})^{(\geq 2)} \to \mathscr{P}, \quad \text{such that } \partial(\gamma) + \gamma \star \gamma = 0.$$

Infinity-morphisms of homotopy operads: morphism of dg cooperads

$$\mathrm{B}\mathscr{P} = \big(\mathscr{T}^c(s\mathscr{P}), d_1 + d_\gamma\big) \to \mathrm{B}\mathscr{Q} = \big(\mathscr{T}^c(s\mathscr{Q}), d_1' + d_\nu\big),$$

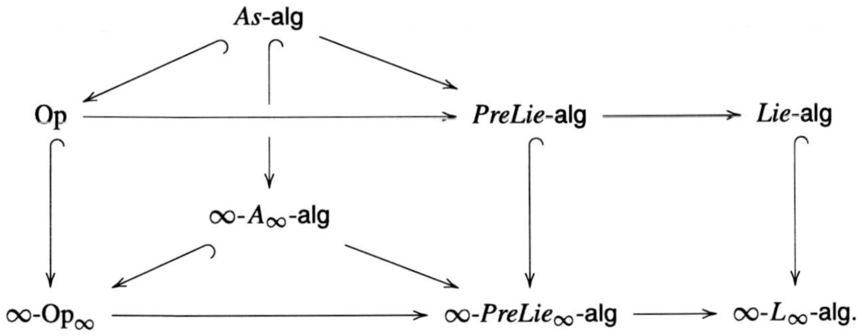

HOMOTOPY REPRESENTATION OF OPERAD.

$$\begin{aligned}\mathrm{Hom}_{\infty\text{-}\mathsf{Op}}(\mathscr{P}, \mathrm{End}_A) &\cong \mathrm{Hom}_{\mathsf{dg\ Op}}(\Omega \mathrm{B}\mathscr{P}, \mathrm{End}_A) \\ &= \{\Omega\mathrm{B}\mathscr{P}\text{-}algebra\ structures\ on\ A\}.\end{aligned}$$

10.7 Exercises

Exercise 10.7.1 (Homotopy $\mathscr{P}$-algebra concentrated in degree 0). Let A be a $\mathbb{K}$-module. We consider it as a dg module concentrated in degree 0 with trivial differential. Prove that a $\mathscr{P}_\infty$-algebra structure on A is a $\mathscr{P}$-algebra structure.

Exercise 10.7.2 (Universal enveloping algebra of an L_∞-algebra [LM95]). In Proposition 10.1.8, we introduced a functor from A_∞-algebras to L_∞-algebras, which is the pullback functor f^* associated to the morphism of operads $f : \Omega Lie^{¡} \to \Omega Ass^{¡}$, see Sect. 5.2.12.

Show that this functor admits a left adjoint functor provided by the *universal enveloping algebra of an* L_∞*-algebra* $(A, d_A, \{\ell_n\}_{n\geq 2})$:

$$U(A) := Ass_\infty(A)/I,$$

where I is the ideal generated, for $n \geq 2$, by the elements

$$\sum_{\sigma\in\mathbb{S}_n} \mathrm{sgn}(\sigma)\varepsilon\big(\mu_n^c; a_{\sigma^{-1}(1)}, \ldots, a_{\sigma^{-1}(n)}\big) - \ell_n(a_1, \ldots, a_n),$$

where ε is the sign induced by the permutation of the graded elements $a_1, \ldots, a_n \in A$.

HINT. It is a direct consequence of Sect. 5.2.12 and Exercise 5.11.26, where $U(A) = f_!(A)$.

Exercise 10.7.3 (Homotopy pre-Lie-algebra). Make explicit the notion of homotopy pre-Lie algebra, see Sect. 13.4, together with the convolution algebra $\mathfrak{g}_{preLie, A}$, which controls it.

HINT. The Koszul dual operad of *preLie* is *Perm*, which admits a simple presentation, see Sect. 13.4.6.

Exercise 10.7.4 (Equivalent Maurer–Cartan equation ★). Let $\mathscr{C}$ be a coaugmented cooperad, with coaugmentation $\eta : \mathrm{I} \to \mathscr{C}$, and let (A, d_A) be a dg module. To any morphism $\alpha : \mathscr{C} \to \mathrm{End}_A$ of $\mathbb{S}$-modules, such that $\alpha \circ \eta = 0$, we associate the morphism of $\mathbb{S}$-modules $\tilde{\alpha} : \mathscr{C} \to \mathrm{End}_A$ defined by $\tilde{\alpha} \circ \eta(\mathrm{id}) := d_A$ and $\tilde{\alpha} := \alpha$ otherwise. If α has degree -1, then $\tilde{\alpha}$ has also degree -1.

1. Prove that α satisfies the Maurer–Cartan equation $\partial(\alpha) + \alpha \star \alpha = 0$ if and only if $\tilde{\alpha}$ squares to zero, $\tilde{\alpha} \star \tilde{\alpha} = 0$.

Let (B, d_B) be another chain complex and let $\varphi \in \mathrm{Tw}(\mathscr{P}^{\text{¡}}, \mathrm{End}_A)$ and $\psi \in \mathrm{Tw}(\mathscr{P}^{\text{¡}}, \mathrm{End}_B)$ be two $\mathscr{P}_\infty$-algebra structures on A and B respectively.

2. Show that any morphism of dg $\mathbb{S}$-modules $f : \mathscr{P}^{\text{¡}} \to \mathrm{End}^A_B$ is an ∞-morphism if and only if $f * \tilde{\varphi} = \tilde{\psi} \circledast f$.

Let $f : \mathscr{P}^{\text{¡}} \to \mathrm{End}^A_B$ and $g : \mathscr{P}^{\text{¡}} \to \mathrm{End}^B_C$ be two ∞-morphisms.

3. Show directly that the composite

$$g \circledcirc f = \mathscr{P}^{\text{¡}} \xrightarrow{\Delta} \mathscr{P}^{\text{¡}} \circ \mathscr{P}^{\text{¡}} \xrightarrow{g \circ f} \mathrm{End}^B_C \circ \mathrm{End}^A_B \to \mathrm{End}^A_C$$

of Proposition 10.2.4 is an ∞-morphism.

Exercise 10.7.5 (Action of the convolution algebra ★). Let A and B be two dg modules and let $\mathscr{P}$ be a Koszul operad. We denote by $\mathfrak{g}_B := \mathrm{Hom}_{\mathbb{S}}(\mathscr{P}^{\text{¡}}, \mathrm{End}_B)$ the convolution pre-Lie algebra and we consider $\mathfrak{g}^A_B = \mathrm{Hom}_{\mathbb{S}}(\mathscr{P}^{\text{¡}}, \mathrm{End}^A_B)$, as in Sect. 10.2.3. We defined the action $\psi \circledast f$, for $\psi \in \mathfrak{g}_B$ and $f \in \mathfrak{g}^A_B$ and the action $\psi \circledcirc \xi$ on $\mathfrak{g}_B$ by the following composite

$$\psi \circledcirc \xi := \mathscr{P}^{\text{¡}} \xrightarrow{\Delta} \mathscr{P}^{\text{¡}} \circ \mathscr{P}^{\text{¡}} \xrightarrow{\psi \circ \xi} \mathrm{End}_B \circ \mathrm{End}_B \xrightarrow{\gamma_{\mathrm{End}_B}} \mathrm{End}_B.$$

1. Show that $\circledcirc$ defines an associative algebra structure on $\mathfrak{g}_B$, where the unit is the composite of the coaugmentation of $\mathscr{P}^{\text{¡}}$ followed by the unit of End_B: $\mathscr{P}^{\text{¡}} \to \mathrm{I} \to \mathrm{End}_A$.
2. Show that $\circledast$ defines a left module action of the associative algebra $(\mathfrak{g}_B, \circledcirc)$ on $\mathfrak{m}$.
3. In the same way, show that the action $*$ of $\mathfrak{g}_A := \mathrm{Hom}_{\mathbb{S}}(\mathscr{P}^{\text{¡}}, \mathrm{End}_A)$ on $\mathfrak{m}$ is a right pre-Lie action.

Exercise 10.7.6 (Kleisli category [HS10]).

1. Let $\alpha : \mathscr{C} \to \mathscr{P}$ be an operadic twisting morphism between a dg cooperad $\mathscr{C}$ and a dg operad $\mathscr{P}$. Show that one can define a comonad K_α in the category of $\mathscr{P}$-algebras by setting

$$K_\alpha(A) := \mathscr{P} \circ_\alpha \mathscr{C} \circ_\alpha A.$$

(We refer to Sect. 11.3 for more details about the construction $\mathscr{P} \circ_\alpha \mathscr{C} \circ_\alpha A = \Omega_\alpha \mathrm{B}_\alpha A$. Notice that the comonad $K_\alpha = \Omega_\alpha \mathrm{B}_\alpha$ is the comonad arising from the bar–cobar adjunction.)

2. Let C be a category and let (K, Δ, ε) be a comonad on C. Show that one can define a category C_K on the following data. The objects of the category C_K are the ones of the category C. The morphisms in the category C_K between two objects A, A' are the morphisms in the category C between $K(A)$ and A':
$$\mathrm{Hom}_{\mathsf{C}_K}\big(A, A'\big) := \mathrm{Hom}_{\mathsf{C}}\big(K(A), A'\big).$$
This category is called the *Kleisli category* C_K associated to the comonad K.
3. Prove that there is an isomorphism of categories
$$\mathsf{C}_{K_\kappa} \cong \infty\text{-}\mathscr{P}\text{-}\mathsf{alg}.$$
4. Prove that there is an isomorphism of categories
$$\mathsf{C}_{K_\iota} \cong \infty\text{-}\mathscr{P}_\infty\text{-}\mathsf{alg}.$$

For more details, we refer to the paper [HS10] of K. Hess and J. Scott.

Exercise 10.7.7 (Inverse of ∞-isomorphisms ★). Make explicit the inverse of ∞-isomorphisms given in Theorem 10.4.1 in the particular cases where the operad $\mathscr{P}$ is the ns operad *As*, the operad *Com*, and the operad *PreLie*.

Exercise 10.7.8 (Homotopy Transfer Theorem, Solution II). This exercise proposes another proof to the Homotopy Transfer Theorem.

1. Let $\mathscr{C}$ be a dg cooperad and let $\mathscr{P}$ and $\mathscr{Q}$ be two dg operads. Show that any ∞-morphism $\Phi : \mathscr{P} \rightsquigarrow \mathscr{Q}$ between the dg operads $\mathscr{P}$ and $\mathscr{Q}$ naturally induces an ∞-morphism
$$\Phi_* : \mathrm{Hom}(\mathscr{C}, \mathscr{P}) \rightsquigarrow \mathrm{Hom}(\mathscr{C}, \mathscr{Q})$$
between the associated convolution operads.
By Proposition 10.5.6, the ∞-morphism Φ_* induces an ∞-morphism between the dg Lie algebras $\prod_n \mathrm{Hom}(\mathscr{C}(n), \mathscr{P}(n))$ and $\prod_n \mathrm{Hom}(\mathscr{C}(n), \mathscr{Q}(n))$.
2. Show that this ∞-morphism passes to invariant elements, i.e. it induces an ∞-morphism ϕ between the convolution dg Lie algebras $\mathrm{Hom}_{\mathbb{S}}(\mathscr{C}, \mathscr{P})$ and $\mathrm{Hom}_{\mathbb{S}}(\mathscr{C}, \mathscr{Q})$.
Hence, the ∞-morphism $\Psi : \mathrm{End}_W \to \mathrm{End}_V$ of operads of Proposition 10.3.2 induces a natural ∞-morphism
$$\psi_\bullet : \mathfrak{g}_W = \mathrm{Hom}_{\mathbb{S}}\big(\mathscr{P}^{\text{i}}, \mathrm{End}_W\big) \to \mathfrak{g}_V = \mathrm{Hom}_{\mathbb{S}}\big(\mathscr{P}^{\text{i}}, \mathrm{End}_V\big)$$
between the associated convolution dg Lie algebras.

3. Let α be a Maurer–Cartan element in $\mathfrak{g}_W$, which vanishes on the coaugmentation of $\mathscr{P}^{\text{i}}$. Show that $\sum_{n=1}^{\infty} \frac{1}{n!}\psi(\alpha, \ldots, \alpha)$ defines a Maurer–Cartan element in $\mathfrak{g}_V$, which vanishes on the coaugmentation of $\mathscr{P}^{\text{i}}$.
4. Compare this proof of the Homotopy Transfer Theorem with the one given in Proposition 10.3.3.

Chapter 11
Bar and Cobar Construction of an Algebra over an Operad

La mathématique est une science dangereuse: elle dévoile les supercheries et les erreurs de calcul.

Galileo Galilei

In the algebra case, a twisting morphism is defined as a particular map from a dg coassociative coalgebra to a dg associative algebra. Starting with dg associative algebras, why should one consider the category of dg coassociative coalgebras? The conceptual explanation is given by the Koszul duality theory for operads: the operad *As* is Koszul and its Koszul dual operad is itself. In order to generalize the notion of twisting morphism to dg $\mathscr{P}$-algebras, one needs to work with dg $\mathscr{P}^{¡}$-coalgebras. Such a phenomenon has already been noticed in the literature. For instance, in rational homotopy theory the case $\mathscr{P} = Lie$ and $\mathscr{P}^{¡} = Com^c$ was treated by Quillen in [Qui69] and the case $\mathscr{P} = Com$ and $\mathscr{P}^{¡} = Lie^c$ was treated by Sullivan in [Sul77].

In this chapter, we extend to dg $\mathscr{P}$-algebras the notions of twisting morphism, bar and cobar constructions introduced in the context of dga algebras and dga coalgebras in Chap. 2. When the operad $\mathscr{P}$ is Koszul, this allows us to define functorial quasi-free resolutions for $\mathscr{P}$-algebras and $\mathscr{P}_\infty$-algebras. They will be used in the next chapter to compute homology groups.

Another application is the *rectification* of homotopy $\mathscr{P}$-algebras, which states that any homotopy $\mathscr{P}$-algebra is naturally and universally ∞-quasi-isomorphic to a dg $\mathscr{P}$-algebra. This proves that the homotopy category of dg $\mathscr{P}$-algebras is equivalent to the homotopy category of $\mathscr{P}_\infty$-algebras with their ∞-morphisms.

This chapter essentially follows [Pro86, GJ94, Liv98b, DTT07, HM10, Mil11].

In this chapter, we work over a ground field $\mathbb{K}$ of characteristic 0. Notice that all the constructions and some of the results hold true without this hypothesis. All the coalgebras over a cooperad considered here are conilpotent by assumption.

J.-L. Loday, B. Vallette, *Algebraic Operads*,
Grundlehren der mathematischen Wissenschaften 346,
DOI 10.1007/978-3-642-30362-3_11,

11.1 Twisting Morphism for $\mathscr{P}$-Algebras

One can work at two different levels: the level of operads (Chap. 6) or the level of algebras. In Chap. 2, we defined the notion of twisting morphism between a dga coalgebra and a dga algebra and in Sect. 6.4, we defined the notion of operadic twisting morphism between a dg cooperad and a dg operad.

In this section, we go much further at the algebra level. Associated to an operadic twisting morphism $\alpha : \mathscr{C} \to \mathscr{P}$ from a dg cooperad $\mathscr{C}$ to a dg operad $\mathscr{P}$, we define the notion of *twisting morphism between a dg conilpotent $\mathscr{C}$-coalgebra and a dg $\mathscr{P}$-algebra with respect to α.*

When $\mathscr{P}$ is a quadratic operad, we consider the canonical twisting morphism $\kappa : \mathscr{P}^{¡} \to \mathscr{P}$. If, moreover, $\mathscr{P}$ is binary, we give a Lie theoretic interpretation of twisting morphisms between dg $\mathscr{P}^{¡}$-coalgebras and dg $\mathscr{P}$-algebras.

11.1.1 Definition

Let $\alpha : \mathscr{C} \to \mathscr{P}$ be an operadic twisting morphism from a dg cooperad $\mathscr{C}$ to a dg operad $\mathscr{P}$ (see Sect. 6.4). Let A be a dg $\mathscr{P}$-algebra and let C be a dg $\mathscr{C}$-coalgebra. We consider the following unary operator $\star_\alpha$ of degree -1 on $\mathrm{Hom}(C, A)$:

$$\star_\alpha(\varphi) : C \xrightarrow{\Delta_C} \mathscr{C} \circ C \xrightarrow{\alpha \circ \varphi} \mathscr{P} \circ A \xrightarrow{\gamma_A} A, \quad \text{for } \varphi \in \mathrm{Hom}(C, A).$$

A *twisting morphism with respect to* α is a linear map $\varphi : C \to A$ of degree 0 which is a solution to the *Maurer–Cartan* equation

$$\partial(\varphi) + \star_\alpha(\varphi) = 0.$$

We denote the space of twisting morphisms with respect to α by $\mathrm{Tw}_\alpha(C, A)$.

Let us recall the three main examples of operadic twisting morphisms between a dg cooperad and a dg operad, to which we will apply this definition. One can either work with the universal twisting morphism $\pi : \mathrm{B}\mathscr{P} \to \mathscr{P}$, associated to any augmented dg operad $\mathscr{P}$, or with the universal twisting morphism $\iota : \mathscr{C} \to \Omega\mathscr{C}$, associated to any conilpotent dg cooperad. When $\mathscr{P}$ is a quadratic operad, we can also consider the twisting morphisms with respect to the canonical twisting morphism $\kappa : \mathscr{P}^{¡} \to \mathscr{P}$, defined in Sect. 7.4.

11.1.2 Lie Theoretical Interpretation

In the binary quadratic case, the term $\star_\alpha(\varphi)$ becomes quadratic. It allows us to interpret the previous equation as a Maurer–Cartan equation in a dg Lie algebra as follows.

Proposition 11.1.1. *Let $\mathscr{P}$ be a binary quadratic operad. For any conilpotent dg $\mathscr{P}^{¡}$-coalgebra C and any dg $\mathscr{P}$-algebra A, the space $s\,\mathrm{Hom}(C, A)$ is a dg Lie algebra.*

Proof. Let us give a proof when $\mathscr{P}$ is finitely generated. Recall from Sect. 7.2.2 that C is a $\mathscr{P}^{¡}$-coalgebra if and only if sC is a $\mathscr{P}^{¡} \underset{\mathrm{H}}{\otimes} \mathrm{End}^c_{s\mathbb{K}}$-coalgebra, that is a $\mathscr{P}^{!*}$-coalgebra. Proposition 7.6.5 provides a morphism of operads $Lie \to \mathscr{P}^! \underset{\mathrm{H}}{\otimes} \mathscr{P}$. Since $\mathrm{Hom}(sC, A) \cong s\,\mathrm{Hom}(C, A)$ is an algebra over $\mathscr{P}^! \underset{\mathrm{H}}{\otimes} \mathscr{P}$, it carries a Lie algebra structure. The result holds in full generality since we are working with a space of maps, see formula below. □

When $\mathscr{P} = \mathscr{P}(E, R)$ is a binary quadratic operad, the Lie bracket is explicitely given by the composite $[f, g]$:

$$sC \xrightarrow{s\Delta_C} s\mathscr{P}^{¡}(C) \twoheadrightarrow s\big(sE(2) \otimes C^{\otimes 2}\big)^{\mathbb{S}_2} \to E(2) \otimes (sC)^{\otimes 2}$$
$$\xrightarrow{\mathrm{Id}_E \otimes f \otimes g - (-1)^{|f||g|}\mathrm{Id}_E \otimes g \otimes f} E(2) \otimes A^{\otimes 2} \xrightarrow{\gamma_A} A,$$

for $f, g \in \mathrm{Hom}(sC, A)$.

Corollary 11.1.2. *When $\mathscr{P}$ is a binary quadratic operad, the equation $\partial(\varphi) + \star_\kappa(\varphi) = 0$ in $\mathrm{Hom}(C, A)$ is equivalent to the Maurer–Cartan equation $\partial(s^{-1}\varphi) + \frac{1}{2}[s^{-1}\varphi, s^{-1}\varphi] = 0$ in the dg Lie algebra $s\,\mathrm{Hom}(C, A)$, under the desuspension isomorphism*

$$\mathrm{Hom}(C, A) \xrightarrow{\cong} s^{-1}\mathrm{Hom}(sC, A),$$
$$\varphi \mapsto \big(s^{-1}\varphi : sc \mapsto (-1)^{|\varphi|}\varphi(c)\big).$$

Proof. When $\mathscr{P} = \mathscr{P}(E, R)$ is a binary quadratic operad, the above formula for the Lie bracket on $s\,\mathrm{Hom}(C, A) \cong \mathrm{Hom}(sC, A)$ allows us to compute $\frac{1}{2}[s^{-1}\varphi, s^{-1}\varphi]$ as the following composite

$$sC \xrightarrow{s\Delta_C} s\mathscr{P}^{¡}(C) \twoheadrightarrow E(sC) \xrightarrow{\mathrm{Id}_E \circ s^{-1}\varphi} E(A) \rightarrowtail \mathscr{P}(A) \xrightarrow{\gamma_A} A.$$

The map between $\mathscr{P}^{¡}(C)$ and $\mathscr{P}(A)$ is equal to the composite $\kappa(\varphi)$, up to the degree shift. □

Moreover, when $\mathscr{P}$ is a nonsymmetric operad, Proposition 7.7.2 shows that the Lie algebra structure on $s\,\mathrm{Hom}(C, A)$ comes from an associative algebra structure, whose product is denoted by $\star$. In this case, the Maurer–Cartan equation $\partial(\varphi) + \star_\kappa(\varphi) = 0$ reads

$$\partial(s^{-1}\varphi) + \frac{1}{2}\big[s^{-1}\varphi, s^{-1}\varphi\big] \equiv \partial\big(s^{-1}\varphi\big) + \big(s^{-1}\varphi\big) \star \big(s^{-1}\varphi\big) = 0.$$

11.1.3 Recollection with the Classical Notion

In the case of the operadic twisting morphism $\kappa : As^{\text{!`}} \to As$, we recover the classical notion of twisting morphisms between coaugmented dga coalgebras and augmented dga algebras of Chap. 2.

Recall that C is a dg $As^{\text{!`}}$-coalgebra if and only if $sC$ is a dg $As^{\text{!`}} \underset{\mathrm{H}}{\otimes} \mathrm{End}^c_{s\mathbb{K}}$-coalgebra. Since the cooperads $As^{\text{!`}} \underset{\mathrm{H}}{\otimes} \mathrm{End}^c_{s\mathbb{K}} \cong As^*$ are isomorphic, sC is a dga coalgebra without counit. A dg As-algebra A is a dga algebra without unit. The dga algebra structure (without unit) on $s\,\mathrm{Hom}(C, A) \cong \mathrm{Hom}(sC, A)$ described above is equal to the dg convolution algebra of Proposition 2.1.1.

Recall the equivalences of categories between dga coalgebras without counit sC and coaugmented dga coalgebras $sC_+ := sC \oplus \mathbb{K}1$ and between dga algebras without unit A and augmented dga algebras $A_+ := A \oplus \mathbb{K}1$. Under these equivalences, the twisting morphisms with respect to κ are nothing but the classical twisting morphisms of dga (co)algebras:

$$\mathrm{Tw}_\kappa(C, A) \cong \mathrm{Tw}(sC_+, A_+).$$

11.2 Bar and Cobar Construction for $\mathscr{P}$-Algebras

Are the covariant functor $\mathrm{Tw}_\alpha(C, -)$ and the contravariant functor $\mathrm{Tw}_\alpha(-, A)$ representable? The bar and cobar constructions of this section provide an affirmative answer to that question. They are the direct generalizations of the constructions of Sect. 2.2.

11.2.1 Relative Composite Product

In this section, we generalize to operads and modules over an operad the notion of relative tensor product of modules over an algebra, see Exercise 1.8.14.

Let $\mathscr{P}$ be an operad. Let $\rho : M \circ \mathscr{P} \to M$ be a right $\mathscr{P}$-module structure on the $\mathbb{S}$-module M and $\lambda : \mathscr{P} \circ N \to N$ be a left $\mathscr{P}$-module structure on N, see Sect. 5.2.2.

The *relative composite product* $M \circ_{\mathscr{P}} N$ *of* M *and* N *over* $\mathscr{P}$ is defined by the following coequalizer, i.e. the cokernel of the difference of the two maps,

$$M \circ \mathscr{P} \circ N \underset{\mathrm{Id}_M \circ \lambda}{\overset{\rho \circ \mathrm{Id}_N}{\rightrightarrows}} M \circ N \twoheadrightarrow M \circ_{\mathscr{P}} N.$$

When $M = M' \circ \mathscr{P}$ is a free $\mathscr{P}$-module, the relative composite product $M \circ_{\mathscr{P}} N$ is naturally isomorphic to $(M' \circ \mathscr{P}) \circ_{\mathscr{P}} N \cong M' \circ N$. The same statement holds on the right-hand side. This construction extends naturally to the differential graded case.

We dualize the arguments to define the relative composite product for comodules over a cooperad. Let $\mathcal{C}$ be a cooperad, $\rho : M \to M \circ \mathcal{C}$ be a right $\mathcal{C}$-comodule and $\lambda : N \to \mathcal{C} \circ N$ be a left $\mathcal{C}$-comodule. The *relative composite product* $M \circ^{\mathcal{C}} N$ *of* M *and* N *under* $\mathcal{C}$ is defined by the following equalizer

$$M \circ \mathcal{C} \circ N \underset{\mathrm{Id}_M \circ \lambda}{\overset{\rho \circ \mathrm{Id}_N}{\leftleftarrows}} M \circ N \longleftarrow M \circ^{\mathcal{C}} N.$$

In the same way, we have $(M' \circ \mathcal{C}) \circ^{\mathcal{C}} N \cong M' \circ N$, for $M = M' \circ \mathcal{C}$ being any cofree $\mathcal{C}$-comodule.

11.2.2 Definition of the Bar Construction

To any twisting morphism $\alpha : \mathcal{C} \to \mathcal{P}$ from a dg cooperad $\mathcal{C}$ to a dg operad $\mathcal{P}$, we associate a functor

$$\mathrm{B}_\alpha : \{\text{dg } \mathcal{P}\text{-algebras}\} \longrightarrow \{\text{conil. dg } \mathcal{C}\text{-coalgebras}\}.$$

Let A be a $\mathcal{P}$-algebra. We consider the cofree $\mathcal{C}$-coalgebra $\mathcal{C}(A)$ on A and we denote by d_1 the square zero coderivation $d_{\mathcal{C}} \circ \mathrm{Id}_A$. Since it is a cofree $\mathcal{C}$-coalgebra, there is a unique coderivation d_2 which extends the degree -1 map

$$\mathcal{C}(A) = \mathcal{C} \circ A \xrightarrow{\alpha \circ \mathrm{Id}_A} \mathcal{P} \circ A \xrightarrow{\gamma_A} A,$$

by Proposition 6.3.8. The coderivation d_2 is equal to the composite

$$\mathcal{C} \circ A \xrightarrow{\Delta_{(1)} \circ \mathrm{Id}_A} (\mathcal{C} \circ_{(1)} \mathcal{C}) \circ A$$
$$\xrightarrow{(\mathrm{Id}_{\mathcal{C}} \circ_{(1)} \alpha) \circ \mathrm{Id}_A} (\mathcal{C} \circ_{(1)} \mathcal{P}) \circ A \cong \mathcal{C} \circ (A; \mathcal{P} \circ A) \xrightarrow{\mathrm{Id}_{\mathcal{C}} \circ (\mathrm{Id}_A; \gamma_A)} \mathcal{C} \circ A.$$

Lemma 11.2.1. *There is a natural isomorphism*

$$\big(\mathcal{C}(A), d_1 + d_2\big) \cong \big((\mathcal{C} \circ_\alpha \mathcal{P}) \circ_{\mathcal{P}} A, d_\alpha\big).$$

Hence, the coderivation $d_1 + d_2$ *is a square zero coderivation,* $(d_1 + d_2)^2 = 0$.

Proof. The underlying modules $\mathcal{C} \circ A \cong (\mathcal{C} \circ \mathcal{P}) \circ_{\mathcal{P}} A$ are naturally isomorphic. The formula for the twisted differential d_α given in Sect. 6.4.5 shows that it is equal to $d_1 + d_2$. Actually, the maps d_α^r and d_2 are defined as the unique coderivation extending the same kind of map. □

The dg $\mathcal{C}$-coalgebra $\mathrm{B}_\alpha A := (\mathcal{C}(A), d_1 + d_2)$ is called the *bar construction of* A *with respect to* α.

We extend this functor to dg $\mathscr{P}$-algebras as follows. When (A, d_A) is a dg $\mathscr{P}$-algebra, the differential d_A induces a square zero coderivation on $\mathscr{C}(A)$: it is the unique coderivation on $\mathscr{C}(A)$ which extends

$$\mathscr{C}(A) \xrightarrow{\varepsilon(A)} \mathrm{I}(A) \cong A \xrightarrow{d_A} A.$$

It is explicitly given by $\mathrm{Id}_{\mathscr{C}} \circ' d_A$. Now we consider the square zero coderivation $d_1 := d_{\mathscr{C} \circ A} = d_{\mathscr{C}} \circ \mathrm{Id}_A + \mathrm{Id}_{\mathscr{C}} \circ' d_A$. Since d_A is a derivation, d_1 and d_2 anticommute, that is $d_1 \circ d_2 + d_2 \circ d_1 = 0$. And the coderivation $d_1 + d_2$ still squares to zero.

In this case, the total complex is called the *bar construction of the dg $\mathscr{P}$-algebra (A, d_A) with respect to α*, which is denoted by

$$\mathrm{B}_\alpha A := \big(\mathscr{C}(A), d = d_1 + d_2\big).$$

The isomorphism of chain complexes $\mathrm{B}_\alpha A \cong (\mathscr{C} \circ_\alpha \mathscr{P}) \circ_{\mathscr{P}} A$ still holds in this case. When $\mathscr{P}$ is a quadratic operad, we consider the bar construction for dg $\mathscr{P}$-algebras associated to the canonical twisting morphism $\kappa : \mathscr{P}^{¡} \to \mathscr{P}$ denoted by $\mathrm{B}_\kappa A := (\mathscr{P}^{¡}(A), d_1 + d_2)$. It is equal to the relative composite product of the right Koszul complex with A:

$$\mathrm{B}_\kappa A \cong \big(\mathscr{P}^{¡} \circ_\kappa \mathscr{P}\big) \circ_{\mathscr{P}} A.$$

11.2.3 Recollection with the Classical Notion

In the case of the Koszul operad *As*, we recover the classical bar construction of Eilenberg–Mac Lane, introduced in Sect. 2.2.1, with the twisting morphism $\kappa : As^{¡} \to As$, as follows.

Proposition 11.2.2. *For any augmented dga algebra A, the bar construction $\mathrm{B}A$ of Sect. 2.2.1 is related to $\mathrm{B}_\kappa A$ by the following isomorphism of dga coaugmented coalgebras*

$$\mathrm{B}A \cong \mathbb{K} \oplus s\mathrm{B}_\kappa \bar{A},$$

under the isomorphism of categories $C \mapsto sC_+ = sC \oplus \mathbb{K}1$ between dg $As^{¡}$-coalgebras and coaugmented dga coalgebras.

Proof. The bar construction $\mathrm{B}A$ is given by the cofree coalgebra $T^c(s\bar{A})$. The counit splits and $\overline{T}^c(s\bar{A})$ is isomorphic to $s\,As^{¡}(\bar{A})$, since $As^{¡} \underset{\mathrm{H}}{\otimes} \mathrm{End}^c_{s\mathbb{K}} \cong As^*$. The coderivation d_2 consists in extracting an element of $As^{¡}(2)$ from $As^{¡}$, transform it into an element of $As(2)$ via κ and make it act on $\bar{A}$. This composite is equal to the coderivation d_2 defining $\mathrm{B}A$. □

Notice that, in the definition of the bar construction given in this section, there is no extra suspension of the algebra A. The sign and the degree shift are all encoded into the dg cooperad $\mathscr{C}$, which is often $\mathscr{P}^{¡}$. The bar construction B_κ produces a dg $\mathscr{P}^{¡}$-coalgebra. To recover the classical cases, like the ones arising in rational homotopy theory (Sect. 11.3.5), one can equivalently consider $s\mathrm{B}_\kappa A$ which is a dg $\mathscr{P}^{!*}$-coalgebra, with underlying space $\mathscr{P}^{!*}(sA)$. Since these two definitions are equivalent, we prefer the first one which is more intrinsic to the Koszul duality theory of operads.

11.2.4 Bar Construction and Quasi-isomorphisms

Proposition 11.2.3. *Let $\alpha : \mathscr{C} \to \mathscr{P}$ be an operadic twisting morphism and let $f : A \xrightarrow{\sim} A'$ be a quasi-isomorphism of dg $\mathscr{P}$-algebras. The induced morphism $\mathrm{B}_\alpha f : \mathrm{B}_\alpha A \xrightarrow{\sim} \mathrm{B}_\alpha A'$ is a quasi-isomorphism of dg $\mathscr{C}$-coalgebras.*

Proof. We consider the filtration on $\mathrm{B}_\alpha A$ defined by $F_p\mathrm{B}_\alpha A := \mathscr{C}_{\leq p}(A)$, where $\mathscr{C}_{\leq p}$ is the sub-$\mathbb{S}$-module of $\mathscr{C}$ made up of elements of homological degree less than p: $\mathscr{C}_{\leq p}(n) := \bigoplus_{d\leq p} \mathscr{C}_d(n)$. This filtration is stable under the differential of the bar construction: $\mathrm{Id}_{\mathscr{C}} \circ' d_A : F_p \to F_p$, $d_{\mathscr{C}} \circ \mathrm{Id}_A : F_p \to F_{p-1}$ and $d_2 : F_p \to F_{\leq p-1}$ since $\mathscr{C}$ and $\mathscr{P}$ are nonnegatively graded. Hence, the first term of the associated spectral sequence is equal to $E^0_{pq}\mathrm{B}_\alpha A \cong (\mathscr{C}_p(A))_{p+q}$ with $d^0 = \mathrm{Id}_{\mathscr{C}} \circ' d_A$. The map $E^0_{p\bullet}(f)$ is a quasi-isomorphism by Künneth formula. Since this filtration is increasing, bounded below and exhaustive it converges to the homology of the bar construction by Theorem 1.5.1. The map $E^\infty_{\bullet\bullet}(f)$ being an isomorphism, it proves that $\mathrm{B}_\alpha f$ is a quasi-isomorphism. □

11.2.5 Definition of the Cobar Construction

Dually, to any operadic twisting morphism $\alpha : \mathscr{C} \to \mathscr{P}$, we associate a functor in the opposite direction

$$\Omega_\alpha : \{\text{conil. dg } \mathscr{C}\text{-coalgebras}\} \longrightarrow \{\text{dg } \mathscr{P}\text{-algebras}\}.$$

Let C be a conilpotent $\mathscr{C}$-coalgebra, the underlying module of $\Omega_\alpha C$ is the free $\mathscr{P}$-algebra on C, that is $\mathscr{P}(C)$. It is endowed with a first square zero derivation $d_1 := -d_{\mathscr{P}} \circ C$. By Proposition 6.3.6, there is a unique derivation $-d_2$ which extends

$$C \xrightarrow{\Delta_C} \mathscr{C} \circ C \xrightarrow{\alpha \circ \mathrm{Id}_C} \mathscr{P} \circ C.$$

It is explicitly given by

$$\begin{aligned} \mathscr{P} \circ C &\xrightarrow{\mathrm{Id}_{\mathscr{P}} \circ' \Delta_C} \mathscr{P} \circ (C; \mathscr{C} \circ C) \\ &\xrightarrow{\mathrm{Id}_{\mathscr{P}} \circ (\mathrm{Id}_C; \alpha \circ \mathrm{Id}_C)} \mathscr{P} \circ (C; \mathscr{P} \circ C) \cong (\mathscr{P} \circ_{(1)} \mathscr{P})(C) \xrightarrow{\gamma_{(1)} \circ \mathrm{Id}_C} \mathscr{P} \circ C. \end{aligned}$$

Lemma 11.2.4. *There is a natural isomorphism*

$$(\mathscr{P}(C), -d_1 - d_2) \cong ((\mathscr{P} \circ_\alpha \mathscr{C}) \circ^{\mathscr{C}} C, d_\alpha).$$

Hence, the derivation $d_1 + d_2$ is a square zero derivation, $(d_1 + d_2)^2 = 0$.

Proof. The underlying modules $\mathscr{P} \circ C \cong (\mathscr{P} \circ \mathscr{C}) \circ^{\mathscr{C}} C$ are naturally isomorphic. The explicit formula for the twisted differential d_α given in Sect. 6.4.5 shows that it is equal to the one given above for d_2. Both of them are defined as the unique derivation extending the same kind of map. □

The dg $\mathscr{P}$-algebra $\Omega_\alpha C := (\mathscr{P}(C), d_1 + d_2)$ is called the *cobar construction of C with respect to α*.

We extend this functor to dg $\mathscr{C}$-coalgebras as follows. When (C, d_C) is a dg $\mathscr{C}$-coalgebra, the differential d_C induces a square zero derivation $\mathrm{Id}_{\mathscr{P}} \circ' d_C$ on $\mathscr{P}(C)$. By definition, the derivations $d_{\mathscr{P}} \circ \mathrm{Id}_C$ and $\mathrm{Id}_{\mathscr{P}} \circ' d_C$ anticommute. We denote their sum by $d_1 := -d_{\mathscr{P}} \circ \mathrm{Id}_C + \mathrm{Id}_{\mathscr{P}} \circ' d_C$. Since d_C is a coderivation, d_1 and d_2 anticommute, that is $d_1 \circ d_2 + d_2 \circ d_1 = 0$ and $d_1 + d_2$ is still a square zero derivation.

The total complex of this bicomplex is called the *cobar construction of the dg $\mathscr{C}$-coalgebra (C, d_C) with respect to α*. We denote it by

$$\Omega_\alpha C := \big(\mathscr{P}(C), d = d_1 + d_2\big).$$

The isomorphism of chain complexes $\Omega_\alpha C \cong (\mathscr{P} \circ_\alpha \mathscr{C}) \circ^{\mathscr{C}} C$ still holds in this case. Given a quadratic operad $\mathscr{P}$, we can consider the cobar construction Ω_κ, for conilpotent dg $\mathscr{P}^{\text{!`}}$-coalgebras, associated to the twisting morphism $\kappa : \mathscr{P}^{\text{!`}} \to \mathscr{P}$. It is given by the left Koszul complex

$$\Omega_\kappa C \cong \big(\mathscr{P} \circ_\kappa \mathscr{P}^{\text{!`}}\big) \circ^{\mathscr{P}^{\text{!`}}} C.$$

11.2.6 Recollection with the Classical Notion

In the case of the operad As, with the twisting morphism $\kappa : As^{\text{!`}} \to As$, we recover the cobar construction of J.F. Adams [Ada56], introduced in Sect. 2.2.2.

In this case, since the twisting morphism κ vanishes outside $As^{\text{!`}}(2)$, the cobar construction also applies to any dg $As^{\text{!`}}$-coalgebras, non-necessarily conilpotent.

Proposition 11.2.5. *Let C be a coaugmented dga coalgebra. Its cobar construction is isomorphic to the augmented dg algebra*

$$\Omega C \cong \mathbb{K} \oplus \Omega_\kappa\big(s^{-1}\overline{C}\big),$$

where $s^{-1}\overline{C}$ is a dg $As^{\text{!`}}$-coalgebra.

Proof. The underlying module of $\mathbb{K} \oplus \Omega_\kappa(s^{-1}\overline{C}) \cong \mathbb{K} \oplus As(s^{-1}\overline{C})$ is the counital cofree coalgebra $T^c(s^{-1}\overline{C}) \cong \mathbb{K} \oplus \overline{T}^c(s^{-1}\overline{C})$. The derivation d_2 defining Ω_κ is the unique derivation which extends the coproduct of C. Therefore it agrees with the derivation d_2 of the classical bar construction ΩC. □

11.2.7 Cobar Construction and Quasi-isomorphisms

A non-negatively graded dg $\mathscr{C}$-coalgebra C is called *connected* if $C_0 = 0$.

Proposition 11.2.6. *Let $\alpha : \mathscr{C} \to \mathscr{P}$ be an operadic twisting morphism which vanishes on $\mathscr{C}(1)$. Let $f : C \to C'$ be a quasi-isomorphism between connected dg $\mathscr{C}$-coalgebras. The induced morphism $\Omega_\alpha f : \Omega_\alpha C \xrightarrow{\sim} \Omega_\alpha C'$ between the cobar constructions is a quasi-isomorphism.*

Proof. We consider the following filtration on the cobar construction

$$F_p \Omega_\alpha C := \bigoplus_{n \geq -p} \mathscr{P}(n) \otimes_{\mathbb{S}_n} C^{\otimes n}.$$

This increasing filtration is preserved by the differential of the cobar construction: $d_1 : F_p \to F_p$ and $d_2 : F_p \to F_{\leq p-1}$. Hence the first term of the associated spectral sequence is equal to $E^0_{pq} = (\mathscr{P}(-p) \otimes_{\mathbb{S}_{-p}} C^{\otimes -p})_{p+q}$, with $d^0 = d_1$. By Künneth formula, $E^0_{p\bullet}(\Omega_\alpha f) = \mathrm{Id}_{\mathscr{P}} \otimes f^{\otimes -p}$ is a quasi-isomorphism. Since C (respectively C') is a connected dg $\mathscr{C}$-coalgebra, the degree of the elements of $C^{\otimes -p}$ is at least $-p$. Therefore $(F_p \Omega_\alpha C)_d = 0$ for $p < -d$. The filtration being exhaustive and bounded below, this spectral sequence converges to the homology of the cobar construction by Theorem 1.5.1, which concludes the proof. □

This result is the generalization of Proposition 2.2.5. Notice that C is a connected dg $\mathscr{C}$-coalgebra if and only if sC is a 2-connected dg $\mathscr{C} \otimes \mathrm{End}^c_{s\mathbb{K}}$-coalgebra. So the hypotheses are the same. Without this assumption, the result does not hold anymore, as Proposition 2.4.3 shows.

11.3 Bar–Cobar Adjunction for $\mathscr{P}$-Algebras

The bar and the cobar constructions are shown to form a pair of adjoint functors which represent the two functors associated to twisting morphisms with respect to the operadic twisting morphism $\alpha : \mathscr{C} \to \mathscr{P}$.

$$\Omega_\alpha : \{\text{conil. dg } \mathscr{C}\text{-coalgebras}\} \rightleftharpoons \{\text{dg } \mathscr{P}\text{-algebras}\} : \mathrm{B}_\alpha.$$

From this adjunction arise universal twisting morphisms. When the twisting morphism α is a Koszul morphism, the adjunction unit and counit are quasi-isomorphisms.

11.3.1 The Adjunction

Proposition 11.3.1. *Let $\alpha : \mathscr{C} \to \mathscr{P}$ be an operadic twisting morphism. For any dg $\mathscr{P}$-algebra A and any conilpotent dg $\mathscr{C}$-coalgebra C, there are natural bijections*

$$\mathrm{Hom}_{\mathsf{dg}\ \mathscr{P}\text{-}\mathsf{alg}}(\Omega_\alpha C, A) \cong \mathrm{Tw}_\alpha(C, A) \cong \mathrm{Hom}_{\mathsf{dg}\ \mathscr{C}\text{-}\mathsf{coalg}}(C, \mathrm{B}_\alpha A).$$

Proof. Since $\mathscr{P}(C)$ is a free $\mathscr{P}$-algebra, any morphism of $\mathscr{P}$-algebras $\Phi : \mathscr{P}(C) \to A$ is characterized by its restriction $\varphi : C \to A$ on C, under the formula $\Phi = \gamma_A(\mathrm{Id}_{\mathscr{P}} \circ \varphi)$. The map Φ is a morphism of *dg* $\mathscr{P}$-algebras if and only if $\Phi \circ (d_1 + d_2) = d_A \circ \Phi$. Since d_1 and d_2 are derivations on the free $\mathscr{P}$-algebra $\mathscr{P}(C)$, it is enough to check this relation on the space of generators C of $\mathscr{P}(C)$, that is $(\Phi \circ (d_1 + d_2))|_C = d_A \circ \varphi$. We have $(\Phi \circ d_2)|_C = - \star_\alpha (\varphi)$ and $\partial(\varphi) = d_A \circ \varphi - (\Phi \circ d_1)|_C$. So the previous equation is equal to $\partial(\varphi) + \star_\alpha \varphi = 0$, which is the Maurer–Cartan equation. Therefore a map $\Phi : \Omega_\alpha C \to A$ is a morphism of dg $\mathscr{P}$-algebras if and only if its restriction $\varphi : C \to A$ is a twisting morphism with respect to α.

The bijection on the right-hand side can be made explicit using the same arguments. Any map $\varphi : C \to A$ extends to a unique morphism of $\mathscr{C}$-coalgebras by the formula $(\mathrm{Id}_{\mathscr{C}} \circ \varphi)(\Delta_C) : C \to \mathscr{C} \circ C \to \mathscr{C} \circ A$, see Proposition 10.2.1. This morphism commutes with the differential if and only if φ is a twisting morphism with respect to α. □

11.3.2 Universal Twisting Morphisms for $\mathscr{P}$-Algebras

As a corollary to the preceding adjunction, we define several canonical morphisms. When $C = \mathrm{B}_\alpha A$, associated to the identity $\mathrm{Id}_{B_\alpha A}$ on the right-hand side, there is a universal twisting morphism $\pi_\alpha(A) : \mathrm{B}_\alpha A \to A$ and a morphism of dg $\mathscr{P}$-algebras $\varepsilon_\alpha(A) : \Omega_\alpha \mathrm{B}_\alpha A \to A$, which is the *counit* of the adjunction. (For simplicity, we will often forget the letter A in the notation.) The former is equal to the composite with the counit of $\mathscr{C}$

$$\pi_\alpha : \mathrm{B}_\alpha A = \mathscr{C} \circ A \xrightarrow{\varepsilon \circ \mathrm{Id}_A} \mathrm{I} \circ A = A,$$

and the latter is equal to the composite

$$\varepsilon_\alpha : \Omega_\alpha \mathrm{B}_\alpha A = \mathscr{P} \circ \mathscr{C} \circ A \xrightarrow{\mathrm{Id}_{\mathscr{P}} \circ \pi_\alpha} \mathscr{P} \circ A \xrightarrow{\gamma_A} A.$$

Dually, let $A = \Omega_\alpha C$. Associated to the identity $\mathrm{Id}_{\Omega_\alpha C}$ on the left hand side, there is a universal twisting morphism $\iota_\alpha : C \to \Omega_\alpha C$ and a morphism of dg $\mathscr{C}$-coalgebras $\upsilon_\alpha : C \to \mathrm{B}_\alpha \Omega_\alpha C$, which is the *unit* of the adjunction. The former is equal to the composite with the unit of $\eta : \mathrm{I} \to \mathscr{P}$ of the operad

$$\iota_\alpha : C \cong \mathrm{I} \circ C \xrightarrow{\eta \circ \mathrm{Id}_C} \mathscr{P} \circ C = \Omega_\alpha C,$$

and the latter is equal to the composite

$$\upsilon_\alpha : C \xrightarrow{\Delta_C} \mathscr{C} \circ C \cong \mathscr{C} \circ \mathrm{I} \circ C \xrightarrow{\mathrm{Id}_{\mathscr{C}} \circ \eta \circ \mathrm{Id}_C} \mathscr{C} \circ \mathscr{P} \circ C = \mathrm{B}_\alpha \Omega_\alpha C.$$

Both π_α and ι_α satisfy the following universal property.

Proposition 11.3.2. *Any twisting morphism $\varphi : C \to A$ with respect to α factors uniquely through the universal twisting morphisms π_α and ι_α:*

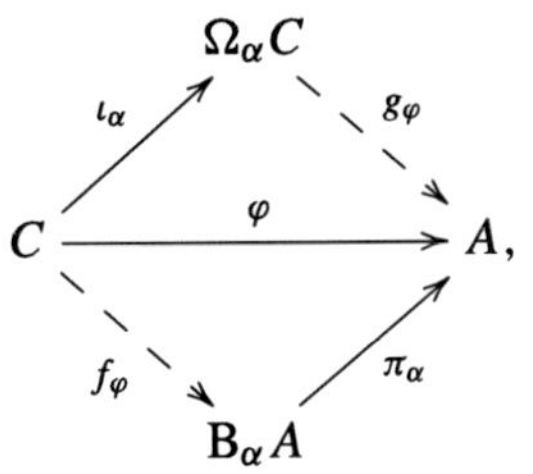

where the map $g : \Omega_\alpha C \to A$ is a morphism of dg $\mathscr{P}$-algebras and where the map $f : C \to \mathrm{B}_\alpha A$ is a morphism of dg $\mathscr{C}$-coalgebras.

11.3.3 Bar–Cobar Resolutions

Theorem 11.3.3. *Let $\alpha : \mathscr{C} \to \mathscr{P}$ be an operadic twisting morphism. Suppose that $\mathscr{C}$ and $\mathscr{P}$ are connected weight graded and that α preserves this weight grading.*

The twisting morphism α is a Koszul morphism if and only if the counit of the adjunction $\varepsilon_\alpha(A) : \Omega_\alpha \mathrm{B}_\alpha A \xrightarrow{\sim} A$ is a quasi-isomorphism of dg $\mathscr{P}$-algebras, for every dg $\mathscr{P}$-algebra A.

Proof.

($\Rightarrow$) The underlying module of $\Omega_\alpha \mathrm{B}_\alpha A$ is isomorphic to $\mathscr{P} \circ \mathscr{C} \circ A$. We consider the increasing filtration F_p made up of the elements of $\mathscr{P} \circ \mathscr{C} \circ A$ whose total weight in $\mathscr{P}$ and $\mathscr{C}$ is less than p. On A, we consider the trivial filtration, where $\mathrm{F}'_p A := A$, for any $p \in \mathbb{N}$. The counit of adjunction preserves this weight grading. The first page E^0 is isomorphic to the chain complex $\mathscr{P} \circ_\alpha \mathscr{C} \circ A$. Since α is a Koszul morphism, the spectral sequence collapses at rank one, $E^1 \cong H(A)$, by the operadic Künneth Formula (Proposition 6.2.3). Therefore $E^1 \varepsilon_\alpha$ is an isomorphism. The respective filtrations are bounded below and exhaustive; so we conclude by the classical convergence theorem of spectral sequences (Theorem 1.5.1).

($\Leftarrow$) We apply the result to the trivial $\mathscr{P}$-algebra V^{tr}, where V is any dg module concentrated in degree 0. In this case, the bar–cobar construction is quasi-isomorphic to $\Omega_\alpha \mathrm{B}_\alpha V^{\mathrm{tr}} \cong \mathscr{P} \circ_\alpha \mathscr{C} \circ V$. Hence, we get $H(\mathscr{P} \circ_\alpha \mathscr{C}) \circ V \cong V$, for any $\mathbb{K}$-module V. We conclude that $\mathscr{P} \circ_\alpha \mathscr{C}$ is acyclic by the Schur's lemma (Corollary 6.2.4). □

Theorem 11.3.4. *Let $\alpha : \mathscr{C} \to \mathscr{P}$ be an operadic twisting morphism, which vanishes on $\mathscr{C}(1)$.*

The twisting morphism α is a Koszul morphism if and only if the unit of the adjunction $\upsilon_\alpha(C) : C \xrightarrow{\sim} \mathrm{B}_\alpha \Omega_\alpha C$ is a quasi-isomorphism of dg $\mathscr{C}$-coalgebras, for every dg conilpotent $\mathscr{C}$-coalgebra C.

Proof.

($\Rightarrow$) The underlying module of $\mathrm{B}_\alpha \Omega_\alpha C$ is isomorphic to $\mathscr{C} \circ_\alpha \mathscr{P} \circ C$. We consider the increasing filtration $\mathrm{F}_p := \bigoplus_{n \le p} (\mathscr{C} \circ \mathscr{P})(n) \otimes_{\mathbb{S}_n} C^{\otimes n}$. It is increasing, bounded below and exhaustive. We conclude with the same arguments as in the previous proof.

($\Leftarrow$) We apply the result to the trivial $\mathscr{C}$-coalgebra W^{tr}, where W is any dg module concentrated in degree 0. □

When these theorems apply, we call the resulting functorial resolutions provided by the counit (respectively by the unit) of the bar–cobar adjunction, the *bar–cobar resolution* of A (respectively of C).

Corollary 11.3.5. *Let $\mathscr{P}$ be a Koszul operad. For every dg $\mathscr{P}$-algebra A, the counit of the adjunction*

$$\varepsilon_\kappa : \Omega_\kappa \mathrm{B}_\kappa A = \mathscr{P} \circ_\kappa \mathscr{P}^{\text{!`}}(A) \xrightarrow{\sim} A$$

is a quasi-isomorphism of dg $\mathscr{P}$-algebras.

Dually, for every conilpotent dg $\mathscr{P}^{\text{!`}}$-coalgebra C, the unit of the adjunction

$$\upsilon_\kappa : C \xrightarrow{\sim} \mathrm{B}_\kappa \Omega_\kappa C = \mathscr{P}^{\text{!`}} \circ_\kappa \mathscr{P}(C)$$

is a quasi-isomorphism of dg $\mathscr{P}^{\text{!`}}$-coalgebras.

Proof. It is a direct corollary of Theorem 11.3.3. □

When $\mathscr{P}$ is not a Koszul operad, one can still apply Theorem 11.3.3 to the Koszul morphism $\pi : \mathrm{B}\mathscr{P} \to \mathscr{P}$ to get functorial quasi-free resolutions

$$\varepsilon_\pi : \Omega_\pi \mathrm{B}_\pi A = \mathscr{P} \circ \mathrm{B}\mathscr{P} \circ A \xrightarrow{\sim} A,$$

for any dg $\mathscr{P}$-algebra A.

11.3.4 Recollection with the Classical Case

In the case of the Koszul operad As with its Koszul morphism $\kappa : As^{\text{!`}} \to As$, we recover the classical bar–cobar resolutions for (co)augmented dga (co)algebras of Proposition 2.3.2.

Proposition 11.3.6. *For any augmented dga algebra A, there is an isomorphism of augmented dga algebras*

$$\Omega \mathrm{B} A \cong \mathbb{K} \oplus \Omega_\kappa \mathrm{B}_\kappa \bar{A} \xrightarrow{\sim} A.$$

Dually, for any conilpotent dga coalgebra C, there is an isomorphism of coaugmented dga coalgebras

$$C \xrightarrow{\sim} \mathrm{B}\Omega C \cong \mathbb{K} \oplus s\mathrm{B}_\kappa \Omega_\kappa \left(s^{-1}\overline{C}\right).$$

Proof. It is a direct corollary of Propositions 11.2.2 and 11.2.5. □

11.3.5 Rational Homotopy Theory

When $\mathscr{P}$ is the Koszul operad *Lie* encoding Lie algebras, the bar construction B_κ is the functor "$\mathscr{C}$" in Quillen [Qui69] and the cobar construction Ω_κ is his functor "$\mathscr{L}$" (see Sect. 22 of [FHT01]).

When $\mathscr{P}$ is the Koszul operad *Com* encoding commutative algebras, the bar construction B_κ was introduced and used by Sullivan in [Sul77] to define algebraic models which compute rational homotopy groups of topological spaces. The bar construction B_κ is the functor Γ and the cobar construction Ω_κ is the functor $\mathscr{A}$ introduced in [SS85] by Schlessinger and Stasheff. These authors also made the induced bar–cobar resolution explicit and used it to compute Harrison and André–Quillen cohomology for commutative algebras. For more details on this subject, we refer the reader to the next chapter.

In the case of the Koszul operad *Leib* encoding Leibniz algebras, see Sect. 13.5, the bar and the cobar constructions were made explicit in [Liv98b] by Livernet.

11.3.6 Koszul Duality Theory for Quadratic $\mathscr{P}$-Algebras

In the light of Chaps. 2 and 3, one can define the notion of *Koszul morphisms for $\mathscr{P}$-algebras* and then prove a *Koszul duality theory for $\mathscr{P}$-algebras*. The main difficulty here is to define the proper analog of the twisted tensor product, i.e. the Koszul complex, for $\mathscr{P}$-algebras. This problem has been solved by Joan Millès in [Mil10]. This theory allows one to simplify the bar–cobar resolutions for quadratic $\mathscr{P}$-algebras. It provides smaller and minimal resolutions for some quadratic $\mathscr{P}$-algebras, naturally called *Koszul $\mathscr{P}$-algebras*.

11.4 Homotopy Theory of $\mathscr{P}$-Algebras

Let $\mathscr{P}$ be a Koszul operad. We apply the previous results to the Koszul morphisms $\kappa : \mathscr{P}^{\text{!`}} \to \mathscr{P}$ and $\iota : \mathscr{P}^{\text{!`}} \to \Omega \mathscr{P}^{\text{!`}} = \mathscr{P}_\infty$.

$$\begin{array}{ccc} \mathscr{P} & \xleftarrow{\ \kappa\ } & \mathscr{P}^{¡} \\ \uparrow{\scriptstyle\sim} & \swarrow_{\iota} & \\ \mathscr{P}_\infty = \Omega\mathscr{P}^{¡}. & & \end{array}$$

The induced bar construction $\mathrm{B}_\iota A \cong \mathscr{P}^{¡}(A)$ coincides with the third definition of $\mathscr{P}_\infty$-algebras. The counit of adjunction gives a functorial quasi-free resolution for $\mathscr{P}_\infty$-algebras. Combining the two adjunctions associated to κ and ι, we define a pair of adjoint functors between dg $\mathscr{P}$-algebras and $\mathscr{P}_\infty$-algebras with ∞-morphisms. The unit of this adjunction allows us to *rectify* universally $\mathscr{P}_\infty$-algebras: any $\mathscr{P}_\infty$-algebra is universally ∞-quasi-isomorphic to a dg $\mathscr{P}$-algebra. As a corollary, it proves that the homotopy categories of the dg $\mathscr{P}$-algebras and $\mathscr{P}_\infty$-algebras, *with* ∞-morphisms, are equivalent.

11.4.1 Third Definition of Homotopy $\mathscr{P}$-Algebras

Let us consider the operadic twisting morphism $\iota : \mathscr{P}^{¡} \to \Omega\mathscr{P}^{¡}$. By Proposition 11.3.1, it induces the following adjunction

$$\Omega_\iota : \{\text{conil. dg } \mathscr{P}^{¡}\text{-coalgebras}\} \rightleftharpoons \{\mathscr{P}_\infty\text{-algebras}\} : \mathrm{B}_\iota.$$

The bar construction $\mathrm{B}_\iota A = \mathscr{P}^{¡}(A)$ is the third equivalent definition of a $\mathscr{P}_\infty$-algebra, see Sect. 10.1.8. The ∞-morphisms between two $\mathscr{P}_\infty$-algebras were actually defined in Sect. 10.2.2 as the morphisms of dg $\mathscr{P}^{¡}$-coalgebras between their images under the bar construction B_ι. This result translates exactly into the following proposition.

Proposition 11.4.1. *The bar construction* B_ι, *associated to the operadic twisting morphism* $\iota : \mathscr{P}^{¡} \to \Omega\mathscr{P}^{¡}$, *extends to an isomorphism of categories*

$$\widetilde{\mathrm{B}}_\iota : \infty\text{-}\mathscr{P}_\infty\text{-alg} \xrightarrow{\cong} \text{quasi-free } \mathscr{P}^{¡}\text{-coalg}$$

between the category of $\mathscr{P}_\infty$*-algebras with their* ∞*-morphisms and the full subcategory of* $\mathscr{P}^{¡}$*-coalgebras composed by quasi-free ones.*

11.4.2 Bar–Cobar Resolution of $\mathscr{P}_\infty$-Algebras

Proposition 11.4.2. *Let A be a* $\mathscr{P}_\infty$*-algebra. The bar–cobar construction associated to the operadic twisting morphism* $\iota : \mathscr{P}^{¡} \to \Omega\mathscr{P}^{¡}$ *provides a quasi-free*

$\mathscr{P}_\infty$-algebra quasi-isomorphic to A:

$$\Omega_\iota \mathrm{B}_\iota A \cong \mathscr{P}_\infty\big(\mathscr{P}^{i}(A)\big) \xrightarrow{\sim} A.$$

Proof. We proved in Lemma 6.5.9 that $\iota : \mathscr{P}^{\mathrm{i}} \to \Omega\mathscr{P}^{\mathrm{i}}$ is a Koszul morphism. So this proposition is a direct corollary of Theorem 11.3.3. □

This result gives a functorial quasi-free resolution to any $\mathscr{P}_\infty$-algebra. It will be used in the next chapter to compute the (co)homology theories of $\mathscr{P}_\infty$-algebras, see Sect. 12.4.4. In fact, the assumption that $\mathscr{P}$ is Koszul is not necessary, one can apply it to any quadratic operad and to any $\Omega\mathscr{P}^{\mathrm{i}}$-algebra.

Notice that the morphism between $\Omega_\iota \mathrm{B}_\iota A$ and A is a "strict" morphism of $\mathscr{P}_\infty$-algebras as defined in Sect. 10.2.1.

11.4.3 Rectification

To sum up, the Koszul morphisms κ and ι induce the following diagram.

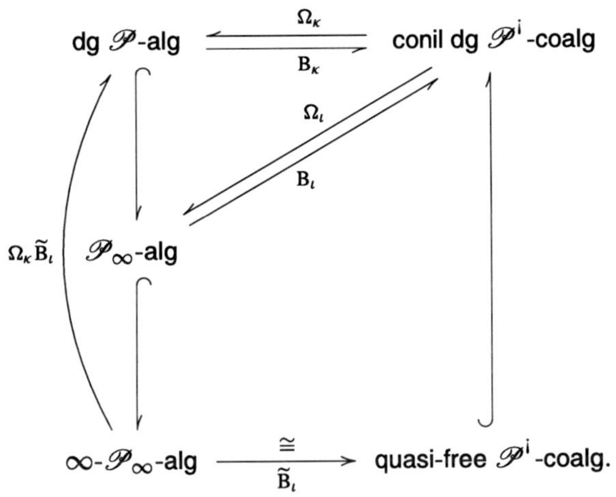

Let us denote by j : dg $\mathscr{P}$-alg $\rightarrowtail$ $\mathscr{P}_\infty$-alg and by i : dg $\mathscr{P}$-alg $\rightarrowtail$ $\infty-\mathscr{P}_\infty$-alg the two "inclusions". Notice that the bar constructions B_ι and $\widetilde{\mathrm{B}}_\iota$ extend the bar construction B_κ in the following sense: $\mathrm{B}_\kappa = \mathrm{B}_\iota \circ j = \widetilde{\mathrm{B}}_\iota \circ i$.

Proposition 11.4.3. *Let $\mathscr{P}$ be a Koszul operad. The functors*

$$\Omega_\kappa \widetilde{\mathrm{B}}_\iota : \infty-\mathscr{P}_\infty\text{-alg} \rightleftharpoons \text{dg } \mathscr{P}\text{-alg} : i$$

form a pair of adjoint functors.

Proof. We use the characterization of adjoint functors given in terms of the unit and the counit of adjunction, see Proposition B.2.1.

To any $\mathcal{P}_\infty$-algebra A, we apply the unit of the adjunction associated to the twisting morphism κ to the dg $\mathcal{P}^{\text{i}}$-coalgebra $\mathrm{B}_\iota A$:

$$\upsilon_\kappa(\mathrm{B}_\iota A) : \mathrm{B}_\iota A \to \mathrm{B}_\kappa \Omega_\kappa(\mathrm{B}_\iota A) = \mathrm{B}_\iota\big(i\Omega_\kappa \widetilde{\mathrm{B}}_\iota(A)\big).$$

Equivalently, this defines an ∞-morphism denoted by $\upsilon(A) : A \rightsquigarrow i\Omega_\kappa \widetilde{\mathrm{B}}_\iota(A)$, which in turn induces a transformation of functors $\upsilon : \mathrm{Id} \longrightarrow i\Omega_\kappa \widetilde{\mathrm{B}}_\iota$.

In the other way round, the unit of the adjunction associated to κ gives, for any dg $\mathcal{P}$-algebra A', the following morphism of dg $\mathcal{P}$-algebras

$$\varepsilon_\kappa(A') : \Omega_\kappa \widetilde{\mathrm{B}}_\iota i(A') = \Omega_\kappa \mathrm{B}_\kappa(A') \to A'.$$

This induces the transformation of functors $\varepsilon : \Omega_\kappa \widetilde{\mathrm{B}}_\iota i \longrightarrow \mathrm{Id}$.

For any $\mathcal{P}_\infty$-algebra A, the composite $\varepsilon(\Omega_\kappa \widetilde{\mathrm{B}}_\iota(A)) \circ \Omega_\kappa \widetilde{\mathrm{B}}_\iota(\upsilon(A))$ is equal to

$$\Omega_\kappa(\mathrm{B}_\iota A) \xrightarrow{\Omega_\kappa(\upsilon_\kappa(\mathrm{B}_\iota A))} \Omega_\kappa\big(\mathrm{B}_\kappa \Omega_\kappa(\mathrm{B}_\iota A)\big) \xrightarrow{\varepsilon_\kappa(\Omega_\kappa(\mathrm{B}_\iota A))} \Omega_\kappa(\mathrm{B}_\iota A)$$

which is the identity by the adjunction associated to κ.

In the same way, for any dg $\mathcal{P}$-algebra A', the composite $i(\varepsilon(A')) \circ \upsilon(i(A'))$ is an ∞-morphism, whose image under the functor $\widetilde{\mathrm{B}}_\iota$ is equal to the following morphism of dg $\mathcal{P}^{\text{i}}$-coalgebras

$$\mathrm{B}_\kappa A' \xrightarrow{\upsilon_\kappa(\mathrm{B}_\kappa A')} (\mathrm{B}_\kappa \Omega_\kappa)\mathrm{B}_\kappa A' \xrightarrow{\mathrm{B}_\kappa(\varepsilon_\kappa(A'))} \mathrm{B}_\kappa A'.$$

Once again, it is the identity by the adjunction associated to κ. □

We use this adjunction to *rectify* $\mathcal{P}_\infty$-algebras in a natural way.

Theorem 11.4.4 (Rectification). *Let $\mathcal{P}$ be a Koszul operad. Any homotopy $\mathcal{P}$-algebra A is naturally ∞-quasi-isomorphic to the dg $\mathcal{P}$-algebra $\Omega_\kappa \mathrm{B}_\iota A$:*

$$A \overset{\sim}{\rightsquigarrow} \Omega_\kappa \mathrm{B}_\iota A.$$

Proof. To any $\mathcal{P}_\infty$-algebra A, we apply the unit of the adjunction defined in Proposition 11.4.3: $A \rightsquigarrow i\Omega_\kappa \widetilde{\mathrm{B}}_\iota A$, whose right-hand side is equal to the dg $\mathcal{P}$-algebra $\Omega_\kappa \mathrm{B}_\iota A$. Since it is defined by the unit $\upsilon_\kappa(\mathrm{B}_\iota A)$ of the adjunction associated to the twisting morphism κ applied to $\mathrm{B}_\iota A$, its first component is equal to the chain morphism $A \cong \mathrm{I} \circ \mathrm{I}(A) \to \mathcal{P} \circ_\kappa \mathcal{P}^{\text{i}}(A) \cong \Omega_\kappa \mathrm{B}_\iota A$.

We filter the right-hand side by the total weight of $\mathcal{P} \circ \mathcal{P}^{\text{i}}$. This defines an increasing, bounded below and exhaustive filtration, whose first page is isomorphic to $\mathrm{E}^0 \cong (q\mathcal{P} \circ_\kappa q\mathcal{P}^{\text{i}}) \circ A$. By the operadic Künneth formula (Proposition 6.2.3), the second page is isomorphic to $\mathrm{E}^1 \cong \mathrm{H}_\bullet(A)$. We conclude by the classical convergence theorem of spectral sequences (Theorem 1.5.1). □

Together with Theorem 10.3.10, this theorem proves that in the homotopy class of any $\mathscr{P}_\infty$-algebra, there are a minimal $\mathscr{P}_\infty$-algebra and a dg $\mathscr{P}$-algebra

$$(H, 0, \text{transferred structure}) \sim (A, d_A, \varphi) \sim (\Omega_\kappa \widetilde{\mathrm{B}}_\iota A, d, \mathscr{P}\text{-algebra structure}).$$

Proposition 11.4.5 (Universal property ($\mathscr{U}$)). *Any ∞-morphism $F : A \rightsquigarrow V$ from a $\mathscr{P}_\infty$-algebra A to a dg $\mathscr{P}$-algebra V factors uniquely through $\upsilon(A) : A \overset{\sim}{\rightsquigarrow} \Omega_\kappa \mathrm{B}_\iota A$*

$$\begin{array}{ccc} A & \overset{\upsilon(A)}{\rightsquigarrow} & \Omega_\kappa \mathrm{B}_\iota A \\ & {\scriptstyle F}\searrow & \downarrow {\scriptstyle \exists!\,\overline{F}} \\ & & V, \end{array} \qquad (\mathscr{U})$$

where $\overline{F} : \Omega_\kappa \mathrm{B}_\iota A \to V$ is a morphism of dg $\mathscr{P}$-algebras.

Proof. Since this rectification of $\mathscr{P}_\infty$-algebras is produced by a left adjoint functor, it satisfies this universal property. □

Proposition 11.4.6 (Universality of the rectification). *Under the property ($\mathscr{U}$), any $\mathscr{P}_\infty$-algebra admits an ∞-quasi-isomorphic dg $\mathscr{P}$-algebra, which is unique up to unique isomorphism.*

Proof. By the left adjoint property (Sect. B.2.2), there is a unique dg $\mathscr{P}$-algebra associated, in a natural way, to any $\mathscr{P}_\infty$-algebra, which satisfies the property ($\mathscr{U}$). Moreover, it is unique up to unique isomorphism. □

Notice moreover that, for any $\mathscr{P}_\infty$-algebra A, any such dg $\mathscr{P}$-algebra $\mathscr{F}(A)$ is actually ∞-quasi-isomorphic to it since one of them is

$$\begin{array}{ccc} & & \mathscr{F}(A) \\ & \overset{\sim}{\nearrow} & \uparrow \wr \\ A & \underset{\upsilon(A)}{\overset{\sim}{\rightsquigarrow}} & \Omega_\kappa \mathrm{B}_\iota A. \end{array}$$

11.4.4 Equivalence of Homotopy Categories

Recall that the *homotopy category* Ho(dg $\mathscr{P}$-alg) is obtained by localizing the category of dg $\mathscr{P}$-algebras with respect to the class of quasi-isomorphisms, see Appendix B.6.1. Respectively the homotopy category Ho(∞-$\mathscr{P}_\infty$-alg) is obtained by localizing the category of $\mathscr{P}_\infty$-algebras with respect to the class of ∞-quasi-isomorphisms Ho(∞-$\mathscr{P}_\infty$-alg). The two previous results show that the adjunction (Proposition 11.4.3) induces an equivalence between these two homotopy categories.

A morphism of dg $\mathscr{P}^{\text{i}}$-coalgebras $g : C \to D$ is called a *weak equivalence* if its image $\Omega_\kappa g : \Omega_\kappa C \xrightarrow{\sim} \Omega_\kappa D$ under the cobar construction Ω_κ is a quasi-isomorphism of dg $\mathscr{P}$-algebras.

Proposition 11.4.7. *Let $\mathscr{P}$ be a Koszul operad and let $f : A \rightsquigarrow A'$ be an ∞-morphism of $\mathscr{P}_\infty$-algebras. It is an ∞-quasi-isomorphism if and only if its image $\widetilde{\mathrm{B}}_\iota f : \widetilde{\mathrm{B}}_\iota A \to \widetilde{\mathrm{B}}_\iota A'$ under the bar construction $\widetilde{\mathrm{B}}_\iota$ is a weak equivalence of dg $\mathscr{P}^{\text{i}}$-coalgebras.*

Proof. Since the unit υ of the adjunction (Proposition 11.4.3) is a transformation of functors, the following diagram of ∞-morphisms commutes

$$\begin{array}{ccc} \Omega_\kappa \widetilde{\mathrm{B}}_\iota A & \xrightarrow{\Omega_\kappa \widetilde{\mathrm{B}}_\iota f} & \Omega_\kappa \widetilde{\mathrm{B}}_\iota A' \\ \sim \Big\uparrow \upsilon(A) & & \sim \Big\uparrow \upsilon(A') \\ A & \underset{f}{\rightsquigarrow} & A'. \end{array}$$

The ∞-morphisms $\upsilon(A)$ and $\upsilon(A')$ are ∞-quasi-isomorphisms by the Rectification Theorem 11.4.4. The restriction of this diagram to the first components of the respective ∞-morphisms gives the following diagram of chain morphisms

$$\begin{array}{ccc} \Omega_\kappa \widetilde{\mathrm{B}}_\iota A & \xrightarrow{\Omega_\kappa \widetilde{\mathrm{B}}_\iota f} & \Omega_\kappa \widetilde{\mathrm{B}}_\iota A' \\ \sim \Big\uparrow & & \sim \Big\uparrow \\ A & \xrightarrow[f_{(0)}]{} & A'. \end{array}$$

Therefore $\Omega_\kappa \widetilde{\mathrm{B}}_\iota f$ is a quasi-isomorphism if and only if $f_{(0)}$ is a quasi-isomorphism, which concludes the proof. □

As in Sect. 2.4, one can prove that weak equivalences of dg $\mathscr{P}^{\text{i}}$-coalgebras form a strict sub-class of quasi-isomorphisms.

Theorem 11.4.8 (Equivalence of homotopy categories). *Let $\mathscr{P}$ be a Koszul operad. The homotopy category of dg $\mathscr{P}$-algebras and the homotopy category of $\mathscr{P}_\infty$-algebras with the ∞-morphisms are equivalent*

$$\mathsf{Ho}(\mathsf{dg}\ \mathscr{P}\text{-}\mathsf{alg}) \cong \mathsf{Ho}(\infty\text{-}\mathscr{P}_\infty\text{-}\mathsf{alg}).$$

Proof. We consider the adjunction of Proposition 11.4.3. The functor i sends quasi-isomorphisms between dg $\mathscr{P}$-algebras to ∞-quasi-isomorphisms by definition. The functor $\Omega_\kappa \widetilde{\mathrm{B}}_\iota$ sends ∞-quasi-isomorphisms of $\mathscr{P}_\infty$-algebras to quasi-isomorphisms by Proposition 11.4.7. So this pair of functors induces a pair of adjoint functors between the homotopy categories.

Since the counit of adjunction is equal to $\varepsilon_\kappa(A) : \Omega_\kappa \widetilde{\mathrm{B}}_\iota i(A) = \Omega_\kappa \mathrm{B}_\kappa(A) \to A$ for any dg $\mathscr{P}$-algebra A, it is a quasi-isomorphism by Corollary 11.3.5. This result and Theorem 11.4.4 prove that the unit and the counit of adjunction induce isomorphisms on the level of the homotopy categories, which concludes the proof. □

These two homotopy categories $\mathsf{Ho}(\mathsf{dg}\ \mathscr{P}\text{-}\mathsf{alg})$ and $\mathsf{Ho}(\infty\text{-}\mathscr{P}_\infty\text{-}\mathsf{alg})$ are also equivalent to the homotopy category $\mathsf{Ho}(\infty\text{-}\mathscr{P}\text{-}\mathsf{alg})$ of dg $\mathscr{P}$-algebras with the ∞-morphisms. This theorem proves that the homotopy theory of dg $\mathscr{P}$-algebras is "the same as" the homotopy theory of $\mathscr{P}_\infty$-algebras or of $\mathscr{P}$-algebras but taken with the ∞-morphisms. The main gain lies in the fact that ∞-quasi-isomorphisms are "invertible" on the opposite to quasi-isomorphisms, see Theorem 10.4.4 and the next proposition.

11.4.5 Quasi-isomorphisms vs ∞-Quasi-isomorphisms

Theorem 11.4.9. *Let $\mathscr{P}$ be a Koszul operad and let A and B be two dg $\mathscr{P}$-algebras. The following assertions are equivalent:*

1. *there exists a zig-zag of quasi-isomorphisms of dg $\mathscr{P}$-algebras*

$$A \xleftarrow{\sim} \bullet \xrightarrow{\sim} \bullet \xleftarrow{\sim} \bullet \cdots \bullet \xrightarrow{\sim} B,$$

2. *there exist two quasi-isomorphisms of dg $\mathscr{P}$-algebras*

$$A \xleftarrow{\sim} \bullet \xrightarrow{\sim} B,$$

3. *there exists an ∞-quasi-isomorphism of dg $\mathscr{P}$-algebras*

$$A \overset{\sim}{\rightsquigarrow} B.$$

Proof.

(2) $\Rightarrow$ (1) Obvious.

(1) $\Rightarrow$ (3) The quasi-isomorphism $A \xleftarrow{\sim} \bullet$ of dg $\mathscr{P}$-algebras is a particular case of ∞-quasi-isomorphism. Theorem 10.4.4 shows that there is an ∞-quasi-isomorphism $A \overset{\sim}{\rightsquigarrow} \bullet$. Its composite with $\bullet \xrightarrow{\sim} B$ provides the required ∞-quasi-isomorphism $A \overset{\sim}{\rightsquigarrow} B$.

(3) $\Rightarrow$ (2) Since A and B are dg $\mathscr{P}$-algebras, they satisfy $\mathrm{B}_\iota A \cong \mathrm{B}_\kappa A$. By Proposition 11.4.7, an ∞-quasi-isomorphism $A \overset{\sim}{\rightsquigarrow} B$ induces a quasi-isomorphism $\Omega_\kappa \mathrm{B}_\kappa A \xrightarrow{\sim} \Omega_\kappa \mathrm{B}_\kappa B$ of dg $\mathscr{P}$-algebras. Finally, the bar–cobar resolution (Corollary 11.3.5) proves

$$A \xleftarrow{\sim} \Omega_\kappa \mathrm{B}_\kappa A \xrightarrow{\sim} \Omega_\kappa \mathrm{B}_\kappa B \xrightarrow{\sim} B.$$

□

11.4.6 Formality

Let $\mathscr{P}$ be an operad that we consider as a dg operad with trivial differential. A dg $\mathscr{P}$-algebra A is called *formal* when A and the graded $\mathscr{P}$-algebra $H(A)$ of Proposition 6.3.5 satisfy the above theorem.

Proposition 11.4.10. *Let $\mathscr{P}$ be a Koszul operad and let A be a dg $\mathscr{P}$-algebra. If the operadic Massey products on $H(A)$ vanish, then A is formal.*

Proof. Since we work over a field, $H(A)$ can be considered as a deformation retract of A. By Theorem 10.3.10 the quasi-isomorphism i can be extended into a ∞-quasi-isomorphism $i_\infty : H(A) \rightsquigarrow A$. It is an ∞-quasi-isomorphism of $\mathscr{P}$-algebras since the operadic Massey products are trivial. So, assertion (3) of Theorem 11.4.9 is fulfilled. As a consequence A is formal. □

11.5 Résumé

11.5.1 Twisting Morphism for $\mathscr{P}$-Algebras

Let $\alpha : \mathscr{C} \to \mathscr{P}$ be an operadic twisting morphism, let C be a dg $\mathscr{C}$-coalgebra C, and let A be a dg $\mathscr{P}$-algebra A.

Twisting morphism with respect to α, $\mathrm{Tw}_\alpha(C, A)$*:* solution $\varphi \in \mathrm{Hom}(C, A)$ to the Maurer–Cartan equation

$$\partial(\varphi) + \star_\alpha(\varphi) = 0.$$

11.5.2 Bar and Cobar Construction for $\mathscr{P}$-Algebras

There is a pair of adjoint functors

$$\Omega_\alpha : \{\text{conil. dg } \mathscr{C}\text{-coalgebras}\} \rightleftharpoons \{\text{dg } \mathscr{P}\text{-algebras}\} : \mathrm{B}_\alpha,$$

which represent the twisting morphism bifunctor

$$\mathrm{Hom}_{\mathsf{dg}\ \mathscr{P}\text{-}\mathsf{alg}}(\Omega_\alpha C, A) \cong \mathrm{Tw}_\alpha(C, A) \cong \mathrm{Hom}_{\mathsf{dg}\ \mathscr{C}\text{-}\mathsf{coalg}}(C, \mathrm{B}_\alpha A).$$

BAR–COBAR RESOLUTION.

$$\alpha \in \mathrm{Kos}(\mathscr{C}, \mathscr{P}) \iff \varepsilon_\alpha : \Omega_\alpha \mathrm{B}_\alpha A \xrightarrow{\sim} A \iff \upsilon_\alpha : C \xrightarrow{\sim} \mathrm{B}_\alpha \Omega_\alpha C.$$

11.5.3 Homotopy Theory of Homotopy $\mathscr{P}$-Algebras

Let $\mathscr{P}$ be a Koszul operad. The twisting morphisms κ and ι induce the following diagrams.

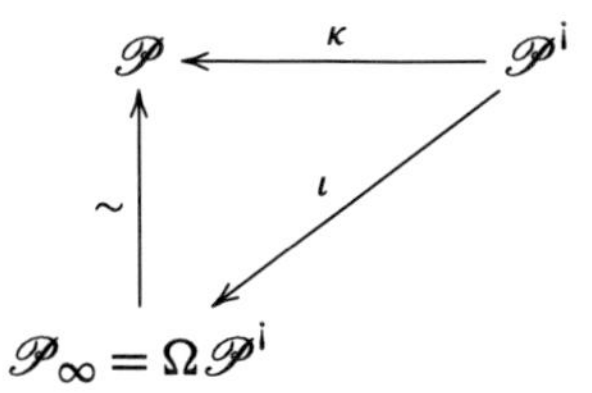

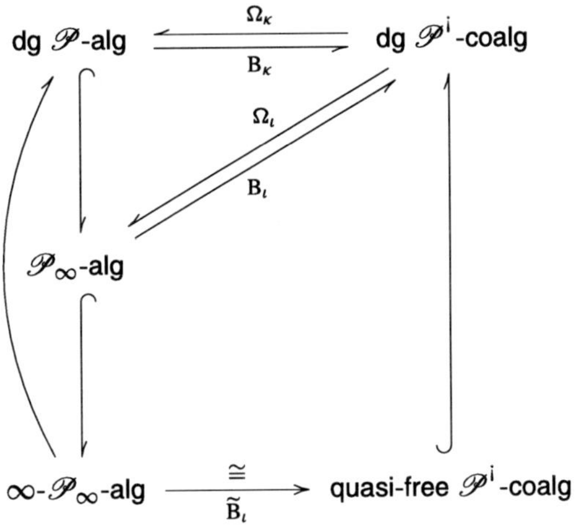

BAR–COBAR RESOLUTION OF $\mathscr{P}$-ALGEBRAS.

$$\varepsilon_\kappa : \Omega_\kappa \mathrm{B}_\kappa A \cong \mathscr{P}\big(\mathscr{P}^{\text{i}}(A)\big) \xrightarrow{\sim} A.$$

BAR–COBAR RESOLUTION OF $\mathscr{P}_\infty$-ALGEBRAS.

$$\varepsilon_\iota : \Omega_\iota \mathrm{B}_\iota A \cong \mathscr{P}_\infty\big(\mathscr{P}^{\text{i}}(A)\big) \xrightarrow{\sim} A.$$

Proposition. *The following functors are adjoint to each other.*

$$\Omega_\kappa \widetilde{\mathrm{B}}_\iota : \infty\text{-}\mathscr{P}_\infty\text{-alg} \rightleftharpoons \text{dg } \mathscr{P}\text{-alg} : i.$$

Theorem. *The counit of this adjunction*

$$A \overset{\sim}{\rightsquigarrow} \Omega_\kappa \mathrm{B}_\iota A$$

provides a natural and universal rectification *for homotopy $\mathscr{P}$-algebras.*

HOMOTOPY CLASS OF A $\mathscr{P}_\infty$-ALGEBRA.

$$(H, 0, \textit{transferred structure}) \sim (A, d_A, \varphi) \sim (\Omega_\kappa \widetilde{\mathrm{B}}_\iota A, d, \mathscr{P}\textit{-algebra structure}).$$

Theorem.

$$\mathrm{Ho}(\mathscr{P}\text{-alg}) \cong \mathrm{Ho}(\infty\text{-}\mathscr{P}\text{-alg}) \cong \mathrm{Ho}(\infty\text{-}\mathscr{P}_\infty\text{-alg}).$$

Theorem. *The following assertions are equivalent*

- $\exists\, A \overset{\sim}{\leftarrow} \bullet \overset{\sim}{\rightarrow} \bullet \overset{\sim}{\leftarrow} \bullet \cdots \bullet \overset{\sim}{\rightarrow} B$,
- $\exists\, A \overset{\sim}{\leftarrow} \bullet \overset{\sim}{\rightarrow} B$,
- $\exists\, A \overset{\sim}{\rightsquigarrow} B$.

11.6 Exercises

Exercise 11.6.1 (Bar–cobar adjunction). Prove that $\mathrm{Tw}_\alpha(C,-)$, $\mathrm{Tw}_\alpha(-,A)$, B_α and Ω_α are indeed functors. Show that the bijections given in Proposition 11.3.1 are natural.

Exercise 11.6.2 (Bar–cobar resolutions). When $\mathscr{P}$ is a Koszul operad, prove that the two resolutions $\varepsilon : \mathscr{P} \circ \mathscr{P}^{\text{!`}}(A) \xrightarrow{\sim} A$ and $\varepsilon_\pi : \mathscr{P} \circ \mathrm{B}\mathscr{P}(A) \xrightarrow{\sim} A$ are homotopy equivalent.

Exercise 11.6.3 (Quasi-isomorphisms vs weak equivalences). Let $\mathscr{P}$ be a Koszul operad such that $\mathscr{P}(1) = \mathrm{I}$. Prove that weak equivalences of dg $\mathscr{P}^{\text{!`}}$-coalgebras form a strict sub-class of quasi-isomorphisms.

Exercise 11.6.4 ($\mathit{Lie}^{\langle k\rangle}$-interpretation of twisting morphisms with respect to k-ary (co)operads ★). Let $\mathscr{P} = \mathscr{P}(E,R)$ be a quadratic operad generated by k-ary operations, with the integer k fixed. Interpret the Maurer–Cartan equation $\partial(\varphi) - \star_\kappa(\varphi) = 0$ in terms of a Maurer–Cartan equation

$$\partial(s^{-1}\varphi) + \frac{1}{k!}\big[s^{-1}\varphi, \ldots, s^{-1}\varphi\big] = 0$$

in a dg $\mathit{Lie}^{\langle k\rangle}$-algebra, see Sect. 13.11.3 for a definition.

Exercise 11.6.5 (Relative composite product ★). Prove the isomorphism $(M' \circ \mathscr{P}) \circ_{\mathscr{P}} N \cong M' \circ N$ for the relative composite product where $M = M' \circ \mathscr{P}$ is a free $\mathscr{P}$-module.

Dually, prove the isomorphism $(M' \circ \mathscr{C}) \circ^{\mathscr{C}} N \cong M' \circ N$ for the relative composite product where $M = M' \circ \mathscr{C}$ is a cofree $\mathscr{C}$-comodule.

Chapter 12
(Co)Homology of Algebras over an Operad

Les mathématiques ne sont pas une moindre immensité que la mer.

Victor Hugo

Given an algebra A over a quadratic operad $\mathscr{P}$, one can construct a chain complex $C_\bullet^{\mathscr{P}}(A) := (\mathscr{P}^{\text{!`}}(A), d)$ out of the Koszul dual cooperad $\mathscr{P}^{\text{!`}}$, whence homology groups $H_\bullet^{\mathscr{P}}(A)$. On the other hand, there is a general theory called André–Quillen (co)homology theory, which provides homological invariants for an algebra over an operad. It plays a role in many classification problems, like for instance deformation theory. We show that, when $\mathscr{P}$ is Koszul, the operadic homology coincides with André–Quillen homology, thus providing a small explicit complex to compute it.

In this chapter, we introduce the André–Quillen cohomology and homology of algebras over an operad. The classical method in homological algebra to define (co)homology theories is to use the notion of derived functors between abelian categories. Since the various categories involved here are not abelian, we have to use the general framework of model categories, Quillen adjunctions and total derived functors, recalled in Appendix B.6. Thanks to this conceptual approach, we can compute the associated (co)homology groups with any cofibrant resolution. We use the resolutions provided by the Koszul duality theory and made explicit in the previous chapter.

In the first section, we consider the *Quillen homology of algebras with trivial coefficients*. It is defined conceptually by deriving the non-abelian functor of indecomposable elements. When the underlying operad is Koszul, the Quillen homology is equal to the homology of the bar construction of the algebra. This gives yet another way to prove that an operad is Koszul: one has just to check whether the homology of the bar construction of the free algebra is trivial, see Ginzburg and Kapranov [GK94] and Getzler and Jones [GJ94].

The *deformation theory* of algebras over an operad is the subject of the second section. It is governed by the convolution dg Lie algebra

$$\mathfrak{g} := \big(\mathrm{Hom}_{\mathbb{S}}(\mathscr{P}^{\text{!`}}, \mathrm{End}_A), [\,,\,], \partial\big),$$

J.-L. Loday, B. Vallette, *Algebraic Operads*,
Grundlehren der mathematischen Wissenschaften 346,
DOI 10.1007/978-3-642-30362-3_12, © Springer-Verlag Berlin Heidelberg 2012

introduced in Sect. 10.1.2. First, the set of $\mathscr{P}$-algebra structures on a space A is in one-to-one correspondence with the set of Maurer–Cartan elements of $\mathfrak{g}$. Then, given such an element, one can twist this dg Lie algebra to obtain another one. Maurer–Cartan elements in this twisted dg Lie algebra correspond to deformations of the original structure. The underlying cochain complex of the twisted dg Lie algebra is called the deformation complex. Its cohomology groups are isomorphic to the André–Quillen cohomology of the $\mathscr{P}$-algebra A with coefficients into itself. The particular case of associative algebras, as treated by Gerstenhaber in [Ger63], serves as a paradigm for this general theory. This treatment applies as well to study the deformation theory of $\mathscr{P}_\infty$-algebras.

In the third section, we give the complete definition of *André–Quillen cohomology and homology* of algebras over an operad after Hinich [Hin97] and Goerss and Hopkins [GH00]. It follows the method used by Quillen in [Qui70] and André in [And74] defining the (co)homology theory of commutative algebras, but extended to any category of algebras over an operad. It involves several general constructions like the *enveloping algebra*, the *module of Kähler differential forms* and the *cotangent complex* for example. This latter plays a key role since it represents the André–Quillen cohomology theory.

In the last section, we define an *operadic cohomology theory with coefficients* for any algebra over an operad by an explicit cochain complex. When the operad is Koszul, this is proved to coincide with the André–Quillen cohomology. So this provides a cochain complex which computes the André–Quillen cohomology groups.

This chapter mainly comes from Ginzburg–Kapranov [GK94], Getzler–Jones [GJ94], Hinich [Hin97], Balavoine [Bal97, Bal98], Goerss–Hopkins [GH00], Millès [Mil11], and Fresse [Fre09a].

12.1 Homology of Algebras over an Operad

In this section, we introduce the homology of an algebra over an operad, with trivial coefficients. When the governing operad $\mathscr{P}$ is quadratic, we define the operadic homology of $\mathscr{P}$-algebras by an explicit chain complex made out of the Koszul dual cooperad. For any augmented operad $\mathscr{P}$, we define the Quillen homology of $\mathscr{P}$-algebras by a total derived functor. We show that the operadic homology is equal to the Quillen homology if and only if the operad is Koszul.

This section comes from Ginzburg–Kapranov [GK94] and from Getzler–Jones [GJ94], see also Livernet [Liv98a].

12.1.1 Operadic Chain Complex of a $\mathscr{P}$-Algebra

For any homogeneous quadratic operad $\mathscr{P} = \mathscr{P}(E, R)$ and any $\mathscr{P}$-algebra A, we consider the chain complex given by the bar construction of A

$$C_\bullet^{\mathscr{P}}(A) := \mathrm{B}_\kappa A = \left(\mathscr{P}^{\mathrm{i}} \circ_\kappa \mathscr{P}\right) \circ_{\mathscr{P}} A \cong \left(\mathscr{P}^{\mathrm{i}}(A), d\right).$$

By definition, it forms a dg $\mathscr{P}^{¡}$-coalgebra. Recall from Sect. 11.2.2 that the differential map $d = d_2$ is the unique coderivation which extends the composite of the twisting morphism $\kappa : \mathscr{P}^{¡} \to \mathscr{P}$ with the composition product γ_A of A. It is explicitly given by the following composite

$$\begin{aligned}\mathscr{P}^{¡}(A) = \mathscr{P}^{¡} \circ A &\xrightarrow{\Delta_{(1)} \circ \mathrm{Id}_A} \left(\mathscr{P}^{¡} \circ_{(1)} \mathscr{P}^{¡}\right) \circ A \\ &\xrightarrow{(\mathrm{Id} \circ_{(1)} \kappa) \circ \mathrm{Id}_A} \left(\mathscr{P}^{¡} \circ_{(1)} \mathscr{P}\right) \circ A \rightarrowtail \mathscr{P}^{¡} \circ \mathscr{P} \circ A \\ &\xrightarrow{\mathrm{Id} \circ \gamma_A} \mathscr{P}^{¡} \circ A = \mathscr{P}^{¡}(A).\end{aligned}$$

When E and A are concentrated in degree 0, the homological degree of $\mathscr{P}^{¡\,(n)}$ is n and the chain complex has the following form

$$C_\bullet^{\mathscr{P}}(A) : \quad \cdots \to \mathscr{P}^{¡\,(3)}(A) \to \mathscr{P}^{¡\,(2)}(A) \to \mathscr{P}^{¡\,(1)}(A) \to A.$$

In low degree, we have explicitly,

$$\cdots \to s^2 R(A) \to sE(A) \to A.$$

This definition extends in two directions: to dg $\mathscr{P}$-algebras (A, d_A) and to inhomogeneous quadratic operads $\mathscr{P}$, satisfying the conditions (ql_1) and (ql_2) of Sect. 7.8. In this case, the differential map in the bar construction is the sum of several terms, as explained in Sect. 11.2.2.

Proposition 12.1.1. *If the operad $\mathscr{P} = \mathscr{P}(E, R)$ is binary and quadratic, then the Koszul complex $C_\bullet^{\mathscr{P}}(A)$ of the $\mathscr{P}$-algebra A is*

$$C_n^{\mathscr{P}}(A) = \mathscr{P}^{i}(n+1) \otimes_{\mathbb{S}_{n+1}} A^{\otimes n+1},$$

with differential

$$\begin{aligned}&d\big(\delta \otimes (a_1, \ldots, a_{n+1})\big) \\ &\quad = \sum \xi \otimes \big(a_{\sigma^{-1}(1)}, \ldots, a_{\sigma^{-1}(i-1)}, \mu(a_{\sigma^{-1}(i)}, a_{\sigma^{-1}(i+1)}), a_{\sigma^{-1}(i+2)}, \ldots, a_{\sigma^{-1}(n+1)}\big),\end{aligned}$$

for

$$\Delta_{(1)}(\delta) = \sum (\xi; \mathrm{id}, \ldots, \mathrm{id}, \mu, \mathrm{id}, \ldots, \mathrm{id}; \sigma),$$
$$\delta \in \mathscr{P}^{i}(n+1), \quad \xi \in \mathscr{P}^{i}(n), \quad \mu \in \mathscr{P}^{i}(2) = E, \quad \textit{and} \quad \sigma \in \mathbb{S}_{n+1}.$$

Proof. Since $\mathscr{P}$ is binary, the infinitesimal decomposition map splits the $n+1$-cooperation δ of $\mathscr{P}^{¡}$ into an n-cooperation and a binary cooperation of $\mathscr{P}^{¡}$. By κ this latter cooperation is viewed as a binary operation of $\mathscr{P}$. The sum is over all these possibilities of splitting. □

EXAMPLE. In the case of the operad *As* encoding associative algebras, we recover the Hochschild chain complex introduced in Sect. 9.1.6.

12.1.2 Operadic Homology of a $\mathscr{P}$-Algebra

By definition, the homology of the chain complex $C_\bullet^{\mathscr{P}}(A)$ is called the *operadic homology* of the $\mathscr{P}$-algebra A and we denote the homology groups by $H_\bullet^{\mathscr{P}}(A)$.

Proposition 12.1.2. *For any quadratic operad $\mathscr{P}$ and any dg $\mathscr{P}$-algebra A, the operadic homology $H_\bullet^{\mathscr{P}}(A)$ of A is a graded $\mathscr{P}^{¡}$-coalgebra. Equivalently $H_{\bullet+1}^{\mathscr{P}}(A)$ is a graded $\mathscr{P}^!$-coalgebra.*

Proof. Since the chain complex $C_\bullet^{\mathscr{P}}(A)$ is a dg $\mathscr{P}^{¡}$-coalgebra, its homology is a graded $\mathscr{P}^{¡}$-coalgebra. The second statement follows from Sect. 7.2.3. □

Let $\mathscr{P}$ be an augmented operad $\mathscr{P} = \mathrm{I} \oplus \overline{\mathscr{P}}$ and let A be a $\mathscr{P}$-algebra. The space of *indecomposable elements of A* is the cokernel of

$$\gamma_A : \overline{\mathscr{P}}(A) \to A.$$

We denote it by $\mathrm{Indec}(A)$. For instance, the indecomposable elements of the free $\mathscr{P}$-algebra $\mathscr{P}(V)$ is equal to the generating space V. When $\mathscr{P}$ is binary, we have $\mathrm{Indec}(A) = A/A^2$, where $A^2 := \gamma(E(2) \otimes A^{\otimes 2})$, since any product of elements of A is an iterated composition of binary operations.

Proposition 12.1.3. *For any quadratic operad $\mathscr{P}$ and for any $\mathscr{P}$-algebra A, we have $H_0^{\mathscr{P}}(A) = \mathrm{Indec}(A)$.*

Proof. It comes from the explicit form of the operadic complex given above. □

This homology theory permits us to rephrase the Koszul property of an operad.

Theorem 12.1.4. *Let $\mathbb{K}$ be a field of characteristic zero. A quadratic operad $\mathscr{P}$ is Koszul if and only if, for any vector space V, the operadic homology of the free $\mathscr{P}$-algebra $\mathscr{P}(V)$ is equal to*

$$H_n^{\mathscr{P}}\big(\mathscr{P}(V)\big) = \begin{cases} 0 & \text{if } n \geq 1, \\ V & \text{if } n = 0. \end{cases}$$

Proof. The bar construction of a free $\mathscr{P}$-algebra $\mathscr{P}(V)$ is equal to

$$C_\bullet^{\mathscr{P}}\big(\mathscr{P}(V)\big) = \mathrm{B}_\kappa \mathscr{P}(V) = \big(\mathscr{P}^{¡} \circ_\kappa \mathscr{P}\big) \circ V.$$

We conclude with Corollary 6.2.4, which states that $H_\bullet((\mathscr{P}^{¡} \circ_\kappa \mathscr{P}) \circ V) = V$ if and only if $H_\bullet(\mathscr{P}^{¡} \circ_\kappa \mathscr{P}) = \mathrm{I}$. □

This result provides a method to prove that an operad is Koszul. Several examples are made explicit in Chap. 13.

REMARK. Recall that the Koszul duality theory for operads holds over a ring of positive characteristic [Fre04]. In this case, if the operadic homology of any free $\mathscr{P}$-algebra is concentrated in degree 0, then the operad is Koszul (see Proposition 5.3.5. of loc. cit.). But the converse does not hold true because of the torsion phenomenon. For instance Harrison homology of free commutative algebras is not always trivial in positive characteristic, see Barr [Bar68] and Harrison [Har62]. But the operad *Com* is Koszul in positive characteristic, by the rewriting method (Sect. 8.3), for instance.

12.1.3 Quillen Homology of a $\mathscr{P}$-Algebra

To give a conceptual definition of the homology of a $\mathscr{P}$-algebra, we apply to operads the same method as the one which defines the Tor functors of modules over an associative algebra, see Appendix B.7.2. However, since we work in a non-additive setting, we need to use the formalism of model categories: total derived functors and cofibrant objects, see Appendix B.6. We show that it coincides with the previous operadic homology when the operad $\mathscr{P}$ is Koszul.

Let $f : \mathscr{P} \to \mathscr{Q}$ be a morphism of operads. The pullback along f defines a functor $f^* : \mathsf{dg}\ \mathscr{Q}\text{-}\mathsf{alg} \to \mathsf{dg}\ \mathscr{P}\text{-}\mathsf{alg}$.

Lemma 12.1.5. *The functor* $f_! : \mathsf{dg}\ \mathscr{P}\text{-}\mathsf{alg} \to \mathsf{dg}\ \mathscr{Q}\text{-}\mathsf{alg}$, *given by the relative composite product* $f_!(A) := \mathscr{Q} \circ_{\mathscr{P}} A$ *is left adjoint to* f^*.

$$f_! : \mathsf{dg}\mathscr{P}\text{-}\mathsf{alg} \rightleftharpoons \mathsf{dg}\mathscr{Q}\text{-}\mathsf{alg} : f^*.$$

Proof. The proof is left to the reader as a good exercise. □

The notation $f_!$ is common in the literature and should not be confused with any Koszul dual.

In general, neither the category of dg $\mathscr{P}$-algebras nor the functor $f_!$ are additive, therefore we cannot try to derive it in the classical sense. Instead, we have to consider its total derived functor à la Quillen, see Appendix B.7.3. Under some assumptions, for instance when the ground ring $\mathbb{K}$ is a characteristic 0 field, the two categories of dg algebras can be endowed with model category structures, see Proposition B.6.5, and the aforementioned adjunction is a Quillen functor. So the two functors can be derived to induce the following adjunction

$$\mathbb{L}f_! : \mathsf{Ho}(\mathsf{dg}\mathscr{P}\text{-}\mathsf{alg}) \rightleftharpoons \mathsf{Ho}(\mathsf{dg}\mathscr{Q}\text{-}\mathsf{alg}) : \mathbb{R}f^*.$$

We apply these results to the augmentation map $\varepsilon : \mathscr{P} \to \mathrm{I}$ of an augmented operad $\mathscr{P}$. In this case, the functor $\varepsilon^* : \mathsf{dg}\ \mathsf{Mod}_{\mathbb{K}} \to \mathsf{dg}\ \mathscr{P}\text{-}\mathsf{alg}$ amounts just to restricting the action to the scalars. Its left adjoint $\varepsilon_! : \mathsf{dg}\ \mathscr{P}\text{-}\mathsf{alg} \to \mathsf{dg}\ \mathsf{Mod}_{\mathbb{K}}$ is equal to $\varepsilon_!(A) := \mathrm{I} \circ_{\mathscr{P}} A = \mathrm{Indec}(A)$, the space of indecomposable elements of the $\mathscr{P}$-algebra A.

The *Quillen homology of a dg $\mathcal{P}$-algebra A* is defined as being the homology $H_\bullet(\mathbb{L}\,\mathrm{Indec}(A))$ of the total left derived functor of the indecomposable elements functor.

As usual, one computes it as the homology $H_\bullet(\mathrm{Indec})(R)$, for any cofibrant replacement $R \xrightarrow{\sim} A$ of the dg $\mathcal{P}$-algebra A. Recall from Sect. 11.3.3 that, when the operad $\mathcal{P}$ is Koszul, the bar–cobar construction $\Omega_\kappa \mathrm{B}_\kappa A$, associated to the Koszul morphism κ, provides a functorial quasi-free resolution for dg $\mathcal{P}$-algebras A.

Theorem 12.1.6. *Let $\mathcal{P}$ be a Koszul operad over a field $\mathbb{K}$ of characteristic 0. For any dg $\mathcal{P}$-algebra A, the operadic homology of A is equal to its Koszul homology*

$$H_\bullet\big(\mathbb{L}\,\mathrm{Indec}(A)\big) = H_\bullet^{\mathcal{P}}(A).$$

Proof. It is enough to prove that the resolution

$$\Omega_\kappa \mathrm{B}_\kappa A = \big(\mathcal{P} \circ_\kappa \mathcal{P}^{¡} \circ_\kappa \mathcal{P}\big) \circ_{\mathcal{P}} A \xrightarrow{\sim} A$$

is cofibrant. One considers the filtration on $\mathcal{P}^{¡}(A)$ given by the weight of $\mathcal{P}^{¡}$. It is straightforward to see that the boundary map lowers the weight filtration. So this dg $\mathcal{P}$-algebra is quasi-free and triangulated, according to the terminology of Appendix B.6.6. Therefore, it is cofibrant. □

REMARK. Recall from Sect. 11.3.3 that

$$\Omega_\pi \mathrm{B}_\pi A = (\mathcal{P} \circ_\pi \mathrm{B}\mathcal{P} \circ_\pi \mathcal{P}) \circ_{\mathcal{P}} \circ A \xrightarrow{\sim} A$$

is always a quasi-free resolution of A. Among graded $\mathcal{P}$-algebras A, i.e. with trivial differential, the dg $\mathcal{P}$-algebra $\Omega_\pi B_\pi A$ provides a functorial cofibrant replacement. To prove it, it is enough to consider the filtration on $(\mathrm{B}\mathcal{P})(A)$ given by the numbers of elements of $\overline{\mathcal{P}}$, see Appendix B.6.6. This resolution gives an explicit chain complex which computes the Quillen homology of graded $\mathcal{P}$-algebras, even when the operad $\mathcal{P}$ fails to be Koszul.

Moreover, the converse of the above theorem is also true.

Proposition 12.1.7. *Let $\mathcal{P}$ be a quadratic operad over a field $\mathbb{K}$ of characteristic 0. If the operadic homology of any graded $\mathcal{P}$-algebra A is equal to its Koszul homology*

$$H_\bullet\big(\mathbb{L}\,\mathrm{Indec}(A)\big) = H_\bullet^{\mathcal{P}}(A),$$

then the operad $\mathcal{P}$ is Koszul.

Proof. We apply the assumption to the free $\mathcal{P}$-algebra $\mathcal{P}(V)$ and we compute its operadic homology with the aforementioned bar–cobar construction $\Omega_\pi B_\pi \mathcal{P}(V)$. It is equal to $H_\bullet(\mathrm{B}\mathcal{P} \circ_\pi \mathcal{P})(V) = V$. We conclude with Theorem 12.1.4. □

12.1.4 Operadic Homology of a $\mathscr{P}_\infty$-Algebra

In the same way, we define the *operadic homology of a* $\mathscr{P}_\infty$*-algebra* A by the homology of its bar construction

$$H_\bullet^{\mathscr{P}_\infty}(A) := H_\bullet(\mathrm{B}_\iota A).$$

Recall from Sect. 11.2.2, that the bar construction $\mathrm{B}_\iota A$ of a $\mathscr{P}_\infty$-algebra A is the dg $\mathscr{P}^{\text{i}}$-coalgebra

$$\mathrm{B}_\iota A = \left(\mathscr{P}^{\text{i}} \circ_\iota \mathscr{P}_\infty\right) \circ_{\mathscr{P}_\infty} A = \left(\mathscr{P}^{\text{i}}(A), d\right).$$

Over a field of characteristic 0 and over nonnegatively graded dg modules, it computes the Quillen homology of $\mathscr{P}_\infty$-algebras, which is defined as the homology of the derived functor of the indecomposable functor. This uses the model category structure on the category $\mathscr{P}_\infty$-alg of $\mathscr{P}_\infty$-algebras with "strict" morphisms. Notice that we do not need $\mathscr{P}$ to be a Koszul operad to obtain these results.

12.2 Deformation Theory of Algebra Structures

In this section, we use the convolution dg Lie algebra

$$\mathfrak{g} := \left(\mathrm{Hom}_{\mathbb{S}}\left(\mathscr{P}^{\text{i}}, \mathrm{End}_A\right), [\,,\,], \partial\right)$$

introduced in Sect. 10.1.2 to study the deformation theory of $\mathscr{P}$-algebra and $\mathscr{P}_\infty$-algebra structures. We use the general framework of deformation theory associated to a dg Lie recalled in Sect. 13.2.14. First, we study the case of $\mathscr{P}$-algebra structures: moduli space, deformation complex, deformation functor, infinitesimal and formal deformations. Since all the proofs are based on general arguments using "Lie calculus" in the convolution dg Lie algebra, we show that they extend mutatis mutandis to the $\mathscr{P}_\infty$-algebra case.

The deformation theory of algebraic structures was initiated by Murray Gerstenhaber on the level of associative algebras in [Ger63, Ger64]. It was extended to Lie algebras by Nijenhuis and Richardson [NR66, NR67]. The case of algebras over a binary quadratic operads was treated by Balavoine in [Bal97].

For more details and applications on the deformation theory of algebraic structures, we refer the reader to the book of Kontsevich and Soibelman [KS10] and to the survey of Keller [Kel05].

12.2.1 Moduli Space of $\mathscr{P}$-Algebra Structures

Recall that in Sect. 10.1.2, we have associated a convolution dg Lie algebra

$$\mathfrak{g} = \mathfrak{g}_{\mathscr{P},A} := \left(\mathrm{Hom}_{\mathbb{S}}\left(\mathscr{P}^{\text{i}}, \mathrm{End}_A\right), [\,,\,], \partial\right)$$

to any a pair $(\mathscr{P}, A)$, where $\mathscr{P} = \mathscr{P}(E, R)$ is a quadratic operad and where A is a dg $\mathbb{K}$-module. When A is concentrated in degree 0, we consider the following cohomological degree on $\mathfrak{g}$ induced by the weight grading

$$\underset{0}{\operatorname{Hom}(A, A)} \xrightarrow{\partial} \underset{1}{\operatorname{Hom}_{\mathbb{S}}\big(\mathscr{P}^{¡(1)}, \operatorname{End}_A\big)} \xrightarrow{\partial} \underset{2}{\operatorname{Hom}_{\mathbb{S}}\big(\mathscr{P}^{¡(2)}, \operatorname{End}_A\big)} \cdots,$$

where $\operatorname{Hom}_{\mathbb{S}}(\mathrm{I}, \operatorname{End}_A)$ is identified with $\operatorname{Hom}(A, A)$. The coboundary map ∂ is null when the operad $\mathscr{P}$ is homogeneous quadratic and is equal to the pullback of the differential of $\mathscr{P}^{¡}$ in the inhomogeneous case. Under this convention of grading, $\mathfrak{g}$ becomes a cohomological dg Lie algebra. Its set $\mathrm{MC}(\mathfrak{g})$ of degree 1 elements φ, which satisfy the Maurer–Cartan equation $\partial(\varphi) + \frac{1}{2}[\varphi, \varphi] = 0$, is in one-to-one correspondence with the set of $\mathscr{P}$-algebra structures on A by Proposition 10.1.4.

A finer object to study is the coset of $\mathscr{P}$-algebra structures on A modulo isomorphisms. It is called the *moduli space* of $\mathscr{P}$-algebra structures on A and denoted by $\mathscr{P}\text{-alg}(A)/\sim$. Let us now apply the general deformation theory associated to a dg Lie algebra, as recalled in Sect. 13.2.14. The degree 0 Lie subalgebra is here equal to $\mathfrak{g}^0 = \operatorname{Hom}(A, A) = \mathfrak{gl}(A)$. Its Lie group, under proper hypotheses, is the general Lie group $GL(A)$. It acts on the set of Maurer–Cartan elements to give the moduli space

$$\mathscr{MC}(\mathfrak{g}_{\mathscr{P},A}) := \mathrm{MC}(\mathfrak{g}_{\mathscr{P},A})/GL(A).$$

Proposition 12.2.1. *Over the field $\mathbb{C}$ of complex numbers, for any quadratic operad $\mathscr{P}$ and any finite dimensional module A, there is a natural bijection of moduli spaces*

$$\mathscr{MC}(\mathfrak{g}_{\mathscr{P},A}) \cong \mathscr{P}\text{-}alg(A)/\sim.$$

Proof. By definition of the gauge group action, two Maurer–Cartan elements φ and ψ are equivalent if and only if there exists a $\lambda \in \operatorname{Hom}(A, A)$ such that $\psi = e^{\mathrm{ad}_\lambda}(\varphi)$ in $\operatorname{Hom}_{\mathbb{S}}(\mathscr{P}^{¡(1)}, \operatorname{End}_A)$. A short computation shows that this is equivalent to $\psi(\mu) = e^\lambda \circ \psi(\mu) \circ (e^{-\lambda}, \dots, e^{-\lambda})$ in $\operatorname{Hom}(A^{\otimes m}, A)$, for any $\mu \in \mathscr{P}^{¡(1)}(m)$. It means that there exists $f = e^\lambda \in GL(A)$ such that $\psi(\mu) \circ (f, \dots, f) = f \circ \varphi(\mu)$, that is f is an isomorphism of $\mathscr{P}$-algebras. □

The set of $\mathscr{P}$-algebra structures on a module A carries the following geometrical structure.

Proposition 12.2.2. *When the spaces $\mathscr{P}^{¡(1)} = sE$ and A are finite dimensional, the set of $\mathscr{P}$-algebra structures* $\mathrm{MC}(\mathfrak{g})$ *is an algebraic variety in the affine space* $\mathfrak{g}^1 = \operatorname{Hom}_{\mathbb{S}}(\mathscr{P}^{¡(1)}, \operatorname{End}_A)$, *which is an intersection of quadrics.*

Proof. Under these assumptions, the vector space of $\mathfrak{g}^1 = \operatorname{Hom}_{\mathbb{S}}(sE, \operatorname{End}_A)$ is finite dimensional. We conclude with Proposition 13.2.7. □

As a quotient, the moduli space $\mathscr{MC}(\mathfrak{g}_{\mathscr{P},A})$ might be a singular space. In general, it forms an algebraic stack.

12.2.2 Deformation Complex of a $\mathscr{P}$-Algebra

Given any $\mathscr{P}$-algebra structure $\varphi \in \mathrm{MC}(\mathfrak{g})$ on A, we associate its *deformation complex* given by the cochain complex

$$\left(\mathrm{Hom}_{\mathbb{S}}\left(\mathscr{P}^{\mathrm{i}}, \mathrm{End}_A\right), \partial_\varphi\right),$$

where $\partial_\varphi(f) := \partial(f) + [\varphi, f]$ is the twisted differential introduced in Sect. 6.4.4.

The homology groups of the deformation complex are often called the *tangent homology*. Later in Sect. 12.2.6, we will explain why the homology of the deformation complex can be interpreted as the tangent space at point φ of the algebraic stack $\mathscr{MC}(\mathfrak{g}_{\mathscr{P},A})$.

12.2.3 Intrinsic Lie Bracket

Before analyzing the associated cohomology groups, let us recall that, by its very definition, the deformation complex carries a Lie bracket. It is called the *intrinsic Lie bracket* by Stasheff in [SS85, Sta93].

Proposition 12.2.3. *The deformation complex forms a dg Lie algebra*

$$\mathfrak{g}^\varphi = \mathfrak{g}^\varphi_{\mathscr{P},A} := \left(\mathrm{Hom}_{\mathbb{S}}\left(\mathscr{P}^{i}, \mathrm{End}_A\right), [\,,\,], \partial_\varphi\right).$$

Proof. See Proposition 6.4.6. □

Proposition 12.2.4. *For any $\mathscr{P}$-algebra structure $\varphi \in \mathrm{MC}(\mathfrak{g})$ on A, the following equivalence holds*

$$\alpha \in \mathrm{MC}\left(\mathfrak{g}^\varphi\right) \iff \alpha + \varphi \in \mathrm{MC}(\mathfrak{g}).$$

Proof. Under $\partial(\varphi) + \frac{1}{2}[\varphi, \varphi] = 0$, the equation $\partial(\alpha) + [\varphi, \alpha] + \frac{1}{2}[\alpha, \alpha] = 0$ is equivalent to $\partial(\alpha) + \partial(\varphi) + \frac{1}{2}[\alpha + \varphi, \alpha + \varphi] = 0$. □

The sum $\alpha + \varphi$ of two $\mathscr{P}$-algebra structures on A is not in general a $\mathscr{P}$-algebra structure. This proposition shows that $\alpha + \varphi$ is indeed a $\mathscr{P}$-algebra structure on A if and only if α is a Maurer–Cartan element in the twisted convolution dg Lie algebra $\mathfrak{g}^\varphi$.

Equivalently, the algebraic variety $\mathrm{MC}(\mathfrak{g}^\varphi)$ is the translation under $-\varphi$ of $\mathrm{MC}(\mathfrak{g})$ in the affine space $\mathfrak{g}^1$.

12.2.4 Deformation Complex of Associative Algebras

The convolution dg Lie algebra associated to the nonsymmetric operad As has already been made explicit in Sect. 10.1.5. In the case where the dg module A is concentrated in degree 0, it coincides with the Hochschild cochain complex of A with coefficients into itself, endowed with the Lie bracket defined by Gerstenhaber in [Ger63].

When the nonsymmetric operad $\mathcal{P} = As$, the Koszul dual ns cooperad $As^{¡}$ is one-dimensional in each arity, so the underlying graded module of the deformation complex is

$$\mathfrak{g}_{As,A} = \operatorname{Hom}\big(As^{¡}, \operatorname{End}_A\big) \cong \big\{\operatorname{Hom}\big(A^{\otimes n}, A\big)\big\}_{n\in\mathbb{N}^*},$$

where $\operatorname{Hom}(A^{\otimes n}, A)$ lies in cohomological degree $n-1$. The intrinsic Lie bracket is equal to

$$[f,g] = \sum_{i=1}^{n} (-1)^{(i-1)(m-1)} f \circ_i g - (-1)^{(n-1)(m-1)} \sum_{j=1}^{m} (-1)^{(j-1)(n-1)} g \circ_j f$$

for $f \in \operatorname{Hom}(A^{\otimes n}, A)$ and $g \in \operatorname{Hom}(A^{\otimes m}, A)$. Since the operad As and the dg module A are concentrated in degree 0, the cohomological degree considered here is the opposite of the homological degree considered in Sect. 10.1.5. Therefore the signs are the same and this Lie bracket is equal to the one defined by Gerstenhaber in [Ger63].

An element $\mu \in \operatorname{Hom}(A^{\otimes 2}, A)$ satisfies the Maurer–Cartan equation if and only if it is associative

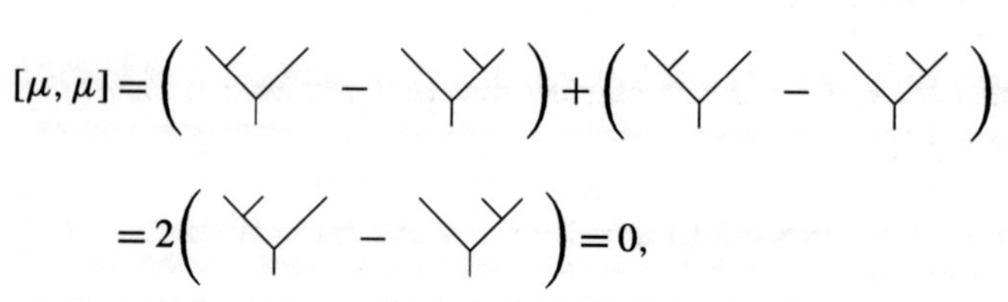

when the characteristic of the ground field is not equal to 2.

Given such an associative product μ, the twisted differential ∂_μ on

$$\underset{0}{\operatorname{Hom}(A,A)} \xrightarrow{\partial_\mu} \underset{1}{\operatorname{Hom}\big(A^{\otimes 2},A\big)} \xrightarrow{\partial_\mu} \underset{2}{\operatorname{Hom}\big(A^{\otimes 3},A\big)} \cdots,$$

is equal to

$$\partial_\mu(f) := [\mu, f]$$

$$= $$ 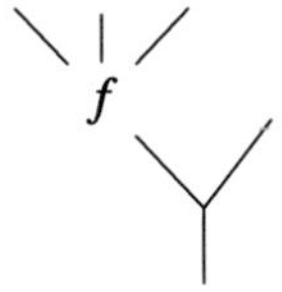 $$+ (-1)^{n-1}$$ 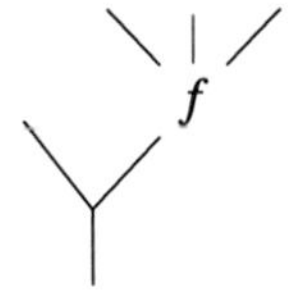

$$- (-1)^{(n-1)} \sum_{j=1}^{n} (-1)^{j-1}$$ 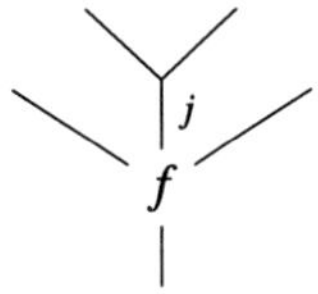

for $f \in \mathrm{Hom}(A^{\otimes n}, A)$. Up to a factor $(-1)^{n-1}$, we recover the differential map

$$\begin{aligned}(df)(a_1, \dots, a_{n+1}) &= a_1.f(a_2, \dots, a_{n+1}) \\ &\quad + \sum_{j=1}^{n} (-1)^j f(a_1, \dots, a_j.a_{j+1}, \dots, a_{n+1}) \\ &\quad + (-1)^{n+1} f(a_1, \dots, a_n).a_{n+1}\end{aligned}$$

of the Hochschild cochain complex $C^\bullet(A, A)$, see Sect. 9.1.7.

In the case of the operad *Lie*, we recover the Chevalley–Eilenberg cochain complex together with the Nijenhuis–Richardson bracket. For more details, we refer the reader to Exercise 12.6.2.

12.2.5 Deformation of $\mathscr{P}$-Algebra Structures

Let $\mathfrak{R}$ be a local commutative algebra (or ring) with maximal ideal $\mathfrak{m}$ and with residue field $\mathbb{K}$, i.e. $\mathfrak{R} \cong \mathbb{K} \oplus \mathfrak{m}$. Later on we will work in detail on the case of the algebra of dual numbers $\mathfrak{R} = \mathbb{K}[t]/(t^2)$ and the case of the algebra of formal power series $\mathfrak{R} = \mathbb{K}[[t]]$.

Extending the scalars, one can consider operads and algebras in the symmetric monoidal category $(\mathsf{Mod}_{\mathfrak{R}}, \otimes_{\mathfrak{R}}, \mathfrak{R})$ of $\mathfrak{R}$-modules. Any $\mathbb{S}$-module M in the category of $\mathbb{K}$-modules extends to an $\mathbb{S}$-module

$$M \otimes_{\mathbb{K}} \mathfrak{R} := \big(M(0) \otimes_{\mathbb{K}} \mathfrak{R}, M(1) \otimes_{\mathbb{K}} \mathfrak{R}, \dots\big)$$

in the category of $\mathfrak{R}$-modules. When $(\mathscr{P}, \gamma, \iota)$ is a $\mathbb{K}$-linear operad, then $\mathscr{P} \otimes_{\mathbb{K}} \mathfrak{R}$ is endowed with the following $\mathfrak{R}$-linear operad structure

$$(\mu \otimes r; \nu_1 \otimes r_1, \dots, \nu_k \otimes r_k) \mapsto \gamma(\mu; \nu_1, \dots, \nu_k) \otimes r r_1 \dots r_k.$$

For instance, any $\mathbb{K}$-linear quadratic operad $\mathscr{P} = \mathscr{P}(E, R)$ extends to an $\mathfrak{R}$-linear quadratic operad $\mathscr{P} \otimes_{\mathbb{K}} \mathfrak{R} \cong \mathscr{P}(E \otimes_{\mathbb{K}} \mathfrak{R}, R \otimes_{\mathbb{K}} \mathfrak{R})$ since the composition map of operads is $\mathfrak{R}$-linear in the latter case. From now on and as usual, we simplify the notation $\otimes_{\mathbb{K}}$ into $\otimes$.

In the same way, the extension of scalars provides an $\mathfrak{R}$-module $A \otimes \mathfrak{R}$, whose endomorphism operad is equal to

$$\begin{aligned} \mathrm{End}_{A\otimes\mathfrak{R}} &= \left\{\mathrm{Hom}_{\mathfrak{R}}\left((A\otimes\mathfrak{R})^{\otimes_{\mathfrak{R}} n}, A\otimes\mathfrak{R}\right)\right\}_{n\in\mathbb{N}} \\ &\cong \left\{\mathrm{Hom}_{\mathbb{K}}\left(A^{\otimes n}, A\right)\otimes\mathfrak{R}\right\}_{n\in\mathbb{N}} \cong \mathrm{End}_A\otimes\mathfrak{R} \end{aligned}$$

by $\mathfrak{R}$-linearity.

Lemma 12.2.5. *The convolution dg Lie algebra $\mathfrak{g}_{\mathscr{P}\otimes\mathfrak{R},A\otimes\mathfrak{R}}$ is isomorphic to the $\mathfrak{R}$-linear extension of the convolution dg Lie algebra $\mathfrak{g}_{\mathscr{P},A}$*

$$\mathfrak{g}_{\mathscr{P}\otimes\mathfrak{R},A\otimes\mathfrak{R}} \cong \mathfrak{g}_{\mathscr{P},A}\otimes\mathfrak{R} = \left(\mathrm{Hom}_{\mathbb{S}}\left(\mathscr{P}^{\text{!`}}, \mathrm{End}_A\right)\otimes\mathfrak{R}, [\,,], \partial\right).$$

Proof. We have seen that the operad $\mathscr{P}\otimes\mathfrak{R}$ is equal to the $\mathfrak{R}$-linear quadratic operad $\mathscr{P}(E\otimes\mathfrak{R}, R\otimes\mathfrak{R})$. Its Koszul dual cooperad $\mathscr{C}(sE\otimes\mathfrak{R}, s^2R\otimes\mathfrak{R})$ is isomorphic to $\mathscr{C}(sE, s^2R)\otimes\mathfrak{R} = \mathscr{P}^{\text{!`}}\otimes\mathfrak{R}$ by $\mathfrak{R}$-linearity.

By definition, the convolution dg Lie algebra $\mathfrak{g}_{\mathscr{P}\otimes\mathfrak{R},A\otimes\mathfrak{R}}$ is made up of R-linear $\mathbb{S}_n$-equivariant morphisms from $(\mathscr{P}\otimes\mathfrak{R})^{\text{!`}}(n)$ to $\mathrm{Hom}_{\mathfrak{R}}((A\otimes\mathfrak{R})^{\otimes_{\mathfrak{R}} n}, A\otimes\mathfrak{R})$. Such morphisms are equivalent to $\mathbb{K}$-linear $\mathbb{S}_n$-equivariant morphisms from $\mathscr{P}^{\text{!`}}(n)$ to $\mathrm{Hom}(A^{\otimes n}, A)\otimes\mathfrak{R}$, which concludes the proof. □

Recall that the tensor product of a dg Lie algebra with a commutative algebra is endowed with a dg Lie algebra structure, where $[\alpha\otimes X, \beta\otimes Y] := [\alpha, \beta]\otimes XY$.

In the other way round, one recovers $\mathfrak{g}$ from $\mathfrak{g}\otimes\mathfrak{R}\cong\mathfrak{g}\oplus\mathfrak{g}\otimes\mathfrak{m}$ by reducing modulo $\mathfrak{m}$: $\mathfrak{g}\cong(\mathfrak{g}\otimes_{\mathbb{K}}\mathfrak{R})\otimes_{\mathfrak{R}}\mathbb{K}$.

Let $\varphi\in\mathrm{MC}(\mathfrak{g})$ be a $\mathscr{P}$-algebra structure on A. An $\mathfrak{R}$-*deformation* of φ is an $\mathfrak{R}$-linear $\mathscr{P}\otimes\mathfrak{R}$-algebra structure on $A\otimes\mathfrak{R}$, that is $\Phi\in\mathrm{MC}(\mathfrak{g}\otimes\mathfrak{R})$, which reduces to φ modulo $\mathfrak{m}$. We denote by $\mathrm{Def}_\varphi(\mathfrak{R})$ the set of $\mathfrak{R}$-deformations of φ.

When $\mathfrak{m}^2=0$, the associated deformations are called *infinitesimal deformations*. The paradigm of infinitesimal deformation is the one-parameter case given by the algebra of dual numbers $D=\mathbb{K}[t]/(t^2)$.

When the local algebra $\mathfrak{R}$ is *complete*, i.e. when $\mathfrak{R}$ is equal to the limit $\mathfrak{R} = \varprojlim \mathfrak{R}/\mathfrak{m}^n$, the associated deformations are called *formal deformations*. The one-parameter case is given by the algebra of formal power series $\mathbb{K}[[t]]$.

Any $\mathscr{P}$-algebra structure $\varphi\in\mathrm{MC}(\mathfrak{g})$ admits a trivial $\mathfrak{R}$-deformation $\varphi+0\in\mathrm{MC}(\mathfrak{g}\otimes\mathfrak{R})$, which corresponds to the trivial $\mathfrak{R}$-linear extension of the $\mathscr{P}$-algebra structure. The following result shows that the $\mathfrak{R}$-deformations of φ are controlled by the dg Lie algebra $\mathfrak{g}^\varphi\otimes\mathfrak{m}$.

Proposition 12.2.6. *For any $\mathscr{P}$-algebra structure φ on A, its set of $\mathfrak{R}$-deformations is naturally in bijection with the set of Maurer–Cartan elements in the dg Lie algebra $\mathfrak{g}^\varphi\otimes\mathfrak{m}$:*

$$\mathrm{Def}_\varphi(\mathfrak{R})\cong\mathrm{MC}\left(\mathfrak{g}^\varphi\otimes\mathfrak{m}\right).$$

Proof. Since $\varphi + 0 \in \mathrm{MC}(\mathfrak{g} \otimes \mathfrak{R})$, Proposition 12.2.4 shows that $\Phi = \varphi + \overline{\Phi}$, with $\overline{\Phi} \in \mathfrak{g}^1 \otimes \mathfrak{m}$, is in $\mathrm{MC}(\mathfrak{g} \otimes \mathfrak{R})$ if and only if $\overline{\Phi}$ belongs to $\mathrm{MC}((\mathfrak{g} \otimes \mathfrak{R})^{\varphi+0})$. This condition is equivalent to $\overline{\Phi} \in \mathrm{MC}(\mathfrak{g}^{\varphi} \otimes \mathfrak{m})$. □

An $\mathfrak{R}$-deformation Φ is *equivalent* to another $\mathfrak{R}$-deformation Ψ if there exists an $\mathfrak{R}$-linear isomorphism of $\mathscr{P} \otimes \mathfrak{R}$-algebras $(A \otimes \mathfrak{R}, \Phi) \to (A \otimes \mathfrak{R}, \Psi)$, whose restriction modulo $\mathfrak{m}$ is equal to the identity of (A, φ). Notice that it forms an equivalence relation denoted $\Phi \sim \Psi$. The associated deformation functor is denoted by

$$\mathscr{D}ef_{\varphi}(\mathfrak{R}) := \mathrm{Def}_{\varphi}(\mathfrak{R})/\sim .$$

12.2.6 Infinitesimal Deformations and Tangent Homology

In this section, we consider the case of the algebra of dual numbers $\mathfrak{R} = D = \mathbb{K}[t]/(t^2)$.

In this case, a (one-parameter) infinitesimal deformation of φ is an element of the form $\Phi = \varphi + \varphi_1 t$ in $\mathfrak{g}^1 \oplus \mathfrak{g}^1 \otimes t$. By Proposition 12.2.6, this latter element is an infinitesimal deformation if and only if

$$\partial(\varphi_1)t + [\varphi, \varphi_1]t + \frac{1}{2}[\varphi_1, \varphi_1]t^2 = \big(\partial(\varphi_1) + [\varphi, \varphi_1]\big)t = \partial_{\varphi}(\varphi_1)t = 0.$$

So the set of infinitesimal deformations of φ is canonically in bijection with the set of 1-cocycles in $\mathfrak{g}^{\varphi}$.

Two infinitesimal deformations $\Phi = \varphi + \varphi_1 t$ and $\Psi = \varphi + \psi_1 t$ are equivalent if and only if there exists an isomorphism of $\mathscr{P} \otimes D$-algebras, whose restriction to A reads

$$f = \mathrm{Id}_A + f_1 t : A \to A \oplus A \otimes t.$$

Notice that an $\mathfrak{R}$-linear endomorphism $f = \mathrm{Id}_A + f_1 t : A \to A \oplus A \otimes t$ of $A \otimes \mathfrak{R}$ is always an isomorphism with inverse $\mathrm{Id}_A - f_1 t$. So the condition for the deformations Φ and Ψ depends only on the endomorphism $f_1 \in \mathrm{Hom}(A, A)$ as follows.

Theorem 12.2.7. *There are canonical bijections*

$$\mathrm{Def}_{\varphi}\big(\mathbb{K}[t]/(t^2)\big) \cong Z^1(\mathfrak{g}^{\varphi}) \quad \textit{and} \quad \mathscr{D}ef_{\varphi}\big(\mathbb{K}[t]/(t^2)\big) \cong H^1(g^{\varphi}).$$

Proof. An endomorphism $f_1 \in \mathrm{Hom}(A, A)$ induces a morphism of $\mathscr{P} \otimes D$-algebras $f = \mathrm{Id}_A + f_1 t$ if and only if the following diagram commutes

$$\begin{array}{ccc}
\mu \in sE = \mathscr{P}^{\text{¡}(1)} & \xmapsto{\ \ \Phi\ \ } & \varphi(\mu) + \varphi_1(\mu)t \in \mathrm{End}_A \otimes D \\
\Big\downarrow{\scriptstyle \Psi} & & \Big\downarrow{\scriptstyle f_*} \\
\varphi(\mu) + \psi_1(\mu)t \in \mathrm{End}_A \otimes D & \xmapsto{\ f^*\ } & \begin{array}{c} \varphi(\mu) + (\varphi_1(\mu) + f_1 \circ \varphi(\mu))t \\ = \varphi(\mu) + (\psi_1(\mu) + \sum \varphi(\mu) \circ (\mathrm{Id}_A \otimes \cdots \otimes f_1 \otimes \cdots \otimes \mathrm{Id}_A))t \end{array}
\end{array}$$

for any $\mu \in sE$ and where the last sum runs over all the possible positions of f_1. If $\mu \in sE(m)$, then $\varphi(\mu) \in \mathrm{Hom}(A^{\otimes m}, A)$. Therefore, it is equivalent to

$$\varphi_1(\mu) - \psi_1(\mu) = -f_1 \circ \varphi(\mu) + \sum \varphi(\mu) \circ (\mathrm{Id}_A \otimes \cdots \otimes f_1 \otimes \cdots \otimes \mathrm{Id}_A).$$

It remains to show that the right-hand side is equal to $\partial_\varphi(f_1)(\mu)$. By definition, this latter element is equal to

$$\partial_\varphi(f_1) : \mathscr{P}^{¡\,(1)} \xrightarrow{\Delta_{(1)}} \mathrm{I} \circ \mathscr{P}^{¡\,(1)} \oplus \mathscr{P}^{¡\,(1)} \circ_{(1)} \mathrm{I} \xrightarrow{-f_1 \circ \varphi + \varphi \circ_{(1)} f_1} \mathrm{End}_A.$$

This concludes the proof. □

When t tends to 0, any infinitesimal deformation $\varphi + \varphi_1 t$ tends to φ in the direction given by φ_1. With the aforementioned results, this explains why $Z^1(\mathfrak{g}^\varphi)$ can be interpreted as the tangent space at φ of the algebraic variety $\mathrm{MC}(\mathfrak{g})$ and why $H^1(\mathfrak{g}^\varphi)$ can be interpreted as the tangent space at φ of the algebraic stack $\mathscr{MC}(\mathfrak{g})$. For more details on this interpretation, we refer the reader to [CFK01].

12.2.7 Formal Deformations

Let us now work with $\mathfrak{R} = \mathbb{K}[[t]]$ and (one-parameter) formal deformations $\Phi = \varphi + \sum_{n\geq 1} \varphi_n t^n$, where each $\varphi_n \in \mathfrak{g}^1$, for $n \geq 1$. Denoting $\overline{\Phi} := \sum_{n\geq 1} \varphi_n t^n$, Proposition 12.2.6 shows that Φ is a formal deformation if and only if

$$\partial(\overline{\Phi}) + [\varphi, \overline{\Phi}] + \frac{1}{2}[\overline{\Phi}, \overline{\Phi}] = 0.$$

By t-linearity, this latter equation is equivalent to the following system

$$\partial(\varphi_n) + [\varphi, \varphi_n] + \frac{1}{2}\sum_{k=1}^{n-1}[\varphi_k, \varphi_{n-k}] = 0, \quad \text{for } n \geq 1.$$

For $n = 1$, it gives $\partial(\varphi_1) + [\varphi, \varphi_1] = 0$, that is φ_1 is a 1-cocycle of $\mathfrak{g}^\varphi$. More generally, the first nontrivial element φ_k of a formal deformation $\varphi + \sum_{n\geq k} \varphi_n t^n$ of φ is a 1-cocycle of $\mathfrak{g}^\varphi$. For $n \geq 2$, it can be rewritten as

$$\partial_\varphi(\varphi_n) = -\frac{1}{2}\sum_{k=1}^{n-1}[\varphi_k, \varphi_{n-k}].$$

Theorem 12.2.8 (Obstructions). *If $H^2(\mathfrak{g}^\varphi) = 0$, then any 1-cocycle of $\mathfrak{g}^\varphi$ extends to a formal deformation of φ.*

Proof. It is enough to prove that, for any $n \geq 2$, if $\varphi \in \mathrm{MC}(\mathfrak{g})$ and if $\varphi_1, \ldots, \varphi_{n-1} \in \mathfrak{g}^1$ satisfy the above equations up to $n-1$, then the element $\sum_{k=1}^{n-1}[\varphi_k, \varphi_{n-k}]$ is a 2-cocycle in $\mathfrak{g}^{\varphi}$.

Since ∂_{φ} is a derivation with respect to the Lie bracket, we have

$$\begin{aligned}\partial_{\varphi}\left(\sum_{k=1}^{n-1}[\varphi_k, \varphi_{n-k}]\right) &= \sum_{k=1}^{n-1}\big[\partial_{\varphi}(\varphi_k), \varphi_{n-k}\big] - \big[\varphi_k, \partial_{\varphi}(\varphi_{n-k})\big]\\ &= 2\sum_{k=1}^{n-1}\big[\partial_{\varphi}(\varphi_k), \varphi_{n-k}\big].\end{aligned}$$

Using the equations satisfied by the element φ_k, we get

$$\partial_{\varphi}\left(\sum_{k=1}^{n-1}[\varphi_k, \varphi_{n-k}]\right) = -\sum_{k=1}^{n-1}\sum_{l=1}^{k-1}\big[[\varphi_l, \varphi_{k-l}], \varphi_{n-k}\big] = \sum_{\substack{a+b+c=n,\\ a,b,c\geq 1}}\big[[\varphi_a, \varphi_b], \varphi_c\big],$$

which vanishes by the Jacobi identity. □

In other words, the second cohomology group $H^2(\mathfrak{g}^{\varphi})$ of the deformation complex carries the obstructions to formal deformations.

An equivalence between two formal deformations is a $\mathbb{K}[[t]]$-linear isomorphism of $\mathscr{P} \otimes \mathbb{K}[[t]]$-algebras, whose first component is the identity of A. It amounts to a family of $\mathbb{K}$-linear maps

$$f = \mathrm{Id}_A + \sum_{n\geq 1} f_n t^n : A \to A \oplus A \otimes t \oplus A \otimes t^2 \oplus \cdots.$$

Since the first component is invertible, the induced $\mathbb{K}[[t]]$-linear endomorphism of $A \otimes \mathbb{K}[[t]]$ is always invertible, see Exercise 12.6.4. So it is an equivalence of deformations if and only if it is a morphism of $\mathscr{P} \otimes \mathbb{K}[[t]]$-algebras.

Theorem 12.2.9 (Rigidity). *If $H^1(g^{\varphi}) = 0$, then any formal deformation of φ is equivalent to the trivial one.*

Proof. Let $\Phi = \varphi + \sum_{n\geq 1} \varphi_n t^n$ be a formal deformation of φ. The same kind of commutative diagram as in the proof of Theorem 12.2.7 shows that $f = \mathrm{Id}_A + \sum_{n\geq 1} f_n t^n$ is an equivalence between the deformation Φ and the trivial deformation $\varphi + 0$ if and only if for any $\mu \in sE(m)$

$$\sum_{k,l\geq 0} f_k \circ \varphi_l(\mu) t^{k+l} = \sum_{i_1,\ldots,i_m\geq 0} \varphi(\mu) \circ (f_{i_1} \otimes \cdots \otimes f_{i_m}) t^{i_1+\cdots+i_m},$$

where $f_0 = \mathrm{Id}_A$ and $\varphi_0 = \varphi$ by convention. Therefore, f is an equivalence if and only if for any $n \geq 1$ and for any $\mu \in sE(m)$:

$$\sum_{k+l=n} f_k \circ \varphi_l(\mu) = \sum_{i_1+\cdots+i_m=n} \varphi(\mu) \circ (f_{i_1} \otimes \cdots \otimes f_{i_m}).$$

This latter relation can be rewritten as

$$\begin{aligned}\Theta_n(\mu) &:= \sum_{k=0}^{n-1} f_k \circ \varphi_{n-k}(\mu) - \sum_{\substack{i_1+\cdots+i_m=n\\ i_1,\ldots,i_m<n}} \varphi(\mu)\circ (f_{i_1}\otimes\cdots\otimes f_{i_m})\\ &= -f_n\circ\varphi(\mu) + \sum \varphi(\mu)\circ(\mathrm{Id}_A\otimes\cdots\otimes f_n\otimes\cdots\otimes \mathrm{Id}_A)\\ &= \partial_\varphi(f_n)(\mu).\end{aligned}$$

We now prove by induction on $n \geq 1$ that $\Theta_n \in \mathrm{Hom}_{\mathbb{S}}(\mathscr{P}^{\text{!`}(1)}, \mathrm{End}_A)$ is a 1-cocycle, i.e. $\partial_\varphi(\Theta_n) = 0$. Since $H^1(\mathfrak{g}^\varphi) = 0$, this will imply that there exists an element $f_n \in \mathrm{Hom}(A, A)$ such that $\partial_\varphi(f_n) = \Theta_n$, and thus concludes the proof.

For $n = 1$, we have $\Theta_1 = \varphi_1$, which is a 1-cocycle since Φ is a deformation of φ, see above.

We suppose that the result is true up to $n-1$ and we prove it at rank n. It is a straightforward computation and we leave it to the reader. □

12.2.8 Deformation Functor

To study the general case, we have to consider the deformations $\mathrm{Def}_\varphi(\mathfrak{R})$ over *Artin* local algebras $\mathfrak{R}$, see Sect. 13.2.14. In this case, the dg Lie algebra $\mathfrak{g}^\varphi \otimes \mathfrak{m}$ becomes nilpotent. So the Lie subalgebra $\mathfrak{g}^0$ admits a Lie group G, called the *gauge group*, which acts on the set of Maurer–Cartan elements. This yields a Deligne groupoid and a moduli space $\mathscr{MC}(\mathfrak{g}^\varphi\otimes\mathfrak{m}) := \mathrm{MC}(\mathfrak{g}^\varphi\otimes\mathfrak{m})/G$.

Theorem 12.2.10. *Let φ be a $\mathscr{P}$-algebra structure on A. For any Artin local algebra $\mathfrak{R} = \mathbb{K}\oplus\mathfrak{m}$, there is a natural bijection, respectively equivalence of Deligne groupoids*

$$\mathscr{D}ef_\varphi(\mathfrak{R}) = \mathrm{Def}_\varphi(\mathfrak{R})/\sim\, \cong \mathscr{MC}\big(\mathfrak{g}^\varphi\otimes\mathfrak{m}\big) := \mathrm{MC}\big(\mathfrak{g}^\varphi\otimes\mathfrak{m}\big)/G.$$

Proof. We have already proved in Proposition 12.2.6 the bijection between the deformations and the Maurer–Cartan elements. Let us now prove that the equivalence relation corresponds to the gauge group action.

Let $\Phi = \varphi + \overline{\Phi}$ and $\Psi = \varphi + \overline{\Psi}$ be two $\mathfrak{R}$-deformations of φ. Let $\lambda \in \mathrm{Hom}(A, A)\otimes\mathfrak{m}$ such that $e^\lambda.\overline{\Phi} = \overline{\Psi}$ in $\mathrm{Hom}_{\mathbb{S}}(\mathscr{P}^{\text{!`}(1)}, \mathrm{End}_A)\otimes\mathfrak{m}$. Since λ is a nilpotent element in the associative algebra $\mathrm{Hom}(A, A)\otimes\mathfrak{m}$, this equality is equivalent to

$$\begin{aligned}\big(e^\lambda.\overline{\Phi}\big)(\mu) &= e^\lambda\circ\overline{\Phi}(\mu)\circ\big(e^{-\lambda},\ldots,e^{-\lambda}\big) + e^\lambda\circ\varphi(\mu)\circ\big(e^{-\lambda},\ldots,e^{-\lambda}\big) - \varphi(\mu)\\ &= \overline{\Psi}(\mu),\end{aligned}$$

for any $\mu\in\mathscr{P}^{\text{!`}(1)}$. Therefore, the two Maurer–Cartan elements $\overline{\Phi}$ and $\overline{\Psi}$ are gauge equivalent if and only if there exists $\lambda\in\mathrm{Hom}(A, A)\otimes\mathfrak{m}$ such that $e^\lambda\circ\Phi(\mu) =$

$\Psi(\mu) \circ (e^\lambda, \dots, e^\lambda)$. In this case, the two deformations Φ and Ψ are equivalent under the $\mathfrak{R}$-linear isomorphism $f = e^\lambda$, whose reduction modulo $\mathfrak{m}$ is equal to the identity on A. In the other way round, an $\mathfrak{R}$-linear isomorphism $f = \mathrm{Id}_A + \bar{f}$ between $(A \otimes \mathfrak{R}, \Phi)$ and $(A \otimes \mathfrak{R}, \Psi)$ is a morphism of $\mathscr{P} \otimes \mathfrak{R}$-algebras if and only if $f \circ \Phi(\mu) = \Psi(\mu) \circ (f, \dots, f)$, for any $\mu \in \mathscr{P}^{\text{¡}(1)}$. The one-to-one correspondence $f \leftrightarrow \lambda$ is given by $f = e^\lambda$ and $\lambda = \log(\mathrm{Id}_A + \bar{f})$ in the nilpotent algebra $\mathrm{Hom}(A, A) \otimes \mathfrak{m}$.

We leave it to the reader to verify that this construction is natural. Notice that here, we actually consider a functor from the category of Artin local algebras to groupoids. □

This result shows that the dg Lie algebra $\mathfrak{g}^\varphi$ faithfully encodes the deformation problem of the $\mathscr{P}$-algebra structure φ on A, according to the philosophy recalled in Sect. 13.2.14.

Notice that the algebra of formal power series is not an Artin algebra. One approximates its deformations by the following limit $\mathscr{D}ef_\varphi(\mathbb{K}[[t]]) \cong \varprojlim \mathscr{D}ef_\varphi(\mathbb{K}[t]/(t^n))$, where $\mathbb{K}[t]/(t^n)$ is an Artin local algebra.

12.2.9 Obstruction Theory for $\mathscr{P}_\infty$-Algebra Structures

At this point, the reader might ask: how do we treat the case of *dg* $\mathscr{P}$-algebras? what is the use of the higher components $\mathrm{Hom}_{\mathbb{S}}(\mathscr{P}^{\text{¡}(n)}, \mathrm{End}_A)$, for $n \geq 2$? and why have we not used the hypothesis that $\mathscr{P}$ is a Koszul operad? The answer is: when the operad $\mathscr{P}$ is Koszul, the *full* convolution dg Lie algebra

$$\mathfrak{g} = \big(\mathrm{Hom}_{\mathbb{S}}\big(\mathscr{P}^{\text{¡}}, \mathrm{End}_A\big), [,], \partial\big)$$

controls the deformations of the *$\mathscr{P}_\infty$-algebra structures* on the chain complex (A, d_A).

When (A, d_A) is a chain complex, we consider the convolution dg Lie algebra $\mathfrak{g}$, with the usual homological degree. Proposition 10.1.1 shows that a $\mathscr{P}_\infty$-algebra structure on (A, d_A) is equivalent to a twisting morphism, i.e. a degree -1 Maurer–Cartan element $\partial\varphi + \frac{1}{2}[\varphi, \varphi] = 0$ in $\mathfrak{g}$, with vanishing component on I.

From that point, everything works mutatis mutandis as before in the $\mathscr{P}$-algebra case since we only used the properties of the convolution dg Lie algebra. For instance, the following lemma states that the convolution dg Lie algebra $\mathfrak{g}$, which controls the $\mathscr{P}_\infty$-algebra structures, shares similar properties with the convolution dg Lie algebra $\mathfrak{g}^\varphi \otimes \mathbb{K}[[t]]$ which controls formal deformations of $\mathscr{P}$-algebras.

Lemma 12.2.11. *The convolution dg Lie algebra*

$$\mathfrak{g} = \big(\mathrm{Hom}_{\mathbb{S}}\big(\mathscr{P}^{i}, \mathrm{End}_A\big), [,], \partial\big)$$

is a weight graded Lie algebra, $\mathfrak{g} = \prod_{n\geq 0} \mathfrak{g}^{(n)}$. *Its differential map is the sum of two anti-commuting square-zero derivations,* $\partial = \partial_0 + \partial_1$, *such that the first one preserves this grading,* $\partial_0 : \mathfrak{g}^{(n)} \to \mathfrak{g}^{(n)}$, *and such that the second one raises it by one,* $\partial_1 : \mathfrak{g}^{(n-1)} \to \mathfrak{g}^{(n)}$.

Proof. Since the Koszul dual cooperad is a weight graded cooperad, the convolution Lie algebra is a weight graded Lie algebra $\mathfrak{g}^{(n)} := \mathrm{Hom}_{\mathbb{S}}(\mathscr{P}^{¡\,(n)}, \mathrm{End}_A)$, for $n \geq 0$. The differential is the sum of the following two terms:

$$\partial_\varphi(f) = d_{\mathrm{End}_A} \circ f - (-1)^{|f|} f \circ d_{\mathscr{P}^{¡}}.$$

The component $\partial_0(f) := d_{\mathrm{End}_A} \circ f$ preserves the weight grading and the component $\partial_1(f) := -(-1)^{|f|} f \circ d_{\mathscr{P}^{¡}}$ raises it by 1. □

As in Sect. 12.2.7, this result allows us to settle an obstruction theory for $\mathscr{P}_\infty$-algebra structures: we give below a criterion to say when a family of operations can be extended to a full $\mathscr{P}_\infty$-algebra structure.

Theorem 12.2.12. *Let* $\mathscr{P}$ *be a Koszul operad and let* (A, d_A) *be a dg module. Suppose that we are given an element* $\varphi_1 \in \mathrm{Hom}_{\mathbb{S}}(\mathscr{P}^{i\,(1)}, \mathrm{End}_A)$, *such that* d_A *is a derivation with respect to image operations, that is* $\partial_0(\varphi_1) := d_{\mathrm{End}_A} \circ \varphi_1 = 0$.

If $H_{-2}(\mathrm{Hom}_{\mathbb{S}}(\mathscr{P}^{i\,(n)}, \mathrm{End}_A), \partial_0) = 0$ *for* $n \geq 2$, *then* φ_1 *extends to a* $\mathscr{P}_\infty$-*algebra structure on* A.

Proof. Using Lemma 12.2.11, the proof is similar to that of Theorem 12.2.8. A $\mathscr{P}_\infty$-algebra structure φ on A is a Maurer–Cartan element in $\mathfrak{g}$ of degree -1, which vanishes on I, that is $\varphi = \sum_{n\geq 1} \varphi_n$, where $\varphi_n \in \mathfrak{g}_{-1}^{(n)}$. The Maurer–Cartan equation splits with respect to the weight grading; it is therefore equivalent to

$$\xi_n := \partial_1(\varphi_{n-1}) + \frac{1}{2}\sum_{k=1}^{n-1} [\varphi_k, \varphi_{n-k}] = -\partial_0(\varphi_n),$$

in $\mathfrak{g}_{-2}^{(n)}$, for any $n \geq 1$. As in the proof of Theorem 12.2.8, one proves by induction on n that ξ_n is a ∂_0-cycle and, since $H_{-2}(\mathrm{Hom}_{\mathbb{S}}(\mathscr{P}^{¡\,(n)}, \mathrm{End}_A), \partial_0) = 0$, there exists $\varphi_n \in \mathfrak{g}_{-1}^{(n)}$ such that $\xi_n = -\partial_0(\varphi_n)$. The induction is founded by the hypothesis of the theorem:

$$\partial_0(\xi_1) = \partial_0\left(\frac{1}{2}[\varphi_1, \varphi_1]\right) = \big[\partial_0(\varphi_1), \varphi_1\big] = 0.$$

We leave the computations to the reader as a good exercise. The details can be found in [GCTV09, Appendix C]. □

This method was used in the case of homotopy Lie algebras in [BFLS98]. In [GCTV09], it plays also a crucial role in the proof of the Lian–Zuckerman conjecture, which states that the BRST complex of a Vertex Operator Algebra is endowed with a homotopy Batalin–Vilkovisky algebra structure. Notice that this general method extends to Koszul properads as well, see [GCTV09, Appendix C], and allows one to endow the differential forms of a closed oriented manifold with a homotopy Frobenius bialgebra structure, see [Wil07, HM10].

12.2.10 Deformation Theory of $\mathscr{P}_\infty$-Algebra Structures

In this section, we study the possible deformations of a given $\mathscr{P}_\infty$-algebra structure. The same "Lie calculus" holds and allows us to prove exactly the same results in the $\mathscr{P}$-algebra case. The only new point comes with the introduction of ∞-morphisms in the definition of the equivalence relation. From that point, everything works mutatis mutandis.

Given a $\mathscr{P}_\infty$-algebra structure φ on A, we consider, as in Sect. 2.1.3, the twisted differential $\partial_\varphi(f) := \partial(f) + [\varphi, f]$, which squares to 0. The associated twisted chain complex

$$\left(\mathrm{Hom}_{\mathbb{S}}\left(\mathscr{P}^{\text{¡}}, \mathrm{End}_A\right), \partial_\varphi\right)$$

defines the *deformation complex of the* $\mathscr{P}_\infty$*-algebra* (A, φ). Its homology groups are again called the *tangent homology* and it is endowed with the same intrinsic Lie bracket as before. Notice that when (A, φ) is concentrated in degree 0 and when φ is a $\mathscr{P}$-algebra structure on A, one recovers the deformation complex of the previous sections.

Proposition 12.2.13. *Let $\mathscr{P}$ be a Koszul operad and let (A, d_A) be a chain complex. For any $\mathscr{P}_\infty$-algebra structure $\varphi \in \mathrm{MC}(\mathfrak{g})$ on A, the following equivalence holds*

$$\alpha \in \mathrm{MC}\left(\mathfrak{g}^\varphi\right) \iff \alpha + \varphi \in \mathrm{MC}(\mathfrak{g}).$$

Proof. Same as Proposition 12.2.4. □

Let $(\mathfrak{R}, \mathfrak{m})$ be a local commutative algebra. An $\mathfrak{R}$*-deformation* of a $\mathscr{P}_\infty$-algebra (A, φ) is an $\mathfrak{R}$-linear $\mathscr{P}_\infty \otimes \mathfrak{R}$-algebra structure on $A \otimes \mathfrak{R}$ which reduces to φ modulo $\mathfrak{m}$. The set of $\mathfrak{R}$-deformations of φ is denoted by $\mathrm{Def}_\varphi(\mathfrak{R})$. By the preceding proposition and Lemma 12.2.5, we still have a natural bijection

$$\mathrm{Def}_\varphi(\mathfrak{R}) \cong \mathrm{MC}\left(\mathfrak{g}^\varphi \otimes \mathfrak{m}\right).$$

The only new point comes with the following definition of the equivalence relation. First we define the notion of an $\mathfrak{R}$-linear ∞-morphism between two $\mathscr{P}_\infty \otimes \mathfrak{R}$-algebras $(A \otimes \mathfrak{R}, \Phi)$ and $(A \otimes \mathfrak{R}, \Psi)$ by $\mathfrak{R}$-linear extension of the arguments of Sect. 10.2.

One can extend Theorem 10.2.3 as follows. Since here the source and target spaces are the same, $(A \otimes \mathfrak{R}, \Phi)$, the operator $*$ is replaced by the $\mathfrak{R}$-linear extension of the preLie product in $\mathrm{Hom}_{\mathbb{S}}(\mathscr{P}^{\mathrm{i}}, \mathrm{End}_A) \otimes \mathfrak{R}$, still denoted by $\star$:

$$f \star g := \mathscr{P}^{\mathrm{i}} \xrightarrow{\Delta_{(1)}} \mathscr{P}^{\mathrm{i}} \circ_{(1)} \mathscr{P}^{\mathrm{i}} \xrightarrow{f \circ_{(1)} g} (\mathrm{End}_A \otimes \mathfrak{R}) \circ_{(1)} (\mathrm{End}_A \otimes \mathfrak{R}) \to \mathrm{End}_A \otimes \mathfrak{R},$$

where the last arrow stands for the composition of morphisms and the product in the algebra $\mathfrak{R}$. In the same way, we replace the operator $\circledast$ by the $\mathfrak{R}$-linear extension of the associative product, still denoted $\circledcirc$:

$$f \circledcirc g := \mathscr{P}^{\mathrm{i}} \xrightarrow{\Delta_{(1)}} \mathscr{P}^{\mathrm{i}} \circ \mathscr{P}^{\mathrm{i}} \xrightarrow{f \circ g} (\mathrm{End}_A \otimes \mathfrak{R}) \circ (\mathrm{End}_A \otimes \mathfrak{R}) \to \mathrm{End}_A \otimes \mathfrak{R}.$$

Finally an $\mathfrak{R}$-linear ∞-morphism is equivalently given by an element $f \in \mathrm{Hom}_{\mathbb{S}}(\mathscr{P}^{\mathrm{i}}, \mathrm{End}_A) \otimes \mathfrak{R}$ of degree 0, such that

$$f \star \Phi - \Psi \circledcirc f = \partial(f),$$

in $\mathfrak{g}_{-1} \otimes \mathfrak{R}$.

An $\mathfrak{R}$-deformation Φ is *equivalent* to another $\mathfrak{R}$-deformation Ψ if there exists an $\mathfrak{R}$-linear ∞-isomorphism of $\mathscr{P}_\infty \otimes \mathfrak{R}$-algebras $(A \otimes \mathfrak{R}, \Phi) \rightsquigarrow (A \otimes \mathfrak{R}, \Psi)$, whose restriction modulo $\mathfrak{m}$ is equal to the identity of A. By definition 10.2.4, it means that the first component $\mathrm{Id}_A / \bar{f}$ of f,

$$f_{(0)} : \mathrm{I} \to \mathrm{Hom}(A, A) \otimes \mathfrak{R}; \qquad \mathrm{id} \mapsto \mathrm{Id}_A + \bar{f},$$

is invertible in $\mathrm{Hom}(A, A) \otimes \mathfrak{R}$. We leave it to the reader to prove that such an $\mathfrak{R}$-linear ∞-isomorphism admits an inverse of the same form; see Exercise 12.6.5 for more details.

In this case, this equivalence relation is denoted by $\Phi \sim \Psi$ and the associated deformation functor is denoted by

$$\mathscr{D}ef_\varphi(\mathfrak{R}) := \mathrm{Def}_\varphi(\mathfrak{R}) / \sim.$$

Under these definitions, all the results proved in the case of $\mathscr{P}$-algebras extend to the case of $\mathscr{P}_\infty$-algebras as follows.

Theorem 12.2.14. *Let $\mathscr{P}$ be a Koszul operad and let φ be a $\mathscr{P}_\infty$-algebra structure on (A, d_A).*

- [Infinitesimal deformations] *There are canonical bijections*

$$\mathrm{Def}_\varphi(\mathbb{K}[t]/(t^2)) = Z_{-1}(\mathfrak{g}^\varphi) \quad \textit{and} \quad \mathscr{D}ef_\varphi(\mathbb{K}[t]/(t^2)) = H_{-1}(\mathfrak{g}^\varphi).$$

- [Obstructions] *If $H_{-2}(\mathfrak{g}^\varphi) = 0$, then any (-1)-cycle of $\mathfrak{g}^\varphi$ extends to a formal deformation of φ.*
- [Rigidity] *If $H_{-1}(g^\varphi) = 0$, then any formal deformation of φ is equivalent to the trivial one.*

◇ [Deformation functor] *For any Artin local algebra* $(\mathfrak{R}, \mathfrak{m})$, *there is a natural bijection, respectively equivalence of Deligne groupoids, between*

$$\mathscr{D}ef_{\varphi}(\mathfrak{R}) = \mathrm{Def}_{\varphi}(\mathfrak{R})/\sim \cong \mathscr{MC}(\mathfrak{g}^{\varphi} \otimes \mathfrak{m}) := \mathrm{MC}(\mathfrak{g}^{\varphi} \otimes \mathfrak{m})/G.$$

Proof. Mutatis mutandis, this proof is similar the previous proofs in the $\mathscr{P}$-algebra case. Therefore it is left to the reader as a good exercise. □

12.2.11 Deformation Theory of Morphisms of Operads

The reader may have noticed that we never used the particular form of the operad End_A in the preceding sections. Actually the aforementioned methods apply as well to study the deformation theory of morphisms from the operad $\mathscr{P}$, respectively $\mathscr{P}_\infty$, to any operad $\mathscr{Q}$. One just has to replace End_A by $\mathscr{Q}$ everywhere.

For instance, an operad $\mathscr{Q}$ is called *multiplicative* when it is endowed with a morphism of operads $\varphi : Ass \to \mathscr{Q}$, where Ass is the operad encoding associative algebras. The associated deformation complex of φ is often called the *Hochschild cohomology of* $\mathscr{Q}$. This complex plays a crucial role in the solutions of the Deligne conjecture, see [GV95, MS02].

One can consider the operad $\mathscr{Q} = Pois$ encoding Poisson algebras, see Sect. 13.3. It comes equipped with a canonical morphism $Ass \to Pois$ of operads. The associated deformation complex is related to the rational homology of the long knots and to Vassiliev invariants, as explained in [Tou04, Sin06, Sin09, LTV10].

12.2.12 Beyond the Koszul Case

When an operad $\mathscr{P}$ admits a quasi-free resolution $\mathscr{T}(\mathscr{C}) \xrightarrow{\sim} \mathscr{P}$, not necessarily quadratic, there exists a convolution algebra $\mathrm{Hom}_{\mathbb{S}}(\mathscr{C}, \mathrm{End}_A)$, which carries an L_∞-algebra in general. It is a strict dg Lie algebra only when the resolution is quadratic. In the same way as before, this L_∞-algebra controls the deformations of homotopy $\mathscr{P}$-algebra structures. For more details, we refer the reader to Van der Laan [vdL02], Kontsevich–Soibelman [KS10] and Merkulov–Vallette [MV09a, MV09b].

12.3 André–Quillen (Co)Homology of Algebras over an Operad

This section is the generalization to algebras over an operad of the conceptual definition of the (co)homology of commutative algebras given by D. Quillen in [Qui70] and M. André in [And74].

First, we define the notion of module over an algebra over an operad, which provides the space of coefficients of this cohomology theory. We consider the space of derivations from an algebra to such a module. It gives rise to a Quillen adjunction, that we derive to define the André–Quillen cohomology. Since the functor of derivations is represented on the left by the module of Kähler differential forms, the André–Quillen cohomology theory is represented by an object called the cotangent complex.

The category of modules over an algebra A is equivalent to the category of left modules over the enveloping algebra of A. We define the André–Quillen homology of A with coefficients into right modules over the enveloping algebra of A.

Throughout the section, we study the examples of the operads *As*, *Com*, and *Lie*. In this way, we recover, up to a degree shift, the Hochschild (co)homology of associative algebras, the Chevalley–Eilenberg (co)homology of Lie algebras and the Harrison (co)homology of commutative algebras.

This section comes from V. Ginzburg and M.M. Kapranov [GK94], V. Hinich [Hin97], D. Balavoine [Bal98], P. Goerss and M. Hopkins [GH00], J. Millès [Mil11] and B. Fresse [Fre09a].

12.3.1 Module over a $\mathscr{P}$-Algebra

Let $(\mathscr{P}, \gamma, \eta)$ be an operad and let (A, γ_A) be a $\mathscr{P}$-algebra. Recall from Sect. 6.1.1 that to any $\mathbb{K}$-module M, we associate the following linearized version of the composite product

$$\mathscr{P} \circ (A; M) := \bigoplus_{n \geq 0} \mathscr{P}(n) \otimes_{\mathbb{S}_n} \Bigg(\bigoplus_{1 \leq i \leq n} A \otimes \cdots \otimes A \otimes \underbrace{M}_{i\text{th position}} \otimes A \otimes \cdots \otimes A \Bigg).$$

An *A-module over $\mathscr{P}$*, or *A-module* when the operad is understood, is a $\mathbb{K}$-module M endowed with two maps

$$\gamma_M : \mathscr{P} \circ (A; M) \to M \quad \text{and} \quad \eta_M : M \to \mathscr{P} \circ (A; M),$$

satisfying the following commutative diagrams

$$\begin{array}{ccc}
 & & \mathscr{P} \circ (\mathscr{P}(A); \mathscr{P} \circ (A; M)) \xrightarrow{\mathrm{Id} \circ (\gamma_A; \gamma_M)} \mathscr{P} \circ (A; M) \\
 & \overset{\cong}{\nearrow} & \downarrow \gamma_M \\
(\mathscr{P} \circ \mathscr{P}) \circ (A; M) & & \\
\downarrow \gamma \circ (\mathrm{Id}; \mathrm{Id}) & & \\
\mathscr{P} \circ (A; M) & \xrightarrow{\gamma_M} & M
\end{array}$$

and

$$\begin{array}{ccc} M & \xrightarrow{\eta_M} & \mathscr{P}\circ(A;M) \\ & \searrow^{=} & \downarrow{\scriptstyle \gamma_M} \\ & & M. \end{array}$$

A morphism of A-modules is a morphism of $\mathbb{K}$-modules $f : M \to N$ which commutes to the structure maps γ_M, γ_N and η_M, η_N. The associated category of A-modules over $\mathscr{P}$ is denoted by $\mathsf{Mod}_A^{\mathscr{P}}$.

The algebra A itself is the first example of an A-module.

Proposition 12.3.1.

- ◇ *In the case of the (nonsymmetric) operad $\mathscr{P} = As$ and of an associative algebra A, an A-module over As is an A-bimodule $A \otimes M \to M$, $M \otimes A \to M$ over A.*
- ◇ *In the case of the operad $\mathscr{P} = Com$ and of a commutative algebra A, an A-module over Com is a classical A-module $A \otimes M \to M$.*
- ◇ *In the case of the operad $\mathscr{P} = Lie$ and of a Lie algebra $\mathfrak{g}$, an $\mathfrak{g}$-module over Lie is a classical Lie module $\mathfrak{g} \otimes M \to M$ over $\mathfrak{g}$.*

Proof. When the operad $\mathscr{P}(E, R)$ admits a presentation by generators and relations, the defining action $\gamma_M : \mathscr{P}\circ(A; M) \to M$ of an A-module includes an action of the generating operations $E \circ (A, M) \rightarrowtail \mathscr{P}\circ(A; M) \to M$. The relations that they have to satisfy is given by the space R. In the other way round, one can reconstruct the action of any element of $\mathscr{P}(A; M)$ thanks to the composition map of the operad $\mathscr{P}$. These arguments, applied to the three above cases, give the classical notions of modules. The details are left to the reader as an easy but interesting exercise. □

Recall that the notion of Lie module over a Lie algebra $\mathfrak{g}$ is equivalent to the notion of left module over the associative universal enveloping algebra $U(\mathfrak{g})$, see Sect. 1.1.11.

12.3.2 Abelian Extension of A by M

An *extension of A by M*

$$0 \to M \rightarrowtail B \twoheadrightarrow A \to 0$$

is a short exact sequence in the category of $\mathscr{P}$-algebras. A *split* extension is an extension

$$0 \longrightarrow M \overset{i}{\hookrightarrow} B \underset{\overset{s}{\curvearrowleft}}{\overset{p}{\twoheadrightarrow}} A \longrightarrow 0,$$

endowed with a $\mathbb{K}$-linear map s satisfying $ps = \mathrm{Id}_A$ and so

$$\mathrm{B} \cong A \oplus M.$$

Suppose now that the operad $\mathscr{P}$ is augmented $\mathscr{P} \cong \overline{\mathscr{P}} \oplus \mathrm{I}$. An *abelian extension* is a split extension such that the $\mathscr{P}$-algebra structure on M is trivial, i.e. $\overline{\mathscr{P}}(M) \to M$ is zero.

If $\mathscr{P}$ happens to be binary, then it is equivalent to requiring that any product of two elements of M is zero. In this case, it is sometimes called a (split) *square-zero extension* in the literature.

When M is an A-module, we consider the following $\mathscr{P}$-algebra structure on $A \oplus M$:

$$\mathscr{P}(A \oplus M) \twoheadrightarrow \mathscr{P}(A) \oplus \mathscr{P} \circ (A; M) \xrightarrow{\gamma_A + \gamma_M} A \oplus M.$$

Such a $\mathscr{P}$-algebra structure is denoted by $A \ltimes M$. It is the archetype of abelian extensions of A by M,

$$0 \to M \rightarrowtail A \ltimes M \twoheadrightarrow A \to 0,$$

as the following proposition shows.

Proposition 12.3.2. *Let $\mathscr{P}$ be an augmented operad, let A be a $\mathscr{P}$-algebra and let M be a $\mathbb{K}$-module. The data of an A-module on M is equivalent to the data of an abelian extension of A by M.*

Proof. ($\Rightarrow$) The commutative diagram in the definition of the $\mathscr{P}$-algebra structure on $A \ltimes M$ is equivalent to

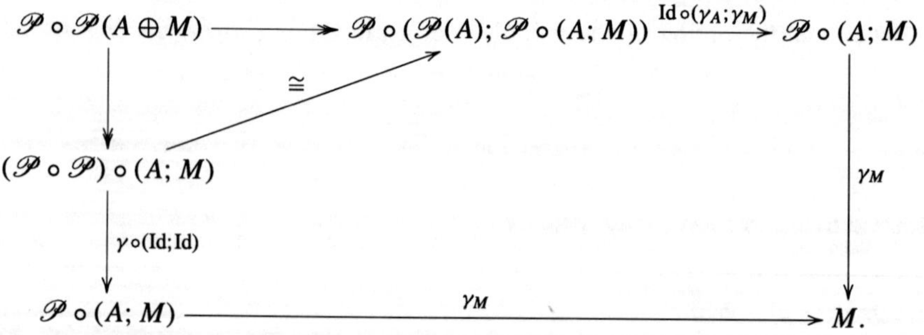

The other one is

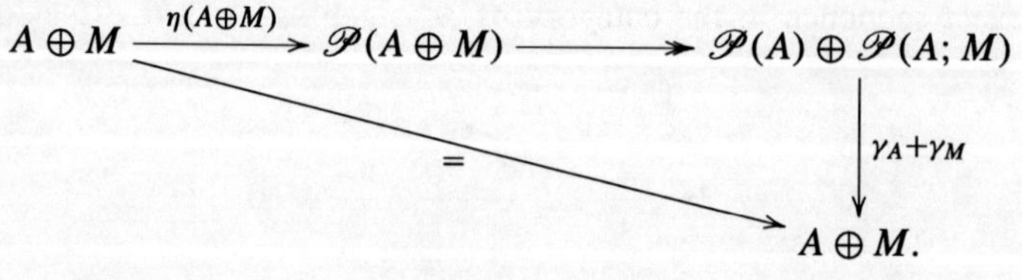

Since A is already a $\mathscr{P}$-algebra, it is equivalent to the second commutative diagram satisfied by an A-module M. So $A \ltimes M$ is a $\mathscr{P}$-algebra, which induces an abelian extension of A by M. (Notice that we do not need the hypothesis $\mathscr{P}$ to be augmented here.)

($\Leftarrow$) In the other way round, let $0 \to M \rightarrowtail B \twoheadrightarrow A \to 0$ be an abelian extension. Under the isomorphism $B \cong A \oplus M$, the $\mathscr{P}$-algebra structure map on B is given by the composite

$$\mathscr{P} \circ (A; M) \to \mathscr{P}(B) \xrightarrow{\gamma_B} B \to M.$$

We leave it to the reader to check that it satisfies the axioms of an A-module. □

12.3.3 Free A-Module over $\mathscr{P}$

The forgetful functor

$$\mathscr{U} : \mathsf{Mod}_A^{\mathscr{P}} \to \mathsf{Mod}_{\mathbb{K}},$$

which forgets the A-module structure, admits a left adjoint

$$A \otimes^{\mathscr{P}} - : \mathsf{Mod}_{\mathbb{K}} \to \mathsf{Mod}_A^{\mathscr{P}}$$

as follows.

Let N be a $\mathbb{K}$-module. We consider the following coequalizer in the category of $\mathbb{K}$-modules

$$\mathscr{P} \circ \big(\mathscr{P}(A); N\big) \underset{\widetilde{\gamma_A}}{\overset{\tilde{\gamma}}{\rightrightarrows}} \mathscr{P} \circ (A; N) \longrightarrow A \otimes^{\mathscr{P}} N,$$

where

$$\tilde{\gamma} : \mathscr{P} \circ \big(\mathscr{P}(A); N\big) \to (\mathscr{P} \circ \mathscr{P}) \circ (A; N) \xrightarrow{\gamma \circ (\mathrm{Id}, \mathrm{Id})} \mathscr{P} \circ (A; N)$$

and

$$\widetilde{\gamma_A} : \mathscr{P} \circ \big(\mathscr{P}(A); N\big) \xrightarrow{\mathrm{Id} \circ (\gamma_A, \mathrm{Id})} \mathscr{P} \circ (A; N).$$

Notice that the notation $A \otimes^{\mathscr{P}} N$ stands for a quotient of $\mathscr{P}(A; N)$, which is not in general isomorphic to the tensor product $A \otimes N$. However, in the case of the operad *Com*, we will see that $A \otimes^{Com} N \cong (A \oplus \mathbb{K}) \otimes N$.

Lemma 12.3.3. *The composite*

$$\begin{aligned}\mathscr{P}\circ\big(A;\mathscr{P}\circ(A;N)\big)&\to(\mathscr{P}\circ\mathscr{P})\circ(A;N)\\&\xrightarrow{\gamma\circ(\mathrm{Id};\mathrm{Id})}\mathscr{P}\circ(A;N)\end{aligned}$$

passes to the quotient $\mathscr{P}\circ(A;N)\twoheadrightarrow A\otimes^{\mathscr{P}}N$ *and induces on* $A\otimes^{\mathscr{P}}N$ *an A-module structure.*

Proof. In this proof, we use the shorter notations $\mathscr{P}(A;N)$ instead of $\mathscr{P}\circ(A;N)$ and $\mathscr{P}\mathscr{P}$ instead of $\mathscr{P}\circ\mathscr{P}$.

The following diagram

$$\begin{array}{ccccc}
\mathscr{P}(A;\mathscr{P}(\mathscr{P}(A);N)) & \longrightarrow & (\mathscr{P}\mathscr{P})(\mathscr{P}(A);N) & \xrightarrow{\gamma(\mathrm{Id};\mathrm{Id})} & \mathscr{P}(\mathscr{P}(A);N)\\
\downarrow & & \downarrow & & \downarrow\\
\mathscr{P}(A;(\mathscr{P}\mathscr{P})(A;N)) & & ((\mathscr{P}\mathscr{P})\mathscr{P})(A;N) & \searrow^{(\gamma\,\mathrm{Id})(\mathrm{Id}:\mathrm{Id})} & \downarrow\\
\downarrow{\scriptstyle \mathrm{Id}(\mathrm{Id};\gamma(\mathrm{Id};\mathrm{Id}))} & & \downarrow\cong & & \\
 & & (\mathscr{P}(\mathscr{P}\mathscr{P}))(A;N) & & (\mathscr{P}\mathscr{P})(A;N)\\
 & & \downarrow{\scriptstyle (\mathrm{Id}\,\gamma)(\mathrm{Id}:\mathrm{Id})} & & \downarrow{\scriptstyle \gamma(\mathrm{Id};\mathrm{Id})}\\
\mathscr{P}(A;\mathscr{P}(A;N)) & \longrightarrow & (\mathscr{P}\mathscr{P})(A;N) & \xrightarrow{\gamma(\mathrm{Id};\mathrm{Id})} & \mathscr{P}(A;N)
\end{array}$$

commutes by the associativity of the composition map γ of the operad $\mathscr{P}$. The following diagram

$$\begin{array}{ccccc}
\mathscr{P}(A;\mathscr{P}(\mathscr{P}(A);N)) & \longrightarrow & (\mathscr{P}\mathscr{P})(\mathscr{P}(A);N) & \xrightarrow{\gamma(\mathrm{Id};\mathrm{Id})} & \mathscr{P}(\mathscr{P}(A);N)\\
\downarrow{\scriptstyle \mathrm{Id}(\mathrm{Id};\mathrm{Id}(\gamma_A;\mathrm{Id}))} & & \downarrow & & \downarrow{\scriptstyle \mathrm{Id}(\gamma_A;\mathrm{Id})}\\
\mathscr{P}(A;\mathscr{P}(A;N)) & \longrightarrow & (\mathscr{P}\mathscr{P})(A;N) & \xrightarrow{\gamma(\mathrm{Id};\mathrm{Id})} & \mathscr{P}(A;N)
\end{array}$$

commutes by functoriality of the composite product $\circ$. It shows that the composite of the lemma passes to the quotient $A\otimes^{\mathscr{P}}N$ and induces a morphism $\bar{\gamma}:\mathscr{P}\circ(A;A\otimes^{\mathscr{P}}N)\to A\otimes^{\mathscr{P}}N$.

Let us now show that $\bar{\gamma}$ defines an A-module structure on $A\otimes^{\mathscr{P}}N$. The top-right composite in the commutative diagram defining an A-module lifts into

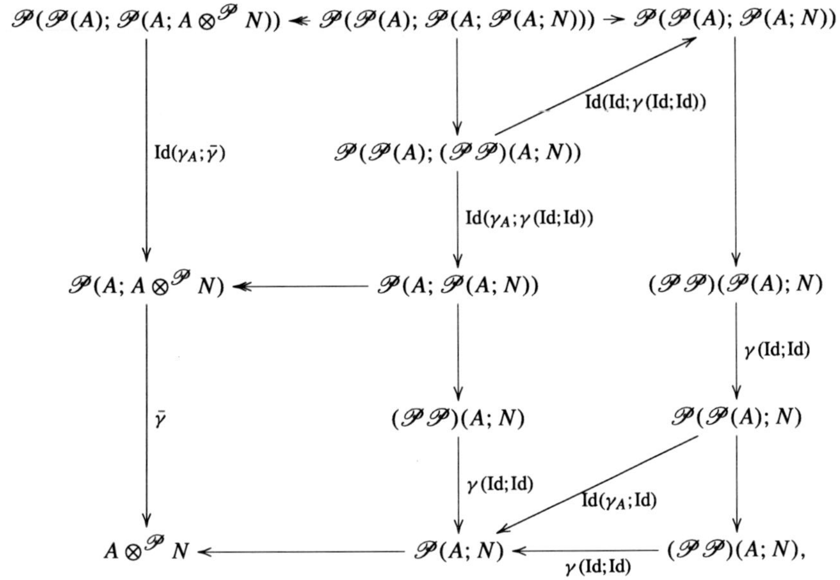

where the bottom-right triangle is not commutative itself, but only after the composite with $\mathscr{P}(A; N) \twoheadrightarrow A \otimes^{\mathscr{P}} N$. The bottom-left composite in the commutative diagram defining an A-module lifts into

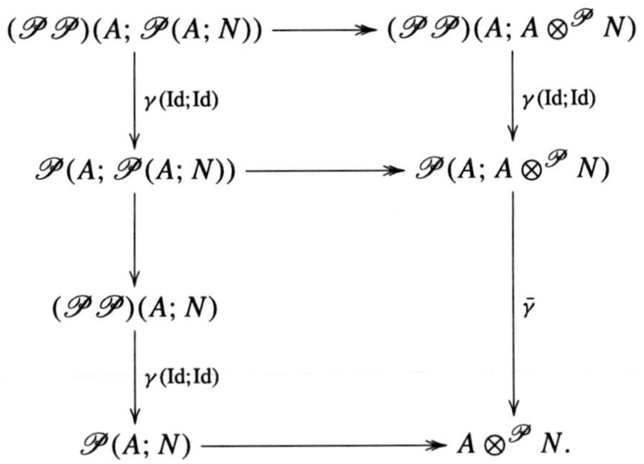

The top-right composite of the former diagram is equal to the bottom-left composite of the latter diagram by associativity of the composition map γ of the operad $\mathscr{P}$.

We define the second structure map $\bar{\eta} : A \otimes^{\mathscr{P}} N \to \mathscr{P}(A; A \otimes^{\mathscr{P}} N)$ of the A-module structure on $A \otimes^{\mathscr{P}} N$ by

$$\begin{array}{ccccc}
\mathscr{P}(A;N) & \xrightarrow{\cong} & \mathscr{P}(A;\mathrm{I}(A;N)) & \xrightarrow{\mathrm{Id}(\mathrm{Id};\eta(\mathrm{Id};\mathrm{Id}))} & \mathscr{P}(A;\mathscr{P}(A;N)) \\
\downarrow & & & & \downarrow \\
A\otimes^{\mathscr{P}} N & & \xrightarrow{\quad\bar{\eta}\quad} & & \mathscr{P}(A;A\otimes^{\mathscr{P}} N).
\end{array}$$

The last verifications are left to the reader. □

The following proposition shows that for any $\mathbb{K}$-module N, the data $(A\otimes^{\mathscr{P}} N,\bar{\gamma},\bar{\iota})$ is the *free A-module on N*.

Theorem 12.3.4. *For any operad $\mathscr{P}$ and any $\mathscr{P}$-algebra A, the two functors*

$$A\otimes^{\mathscr{P}} - : \mathsf{Mod}_{\mathbb{K}} \rightleftharpoons \mathsf{Mod}_A^{\mathscr{P}} : \mathscr{U}$$

form an adjoint pair, i.e. there are natural linear isomorphisms

$$\mathrm{Hom}_{\mathsf{Mod}_A^{\mathscr{P}}}\big(A\otimes^{\mathscr{P}} N, M\big)\cong \mathrm{Hom}_{\mathsf{Mod}_{\mathbb{K}}}\big(N,\mathscr{U}(M)\big).$$

for any A-module M and any $\mathbb{K}$-module N.

Proof. To any morphism $f : A\otimes^{\mathscr{P}} N\to M$ of A-modules, we associate its "restriction" defined by the following composite

$$\bar{f} : N\cong \mathrm{I}\circ(A;N)\xrightarrow{\iota\circ(\mathrm{Id};\mathrm{Id})}\mathscr{P}\circ(A;N)\twoheadrightarrow A\otimes^{\mathscr{P}} N\xrightarrow{f} M.$$

In the other way round, given any morphism of $\mathbb{K}$-modules $g : N\to\mathscr{U}(M)$, we extend it to a morphism of $\mathbb{K}$-modules $\tilde{g} : A\otimes^{\mathscr{P}} N\to M$ as follows. Let $\overline{\mu(a_1,\dots,n,\dots,a_k)}$ be an element of $A\otimes^{\mathscr{P}} N$ represented by $\mu(a_1,\dots,n,\dots,a_k)\in\mathscr{P}\circ(A;N)$. Its image under $\tilde{g}$ is equal to

$$\tilde{g}\big(\overline{\mu(a_1,\dots,n,\dots,a_k)}\big) := \gamma_M\big(\mu(a_1,\dots,g(n),\dots,a_k)\big).$$

We leave it to the reader to verify that $\tilde{g}$ is actually a morphism of A-modules and that the two aforementioned maps are natural linear isomorphisms. □

Notice that

$$\begin{aligned}
\mathscr{P}\circ(A;N) &:= \bigoplus_{k\geq 0}\mathscr{P}(k)\otimes_{\mathbb{S}_k}\Big(\bigoplus_{1\leq i\leq k} A\otimes\cdots\otimes A\otimes \underbrace{N}_{i\text{th position}}\otimes A\otimes\cdots\otimes A\Big)\\
&\cong \bigoplus_{k\geq 0}\mathscr{P}(k+1)\otimes_{\mathbb{S}_k}\big(A^{\otimes k}\otimes N\big).
\end{aligned}$$

We denote its elements simply by $\mu(a_1,\dots,a_k;n)$.

Proposition 12.3.5. *Let $\mathscr{P}$ be an operad and let (A, γ_A) be a $\mathscr{P}$-algebra. The free A-module $A \otimes^{\mathscr{P}} N$ over N is given by the quotient of $\mathscr{P} \circ (A; N)$ under the relation*

$$\mu\big(a_1, \ldots, a_{i-1}, \gamma_A\big(\nu(a_i, \ldots, a_{i+l-1})\big), a_{i+l}, \ldots, a_k; n\big) \sim (\mu \circ_i \nu)(a_1, \ldots, a_k; n),$$

for any $k \geq 1$, $i + l \leq k + 1$, $\mu \in \mathscr{P}(k+1)$, $\nu \in \mathscr{P}(l)$, $a_1, \ldots, a_k \in A$ and $n \in N$.

Proof. It is a direct consequence of the partial definition 5.3.4 of an operad. □

EXAMPLES. To treat the examples *As*, *Com* and *Lie* below, one can either use Proposition 12.3.1, and conclude by the well-known respective notions of free module, or use Proposition 12.3.5 to compute the coequalizer $A \otimes^{\mathscr{P}} N$, each time.

◇ In the case of the (nonsymmetric) operad $\mathscr{P} = As$ and of an associative (non-necessarily unital) algebra A, the underlying $\mathbb{K}$-module of the free A-module over *As* is given by

$$A \otimes^{As} N \cong N \oplus A \otimes N \oplus N \otimes A \oplus A \otimes N \otimes A \cong (A \oplus \mathbb{K}) \otimes N \otimes (A \oplus \mathbb{K}).$$

To prove it, it is enough to notice that the composition of the following elements gives

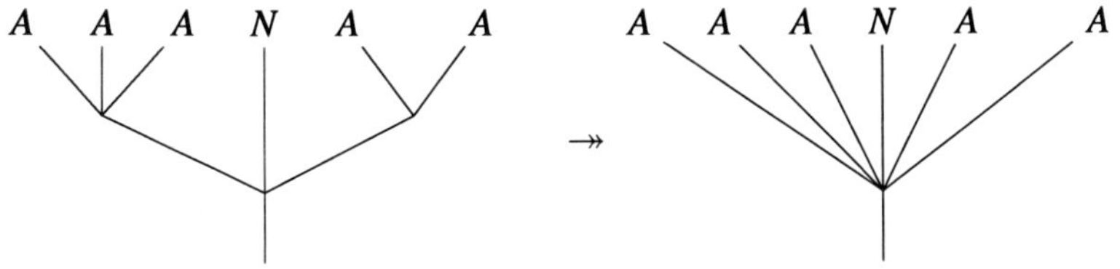

in the operad *As*. Therefore $\overline{a_1 \otimes a_2 \otimes a_3 \otimes \nu \otimes a_4 \otimes a_5}$ is equal to $\overline{a_1.a_2.a_3 \otimes \nu \otimes a_4.a_5}$ in $A \otimes^{As} N$.

◇ In the case of the operad $\mathscr{P} = Com$ and of a commutative algebra A, the underlying $\mathbb{K}$-module of the free A-module over *Com* is given by

$$A \otimes^{Com} N \cong N \oplus A \otimes N \cong (A \oplus \mathbb{K}) \otimes N.$$

◇ In the case of the operad $\mathscr{P} = Lie$ and of a Lie algebra $\mathfrak{g}$, the underlying $\mathbb{K}$-module of the free $\mathfrak{g}$-module over *Lie* is given by

$$\mathfrak{g} \otimes^{Lie} N \cong U(\mathfrak{g}) \otimes N,$$

where $U(\mathfrak{g})$ is the universal enveloping algebra of the Lie algebra $\mathfrak{g}$.

This last example will be generalized in the next section.

Proposition B.4.1, applied to the above adjunction, endow the composite $\mathscr{U}(A \otimes^{\mathscr{P}} -)$ with a monad structure on $\mathsf{Mod}_{\mathbb{K}}$.

Proposition 12.3.6. *The category of A-modules over $\mathcal{P}$ is isomorphic to the category of algebras over the monad $\mathcal{U}(A \otimes^{\mathcal{P}} -)$.*

Proof. This result follows from Beck's Theorem, see [ML98, Chap. VI, Sect. 7] and the study of reflexive coequalizers given in Goerss–Hopkins [GH00]. □

12.3.4 Enveloping Algebra of a $\mathcal{P}$-Algebra

We consider the free A-module $A \otimes^{\mathcal{P}} \mathbb{K}$ over the ground ring $\mathbb{K}$. It can be endowed with the following binary operation

$$\mu : (A \otimes^{\mathcal{P}} \mathbb{K}) \otimes (A \otimes^{\mathcal{P}} \mathbb{K}) \cong (A \otimes^{\mathcal{P}} (A \otimes^{\mathcal{P}} \mathbb{K})) \to A \otimes^{\mathcal{P}} \mathbb{K},$$

where the first map comes from the general isomorphism $(A \otimes^{\mathcal{P}} \mathbb{K}) \otimes N \cong A \otimes^{\mathcal{P}} N$ and where the second map is given by the composition product of the monad $\mathcal{U}(A \otimes^{\mathcal{P}} -)$. We denote by

$$u := \bar{\eta} : \mathbb{K} \to A \otimes^{\mathcal{P}} \mathbb{K}$$

the second structure map of the free A-module over $\mathbb{K}$.

The triple

$$U_{\mathcal{P}}(A) := \left(A \otimes^{\mathcal{P}} \mathbb{K}, \mu, u\right)$$

is called *enveloping algebra of the $\mathcal{P}$-algebra A* .

Lemma 12.3.7. *The enveloping algebra $U_{\mathcal{P}}(A)$ of a $\mathcal{P}$-algebra A is a unital associative algebra.*

Proof. The associativity of the product μ is a direct consequence of the associativity of the monad $\mathcal{U}(A \otimes^{\mathcal{P}} -)$. □

Proposition 12.3.5 shows that the elements of the form $\nu(a_1, \ldots, a_k; 1)$ provide representatives for the enveloping algebra $U_{\mathcal{P}}(A)$ under the relation $\sim$. Under this presentation, the product μ of the enveloping algebra is equal to

$$\mu\big(\nu(a_1, \ldots, a_k; 1), \xi(a_{k+1}, \ldots, a_{k+l}; 1)\big) = (\nu \circ_{k+1} \xi)(a_1, \ldots, a_{k+l}; 1).$$

The relation (I) of Sect. 5.3.4 satisfied by the partial products $\circ_i$ of an operad gives another proof of the associativity of μ.

EXAMPLES.

◇ The enveloping algebra $U_{As}(A)$ of an associative (non-necessarily unital) algebra A is equal to the classical enveloping algebra of the augmentation $A_+ := A \oplus \mathbb{K}$ of A:

$$U_{As}(A) \cong A_+ \otimes A_+^{\mathrm{op}},$$

where A_+^{op} is the opposite algebra associated to A_+.

◇ The enveloping algebra $U_{Com}(A)$ of a commutative algebra A is equal to the augmentation of A:

$$U_{Com}(A) \cong A_+ := A \oplus \mathbb{K}.$$

◇ The enveloping algebra $U_{Lie}(\mathfrak{g})$ of a Lie algebra $\mathfrak{g}$ is equal to the classical universal enveloping algebra:

$$U_{Lie}(\mathfrak{g}) \cong U(\mathfrak{g}).$$

Proposition 12.3.8. *The category of A-modules over $\mathscr{P}$ is isomorphic to the category of left modules over the enveloping algebra $U_{\mathscr{P}}(A)$.*

Proof. Using the isomorphism of $\mathbb{K}$-modules $(A \otimes^{\mathscr{P}} \mathbb{K}) \otimes M \cong \mathscr{U}(A \otimes^{\mathscr{P}} M)$, we prove that the category of left modules over the enveloping algebra $U_{\mathscr{P}}(A)$ is isomorphic to the category of algebras over the monad $\mathscr{U}(A \otimes^{\mathscr{P}} -)$. We conclude with Proposition 12.3.6. □

The aforementioned isomorphism $A \otimes^{\mathscr{P}} N \cong U_{\mathscr{P}}(A) \otimes N$ gives two ways of making the free A-module on N explicit.

Proposition 12.3.9. *For any operad $\mathscr{P}$, the enveloping algebra construction provides a functor*

$$U_{\mathscr{P}} : \mathscr{P}\textsf{-alg} \to uAs\textsf{-alg}.$$

Proof. Let $f : B \to A$ be a morphism of $\mathscr{P}$-algebras. The map $U_{\mathscr{P}}(f) : U_{\mathscr{P}}(B) \to U_{\mathscr{P}}(A)$ induced by $\mathscr{P} \circ (B; \mathbb{K}) \xrightarrow{\mathrm{Id} \circ (f;\mathrm{Id})} \mathscr{P} \circ (A; \mathbb{K})$ is a morphism of unital associative algebras. □

12.3.5 Relative Free Module

Any morphism $f : B \to A$ of $\mathscr{P}$-algebras induces a functor

$$f^* : \textsf{Mod}_A^{\mathscr{P}} \to \textsf{Mod}_B^{\mathscr{P}},$$

called *restriction of scalars*. We show that it admits a left adjoint

$$f_! : \textsf{Mod}_B^{\mathscr{P}} \to \textsf{Mod}_A^{\mathscr{P}},$$

as follows.

By Proposition 12.3.9, the map $U_{\mathcal{P}}(f) : U_{\mathcal{P}}(B) \to U_{\mathcal{P}}(A)$ is a morphism of unital associative algebras. It endows $U_{\mathcal{P}}(A)$ with a right $U_{\mathcal{P}}(B)$-module structure. For any B-module M, we define its *free relative A-module* $f_!(M) := A \otimes_B^{\mathcal{P}} M$ by the classical relative tensor product

$$A \otimes_B^{\mathcal{P}} M := U_{\mathcal{P}}(A) \otimes_{U_{\mathcal{P}}(B)} M,$$

which is defined by the following coequalizer

$$U_{\mathcal{P}}(A) \otimes U_{\mathcal{P}}(B) \otimes M \rightrightarrows U_{\mathcal{P}}(A) \otimes M \twoheadrightarrow U_{\mathcal{P}}(A) \otimes_{U_{\mathcal{P}}(B)} M.$$

The notation $f_!$ is classical and should not be confused with any Koszul dual construction.

Proposition 12.3.10. *For any operad $\mathcal{P}$ and any morphism $f : B \to A$ of $\mathcal{P}$-algebras, the two functors*

$$f_! : \mathsf{Mod}_B^{\mathcal{P}} \rightleftarrows \mathsf{Mod}_A^{\mathcal{P}} : f^*$$

form an adjoint pair, i.e. there are natural linear isomorphisms

$$\mathrm{Hom}_{\mathsf{Mod}_A^{\mathcal{P}}}\big(f_!(N), M\big) \cong \mathrm{Hom}_{\mathsf{Mod}_B^{\mathcal{P}}}\big(N, f^*(M)\big),$$

for any A-module M and any B-module N.

Proof. It is a direct corollary of Proposition 12.3.8. □

EXAMPLES. Using the classical relative tensor product $\otimes_B$, we recover the classical cases as follows.

◇ In the case of the (nonsymmetric) operad $\mathcal{P} = As$, the underlying $\mathbb{K}$-module of the free A-module over the algebra B is given by

$$A \otimes_B^{As} N \cong (A \oplus \mathbb{K}) \otimes_B N \otimes_B (A \oplus \mathbb{K}),$$

where the action of B on $\mathbb{K}$ is null.

◇ In the case of the operad $\mathcal{P} = Com$, the underlying $\mathbb{K}$-module of the free A-module over the algebra B is given by

$$A \otimes_B^{As} N \cong (A \oplus \mathbb{K}) \otimes_B N,$$

where the action of B on $\mathbb{K}$ is null.

◇ In the case of the operad $\mathcal{P} = Lie$, the underlying $\mathbb{K}$-module of the free $\mathfrak{g}$-module over $\mathfrak{g}'$ is given by

$$\mathfrak{g} \otimes_{\mathfrak{g}'}^{Lie} N \cong U(\mathfrak{g}) \otimes_{\mathfrak{g}'} N,$$

where $U(\mathfrak{g})$ is the universal enveloping algebra of the Lie algebra $\mathfrak{g}$.

12.3.6 The Category of $\mathscr{P}$-Algebras over a Fixed $\mathscr{P}$-Algebra

Let A be a fixed $\mathscr{P}$-algebra. We consider the category $\mathscr{P}\text{-}\mathsf{alg}/A$ of $\mathscr{P}$-algebras over A. Its objects are the morphisms of $\mathscr{P}$-algebras $f : B \to A$ and its morphisms are the commutative triangles

$$\begin{array}{ccc} B & \longrightarrow & B' \\ & {\scriptstyle f}\searrow \quad \swarrow{\scriptstyle f'} & \\ & A & \end{array}$$

of $\mathscr{P}$-algebras.

For instance, when M is an A-module, the abelian extension $A \ltimes M \to A$ is a $\mathscr{P}$-algebra over A.

12.3.7 Derivations

Let $f : B \to A$ be a $\mathscr{P}$-algebra over A and let M be an A-module. An A-*derivation* from B to M is a morphism of $\mathbb{K}$-modules $d : B \to M$ satisfying the following commutative diagram

$$\begin{array}{ccccc} \mathscr{P}(B) \cong \mathscr{P} \circ B & \xrightarrow{\mathrm{Id} \circ' d} & \mathscr{P} \circ (B; M) & \xrightarrow{\mathrm{Id} \circ (f; \mathrm{Id})} & \mathscr{P} \circ (A; M) \\ \downarrow{\scriptstyle \gamma_B} & & & & \downarrow{\scriptstyle \gamma_M} \\ B & & \xrightarrow{\qquad d \qquad} & & M. \end{array}$$

Recall that the infinitesimal composite of morphisms $\mathrm{Id} \circ' d$ was defined in Sect. 6.1.3. Notice that the linear combination of A-derivations is again an A-derivation. The module of A-derivations from B to M is denoted by $\mathrm{Der}_A(B, M)$. It induces a bifunctor from the category $(\mathscr{P}\text{-}\mathsf{alg}/A)^{\mathrm{op}} \times \mathsf{Mod}_A^{\mathscr{P}}$, which is representable on the right-hand side by the functor of abelian extensions.

Proposition 12.3.11. *For any operad $\mathscr{P}$ and any $\mathscr{P}$-algebra A, there are natural linear isomorphisms*

$$\mathrm{Der}_A(B, M) \cong \mathrm{Hom}_{\mathscr{P}\text{-}\mathsf{alg}/A}(B, A \ltimes M),$$

for any A-module M and any $\mathscr{P}$-algebra $B \to A$ over A.

Proof. Let us denote $f : B \to A$ the morphism of $\mathscr{P}$-algebras. Any morphisms in $\mathscr{P}\text{-}\mathsf{alg}/A$ from B to $A \ltimes M$ is of the form

$$\begin{array}{ccc} B & \xrightarrow{f+d} & A \ltimes M = A \oplus M \\ & {\scriptstyle f}\searrow \quad \swarrow & \\ & A. & \end{array}$$

The map $f + d$ is a morphism of $\mathcal{P}$-algebras if and only if the map d is an A-derivation. □

EXAMPLES. In the examples of the operads *As*, *Com* and *Lie*, we recover the classical notions of derivations.

12.3.8 Module of Kähler Differential Forms

To represent the module of A-derivations on the left-hand side, we introduce the module of Kähler differential forms as follows.

To any $\mathcal{P}$-algebra A, we associate the coequalizer

$$A \otimes^{\mathcal{P}} \mathcal{P}(A) \underset{\widetilde{\gamma_A}}{\overset{\widetilde{\gamma_{(1)}}}{\rightrightarrows}} A \otimes^{\mathcal{P}} A \twoheadrightarrow \Omega_{\mathcal{P}} A,$$

where the map $\widetilde{\gamma_{(1)}}$ comes from

$$\mathcal{P} \circ \big(A; \mathcal{P}(A)\big) \to \mathcal{P} \circ \big(A; \mathcal{P} \circ (A; A)\big) \to (\mathcal{P} \circ_{(1)} \mathcal{P}) \circ (A; A) \xrightarrow{\gamma_{(1)} \circ (\mathrm{Id}; \mathrm{Id})} \mathcal{P} \circ (A; A)$$

and where the map $\widetilde{\gamma_A}$ comes from

$$\mathcal{P} \circ \big(A; \mathcal{P}(A)\big) \xrightarrow{\mathrm{Id} \circ (\mathrm{Id}; \gamma_A)} \mathcal{P} \circ (A; A).$$

The A-module structure on $A \otimes^{\mathcal{P}} A$ passes to the quotient $\Omega_{\mathcal{P}} A$. We call this A-module the *module of Kähler differential forms*.

Lemma 12.3.12. *The module of Kähler differential forms $\Omega_{\mathcal{P}} A$ of a $\mathcal{P}$-algebra A is given by the quotient of $\mathcal{P}(A; A)$, whose elements are denoted by $\mu(b_1, \ldots, db_i, \ldots, b_k)$, under the relations*

$$\mu\big(a_1, \ldots, a_{i-1}, \gamma_A\big(\nu(a_i, \ldots, a_{i+l-1})\big), a_{i+l}, \ldots, da_j, \ldots, a_k\big) \approx (\mu \circ_i \nu)(a_1, \ldots, da_j, \ldots, a_k),$$

and

$$\mu\big(a_1, \ldots, a_{i-1}, d\gamma_A\big(\nu(a_i, \ldots, a_{i+l-1})\big), a_{i+l}, \ldots, a_k\big) \approx \sum_{j=0}^{l-1} (\mu \circ_i \nu)(a_1, \ldots, da_{i+j}, \ldots, a_k).$$

Proof. The first relation stands for the free module relation $\sim$ introduced in Proposition 12.3.5. The second relation is equivalent to the above coequalizer. □

When $A = \mathscr{P}(V)$ is a free $\mathscr{P}$-algebra, its module of Kähler differential forms is equal to

$$\Omega_{\mathscr{P}}\mathscr{P}(V) \cong \mathscr{P}(V, V).$$

EXAMPLES.

◇ In the case of the operad $\mathscr{P} = Com$, the operadic module of Kähler differential forms agrees with the classical module of Kähler differential forms

$$\Omega_{Com}A \cong \Omega^1_A,$$

where Ω^1_A is the free A_+-module on A modulo the derivation relation

$$a \otimes d(b.c) = (a.b) \otimes dc + (a.c) \otimes db,$$

see for instance [Lod98, Chap. 1] for more details.

◇ In the case of the (nonsymmetric) operad $\mathscr{P} = As$, the module of Kähler differential forms of an associative algebra agrees with the notion introduced in noncommutative geometry [Con85, Kar87]. In this case, it is equal to

$$\Omega_{As}A \cong A_+ \otimes A \otimes A_+ / \equiv,$$

that is the free A_+-bimodule over A, modulo the relation

$$a \otimes d(bb') \otimes c \equiv (ab) \otimes db' \otimes c + a \otimes db \otimes (b'c).$$

◇ In the case of the operad $\mathscr{P} = Lie$, the module of Kähler differential forms of a Lie algebra $\mathfrak{g}$ is isomorphic to the augmentation ideal of its enveloping algebra

$$\Omega_{Lie}(\mathfrak{g}) \cong \overline{U}(\mathfrak{g}),$$

with its left $U(\mathfrak{g})$-module structure, see [Fre09a, Sect. 4.4].

Proposition 12.3.13. *For any operad $\mathscr{P}$ and any $\mathscr{P}$-algebra A, there are natural linear isomorphisms*

$$\mathrm{Hom}_{\mathsf{Mod}_B^{\mathscr{P}}}\big(\Omega_{\mathscr{P}}B, f^*(M)\big) \cong \mathrm{Der}_A(B, M),$$

for any A-module M and any $\mathscr{P}$-algebra $f : B \to A$ over A.

Proof. We use the description of $\Omega_{\mathscr{P}}B$ given in Lemma 12.3.12. To any map $d : B \to M$, we associate the map $D : \mathscr{P}(B; B) \to M$ defined by

$$\mu(b_1, \dots, db_i, \dots, b_k) \mapsto \gamma_M\big(\mu\big(f(b_1), \dots, d(b_i), \dots, f(b_k)\big)\big).$$

Since d is an A-derivation, it factors through the quotient under the relations $\approx$ to define a morphism of B-modules $\widetilde{D} : \Omega_{\mathscr{P}}B \to f^*(M)$. In the other way round, the restriction of any map $\Omega_{\mathscr{P}}B \to f^*(M)$ to $\mathrm{I} \circ (db)$ gives a map $B \to M$. If the former is a morphism of B-modules, then the latter is an A-derivation. □

12.3.9 Quillen Adjunction

The two aforementioned propositions do not induce directly a pair of adjoint functors because the underlying category of the left-hand side "representation" depends on the $\mathcal{P}$-algebra B. The final adjunction, which we are interested in, comes after considering the adjunction of Proposition 12.3.10 between the categories of B-modules and A-modules.

Theorem 12.3.14. *Let $\mathcal{P}$ be an operad and let A be a $\mathcal{P}$-algebra. The following two functors*

$$A \otimes^{\mathcal{P}}_{-} \Omega_{\mathcal{P}} - : \mathcal{P}\text{-}\mathsf{alg}/A \rightleftharpoons \mathsf{Mod}^{\mathcal{P}}_{A} : A \ltimes -$$

form an adjoint pair. They represent the bifunctor of A-derivations, i.e. there exist natural linear isomorphisms

$$\mathrm{Hom}_{\mathsf{Mod}^{\mathcal{P}}_{A}}\big(A \otimes^{\mathcal{P}}_{B} \Omega_{\mathcal{P}} B, M\big) \cong \mathrm{Der}_A(B, M) \cong \mathrm{Hom}_{\mathcal{P}\text{-}\mathsf{alg}/A}(B, A \ltimes M),$$

for any A-module M and any $\mathcal{P}$-algebra $B \to A$ over A.

Proof. The proof follows directly from Propositions 12.3.11, 12.3.13 and 12.3.10. □

All the constructions of this section can be generalized in a straightforward way from the category of $\mathbb{K}$-modules to the category of dg modules. All the propositions, like the adjunctions, generalize to the various categories of dg modules as follows. Either one considers morphisms preserving the grading and then the linear isomorphisms of the various results extend to isomorphisms of dg modules. Or one can consider morphisms of dg modules and then the linear isomorphisms extend only to grading preserving linear isomorphisms.

Let us assume, from now on, that the ground ring $\mathbb{K}$ is a field of characteristic 0. One can extend Proposition B.6.5 and show that the category of dg $\mathcal{P}$-algebras over a fixed dg $\mathcal{P}$-algebra A, still denoted $\mathcal{P}\text{-}\mathsf{alg}/A$, can be endowed with a model category structure such that the weak equivalences (resp. fibrations) are the quasi-isomorphisms (resp. degree-wise epimorphisms).

In the same way, the category of dg A-modules over $\mathcal{P}$, still denoted $\mathsf{Mod}^{\mathcal{P}}_{A}$, is endowed with a model category structure such that the weak equivalences (resp. fibrations) are the quasi-isomorphisms (resp. degree-wise epimorphisms).

For these two results, we refer the reader to Hinich [Hin97]. For more refined results, for instance with weaken assumptions on the ground ring, we refer the reader to the monograph [Fre09a].

Proposition 12.3.15. *Let $\mathcal{P}$ be a dg operad and let A be a dg $\mathcal{P}$-algebra. The adjoint pair of functors*

$$A \otimes^{\mathcal{P}}_{-} \Omega_{\mathcal{P}} - : \mathcal{P}\text{-}\mathsf{alg}/A \rightleftharpoons \mathsf{Mod}^{\mathcal{P}}_{A} : A \ltimes -$$

is a Quillen adjunction.

Proof. For the notion of Quillen adjunction, see Appendix B.7.3. It is enough to show that the right adjoint functor $A \ltimes -$ preserves the fibrations and the acyclic fibrations. Since $A \ltimes M = A \oplus M$, the proof is straightforward from the definition of the two model category structures considered here. □

Therefore, we can consider the associated total derived functors

$$\mathbb{L}\big(A \otimes_{-}^{\mathscr{P}} \Omega_{\mathscr{P}} -\big) \,:\, \mathsf{Ho}(\mathscr{P}\text{-}\mathsf{alg}/A) \rightleftharpoons \mathsf{Ho}\big(\mathsf{Mod}_A^{\mathscr{P}}\big) \,:\, \mathbb{R}(A \ltimes -)\,,$$

which form an adjoint pair between the homotopy categories.

12.3.10 Cotangent Complex

The *cotangent complex of* A is the image under the total derived functor $\mathbb{L}(A \otimes_{-}^{\mathscr{P}} \Omega_{\mathscr{P}} -)$ of the dg $\mathscr{P}$-algebra $A \to A$ over A in the homotopy category of dg A-modules.

To make explicit one representative, we consider a cofibrant replacement $R \xrightarrow{\sim} A$ in the category of dg $\mathscr{P}$-algebras. It provides a cofibrant replacement of A in the category of dg $\mathscr{P}$-algebras over A. The class of

$$\mathbb{L}_{R/A} := A \otimes_R^{\mathscr{P}} \Omega_{\mathscr{P}} R$$

in $\mathsf{Ho}(\mathsf{Mod}_A^{\mathscr{P}})$ is a representative of the cotangent complex.

12.3.11 André–Quillen (Co)Homology

The *André–Quillen cohomology of a dg $\mathscr{P}$-algebra A with coefficients in a dg A-module M* is defined by

$$H_{AQ}^{\bullet}(A, M) := H_{\bullet}\big(\mathrm{Hom}_{\mathsf{Ho}(\mathsf{Mod}_A^{\mathscr{P}})}(\mathbb{L}_{R/A}, M)\big).$$

Proposition 12.3.15 shows that these homology groups are independent of the cofibrant resolution $R \xrightarrow{\sim} A$ of the $\mathscr{P}$-algebra A.

Theorem 12.3.14 and Proposition 12.3.15 show that the André–Quillen cohomology is actually equal to the homology of the "derived" functor of A-derivations:

$$H_{AQ}^{\bullet}(A, M) \cong H_{\bullet}\big(\mathrm{Der}_A(R, M)\big).$$

Dually, the *André–Quillen homology of a dg $\mathscr{P}$-algebra A with coefficients in a dg right $U_{\mathscr{P}}(A)$-module M* is defined by

$$H_{\bullet}^{AQ}(A, M) := H_{\bullet}(M \otimes_{U_{\mathscr{P}}(A)} \mathbb{L}_{R/A}).$$

The André–Quillen homology can be seen as the homology of the derived functor of the functor

$$\big(B \xrightarrow{f} A\big) \mapsto f^*(M) \otimes_{U_{\mathcal{P}}(B)} \Omega_{\mathcal{P}} B,$$

see [Fre09a, Chap. 13] for more details.

Notice that the coefficients for the André–Quillen cohomology live in the category of *left* $U_{\mathcal{P}}(A)$-modules, whereas the coefficients for the André–Quillen homology live in the category of *right* $U_{\mathcal{P}}(A)$-modules. In the case the three operads *As*, *Com* and *Lie*, the two categories of left and right $U_{\mathcal{P}}(A)$-modules are equivalent. The first example of an operad for which it is not the case, was given in [LP93] by the operad *Leib* encoding Leibniz algebras, see Exercise 12.6.10.

12.4 Operadic Cohomology of Algebras with Coefficients

To an algebra A over a quadratic operad $\mathcal{P}$ and to an A-module M, one can associate the operadic cochain complex

$$C^\bullet_{\mathcal{P}}(A, M) := \big(\mathrm{Hom}\big(\mathcal{P}^{¡}(A), M\big), \partial_{\pi_\kappa}\big).$$

When the underlying operad is Koszul, this cochain complex computes the André–Quillen cohomology groups. It shows, for instance, that the first cohomology group $H^1_{\mathcal{P}}(A, M)$ is in one-to-one correspondence with the coset of abelian extension A by M. All these definitions apply as well to $\mathcal{P}_\infty$-algebras to define the André–Quillen (co)homology of homotopy $\mathcal{P}$-algebras.

12.4.1 Operadic Cochain Complex of a $\mathcal{P}$-Algebra with Coefficients

Let $\mathcal{P} = \mathcal{P}(E, R)$ be a quadratic operad, let A be a $\mathcal{P}$-algebra, and let M be an A-module. Suppose first that $\mathcal{P}$ is homogeneous quadratic and that A and M are concentrated in degree 0.

We define the *operadic cochain complex* by

$$C^\bullet_{\mathcal{P}}(A, M) := \big(\mathrm{Hom}\big(\mathcal{P}^{¡}(A), M\big), \partial_{\pi_\kappa}\big),$$

where the differential map ∂_{π_κ} is given by

$$\partial_{\pi_\kappa}(g) := \bar{\partial}(g) - (-1)^{|g|} g d.$$

The notation d stands for the differential of the bar construction $\mathrm{B}_\kappa A := (\mathcal{P}^{¡}(A), d)$ of A, see Sect. 11.2.2 and the map $\bar{\partial}(g)$ is equal to the following composite

$$\mathcal{P}^{¡}(A) \xrightarrow{\Delta} \mathcal{P}^{¡} \circ \mathcal{P}^{¡}(A) \xrightarrow{\kappa \circ (\pi_\kappa; g)} \mathcal{P} \circ (A; M) \xrightarrow{\gamma_M} M.$$

The map $\pi_\kappa : \mathscr{P}^{\text{i}}(A) \twoheadrightarrow \mathrm{I} \circ A \cong A$ is the universal twisting morphism, see Sect. 11.3.2.

Considering the cohomological degree, which is the opposite of the homological degree, we get the following cochain complex

$$\begin{array}{ccccc}\mathrm{Hom}(A, M) & \longrightarrow & \mathrm{Hom}_{\mathbb{S}}\big(\mathscr{P}^{\text{i}\,(1)}(A), M\big) & \longrightarrow & \mathrm{Hom}_{\mathbb{S}}\big(\mathscr{P}^{\text{i}\,(2)}(A), M\big) \cdots . \\ 0 & & 1 & & 2\end{array}$$

With the presentation $\mathscr{P}(E, R)$ of the operad $\mathscr{P}$, it gives explicitly

$$\mathrm{Hom}(A, M) \longrightarrow \mathrm{Hom}_{\mathbb{S}}\big(E(A), M\big) \longrightarrow \mathrm{Hom}_{\mathbb{S}}\big(R(A), M\big) \cdots .$$

It can be checked by hand that $(\partial_{\pi_\kappa})^2 = 0$, but it is also a direct consequence of the next section. The associated cohomology groups form the *operadic cohomology of A with coefficients into M.*

This definition extends in two directions. One can consider inhomogeneous quadratic operads $\mathscr{P}$, satisfying the conditions (ql_1) and (ql_2) of 7.8. Or one can consider dg $\mathscr{P}$-algebras (A, d_A) and dg A-module M. (In this last case, the degree is not given by the weight of the Koszul dual cooperad anymore.) In these cases, the differential map is the sum of several terms

$$\partial_{\pi_\kappa}(g) := \bar{\partial}(g) + d_M g - (-1)^{|g|} g(d_2 + d_{\mathscr{P}^{\text{i}}} \circ \mathrm{Id} + \mathrm{Id} \circ' d_A).$$

Proposition 12.4.1. *Let $\mathscr{P}$ be a quadratic operad and let A and M be respectively a $\mathscr{P}$-algebra and an A-module. The operadic cohomology of A with coefficients into M is a subspace of the tangent homology of the $\mathscr{P}$-algebra $A \ltimes M$.*

Proof. Under the inclusion

$$\mathrm{Hom}\big(\mathscr{P}^{\text{i}}(A), M\big) \cong \mathrm{Hom}_{\mathbb{S}}\big(\mathscr{P}^{\text{i}}, \mathrm{End}^A_M\big) \subset \mathrm{Hom}_{\mathbb{S}}\big(\mathscr{P}^{\text{i}}, \mathrm{End}_{A\oplus M}\big) = \mathfrak{g}_{A\ltimes M},$$

the chain complex

$$\big(\mathrm{Hom}\big(\mathscr{P}^{\text{i}}(A), M\big), \partial_{\pi_\kappa}\big) \subset \big(\mathrm{Hom}_{\mathbb{S}}\big(\mathscr{P}^{\text{i}}, \mathrm{End}_{A\oplus M}\big), \partial_\varphi\big)$$

is a chain subcomplex of the deformation complex. □

Since a $\mathscr{P}$-algebra A is an example of a module over itself, we can consider the operadic cohomology $H^\bullet_{\mathscr{P}}(A, A)$ of A with coefficients in A. These (co)homology groups are isomorphic to the tangent homology of the $\mathscr{P}$-algebra A, that is the homology of the deformation complex of A, see Sect. 12.2.2.

EXAMPLES. We recover the classical cases, up to a degree shift.

◇ In the case of the operad *As*, the operadic (co)homology is isomorphic to the Hochschild (co)homology of associative algebras, see Sect. 9.1.7.

◇ In the case of the operad *Com*, the operadic (co)homology is isomorphic to the Harrison (co)homology of commutative algebras, see Sect. 13.1.7 (recall that we work over a field $\mathbb{K}$ of characteristic 0).
◇ In the case of the operad *Lie*, the operadic (co)homology is isomorphic to the Chevalley–Eilenberg (co)homology of Lie algebras, see Sect. 13.2.8.

Proposition 12.4.2. *For any $\mathscr{P}$-algebra A and any A-module M, both concentrated in degree 0, there is a natural isomorphism*

$$H^0_{\mathscr{P}}(A, M) \cong \mathrm{Der}_A(A, M).$$

Proof. A map $d : A \to M$ satisfies $\partial_{\pi_\kappa}(d) = 0$ if and only if $d \in \mathrm{Der}_A(A, M)$. □

12.4.2 Operadic Cohomology vs André–Quillen Cohomology

The goal of this section is to prove the following theorem.

Theorem 12.4.3. *Working over nonnegatively graded chain complexes, if $\mathscr{P}$ is a Koszul operad, then the André–Quillen cohomology is isomorphic to the operadic cohomology*

$$H^\bullet_{AQ}(A, M) \cong H^\bullet_{\mathscr{P}}(A, M) \cong H_\bullet\big(\mathrm{Hom}(\mathscr{P}^{¡}(A), M), \partial_{\pi_\kappa}\big),$$

and a representative of the cotangent complex of any dg $\mathscr{P}$-algebra A is given by

$$\mathbb{L}_{\Omega_\kappa B_\kappa A/A} \cong \big(A \otimes^{\mathscr{P}} \mathscr{P}^{¡}(A), d_{\pi_\kappa}\big).$$

We suppose that the operad $\mathscr{P}$ is equipped with an operadic twisting morphism $\alpha : \mathscr{C} \to \mathscr{P}$. We suppose that the dg $\mathscr{P}$-algebra A admits a twisting morphism $\varphi : C \to A$, where C is a conilpotent dg $\mathscr{C}$-coalgebra, such that the induced morphism of dg $\mathscr{P}$-algebras

$$f_\varphi : \Omega_\alpha C := \mathscr{P}(C) \xrightarrow{\sim} A$$

is a quasi-isomorphism, see Sect. 11.3.2. Let us assume that the dg $\mathscr{P}$-algebra $\Omega_\alpha C$ is cofibrant. Since it is a quasi-free dg $\mathscr{P}$-algebra, it is cofibrant when working over nonnegatively graded chain complexes by Proposition B.6.6.

As in the last section, we denote by $\alpha \circ (\varphi; \mathrm{Id}) : \mathscr{C}(C) \to \mathscr{P} \circ (A; C)$ the composite

$$\mathscr{C}(C) \xrightarrow{\alpha \circ' \mathrm{Id}} \mathscr{P} \circ (C; C) \xrightarrow{\mathrm{Id} \circ (\varphi; \mathrm{Id})} \mathscr{P} \circ (A; C).$$

Lemma 12.4.4. *The above cofibrant resolution $\Omega_\alpha C$ of the dg $\mathscr{P}$-algebra A gives the following representative of the cotangent complex*

$$\mathbb{L}_{\Omega_\alpha C/A} \cong \big(A \otimes^{\mathscr{P}} C, d_\varphi\big),$$

where the differential map d_φ is the sum of $d_{A\otimes^{\mathscr{P}}C}$ with the differential coming from the composite

$$\mathscr{P}\circ(A;C)\xrightarrow{\mathrm{Id}\circ(\mathrm{Id};\Delta_C)}\mathscr{P}\circ\big(A;\mathscr{C}(C)\big)\xrightarrow{\mathrm{Id}\circ(\mathrm{Id};\alpha\circ(\varphi;\mathrm{Id}))}\mathscr{P}\circ\big(A;\mathscr{P}\circ(A;C)\big)$$
$$\to(\mathscr{P}\circ_{(1)}\mathscr{P})\circ(A;C)\xrightarrow{\gamma_{(1)}\circ(\mathrm{Id};\mathrm{Id})}\mathscr{P}\circ(A;C).$$

Proof. The above composite passes to the quotient which defines the free A-module $A\otimes^{\mathscr{P}}C$ on C. Tracing through the various coequalizers, we get an isomorphism of graded $\mathbb{K}$-modules

$$A\otimes^{\mathscr{P}}_{\mathscr{P}(C)}\Omega_{\mathscr{P}}\mathscr{P}(C)\cong A\otimes^{\mathscr{P}}C.$$

Under this isomorphism, the differential on $A\otimes^{\mathscr{P}}_{\mathscr{P}(C)}\Omega_{\mathscr{P}}\mathscr{P}(C)$ is sent to the expected differential. □

FUNDAMENTAL EXAMPLE. We consider the case of the operad $\mathscr{P}=As$, the cooperad $\mathscr{C}=As^{\text{!`}}$ and the Koszul morphism $\alpha=\kappa:As^{\text{!`}}\to As$. In this case, we get the following representative of the cotangent complex

$$A\otimes^{As}C\cong A_+\otimes_\varphi C\otimes_\varphi A_+,$$

which is nothing but the twisted bitensor product introduced in Exercise 2.7.6.

Let M be a dg A-module. We consider the following twisted differential $\partial_\varphi:=\partial+\bar{\partial}_\varphi$ on the mapping space $\mathrm{Hom}(C,M)$, where

$$\bar{\partial}_\varphi(g):=C\xrightarrow{\Delta_C}\mathscr{C}(C)\xrightarrow{\alpha\circ(\varphi;g)}\mathscr{P}\circ(A;M)\xrightarrow{\gamma_M}M.$$

Proposition 12.4.5. *Under the above given data, any cofibrant resolution $f_\varphi:\Omega_\alpha C\xrightarrow{\sim}A$ gives rise to a natural isomorphisms of dg modules*

$$\big(\mathrm{Der}_A(\Omega_\alpha C,M),\partial\big)\cong\big(\mathrm{Hom}(C,M),\partial_\varphi\big).$$

Proof. It is a consequence of Lemma 12.4.4. □

Hence the André–Quillen cohomology of a $\mathscr{P}$-algebra A with coefficients in an A-module M can be computed using this resolution:

$$H^\bullet_{AQ}(A,M)\cong H_\bullet\big(\mathrm{Hom}(C,M),\partial_\varphi\big).$$

When the operad $\mathscr{P}$ is Koszul, we take $\mathscr{C}=\mathscr{P}^{\text{!`}}$ to be the Koszul dual cooperad, $\alpha=\kappa:\mathscr{P}^{\text{!`}}\to\mathscr{P}$ to be the Koszul morphism and $C=B_\kappa A$. This data induces a universal twisting morphism $\pi_\kappa:C:=B_\kappa A\to A$, see Proposition 11.3.2, and provides a cofibrant replacement functor

$$\Omega_\kappa B_\kappa A \xrightarrow{\sim} A,$$

by Theorem 11.3.5.

Proof of Theorem 12.4.3. It is corollary of Lemma 12.4.4 and Proposition 12.4.5. □

12.4.3 Extensions and $H^1_{\mathcal{P}}(A, M)$

Recall that we defined the notion of abelian extension

$$0 \to M \rightarrowtail B \twoheadrightarrow A \to 0$$

of a $\mathcal{P}$-algebra A by an A-module M in Sect. 12.3.2.

Two abelian extensions are *equivalent* when there exists a morphism of $\mathcal{P}$-algebras $f : B \to B'$ such that the following diagram is commutative

$$\begin{array}{ccccccccc} 0 & \longrightarrow & M & \hookrightarrow & B & \twoheadrightarrow & A & \longrightarrow & 0 \\ & & \| & & \downarrow f & & \| & & \\ 0 & \longrightarrow & M & \hookrightarrow & B' & \twoheadrightarrow & A & \longrightarrow & 0. \end{array}$$

Since such an f is an isomorphism, this defines an equivalence relation among extensions of A by M, whose coset is denoted by $\mathcal{E}xt(A, M)$. Using the cohomological degree convention of the previous section, we have the following interpretation.

Theorem 12.4.6. *Let $\mathcal{P}$ be a quadratic operad. Let A and M be respectively a $\mathcal{P}$-algebra and an A-module, both concentrated in degree* 0. *There is a canonical bijection*

$$H^1_{\mathcal{P}}(A, M) \cong \mathcal{E}xt(A, M).$$

Proof. To any element $\varphi \in \mathrm{Hom}_{\mathbb{S}}(\mathcal{P}^{¡(1)}, \mathrm{End}^A_M)$, we associate a map $s^{-1}\varphi :$ $E(A) \cong s^{-1}\mathcal{P}^{¡(1)}(A) \to M$. When φ is a 1-cocycle, that is $\partial_{\pi_\kappa}(\varphi) = 0$, it induces a map $\Phi : \mathcal{P}(A) \to M$. From this data, we consider the following $\mathcal{P}$-algebra structure on $A \oplus M$:

$$\mathcal{P} \circ (A; M) \twoheadrightarrow \mathcal{P}(A) \oplus \mathcal{P} \circ (A; M) \xrightarrow{\gamma_A \oplus (\Phi + \gamma_M)} A \oplus M.$$

We denote the associated abelian extension by $A \ltimes_\varphi M$.

Let φ' be another 1-cocycle such that $\varphi' = \varphi + \partial_{\pi_\kappa}(\alpha)$, with $\alpha \in \mathrm{Hom}(A, M)$. Then the morphism of $\mathcal{P}$-algebras $A \ltimes_{\varphi'} M \to A \ltimes_\varphi M$ defined by $(a, m) \mapsto$ $(a, m + \alpha(a))$ yields an equivalence of extensions.

In the other way round, let $0 \to M \rightarrowtail B \twoheadrightarrow A \to 0$ be an abelian extension. Under the isomorphism $B \cong A \oplus M$, the $\mathcal{P}$-algebra structure on $A \oplus M$ induces a 1-cocycle $\mathcal{P}^{\text{i}\,(1)}(A) \cong sE(A) \to M$. This defines a map, which passes to the respective quotients. □

Notice the parallel with Theorem 12.2.7, which relates infinitesimal deformations and the cohomology group H^1 of the deformation complex. In the present case, we deform the $\mathcal{P}$-algebra structure $A \ltimes M$ in a certain way.

An interpretation of $H^2_{\mathcal{P}}(A, M)$ in terms of crossed modules is given in [BMR04].

12.4.4 André–Quillen Cohomology of $\mathcal{P}_\infty$-Algebras

One can apply the definition of the André–Quillen cohomology and homology to $\mathcal{P}_\infty$-algebras as well. In this case, we use the functorial cofibrant replacement

$$\Omega_\iota B_\iota A = \mathcal{P}_\infty\big(\mathcal{P}^{\text{i}}(A)\big) \xrightarrow{\sim} A$$

of $\mathcal{P}_\infty$-algebras given in Proposition 11.4.2.

Proposition 12.4.7. *With this resolution, the André–Quillen cohomology of $\mathcal{P}_\infty$-algebras is computed by*

$$H^\bullet_{AQ}(A, M) \cong H^\bullet_{\mathcal{P}_\infty}(A, M) := H_\bullet\big(\operatorname{Hom}\big(\mathcal{P}^{i}(A), M\big), \partial_{\pi_\iota}\big).$$

The same interpretation in terms of tangent homology as in Proposition 12.4.1 holds in the case of $\mathcal{P}_\infty$-algebras. Notice that the operad $\mathcal{P}$ need not be Koszul; this treatment holds for any operad of the form $\Omega\mathcal{C}$.

In the case of the operad A_∞, one recovers the cohomology theory defined by Markl in [Mar92]. For more details, see [Mil10, Sect. 3].

12.4.5 André–Quillen Cohomology and Classical Ext Functor

Since the notion of A-module is equivalent to the notion of left module over the enveloping algebra $U_{\mathcal{P}}(A)$, one can study the following question: is the André–Quillen cohomology equal to the classical Ext-functor

$$H^\bullet_{\mathcal{P}_\infty}(A, M) \cong \operatorname{Ext}^\bullet_{U_{\mathcal{P}}(A)}(\Omega_{\mathcal{P}} A, M)?$$

It is known to be the case for the Koszul operads *As* and *Lie* and it is not the case for the Koszul operad *Com*. A general answer is given by the following result. An operad satisfies the *PBW property*, when the graded module associated to the natural filtration of $U_{\mathscr{P}}(A)$ is isomorphic to the enveloping algebra $U_{\mathscr{P}}(A^{tr})$ of the trivial algebra on A: $\mathrm{gr}U_{\mathscr{P}}(A) \cong U_{\mathscr{P}}(A^{tr})$, for any $\mathscr{P}$-algebra A.

Proposition 12.4.8. [Mil10] *To any Koszul operad $\mathscr{P}$ satisfying the PBW property, there exists a chain complex of obstructions $\mathbb{O}_{\mathscr{P}}$, such that $\mathbb{O}_{\mathscr{P}}$ is acyclic if and only if the André–Quillen cohomology of $\mathscr{P}$-algebras is isomorphic to the* Ext*-functor over the enveloping algebra.*

For instance, the author proves that $\mathbb{O}_{As}$ and $\mathbb{O}_{Lie}$ are acyclic. He exhibits non-trivial homology groups in $\mathbb{O}_{Com}$.

Moreover he proves that $\mathbb{O}_{\Omega\mathscr{C}}$ is acyclic, for any cooperad $\mathscr{C}$, showing that the André–Quillen cohomology of $\mathscr{P}_\infty$-algebra is an Ext-functor. This induces the following result.

Theorem 12.4.9. [Mil10] *Let $\mathscr{P}$ be a Koszul operad, let A be a $\mathscr{P}$-algebra and let M be an A-module. There are natural isomorphisms*

$$H^\bullet_{AQ}(A, M) \cong H^\bullet_{\mathscr{P}_\infty}(A, M) \cong \mathrm{Ext}^\bullet_{U_{\mathscr{P}_\infty}(A)}(\Omega_{\mathscr{P}_\infty} A, M).$$

12.4.6 Other Cohomology Theories

Cotriple Cohomology Since an operad $\mathscr{P}$ is a particular kind of monad, one can consider the *triple cohomology* of $\mathscr{P}$-algebras as defined by Barr and Beck in [BB69]. The André–Quillen cohomology of $\mathscr{P}$-algebras was proved to be isomorphic to the triple cohomology by Fresse in [Fre09a, Sect. 13].

Gamma Cohomology When the operad $\mathscr{P}$ fails to have nice enough properties so as to induce a model category structure on the category of $\mathscr{P}$-algebras, one cannot apply the aforementioned arguments. However, it was proved directly by Hoffbeck in [Hof10a] that the various derived functors are well defined. This leads to the definition of the Γ-*cohomology* of $\mathscr{P}$-algebras. In the case of the operad *Com*, one recovers the Γ-cohomology of Robinson and Whitehouse [RW02, Rob03]. The first Γ-cohomology groups are interpreted as obstructions to lifting homotopy maps of $\mathscr{P}$-algebras in [Liv99, Hof10b].

12.5 Résumé

12.5.1 Homology of Algebras over an Operad

Operadic homology:

$$H^{\mathscr{P}}_\bullet(A) := H_\bullet(\mathrm{B}_\kappa A) \cong H_\bullet\big(\mathscr{P}^{¡}(A), d\big) : \text{graded } \mathscr{P}^{¡}\text{-coalgebra},$$

$$C^{\mathscr{P}}_\bullet(A) : \quad \cdots \to \mathscr{P}^{¡\,(3)}(A) \to \mathscr{P}^{¡\,(2)}(A) \to \mathscr{P}^{¡\,(1)}(A) \to A.$$

Theorem.

$$H_0(A) \cong \text{Indec}(A) \quad \& \quad \mathscr{P} \textit{ Koszul} \quad \Longleftrightarrow \quad H_\bullet^{\mathscr{P}}\big(\mathscr{P}(V)\big) \cong V.$$

Quillen homology: $H_\bullet(\mathbb{L}\,\text{Indec}(A))$.

Theorem.

$$\mathscr{P} \textit{ Koszul} \quad \Longleftrightarrow \quad \textit{operadic homology} \cong \textit{Quillen homology}.$$

12.5.2 Deformation Theory of Algebra Structures

Convolution dg Lie algebra:

$$\mathfrak{g} = \mathfrak{g}_{\mathscr{P},A} := \big(\text{Hom}_{\mathbb{S}}\big(\mathscr{P}^{\text{i}}, \text{End}_A\big), [\,,\,], \partial\big)$$

Proposition.

$$\begin{aligned} \mathscr{P}\textit{-algebra structure on } A \quad &\overset{1-1}{\longleftrightarrow} \quad \varphi \in \text{MC}(\mathfrak{g}), \\ \mathscr{MC}(\mathfrak{g}) := \text{MC}(\mathfrak{g})/GL(A) &\cong \mathscr{P}\textit{-alg}(A)/\textit{iso}. \end{aligned}$$

Twisted convolution dg Lie algebra: Given a $\mathscr{P}$-algebra structure $\varphi \in \text{MC}(\mathfrak{g})$ on A, we consider the twisting dg Lie algebra

$$\mathfrak{g}^{\varphi} = \mathfrak{g}^{\varphi}_{\mathscr{P},A} := \big(\text{Hom}_{\mathbb{S}}\big(\mathscr{P}^{\text{i}}, \text{End}_A\big), [\,,\,], \partial_\varphi\big)$$

called the *deformation complex*.

Tangent homology: $H^\bullet(\mathfrak{g}^\varphi)$.

$\mathfrak{R}$-deformation of φ: $\mathscr{P} \otimes \mathbb{R}$-algebra structure on $A \otimes \mathfrak{R}$ which reduces to φ modulo $\mathfrak{m}$.

Theorem. *$\mathfrak{R}$ Artin ring*

$$\mathscr{MC}\big(\mathfrak{g}^\varphi \otimes \mathfrak{m}\big) := \text{MC}\big(\mathfrak{g}^\varphi \otimes \mathfrak{m}\big)/G \cong \mathscr{D}ef_{\varphi}(\mathfrak{R}) = \text{Def}_\varphi(\mathfrak{R})/\sim.$$

Infinitesimal deformation: $\mathfrak{R} = \mathbb{K}[t]/(t^2)$.

Theorem.

$$\mathscr{D}\text{ef}_\varphi\big(\mathbb{K}[t]/(t^2)\big) \cong H^1\big(g^\varphi\big).$$

Formal deformation: $\mathfrak{R} = \mathbb{K}[[t]]$.

Theorem (Obstructions). *If $H^2(\mathfrak{g}^\varphi)=0$, then any 1-cocycle of $\mathfrak{g}^\varphi$ extends to a formal deformation of φ.*

(Rigidity). *If $H^1(g^\varphi)=0$, then any formal deformation of φ is equivalent to the trivial one.*

The full convolution dg Lie algebra $\mathfrak{g}=\mathfrak{g}_{\mathcal{P},A}$ encodes the $\mathcal{P}_\infty$-algebra structures on A.

Theorem (Obstructions for $\mathcal{P}_\infty$-algebras).. *If $H_{-2}(\mathrm{Hom}_{\mathbb{S}}(\mathcal{P}^{¡\,(n)}, \mathrm{End}_A), \partial_0)=0$ for $n\geq 2$, then any weight one ∂_0-cocycle extends to a $\mathcal{P}_\infty$-algebra structure on A.*

By considering ∞-isomorphisms, the above results extend to $\mathcal{P}_\infty$-algebras.

12.5.3 André–Quillen (Co)Homology of Algebras over an Operad

A-module over $\mathcal{P}$: $\mathsf{Mod}_A^{\mathcal{P}}$.
Free A-module: $A\otimes^{\mathcal{P}} N$.
Enveloping algebra: $U_{\mathcal{P}}(A):=A\otimes^{\mathcal{P}}\mathbb{K}$.

Proposition. $\mathsf{Mod}_A^{\mathcal{P}}\cong U_{\mathcal{P}}(A)\text{-}\mathsf{Mod}$.

Module of Kähler differential forms: $\Omega_{\mathcal{P}}A$.

Theorem.

$$\mathrm{Hom}_{\mathsf{Mod}_A^{\mathcal{P}}}\big(A\otimes_B^{\mathcal{P}}\Omega_{\mathcal{P}}B, M\big)\cong \mathrm{Der}_A(B,M)\cong \mathrm{Hom}_{\mathcal{P}\text{-}\mathsf{alg}/A}(B, A\ltimes M).$$

It forms a Quillen adjunction that we derive to give the

Cotangent complex:

$$\mathbb{L}_{R/A}:=A\otimes_R^{\mathcal{P}}\Omega_{\mathcal{P}}R$$

in $\mathsf{Ho}\big(\mathsf{Mod}_A^{\mathcal{P}}\big)$ for a cofibrant resolution $R\xrightarrow{\sim}A$.
André–Quillen cohomology of a $\mathcal{P}$-algebra A with coefficients into a *left* $U_{\mathcal{P}}(A)$-module M:

$$H^\bullet_{\mathcal{P}}(A,M):=H_\bullet\big(\mathrm{Hom}_{\mathsf{Ho}(\mathsf{Mod}_A^{\mathcal{P}})}(\mathbb{L}_{R/A},M)\big)\cong H_\bullet\big(\mathrm{Der}_A(R,M)\big).$$

André–Quillen homology of a $\mathscr{P}$-algebra A with coefficients into a *right* $U_{\mathscr{P}}(A)$-module M:

$$H_\bullet^{\mathscr{P}}(A, M) := H_\bullet(M \otimes_{U_{\mathscr{P}}(A)} \mathbb{L}_{R/A}).$$

12.5.4 Operadic (Co)Homology of Algebras over an Operad

When $\mathscr{P}$ is a Koszul operad:

$$H^\bullet_{\mathscr{P}}(A, M) \cong H_\bullet\big(\mathrm{Hom}\big(\mathscr{P}^{\mathrm{i}}(A), M\big), \partial_{\pi_\kappa}\big),$$
$$H^0_{\mathscr{P}}(A, M) \cong \mathrm{Der}_A(A, M) \quad \& \quad H^1_{\mathscr{P}}(A, M) \cong \mathscr{E}xt(A, M).$$

12.6 Exercises

Exercise 12.6.1 (Chevalley–Eilenberg homology of the free Lie algebra). Compute the Chevalley–Eilenberg homology of the free Lie algebra.

HINT. The Chevalley–Eilenberg homology (Sect. 13.2.7) is equal to the operadic homology (and to the Quillen homology).

Exercise 12.6.2 (Deformation complex of Lie algebras). Make explicit the deformation complex of Lie algebras, together with its intrinsic Lie bracket, in the same way as Sect. 12.2.4 for associative algebras.

Show that one recovers the Chevalley–Eilenberg cochain complex of a Lie algebra Sect. 13.2.8, up to a degree shift, endowed with the Nijenhuis–Richardson bracket [NR66, NR67].

HINT. Use the description of the convolution dg Lie algebra associated to the operad *Lie* given in Sect. 10.1.6.

Notice that the interpretation of the cohomology groups H^1 and H^2 given in Theorems 12.2.7, 12.2.8 and 12.2.9 in terms of infinitesimal and formal deformations coincide with the results of [NR66, NR67] in the case of Lie algebras.

Exercise 12.6.3 (Equivalence of deformations). Using the fact that any $\mathfrak{R}$-linear endomorphism of $A \otimes \mathfrak{R}$ is completely characterized by its restriction on A:

$$A \to A \otimes \mathfrak{R} \cong A \oplus A \otimes \mathfrak{m},$$

prove that the equivalence of $\mathfrak{R}$-deformations defined in Sect. 12.2.5 is an equivalence relation.

Exercise 12.6.4 (Invertibility). Let A be a $\mathbb{K}$-module. Show that a $\mathbb{K}[[t]]$-linear endomorphism f of $A \otimes \mathbb{K}[[t]]$ is an isomorphism if and only if its first component f_0 (or reduction modulo t) is a $\mathbb{K}$-linear isomorphism of A.

$$f_{|A} = f_0 + \sum_{n\geq 1} f_n t^n : A \to A \oplus A \otimes t \oplus A \otimes t^2 \oplus \cdots .$$

Extend this result (with the same kind of proof), to any $\mathfrak{R}$-extension $A \otimes \mathfrak{R}$ of A, where $\mathfrak{R}$ is a local complete ring, i.e. $\mathfrak{R} = \varprojlim \mathfrak{R}/\mathfrak{m}^n$.

What happens when $\mathfrak{R}$ is an Artin local ring?

Exercise 12.6.5 (Equivalence and ∞-isomorphism). Let $(\mathfrak{R}, \mathfrak{m})$ be a local complete ring. Using the result of the previous exercise, show that an $\mathfrak{R}$-linear ∞-morphism between two $\mathscr{P}_\infty \otimes \mathfrak{R}$-algebras $(A \otimes \mathfrak{R}, \Phi)$ and $(A \otimes \mathfrak{R}, \Psi)$ is invertible if and only if the reduction modulo $\mathfrak{m}$ of its first component, $I = \mathscr{P}^{¡(0)} \to \mathrm{Hom}(A, A)$, is invertible in $\mathrm{Hom}(A, A)$.

HINT. Use the characterization of invertible ∞-morphism given in Theorem 10.4.1.

From this, conclude that the relation $\sim$ between $\mathfrak{R}$-deformations of $\mathscr{P}_\infty$-algebra structures, defined in Sect. 12.2.10, is an equivalence relation.

Exercise 12.6.6 (Examples of deformation complexes). Make explicit the deformation complexes of A_∞-algebras, L_∞-algebras, C_∞-algebras, G_∞-algebras and BV_∞-algebras.

Explain the relationship between them.

HINT. The last two cases are described in detail in [GCTV09].

Exercise 12.6.7 (Obstructions to $\mathscr{P}_\infty$-algebra structure). Fulfill the missing computations in the proof of Theorem 12.2.12.

Exercise 12.6.8 (Independence of the deformation functor). Show that the deformation theory of $\mathscr{P}_\infty$-algebras can equivalently be studied with the bar–cobar construction as follows.

When $\mathscr{P}$ is a Koszul operad, prove that the convolution dg Lie algebra $\mathrm{Hom}_{\mathbb{S}}(\mathscr{P}^{¡}, \mathrm{End}_A)$ is quasi-isomorphic to the convolution dg Lie algebra $\mathrm{Hom}_{\mathbb{S}}(\mathrm{B}\mathscr{P}, \mathrm{End}_A)$ associated to the bar construction of $\mathscr{P}$. Conclude that the associated deformation functors are isomorphic.

HINT. Do not forget to use Theorem 13.2.10.

Exercise 12.6.9 (Modules over a $\mathscr{P}$-algebra). Write the details of the proof of Proposition 12.3.1, which states that for the three graces *As*, *Com* and *Lie*, the respective notions of module over a $\mathscr{P}$-algebra are equivalent to the classical notions of modules, i.e. bimodule, module and Lie module respectively.

Give a similar equivalent definition for the notion of A-module over an operad $\mathscr{P}$, in the cases where $\mathscr{P} = \mathit{PreLie}$, see Sect. 13.4 and [Dzh99], $\mathscr{P} = \mathit{Perm}$, see Sect. 13.4.6 and [Mil11], $\mathscr{P} = \mathit{Leib}$, see Sect. 13.5 and [LP93], $\mathscr{P} = \mathit{Zinb}$, see Sect. 13.5.2 and [Bal98], $\mathscr{P} = \mathit{Dias}$, see Sect. 13.6.7 and [Fra01], $\mathscr{P} = \mathit{Pois}$, see Sect. 13.3 and [Fre06].

HINT. The references given above provide solutions.

Exercise 12.6.10 (Left and right modules over $U_{Leib}(A)$ ★). Give a Leibniz algebra A for which the category of left modules over the enveloping algebra $U_{Leib}(A)$ is not equivalent to the category of right modules over $U_{Leib}(A)$.

See Sect. 13.5 for the definition of Leibniz algebras.

Exercise 12.6.11 (Explicit module-enveloping algebra-Kähler ★). Let $\mathscr{P}$ be an operad given with the following presentation $\mathscr{P} = \mathscr{P}(E, R) = \mathscr{T}(E)/(R)$ by generators and relations. Let A be a $\mathscr{P}$-algebra.

Make explicit the notions of free A-module $A \otimes^{\mathscr{P}} N$, enveloping algebras $U_{\mathscr{P}}(A)$ and module of Kähler differential forms $\Omega_{\mathscr{P}} A$ in terms of E and R.

Using this results, recover the examples of the operads *As*, *Com* and *Lie*.

Exercise 12.6.12 (Examples of André–Quillen (co)homology ★). Make explicit the André–Quillen (co)homology of algebras over the operads $Dias$ [Fra01], $Leib$ [LP93] and $Perm$ [Mil10].

Exercise 12.6.13 (Operadic homology with coefficients ★). When the operad $\mathscr{P}$ is Koszul, using the representation of the cotangent complex obtained in Theorem 12.4.3, make explicit a chain complex which computes the André–Quillen homology with coefficients.

Since $\mathbb{K}$ is a right $U_{\mathscr{P}}(A)$-module, show that, in this case, we recover the operadic (or equivalently the Quillen) homology of the $\mathscr{P}$-algebra A, as defined in Sect. 12.1.

Exercise 12.6.14 ((Co)homology of operads ★). Let $(\mathscr{P}, \gamma)$ be an operad. An *infinitesimal $\mathscr{P}$-bimodule* is an $\mathbb{S}$-module M equipped with a left action $\lambda : \mathscr{P} \circ (\mathscr{P}; M) \to M$ and a right action $\rho : M \circ \mathscr{P} \to$ such that the following diagrams commute

$$
\begin{CD}
M \circ \mathscr{P} \circ \mathscr{P} @>{\mathrm{Id} \circ \gamma}>> M \circ \mathscr{P} \\
@V{\rho \circ \mathrm{Id}}VV @VV{\rho}V \\
M \circ \mathscr{P} @>{\rho}>> M
\end{CD}
$$

$$
\begin{CD}
\mathscr{P} \circ (\mathscr{P} \circ \mathscr{P}; \mathscr{P} \circ (\mathscr{P}; M)) @>{\mathrm{Id} \circ (\gamma; \lambda)}>> \mathscr{P} \circ (\mathscr{P}; M) \\
@V{\cong}VV @VV{\lambda}V \\
(\mathscr{P} \circ \mathscr{P}) \circ (\mathscr{P}; M) @. M \\
@V{\gamma \circ (\mathrm{Id}; \mathrm{Id})}VV @| \\
\mathscr{P} \circ (\mathscr{P}; M) @>{\lambda}>> M
\end{CD}
$$

$$
\begin{CD}
\mathscr{P} \circ (\mathscr{P} \circ \mathscr{P}; M \circ \mathscr{P}) @>{\mathrm{Id} \circ (\gamma; \rho)}>> \mathscr{P} \circ (\mathscr{P}; M) \\
@V{\cong}VV @VV{\lambda}V \\
\mathscr{P} \circ (\mathscr{P}; M) \circ \mathscr{P} @. M \\
@V{\lambda \circ \mathrm{Id}}VV @| \\
M \circ \mathscr{P} @>{\rho}>> M.
\end{CD}
$$

Show that such a data is equivalent to an operad structure on $\mathscr{P} \oplus M$, where the composite of at least two elements coming from M vanishes. This operad is called an *abelian extension* of $\mathscr{P}$ by M and is denoted by $\mathscr{P} \ltimes M$.

Give an equivalent definition of the notion of infinitesimal $\mathscr{P}$-bimodule in terms of two infinitesimal actions of the form

$$\bar{\rho} : M \circ_{(1)} \mathscr{P} \to M \qquad \text{and} \qquad \bar{\lambda} : \mathscr{P} \circ_{(1)} M \to M.$$

Show that the free infinitesimal $\mathscr{P}$-bimodule over N is given by the $\mathbb{S}$-module $\mathscr{P} \circ (\mathscr{P}; N) \circ \mathscr{P}$.

Let $f : \mathscr{Q} \to \mathscr{P}$ be a morphism of operads and let M be an infinitesimal $\mathscr{P}$-bimodule. A *$\mathscr{P}$-derivation from $\mathscr{Q}$ to M* is a morphism of $\mathbb{S}$-modules $d : \mathscr{Q} \to M$ satisfying

$$
\begin{CD}
\mathscr{Q} \circ_{(1)} \mathscr{Q} @>{\gamma_{(1)}^{\mathscr{Q}}}>> \mathscr{Q} \\
@V{f \circ_{(1)} d \oplus d \circ_{(1)} f}VV @VV{d}V \\
\mathscr{P} \circ_{(1)} M \oplus M \circ_{(1)} \mathscr{P} @>{\bar{\lambda} + \bar{\rho}}>> M.
\end{CD}
$$

Show that there exist natural isomorphisms

$$\mathrm{Der}_{\mathscr{P}}(\mathscr{Q}, M) \cong \mathrm{Hom}_{\mathsf{op}/\mathscr{P}}(\mathscr{Q}, \mathscr{P} \ltimes M),$$

where $\mathsf{op}/\mathscr{P}$ denotes the category of operads over the fixed operad $\mathscr{P}$.

To any operad $\mathscr{Q}$, make explicit a suitable notion of *infinitesimal Kähler $\mathscr{Q}$-bimodule of differential forms* $\Omega(\mathscr{Q})$, such that

$$\mathrm{Hom}_{\mathsf{inf}\text{-}\mathscr{Q}\mathsf{biMod}}\big(\Omega(\mathscr{Q}), f^*(M)\big) \cong \mathrm{Der}_{\mathscr{P}}(\mathscr{Q}, M),$$

where $f^*(M)$ stands for the infinitesimal $\mathscr{P}$-bimodule structure on M obtained under f.

Make explicit the left adjoint $f_! : \mathsf{inf}\text{-}\mathscr{Q}\mathsf{biMod} \to \mathsf{inf}\text{-}\mathscr{P}\mathsf{biMod}$ to $f^* : \mathsf{inf}\text{-}\mathscr{P}\mathsf{biMod} \to \mathsf{inf}\text{-}\mathscr{Q}\mathsf{biMod}$.

Show that

$$\mathrm{Hom}_{\mathsf{inf}\text{-}\mathscr{P}\mathsf{biMod}}\big(f_!\big(\Omega(\mathscr{Q}), M\big)\big) \cong \mathrm{Der}_{\mathscr{P}}(\mathscr{Q}, M) \cong \mathrm{Hom}_{\mathsf{op}/\mathscr{P}}(\mathscr{Q}, \mathscr{P} \ltimes M)$$

forms a Quillen adjunction. Derive it to define the *André–Quillen cohomology of the operad $\mathscr{P}$ with coefficients into M*.

Let $f : \mathscr{R} \xrightarrow{\sim} \mathscr{P}$ be a cofibrant resolution of the operad $\mathscr{P}$. Prove that the André–Quillen cohomology of $\mathscr{P}$ is represented by the *cotangent complex of $\mathscr{P}$*, whose representative is given by

$$\mathbb{L}_{\mathscr{R}/\mathscr{P}} := f_!\big(\Omega(\mathscr{R})\big).$$

Using either the bar–cobar resolution $\Omega \mathrm{B}\mathscr{P}$ of $\mathscr{P}$ or the Koszul resolution $\Omega\mathscr{P}^{¡}$ of $\mathscr{P}$, make this cotangent complex explicit.

Let $\mathscr{P}$ is a Koszul operad and let A be a $\mathscr{P}$-algebra $\mathscr{P} \to \mathrm{End}_A$. Show that the André–Quillen cohomology of $\mathscr{P}$ with coefficients in End_A is isomorphic to the tangent homology of A, defined in Sect. 12.2.2.

This exercise comes from [MV09b, Sect. 2]; see also [Rez96, BJT97].

Chapter 13
Examples of Algebraic Operads

Autrement dit, voici un livre qui ne peut se conclure que sur un 'et coetera'.

Umberto Eco
"Vertige de la liste", Flammarion, 2009

In Chap. 9, we studied in detail the operad *Ass* encoding the associative algebras. It is a paradigm for nonsymmetric operads, symmetric operads, cyclic operads. In this chapter we present several other examples of operads. First, we present the two other "graces", the operads *Com* and *Lie* encoding respectively the commutative (meaning commutative and associative) algebras, and the Lie algebras. Second, we introduce more examples of binary quadratic operads: Poisson, Gerstenhaber, pre-Lie, Leibniz, Zinbiel, dendriform, magmatic, several variations like Jordan algebra, divided power algebra, Batalin–Vilkovisky algebra. Then we present various examples of operads involving higher ary-operations: homotopy algebras, infinite-magmatic, brace, multibrace, Jordan triples, Lie triples. The choice is dictated by their relevance in various parts of mathematics: differential geometry, noncommutative geometry, harmonic analysis, algebraic combinatorics, theoretical physics, computer science. Of course, this list does not exhaust the examples appearing in the existing literature. The reader may have a look at the cornucopia of types of algebras [Zin12] to find more examples.

In many cases the (co)homology theory of a given type of algebras was devised before the operad theory told us how to construct an explicit chain complex. Sometimes the equivalence between the two is not immediate, the *Com* case for instance. In general we first describe the known chain complex and then, we relate it with the operadic chain complex. We leave it to the reader to figure out the notion of module and the cohomology theory with coefficients. In fact, in many cases it already exists in the literature.

We make this chapter as self-contained as possible concerning the definitions and the statements. So there is some redundancy with other parts of this book. In most cases we refer to the "general theorems" proved in the previous chapters or to the literature. We work in the category of vector spaces over a field $\mathbb{K}$. For symmetric

J.-L. Loday, B. Vallette, *Algebraic Operads*,
Grundlehren der mathematischen Wissenschaften 346,
DOI 10.1007/978-3-642-30362-3_13, © Springer-Verlag Berlin Heidelberg 2012

operads the assumption "$\mathbb{K}$ is a characteristic zero field" is, in general, required. However all the definitions make sense in the category of modules over a commutative ring, like $\mathbb{Z}$ or a polynomial ring for instance. For many examples, the interesting notion takes place in the category of graded vector spaces (supercase), or even in the category of differential graded vector spaces. Most of the time we leave it to the reader to make explicit the definitions in these cases since they are straightforward.

Many of the operads presented in this chapter are Koszul operads. Therefore all the results proved for Koszul operads in the preceding chapters can be applied. In some specific cases, they have already been proved and published in the literature. In general we leave it to the reader to state the exact statements, though they are not always as straightforward as one could hope.

In this book we have presented four types of operads: symmetric operads, nonsymmetric operads, shuffle operads, unital associative algebras. In Sect.13.14 we present, very briefly, other kinds of operads from the combinatorial point of view.

13.1 Commutative Algebras and the Operad *Com*

We study the operad *Com* which encodes the (nonunital) commutative algebras (associativity is understood). It is a binary, quadratic, Koszul symmetric operad, which comes from a set-theoretic operad. Its Koszul dual operad is: $Com^{!} = Lie$ encoding Lie algebras. The operadic homology is Harrison homology in characteristic zero. The notion of homotopy commutative algebra in the operadic sense coincides with the notion of C_∞-algebra due to T. Kadeishvili [Kad82]. We compare commutative algebras with divided power algebras and in the last subsection we "split" the commutative operation.

13.1.1 Commutative Algebra

By definition a *commutative algebra* over the field $\mathbb{K}$ is an associative algebra A which satisfies the following *commutativity symmetry*:

$$xy = yx$$

for any $x, y \in A$. Let $\tau : V \otimes V \to V \otimes V$ be the switching map (cf. Sect. 1.5.2). Denoting by μ the associative binary operation, the commutativity relation reads

$$\mu^\tau = \mu.$$

There is an obvious notion of morphism between commutative algebras and we denote by *Com*-alg the category of commutative algebras.

Here we work in the symmetric monoidal category of vector spaces over $\mathbb{K}$, but, because of the form of the relations, we could as well work in the symmetric monoidal category of sets, resp. topological sets, resp. simplicial sets. Then we would obtain the notion of commutative monoid, resp. topological commutative monoid, resp. simplicial commutative monoid. If we replace the category Vect by the category gVect of sign-graded vector spaces, the commutativity relation becomes

$$yx = (-1)^{|x||y|}xy.$$

One can either work with nonunital commutative algebras or with unital commutative algebras. In this latter case we denote the operad by $uCom$.

It is sometimes necessary to work with algebras equipped with a binary operation which satisfies the commutativity symmetry relation, but which is not associative. We propose to call them *commutative magmatic* algebras and to denote the associated operad by *ComMag*, cf. Sect. 13.8.4.

13.1.2 Free Commutative Algebra

The free commutative algebra over the vector space V is known to be the reduced symmetric module $\overline{S}(V) = \bigoplus_{n\geq 1}(V^{\otimes n})_{\mathbb{S}_n}$ equipped with the concatenation product. It is called the *reduced symmetric algebra.* If V is spanned by the basis elements $x_1, \dots, x_n$, then $\overline{S}(V)$ is the algebra of polynomials in $x_1, \dots, x_n$ modulo the constants: $\mathbb{K}[x_1, \dots, x_n]/\mathbb{K}1$. The composition of polynomials $\gamma(V) : \overline{S}(\overline{S}(V)) \to \overline{S}(V)$ is given by substitution of polynomials: if $P(X_1, \dots, X_k)$ is a polynomial in the variables X_is and if each X_i is a polynomial in the variables x_js, then $P(X_1(x_1, \dots, x_n), \dots, X_k(x_1, \dots, x_n))$ is a polynomial in the variables x_js called the composite. This composition is obviously associative and commutative.

In the unital case the free algebra is the symmetric algebra: $uCom(V) = S(V)$.

13.1.3 The Operad *Com*

We denote by *Com* the operad encoding the category *Com*-alg of commutative algebras. From the polynomial description of the free commutative algebra it follows that the space of n-ary operations is $Com(n) \cong \mathbb{K}$ equipped with the trivial action of the symmetric group. The n-ary operation $\mu_n \in Com(n) \cong \mathbb{K}$, $\mu_n \mapsto 1$, corresponding to the generator is

$$\mu_n(x_1, \dots, x_n) := x_1 \dots x_n.$$

In the nonunital case we have $Com(0) = 0$ and in the unital case $uCom(0) = \mathbb{K}1$.

Since the composition γ in the operad *Com* is given by the substitution of polynomials, we get

$$\gamma(\mu_k; \mu_{i_1}, \dots, \mu_{i_k}) = \mu_{i_1+\dots+i_k}.$$

The generating series of the operad *Com* is given by

$$f^{Com}(x) = \sum_{n\geq 1} \frac{x^n}{n!} = \exp(x) - 1.$$

As a quadratic operad *Com* is presented by the quadratic data (E_{Com}, R_{Com}) where E_{Com} is the $\mathbb{S}$-module concentrated in arity 2: $E_{Com}(2) = \mathbb{K}\mu$ being the trivial representation. Let us think of this generator as the operation xy, which satisfies $xy = yx$. This operation determines the space $\mathcal{T}(E_{Com})^{(2)}$, which is three-dimensional spanned by $x(yz)$, $y(zx)$, $z(xy)$ or equivalently $x(yz)$, $(xz)y$, $(xy)z$ (compare with the shuffle trees). The space of relations R_{Com} is the sub-$\mathbb{S}_3$-space of $\mathcal{T}(E_{Com})^{(2)}$ generated by the associator $(xy)z - x(zy)$. Therefore it is spanned (for instance) by the two elements $x(yz) - y(zx)$, $y(zx) - z(xy)$. Observe that the quotient $\mathcal{T}(E_{Com})^{(2)}/R_{Com}$ is the trivial representation of $\mathbb{S}_3$ as expected.

13.1.4 Relationship with Other Types of Algebras

The operad *Com* is obviously related to many operads: *Ass*, *Pois*, *Gerst*, *Zinb*, *BV*, *ComMag* and many others, see below.

By forgetting the commutativity property, a commutative algebra can be considered as an associative algebra. So there is an "inclusion" functor:

$$\mathit{Com}\text{-alg} \to \mathit{Ass}\text{-alg}$$

giving rise to a morphism of operads $Ass \twoheadrightarrow Com$. In arity n the representation $Ass(n)$ is the regular representation of $\mathbb{S}_n$, which is isomorphic to the sum of its isotypic components. The isotypic component corresponding to the trivial representation is one-dimensional and the map

$$\mathbb{K}[\mathbb{S}_n] = Ass(n) \twoheadrightarrow Com(n) = \mathbb{K},$$

is precisely the projection onto this component. It is the augmentation map $\sigma \mapsto 1$ for $\sigma \in \mathbb{S}_n$. The relationship with other operads will be treated in the relevant sections.

13.1.5 Koszul Dual of *Com*

We show that the dual of *Com* is *Lie* and that both operads are Koszul operads.

Proposition 13.1.1. *The Koszul dual operad of the operad Com encoding commutative algebras is the operad Lie encoding Lie algebras:*

$$Com^! = Lie.$$

Proof. Since *Com* is generated by a symmetric operation, $Com^!$ is generated by an antisymmetric operation. Let us denote this operation by $[x, y]$. In arity 3 the operations $[[x, y], z]$, $[[y, z], x]$, $[[z, x], y]$ form a linear basis of the weight 2 part of the free operad generated by $[-, -]$. As a representation of $\mathbb{S}_3$ we get a copy of the signature representation and a copy of the hook representation:

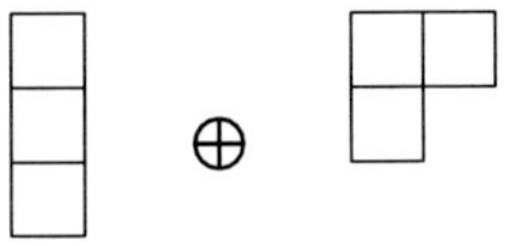

The module of relations R_{Com} defining *Com* is two-dimensional and spanned, for instance, by $(xy)z - x(yz)$ and $(yz)x - y(zx)$. It follows that ${R_{Com}}^{\perp}$ is one-dimensional; so ${R_{Com}}^{\perp}$ is the signature representation. It is spanned by the sum of the three generators. Hence the relation defining $Com^!$ is

$$\big[[x, y], z\big] + \big[[y, z], x\big] + \big[[z, x], y\big] = 0.$$

This is the Jacobi relation, hence $Com^! = Lie$. □

Proposition 13.1.2. *The operad Com is a Koszul operad.*

Proof. There are several ways to prove that *Com* (and therefore *Lie*) is Koszul. For instance, it suffices by Proposition 7.4.5 to show that $H_\bullet^{Com}(Com(V)) = V$ concentrated in degree 1. We will show below that the operadic homology of *Com* is Harrison homology. The computation of Harrison homology of the symmetric algebra is well known. It follows from the computation of the Hochschild homology of the symmetric algebra (cf. [Lod98] for instance).

Another proof would consist in computing the Lie algebra homology of the free Lie algebra (cf. Sect. 13.2.7). The poset method (cf. Sect. 8.7) does not need the characteristic zero hypothesis. The rewriting method (cf. Sect. 8.5.3) also works pretty well. □

13.1.6 Comparison of *Ass* with *Com* and *Lie*

Since a commutative algebra is a particular case of associative algebra, there is a well-defined morphism of operads $Ass \twoheadrightarrow Com$. Taking the Koszul dual of this operad morphism, we get of morphism of cooperads:

$$Ass^{¡} = H^0(\mathrm{B}\,Ass) \to H^0(\mathrm{B}\,Com) = Com^{¡}.$$

Translated to the Koszul dual operads we get

$$Lie = Com^! \to Ass.$$

Proposition 13.1.3. *The Koszul dual of the functor Com*-alg $\to$ *Ass*-alg *is the functor Ass*-alg $\to$ *Lie*-alg, $A \mapsto A_{Lie}$.

Proof. For the purpose of this proof, we denote by $\mu \in Ass(2)$ the generating operation of *Ass* and by $\nu \in Com(2)$ the generating operation of *Com*. We know that both μ and $\mu^{(12)}$ map to ν. As a consequence, the cooperad morphism $Ass^{\text{¡}} \to Com^{\text{¡}}$ maps both μ^c and $(\mu^{(12)})^c$ to ν^c. Denoting by m the linear dual of μ^c in $Ass^! = \mathcal{S} \underset{\mathrm{H}}{\otimes} (Ass^{\text{¡}})^*$ and by c ("crochet" in French) the linear dual of ν^c in $Lie = Com^! = \mathcal{S} \underset{\mathrm{H}}{\otimes} (Com^{\text{¡}})^*$, we see that the map $Lie(2) \to Ass(2)$ sends c to $m - m^{(12)}$. In other words, we get $[x, y] = xy - yx$ as expected. □

13.1.7 Harrison (Co)Homology

Let us recall the definition of *Harrison homology* [Har62] of a (nonunital) commutative algebra A. Let $C_\bullet(A)$ be the Hochschild complex of A, that is $C_n(A) = A^{\otimes n}$ and the Hochschild boundary map is b', cf. Sect. 9.1.6. We denote by $C_\bullet^{Harr}(A)$ the quotient of $C_\bullet(A)$ by the nontrivial signed shuffles, that is by the $(p_1, \dots, p_r)$-shuffles for $p_i \geq 1$, $p_1 + \cdots + p_r = n$, and $r \geq 2$. For instance

$$C_2^{Harr}(A) = A^{\otimes 2}/\{\text{nontrivial shuffles}\} = A^{\otimes 2}/\{a \otimes b - b \otimes a\} = S^2(A).$$

Thinking of $C_\bullet(A)$ as the graded shuffle algebra over the space A of Sect. 1.3.2, $C_\bullet^{Harr}(A)$ is the space of indecomposables. Since A is commutative, the boundary operator b' passes to the quotient and we get a chain complex called the *Harrison complex* of A (for a conceptual explanation of this compatibility between b' and the shuffles, see [Lod89]). Its homology is Harrison homology $H_\bullet^{Harr}(A)$. Observe that we are working here in the nonunital framework.

Proposition 13.1.4. *In characteristic zero, the operadic homology of Com-algebras is the Harrison homology of commutative algebras, up to a shift of degree.*

Proof. Recall from Proposition 12.1.1 that the operadic chain complex of A is such that $C_{n-1}^{Com}(A) := Lie(n)^* \otimes_{\mathbb{S}_n} A^{\otimes n}$. By Theorem 1.3.6, we know that $Lie(n)^*$ is isomorphic to the quotient of the regular representation $\mathbb{K}[\mathbb{S}_n]$ by the nontrivial shuffles. Hence we get $Lie(n)^* \otimes_{\mathbb{S}_n} A^{\otimes n} = C_n^{Harr}(A)$. By Proposition 12.1.1 it follows that the operadic differential corresponds to the Hochschild boundary map under this isomorphism, since the dual of the operad map $Ass \to Com$ is the operad map $Lie \to Ass$. □

Corollary 13.1.5. *The operadic homology $H_{\bullet+1}^{Com}(A)$ of a commutative algebra A inherits a graded Lie-coalgebra structure.*

Proof. This is an immediate consequence of Proposition 13.1.1. □

13.1.8 Homotopy Commutative Algebra, Alias C_∞-Algebra

From the general theory of homotopy algebras, cf. Chap. 10, we know that a *homotopy commutative algebra*, also called strong homotopy commutative algebra, is an algebra over the dg operad $Com_\infty := \Omega Com^{¡}$. From the computation of the dual operad of *Com* we know that the cooperad $Com^{¡}$ is, up to suspension, the cooperad Lie^c encoding Lie coalgebras.

As for homotopy associative algebras, the notion of homotopy commutative algebra appeared in the literature in [Kad88] before the Koszul duality theory for operads was set up. It is called *C_∞-algebra*.

By definition a *C_∞-algebra* is a graded vector space $A = \{A_k\}_{k\in\mathbb{Z}}$ equipped with a differential map $d : Lie^c(sA) \to Lie^c(sA)$ (so $|d| = -1$ and $d \circ d = 0$) which is a coderivation (cf. Sect. 1.2.7). Recall that Lie^c is the cofree Lie coalgebra functor. These two notions coincide since this last definition is the fourth one in the Rosetta Stone (Sect. 10.1.9).

Proposition 13.1.6. *A C_∞-algebra structure on a dg module (A, d_A) is an A_∞-algebra $(A, \{m_n\}_{n\geq 2})$ such that each map $m_n : A^{\otimes n} \to A$ is a Harrison cochain, i.e. m_n vanishes on the sum of all (p,q)-shuffles for $p+q=n$, $p \geq 1$.*

Proof. The aforementioned morphism of cooperad $Ass^{¡} \to Com^{¡}$ shows that a C_∞-algebra induces an A_∞-algebra:

$$\begin{array}{ccc} Ass^{¡} & \longrightarrow & Com^{¡} \\ & \searrow & \downarrow \\ & & \mathrm{End}_A. \end{array}$$

Since $Com^{¡} \cong \mathrm{End}^c_{s^{-1}\mathbb{K}} \otimes_H Lie^c$, we conclude with Ree's theorem 1.3.6, which states that the kernel of $Ass^c \to Lie^c$ is spanned by the sums of the nontrivial shuffles. □

Notice that, in the above definition of a C_∞-algebra, only the associativity relation of a commutative algebra has been relaxed up to homotopy. The symmetry of the binary product remains strict. A differential graded commutative algebra, dgc algebra for short, is a C_∞-algebra whose higher operations vanish: $c_n = 0$ for $n \geq 3$. The linear dual notion of *C_∞-coalgebra*, defined as a square-zero derivation on the free Lie algebra, is ubiquitous in rational homotopy theory through Quillen's constructions of [Qui69].

EXAMPLE. The Lie algebra L_1 of polynomial vector fields over the line $\mathbb{K}^1$, mentioned in Sect. 1.4.3 (b) has a computable cohomology. It can be shown that $\dim H^n_{CE}(L_1) = 2$ for any $n \geq 2$. But, as a C_∞-algebra it is generated by $H^1_{CE}(L_1)$ as shown by Millionshchikov in [Mil09].

13.1.9 Homotopy Transfer Theorem for C_∞-Algebras

The structure of differential graded commutative (dgc) algebra is not stable under homotopy equivalence, but the structure of C_∞-algebra is. The homotopy transfer theorem for commutative algebras takes the same form as the transfer theorem 9.4.7 for associative algebras mutatis mutandis. See for instance [CG08].

Proposition 13.1.7. *Let*

$$h \circlearrowright (A, d_A) \underset{i}{\overset{p}{\rightleftarrows}} (V, d_V)$$

be a homotopy retract, i.e. $\mathrm{Id}_A - ip = dh + hd$ *and* i *is a quasi-isomorphism, and let* $\{\mu_n : A^{\otimes n} \to A\}_{n\geq 2}$ *be a* C_∞*-algebra structure on* A*. The transferred* A_∞*-algebra structure on* V *given by the tree-formula of Theorem* 9.4.7

$$m_n := \sum_{t \in PT_{n+2}} \pm m_t,$$

where, for any planar tree t*, the* n*-ary operation* m_t *is obtained by putting* i *on the leaves,* μ_k *on the* k*-ary vertices,* h *on the internal edges and* p *on the root, is a* C_∞*-algebra structure. The* ∞*-*A_∞*-morphism* $i_\infty : V \stackrel{\sim}{\rightsquigarrow} A$*, extending* i*, given by the tree formula of Theorem* 9.4.7 *defines an* ∞*-*C_∞*-quasi-isomorphism.*

Proof. This statement follows from the general homotopy transfer theorem 10.3.1, with the explicit formulas of Theorems 10.3.3 and 10.3.6. The relation between the A_∞-algebra structures and the C_∞-algebra structures comes from the morphism of cooperads $Lie^{¡} \to Ass^{¡}$. □

As an application, this proposition allows one [CG08] to transfer the dgc algebra structure of the De Rham cochain complex of the differential forms $\Omega^\bullet\Delta^n$ on the n-simplex Δ^n to the normalized simplicial cochain complex $N^\bullet\Delta^n$ through the Dupont contraction [Dup76, Dup78]. This result has applications in deformation theory [Get09] and renormalization theory [Mnë09].

13.1.10 E-Infinity Algebra

In algebraic topology, one also needs to relax the symmetry of the commutative product "up to homotopy". By definition, an E_∞*-operad* is a dg operad $\mathscr{E}$ which is a model for *Com*, that is endowed with a quasi-isomorphism of dg operads $\mathscr{E} \xrightarrow{\sim} Com$. Moreover, one often requires the underlying $\mathbb{S}$-module of $\mathscr{E}$ to be projective, i.e. $\mathscr{E}(n)$ is a projective $\mathbb{K}[\mathbb{S}_n]$-module, for any $n \geq 1$. The property ensures that the category of algebras over $\mathscr{E}$ has nice homotopy behavior, namely it admits a model category structure with quasi-isomorphisms as weak equivalences, see Sect. B.6.7. Notice that this extra assumption is always satisfied over a field $\mathbb{K}$ of characteristic 0; in that case Com_∞ is an E_∞-operad. In this framework, a projective quasi-free model is cofibrant.

The notation E_∞ stands for "everything up to homotopy" since everything, associativity and commutativity, has been relaxed up to homotopy. An *E_∞-algebra* is an algebra over an *E_∞-operad*. Observe that, in this terminology, two E_∞-algebras may be defined over different E_∞-operads $\mathscr{E}'$ and $\mathscr{E}''$. However this is not too much of a problem since there always exists another model $\mathscr{E}$ with quasi-isomorphisms

$$\begin{array}{ccc} \mathscr{E} & \overset{\sim}{\dashrightarrow} & \mathscr{E}'' \\ {\scriptstyle\sim}\downarrow & \searrow{\scriptstyle\sim} & \downarrow{\scriptstyle\sim} \\ \mathscr{E}' & \overset{\sim}{\longrightarrow} & \mathit{Com}. \end{array}$$

By the end, it is enough to consider a cofibrant E_∞-operad $\mathscr{E}$, see Proposition B.6.3. So, by composition, the two E_∞-algebras can be considered as algebras over the same operad $\mathscr{E}$.

Recall that, when the characteristic of the ground field is 0, rational homotopy theory of Quillen [Qui69] and Sullivan [Sul77], tells us that the *rational* homotopy type of a simply-connected CW-complex, satisfying mild assumptions, is faithfully determined by the algebraic structure of some dg commutative algebra (resp. dg Lie algebra). The main interest in E_∞-algebra structures lies in the following results of Mandell [Man06, Man01], which generalize the philosophy of rational homotopy theory to modules over $\mathbb{Z}$ and in characteristic p. It asserts that the singular cochain complex of a simply-connected CW-complex, satisfying finiteness and completeness conditions, carries an E_∞-algebra structure, which faithfully determines its homotopy type, respectively its p-adic homotopy type.

13.1.11 Barratt–Eccles Operad

An example of an E_∞-operad is given by the *Barratt–Eccles operad* $\mathscr{E}$, introduced in [BF04] and dubbed after [BE74]. For any $k \geq 1$, one defines $\mathscr{E}(k)$ to be the normalized bar construction of the symmetric group $\mathbb{S}_k$. In other words, the space $\mathscr{E}_d(k)$ is the quotient of the free $\mathbb{K}$-module ($\mathbb{K}$ might be $\mathbb{Z}$) on the set of $d+1$-tuples $(\omega_0, \dots, \omega_d)$ of elements of $\mathbb{S}_k$ by the degenerate tuples such that $\omega_i = \omega_{i+1}$, for at least one integer i, $0 \leq i \leq d-1$. The differential map is the simplicial one given by

$$d_{\mathscr{E}}(\omega_0, \dots, \omega_d) := \sum_{i=0}^{d} (-1)^i (\omega_0, \dots, \widehat{\omega_i}, \dots, \omega_d).$$

The action of the symmetric group is diagonal

$$(\omega_0, \dots, \omega_d)^\sigma := \big(\omega_0^\sigma, \dots, \omega_d^\sigma\big).$$

The composite product of this operad is given by the combinatorics of the permutations. We refer the reader to Sect. 1.1.3 of [BF04] for details. Notice that the Barratt–Eccles operad is a Hopf operad with the coproduct

$$\Delta(\omega_0, \ldots, \omega_d) := \sum_{l=0}^{d} (\omega_0, \ldots, \omega_l) \otimes (\omega_{l+1}, \ldots, \omega_d).$$

One defines a quasi-isomorphism of dg operads $\mathcal{E} \xrightarrow{\sim} Com$ by the augmentation $(\omega_0, \ldots, \omega_d) \mapsto 0$ for $d > 0$ and $(\omega_0, \ldots, \omega_d) \mapsto 1$ for $d = 0$.

The suboperad of $\mathcal{E}$, made of the elements of degree 0, $\mathcal{E}_0(k) = \mathbb{K}[\mathbb{S}_k]$, is the symmetric operad Ass, inducing the following diagram of dg operads

$$\begin{array}{ccc} Ass & \rightarrowtail & \mathcal{E} \\ & \searrow & \downarrow \sim \\ & & Com. \end{array}$$

Actually the Barratt–Eccles operad comes equipped with a filtration, defined by the number of descents of permutations, such that

$$Ass = \mathcal{E}_1 \subset \mathcal{E}_2 \subset \cdots \subset \mathcal{E}_n \subset \cdots \subset \operatorname{colim}_n \mathcal{E}_n = \mathcal{E}.$$

This gives intermediate ways to relax the notion of commutative algebra up to homotopy. In [Fre11b], Fresse proved that, in a way, the operads $\mathcal{E}_n$ are Koszul, see Sect. 13.3.17 for more details.

For each $n \geq 1$, the operad $\mathcal{E}_n$ is quasi-isomorphic over $\mathbb{Z}$ to the chains of the little n-discs operad [Smi89, Kas93, Ber96]. Such a dg operad is called an *E_n-operad.*

13.1.12 Divided Power Algebras

Let $\mathbb{K}$ be a field, possibly of finite characteristic, or $\mathbb{Z}$. A *divided power algebra* A is an augmented commutative algebra $A = \mathbb{K}1 \oplus \bar{A}$ equipped with operations $\gamma_i : \bar{A} \to \bar{A}$, $i \geq 1$, called *divided power operations*, such that the following relations hold:

$$\gamma_1(x) = x, \qquad \gamma_i(\lambda x) = \lambda^i \gamma(x) \quad \text{for } \lambda \in \mathbb{K},$$

$$\gamma_i(x + y) = \sum_{j=0}^{i} \gamma_j(x)\gamma_{i-j}(y),$$

$$\gamma_i(xy) = i!\gamma_i(x)\gamma_i(y),$$

$$\gamma_i(x)\gamma_j(x) = \frac{(i+j)!}{i!j!}\gamma_{i+j}(x),$$

$$\gamma_i\big(\gamma_j(x)\big) = \frac{(ij)!}{i!(j!)^i}\gamma_{ij}(x),$$

for any $x, y \in \bar{A}$ and where, by convention $\gamma_0(x) = 1$. Observe that, if $\mathbb{K}$ is a characteristic zero field, then we have $\gamma_n(x) = \frac{x^{*n}}{n!}$, so there is no extra data. In characteristic p the divided power algebra is completely determined by the associative and commutative product and by $\gamma(x) := \gamma_p(x)$.

We have mentioned in Sect. 5.2.9 that any symmetric operad $\mathscr{P}$ gives rise to an operad $\Gamma\mathscr{P}$ encoding the $\mathscr{P}$-algebras with divided powers. For *Com* we get Γ*Com*. One can check, cf. [Fre00], that a Γ*Com*-algebra is a commutative algebra with divided powers as defined above. In the literature, the divided power algebra $\Gamma Com(V)$ over the vector space V is denoted $\Gamma(V)$.

EXAMPLE. On the space $\mathbb{K}[x]$ of polynomials in one variable, one can define two different commutative algebra structures:

$$x^n x^m := x^{n+m},$$
$$x^n * x^m := \binom{n+m}{n} x^{n+m},$$

see Exercise 1.8.5. We denote the first algebra by $\mathbb{K}[x]$ and the second one by $\Gamma(\mathbb{K}x)$. In the second case, we observe that x^n is not obtainable from x under the product $*$, unless $n!$ is invertible, since

$$x^{*n} := \underbrace{x * x * \ldots * x}_{n} = n! x^n.$$

The unary operation $\gamma_n(x) = x^n$ obviously satisfies the formal properties of $\frac{x^{*n}}{n!}$ even when $n!$ is not invertible in $\mathbb{K}$, e.g. when $\mathbb{K}$ is of finite characteristic. This is the free Γ*Com*-algebra on one generator x.

13.1.13 Splitting Associativity

The commutative algebra $\Gamma(\mathbb{K}x)$, i.e. the polynomials in one variable equipped with the product $*$, has the following property:

the associative product $*$ splits into the sum

$$x^n * x^m = x^n \prec x^m + x^m \prec x^n,$$

where

$$x^n \prec x^m := \binom{n+m-1}{n-1} x^{n+m}.$$

It is a consequence of the binomial formula $\binom{n+m}{n} = \binom{n+m-1}{n-1} + \binom{n+m-1}{m-1}$. It is immediate to check that the binary product $x^n \prec x^m$ satisfies the following relation

$$(x^n \prec x^m) \prec x^p = x^n \prec (x^m \prec x^p) + x^n \prec (x^p \prec x^m).$$

This is a particular case of a more general phenomenon that will be dealt with in Sect. 13.5.

13.2 Lie Algebras and the Operad *Lie*

The category of Lie algebras determines the operad *Lie*. It is a binary quadratic operad which is Koszul. Its Koszul dual operad is *Com*. However, unlike *Com*, it does not come from a set-theoretic operad. We mentioned various characterizations of $Lie(V)$ as a subspace of $T(V)$ in Sect. 1.1.9. The representation $Lie(n)$ is a rather complicated one, which has been treated in several papers, cf. for instance [Reu93]. The operadic (co)homology theory is the classical Chevalley–Eilenberg theory. Homotopy Lie algebras, or L_∞-algebras, play a prominent role in many parts of mathematics, for instance in the deformation quantization of Poisson manifolds by Maxim Kontsevich [Kon03].

13.2.1 Lie Algebra

Let us recall that a *Lie algebra* $\mathfrak{g}$ is a vector space equipped with a binary operation called the *bracket* (or Lie bracket), which is skew-symmetric:

$$[x,y] = -[y,x]$$

and which satisfies the Jacobi identity

$$\big[[x,y],z\big] + \big[[y,z],x\big] + \big[[z,x],y\big] = 0.$$

See Sect. 1.1.9 for comments on this presentation. Its name comes from the Norwegian mathematician Sophus Lie.

It is often helpful to adopt the notation $c(x,y) := [x,y]$. The symmetry property and the Jacobi relation become:

$$c^{(12)} = -c, \qquad c \circ (c \otimes \mathrm{id}) + c \circ (c \otimes \mathrm{id})^{(123)} + c \circ (c \otimes \mathrm{id})^{(321)} = 0.$$

Applying these formulas in the sign graded vector space framework gives the notion of "graded Lie algebra" (also called super Lie algebra).

13.2.2 Free Lie Algebra

There are several characterizations of the free Lie algebra $Lie(V)$ over the vector space V as a subspace of $T(V)$. The following is a helpful result in the study of the free Lie algebra and its associated representations.

Proposition 13.2.1. *Each one of the following statements characterizes* $Lie(V)$ *as a subspace of the tensor algebra* $T(V)$:

a) *The subspace $Lie(V)$ of $T(V)$ is generated by V under the bracket operation.*
b) *Let $D : V^{\otimes n} \to V^{\otimes n}$ (Dynkin bracketing) be the map defined by*

$$D(v_1 \cdots v_n) := \big[\ldots\big[[v_1, v_2], v_3\big], \ldots, v_n\big].$$

The element $\omega \in V^{\otimes n}$ lies in $Lie(V)$ if and only if $D(\omega) = n\omega$.
c) *The space $Lie(V)$ is the image of the Eulerian idempotent defined as*

$$e^{(1)} := \log^{\star}(\mathrm{Id}) : T(V) \to T(V),$$

where $\star$ is the convolution product of the Hopf algebra $T(V)$.
d) *The space $Lie(V)$ is made of the primitive elements in the tensor algebra $T(V)$ (viewed as a cocommutative bialgebra).*

Proof. We mentioned this result in Proposition 1.3.3. □

13.2.3 The Operad *Lie*

From the presentation of a Lie algebra it follows that the operad *Lie* encoding Lie algebras is the quotient

$$Lie = \mathcal{T}(E_{Lie})/(R_{Lie}),$$

where $E_{Lie} = \mathbb{K}c$, c being an antisymmetric operation, and R_{Lie} is the $\mathbb{S}_3$-submodule of $\mathcal{T}(\mathbb{K}c)^{(2)}$ generated by the Jacobiator. Recall that $\mathcal{T}(\mathbb{K}c)^{(2)}$ is three-dimensional spanned by $c \circ_1 c$, $(c \circ_1 c)^{(123)}$, $(c \circ_1 c)^{(132)}$ and the Jacobiator is the sum of these three elements. As a consequence the quotient $Lie(3) = \mathcal{T}(\mathbb{K}c)^{(2)}/R_{Lie}$ is the two-dimensional hook representation.

From the various constructions of $Lie(V)$ it can be deduced that $\dim Lie(n) = (n-1)!$. As a consequence the Poincaré series is

$$f^{Lie}(x) = \sum_{n\geq 1} \frac{x^n}{n} = -\log(1-x).$$

As a complex representation of $\mathbb{S}_n$, it can be shown (cf. [Klj74]) that $Lie(n)$ is isomorphic to the induced representation $\mathrm{Ind}_{\mathbb{Z}/n\mathbb{Z}}^{\mathbb{S}_n}(\rho)$, where (ρ) is the one-dimensional representation of the cyclic group given by an irreducible nth root of unity.

13.2.4 A Basis for *Lie*(*n*)

For the operads *Ass* and *Com* we do know about a basis of the space of n-ary operations which behaves well with respect to the operadic composition and the action of the symmetric group. For $Lie(n)$ the situation is more complicated. This can be readily seen for $n = 3$. Indeed, $Lie(3)$ is the unique two-dimensional irreducible representation of $\mathbb{S}_3$ (hook representation $S^{2,1}$). It can be constructed as the hyperplane of $\mathbb{K}u \oplus \mathbb{K}v \oplus \mathbb{K}w$ orthogonal to the vector $u + v + w$. The three vectors $u - v$, $v - w$, $w - u$ lie in this hyperplane and any pair forms a basis. A bijection with $Lie(3)$ can be obtained by

$$\begin{cases} c \circ_1 c = u - v, \\ (c \circ_1 c)^{(123)} = v - w, \\ (c \circ_1 c)^{(321)} = w - u, \end{cases}$$

where c is the Lie bracket and (123), resp. (321) are cyclic permutations.

Therefore in order to bypass this difficulty, in most cases the way to handle $Lie(n)$ is to view it as a subspace of $Ass(n) = \mathbb{K}[\mathbb{S}_n]$, see Proposition 13.2.1. The construction of the Harrison chain complex is a typical example.

However there are several ways to construct an explicit basis of $Lie(n)$ as elements of $Lie(\mathbb{K}x_1 \oplus \cdots \oplus \mathbb{K}x_n)$. Here are two of them.

Dynkin Elements

Consider the element

$$[[\ldots[[x_1, x_2], x_3], \ldots], x_n].$$

Let the symmetric group $\mathbb{S}_{n-1}$ act on the index set $\{2, \ldots, n\}$. The $(n-1)!$ elements given by the action of $\mathbb{S}_{n-1}$ span $Lie(n)$, cf. [Reu93]. The behavior of this basis under operadic composition is rather complicated (cf. Sect. 8.3).

Indexed Trees [MR96]

Let t be a planar binary tree with n leaves (cf. Appendix C.1.1). Index the leaves by $\{1, \ldots, n\}$ such that, for any given vertex, the left-most (resp. right-most) index is the smallest (resp. largest) index among the indices involved by this vertex. Any such indexed tree determines an element in $Lie(n)$ by bracketing. It is proved in loc. cit. that one gets a basis of $Lie(n)$. For instance (we write just i in place of x_i):

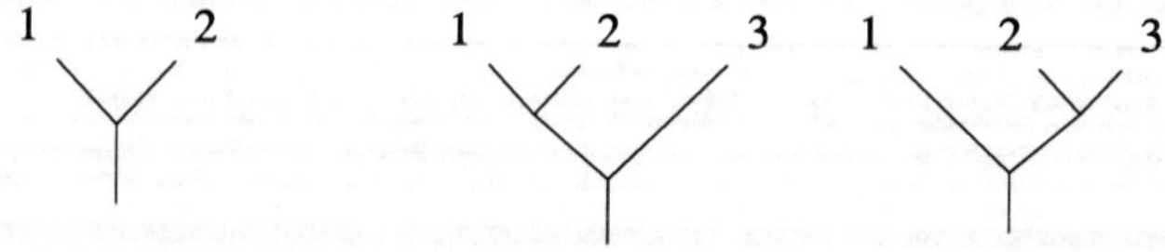

$[1,2] \in Lie(2)$, $\quad [[1,2],3]$, $\quad [1,[2,3]] \in Lie(3)$,

in $Lie(4)$:

$$[[[1,2],3],4], \quad [[1,[2,3]],4], \quad [[1,2],[3,4]], \quad [[1,3],[2,4]],$$
$$[1,[[2,3],4]], \quad [1,[2,[3,4]]].$$

This basis behaves well with respect to partial composition. Indeed, if t and s are two basis elements, then so is $t \circ_i s$. It is obtained by grafting the root of s on the leaf of t with label i and by shifting the indices accordingly (standardization). Equivalently, the tree $[t, s]$ is obtained by grafting the trees t and s and standardizing the labels of s.

This basis is the PBW basis obtained in Sect. 8.3. It appears quite often in the literature when one wants to prove that some representation of the symmetric group, coming either from algebraic topology or algebraic combinatorics, is indeed isomorphic to $Lie(n)$, see for instance [Coh76, RW02, Tou06, ST09].

13.2.5 Relationship of *Lie* with Other Types of Algebras

The most important one is the "forgetful" functor:

$$Ass\text{-alg} \xrightarrow{-} Lie\text{-alg}, \qquad A \mapsto A_{Lie}$$

induced by the inclusion $Lie \rightarrowtail Ass$ and already mentioned (cf. Sect. 1.1.9). It is called forgetful because, when dealing with the presentation of *Ass* by a symmetric operation and an antisymmetric operation (cf. Proposition 9.1.1), then it simply consists in forgetting the symmetric operation.

There is of course an obvious relationship with the operad *Pois* encoding Poisson algebras, namely $Lie \rightarrowtail Pois$, cf. Sect. 13.3.

In order for the "bracketing process" to give a Lie algebra, the starting binary operation $x \cdot y$ need not be associative. In fact it suffices that the Jacobi identity holds for $[x, y] := x \cdot y - y \cdot x$. It is easy to check that the "Jacobiator" (left part of the Jacobi identity) is made of 12 monomials corresponding to the two ways of parenthesizing the six permutations of xyz. A vector space equipped with a binary operation satisfying this property is called a *Lie-admissible algebra*.

A pre-Lie algebra (cf. Sect. 1.4) is an example of a Lie-admissible algebra which is not an associative algebra. So there are functors:

$$Ass\text{-alg} \to preLie\text{-alg} \to LieAdm\text{-alg} \xrightarrow{-} Lie\text{-alg},$$

giving rise to operadic morphisms

$$Lie \to LieAdm \to preLie \to Ass.$$

13.2.6 Koszul Dual of *Lie*

We know that the Koszul dual operad of *Com* is *Lie*, therefore, by Proposition 7.2.2, the Koszul dual operad of *Lie* is *Com*:

$$Lie^{!} = Com$$

and *Lie* is a Koszul operad. By Theorem 7.6.7 the Poincaré series of *Com* and *Lie* are inverse to each other under composition. This is no surprise since exp and log are the respective Poincaré series. More interesting is the relationship for the Frobenius characteristic, see [Mac95, p. 120].

We refer to Sect. 8.3 for a direct proof of the Koszulity of the *Lie* operad.

13.2.7 Chevalley–Eilenberg Homology

Chevalley and Eilenberg defined the homology of a Lie algebra $\mathfrak{g}$ as follows [CE48]. The module of n-chains is $C_n^{CE}(\mathfrak{g}) := \Lambda^n(\mathfrak{g})$, the nth exterior power of the space $\mathfrak{g}$ (graded symmetric power of $\mathfrak{g}$ placed in degree 1). The boundary map is induced by the bracket:

$$d(x_1 \wedge \cdots \wedge x_n) = \sum_{i<j} (-1)^{i+j-1}[x_i, x_j] \wedge x_1 \wedge \cdots \wedge \widehat{x_i} \wedge \cdots \wedge \widehat{x_j} \wedge \cdots \wedge x_n.$$

The antisymmetry property of the bracket ensures that the map d is well-defined. The Jacobi identity ensures that $d^2 = 0$.

Proposition 13.2.2. *In characteristic zero, the operadic homology theory of a Lie algebra is the Chevalley–Eilenberg homology theory, up to a shift of degree.*

Proof. Since the Koszul dual operad of *Lie* is *Com*, it follows that the operadic chain complex computing the homology of $\mathfrak{g}$ is such that $C_{n-1}^{Lie}(\mathfrak{g}) = (s\,Com(n)^*) \otimes_{\mathbb{S}_n} \mathfrak{g}^{\otimes n}$. We have to remember that the symmetric group is acting via the signature representation. Since $Com(n)$ is the one-dimensional trivial representation, it follows that

$$C_{n-1}^{Lie}(\mathfrak{g}) = \Lambda^n(\mathfrak{g}).$$

By Proposition 12.1.1 the differential of the operadic chain complex is, up to suspension, the Chevalley–Eilenberg boundary map. □

Corollary 13.2.3. *The operadic homology $H_{\bullet+1}^{Lie}(\mathfrak{g})$ of a Lie algebra $\mathfrak{g}$ is a graded commutative coalgebra.*

Proof. It is a consequence of Theorem 12.1.2 applied to the Koszul operad *Lie* and to the fact that the Koszul dual operad of *Lie* is *Com*, cf. Sect. 13.2.6. □

13.2.8 Lie Homology and Cohomology with Coefficients

In Chap. 12 we have shown that for any Koszul operad one can construct a small chain complex to construct (co)homology with coefficients. In the Lie case it gives precisely the Chevalley–Eilenberg complex (cf. for instance [CE48]).

13.2.9 Homotopy Lie Algebras, Alias L_∞-Algebras

From the general theory of homotopy algebras, cf. Chap. 10, we know that a *homotopy Lie algebra*, also called L_∞*-algebra* (or strong homotopy Lie algebra) in the literature, is an algebra over the operad $Lie_\infty := \Omega Lie^{\text{i}}$. From the computation of the dual operad of *Lie*, we know that the cooperad Lie^{i} is, up to suspension, the cooperad Com^c encoding commutative coalgebras.

Using the Rosetta Stone (Sect. 10.1.9), an L_∞-algebra can equivalently be defined in the following three ways.

By definition, an L_∞*-algebra* is a dg module (A, d_A) equipped with skew-symmetric operations $l_n : A^{\otimes n} \to A$ of degree $n-2$, for $n \geq 2$, which satisfy the relations

$$\sum_{\substack{p+q=n+1\\ p,q>1}} \sum_{\sigma \in Sh_{p,q}^{-1}} \operatorname{sgn}(\sigma)(-1)^{(p-1)q}(l_p \circ_1 l_q)^\sigma = \partial_A(l_n),$$

for $n \geq 1$, where ∂_A is the differential in End_A induced by d_A and where $Sh_{p,q}^{-1}$ denotes the set of (p,q)-unshuffles, cf. Sect. 1.3.2. See Proposition 10.1.7 for more details.

The second definition is given by a square-zero coderivation $d : \overline{S^c}(sA) \to \overline{S^c}(sA)$ of degree -1 on the non-counital cofree cocommutative coalgebra of the suspension of a graded module $A = \{A_k\}_{k \in \mathbb{Z}}$. Hence the coderivation d gives rise, for each $n \geq 1$, to a map

$$l_n : \Lambda^n(A) \to A$$

of degree $n-2$, which corresponds to the n-ary operation. The proof was given in Proposition 10.1.12. An ∞*-morphism* between two L_∞-algebras A and A' is defined as a morphism of dg cocommutative coalgebras between $\overline{S^c}(sA)$ and $\overline{S^c}(sA')$. It is sometimes called an L_∞-morphism in the literature.

A third equivalent definition is given by solutions to the Maurer–Cartan equation in the Nijenhuis–Richardson dg (pre-)Lie algebra $\left(\prod_n s^{-n+1} \operatorname{Hom}(\Lambda^n A, A), \star\right)$, see Proposition 10.1.9. This latter one is the Chevalley–Eilenberg cochain complex with trivial Lie algebra structure on A.

13.2.10 Formal Manifold

The second definition of an L_∞-algebra allows Kontsevich [Kon03, Sect. 4] to describe this notion with the geometrical language of *formal manifolds* as follows.

One can think of the vector space A as a manifold. When it is finite dimensional, the linear dual of $\overline{S^c}(sA)$, is up to suspension, isomorphic to $\overline{S}(A^*)$, the structure sheaf formed by the algebra of formal functions on A which vanish at 0. Under this isomorphism, square-zero coderivations are in one-to-one correspondence with square-zero derivations. A derivation on the free commutative algebra without unit

$\overline{S}(A^*)$ is completely characterized by its restriction to the generators, that is $A^* \to \overline{S}(A^*)$, which corresponds to a vector field in the geometrical language. When it squares to zero, it is called a *homological vector field.*

Proposition 13.2.4. *The data of a homological vector field on a formal manifold is equivalent an L_∞-algebra structure on a finite dimensional chain complex.*

13.2.11 Comparison of L_∞-Algebras with A_∞-Algebras

The morphism of quadratic operads $Lie \rightarrowtail Ass$ induces the following morphism of Koszul dual cooperads $Lie^¡ \rightarrowtail Ass^¡$. Since the cobar construction Ω is a functor, it induces the following morphism of dg operads $Lie_\infty \rightarrowtail Ass_\infty$.

Proposition 13.2.5. *The functor*

$$A_\infty\text{-alg} \to L_\infty\text{-alg}$$

consists in anti-symmetrizing the operations m_n to get the operations l_n.

Proof. See the proof of Proposition 10.1.8. □

Together with Proposition 13.1.6, we get the following functors

$$L_\infty\text{-alg} \leftarrow A_\infty\text{-alg} \leftarrow C_\infty\text{-alg},$$

where a C_∞-algebra is viewed as an A_∞-algebra whose operations vanish on the sum of the shuffles.

13.2.12 The Notion of L_∞-Algebra in the Literature

As for homotopy associative algebras, the notion of homotopy Lie algebra appeared in the literature before the Koszul duality theory of operads was set up.

In the first place, the dual notion of L_∞-coalgebra is ubiquitous in the rational homotopy theory of Sullivan [Sul77] through the minimal model construction, which is a quasi-free dg commutative algebra. It was used in the early 1980s in deformation theory [SS85, GM88], extending the general philosophy of Deligne and Grothendieck which states that a deformation problem should be governed by a dg Lie algebra, see Sect. 13.2.14. This notion was explicitly used in mathematical physics at the beginning of the 1990s, see [LS93], for instance in string field theory by Zwiebach [Zwi93]. The Koszul resolution L_∞ of the operad *Lie* was explicitly given by Hinich and Schechtman in [HS93], before the general paper of

Ginzburg and Kapranov [GK94] on the Koszul duality theory. The notion of ∞-L_∞-morphisms plays a crucial role in the proof the deformation-quantization of Poisson manifolds by Kontsevich [Kon03], as explained below.

Notice that it should not be confused with the Lie algebra structure on the homotopy groups of a topological space [Whi41, Sam53], which is also called "homotopy Lie algebra".

13.2.13 Homotopy Theory of L_∞-Algebras

The general results on homotopy algebras of Chaps. 10 and 11 can be applied to L_∞-algebras. Actually, they were first proved on the level of L_∞-algebras by Sullivan [Sul77] and Kontsevich [Kon03].

Proposition 13.2.6. *For any homotopy retract an L_∞-algebra structure can be transferred so that the resulting L_∞-algebra is homotopy equivalent to the starting one.*

Proof. It is the HTT 10.3.1 applied to the Koszul operad *Lie*. □

An explicit formula, based on labeled trees, is given in Proposition 10.3.5. In the case of the homology of a dg Lie algebra, the induced operations are sometimes called the *Lie–Massey products* [Ret93] in the literature.

The other results are the following ones.

◇ Any L_∞-algebra is ∞-isomorphic to the product (direct sum) of a minimal L_∞-algebra, its homology, with an acyclic trivial L_∞-algebra by Theorem 10.4.3.
◇ For any ∞-quasi-isomorphism between two L_∞-algebras, there exists an ∞-quasi-isomorphism in the other way round, which extends the inverse of the isomorphism on the level of the homotopy groups (the homology of the underlying chain complexes), see Theorem 10.4.4.
◇ One can universally rectify any homotopy Lie algebra by Proposition 11.4.6.
◇ Finally the homotopy category of dg Lie algebras is equivalent to the homotopy category of L_∞-algebras with ∞-morphisms by Theorem 11.4.8.

13.2.14 Deformation Theory

This section deals with deformation theory using dg Lie algebras over a ground field $\mathbb{K}$ of characteristic 0. The philosophy of using dg Lie algebras to encode deformation problems goes back to Deligne and Grothendieck according to [GM88, SS85].

The general idea of deformation theory is to study the possible deformations, up to some equivalence relation, of a given structure, which can be algebraic as in Sect. 12.2 or geometric. There are basically two approaches.

The first one, due to M. Schlessinger [Sch68] and Grothendieck, relies on a functor which associates the coset of equivalent deformations to a ring made up of the possible parameters of deformation. Such a functor is called the *deformation functor*. This method forgets the data of the equivalences themselves to keep track only of the equivalence classes. It might also give rise to singular spaces.

The other approach due to Deligne encodes the two data of the possible deformations and their equivalences into a category. By definition, the objects are the deformations themselves and the morphism sets between two deformations is made of the equivalences between them. Since one works with an equivalence relation, every morphism of the category is an isomorphism. So this category is a groupoid called the *Deligne groupoid*. Deligne groupoids are often obtained as sets of deformations with a group action encoding the equivalences. In this language, two deformation problems are equivalent if the associated Deligne groupoids are equivalent.

Ultimately a deformation problem is encoded by a dg Lie algebra as follows. To any dg Lie algebra $(\mathfrak{g}, [\,,\,], d)$, one can associate the set $\mathrm{MC}(\mathfrak{g})$ of Maurer–Cartan elements:

$$\mathrm{MC}(\mathfrak{g}) := \left\{\alpha \in \mathfrak{g}^1 \mid d\alpha + \frac{1}{2}[\alpha, \alpha] = 0\right\}.$$

Here we use the cohomological convention $|d| = |\alpha| = +1$, which is often the case in the literature, for instance in Sect. 12.2.

Proposition 13.2.7. *When the dimension of $\mathfrak{g}^1$ is finite, the set of $\mathrm{MC}(\mathfrak{g})$ of Maurer–Cartan elements forms an algebraic variety in the affine space $\mathfrak{g}^1$, which is an intersection of quadrics.*

Proof. Choosing a finite basis $\{e_i\}_{1\le i\le m}$ for $\mathfrak{g}^1$ and a finite basis $\{f_j\}_{1\le j\le n}$ for the images of the differential and the bracket in $\mathfrak{g}^2$, we have

$$de_i = \sum_{1\le j\le m} \lambda^i_j f_j \quad \text{and} \quad [e_i, e_{i'}] = \sum_{1\le j\le m} \lambda^{i,i'}_j f_j.$$

Therefore, an element $\alpha = X_1 e_1 + \cdots + X_n e_n$ lies in $\mathrm{MC}(\mathfrak{g})$ if and only if it satisfies the quadratic relations

$$\sum_{1\le i\le n} \lambda^i_j X_i + \frac{1}{2} \sum_{1\le i,i'\le n} \lambda^{i,i'}_j X_i X_{i'} = 0,$$

for any $1 \le j \le m$. □

We denote the quadratic mapping $\mathfrak{g}^1 \to \mathfrak{g}^2$ by $\alpha \mapsto Q(\alpha) := d\alpha + \frac{1}{2}[\alpha, \alpha]$. It satisfies $\mathrm{MC}(\mathfrak{g}) = Q^{-1}(0)$.

Lemma 13.2.8. *The map*

$$\lambda \in \mathfrak{g}^0 \mapsto d\lambda + [-, \lambda] \in \Gamma\big(\mathrm{MC}(\mathfrak{g}), T\,\mathrm{MC}(\mathfrak{g})\big)$$

is a morphism of Lie algebras from $\mathfrak{g}^0$ to the Lie algebra of affine vector fields of $\mathrm{MC}(\mathfrak{g})$.

Proof. Since the set $\mathrm{MC}(\mathfrak{g})$ is the zero locus of the map Q, it is enough to compute its derivative: $d_\alpha Q(\beta) = d\beta + [\alpha, \beta]$. So the tangent space at the point α of $\mathrm{MC}(\mathfrak{g})$ is equal to

$$T_\alpha \mathrm{MC}(\mathfrak{g}) := \{\beta \in \mathfrak{g}^1 \mid d\beta + [\alpha, \beta] = 0\}.$$

For any $\alpha \in \mathrm{MC}(\mathfrak{g})$ and any $\lambda \in \mathfrak{g}^0$, we have

$$\begin{aligned} d_\alpha Q\big(d\lambda + [\alpha, \lambda]\big) &= [d\alpha, \lambda] - [\alpha, d\lambda] + [\alpha, d\lambda] + \big[\alpha, [\alpha, \lambda]\big] \\ &= [d\alpha, \lambda] + \frac{1}{2}\big[[\alpha, \alpha], \lambda\big] = \big[Q(\alpha), \lambda\big] = 0. \end{aligned}$$

This proves that the aforementioned map lands into the vector fields of $\mathrm{MC}(\mathfrak{g})$. We leave it to the reader to verify that this map is a morphism of Lie algebras. □

The idea now is to look for an equivalence relation on Maurer–Cartan elements using the differential equation associated to flow given by elements of $\mathfrak{g}^0$: two elements $\alpha, \beta \in \mathrm{MC}(\mathfrak{g})$ are equivalent if there exist $\lambda \in \mathfrak{g}^0$ and a curve $\xi(t)$ in $\mathrm{MC}(\mathfrak{g})$ such that

$$\frac{d\xi(t)}{dt} = d\lambda + \big[\xi(t), \lambda\big], \qquad \xi(0) = \alpha \quad \text{and} \quad \xi(1) = \beta.$$

To do so, we suppose that the Lie algebra $\mathfrak{g}$ is nilpotent, i.e. the lower central series $\mathfrak{g} \supset [\mathfrak{g}, \mathfrak{g}] \supset [\mathfrak{g}, [\mathfrak{g}, \mathfrak{g}]] \supset \cdots$ vanishes eventually. In this case, the Lie subalgebra $\mathfrak{g}^0$ is the tangent Lie algebra of a Lie group G, called the *gauge group*. Its underlying set is in bijection with $\mathfrak{g}^0$ and explicitly given by elements of the form $\{e^\lambda, \lambda \in \mathfrak{g}^0\}$. The neutral element is e^0 and the group law is given by the Baker–Campbell–Hausdorff formula (cf. Exercise 1.8.13)

$$e^\lambda * e^\mu := e^{\mathrm{BCH}(\lambda,\mu)} = e^{\lambda+\mu+\frac{1}{2}[\lambda,\mu]+\frac{1}{12}[\lambda,[\lambda,\mu]]-\frac{1}{12}[\mu,[\lambda,\mu]]+\cdots}.$$

For more details on this point, we refer to [Ser06]. The one-parameter subgroups $e^{t\lambda}$ of G associated to $\lambda \in \mathfrak{g}^0$ acts on $\mathrm{MC}(\mathfrak{g})$ by the formula

$$e^{t\lambda}.\alpha := e^{t\,\mathrm{ad}_\lambda}(\alpha) + \frac{\mathrm{Id} - e^{t\,\mathrm{ad}_\lambda}}{\mathrm{ad}_\lambda}(d\lambda),$$

where $\mathrm{ad}_\lambda(\alpha) = [\lambda, \alpha]$. Therefore, two elements $\alpha, \beta \in \mathrm{MC}(\mathfrak{g})$ are defined to be equivalent if there exists $\lambda \in \mathfrak{g}^0$ such that $e^\lambda.\alpha = \beta$.

The cosets under this action form the *moduli space* of Maurer–Cartan elements of $\mathfrak{g}$:

$$\mathscr{MC}(\mathfrak{g}) := \mathrm{MC}(\mathfrak{g})/G.$$

One can also consider the associated Deligne groupoid. The objects are the Maurer–Cartan elements and the morphism sets are given by the elements of the gauge group.

Recall that, in an Artin ring $\mathfrak{R}$, any descending chain of ideals stabilizes

$$\cdots = I_{n+1} = I_n \subset I_{n-1} \subset \cdots I_1 \subset I_0.$$

To satisfy the nilpotency condition, one considers the nilpotent dg Lie algebras $\mathfrak{g} \otimes \mathfrak{m}$, where $\mathfrak{m}$ is the maximal ideal of a local Artin ring $\mathfrak{R}$. In the end, the *deformation functor*

$$\begin{aligned} \mathscr{D}ef_{\mathfrak{g}} : \text{local Artin ring} &\to \mathsf{Set} \\ \mathfrak{R} = \mathbb{K} \oplus \mathfrak{m} &\mapsto \mathscr{MC}(\mathfrak{g} \otimes \mathfrak{m}) \end{aligned}$$

faithfully encompasses the global moduli space of Maurer–Cartan elements of $\mathfrak{g}$. The idea of tensoring with a nilpotent maximal ideal is the same as that of distribution in Analysis, as explained below.

Lemma 13.2.9. *By linear dualization, the category of local Artin rings is equivalent to the category of conilpotent coalgebras*

$$\begin{aligned} \text{conilpotent coalgebra} &\to \text{local Artin ring} \\ C &\mapsto C^*. \end{aligned}$$

Proof. The linear dual of a coaugmented coalgebra is a local ring. Under dualization, a conilpotent coalgebra gives an Artin ring. □

So, one can equivalently consider the deformation functor

$$\begin{aligned} \text{conilpotent coalgebra} &\to \mathsf{Set} \\ C = \mathbb{K} \oplus \overline{C} &\mapsto \mathscr{TW}(\overline{C}, \mathfrak{g}), \end{aligned}$$

where $\mathscr{TW}(\overline{C}, \mathfrak{g})$ stands for the moduli space of twisting morphisms, that is Maurer–Cartan elements in the nilpotent convolution Lie algebra $\mathrm{Hom}(\overline{C}, \mathfrak{g})$, see Sect. 11.1.

A deformation problem is encoded by a dg Lie algebra when the possible structures that one wants to consider are in one-to-one correspondence with the set of Maurer–Cartan elements and when the equivalence between them corresponds to the gauge group action. One can consider deformation functors with value in Deligne groupoids, which we do not work out in this book.

Deformation functors are invariant under quasi-isomorphisms of dg Lie algebras.

Theorem 13.2.10. [GM88] *Any quasi-isomorphism $\mathfrak{g} \xrightarrow{\sim} \mathfrak{g}'$ of dg Lie algebras induces an isomorphism between the deformation functors $\mathscr{D}ef_{\mathfrak{g}} \cong \mathscr{D}ef_{\mathfrak{g}'}$.*

Theorem 13.2.11. [Kon03] *Any ∞-quasi-isomorphism $\mathfrak{g} \overset{\sim}{\rightsquigarrow} \mathfrak{g}'$ of dg Lie algebras induces an isomorphism between the deformation functors $\mathscr{D}ef_{\mathfrak{g}} \cong \mathscr{D}ef_{\mathfrak{g}'}$.*

Proof. The rectification theorem 11.4.4 gives the following commutative diagram of ∞-quasi-isomorphisms

$$\begin{array}{ccc} \mathfrak{g} & \overset{\sim}{\rightsquigarrow} & \mathfrak{g}' \\ \sim\uparrow & & \sim\uparrow \\ \Omega_\kappa \mathrm{B}_\iota \mathfrak{g} & \xrightarrow{\sim} & \Omega_\kappa \mathrm{B}_\iota \mathfrak{g}', \end{array}$$

where the three solid arrows are quasi-isomorphisms of dg Lie algebras. The statement is a consequence of the previous proposition. □

So, it is enough to show that two dg Lie algebras are ∞-quasi-isomorphic to prove the equivalence of the associated deformation functors. This method was used in a crucial way by Kontsevich in [Kon03] to prove that any Poisson manifold can be quantized by deformation. More precisely, he extended the Hochschild–Kostant–Rosenberg quasi-isomorphism into an explicit ∞-quasi-isomorphism

$$\Lambda_A \operatorname{Der} A \overset{\sim}{\rightsquigarrow} C^\bullet(A, A)$$

of L_∞-algebras, when $A = C^\infty(M)$ is the algebra of smooth functions on a Poisson manifold M. This result proves the formality of the Hochschild cochain complex as a dg Lie algebra.

For more exhaustive accounts on deformation theory, we refer the reader to [Man99a, Kon03, Kel05, KS10].

13.2.15 Restricted Lie Algebras

Let A be an associative algebra over a characteristic p field $\mathbb{K}$. The bracket $[x, y]$ and the pth power x^p are operations related by the following relation:

$$(x+y)^p = x^p + y^p + \sum_{i=1}^{p-1} s_i(x, y),$$

where the polynomial $s_i(x, y)$ is of degree i in x. It turns out that these polynomials are Lie polynomials in x and y, that is, when $A = T(V)$ they lie in $Lie(V)$. So they are universally determined as Lie polynomials.

By definition, cf. [Jac62], a *restricted Lie algebra* is a pair $(L, [p])$ where L is a Lie algebra over a characteristic $p \neq 0$ field $\mathbb{K}$ and $[p] : L \to L,\ x \mapsto x^{[p]}$ is a map, called the *Frobenius map*, which satisfies the following relations:

$$(\alpha x)^{[p]} = \alpha^p x^{[p]}, \quad \alpha \in \mathbb{K},\ x \in L,$$

$$\big[x, y^{[p]}\big] = \big[\cdots \big[\underbrace{[x, y], y\big] \cdots, y}_{p}\big], \quad x, y \in L,$$

$$(x+y)^{[p]} = x^{[p]} + y^{[p]} + \sum_{i=1}^{p-1} s_i(x, y),$$

where $s_i(x, y)$ is the aforementioned Lie polynomial. A morphism $f : L \to L'$ of restricted Lie algebras is a Lie morphism such that $f(x^{[p]}) = f(x)^{[p]}$.

The relationship with operads is the following. We have mentioned in Sect. 5.2.9 that any symmetric operad $\mathcal{P}$ gives rise to a divided powers operad $\Gamma\mathcal{P}$. Applied to $\mathcal{P} = Lie$ we get the notion of ΓLie-algebra. It is proved in [Fre00] that the notion of ΓLie-algebra coincides with the notion of restricted Lie algebra.

EXAMPLE. Any associative algebra A over a characteristic p field $\mathbb{K}$ has the structure of restricted Lie algebra with Frobenius map given by $a \mapsto a^p$. In the case of $n \times n$-matrices over A this Lie algebra is usually denoted by $\mathfrak{gl}_n(A)$. The Lie subalgebra of trace zero matrices, denoted by $\mathfrak{sl}_n(A)$, is also a restricted Lie algebra.

Let A be a magmatic algebra. The space of derivations D on A is a Lie algebra for the bracket $[D, D'] = D \circ D' - D' \circ D$. It is also a restricted Lie algebra for the Frobenius map $D \mapsto D^p$.

13.3 Poisson Algebras, Gerstenhaber Algebras and Their Operad

Poisson algebras mix a commutative operation and a Lie bracket by means of a distributive law. They appeared in differential geometry as the structure of the space of real valued functions on a symplectic manifold, and more generally Poisson manifold. A Gerstenhaber algebra is a Poisson algebra with the Lie bracket in degree 1. This notion plays a crucial role in current mathematics: the Hochschild cohomology of an associative algebra with coefficients in itself and the polyvector fields of a smooth manifold carry a structure of Gerstenhaber algebra. The Poisson operad is a sort of trivial extension of the operads *Com* and *Lie*.

13.3.1 Poisson Algebra

By definition, a *Poisson algebra* A is a vector space equipped with a symmetric operation $x \cdot y$ and an antisymmetric operation $[x, y]$, which satisfy the following relations:

$$\begin{cases} \big[[x,y],z\big] + \big[[y,z],x\big] + \big[[z,x],y\big] = 0, \\ [x\cdot y, z] = x\cdot[y,z] + [x,z]\cdot y, \\ (x\cdot y)\cdot z = x\cdot(y\cdot z). \end{cases}$$

In other words, one has a commutative operation and a Lie bracket entwined by the Leibniz relation. Its name comes from the French mathematician Siméon Poisson.

13.3.2 Free Poisson Algebra

The free Poisson algebra $Pois(V)$ over V is the reduced tensor module $\overline{T}(V)$ as a vector space. The commutative structure and the Lie structure are obtained as follows. Recall that there is an isomorphism of coalgebras (Eulerian isomorphism) $\overline{T}(V) \cong \overline{S}(Lie(V))$. From this isomorphism we deduce the structure of commutative algebra. The structure of Lie algebra on $Lie(V)$ is obvious. It is extended to $\overline{S}(Lie(V))$ by means of the Leibniz relation (distributive law), and then transferred to $T(V)$.

13.3.3 The Operad *Pois*

The operad encoding Poisson algebras is denoted by *Pois*. By definition, the operad *Pois* fits into the two following sequences of operads

$$Com \rightarrowtail Pois \twoheadrightarrow Lie,$$

$$Lie \rightarrowtail Pois \twoheadrightarrow Com.$$

The operad *Pois* can be constructed out of *Lie* and *Com* by means of a distributive law, see Sect. 8.6,

$$Pois \cong Com \circ Lie.$$

This distributive law $Lie \circ Com \to Com \circ Lie$ is given by the second relation in the presentation of the Poisson operad.

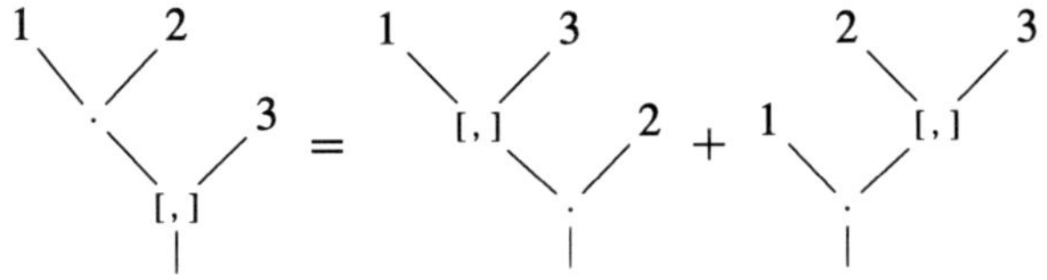

By Proposition 8.6.1, it makes the composite $Com \circ Lie$ into an operad. This is precisely the operad *Pois*.

From this isomorphism and the isomorphism $Ass \cong Com \circ Lie$, it follows that $Pois(n) \cong \mathbb{K}[\mathbb{S}_n]$ (regular representation).

13.3.4 Parametrized Poisson Operad

The Poisson operad and the associative operad are related by a family of operads as follows. Let $q \in \mathbb{K}$ be a parameter. Consider the algebras having a symmetric operation $x \cdot y$ and an antisymmetric operation $[x, y]$ satisfying the relations:

$$\begin{cases} \big[[x,y],z\big]+\big[[y,z],x\big]+\big[[z,x],y\big]=0, \\ \qquad\qquad [x\cdot y,z]=x\cdot[y,z]+[x,z]\cdot y, \\ \qquad\qquad (x\cdot y)\cdot z - x\cdot(y\cdot z)=q[y,[x,z]]. \end{cases}$$

For $q=0$ we get the Poisson operad. We have seen in Proposition 9.1.1 that for $q=1$ we get the associative operad. It is easy to show that, if $(\sqrt{q})^{-1}$ exists in $\mathbb{K}$, then the associated operad is isomorphic to *Ass*.

The parametrized Poisson operad $Pois_q$ can also be presented by a unique operation without symmetry, denoted ab, with relation

$$(ab)c = a(bc) + \frac{q-1}{q+3}\big(-a(cb)+c(ab)+b(ac)-b(ca)\big).$$

Hence, taking $q=0$, a Poisson algebra is defined by a binary operation ab satisfying

$$(ab)c = a(bc) - \frac{1}{3}\big(-a(cb)+c(ab)+b(ac)-b(ca)\big).$$

The relationship with the previous presentation is given by $ab = a\cdot b + [a,b]$.

13.3.5 Koszul Dual of *Pois*

From the duality exchanging *Com* and *Lie* and from the distributive law, one can check that the Poisson operad is Koszul self-dual: $Pois^{!} = Pois$. It is an example of Koszul duality for distributive laws, cf. Sect. 8.6.

More generally the Koszul dual operad of the parametrized Poisson operad is generated by a binary operation ab with relation

$$a(bc) = (ab)c + \frac{q-1}{q+3}\big((ac)b+(bc)a-(ba)c-(ca)b\big).$$

This is precisely the opposite type of $Pois_q$.

13.3.6 Poisson Homology and Eulerian Idempotents

Let A be a Poisson algebra. Since the operad *Pois* is Koszul self-dual, the module of n-chains of A is $C_n(A) := A^{\otimes n}$.

Denote by

$$C_n(A) = C_n^{(1)}(A) \oplus \cdots \oplus C_n^{(n)}(A)$$

the *Eulerian decomposition* (Sect. 1.3.5), cf. [Lod89], of $C_n(A)$ (sometimes called the *Hodge decomposition*). Operadically it is obtained as

$$C_n^{(i)}(A) = \big(Com^c(i) \circ Lie^c\big)(A).$$

Benoit Fresse has shown in [Fre06] that there is a differential map $d^h : C^{(p)}_{p+q}(A) \to C^{(p-1)}_{p+q-1}(A)$ induced by the bracket operation, and a differential map $d^v : C^{(p)}_{p+q}(A) \to C^{(p)}_{p+q-1}(A)$ induced by the commutative operation (dot operation), which define a bicomplex:

$$\begin{array}{ccccccc}
\cdots & & & & & & \\
\downarrow & & & & & & \\
C_3^{(1)} & \longleftarrow & \cdots & & & & \\
\downarrow{\scriptstyle d^v} & & \downarrow{\scriptstyle d^v} & & & & \\
C_2^{(1)} & \xleftarrow{d^h} & C_3^{(2)} & \xleftarrow{d^h} & \cdots & & \\
\downarrow{\scriptstyle d^v} & & \downarrow{\scriptstyle d^v} & & \downarrow & & \\
C_1^{(1)} & \xleftarrow{d^h} & C_2^{(2)} & \xleftarrow{d^h} & C_3^{(3)} & \xleftarrow{d^h} & \cdots
\end{array}$$

In low dimension the equality $d^h d^h = 0$, resp. $d^v d^h + d^h d^v = 0$, resp. $d^v d^v = 0$, is precisely the first, resp. second, resp. third, relation of a Poisson algebra. Fresse has shown that the total complex of this bicomplex is the operadic chain complex $C^{Pois}_{\bullet-1}(A)$. The vertical complex $C^{(1)}_{\bullet}(A) = (Com^c(1) \circ Lie)^c(A)$ is the Harrison complex of the underlying commutative algebra A, and the bottom horizontal complex $\bigoplus_n C_n^{(n)}(A) = \bigoplus_n (Com^c(n) \circ \mathrm{I})(A) = \bigoplus_n \Lambda^n(A)$ is the Chevalley–Eilenberg complex of the Lie algebra A_{Lie}.

13.3.7 Poisson Homology and Straight Shuffles

From the presentation of Poisson algebras by one binary operation ab without symmetry it follows that the boundary map of the chain complex

$$C^{Pois}_{\bullet-1}(A) \qquad \cdots \to A^{\otimes n} \xrightarrow{d_n} A^{\otimes n-1} \to \cdots \to A^{\otimes 3} \xrightarrow{d_3} A^{\otimes 2} \xrightarrow{d_2} A$$

is as follows in low dimension:

$$\begin{aligned}
&d_2(a,b) = ab,\\
&d_3(a,b,c) = 3(ab)\otimes c - 3a \otimes (bc) + a \otimes (cb) - c \otimes (ab) - b \otimes (ac) + b \otimes (ca).
\end{aligned}$$

Making d_n explicit higher up is a challenge, but can be done using the notion of straight shuffles and the computations of the Koszul dual cooperad $Pois^{¡}$, see [GCTV09, Sect. 2].

13.3.8 Poisson Homology of Smooth Algebras

Before the operad theory told us about the correct chain complex to compute the (co)homology of a Poisson algebra, people were using a different complex for Poisson homology with coefficients, defined as follows. Let A be the Poisson algebra and M be a Poisson module. There is a well-defined differential map on $M \otimes_A \Omega_A^\bullet$ induced by the bracket operation. Here $\Omega_A^\bullet$ stands for the space of classical differential forms of the commutative algebra A. When A is smooth as a commutative algebra, it turns out that the bicomplex (with coefficients in M) is quasi-isomorphic to its lower row which is the complex $M \otimes_A \Omega_A^\bullet$ (cf. [Fre06, Sect. 1.4]).

13.3.9 Graded Poisson Algebras

If we denote by m the generator of the operad *Pois* corresponding to the commutative operation, resp. c for the Lie bracket, of a Poisson algebra, then the relations read:

$$c \circ_1 c + (c \circ_1 c)^{(123)} + (c \circ_1 c)^{(321)} = 0,$$

$$c \circ_1 m = m \circ_2 c + (m \circ_1 c)^{(23)},$$

$$m \circ_1 m = m \circ_2 m.$$

A *graded Poisson algebra* is a graded vector space A equipped with two binary operations m and c of degree 0 satisfying the aforementioned relations. When applying these relations on homogeneous elements, the degree comes into play. For instance the second relation gives:

$$[x \cdot y, z] = x \cdot [y, z] + (-1)^{|y||z|}[x, z] \cdot y.$$

13.3.10 Gerstenhaber Algebras

A *Gerstenhaber algebra* [Ger63] is a graded vector space A equipped with a (sign-graded) symmetric associative product $x \cdot y$ of degree 0 on A and a (sign-graded) symmetric operation $\langle x, y\rangle$ of degree $+1$, that is $|\langle x, y\rangle| = |x| + |y| + 1$, which satisfies the Jacobi relation. They are related by the Leibniz relation. In plain words, a Gerstenhaber algebra is a graded Poisson algebra, except that the skew-symmetric degree 0 bracket becomes a symmetric degree 1 binary operation. The assumption that $\langle\,,\rangle$ has homological degree 1 does not change the appearing signs because of the permutation of the variables. For instance, one has $\langle y, x\rangle = (-1)^{|x||y|}\langle x, y\rangle$ and

$$\big\langle\langle x, y\rangle, z\big\rangle + (-1)^{|x|(|y|+|z|)}\big\langle\langle y, z\rangle, x\big\rangle + (-1)^{|z|(|x|+|y|)}\big\langle\langle z, x\rangle, y\big\rangle = 0.$$

One sometimes defines equivalently a Gerstenhaber algebra by Lie bracket $[\,,]$ of degree 0 on sA, the suspension of A, see [Get94, TT00] for instance. It is supposed to satisfy the Leibniz relation with the commutative product, but since they do not act on the same space, we need the preceding refinement to make that relation precise. Moreover, to encode the category of Gerstenhaber algebras with an operad, we need it to act on the same space. A Lie bracket $[\,,]$ on sA is equivalent to a commutative operation $\langle\,,\rangle$ of degree $+1$ on A under the formula

$$\langle -,-\rangle := s^{-1}[-,-] \circ (s\otimes s),$$

which gives $\langle x, y\rangle := (-1)^{|x|}s^{-1}[sx, sy]$ on elements. The commutativity follows from

$$\langle\,,\rangle^{(12)} = \big(s^{-1}[\,,] \circ s\otimes s\big)^{(12)} = -s^{-1}\big([\,,]^{(12)}\big) \circ s\otimes s = s^{-1}[\,,] \circ s\otimes s = \langle\,,\rangle,$$

which, applied to homogeneous elements, reads

$$\begin{aligned}\langle y, x\rangle &= (-1)^{|y|}s^{-1}[sy, sx] = (-1)^{|y|+1+(|x|+1)(|y|+1)}s^{-1}[sx, sy]\\ &= (-1)^{|x||y|}\langle x, y\rangle.\end{aligned}$$

EXAMPLE. Let M be a smooth manifold and let $\Gamma(M, TM)$ denote the space of vector fields, i.e. the sections of the tangent bundle, endowed with the classical Lie bracket. The space of *polyvector fields* $\Gamma(M, \overline{S}(sTM)) \cong \Gamma(M, \Lambda(TM))$ is equipped with a Gerstenhaber algebra structure, whose Lie bracket is obtained from the previous one under the Leibniz rule.

13.3.11 Gerstenhaber Algebra Structure on Hochschild Cohomology

The origin of this algebraic structure lies in the following result.

Proposition 13.3.1. [Ger63] *For any associative algebra A, the Hochschild cohomology of A with coefficients in itself inherits a Gerstenhaber algebra structure whose product is induced by the cup product and whose bracket comes from the pre-Lie product.*

Proof. The Hochschild cochain complex of an associative algebra (A, μ) with coefficients into itself is equal to the convolution dg pre-Lie algebra

$$\mathfrak{g}_{As,A} = \operatorname{Hom}\big(As^{¡}, \operatorname{End}_A\big) \cong \prod_{n\geq 1} s^{-n+1}\operatorname{Hom}\big(A^{\otimes n}, A\big).$$

Under this homological degree convention, the cup product $f \cup g := \mu \circ (f \otimes g)$ endows the desuspension $s^{-1}\mathfrak{g}_{As,A}$ with a degree 0 associative product. Recall that the boundary map is given by $\partial(f) = [\mu, f]$. The relation

$$f \cup g - (-1)^{|f||g|} g \cup f = \partial(f) \star g + (-1)^{|f|} - \partial(f \star g)$$

shows that the induced associative product on homology is commutative. One can write the Leibniz relation in terms $\partial(\{f\}\{g, h\})$, showing that this relation holds on homology. □

This result is a consequence of a more refined algebraic structure on the cochain level, namely that of a multiplicative operad, which induces the brace operations $\{-; -, \dots, -\}$, cf. Sect. 13.11.4. Notice that in the literature, the cohomological degree convention $C^n_{Hoch}(A, A) = s^n \operatorname{Hom}(A^{\otimes n}, A)$ is often used. In this case, the Gerstenhaber algebra operations have opposite degree.

One applies this structure as follows to compute the Hochschild cohomology of smooth algebras. The cycles of $C^1(A, A) = \operatorname{Hom}(A, A)$ are exactly the derivations $HH^1(A, A) \cong \operatorname{Der} A$ of the algebra A, which form a Lie sub-algebra. One extends this Lie algebra structure on the free commutative algebra $\Lambda_A \operatorname{Der} A$ by the Leibniz relation. This endows $\Lambda_A \operatorname{Der} A$ with a Gerstenhaber algebra structure. (This is the algebraic analog of the Gerstenhaber polyvector fields algebra.)

Proposition 13.3.2. [HKR62] *Let $A = C^\infty(M)$ be the algebra of smooth functions on a manifold M or let $A = S(V)$ be a free commutative algebra on a finite vector space V. The natural morphism of Gerstenhaber algebras*

$$\Lambda^\bullet_A \operatorname{Der} A \xrightarrow{\cong} HH^\bullet(A, A)$$

is an isomorphism.

Since the cup product is not commutative on the cochain level, one defines a direct quasi-isomorphism $\Lambda^\bullet_A \operatorname{Der} A \xrightarrow{\sim} C^\bullet(A, A)$ by anti-symmetrization $(d_1, \dots, d_n) \mapsto \sum_{\sigma \in \mathbb{S}_n} \pm d_{\sigma(1)} \cup \dots \cup d_{\sigma(n)}$.

13.3.12 The Operad *Gerst*

A Gerstenhaber algebra is an algebra over the graded operad *Gerst*, which admits the following quadratic presentation. The space of generators is $E_{Gerst} = \mathbb{K}m \oplus \mathbb{K}c$, which is the direct sum of two one-dimensional trivial representations of $\mathbb{S}_2$, one in degree 0 denoted by m and one in degree 1 denoted by c. The space of relations R_{Gerst} is the $\mathbb{K}[\mathbb{S}_3]$-module generated by

$$\begin{cases} c \circ_1 c + (c \circ_1 c)^{(123)} + (c \circ_1 c)^{(321)}, \\ c \circ_1 m - m \circ_2 c - (m \circ_1 c)^{(23)}, \\ m \circ_1 m - m \circ_2 m. \end{cases}$$

Recall that sA is a Lie algebra if and only if $A = s^{-1}(sA)$ is a $Lie \underset{\mathrm{H}}{\otimes} \mathrm{End}_{s^{-1}\mathbb{K}} = \mathscr{S}^{-1}Lie$-algebra. The operad $\mathscr{S}^{-1}Lie$ admits the quadratic presentation $\mathscr{P}(\mathbb{K}c, c \circ_1 c + (c \circ_1 c)^{(123)} + (c \circ_1 c)^{(321)})$. Like the operad *Pois*, the operad *Gerst* is isomorphic to an operad obtained by means of a distributive law:

$$Gerst \cong Com \circ \mathscr{S}^{-1}Lie.$$

13.3.13 Relationship with the Little Discs Operad

We introduced the little discs operad $\mathscr{D}$ in Sect. 5.3.6.

Theorem 13.3.3. [Coh76] *The singular homology $H_\bullet(\mathscr{D})$ of the little discs operad is isomorphic to the operad Gerst.*

Roughly speaking, the little discs operad $\mathscr{D}$ is made of the configurations of n discs inside the unit disc of the plane. The operadic composition is given by shrinking and inserting discs. Since a double loop space $\Omega^2 X$ is an algebra over the little discs operad, this theorem proves that the homology of a double loop space carries a natural structure of Gerstenhaber algebra, where the product is the Pontryagin product and the bracket is the Browder bracket, see [Get94] for details.

13.3.14 Homotopy Gerstenhaber Algebras

As well as for the operad *Pois*, the Koszul duality theory applies to the graded operad *Gerst*. The Koszul resolution $G_\infty := \Omega Gerst^{¡} \xrightarrow{\sim} Gerst$ produces the minimal model of *Gerst*.

Proposition 13.3.4. [GJ94] *The operad Gerst is a Koszul operad, whose Koszul dual operad is given by*

$$Gerst^! \cong \mathscr{S}Gerst = Gerst \underset{\mathrm{H}}{\otimes} \mathrm{End}_{s\mathbb{K}}.$$

As usual, an algebra over the operad G_∞ is called a *G_∞-algebra* or a *homotopy Gerstenhaber algebra*. This notion is different from that of "homotopy G-algebra" [GV95]. To avoid any confusion, we call this later one "GV-algebra", see Sect. 13.11.4. A G_∞-algebra can be equivalently defined by a square-zero coderivation on the cofree "Gerstenhaber coalgebra"

$$G^{¡}(A) \cong s^{-1}Gerst^c(sA) \cong s^{-2}Com^c\big(sLie^c(sA)\big).$$

Up to suspension, the right-hand side is given by the cofree cocommutative coalgebra (without counit) on the cofree Lie coalgebra $\overline{S^c}(sLie^c(A))$. The explicit structure of Gerstenhaber coalgebra on such a space was described in [GCTV09, Sect. 2]. As a corollary, it gives the explicit definition of a G_∞-algebra in terms of generating operations and relations [GCTV09, Proposition 16].

13.3.15 Deligne's Conjecture

Theorems 13.3.1 and 13.3.3 led Deligne [Del93] to ask the following question, which is now called *Deligne's conjecture*.

> *Is there an action of an operad, homotopy equivalent to the operad of singular chains of the little discs operad $\mathcal{D}$, on the Hochschild cochain complex of an associative algebra, which induces the Gerstenhaber algebra structure on cohomology?*

The "homotopy equivalence" refers to the homotopy category of dg operads Appendix B.6; it means that the two operads should be related by a zig-zag of quasi-isomorphisms. This conjecture has now been proved by many authors [Tam99, Vor00, KS00, MS02, BF04, Kau07, Tam07, BB09] for instance. Here we just mention the proof given by Tamarkin using the Koszul model.

Theorem 13.3.5. [Tam99] *There is an action of the operad G_∞ on the Hochschild cochain complex of an associative algebra, which induces the Gerstenhaber algebra on cohomology.*

We also refer the reader to [TT00, Hin03] for more details. To fully answer Deligne's question, Tamarkin proved the formality of the little discs operad.

Theorem 13.3.6. [Tam03] *There is a zig-zag of quasi-isomorphisms of dg operads*

$$C_\bullet(\mathcal{D}) \overset{\sim}{\longleftarrow} \cdot \overset{\sim}{\longrightarrow} H_\bullet(\mathcal{D}).$$

Since $H_\bullet(\mathcal{D}) \cong \mathit{Gerst}$ by Cohen's theorem 13.3.3, the operad G_∞ is homotopy equivalent to the chain operad $C_\bullet(\mathcal{D})$ of the little discs operad.

$$C_\bullet(\mathcal{D}) \overset{\sim}{\longleftarrow} \cdot \overset{\sim}{\longrightarrow} H_\bullet(\mathcal{D}) \cong \mathit{Gerst} \overset{\sim}{\longleftarrow} G_\infty \longrightarrow \mathrm{End}_{C^\bullet(A,A)}.$$

These two results allowed Tamarkin in [Tam99] to give another proof of Kontsevich formality theorem, which implies the deformation-quantization of Poisson manifolds. For a polynomial algebra $A = S(V)$, with V finite-dimensional, he proved the vanishing of the obstructions to extending the Hochschild–Kostant–Rosenberg quasi-isomorphism $\Lambda_A^\bullet \operatorname{Der} A \xrightarrow{\sim} C^\bullet(A, A)$ into an ∞-quasi-isomorphism of G_∞-algebras. Since there is an inclusion $\mathcal{S}^{-1}L_\infty \rightarrowtail G_\infty$, it implies the existence of an ∞-quasi-isomorphism of L_∞-algebras. Contrary to Kontsevich proof, this one is not explicit. In [DTT07], the authors refine these arguments and construct a zig-zag of ∞-quasi-isomorphisms of G_∞-algebras. The method depends on the rectification theorem 11.4.4.

13.3.16 The Operads e_n

The Gerstenhaber operad is defined as the Poisson operad but with a twisting of sign and a degree shift on one generator. More generally, for $n \geq 2$, we define the operads e_n by the following quadratic presentation. The space of generators is $E_{e_n} := \mathbb{K}m \oplus \mathbb{K}c_{n-1}$, which is the direct sum of two one-dimensional representations of $\mathbb{S}_2$, $\mathbb{K}m$ of degree 0 with trivial action and $\mathbb{K}c_{n-1}$ of degree $n-1$ and with $\mathrm{sgn}_{\mathbb{S}_2}^{\otimes n}$-action, that is trivial for n even and $\mathrm{sgn}_{\mathbb{S}_2}$ for n odd. The space of relations R_{e_n} is the $\mathbb{K}[\mathbb{S}_3]$-module generated by

$$\begin{cases} c_{n-1} \circ_1 c_{n-1} + (c_{n-1} \circ_1 c_{n-1})^{(123)} + (c_{n-1} \circ_1 c_{n-1})^{(321)}, \\ c_{n-1} \circ_1 m - m \circ_2 c_{n-1} - (m \circ_1 c_{n-1})^{(23)}, \\ m \circ_1 m - m \circ_2 m. \end{cases}$$

For $n = 1$, the convention is $e_1 := \mathit{Ass}$, the symmetric operad encoding associative algebras. For $n = 2$, we find $e_2 = \mathit{Gerst}$ as expected.

These operads allow one [Coh76] to extend Theorem 13.3.3 to higher dimensions: the homology operads of the little n-discs operads are isomorphic to e_n:

$$H_\bullet(D_n) \cong e_n, \quad \text{for } n \geq 1.$$

There is a sequence of morphisms of operads

$$\mathit{Ass} = e_1 \to e_2 \to \cdots \to e_n \to \cdots \to \operatorname{colim}_n e_n = \mathit{Com},$$

where $i_n : e_n \to e_{n+1}$ is given by $m \mapsto m$ and $c_{n-1} \mapsto 0$. At the limit $n \to \infty$, one recovers the operad Com and the operads e_n interpolate between Ass and Com.

By the same proof as in Proposition 13.3.4, one can show that the operad e_n is Koszul [GJ94] and satisfies

$$e_n^{!} \cong \mathcal{S}^{n-1} e_n = \underbrace{\mathrm{End}_{s\mathbb{K}} \underset{\mathrm{H}}{\otimes} \cdots \underset{\mathrm{H}}{\otimes} \mathrm{End}_{s\mathbb{K}}}_{n-1} \underset{\mathrm{H}}{\otimes} e_n.$$

Let us denote by $\kappa_n : e_n^{¡} \to e_n$ the Koszul morphism, which is given by $sm \mapsto m$ and $sc_{n-1} \mapsto c_{n-1}$. We define the following morphisms of quadratic cooperads

$$j_n : e_n^{¡} \to e_{n+1}^{¡} \quad \text{by} \quad sm \mapsto sm \quad \text{and} \quad sc_{n-1} \mapsto 0,$$

making the following diagram to commute

$$\begin{array}{ccc} e_n^{¡} & \xrightarrow{j_n} & e_{n+1}^{¡} \\ \downarrow{\scriptstyle \kappa_n} & & \downarrow{\scriptstyle \kappa_{n+1}} \\ e_n & \xrightarrow{i_n} & e_{n+1}. \end{array}$$

They correspond to the morphisms $\mathscr{S}e_{n+1} \to e_n$ given by $s^{-1}m \mapsto 0$ and by $s^{-1}c_n \mapsto c_{n-1}$ at the level of the Koszul dual operads. Finally, they induce the following commutative diagram of dg operads

$$\begin{array}{ccccccccc} A_\infty & \longrightarrow \cdots \longrightarrow & \Omega e_n^{¡} & \longrightarrow & \Omega e_{n+1}^{¡} & \longrightarrow \cdots \longrightarrow & C_\infty \\ \downarrow \wr & & \downarrow \wr & & \downarrow \wr & & \downarrow \wr \\ Ass & \longrightarrow \cdots \longrightarrow & e_n & \longrightarrow & e_{n+1} & \longrightarrow \cdots \longrightarrow & Com. \end{array}$$

Therefore, the operads $\Omega e_n^{¡}$ interpolate between A_∞ and C_∞.

13.3.17 Koszul Duality of the Operads $\mathscr{E}_n$

Recall from Sect. 13.1.11 that the Barratt–Eccles dg operad $\mathscr{E}$ comes equipped with a filtration of suboperads

$$Ass = \mathscr{E}_1 \subset \mathscr{E}_2 \subset \cdots \subset \mathscr{E}_n \subset \cdots \subset \mathrm{Colim}_n \mathscr{E}_n = \mathscr{E}.$$

Passing to homology, we get $H_\bullet(\mathscr{E}_n) = e_n$ and so we recover the aforementioned sequence of morphisms

$$Ass = e_1 \to e_2 \to \cdots \to e_n \to \cdots \to \mathrm{colim}_n e_n = Com.$$

Fresse in [Fre11b] has extended the above results to the level of the operads $\mathscr{E}_n$ as follows. By a slight abuse of notation, let us take the following convention for the dg cooperads

$$\mathscr{E}_n^{¡} := \underbrace{\mathrm{End}^c_{s^{-1}\mathbb{K}} \underset{\mathrm{H}}{\otimes} \cdots \underset{\mathrm{H}}{\otimes} \mathrm{End}^c_{s^{-1}\mathbb{K}}}_{n} \underset{\mathrm{H}}{\otimes} \mathscr{E}_n^*.$$

Proposition 13.3.7. [Fre11b] *For any $n \geq 1$, there exists a Koszul morphism $\varkappa_n : \mathscr{E}_n^{¡} \to \mathscr{E}_n$.*

It gives the quasi-free model $\Omega \mathscr{E}_n^{¡} \xrightarrow{\sim} \mathscr{E}_n$. Moreover, there exist morphisms of dg cooperads $\mathscr{E}_n^{¡} \to \mathscr{E}_{n+1}^{¡}$ such that

$$\begin{array}{ccc} \mathscr{E}_n^{¡} & \longrightarrow & \mathscr{E}_{n+1}^{¡} \\ \downarrow \varkappa_n & & \downarrow \varkappa_{n+1} \\ \mathscr{E}_n & \rightarrowtail & \mathscr{E}_{n+1}. \end{array}$$

They imply the following commutative diagram of operads.

$$\begin{array}{ccccccccc}
\Omega Ass^{¡} & \longrightarrow \cdots \longrightarrow & \Omega\mathcal{E}_n{}^{¡} & \longrightarrow & \Omega\mathcal{E}_{n+1}{}^{¡} & \longrightarrow \cdots \longrightarrow & \Omega\mathcal{E}^{¡} \\
\downarrow{\scriptstyle \sim} & & \downarrow{\scriptstyle \sim} & & \downarrow{\scriptstyle \sim} & & \downarrow{\scriptstyle \sim} \\
Ass & \longrightarrow \cdots \longrightarrow & \mathcal{E}_n & \longrightarrow & \mathcal{E}_{n+1} & \longrightarrow \cdots \longrightarrow & \mathcal{E},
\end{array}$$

where $\mathcal{E}^{¡}$ is the colimit of the $\mathcal{E}_n{}^{¡}$.

The quasi-free, thus cofibrant, resolution $\Omega\mathcal{E}_n^{¡} \xrightarrow{\sim} \mathcal{E}_n$ provides a small chain complex which computes the homology of $\mathcal{E}_n$-algebras. This last one is equal to the homology of the n-fold iterated bar construction [Fre11a]. Applied to the normalized singular chain complex of a topological space X, it also computes the homology of its n-fold loop space $\Omega^n X$. The case of the spheres $X = S^m$ is treated in [Fre10].

13.3.18 Lie–Rinehart Algebras

Let A be a commutative algebra and let L be a Lie algebra. We suppose that L is equipped with a module structure over A:

$$A \otimes L \to L, \qquad (a, X) \mapsto aX,$$

and we suppose that there is given a Lie algebra morphism

$$\omega : L \to \operatorname{Der}(A), \qquad X \mapsto \big(a \mapsto X(a)\big)$$

where $\operatorname{Der}(A)$ is the Lie algebra of derivations of the algebra A. The pair (A, L) is called a *Lie–Rinehart algebra* with *anchor* ω, provided the following relations hold:

$$\begin{cases} (aX)(b) = a\big(X(b)\big), & X \in L, a, b \in A, \\ [X, aY] = a[X, Y] + X(a)Y, & X, Y \in L, a \in A. \end{cases}$$

There is an obvious notion of morphism of Lie–Rinehart algebras and therefore a category of Lie–Rinehart algebras.

Any associative algebra R gives rise to a Lie–Rinehart algebra (A, L) as follows: $A = R_{ab} := R/([R, R])$, and $L = R_{Lie}$ is the vector space R considered as a Lie algebra for the usual bracket $[r, s] = rs - sr$.

Proposition 13.3.8. *Any Lie–Rinehart algebra (A, L) gives rise to a Poisson algebra $P = A \oplus L$ such that the two operations $\cdot$ and $[\,,\,]$ take values as follows:*

$$\begin{array}{ll}
A \otimes A \xrightarrow{\cdot} A, & A \otimes A \xrightarrow{[,]} 0, \\
A \otimes L \xrightarrow{\cdot} L, & L \otimes A \xrightarrow{[,]} A, \\
L \otimes L \xrightarrow{\cdot} 0, & L \otimes L \xrightarrow{[,]} L.
\end{array}$$

Conversely, any Poisson algebra P, whose underlying vector space can be split as $P = A \oplus L$ and such that the two operations take values as indicated above, defines a Lie–Rinehart algebra. The two constructions are inverse to each other.

Proof. Let us start with a Lie–Rinehart algebra (A, L). We define the two operations $x \cdot y$ and $[x, y]$ on $A \oplus L$ as follows

$$(a+X)\cdot(b+Y) := ab + (aY + bX),$$
$$[a+X, b+Y] := \big(X(b) - Y(a)\big) + [X, Y],$$

for any $a, b \in A$ and any $X, Y \in L$. Checking the axioms of a Poisson algebra is straightforward.

The rest of the proof is straightforward. □

Let us mention that one can also study Lie–Rinehart algebras operadically by constructing a two-colored Koszul operad whose algebras are the Lie–Rinehart algebras, cf. Sect. 13.14.1.

13.4 Pre-Lie Algebras and Perm-Algebras

The notion of pre-Lie algebra appeared first in the study of Hochschild cohomology and in differential geometry (flat affine connections on a given manifold), but nowadays it is also present in algebraic combinatorics and theoretical physics (renormalization). It also plays a role in operad theory since the convolution algebra of the morphisms from a cooperad to an operad is a pre-Lie algebra. We introduced pre-Lie algebras in Chap. 1, see Sect. 1.4.

13.4.1 Pre-Lie Algebra

By definition a (right) *pre-Lie algebra* is a vector space A equipped with a binary operation $\{x, y\}$ which satisfies the following relation:

$$\big\{\{x, y\}, z\big\} - \big\{x, \{y, z\}\big\} = \big\{\{x, z\}, y\big\} - \big\{x, \{z, y\}\big\}.$$

In plain words, the associator is right-symmetric. For the opposite type (i.e. $\langle x, y\rangle := \{y, x\}$) the associator is left-symmetric.

13.4.2 Rooted Trees and Free Pre-Lie Algebra

For the purpose of this section we consider the set of rooted trees which have no leaves and such that the number of inputs is nonnegative (so 0 input is allowed). We denote by $rT(n)$ the set of such trees endowed with an enumeration of the n vertices (see figures below). The change of enumeration gives an action of the symmetric group. For any space V there is defined a binary operation on $\bigoplus_n \mathbb{K}[rT(n)] \otimes_{\mathbb{S}_n} V^{\otimes n}$ as follows. The parameters $v_1, \dots, v_n$ are labelings of the vertices of the tree. The following pictures represent the elements x, $\{x, y\}$, $\{x, \{y, z\}\}$, $\{\{x, y\}, z\} - \{x, \{y, z\}\}$ respectively:

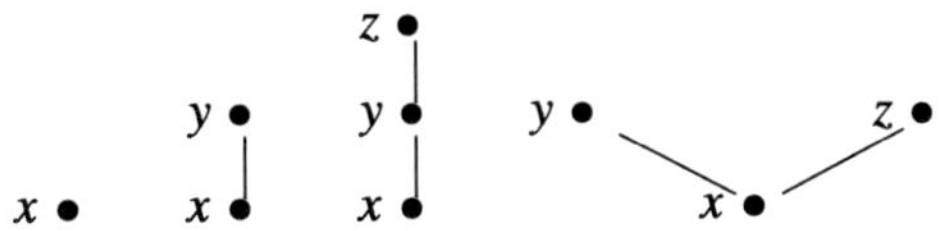

The operation $\{-, -\}$ on two decorated trees is constructed as follows: draw an edge from any vertex of the first tree to the root of the second one, keep the decorations, keep the root of the first tree as the root of the new tree, add all the elements so obtained.

Proposition 13.4.1 (Chapoton–Livernet [CL01]). *The product* $\{-, -\}$ *is pre-Lie and* $\bigoplus_{n\geq 1} \mathbb{K}[rT(n)] \otimes_{\mathbb{S}_n} V^{\otimes n}$ *is the free pre-Lie algebra over* V.

Proof. We refer to the original article [CL01] for the proof of this result. □

13.4.3 The Operad *preLie*

The operad which encodes the pre-Lie algebras is denoted by *preLie* (without "dash"). From the description of the free pre-Lie algebra given above it follows that *preLie*(n) is spanned by the rooted trees with vertices labeled by $\{1, \dots, n\}$.

For $n = 1$, there is only one tree (and no edges), which codes for the identity operation

For $n = 2$, there are two trees coding respectively for the operations $\{x_1, x_2\}$ and $\{x_2, x_1\}$:

For $n = 3$ there are nine trees, six of them code for the operation $\{x_1, \{x_2, x_3\}\}$ and its analogs under the action of $\mathbb{S}_3$ and the other three code for $\{x_1, \{x_2, x_3\}\} - \{\{x_1, x_2\}, x_3\}$ and its analogs:

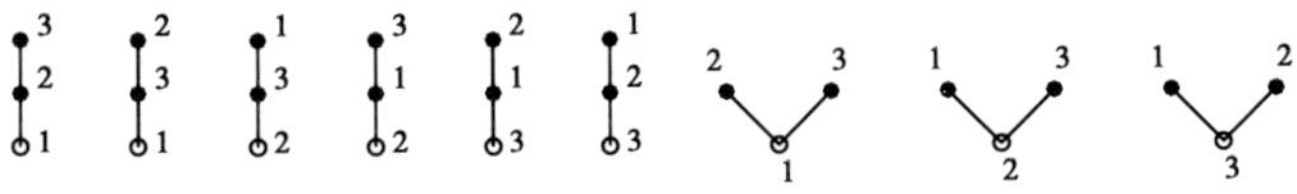

The action of $\mathbb{S}_n$ is by permutation of the labels (recall that the trees are not planar).

13.4.4 Relationship of *preLie* with Other Operads

The terminology "pre-Lie" comes from the following immediate property: the anti-symmetrized operation $[x, y] := \{x, y\} - \{y, x\}$ is a Lie bracket, cf. Lemma 1.4.1. So there is a functor:

$$preLie\text{-}\mathsf{alg} \to Lie\text{-}\mathsf{alg}.$$

Proposition 13.4.2. *The morphism of operads $Lie \to preLie$ induced by $[x, y] := \{x, y\} - \{y, x\}$ is injective.*

Proof. It can either be proved directly, or by using the following commutative diagram of operads (cf. Lemma 13.6.4):

$$\begin{array}{ccc} Lie & \longrightarrow & preLie \\ \downarrow & & \downarrow \\ Ass & \longrightarrow & Dend. \end{array}$$

Since the maps from *Lie* to *Ass* (CMM Theorem) and from *As* to *Dend* (Ronco's Theorem [Ron02]) are known to be injective, so is the map from *Lie* to *preLie*. □

In the case of commutative algebras it is known that the symmetrized product satisfies some relations, for instance the Jordan relation (cf. Sect. 13.10). In the pre-Lie case the situation is completely different. Let us recall that *ComMag* denotes the operad generated by one symmetric binary operation with no relation (cf. Sect. 13.8.4).

Proposition 13.4.3 (Bergeron–Loday [BL11]). *The morphism of operads $ComMag \to preLie$ induced by $x \cdot y := \{x, y\} + \{y, x\}$ is injective.*

Proof. The proof follows from the comparison with the operads *Dend* and *Dup*. We refer the reader to loc. cit. for the details of the proof. □

Proposition 13.4.4. *Let $\mathcal{P}$ be a symmetric operad (resp. a nonsymmetric operad). Composition in the operad $\mathcal{P}$ induces a pre-Lie algebra structure on the space $\bigoplus_n \mathcal{P}(n)$ (resp. $\bigoplus_n \mathcal{P}_n$).*

Proof. We have already mentioned and proved this result, cf. Sects. 5.4.3 and 5.9.15. It is a particular case of a stronger one involving brace algebras, cf. Proposition 13.11.4. □

13.4.5 Symmetric Brace Algebras

Starting with a pre-Lie algebra A one can construct recursively an $(1+n)$-ary operation for all $n \geq 1$ as follows:

$$\begin{aligned} M_{11}(x; y) &:= \{x, y\}, \\ M_{1n}(x; y_1 \ldots y_n) &:= M_{11}\big(M_{1(n-1)}(x; y_1 \ldots y_{n-1}); y_n\big) \\ &\quad - \sum_{1 \le i \le n-1} M_{1(n-1)}\big(x; y_1 \ldots M_{11}(y_i; y_n) \ldots y_{n-1}\big), \end{aligned}$$

for $x, y_1, \ldots, y_n \in V$. So for $n = 1$ this is the pre-Lie bracket. One can show that these multi-ary operations satisfy some relations which make A into a brace algebra (cf. Sect. 13.11.4).

By definition a *symmetric brace algebra* is a brace algebra such that the operation $M_{1n}(x; y_1 \ldots y_n)$ is symmetric in the variables y_is. The aforementioned formulas construct a symmetric brace algebra out of a pre-Lie algebra. On the other hand, any symmetric brace algebra gives rise to a pre-Lie algebra by taking only M_{11} into account.

Proposition 13.4.5 (Guin–Oudom). *The notion of symmetric brace algebra is equivalent to the notion of pre-Lie algebra in characteristic zero.*

Proof. In arity 3 the brace relation reads

$$\big\{\{x, y\}, z\big\} - \big\{x, \{y, z\}\big\} = \{x; y, z\} + \{x : z, y\}.$$

Hence, if 2 is invertible in $\mathbb{K}$ and if the brace is symmetric, then the ternary operation can be written in terms of the binary operation. The same phenomenon holds for the other higher operations. The details of the proof can be found in [OG08, LM05]. □

13.4.6 Koszul Dual Operad of *preLie*: The Operad *Perm*

It is easy to check that the Koszul dual operad of *preLie* is the operad dubbed *Perm* by Chapoton in [Cha01b]. A *Perm*-algebra is a vector space equipped with a binary operation xy satisfying the following relations:

$$(xy)z = x(yz) = (xz)y.$$

In plain words, it is an associative algebra which is commutative on the right-hand side as soon as there are at least three entries. The free *Perm*-algebra over the vector space V is isomorphic to $V \otimes S(V)$. It follows that $\dim \mathit{Perm}(n) = n$. As a representation of the symmetric group, $\mathit{Perm}(n) = \mathbb{K}^n$, where the action is simply the permutation of the coordinates. The basis is given by the operations

$$x_i x_1 x_2 \cdots \widehat{x_i} \cdots x_n, \quad i = 1, \ldots, n.$$

Since $\dim \mathit{Perm}(n) = n$, the generating series is $f^{\mathit{Perm}}(x) = x \exp(x)$. The Koszulity of the operads *preLie* and *Perm* has been proved in [CL01]. It can also be obtained through the poset method and by the rewriting process method. It implies that the generating series of *preLie*, that is $y = f^{\mathit{preLie}}(x)$, satisfies the functional equation $y = x\exp(y)$. It follows that

$$f^{preLie}(x) = \sum_{n\geq 1} \frac{n^{n-1}}{n!} x^n,$$

so $\dim preLie(n) = n^{n-1}$.

13.4.7 Pre-Lie Homology

Let A be a pre-Lie algebra. We construct a chain complex $C_\bullet^{preLie}(A)$ as follows. First we define $C_{n-1}^{preLie}(A) := A \otimes \Lambda^n(A)$. Second we define the boundary map by the formula:

$$d(x_0; x_1, \ldots, x_n) = b'\left(\sum_{\sigma\in\mathbb{S}_n} \operatorname{sgn}(\sigma)(x_0; x_{\sigma(1)}, \ldots, x_{\sigma(n)})\right),$$

where $b'(x_0; x_1, \ldots, x_n) = \sum_{i=0}^{n-1}(-1)^i(x_0; x_1, \ldots, x_i x_{i+1}, \ldots, x_n)$.

Proposition 13.4.6. *The chain complex $C_\bullet^{preLie}(A)$ constructed above is the operadic chain complex of the pre-Lie algebra A.*

Proof. The free Perm-algebra is $Perm(V) = V \otimes S(V)$. Hence the operadic chain complex of a pre-Lie algebra A is such that $C_{n-1}^{preLie}(A) = A \otimes \Lambda^n(A)$. The computation of the boundary map is straightforward from Proposition 12.1.1. □

13.4.8 Homotopy Pre-Lie Algebras

Since we know the Koszul dual operad of *preLie*, the notion of (strong) pre-Lie algebra up to homotopy can be described explicitly. Since $\dim Perm(n) = n$ the space $preLie_\infty(n)$ contains n generating operations. The symmetric group $\mathbb{S}_n$ is acting by permutation. We refer to [Mer05] for details and application of this structure in geometry.

13.4.9 Splitting and *preLie*

According to [Val08] the black Manin product of operads (cf. Sect. 8.8.3) gives the following isomorphisms of operads:

$$preLie \bullet Com = Zinb, \qquad preLie \bullet As = Dend, \qquad preLie \bullet Lie = preLie,$$

where *Zinb* is the operad encoding Zinbiel algebras (cf. Sect. 13.5.2) and *Dend* is the operad encoding dendriform algebras (cf. Sect. 13.6.1). So, in a sense, the functor $preLie \bullet -$ is "splitting" the binary operations when applied to binary operads. This process is similar to the splitting of operations using a Rota–Baxter operator [Uch10].

Dually, there is another way to split the operations of a $\mathcal{P}$-algebra: one can consider the new operad $Perm \bigcirc \mathcal{P}$ obtained by taking the Manin white product with the operad *Perm*.

One can also use the differential of a $\mathcal{P}$-algebra to define *derived products*, like $[d(-), -]$ and $[-, d(-)]$ for a dg Lie algebra. Such constructions play an important role in differential geometry, see [KS96, KS04]. Uchino proved that these two processes are the same [Uch10]. Interpreting the Rota–Baxter operator as the integration operator, this result gives a Koszul duality interpretation $PreLie \overset{!}{\longleftrightarrow} Perm$ of the classical integration–derivation duality.

13.4.10 From Pre-Lie Algebras to Hopf Algebras

Since, by antisymmetrization, a pre-Lie algebra gives rise to a Lie algebra, one can then take the universal enveloping algebra of this Lie algebra to get a Hopf algebra. It turns out that, first, many examples of this type arise in the literature, second, these Hopf algebras are "cofree cocommutative right-sided combinatorial Hopf algebras", studied in [LR10]. For instance, if $preLie(\mathbb{K})$ is the free preLie algebra on one generator, then $U(preLie(\mathbb{K})_{Lie})$ is the Grossman–Larson Hopf algebra, whose graded linear dual is the Connes–Kreimer Hopf algebra (cf. [GL89, CK98, CL01]). This latter Hopf algebra plays a prominent role in renormalization theory, because there is a way to compute the counter-term of some divergent integrals by means of the antipode of this Hopf algebra.

Another example is given by the vector space $L_1 := \oplus_{n\geq 1}\mathbb{K}x_n$ equipped with the pre-Lie product

$$\{x_p, x_q\} := (p+1)x_{p+q}$$

(cf. Sect. 1.4.3). The associated Hopf algebra is dual to the Faà di Bruno Hopf algebra.

13.5 Leibniz Algebras and Zinbiel Algebras

As shown in Bourbaki, the main property of a Lie bracket is to be a derivation for itself. So it is natural to introduce the algebras equipped with a not necessarily anti-symmetric operation satisfying this derivation property. It turned out to be relevant in many mathematical domains.

13.5.1 Leibniz Algebra [Lod93]

By definition a (right) *Leibniz algebra* is a vector space A equipped with a linear map (called bracket)

$$[-,-]: A \otimes A \to A$$

satisfying the Leibniz relation

$$\big[[x,y],z\big] = \big[x,[y,z]\big] + \big[[x,z],y\big].$$

In the sign-graded case (cf. Sect. 1.5.3), this relation reads

$$\big[[x,y],z\big] = \big[x,[y,z]\big] + (-1)^{|y||z|}\big[[x,z],y\big].$$

It is dubbed after George Wilhelm Leibniz (one of our ancestors) because of the form of the relation which says that $[-,z]$ is a derivation. It also appears sometimes in the literature under the name Loday algebra (a terminology due to Yvette Kosmann-Schwarzbach). Obviously a Lie algebra is an example of a Leibniz algebra. It is a Leibniz algebra whose bracket is antisymmetric.

It is shown in loc. cit. that the free Leibniz algebra over V is the reduced tensor module $\overline{T}(V)$ equipped with a binary operation $[-,-]$ such that

$$x_1 \cdots x_n = \big[\ldots\big[[x_1,x_2],x_3\big],\ldots,x_n\big].$$

Here is an example of computation in this algebra (for $x,y,z \in V$):

$$\begin{aligned}[x,yz] &= \big[x,[y,z]\big]\\ &= \big[[x,y],z\big] - \big[[x,z],y\big]\\ &= xyz - xzy.\end{aligned}$$

Though the space $Leib(n)$ is the regular representation: $Leib(n) = \mathbb{K}[\mathbb{S}_n]$, the Leibniz algebras are not encoded by a nonsymmetric operad. This is because, in the Leibniz relation, the variables do not stay in order, so the composition γ involves the symmetric group.

We refer to Sect. 13.6.12 for the relationship of Leibniz algebras with other types of algebras.

The Koszul dual operad of *Leib* is the Zinbiel operad denoted *Zinb* that we now introduce.

13.5.2 Zinbiel Algebras [Lod95, Lod01]

By definition a *Zinbiel algebra* is a vector space A equipped with a binary operation $\prec$ verifying the Zinbiel relation

$$(x \prec y) \prec z = x \prec (y \prec z + z \prec y).$$

It is dubbed after the virtual mathematician G.W. Zinbiel.

In loc. cit. the free Zinbiel algebra over V was shown to be the tensor module $\overline{T}(V)$ equipped with the *half-shuffle* as operation $\prec$:

$$x_1 \cdots x_p \prec x_{p+1} \cdots x_{p+q} = \sum_{\sigma \in Sh(p-1,q)} x_1 x_{\sigma(2)} \cdots x_{\sigma(n)}.$$

As a consequence we deduce the structure of the operad *Zinb* encoding Zinbiel algebras, namely $Zinb(n) = \mathbb{K}[\mathbb{S}_n]$. Using Proposition 7.6.4 it is easy to show that the Koszul dual operad of *Zinb* is *Leib*. This is why Zinbiel algebras were first called "dual Leibniz algebras" (cf. [Lod95]). As an immediate consequence of Proposition 7.6.5 the tensor product of a Leibniz algebra with a Zinbiel algebra has a Lie structure. In fact it can be shown that it is even a pre-Lie algebra, cf. Sect. 13.9.

The Koszul dual of the functor *Lie*-alg $\to$ *Leib*-alg is the functor *Zinb*-alg $\to$ *Com*-alg given by symmetrizing the binary operation. In other words on a Zinbiel algebra the symmetrized operation $xy := x \prec y + y \prec x$ is commutative and associative. For instance the commutative algebra associated to the free Zinbiel algebra is the shuffle algebra (cf. Sect. 1.3.2).

13.5.3 Zinbiel Algebras and Divided Powers

The commutative algebra obtained as the image of a Zinbiel algebra has more properties: it is a divided powers algebra (cf. Sect. 13.1.12). Indeed it is easy to show that the operation

$$x \mapsto x^{\prec n} := \underbrace{\big(x \prec \big(x \prec \cdots \prec (x \prec x) \cdots\big)\big)}_{n}$$

is a divided power operation. Moreover, if we start with a free Zinbiel algebra $Zinb(V) = \overline{T}^{sh}(V)$, then it is free as a divided powers algebra (cf. [Dok09]). If V is one-dimensional generated by x, we get the algebra $\Gamma(\mathbb{K}x)$ mentioned in Exercise 1.8.5. Its Zinbiel operation is given by $x^n \prec x^m = \binom{n+m-1}{m} x^{n+m}$.

13.5.4 Leibniz Homology

The chain complex of a Leibniz algebra was introduced in [Lod93] and was further studied in [LP93]. Up to a shift it turns out to be the same as the operadic chain complex (cf. Proposition 12.1.1), given as follows:

$$C^{Leib}_{\bullet-1}: \quad \cdots \longrightarrow \mathfrak{g}^{\otimes n} \xrightarrow{d} \mathfrak{g}^{\otimes n-1} \longrightarrow \cdots \longrightarrow \mathfrak{g}^{\otimes 2} \to \mathfrak{g},$$

where

$$d(x_1, \ldots, x_n) = \sum_{1 \le i < j \le n} (-1)^j \big(x_1, \ldots, x_{i-1}, [x_i, x_j], \ldots, \widehat{x_j}, \ldots, x_n\big).$$

The homology of this complex is denoted either by $H_\bullet^{Leib}(\mathfrak{g})$ or by $HL_\bullet(\mathfrak{g})$. We remark that it is a lifting of the Chevalley–Eilenberg complex. Indeed, the boundary map in the CE complex can be written:

$$d^{CE}(x_1 \wedge \ldots \wedge x_n) = \sum_{1 \le i < j \le n} (-1)^j \big(x_1 \wedge \ldots \wedge x_{i-1} \wedge [x_i, x_j] \wedge \ldots \wedge \widehat{x_j} \wedge \ldots \wedge x_n\big).$$

Its operadic interpretation is the following. The map of operads $Leib \to Lie$ induces a map of cooperads $Leib^{¡} \to Lie^{¡}$, whence a morphism of chain complexes $C_\bullet^{Leib}(\mathfrak{g}) \to C_\bullet^{Lie}(\mathfrak{g})$. This map is given by quotienting by the action of the symmetric group. Leibniz (co)homology has been studied in various papers, cf. [LP93, Pir94]. Let us just mention one peculiar feature: the Leibniz homology of a finite dimensional semi-simple Lie algebra is trivial, cf. [Pir94, Nto94].

By Theorem 12.1.2 the homology (resp. cohomology) of a Leibniz algebra is a graded Zinbiel coalgebra (resp. graded Zinbiel algebra), and hence a graded commutative coalgebra (resp. graded commutative algebra). This result was proved in [Lod95] by an ad hoc method.

13.5.5 Homotopy Leibniz Algebra, Homotopy Zinbiel Algebra

From the explicit description of the Leibniz operad (resp. Zinbiel operad) it is straightforward to describe the notion of Zinbiel algebra up to homotopy (resp. Leibniz algebra up to homotopy). The structure of $Leib_\infty$-algebra has been used for instance in differential geometry and deformation theory by [AP10] and by [Mer08], where this structure is made explicit.

13.6 Dendriform Algebras and Diassociative Algebras

Periodicity questions in algebraic K-theory led to the construction of "diassociative algebras" [Lod95]. See [Lod97] for a survey on these problems. The explicitation of the homology of a diassociative algebra led to the discovery of the Koszul dual structure, which was later called "dendriform algebra" in [Lod01]. It turned out that this notion found its way into several domains like combinatorial algebra, theoretical physics, algebraic topology. It is not so surprising because, first, the dendriform structure models the "noncommutative shuffles", second, the dendriform operad is closely related to planar binary trees [Lod02], third, the dendriform operations "split associativity". We will see that the dendriform operad comes from a nonsymmetric operad.

13.6.1 Dendriform Algebra

By definition a *dendriform algebra* is a vector space A equipped with two linear maps (binary operations)

$$\prec : A \otimes A \to A \quad \text{and} \quad \succ : A \otimes A \to A$$

called the *left operation* and the *right operation* respectively, satisfying the following three relations

$$\begin{cases} (x \prec y) \prec z = x \prec (y \prec z) + x \prec (y \succ z), \\ (x \succ y) \prec z = x \succ (y \prec z), \\ (x \prec y) \succ z + (x \succ y) \succ z = x \succ (y \succ z). \end{cases}$$

A morphism $f : A \to A'$ of dendriform algebras is a linear map compatible with the two operations:

$$f(a \prec b) = f(a) \prec f(b) \quad \text{and} \quad f(a \succ b) = f(a) \succ f(b) \quad \text{for any } a, b \in A.$$

Let us introduce the operation $*$ given by

$$x * y := x \prec y + x \succ y.$$

Adding the three relations shows that the operation $*$ is associative. So a dendriform algebra is an associative algebra whose product *splits* as the sum of two operations. The axioms imply that the associative algebra A has an extra bimodule structure over itself:

$$\begin{cases} (x \prec y) \prec z = x \prec (y * z), \\ (x \succ y) \prec z = x \succ (y \prec z), \\ (x * y) \succ z = x \succ (y \succ z). \end{cases}$$

As a consequence, a dendriform algebra structure on A gives rise to an abelian extension of associative algebras

$$0 \to A \to A \oplus A \to A \to 0$$

(the product on the kernel is 0 and the product on the cokernel is $*$).

13.6.2 The Free Dendriform Algebra

In order to describe explicitly the free dendriform algebra we need the notion of planar binary trees (pb trees), cf. Appendix C.1.1. Recall that PBT_{n+1} denotes the set of pb trees with n vertices (and so $n+1$ leaves). We first describe the free dendriform algebra on one generator. The free dendriform algebra on a vector space V is easily deduced.

Proposition 13.6.1. *The vector space $\bigoplus_{n>0} \mathbb{K}[PBT_{n+1}]$ equipped with the two binary operations $\prec$ and $\succ$ is defined inductively by the formulas*

$$t \prec s := t^l \vee (t^r * s),$$
$$t \succ s := (t * s^l) \vee s^r,$$

on the trees $t = t^l \vee t^r$ and $s = s^l \vee s^r$, and $| * t = t = t * |$ is a dendriform algebra generated by the tree  *$\in PBT_2$. It is the free dendriform algebra on one generator $Dend(\mathbb{K})$.*

Proof. In the statement of the proposition we use the notation $t = t^l \vee t^r$ meaning that any planar binary tree with at least two leaves can be written uniquely as the grafting of two other trees (cf. Appendix C). For the proof of this proposition we refer to the original article [Lod95, Lod01]. □

Corollary 13.6.2. *The free dendriform algebra on the vector space V is*

$$Dend(V) = \bigoplus_{n>0} \mathbb{K}[PBT_{n+1}] \otimes V^{\otimes n}$$

equipped with the operations

$$(t; v_1 \cdots v_p) \prec (s; v_{p+1} \cdots v_{p+q}) = (t \prec s; v_1 \cdots v_{p+q}),$$
$$(t; v_1 \cdots v_p) \succ (s; v_{p+1} \cdots v_{p+q}) = (t \succ s; v_1 \cdots v_{p+q}).$$

In particular, if $V = \mathbb{K}[X]$ for some set X, then the degree n component of $Dend(V)$ is spanned by $PBT_{n+1} \times X^n$.

As an immediate consequence of this corollary the generating series of the operad *Dend* is

$$f^{Dend}(x) = \frac{1 - x - \sqrt{1-4x}}{2x}.$$

13.6.3 Examples of Computation

Let us denote by Y the generator of $Dend(\mathbb{K}) = Dend(\mathbb{K}\,\mathrm{Y})$, which corresponds to the tree . In low dimension we get:

$\mathrm{Y} \prec \mathrm{Y} =$ 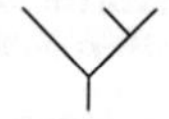 , $\mathrm{Y} \succ \mathrm{Y} =$  ,

$\mathrm{Y} \prec (\mathrm{Y} \prec \mathrm{Y}) =$ 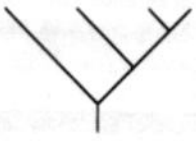 , $\mathrm{Y} \prec (\mathrm{Y} \succ \mathrm{Y}) =$ 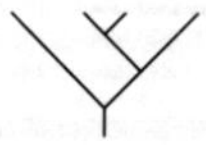 , $\mathrm{Y} \succ \mathrm{Y} \prec \mathrm{Y} =$ 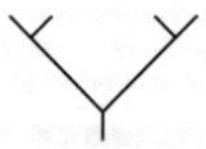 ,

$(\mathrm{Y} \prec \mathrm{Y}) \succ \mathrm{Y} =$ 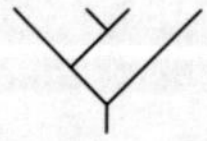 , $(\mathrm{Y} \succ \mathrm{Y}) \succ \mathrm{Y} =$ 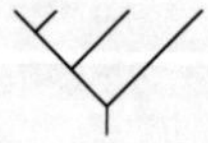 .

Observe that the orientation of the leaves corresponds precisely to the operations involved in the monomial.

We know that on the set of pb trees there is a simple operation: the grafting $t \vee s$. It is not difficult to show that it corresponds to the following operation in $Dend(\mathbb{K})$:

$$t \vee s = t \succ \mathrm{Y} \prec s.$$

13.6.4 Unital Dendriform Algebras

We know that associative algebras can be unitarized, cf. Sect. 1.1.1. Similarly dendriform algebras can be unitarized as follows. By definition a *unital dendriform algebra* is a vector space $A := \mathbb{K}1 \oplus \bar{A}$ such that $\bar{A}$ is a dendriform algebra and the left and right operations are (partially) extended to A by the formulas

$$\begin{cases} 1 \prec a = 0, & a \prec 1 = a, \\ 1 \succ a = a, & a \succ 1 = 0, \end{cases}$$

for any $a \in A$. Observe that one has $1 * a = a = a * 1$, so A is a unital associative algebra, but $1 \prec 1$ and $1 \succ 1$ are *not* defined.

A morphism of unital dendriform algebras is a linear map $f : A \to A'$ which maps 1 to 1, $\bar{A}$ to $\bar{A}'$ and f restricted to A is a dendriform morphism.

Observe that the three relations defining a dendriform algebra are still valid whenever two of the variables are in $\bar{A}$ and the third is equal to 1.

13.6.5 The Operad *Dend*

Since the generating operations have no symmetry and, in the relations, the variables stay in the same order, the category of dendriform algebras is encoded by a nonsymmetric operad. We denote it by *Dend* and we still denote by *Dend*, instead of $Dend \otimes Ass$, the associated symmetric operad if no confusion can arise. As a consequence the operad *Dend* is completely determined by the free dendriform algebra on one generator. By Proposition 13.6.1 we get $Dend_n = \mathbb{K}[PBT_{n+1}]$. Considering *Dend* as a symmetric operad we get $Dend(n) = \mathbb{K}[PBT_{n+1}] \otimes \mathbb{K}[\mathbb{S}_n] = \mathbb{K}[PBT_{n+1} \times \mathbb{S}_n]$. The operadic composition is deduced from the formula given in this proposition.

The operad *Dend* does not come from a set-theoretic operad. However the family of nonempty subsets of PBT_{n+1}, for all n, do form a set-operad. It has been studied in [Lod02].

13.6.6 Koszul Dual Operad of *Dend*

Since *Dend* is a binary quadratic nonsymmetric operad its Koszul dual operad can be computed by applying Theorem 7.7.1. It gives rise to diassociative algebras that we now describe. Historically, diassociative algebras cropped up first, then dendriform algebras appeared in the construction of the homology theory for diassociative algebras, cf. [Lod95].

13.6.7 Diassociative Algebra

By definition a *diassociative algebra* (or *Dias*-algebra) is a vector space A equipped with two linear maps (binary operations)

$$\dashv : A \otimes A \to A \quad \text{and} \quad \vdash : A \otimes A \to A$$

called the *left operation* and the *right operation* respectively, satisfying the following five relations

$$\begin{cases} (x \dashv y) \dashv z = x \dashv (y \dashv z), \\ (x \dashv y) \dashv z = x \dashv (y \vdash z), \\ (x \vdash y) \dashv z = x \vdash (y \dashv z), \\ (x \dashv y) \vdash z = x \vdash (y \vdash z), \\ (x \vdash y) \vdash z = x \vdash (y \vdash z). \end{cases}$$

The operad *Dias* comes from a set-theoretic nonsymmetric operad. The space $Dias_n$ is n-dimensional and one can take, as linear generators, the operations

$$x_1 \vdash \cdots \vdash x_{i-1} \vdash x_i \dashv x_{i+1} \dashv \cdots \dashv x_n, \quad \text{for } i = 1, \ldots, n.$$

In the tensor algebra $T(V)$ the monomial $x_1 \ldots x_n$ is the generic element. Similarly, in the free diassociative algebra $Dias(V)$ it is helpful to denote by

$$x_1 \ldots \check{x}_i \ldots x_n$$

the generic element corresponding to the ith operation, $i = 1, \ldots, n$. The left (resp. right) operation is simply concatenation where one keeps the decoration indicated by the pointer:

$$x_1 x_2 \check{x}_3 \vdash \check{x}_4 x_5 = x_1 x_2 x_3 \check{x}_4 x_5.$$

Let us mention that the Milnor invariants of tangles have been interpreted by Olga Kravchenko and Michael Polyak in [KP11] as a morphism of operads

$$Tangles \to Dias.$$

Proposition 13.6.3. [Lod01] *The operad Dend is a Koszul operad whose Koszul dual operad is*

$$Dend^! = Dias.$$

Proof. The dual basis of $\{\prec, \succ\}$ is denoted $\{\dashv, \vdash\}$. By Theorem 7.7.1 we only need to verify that, in the eight-dimensional space spanned by $(a \circ b) \circ' c$ and $a \circ (b \circ' c)$, where $\circ$ and $\circ' = \prec$ or $\succ$, the relators of *Dend* are dual to the relators of *Dias*. Since the space of relations of *Dend* is of dimension 3, the space of relations of *Dias* is going to be 5-dimensional. Now it suffices to check the orthogonality of the relators. We only do it for two pairs, checking the other cases is similar:

$$\begin{aligned}
&\big\langle (a \prec b) \prec c - a \prec (b \prec c) - a \prec (b \succ c), (a \dashv b) \dashv c - a \dashv (b \dashv c)\big\rangle \\
&\quad = +1 + 0 + 0 + (-1) + 0 + 0 = 0, \\
&\big\langle (a \prec b) \prec c - a \prec (b \prec c) - a \prec (b \succ c), (a \dashv b) \dashv c - a \dashv (b \vdash c)\big\rangle \\
&\quad = +1 + 0 + 0 + 0 + 0 + (-1) = 0.
\end{aligned}$$

The acyclicity of the Koszul complex of *Dend* was first proved in [Lod01]. But it is easier to show the acyclicity of the Koszul complex of *Dias* as done in [Lod08] (and apply Proposition 7.4.4). In fact most of the methods for proving Koszulness (Gröbner basis, poset method, rewriting process) work pretty well in this case. □

13.6.8 Digroups

Because of the form of the relations we can define an object analogous to a diassociative algebra in the category of sets. A *dimonoid* is a set with two binary operations satisfying the five diassociative axioms. Moreover we suppose the existence of an element 1 which is a unit for the bar side:

$$x \dashv 1 = x = 1 \vdash x.$$

There is even an analog of the notion of group as follows. A *digroup* is a dimonoid D such that for any $x \in D$ we are given an element $x^{-1} \in D$ such that

$$x \vdash x^{-1} = 1 = x^{-1} \dashv x.$$

A group is an example of a digroup (for which $\dashv = \vdash$).

13.6.9 Dendriform Homology

From the explicit description of the space $Dias_n$ given in Sect. 13.6.7 it follows that the operadic chain complex of the dendriform algebra A is

$$C^{Dend}_{\bullet-1} \quad \cdots \to \mathbb{K}^n \otimes A^{\otimes n} \xrightarrow{d} \mathbb{K}^{n-1} \otimes A^{\otimes n-1} \to \cdots \to A$$

where the boundary map $d = \sum_{i=1}^{i=n-1} (-1)^i d_i$ is given as follows. First, we define d_i on the set of indices $\{1, \dots, n\}$ by

$$d_i(r) = \begin{cases} r-1 & \text{if } i \le r-1, \\ r & \text{if } i \ge r. \end{cases}$$

Second, we define the operation $\circ_i^r$ as

$$\circ_i^r = \begin{cases} * & \text{if } i < r-1, \\ \succ & \text{if } i = r-1, \\ \prec & \text{if } i = r, \\ * & \text{if } i > r. \end{cases}$$

Recall that $a * b = a \prec b + a \succ b$. Finally the map d_i is given by

$$d_i(r; a_1, \dots, a_n) = \big(d_i(r); a_1, \dots, a_i \circ_i^r a_{i+1}, \dots, a_n\big).$$

From the general result on Koszul dual operad, see Proposition 12.1.2, it follows that the operadic homology $H_{\bullet+1}^{Dend}(A)$ is a graded diassociative coalgebra.

13.6.10 Dendriform Algebra up to Homotopy

Since we know an explicit presentation of $Dias = Dend^!$, and therefore of $Dend^{¡}$ it is easy to describe the notion of $Dend_\infty$-algebra. It has n n-ary generating operations denoted by $m_{n,i}$, $1 \le i \le n$, for any $n \ge 2$. The only relation is given by the explicit expression of the boundary of $m_{n,i}$, which is as follows:

$$\partial(m_{n,i}) = \sum (-1)^{p+qr} m_{p+1+r,\ell}(\underbrace{\mathrm{id}, \cdots, \mathrm{id}}_{p}, m_{q,j}, \underbrace{\mathrm{id}, \cdots, \mathrm{id}}_{r})$$

where, for fixed n and i, the sum is extended to all the quintuples p, q, r, ℓ, j satisfying: $p \ge 0, q \ge 2, r \ge 0, p+q+r = n, 1 \le \ell \le p+1+q, 1 \le j \le q$ and any one of the following:

- $i = q + \ell$, when $1 \le p+1 \le \ell - 1$,
- $i = \ell - 1 + j$, when $p+1 = \ell$,
- $i = \ell$, when $\ell + 1 \le p+1$.

For instance in low dimension we get the following relations:

$$\begin{aligned} \partial(m_{2,1}) &= 0, \\ \partial(m_{2,2}) &= 0, \\ \partial(m_{3,1}) &= m_{2,1} \circ_1 m_{2,1} - m_{2,1} \circ_2 m_{2,1}, \\ \partial(m_{3,2}) &= m_{2,1} \circ_1 m_{2,2} - m_{2,2} \circ_2 m_{2,1}, \\ \partial(m_{3,3}) &= m_{2,2} \circ_1 m_{2,2} - m_{2,2} \circ_2 m_{2,2}. \end{aligned}$$

So, as expected, the dendriform relations are valid only up to homotopy.

The presentation of $Dias_\infty$ algebra is slightly more complicated. It is given, up to sign, in [Lod01].

13.6.11 Relationship with Other Types of Algebras

We already know that *Dend* is related to *As* since the operation $x * y = x \prec y + x \succ y$ is associative. So any dendriform algebra gives rise to an associative algebra (a forgetful functor).

Let us define a *commutative dendriform algebra* as a dendriform algebra in which the following symmetry condition holds: $x \prec y = y \succ x$ for any x and y. This new type of algebras turns out to be the same as Zinbiel algebra which appears as dual of the Leibniz algebra, cf. Sect. 13.5. Recall that the Zinbiel operad is the Manin black product of *preLie* and *Com*. If one replaces *Com* by *Ass*, then it has been shown in [Val08] that

$$preLie \bullet Ass = Dend,$$

in the category of symmetric operads. Composing with the morphism $Lie \to Ass$ and taking into account that *Lie* is neutral for the black product it gives a morphism $preLie \to Dend$. This result can be proved directly as follows.

Lemma 13.6.4. *For any dendriform algebra A, the binary operation*

$$\{a, b\} := a \prec b - b \succ a$$

is a right pre-Lie product.

Proof. It suffices to check that the associator of the operation $\{-,-\}$ is symmetric in the last two variables:

$$\begin{aligned}
\big\{\{a,b\},c\big\} - \big\{a,\{b,c\}\big\} &= (a \prec b - b \succ a) \prec c - c \succ (a \prec b - b \succ a) \\
&\quad - a \prec (b \prec c - c \succ b) + (b \prec c - c \succ b) \succ a \\
&= a \prec (b \succ c) - (b \succ a) \prec c - c \succ (a \prec b) + (c \prec b) \succ a \\
&\quad + a \prec (c \succ b) + (b \prec c) \succ a \\
&= a \prec (b \succ c) - b \succ a \prec c + (c \prec b) \succ a \\
&\quad + a \prec (c \succ b) - c \succ a \prec b + (b \prec c) \succ a.
\end{aligned}$$

□

13.6.12 Butterfly Diagram and More

The following diagram summarizes the relationship between various types of algebras. We use the short notation P to denote the category $\mathscr{P}$-alg:

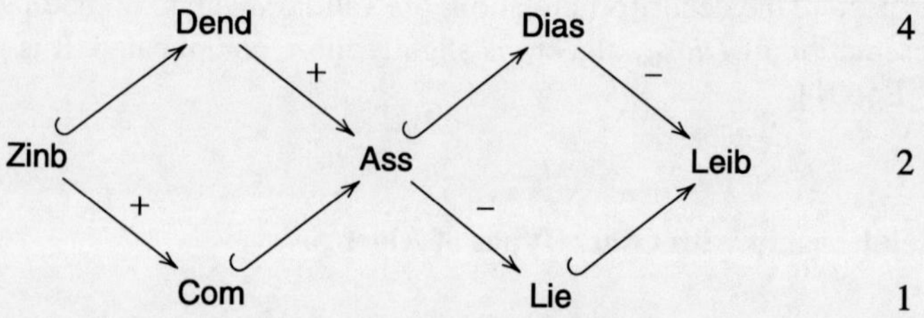

The integer on the right-hand side indicates the dimension of the space $\mathscr{P}(2)$. Koszul duality is given by the symmetry around the vertical axis passing through *As*. This "butterfly diagram" has been completed in [Lod06].

Here is another diagram, due to F. Chapoton [Cha01b], which includes the pre-Lie operad (cf. Lemma 13.6.4):

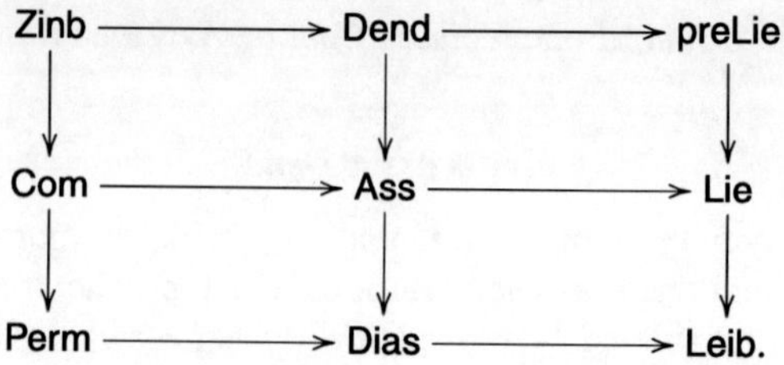

Koszul duality is given by central symmetry. The first column is the middle column made commutative. The first row is the middle row blackproducted with *preLie* and so the last row is the middle row whiteproducted with *Perm* (cf. Theorem 8.8.3). However in several questions (for instance the study of generalized bialgebras) it is better to put brace in place of preLie in this diagram.

13.6.13 Dendriform Variations

There are several variations of dendriform algebras. We already saw the commutative dendriform, i.e. Zinbiel. Here are some more (not an exhaustive list):

◇ *Parametrized dendriform algebras.* Let $\lambda \in \mathbb{K}$ be a parameter. We slightly modify the three relations of a dendriform algebra as follows:

$$\begin{cases} (x \prec y) \prec z = x \prec (y \prec z) + \lambda x \prec (y \succ z), \\ (x \succ y) \prec z = x \succ (y \prec z), \\ \lambda(x \prec y) \succ z + (x \succ y) \succ z = x \succ (y \succ z). \end{cases}$$

So, for $\lambda = 1$ we get the dendriform case. For $\lambda = 0$ we call them *duplicial algebras*, cf. Sect. 13.13.2. It is easy to see that for any λ the operad can be described by planar binary trees, like in the dendriform case.

◇ *Tridendriform algebras* [LR04]. We have seen that the dendriform operad is strongly related to the planar binary trees. In fact there is a way to enlarge this operad so that it involves the planar trees (not just binary) under similar rules, as follows.

We assume that the associative product $*$ splits into the sum of three operations: $x * y = x \prec y + x \succ y + x \cdot y$, and that the associativity relation splits into seven relations (one for each cell of the triangle, cf. loc. cit.):

$$\begin{cases} (x \prec y) \prec z = x \prec (y * z), \\ (x \succ y) \prec z = x \succ (y \prec z), \\ (x * y) \succ z = x \succ (y \succ z), \\ (x \succ y) \cdot z = x \succ (y \cdot z), \\ (x \prec y) \cdot z = x \cdot (y \succ z), \\ (x \cdot y) \prec z = x \cdot (y \prec z), \\ (x \cdot y) \cdot z = x \cdot (y \cdot z). \end{cases}$$

The ns operad *Tridend* can be described in terms of planar trees. We leave it to the reader the pleasure of computing the Koszul dual operad *Trias*, cf. loc. cit. The parametrized version of tridendriform algebras consists in putting

$$x * y := x \prec y + x \succ y + \lambda x \cdot y.$$

The case $\lambda = 0$, studied by Chapoton in [Cha02], see also [BR10], is interesting because it is strongly related to the cell structure of the Stasheff polytope, when the dot operation h has degree one.

◇ *Commutative tridendriform algebras.* They are tridendriform algebras such that $x \prec y = y \succ x$ and $x \cdot y = y \cdot x$. They are strongly related to *quasi-shuffle algebras*, cf. [Lod07].

◇ *Quadrialgebras.* We start with pairs of dendriform operations, that is four operations denoted $\nwarrow, \nearrow, \searrow, \swarrow$, satisfying nine relations, cf. [AL04]. In this case the associative product is split into the sum of four operations. It turns out to be the Manin black product *preLie* ● *Dend* in the category of symmetric operads and the Manin black product *Dend* ■ *Dend* in the category of nonsymmetric operads, cf. Sect. 8.8, [EFG05] and [Val08].

◇ *Locally commutative dendriform algebras.* One can impose a different kind of symmetry property on dendriform operations, namely $x \prec y = y \prec x$ and $x \succ y = y \succ x$. They also imply that the associative operation $x * y$ is commutative. It does not seem that locally commutative dendriform algebras have been studied so far.

For all these operads the poset method is well adapted to prove their Koszulity.

13.7 Batalin–Vilkovisky Algebras and the Operad *BV*

A BV-algebra is a Gerstenhaber algebra with a compatible square-zero operator Δ. Such an algebraic structure appears in differential geometry on polyvector fields, on the Hochschild cohomology of unital cyclic associative algebras, on the homology of free loop spaces, on vertex operator algebras, on conformal field theories and, of course, in mathematical physics with the Batalin–Vilkovisky formalism, see the introduction of [DCV11]. The first example is due to Batalin and Vilkovisky. The general definition is due to Jean-Louis Koszul and appeared in [Kos85].

The introduction of the operad BV with applications to double loop spaces and topological conformal field theories was pioneered by Getzler in [Get94]. The extension of some results from BV-algebras to homotopy BV-algebras was done in [GCTV09].

13.7.1 Definition of BV-Algebras

A Batalin–Vilkovisky algebra, BV-algebra for short, is a graded vector space equipped with a commutative product of degree 0 denoted by $x \cdot y$, and a unary operation of degree 1 denoted by Δ (it is *not* a coproduct), satisfying $\Delta \circ \Delta = 0$. These two operations are supposed to satisfy the following relation, which is ternary (3 variables) and cubical (composite of 3 generating operations):

$$\Delta(- \cdot - \cdot -) - \big(\Delta(- \cdot -) \cdot -\big)^{\mathrm{id}+(123)+(321)} + \big(\Delta(-) \cdot - \cdot -\big)^{\mathrm{id}+(123)+(321)} = 0.$$

With such a structure at hand we can define the binary operation

$$\langle -, - \rangle := \Delta(- \cdot -) - \big(\Delta(-) \cdot -\big) - \big(- \cdot \Delta(-)\big).$$

In plain words: the bracket operation is the obstruction to Δ being a derivation for the commutative product. As a consequence, the operation $\langle -, - \rangle$ is symmetric and of degree one. The aforementioned relation between the commutative product and the operator Δ implies that $\langle -, - \rangle$ satisfies the following Jacobi identity

$$\big\langle \langle -, - \rangle, - \big\rangle + \big\langle \langle -, - \rangle, - \big\rangle^{(123)} + \big\langle \langle -, - \rangle, - \big\rangle^{(132)} = 0,$$

and the Leibniz relation with respect to the commutative product

$$\langle - \cdot -, - \rangle = \big(- \cdot \langle -, - \rangle\big) + \big(\langle -, - \rangle \cdot -\big)^{(23)}.$$

An operator Δ on a commutative algebra, whose associated bracket satisfies the Leibniz relation, is called an operator of *order less than* 2, see [GCTV09, Sect. 2.4].

Therefore a BV-algebra carries a Gerstenhaber algebra structure, see Sect. 13.3.10. So forgetting the operator Δ but keeping the bracket $\langle -, - \rangle$, we get the inclusion of categories

$$BV\text{-alg} \to \mathit{Gerst}\text{-alg}.$$

Proposition 13.7.1. [Get94] *The data of a BV-algebra structure $(A, \cdot, \Delta)$ is equivalent to that of a Gerstenhaber algebra structure $(A, \cdot, \langle -,- \rangle)$ endowed with a square-zero degree 1 unary operator Δ, such that*

$$\langle -,- \rangle := \Delta(- \cdot -) - \big(\Delta(-) \cdot -\big) - \big(- \cdot \Delta(-)\big).$$

Proof. The proof is a straightforward checking and therefore left to the reader. □

Notice that, as a consequence of its definition and independently from the cubical relation, the operator Δ is a derivation for the bracket

$$\Delta\big(\langle -,- \rangle\big) + \big\langle \Delta(-), - \big\rangle + \big\langle -, \Delta(-) \big\rangle = 0.$$

EXAMPLE. Let M be a smooth oriented n-dimensional manifold equipped with a volume form ω. Recall from Sect. 13.3.10, that the space of polyvector fields $\Gamma(M, \Lambda(TM))$ is equipped with a Gerstenhaber algebra structure. The contraction of ω along polyvector fields defines the following isomorphism with the differential forms

$$\begin{aligned} \Gamma\big(M, \Lambda^\bullet(TM)\big) &\to \Omega^{n-\bullet}(M) \\ \pi &\mapsto i_\pi(\omega) := \omega(\pi, -). \end{aligned}$$

The transfer of the de Rham differential map d_{DR}, from the differential forms to polyvector fields, defines the *divergence operator* $\Delta := \mathrm{div}_\omega$, which endows $\Gamma(M, \Lambda(TM))$ with a BV-algebra structure [TT00].

13.7.2 Batalin–Vilkovisky Formalism

Let W be a finite dimensional chain complex. We consider it as a manifold with the structure sheaf of formal functions which vanish at 0:

$$\widehat{S}(W^*) := \prod_{n \geq 1} S^n(W^*),$$

where the symmetric tensor product is the quotient $S^n(W^*) := ((W^*)^{\otimes n})_{\mathbb{S}_n}$. To any finite dimensional chain complex V, we associate its "cotangent bundle" $W := V \oplus sV^*$. (It is equipped with a canonical nondegenerate bilinear form, as any cotangent bundle is endowed with a canonical symplectic manifold structure.) Its commutative algebra of functions $\widehat{S}(V^* \oplus sV)$ carries the following degree -1 operator Δ. Let $\{v_i\}_{1 \leq i \leq n}$ and $\{\nu_i\}_{1 \leq i \leq n}$ denote respectively a basis of V and the dual basis of V^*. To any formal function $\alpha \in \widehat{S}(V^* \oplus sV)$, we associate

$$\Delta(\alpha) := \sum_{i=1}^{n} \frac{\partial \alpha}{\partial v_i} \frac{\partial \alpha}{\partial \nu_i}.$$

Proposition 13.7.2. *The algebra of formal functions* $(\widehat{S}(V^* \oplus sV), \cdot, \Delta)$ *satisfies the relations of a Batalin–Vilkolisky algebra with a differential map and the operator* Δ *of same degree* -1.

Then one considers the quantized version of this algebra, namely

$$\widehat{S}(V^* \oplus sV)[[\hbar]] := \widehat{S}(V^* \oplus sV) \otimes \mathbb{K}[[\hbar]],$$

where the product and the operator Δ are extended by $\hbar$-linearity. It is again a BV-algebra. The Batalin–Vilkovisky formalism [BV81, Sch93] relies on the functions which are solution to the *master equation*:

$$d(\alpha) + \hbar\Delta(\alpha) + \frac{1}{2}\langle \alpha, \alpha \rangle = 0.$$

Following the same ideas as in Chap. 10, S. Merkulov in [Mer10b] showed that solutions to the master equations are in bijective correspondence with some homotopy algebraic structure on V. He further interpreted the passage from classical action to effective action in renormalization theory as a homotopy transfer theorem. In this case, the famous Feynman diagrams correspond bijectively to the graphs appearing in the formulas of the HTT.

REMARK. We have chosen the topological, or homological, definition of a BV-algebra: the underlying differential map and the operator Δ have different signs. In the Batalin–Vilkovisky formalism, the authors use another operad, called the *BD operad*, after Beilinson and Drinfeld [BD04]. This operad can be obtained as the Koszul dual operad of BV and by considering the sub-algebra $\mathbb{K}[[\hbar]]$ as the ground ring.

13.7.3 BV-Algebra Structure on Hochschild Cohomology of a Unital Cyclic Associative Algebra

Since the notion of BV-algebra is a refinement of the notion of Gerstenhaber algebra, the Gerstenhaber algebra structure on the Hochschild cohomology of an associative algebra (cf. Proposition 13.3.1) lifts to a BV-algebra structure. For this extension we need an extra structure on the associative algebra A.

Let A be a *cyclic unital associative algebra*, that is an algebra over the cyclic operad *uAss*, cf. Sect. 13.14.6. By definition, A is a unital associative algebra endowed with a symmetric nondegenerate bilinear form $(-,-) : A \otimes A \to \mathbb{K}$ satisfying the following *invariance* property $(a.b, c) = (a, b.c)$, and thus $(a.b, c) = (-1)^{|a|(|b|+|c|)}(b, c.a)$. In this case, A is finite dimensional and there is an isomorphism $A \to A^*$ of A-bimodules. (In representation theory such an algebraic structure is called a *symmetric algebra*.) This latter isomorphism allows us to transport

Connes' boundary map B (cf. Sect. 10.3.7) from homology to cohomology, giving the BV-operator Δ. This method is the algebraic analog of the aforementioned polyvector fields case.

Proposition 13.7.3. [Tra08, Men09, Gin06] *For any cyclic unital associative algebra A, the Hochschild cohomology of A with coefficients into itself inherits a BV-algebra structure, extending the Gerstenhaber algebra structure of Proposition* 13.3.1.

Proof. When A is a cyclic unital associative algebra, with $\Phi : A \to A^*$ for isomorphism of A-bimodules, there are isomorphisms of chain complexes

$$C^{\bullet}_{Hoch}(A, A) \cong C^{\bullet}_{Hoch}(A, A^*) \cong \left(C_{\bullet}^{Hoch}(A, A)\right)^*.$$

The first one is given by $f \in \mathrm{Hom}(A^{\otimes n}, A) \mapsto \Phi \circ f \in \mathrm{Hom}(A^{\otimes n}, A^*)$ and the second is given by $\mathrm{Hom}(A^{\otimes n}, \mathrm{Hom}(A, \mathbb{K})) \cong \mathrm{Hom}(A^{\otimes n} \otimes A, \mathbb{K})$. We consider the dual of Connes' boundary map $B : C_n^{Hoch}(A, A) \to C_{n+1}^{Hoch}(A, A)$ and transport it, through these isomorphisms, on $C^{\bullet}_{Hoch}(A, A)$ to define a degree 1 operator (with the homological convention), which commutes with the differential map. It induces a square-zero degree 1 operator Δ on cohomology $HH^{\bullet}(A, A)$, which satisfies the axioms of a BV-algebra with the Gerstenhaber algebra. We refer the reader to the aforementioned references for more details. ☐

The explicit formula of Connes' boundary map B (cf. for instance [Lod98, Chap. 2]), induces the following formula for the operator Δ on the cochain level:

$$\left(\Delta(f)(a_1, \ldots, a_{n-1}), a_n\right) = \sum_{i=1}^{n} (-1)^{i(n-1)} \left(f(a_i, \ldots, a_n, a_1, \ldots, a_{i-1}), 1\right).$$

Notice that this result still holds when there is only a quasi-isomorphism $A \to A^*$ of A-bimodules [Gin06]; such an algebra is called a *Calabi–Yau algebra*.

13.7.4 The Operad BV

The previous equivalent definitions of BV-algebras allow us to give several presentations by generators and relations for the operad BV. The first definition gives a presentation with 2 generators m and Δ, under the convention of Sect. 13.3.9, and 3 homogeneous relations, two quadratic and one cubic. The other presentation is made of 3 generators m, c and Δ. In that case, the relations are homogeneous quadratic, except from the one expressing c as the obstruction for Δ to be a derivation for m, which involves quadratic and linear terms. To be able to apply the inhomogeneous Koszul duality theory of Sect. 7.8, we choose this presentation, to which we include in the space of relations, the derivation of Δ with c, so that it satisfies condition (ql_2).

Finally, it gives $BV \cong \mathscr{T}(E_{BV})/(R_{BV})$, with

$$E_{BV} = \mathbb{K}m \oplus \mathbb{K}c \oplus \mathbb{K}\Delta,$$

where $\mathbb{K}\Delta$ is a one-dimensional graded vector space ($\mathbb{S}_1$-module) concentrated in degree 1. The space of relations R_{BV} is the $\mathbb{K}[\mathbb{S}_3]$-module generated by

$$\begin{cases} m \circ_1 m - m \circ_2 m, \\ c \circ_1 c + (c \circ_1 c)^{(123)} + (c \circ_1 c)^{(321)}, \\ \Delta^2, \\ c \circ_1 m - m \circ_2 c - (m \circ_1 c)^{(23)}, \\ c - \Delta \circ_1 m + m \circ_1 \Delta + m \circ_2 \Delta, \\ \Delta \circ_1 c + c \circ_1 \Delta + c \circ_2 \Delta. \end{cases}$$

The last two relations allow us to pull up the operator Δ in the operad BV. More precisely, Proposition 7.8.8 shows that the underlying $\mathbb{S}$-module of the operad BV is given by following composite:

$$BV \cong Com \circ \mathscr{S}^{-1}Lie \circ D,$$

where $D = T(\Delta)/(\Delta^2)$ is the algebra of dual numbers on a degree 1 element.

The operad structure on BV can be extended to a cyclic operad structure.

13.7.5 Relationship with the Framed Little Discs Operad

To extend Theorem 13.3.3 to the operad BV, we need to consider the following refinement of the little discs operad. The framed little discs operad $f\mathscr{D}$ is made of the configurations of n framed discs, that is with a point on the boundary, inside the unit disc of the plane.

Proposition 13.7.4. [Get94] *The singular homology $H_\bullet(f\mathscr{D})$ of the framed little discs operad is an algebraic operad isomorphic to the operad BV.*

Proof. The framing induces the square-zero degree 1 operator Δ on homology. □

Since a double loop space $\Omega^2 X$ on a pointed topological space X endowed with an action of the circle S^1 is an algebra over the framed little discs operad, this theorem proves that the homology of a double loop space carries a natural structure of BV-algebra [Get94].

13.7.6 Relationship with the Riemann Sphere Operad and TCFT

After Segal [Seg04], we consider the moduli space $\mathcal{R}$ of isomorphism classes of connected Riemann surfaces of arbitrary genus with biholomorphic maps from the disjoint union of $n+m$ discs. The gluing along the boundaries endows this topological space with a properad structure, see Sect. 13.14.9. Segal proposed to define a *Conformal Field Theory, CFT* for short, as an algebra over $\mathcal{R}$. In the same way, a *Topological Conformal Field Theory, TCFT* for short, is an algebra over the singular chains $C_\bullet(\mathcal{R})$ of $\mathcal{R}$.

The Riemann spheres R, i.e. Riemann surfaces of genus 0, with n input discs and 1 output disc, form the operadic part of $\mathcal{R}$. The frame little discs operad is a deformation retract of the operad R: look through the outgoing disc to see a configuration of n framed discs inside a unit disc. Therefore Proposition 13.7.4 induces the isomorphism of operads $H_\bullet(R) \cong H_\bullet(f\mathcal{D}) \cong BV$, which leads to the following consequence.

Proposition 13.7.5. [Get94] *The homology of a TCFT carries a natural BV-algebra structure, whose product is given by the degree* 0 *homology class of* $R(2)$ *and whose degree* 1 *operator* Δ *is given by the fundamental class of the circle* $R(1) \sim S^1$.

13.7.7 Homotopy BV-Algebras

The quadratic-linear presentation of the operad BV above in Sect. 13.7.4 satisfies the conditions of the inhomogeneous Koszul duality theory (Sect. 7.8.7). Hence, it provides a quasi-free, but not minimal, resolution

$$BV_\infty := \Omega BV^{¡} \xrightarrow{\sim} BV$$

of the operad BV. Algebras over this cofibrant resolution are called *homotopy BV-algebras*. We refer the reader to Sect. 2.3 of [GCTV09] for the explicit definition in terms of operations and relations.

This definition is a generalization of the notion of *commutative BV_∞-algebra* by O. Kravchenko [Kra00] and it is a particular case of a definition proposed by D. Tamarkin and B. Tsygan in [TT00], which is an algebra over a properad. Contrary to these two other definitions, the one given here comes from a cofibrant operad. Therefore it shares the required homotopy properties, as shown by the following examples.

REMARK. One gets the minimal model of the operad BV by first computing the homotopy $H_\bullet(BV^{¡})$, i.e. the underlying homology, of the dg cooperad $(BV^{¡}, d_\varphi)$. This computation was done in [DCV11, DK09] and the result is related to the homology of the moduli space of curves, which is the operad Gravity, see Sect. 13.11.2. Then, one has to endow $H_\bullet(BV^{¡})$ with a homotopy cooperad structure transferred from the dg cooperad structure on $BV^{¡}$, see [DCV11].

We will need the following formality result, which extends Theorem 13.3.6.

Theorem 13.7.6. [Šev10, GS10] *The framed little discs operad is formal, i.e. there is a zig-zag of quasi-isomorphisms of dg operads*

$$C_\bullet(f\mathscr{D}) \xleftarrow{\sim} \cdot \xrightarrow{\sim} H_\bullet(f\mathscr{D}).$$

Together with the cofibrancy of BV_∞, it implies the following theorem.

Theorem 13.7.7. [GCTV09] *There exists a commuting diagram of quasi-isomorphisms*

$$\begin{array}{ccc} C_\bullet(R) & \longleftarrow & C_\bullet(f\mathscr{D}) \\ \uparrow & & \updownarrow \\ BV_\infty & \longrightarrow & BV \cong H_\bullet(f\mathscr{D}). \end{array}$$
$$BV_\infty \longrightarrow C_\bullet(f\mathscr{D}) \text{ (diagonal)}$$

Proof. The proof relies on the left lifting property of cofibrant objects in the model category of dg operads, see Proposition B.6.3. For the details, we refer the reader to the proof of Theorem 11 of [GCTV09]. □

This theorem and its proof illustrate well how one can use homotopy properties of dg operads to obtain results on algebraic structures.

Corollary 13.7.8. [GCTV09]

◇ *The singular chains $C_\bullet(\Omega^2 X)$ of the double loop space on a topological space X endowed with an action of the circle S^1 carries a homotopy BV-algebra structure, which is homotopy equivalent to that of $C_\bullet(f\mathscr{D})$.*
◇ *Any TCFT carries a homotopy BV-algebra structure, which is homotopy equivalent to the action of the operadic part $C_\bullet(R)$.*

Thus an important part of a TCFT structure is encoded in the algebraic notion of homotopy BV-algebra structure. These two structures lift, on the chain level and up to homotopy, the BV-algebra structures on homology. Then by the Homotopy Transfer Theorem 10.3.10, one extends this BV-algebra structures on homology to a homotopy BV-algebra structure, which defines new homotopy invariants.

13.7.8 Vertex Operator Algebras and Homotopy BV-Algebras

A *vertex operator algebra* is an algebraic structure introduced by Richard Borcherds with motivation in conformal field theory and the monstrous moonshine. We refer to [Kac98, FBZ04] for the full definition. Yi-Zhi Huang [Hua97] proved that a vertex operator algebra is a "partial" algebra over the Riemann sphere operad R. Hence Theorem 13.7.7 motivates the following result.

Theorem 13.7.9 (Lian–Zuckerman conjecture [GCTV09]). *Any topological vertex operator algebra A, with nonnegatively graded conformal weight, carries an explicit homotopy BV-algebra structure, which extends the Lian–Zuckerman operations in conformal weight zero and which induces the Lian–Zuckerman BV-algebra structure on $H_\bullet(A)$.*

Proof. The proof relies on the obstruction theory applied to the convolution dg Lie algebra $\mathfrak{g}_{BV,A} := \mathrm{Hom}_{\mathbb{S}}(BV^{\text{i}}, \mathrm{End}_A)$, see Sect. 10.1.2. Since the Koszul dual dg cooperad BV^{i} is weight-graded, so is $\mathfrak{g}_{BV,A}$. Starting from the Lian–Zuckerman operations in weight 0, one shows that the obstructions, i.e some homology groups, to extending them vanish. The definition of a topological vertex operator algebra actually gives a contracting homotopy for this, which shows that one can explicitly construct the homotopy BV-algebra structure inductively. □

13.7.9 Cyclic Deligne Conjecture

The cyclic Deligne conjecture is the generalization from the Gerstenhaber algebra case to the BV-algebra case, of the Deligne conjecture [Tam99].

Theorem 13.7.10 (Cyclic Deligne conjecture [GCTV09]). *Let A be a cyclic unital associative algebra. There is a homotopy BV-algebra structure on its Hochschild cochain complex, which lifts the BV-algebra structure on Hochschild cohomology.*

Proof. Kaufmann proved in [Kau08] that the operad $C_\bullet(f\mathcal{D})$, made of the singular chains of the framed little discs operad, acts on $C^\bullet(A, A)$. Finally, we conclude with Theorem 13.7.7. □

This conjecture was proved with various topological models in [Kau08] (framed little discs, cacti), in [TZ06] (cyclic Sullivan chord diagram), in [Cos07] (Riemann sphere), and in [KS09] (configuration of points of the cylinder). Since the operad BV_∞ is cofibrant, it provides a canonical model for the cyclic Deligne conjecture.

13.8 Magmatic Algebras

The notion of associative algebra is so important that the algebras of any other type used to be called "nonassociative algebras". This terminology includes the algebras which have one binary operation with no further relation. On the other hand, Bourbaki calls "magma" a set equipped with a binary operation without any further relation. So we adopt the terminology magmatic for this precise type of nonassociative algebras. It is an interesting case because the associated operad is free.

13.8.1 Definition of a Magmatic Algebra

A *magmatic algebra* is a vector space equipped with a binary operation xy with no relation. The free magmatic algebra on one generator x is spanned by the monomials in x with a parenthesizing, for instance:

$$\big((x(xx))(xx)\big).$$

It is often helpful to replace the parenthesizings by the planar binary trees, therefore the free magmatic algebra $Mag(\mathbb{K})$ on one generator x is $Mag(\mathbb{K}) = \bigoplus_{n\geq 1} \mathbb{K}[PBT_n]$. Under this identification the generator is the tree $|$ and the binary product is the grafting: $t_1 t_2 = t_1 \vee t_2$.

As in the associative case we can work either in the unital case (existence of an element 1 such that $1 \cdot x = x = x \cdot 1$, played by the empty tree), or in the nonunital case.

13.8.2 The Operad *Mag*

The category of magmatic algebras can be encoded by a nonsymmetric operad which comes from a set-theoretic operad. We denote it by *Mag*. From the previous discussion we get:

$$Mag_n = \mathbb{K}[PBT_n].$$

As mentioned before the operad *Mag* is the free ns operad on one binary operation that we have met in Sects. 5.5 and 5.9.6:

$$Mag = \mathscr{T}\big(\mathbb{K} \curlyvee\big).$$

It is a nonsymmetric operad, which is binary and quadratic. In the unital case we denote the associated ns operad by *uMag*.

13.8.3 Koszul Dual of *Mag*

The Koszul dual operad of *Mag* is the *nilpotent operad Nil* having one binary operation xy as generator and relations

$$(xy)z = 0 = x(yz).$$

Since *Mag* is free, i.e. $R_{Mag} = 0$, its Koszul dual operad is the quotient of $\mathscr{T}(\mathbb{K}\curlyvee)$ by the operadic ideal generated by $\mathscr{T}(\mathbb{K}\curlyvee)^{(2)}$, that is the composites $\curlyvee \circ_1 \curlyvee$ and $\curlyvee \circ_2 \curlyvee$. It is trivial to show that these operads are Koszul. As a consequence we get, without any calculation, the algebraic expression of the Catalan series. The generating series for the nilpotent operad is simply $f^{Nil}(x) = x + x^2$, and the generating series for the magmatic operad is $f^{Mag}(x) = \sum_{n\geq 1} c_n x^n$, where c_n is the Catalan number (number of planar binary trees). From the functional equation relating the generating series of Koszul dual operads (cf. Theorem 7.6.7) we get

$$\sum_{n\geq 1} c_n x^n = \frac{1-\sqrt{1-4x}}{2},$$

and therefore $c_n = \frac{1}{n+1}\binom{2n}{n}$.

We have mentioned before that the black product with *preLie* amounts to some dichotomization. In the case of *Nil* the following surprising phenomenon occurs: the symmetric operad *preLie* ● *Nil* is not Koszul. The proof can be done by computing the generating series, cf. [Val08].

13.8.4 Commutative Magmatic Algebras

It is sometimes necessary to work with algebras having one commutative operation without any further relation (so not associative for instance). We call it a *commutative magmatic algebra* and we denote the operad by *ComMag*. The operad is described by (non-planar) binary trees. The dimension of the space of n-ary operations is

$$\dim ComMag(n) = (2n-1)!! = 1 \times 3 \times 5 \times 7 \times \cdots \times (2n-1).$$

A basis of $ComMag(n)$ is given by the binary shuffle trees introduced in Sect. 8.2.2.

13.9 Parametrized Binary Quadratic Operads

Associative algebras, Poisson algebras, Leibniz algebras and Zinbiel algebras are examples of types of algebras generated by one operation (without symmetry) with one quadratic relation of the form

$$(xy)z = \sum_{\sigma\in\mathbb{S}_3} a_\sigma \sigma\big(x(yz)\big) \quad \text{for some coefficients } a_\sigma \in \mathbb{K}.$$

We have seen in Proposition 7.6.4 that the Koszul dual operad is generated by one operation with the quadratic relation

$$x(yz) = \sum_{\sigma\in\mathbb{S}_3} \operatorname{sgn}(\sigma) a_{\sigma^{-1}} \sigma\big((xy)z\big).$$

One passes from the type of algebras with relations of the form $(xy)z = \sum(-(--))$ to those with relations of the form $x(yz) = \sum((--)-)$ by taking the "opposite type" given by $x \cdot y := yx$.

If A is an algebra of the first type and B is an algebra of the dual type, then we know by Theorem 7.6.5 that $A \otimes B$ is a Lie algebra. In fact it has a finer structure here.

Proposition 13.9.1. *Under the aforementioned hypothesis the operation* $\{a \otimes b, a' \otimes b'\} := aa' \otimes bb'$ *is a pre-Lie product.*

Proof. The verification is a direct calculation. □

13.10 Jordan Algebras, Interchange Algebras

The symmetrization of an associative product motivates the introduction of the notion of Jordan algebra. The associated operad is binary and cubic (relations involving 3 operations). It is a key tool in the analysis of symmetric spaces. We briefly mention another type of cubic operad: the operad which encodes the interchange algebras.

13.10.1 Definition of Jordan Algebras

A *Jordan algebra* is a vector space equipped with a symmetric binary operation $x \cdot y$ which satisfies the relation:

$$(a^{\cdot 2}) \cdot (b \cdot a) = ((a^{\cdot 2}) \cdot b) \cdot a.$$

A priori operads do not code for such types of algebras because the relation is not multilinear. However, if 3 is invertible in $\mathbb{K}$, the relation is equivalent to its multilinearized form which is the cubic relation:

$$\begin{aligned}&(x \cdot y) \cdot (t \cdot z) + (x \cdot z) \cdot (t \cdot y) + (y \cdot z) \cdot (t \cdot x)\\&\quad = ((x \cdot y) \cdot t) \cdot z + ((x \cdot z) \cdot t) \cdot y + ((y \cdot z) \cdot t) \cdot x\end{aligned}$$

which is symmetric in x, y, z. So, for the purpose of this book we call *Jordan algebra* a vector space equipped with one binary operation satisfying this cubic relation. The associated operad is denoted by *Jord*.

13.10.2 From Associative Algebras to Jordan Algebras

Let A be an associative algebra with product xy. We denote by $x \cdot y := xy + yx$ the symmetrized product. It is well known, and easy to check, that the symmetrized product satisfies the cubic relation of Jordan algebras. More precisely, in arity 3, there are 3 different operations, all of the type $(x \cdot y) \cdot z$, which are linearly independent. In arity 4, there are 15 different operations, 12 of the type $((x \cdot y) \cdot z) \cdot t$ and 3 of the type $(x \cdot y) \cdot (z \cdot t)$ (cf. Sect. 13.8.4). However the 15 operations in arity 4 are not linearly independent, since they satisfy the cubic formula. As a consequence we get a morphism of operads

$$Jord \to Ass.$$

Contrary to the Lie case, this morphism is not injective. In other words there are relations satisfied by the symmetrized associative product which are not consequences of the Jordan relation. The first examples appear in arity 8: the Glennie relations, see page 79 of [ZSSS82].

13.10.3 Jordan Triple

The relation which defines Jordan algebra is not quadratic. However there is a close notion called Jordan triples, whose relation is quadratic. By definition a *Jordan triple* (also called Jordan triple system) is a vector space A equipped with a ternary operation satisfying the symmetry relation

$$(xyz) = (zyx),$$

and the quadratic relation

$$\big(uv(xyz)\big) = \big((uvx)yz\big) - \big(x(yuv)z\big) + \big(xy(uvz)\big).$$

Because of the symmetry relation there are several ways of writing the quadratic relation. The way we wrote it here is close to a Leibniz type relation. The operad of Jordan triples is denoted by JT. The relation with Jordan algebras is as follows. Any Jordan algebra gives rise to a Jordan triple by the formula:

$$(xyz) := x(yz) - y(xz) + (xy)z.$$

Any associative algebra gives a Jordan triple by the formula:

$$(xyz) := xyz + zyx.$$

In the next section we introduce the notion of ternary totally associative algebra, denoted $tAs^{\langle 3\rangle}$-algebra. There is a functor $tAs^{\langle 3\rangle}\text{-alg} \to JT\text{-alg}$, given by $\langle xyz\rangle := (xyz) + (zyx)$. Indeed, the operation $\langle xyz\rangle$ satisfies the symmetry property of the Jordan triples by definition. It satisfies also the quadratic relation because this quadratic relation is already fulfilled for (xyz) if this operation is totally associative.

This functor is similar to the functor *Ass*-alg $\to$ *Lie*-alg. In this Jordan framework the role of commutative algebras is played by the totally associative ternary algebras which satisfy the symmetry relation

$$(xyz) = (zyx).$$

13.10.4 Interchange Algebra

Let $x \cdot y$ and $x * y$ be two binary operations. The following relation

$$(a \cdot b) * (c \cdot d) = (a * c) \cdot (b * d)$$

crops up in many places in mathematics. It is called the *interchange relation*. In many examples the two operations are supposed to be associative. For instance, it is the key formula to show that the higher homotopy groups are abelian (Eckmann–Hilton argument). It also permits us to define the notion of "bicategory". When $\cdot = *$, any idempotent x (i.e. $x * x = x$) gives the identity

$$x * (y * z) = (x * y) * (x * z).$$

This is called the "right-distributivity property". It is related with many structures, like braids and knots for instance.

The operad defined by these two binary operations and the interchange relation is binary cubic and deserves to be studied.

13.11 Multi-ary Algebras

So far we have essentially dealt with examples of operads generated by binary operations. Of course, there is no reason to study only these ones, and in fact there are some very interesting operads generated by operations in any arity. We have already met some, when working with $\mathscr{P}_\infty$-algebras. Here are some others involved in the study of moduli spaces, together with an elementary application to the inversion of power series.

13.11.1 Totally and Partially Associative $\langle k \rangle$-Algebra

By definition a *totally, resp. partially, associative* $\langle k \rangle$*-algebra* is a vector space A equipped with a k-ary operation:

$$(\quad) : A^{\otimes k} \longrightarrow A$$

which satisfies

$$\big((x_1 \cdots x_k) x_{k+1} \cdots x_{2k-1}\big) = \big(x_1 \cdots x_i (x_{i+1} \cdots x_{i+k}) \cdots x_{2k-1}\big) \quad \text{for any } i,$$

$$\text{resp.} \quad \sum_{i=0}^{i=k-1} (-1)^{i(k-1)} \big(x_1 \cdots x_i (x_{i+1} \cdots x_{i+k}) \cdots x_{2k-1}\big) = 0.$$

The category of totally, resp. partially, associative $\langle k\rangle$-algebras is encoded by a nonsymmetric operad denoted by $tAs^{\langle k\rangle}$, resp. $pAs^{\langle k\rangle}$ introduced by Victor Gnedbaye in [Gne97]. The space on n-ary nonsymmetric operations of $tAs^{\langle k\rangle}$ is given by

$$tAs^{\langle k\rangle}_{ik-i-1} = \mathbb{K}, \qquad tAs^{\langle k\rangle}_n = 0 \quad \text{otherwise.}$$

If we put the degree $k-2$ on the generating operation of $pAs^{\langle k\rangle}$, then the ns operads $tAs^{\langle k\rangle}$ and $pAs^{\langle k\rangle}$ are Koszul dual to each other, cf. [MR09]. Moreover the Koszulity can be proved by the rewriting method, cf. Sect. 8.1.

13.11.2 Hypercommutative Algebra and Gravity Algebra

A *hypercommutative algebra* A is a chain complex A equipped with a totally symmetric n-ary operation $(x_1, \ldots, x_n)$ of degree $2(n-2)$ for any $n \geq 2$. They are supposed to satisfy the following generalized associativity relation:

$$\sum_{S_1 \sqcup S_2 = \{1,\ldots,n\}} \big((a, b, x_{S_1}), c, x_{S_2}\big) = \sum_{S_1 \sqcup S_2 = \{1,\ldots,n\}} (-1)^{|c||x_{S_1}|}\big(a, (b, x_{S_1}, c), x_{S_2}\big),$$

for any $n \geq 0$. We denote the associated operad by *HyperCom*.

Consider the moduli space $\mathcal{M}_{g,n+1}$ of smooth projective curves of genus g with $n+1$ marked points. It admits a compactification $\overline{\mathcal{M}}_{g,n+1}$ made of stable curves, which is due to Deligne, Mumford and Knudsen [DM69, Knu83]. The gluing of the latter curves along one point defines an operad structure. Actually, it forms a cyclic and a modular operad (see Sects. 13.14.6 and 13.14.8). Here we think of the first marked point as the output.

Proposition 13.11.1. [KM94, Get95] *The operad formed by the homology $H_\bullet(\overline{\mathcal{M}}_{0,n+1})$ of the genus 0 part of the moduli space of stable curves is isomorphic to the operad encoding hypercommutative algebras.*

Proof. The genus 0 moduli space $\overline{\mathcal{M}}_{0,n+1}$ is a compact manifold of dimension $2(n-2)$. One sends its fundamental class to the generating n-ary operations of the operad *HyperCom*. For the rest of the proof, we refer the reader to the original papers. □

The Gromov–Witten invariants endow the cohomology of any symplectic manifold or any smooth projective variety with a hypercommutative algebra structure. (Notice that in the book [Man99b], this structure is called a $Comm_\infty$-algebra. To avoid confusion with $C_\infty = Com_\infty$-algebra, that is homotopy commutative algebra, we stick to the notation *HyperCom*.)

A *gravity algebra* is a chain complex A endowed with (graded) skew-symmetric operations $[x_1, \ldots, x_n] : A^{\otimes n} \to A$ of degree $n-2$ for any $n \geq 2$. They satisfy the following relations:

$$\sum_{1\le i<j\le k} \pm[[x_i, x_j], x_1, \ldots, \widehat{x_i}, \ldots, \widehat{x_j}, \ldots, x_k, y_1, \ldots, y_l] = \begin{cases} [[x_1, \ldots, x_k], y_1, \ldots, y_l] & \text{for } l > 0, \\ 0 & \text{for } l = 0. \end{cases}$$

We denote the associated operad by *Grav*.

The gluing along one point of two smooth curves does not produce a smooth curve anymore. However, this gluing endows the suspension of the homology $sH_\bullet(\mathcal{M}_{0,n+1})$ with an operad structure.

Proposition 13.11.2. [Get94] *The operad formed by the homology $sH_\bullet(\mathcal{M}_{0,n+1})$ of the genus* 0 *part of the moduli space of smooth curves is isomorphic to the operad encoding gravity algebras.*

Theorem 13.11.3. [Get95] *The operads HyperCom and Grav are Koszul dual to each other.*

Proof. This proof relies on the mixed Hodge structure of the moduli spaces of curves, see [Get95]. □

13.11.3 Lie and Leibniz ⟨k⟩-Algebra

By definition a Leibniz *⟨k⟩-algebra* is a vector space A equipped with a k-ary operation

$$[\quad] : A^{\otimes k} \longrightarrow A$$

satisfying the quadratic relation

$$\big[[x_1 \cdots x_k] y_1 \cdots y_{k-1}\big] = \sum_{i=0}^{i=k-1} \big[x_1 \cdots x_{i-1} [x_i y_1 \cdots y_{k-1}] x_{i+1} \cdots x_k\big].$$

The notion was introduced and studied in [CLP02]. For $k = 2$ it is the notion of Leibniz algebra mentioned in Sect. 13.5.

By definition a Lie *⟨k⟩-algebra* [VV97] is a Leibniz ⟨k⟩-algebra whose k-ary bracket satisfies the following symmetry property:

$$[x_{\sigma(1)} \cdots x_{\sigma(k)}] = \operatorname{sgn}(\sigma)[x_1 \cdots x_k] \quad \text{for any permutation } \sigma.$$

This notion has been used in combinatorics, see for instance [HW95].

For $k = 2$ this is the classical notion of Lie algebras. For $k = 3$ there is an intermediate structure studied in the literature, called *Lie triple systems* [Lis52]. It consists in starting with a Leibniz 3-algebra and imposing the symmetry property:

$$[x_1 x_2 x_3] + [x_2 x_1 x_2] + [x_3 x_1 x_2] = 0.$$

For higher k's this notion is involved in the so-called *Nambu mechanics* [Tak94].

13.11.4 Brace Algebras

A *brace algebra* is a vector space R equipped with a $(1+n)$-ary operation denoted either $\{-;-,\dots,-\}$ or M_{1n} for $n\geq 1$ satisfying the following formulas:

$$\big\{\{x;y_1,\dots,y_n\};z_1,\dots,z_m\big\}=\sum\big\{x;\dots,\{y_1;\dots\},\dots,\{y_n;\dots,\},\dots\big\}.$$

On the right-hand side the dots are filled with the variables z_is (in order) with the convention $\{y_k;\emptyset\}=y_k$. The first nontrivial relation, which relates the 2-ary operation and the 3-ary operation reads

$$\big\{\{x;y\};z\big\}=\{x;z,y\}+\big\{x;\{y;z\}\big\}+\{x;y,z\}.$$

Observe that, as a consequence of this relation, the binary operation $\{-;-\}$ is a pre-Lie product since its associator is right-symmetric:

$$\big\{\{x;y\};z\big\}-\big\{x;\{y;z\}\big\}=\{x;y,z\}+\{x;z,y\}.$$

This notion was implicitly used in [Ger63] and was formally introduced by Gerstenhaber and Voronov in [GV95].

When the $(1+n)$-ary brace operation $\{x;y_1,\dots,y_n\}$ is symmetric in the y variables for any n, then the brace algebra is said to be symmetric. We mentioned the equivalence between symmetric brace algebras and pre-Lie algebras in Sect. 13.4.5.

The pre-Lie structure of $\mathscr{P}(\mathbb{K})$ for a ns operad $\mathscr{P}$ mentioned in Sect. 5.9.15 can be extended as follows.

Proposition 13.11.4. *For any nonsymmetric operad $\mathscr{P}$ the free $\mathscr{P}$-algebra on one generator $\mathscr{P}(\mathbb{K})=\oplus_n\mathscr{P}_n$, is a brace algebra, in particular it is a pre-Lie algebra.*

Proof. The operad structure of $\mathscr{P}$ is determined by the composition maps

$$\gamma_{i_1\cdots i_k}:\mathscr{P}_k\otimes\mathscr{P}_{i_1}\otimes\cdots\otimes\mathscr{P}_{i_k}\to\mathscr{P}_{i_1+\cdots+i_k}.$$

Since $\mathscr{P}(\mathbb{K})=\bigoplus_n\mathscr{P}_n$ we define

$$M_{1n}:\mathscr{P}(\mathbb{K})\otimes\mathscr{P}(\mathbb{K})^{\otimes n}\to\mathscr{P}(\mathbb{K})$$

as follows for $\mu_n\in\mathscr{P}_n$ and $\nu_{i_j}\in\mathscr{P}_{i_j}$:

$$\begin{aligned}
&M_{1n}(\mu_n;\nu_{i_1},\dots,\nu_{i_k})=0\quad\text{if } n<k,\\
&M_{1n}(\mu_n;\nu_{i_1},\dots,\nu_{i_k})=\gamma(\mu_k;\nu_{i_1},\dots,\nu_{i_k})\quad\text{if } n=k,\\
&M_{1n}(\mu_n;\nu_{i_1},\dots,\nu_{i_k})=\sum\gamma(\mu_n;\mathrm{id},\dots,\mathrm{id},\nu_{i_1},\mathrm{id},\dots,\mathrm{id},\nu_{i_k},\dots)\quad\text{if } n>k.
\end{aligned}$$

Observe that the elements on the right-hand side lie in $\mathscr{P}_{i_1+\cdots+i_k+n-k}$.

The brace relations are a consequence of the associativity of γ. □

The operad *Brace* encoding the brace algebras can be described by using the planar rooted trees (see for instance [LR10]). It is related to many other types of algebras. One of the main points is the following.

Proposition 13.11.5 (Ronco [Ron02]). *For any brace algebra R the tensor module $T(R)$ is a unital dendriform algebra. Equipped with the deconcatenation coproduct, this tensor module becomes a cofree Hopf algebra.*

Proof. We refer to the original papers [Ron00, Ron02] for the proofs. Let us state the main formula relating the dendriform structure of $(T(R), \prec, \succ)$ and the brace structure of R. For any $x_1, \dots, x_k \in R$ we introduce Ronco's elements

$$w^i_{\prec}(x_1, \dots, x_k) := \big(x_1 \prec \big(x_2 \prec (\dots \prec x_k)\big)\big),$$
$$w^i_{\succ}(x_1, \dots, x_k) := \big(\big((x_1 \succ x_2) \succ \dots\big) \succ x_k\big).$$

The brace operations and the dendriform operations are related by:

$$\{x; y_1, \dots, y_n\} = \sum_{i=1}^{n+1} (-1)^i w^i_{\prec}(y_1, \dots, y_{i-1}) \succ x \prec w^i_{\succ}(y_i, \dots, y_n).$$ □

REMARK. When $\mathscr{P} = \mathrm{End}(A)$, for A an associative algebra, the Hochschild cochain complex $C^\bullet_{Hoch}(A, A) := \oplus_n \mathrm{Hom}(A^{\otimes n}, A)$ has an associative product and a Lie bracket which make it into a GV-algebra. This structure plays a key role in the proof of the Deligne conjecture, cf. Sect. 13.3.15.

13.11.5 Multibrace Algebra

The notion of brace algebra is a particular case of a more general notion, called multibrace algebra, which is defined as follows.

A *multibrace algebra* is a vector space A equipped with $(p+q)$-ary operations $M_{pq} : A^{\otimes p+q} \to A$, $p \geq 0, q \geq 0$, such that

$$M_{00} = 0, \qquad M_{01} = \mathrm{id} = M_{10}, \qquad M_{0q} = 0 = M_{p0}, \quad \text{for } p > 1, q > 1,$$

$(\mathscr{R}_{ijk})$:

$$\sum_{1 \leq l \leq i+j} M_{lk} \circ \big(M_{i_1 j_1} \dots M_{i_l j_l}; \mathrm{id}^{\otimes k}\big) = \sum_{1 \leq m \leq j+k} M_{im} \circ \big(\mathrm{id}^{\otimes i}; M_{j_1 k_1} \dots M_{j_m k_m}\big)$$

where the left sum is extended to all sets of indices $i_1, \dots, i_l; j_1, \dots, j_l$ such that $i_1 + \dots + i_l = i$; $j_1 + \dots + j_l = j$, and the right sum is extended to all sets of indices $i_1, \dots, i_m; j_1, \dots, j_m$ such that $j_1 + \dots + j_m = j$; $k_1 + \dots + k_m = k$. We denote the associated operad by MB. The notion of differential graded multibrace algebra first appeared under the name B_∞-algebra in [GJ94], see also [GV95]. The plain version was denoted $\mathbf{B}_\infty$ in [LR06]. The relationship with brace algebras is given by the following statement: if, in a given B_∞-algebra, the operations M_{nm} are 0 for $n \geq 2$, then it is a brace algebra for $\{-;-,\ldots,-\} := M_{1,p}$. Our interest for this structure is justified by the following proposition.

Proposition 13.11.6. *For any multibrace algebra R the tensor module $T(R)$ is a unital 2-associative algebra. Equipped with the deconcatenation coproduct, this tensor module becomes a cofree Hopf algebra. Conversely, any cofree Hopf algebra equipped with an isomorphism with the tensor module over its primitive part is of this form.*

Proof. We refer to the original papers [LR06] for the proof. See also [LR10] for more information. □

There is a strong relationship between brace algebras and multibrace algebras. A brace algebra is nothing but a multibrace algebra such that $M_{pq} = 0$ for any $p \geq 2$. Moreover this vanishing condition is equivalent to a "right-sided property" of the Hopf algebra, which reads as follows:

$$\text{for any } q \geq 0 \text{ the space } \bigoplus_{n \geq q} R^{\otimes n} \text{ is a right ideal of } T(R).$$

We refer to [LR10] for more on this subject.

13.11.6 From (Multi)Brace Algebras to Hopf Algebras

If one takes the tensor module over a multibrace algebra, then one gets a bialgebra (in fact a Hopf algebra). The multibrace algebra is the primitive part of the bialgebra. Starting with a brace algebra gives a bialgebra satisfying some special property (rightsidedness), cf. Proposition 13.11.5. We refer to [LR10] and [MV09a] for more information on this subject.

13.11.7 Magmatic-Infinity Algebras

We have seen in Sect. 13.8 that the magmatic operad is the free ns operad on a binary operation. Let us start with one operation in each arity $n \geq 2$. We denote by Mag^∞ the free operad on this set of operations. So a Mag^∞-algebra is simply a vector space equipped with one n-ary operation μ_n for each $n \geq 2$ without any further relation. We recall that this is the beginning of the description of A_∞. We have seen in Sect. 9.2.4 that the free Mag^∞-algebra on one generator is spanned by the planar trees (with at least two inputs). As in the case of Mag, the dual operad is easy to describe since it is a nilpotent operad. The functional equation of the generating series permits us to show that the number of planar trees is given by the little Schroeder numbers. Instead of giving the details, we will show a refinement of this result as follows. Given a generic formal power series, the next proposition make explicit the coefficients of its inverse for composition. It is a well-known result, but the proof that we are going to provide is very simple and makes it clear why the number of planar trees of a given type plays a role in the formula.

Proposition 13.11.7 (Lagrange inversion formula). *et* $f(x) = x + a_1x^2 + \cdots + a_{n-1}x^n + \cdots$ *be a generic formal power series, and let* $g(x) = x + b_1x^2 + \cdots + b_{n-1}x^n + \cdots$ *be its inverse for composition. The coefficient* b_n *is a polynomial in the coefficients* $a_1, \ldots, a_n$ *given by the following formula:*

$$b_1 = -a_1,$$
$$b_2 = 2{a_1}^2 - a_2,$$
$$b_3 = -5{a_1}^3 + 5a_1a_2 - a_3,$$
$$b_4 = 14{a_1}^4 - 21{a_1}^2a_2 + 6a_1a_3 + 3{a_2}^2 - a_4,$$

and more generally

$$b_n = \sum_{i_1+2i_2+\cdots+ki_k=n} (-1)^{i_1+i_2+\cdots+i_k} c_{i_1\cdots i_k} {a_1}^{i_1}{a_2}^{i_2}\ldots {a_k}^{i_k},$$

where the coefficient $c_{i_1\cdots i_k}$ *is the number of planar rooted trees having* $n+2$ *leaves and* i_j *vertices with* $j+1$ *inputs* (*cf.* [Sta97a]).

Proof. [Val07b, Lod06] First we remark that it is sufficient to prove this statement under the hypothesis that the elements a_n are positive integers. Under this assumption, let $\mathscr{P}$ be the free ns operad on a_{n-1} n-ary operations ($n \geq 2$). The space $\mathscr{P}_n$ of n-ary operations is spanned by the compositions of these generating operations. Therefore a linear basis of $\mathscr{P}_n$ is obtained by taking all the planar trees with n leaves and decorating each vertex with j inputs by one of the a_{j-1} operations of arity j. Recall that there are i_j such vertices. Therefore the generating series of $\mathscr{P}$ is (up to signs) $g(x)$.

The dual of $\mathscr{P}$ is the nilpotent operad having a_{n-1} n-ary operations ($n \geq 2$). Because of the nilpotency there is no new operation created by composition. Therefore its generating series is $f(x)$. Since $\mathscr{P}$ is a free operad, it is Koszul and the functional equation between the two generating series (cf. Theorem 7.5.1 and [Val07b, Sect. 4.3]) gives the expected result for $y = 1$ and $x = t$. □

Observe that $c_{i_1\cdots i_k}$ is also the number of cells of the $(k-1)$-dimensional associahedron $\mathscr{K}^{k-1}$, which are of the form $(\mathscr{K}^0)^{i_1} \times \cdots \times (\mathscr{K}^{k-1})^{i_k}$.

13.12 Examples of Operads with 1-Ary Operations, 0-Ary Operation

We first comment briefly on the case of associative algebras viewed as operads.

We have already seen one example of operad mixing unary operation and binary operations: the Batalin–Vilkovisky operad, cf. Sect. 13.7. We briefly mention another one: associative algebras with derivation. In the last subsection we give a brief comment on operads with a 0-ary operation.

13.12.1 Unital Associative Algebras as Operads

We have already mentioned that a unital associative algebra R can be viewed as a ns operad concentrated in arity one: $\mathscr{P}(1) = R$ and $\mathscr{P}(n) = 0$ for $n \geq 2$. Therefore the theorems of the general theory apply, in particular the Homotopy Transfer Theorem. An algebra of this operad $\mathscr{P}$ is simply a left module over R. An algebra over $\Omega\mathscr{P}^{¡}$ is a left dg module over the dg algebra $\Omega R^{¡}$.

Surprisingly the case of the dual numbers $R = \mathbb{K}[\varepsilon]/(\varepsilon^2 = 0)$ gives rise to a structure which generalizes the notion of spectral sequence. We give details on this case in Sect. 10.3.7.

In the case of the symmetric algebra the Koszul duality between $S(V)$ and $\Lambda^c(V)$ gives essentially the Bernstein–Gelfand–Gelfand [BGG78] correspondence.

13.12.2 Associative Algebras with Derivation

Let A be nonunital associative algebra over $\mathbb{K}$. A *derivation* of A is a linear map $D^A : A \to A$ which satisfies the Leibniz relation

$$D^A(ab) = D^A(a)b + aD^A(b)$$

for any $a, b \in A$. There is an obvious notion of morphism. The nonsymmetric operad encoding the category of associative algebras with derivation is denoted by *AsDer*. It admits the following presentation. There are two generating operations, one of arity 1, that we denote by D, and one of arity 2 that we denote by μ. The relations are:

$$\begin{cases} \mu \circ (\mu, \mathrm{id}) = \mu \circ (\mathrm{id}, \mu), \\ \quad D \circ \mu = \mu \circ (D, \mathrm{id}) + \mu \circ (\mathrm{id}, D). \end{cases}$$

The ns operad *AsDer* is determined by a certain family of vector spaces $AsDer_n$, $n \geq 1$, and composition maps

$$\gamma_{i_1,\dots,i_k} : AsDer_k \otimes AsDer_{i_1} \otimes \cdots \otimes AsDer_{i_k} \to AsDer_n$$

where $n = i_1 + \cdots + i_k$. As a vector space $AsDer_n$ is isomorphic to the space of polynomials in n variables:

$$AsDer_n = \mathbb{K}[x_1, \dots, x_n],$$

so it is not finite dimensional. The composition map $\gamma = \gamma_{i_1,\ldots,i_k}$ is given by

$$\begin{aligned}
&\gamma(P; Q_1, \ldots, Q_k)(x_1, \ldots, x_n) \\
&\quad = P(x_1 + \cdots + x_{i_1}, x_{i_1+1} + \cdots + x_{i_1+i_2}, x_{i_1+i_2+1} + \cdots, \ldots) \\
&\qquad \times Q_1(x_1, \ldots, x_{i_1}) Q_2(x_{i_1+1}, \ldots, x_{i_1+i_2}) \cdots .
\end{aligned}$$

Under this identification the operations id, D, μ correspond to $1_1, x_1 \in \mathbb{K}[x_1]$ and to $1_2 \in \mathbb{K}[x_1, x_2]$ respectively. More generally the operation

$$(a_1, \ldots, a_n) \mapsto D^{j_1}(a_1) D^{j_2}(a_2) \cdots D^{j_n}(a_n)$$

corresponds to the monomial $x_1^{j_1} x_2^{j_2} \cdots x_n^{j_n}$.

We refer to [Lod10] for the study of this operad and several generalizations of this case. One of them is particularly interesting: it consists in replacing the polynomials by the rational functions. So in arity one the operation x gets an inverse for composition x^{-1}. Since x corresponds to the derivation D, its inverse corresponds to integration. So the integro-differential calculus can be written within this operad.

13.12.3 Some Examples of Operads with a 0-Ary Operation

So far, we have not worked out the operads with nontrivial 0-ary operation. In the definition of a $\mathscr{P}$-algebra A, any 0-ary operation $u \in \mathscr{P}(0)$ induces a map $\mathbb{K} \to A$, or equivalently a particular element in A.

The main example is unital associative algebras. The nonsymmetric operad *uAs* encoding them is equal to the operad *As* for associative algebras plus one element in arity 0: $uAs(0) = \mathbb{K} u$. The composition product of the ns operad *uAs* is given by the composition product operad in *As* plus some composition involving the unit. Explicitly, the ns operad *uAs* admits the following presentation $uAs = \mathscr{T}(u, \mu)/(R_{uAs})$. The space of relations R_{uAs} is generated by the associator of μ and

$\mu \circ_1 u = \mathrm{id} = \mu \circ_2 u$:

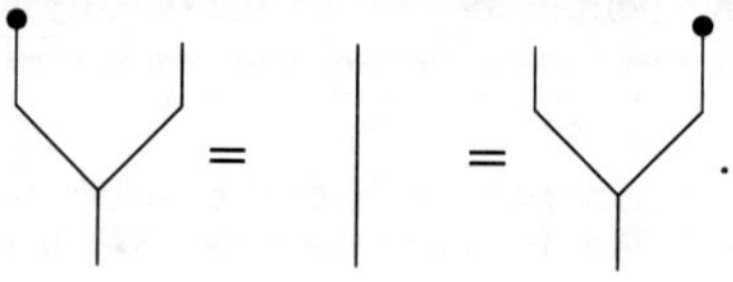

These relations involve binary (2-ary) and constant (0-ary) operations. Therefore this example does not fit into the Koszul duality theory as developed here. We refer the reader to the paper [HM10] of Joey Hirsh and Joan Millès for the suitable generalization. In this case, the Koszul dual cooperad is equipped with a "curvature", which makes it into a non dg object. This theory produces a notion of homotopy unital associative algebra, which is closely related to the ones appearing in symplectic geometry [FOOO09a, FOOO09b, Sei08].

Other types of algebras admit a coherent unit, like commutative algebras, *Perm*-algebras, for instance. In these cases, one has to use operads with a nontrivial 0-ary operation to encode them. There are also cases where it is useful to introduce a unit, though composition is not everywhere defined. The case of dendriform algebras is a paradigm, cf. Sect. 13.6.4. For an elementary introduction to operads with coherent units, see [Lod04b]. Some further study can be found in [Hol06] and [EFG07].

13.13 Generalized Bialgebras and Triples of Operads

We have seen in Sect. 1.3.2 that the free associative algebra is in fact a cocommutative bialgebra. In other words it is naturally equipped with a structure of cocommutative coalgebra satisfying the Hopf compatibility relation. A similar phenomenon occurs in many other situations. For instance, if we replace the Hopf compatibility relation by the unital infinitesimal compatibility relation, then the cooperation is the deconcatenation coproduct (which is coassociative). In this section we give some more examples and we summarize the general results on the "generalized bialgebras". They consist in structure theorems which were stated and proved in [LR06, Lod08].

13.13.1 Unital Infinitesimal Bialgebra

A *unital infinitesimal bialgebra*, u.i. bialgebra for short, is a vector space $\mathscr{H}$ equipped with a unital associative product xy, and a counital coassociative coproduct $\Delta(x)$, which satisfy the *unital infinitesimal compatibility relation*:

$$\Delta(xy) = -x \otimes y + \Delta(x)(1 \otimes y) + (x \otimes 1)\Delta(y).$$

This compatibility relation can be pictured as:

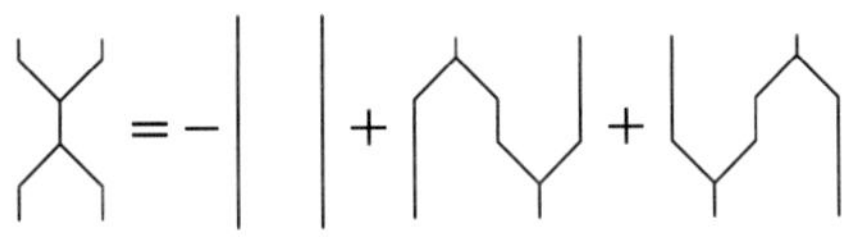

The main example is the following. Let $T(V)$ be the tensor module that we equip with the concatenation product and the deconcatenation coproduct (see Sect. 1.2.6). Then it can easily be proved that concatenation and deconcatenation satisfy the u.i. compatibility relation. It turns out that if the u.i. bialgebra is supposed to be conilpotent, then it is a tensor algebra:

Proposition 13.13.1 (Loday–Ronco [LR06]). *Let $\mathscr{H}$ be a u.i. bialgebra over the field $\mathbb{K}$. Then the following are equivalent:*

a) *$\mathscr{H}$ is conilpotent,*
b) *$\mathscr{H} \cong T(\operatorname{Prim} \mathscr{H})$ as a u.i. bialgebra.*

Here Prim $\mathscr{H}$ is the primitive part of $\mathscr{H}$, that is the kernel of the reduced coproduct. This result is analogous to the Hopf–Borel theorem (a particular case of the CMM theorem), which asserts that, over a characteristic zero field, a commutative cocommutative Hopf algebra which is conilpotent is the symmetric algebra over its primitive part. Observe that, in the associative case, the characteristic zero hypothesis is not necessary. It comes from the fact that the involved operads are nonsymmetric operads and the compatibility relation does not use the symmetric groups.

The Hopf–Borel theorem admits a generalization: the CMM theorem. Similarly the above mentioned structure theorem admits a generalization which involves "duplicial algebras".

13.13.2 Duplicial Structures [Lod08, Chap. 5]

A *duplicial algebra* is a vector space A equipped with two operations $x \prec y$ and $x \succ y$ which satisfy the following axioms:

$$(x \prec y) \prec z = x \prec (y \prec z),$$
$$(x \succ y) \prec z = x \succ (y \prec z),$$
$$(x \succ y) \succ z = x \succ (y \succ z).$$

A *duplicial bialgebra* is a duplicial algebra $\mathscr{H}$ equipped with a coassociative coproduct Δ which satisfy the following n.u.i. compatibility relations:

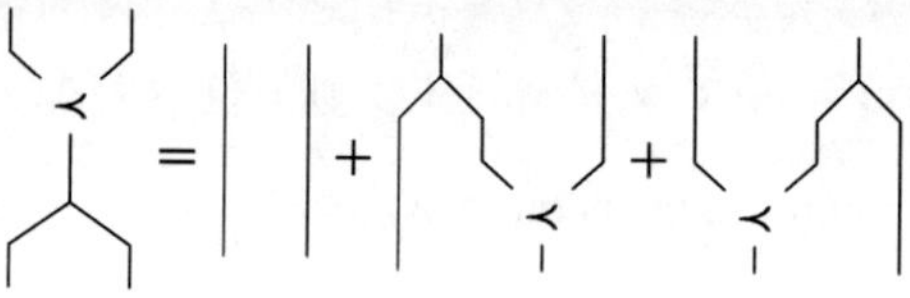

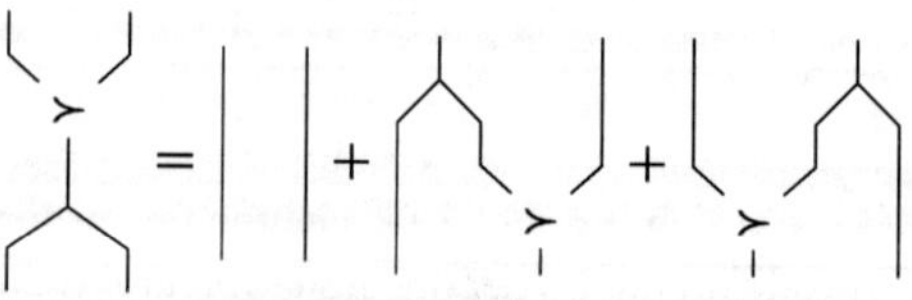

Observe that, here, we work in a nonunital context. It can be shown, by an argument similar to the classical case, that the free duplicial algebra is naturally a duplicial bialgebra. Let us introduce the operation "dot" defined by

$$x \cdot y := x \succ y - x \prec y.$$

It gives rise to a forgetful functor *Dup*-alg $\to$ *Mag*-alg from the category of duplicial algebras to the category of magmatic algebras. Since it is a forgetful functor it admits a left adjoint which we denote by U : *Mag*-alg $\to$ *Dup*-alg.

Theorem 13.13.2 (Loday–Ronco [LR06]). *Let $\mathscr{H}$ be a duplicial bialgebra over a field $\mathbb{K}$. The following conditions are equivalent:*

a) *$\mathscr{H}$ is conilpotent,*
b) *$\mathscr{H} \cong U(\operatorname{Prim}\mathscr{H})$ as a duplicial bialgebra,*
c) *$\mathscr{H} \cong \overline{T}^c(\operatorname{Prim}\mathscr{H})$ as a coassociative coalgebra.*

The key point of the proof is the property of the free duplicial algebra mentioned above: it is naturally a duplicial bialgebra. Its coproduct is constructed as follows. First, we remark that the tensor product of two duplicial algebras A and B can be made into a duplicial algebra by taking

$$(x \otimes y) \prec (x' \otimes y') := (x \prec x') \otimes (y \prec y'),$$
$$(x \otimes y) \succ (x' \otimes y') := (x \succ x') \otimes (y \succ y').$$

The coproduct $\Delta : Dup(V) \to Dup(V) \otimes Dup(V)$ is constructed as follows: it is the unique extension of the map $0 : V \to Dup(V) \otimes Dup(V)$ which satisfies the n.u.i. compatibility relations. Observe that Δ is not 0 since $\Delta(v \prec w) = v \otimes w = \Delta(v \succ w)$ for any $v, w \in V$.

As a corollary of the theorem we get the following isomorphism of functors (or graded vector spaces):

$$Dup \cong As \circ Mag$$

which is the associative analog of $As \cong Com \circ Lie$. In particular the natural map $Mag(V) \to Dup(V)$ is injective. This fact plays a key role in the proof of the property of the symmetrized pre-Lie product, cf. Sect. 13.4.4.

The structure theorem 13.13.2 is an analog of the CMM-PBW theorem in the noncocommutative framework. Several other structure theorems of this type can be proven. The general framework is as follows.

13.13.3 Generalized Bialgebras

Let $\mathscr{A}$ and $\mathscr{C}$ be two algebraic operads. We assume that $\mathscr{A}(0) = 0 = \mathscr{C}(0)$, $\mathscr{A}(1) = \mathbb{K}\,\mathrm{id} = \mathscr{C}(1)$ and that there is a finite number of generating operations in each arity. As a consequence $\mathscr{C}(n)$ and $\mathscr{A}(n)$ are finite dimensional vector spaces.

By definition a *($\mathscr{C}^c$, $\between$, $\mathscr{A}$)-bialgebra*, or *$\mathscr{C}^c$-$\mathscr{A}$-bialgebra* for short, also called *generalized bialgebra*, is a vector space $\mathscr{H}$ which is an $\mathscr{A}$-algebra, a $\mathscr{C}$-coalgebra, and such that the operations of $\mathscr{A}$ and the cooperations of $\mathscr{C}$ acting on $\mathscr{H}$ satisfy some *compatibility relations*, denoted $\between$, read "between" (some equalities involving composition of operations and cooperations valid for any elements of $\mathscr{H}$). This set of relations is, of course, part of the structure. A category of generalized bialgebras is governed by an *algebraic prop* (we simply say a prop). Props can be defined as algebras over some combinatorial monad like in Sect. 13.14. Starting with any presentation of the operad $\mathscr{A}$ and of the cooperad $\mathscr{C}$, this prop is obtained as the quotient of the free prop generated by the generators of $\mathscr{A}$ and $\mathscr{C}$ (considered as multivalued operations), modulo the relations between the operations, the relations between the cooperations and the relations entwining operations and cooperations. The algebras over this quotient are the generalized bialgebras.

A *distributive compatibility relation* between the operation μ and the cooperation δ is a relation of the form

$$\delta\circ\mu=\sum_i\left(\mu_1^i\otimes\cdots\otimes\mu_m^i\right)\circ\omega^i\circ\left(\delta_1^i\otimes\cdots\otimes\delta_n^i\right)\qquad (\between)$$

where

$$\begin{cases}\mu\in\mathscr{A}(n), & \mu_1^i\in\mathscr{A}(k_1),\ldots,\mu_m^i\in\mathscr{A}(k_m),\\ \delta\in\mathscr{C}(m), & \delta_1^i\in\mathscr{C}(l_1),\ldots,\delta_n^i\in\mathscr{C}(l_n),\\ k_1+\cdots+k_m=l_1+\cdots+l_n=r_i, & \\ \omega^i\in\mathbb{K}[S_{r_i}]. & \end{cases}$$

Hence, in a generalized bialgebra with distributive compatibility relations, the composite of an operation and a cooperation can be re-written as cooperations first and then operations. Observe that the identity is both an operation and a cooperation.

13.13.4 Hypothesis (H0)

There is a distributive compatibility relation for any pair (δ,μ) where μ is an operation and δ is a cooperation, which preserves the composition of the operad $\mathscr{A}$ and of the cooperad $\mathscr{C}$.

The distributive compatibility relations induce a *mixed distributive law* in the sense of Fox and Markl [FM97], that is a map

$$\mathscr{P}(m)\otimes\mathscr{C}(n)\to\bigoplus\mathscr{C}(i_1)\otimes\cdots\otimes\mathscr{C}(i_m)\otimes_{S_i}\mathbb{K}[S_N]\otimes_{S_j}\mathscr{P}(j_1)\otimes\cdots\otimes\mathscr{P}(j_n).$$

Here we used the multi-index notation for $i=(i_1,\ldots,i_m)$ and for j, and $S_i:=S_{i_1}\times\cdots\times S_{i_m}$, $N=i_1+\cdots+i_m=j_1+\cdots+j_n$.

Example of a compatibility relation for the pair (δ,μ) with $n=3$, $m=4$ and $r=8$:

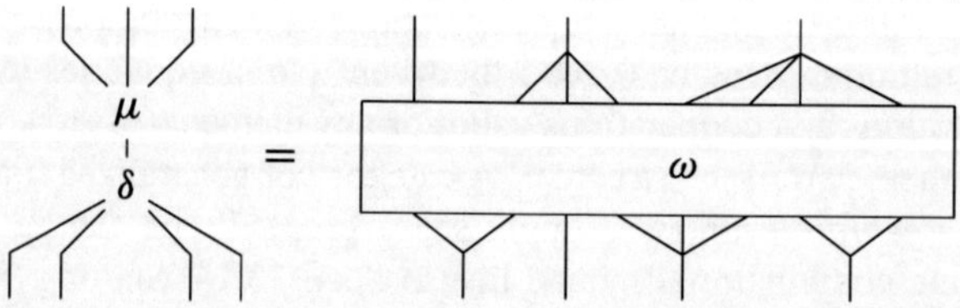

Here we have $l_1 = 1$, $l_2 = 3$, $l_3 = 4$; $k_1 = 2$, $k_2 = 1$, $k_3 = 3$, $k_4 = 2$ and so $r = 1+3+4 = 2+1+3+2$. Observe that, in the general case, the right-hand side term is a sum of such compositions.

When both operads $\mathscr{A}$ and $\mathscr{C}$ are nonsymmetric and, in the compatibility relations, there is no crossing (in particular the only permutations ω are the identity), then we say that this is a *nonsymmetric case* and that $\mathscr{C}^c$-$\mathscr{A}$ is a *nonsymmetric prop*.

13.13.5 The Primitive Operad

Given a type of algebra $(\mathscr{C}, \between, \mathscr{A})$ we assume the following hypothesis:

(H1) The free $\mathscr{A}$-algebra $\mathscr{A}(V)$ is equipped with a $\mathscr{C}^c$-$\mathscr{A}$-bialgebra structure which is functorial in V.

For instance we have seen in Sect. 1.3.2 that $As(V) = \overline{T}(V)$ is a $(Com, Hopf, Ass)$-bialgebra and in Sect. 13.13.1 that it is also a $(As, u.i., As)$-bialgebra (we refer here to the nonunital cases, see Exercise 1.8.11). This hypothesis will prove useful in constructing the operad which encodes the structure of the primitive part of a generalized bialgebra. An operation in $\mathscr{A}(n)$ is said to be *primitive* if, for any cooperation $\delta \in \overline{\mathscr{C}}(m)$, we have $\delta \circ \mu = 0$. This condition amounts to saying that for any generic element $v_1 \ldots v_m \in V^{\otimes m}$ we have $\delta(\mu(v_1 \ldots v_m)) = 0 \in \mathscr{A}(V)^{\otimes m}$. We denote by $\mathscr{P}(n)$ the sub-$\mathbb{S}$-module of $\mathscr{A}(n)$ spanned by the primitive operations and the identity. In [Lod08] we proved the following result.

Theorem 13.13.3. *For any bialgebra type $(\mathscr{C}, \between, \alpha)$ satisfying the hypotheses* (H0) *and* (H1) *the sub-$\mathbb{S}$-module $\mathscr{P}$ is a sub-operad of $\mathscr{A}$.*

It should be noted that, even when $\mathscr{A}$ and $\mathscr{C}$ are presented by a small number of generating operations and relations, it is a challenge to find such a presentation for $\mathscr{P}$. So, in general, we cannot just stay with " types of algebras", we need to use operads. The interest of the primitive operad lies in the following result.

Proposition 13.13.4. *For any bialgebra type $(\mathscr{C}, \between, \mathscr{A})$ satisfying hypotheses* (H0) *and* (H1)*, the primitive part* Prim $\mathscr{H}$ *of a $(\mathscr{C}, \between, \mathscr{A})$-bialgebra $\mathscr{H}$ is a $\mathscr{P}$-algebra.*

In the classical case we know that $\mathscr{P} = Lie$, in the duplicial bialgebra case we can prove that $\mathscr{P} = Mag$.

13.13.6 Triples of Operads

For any generalized bialgebra type $(\mathscr{C}, \between, \mathscr{A})$ satisfying hypotheses (H0) and (H1), we call $(\mathscr{C}, \mathscr{A}, \mathscr{P})$ a triple of operads (the compatibility relations $\between$ are understood, but part of the structure). In the classical case (*Com*, *Ass*, *Lie*) we know that there is a nice structure theorem: the CMM-PBW theorem, cf. 1.3.4. We will find some conditions which ensure that a similar structure theorem holds in more generality.

Since $\mathscr{A}(V)$ is a $\mathscr{C}^c$-$\mathscr{A}$-bialgebra by hypothesis `(H1)` the natural projection map $\varphi(V) : \mathscr{A}(V) \to V$ determines a unique coalgebra map $\mathscr{A}(V) \otimes \mathscr{C}^c(V)$. Let us suppose that there is a natural coalgebra splitting $s(V) : \mathscr{C}^c(V) \to \mathscr{A}(V)$, $\varphi \circ s = \mathrm{id}_{\mathscr{C}}$. This is called hypothesis `(H2epi)`. Then the following structure theorem holds.

Theorem 13.13.5 (Structure Theorem for generalized bialgebras [Lod08]). *Let $\mathscr{C}^c$-$\mathscr{A}$ be a type of generalized bialgebras over a field of characteristic zero. Assume that hypotheses* `(H0)`, `(H1)`, *and* `(H2epi)` *hold. Then, for any $\mathscr{C}^c$-$\mathscr{A}$-bialgebra $\mathscr{H}$, the following statements are equivalent:*

a) *the $\mathscr{C}^c$-$\mathscr{A}$-bialgebra $\mathscr{H}$ is conilpotent,*
b) *there is an isomorphism of bialgebras $\mathscr{H} \cong U(\mathrm{Prim}\,\mathscr{H})$,*
c) *there is an isomorphism of conilpotent coalgebras $\mathscr{H} \cong \mathscr{C}^c(\mathrm{Prim}\,\mathscr{H})$.*

In this statement the functor $U : \mathscr{P}\text{-alg} \to \mathscr{A}\text{-alg}$ is the left adjoint to the forgetful functor $\mathscr{A}\text{-alg} \to \mathscr{P}\text{-alg}$ induced by the inclusion of operads $\mathscr{P} \subset \mathscr{A}$. When $\mathscr{A}$ and $\mathscr{C}$ are nonsymmetric operads and the compatibility relations do not involve the symmetric groups, then $\mathscr{P}$ is also a nonsymmetric operad and the characteristic zero hypothesis is not required.

When the structure theorem holds the triple $(\mathscr{C}, \mathscr{A}, \mathscr{P})$ is called a *good triple of operads*. We have seen that (*Com*, *As*, *Lie*) and (*As*, *Dup*, *Mag*) are good triples of operads. More examples, like (*As*, *Dend*, *Brace*) (due to M. Ronco), and proofs can be found in [Lod08].

13.13.7 Combinatorial Hopf Algebras

The PBW theorem says that a conilpotent cocommutative bialgebra is isomorphic, as a coalgebra, to the cofree coalgebra over its primitive part. This isomorphism is not unique. One can construct a specific iso by using the Eulerian idempotents for instance, but other choices are possible. If, once such a choice has been made, the product is "right-sided", then one can prove that the primitive part has more algebraic structure than being a Lie algebra: it is a pre-Lie algebra. Let us recall the notion of "right-sidedness". Let $\mathscr{H} = (S^c(R), *)$ be a bialgebra structure on the cofree cocommutative coalgebra over R. There is a natural grading $\mathscr{H} = \oplus_n \mathscr{H}^n$ given by $\mathscr{H}^n := S^n(R)$. We say that it is *right-sided* if, for any integer q, the subspace $\bigoplus_{n \geq q} \mathscr{H}^n$ is a right-sided ideal of $\mathscr{H}$.

If the conilpotent bialgebra $\mathscr{H}$ is not cocommutative, then there is no PBW theorem. However, if it happens that it is cofree as a coassociative coalgebra, then we can find a structure theorem. Let us make it explicit in the "right-sided" case (replace S^n by T^n in the previous definition). It can be shown that the primitive part of $\mathscr{H}$ is more than a Lie algebra: it is a brace algebra. Of course the Lie bracket is obtained from the brace structure. It is the antisymmetrization of the binary brace. Moreover one can show that there is also a finer algebra structure on $\mathscr{H} = T^c(R)$ than the associative structure: it is a dendriform algebra. So the good triple which is governing these right-sided cofree coassociative bialgebras is $(As, Dend, Brace)$ (due to M. Ronco [Ron02]).

We have given in [LR10] more examples of this kind and we called them *combinatorial Hopf algebras* because they cover all the Hopf algebras appearing in algebraic combinatorics under this terminology.

One should remark that what we have done here for classical bialgebras, that is extracting finer structures from explicit PBW iso, can also be done in the case of generalized bialgebras. But explicit examples are still to be done.

13.14 Types of Operads

There are several variations of the notion of "symmetric operad". First, one can replace the monoidal category of vector spaces by some other monoidal categories: modules over a commutative ring, sets, simplicial sets, topological spaces, etc. (cf. Sect. 5.3.5). Second, we know that a natural generalization of the notion of monoid is the notion of category: the product of two elements can be performed only when the source of one is equal to the target of the other one. In the operadic framework it gives rise to the notion of "colored operad": operations can be composed only whenever the colors match. Among them we find the operad encoding morphisms between algebras and also the planar algebras devised by Vaughn Jones in his analysis of knot theory. Third, as already mentioned, a symmetric operad can be seen as an algebra over the monad of rooted trees. We have already seen that some other combinatorial objects give rise to other types of operads: planar rooted trees give ns operads, shuffle trees give shuffle operads, ladders give associative algebras. There are many more and we only briefly comment on some of them: cyclic operads, permutads, modular operads, properads. Sometimes Koszul duality theory, as devised in this book, can be extended to these cases, though it is not yet always written in full in the literature, see for instance [vdL03, Val07b].

13.14.1 Colored Operads

A colored operad is to an operad what a category is to a monoid. An operation in a colored operad comes with a color for each input and a color for the output. A composition is going to be possible whenever the colors match. There is a notion of algebra over a colored operad. It is a colored object and the evaluation of an operation on a tuple of elements is possible only if the colors match. This generalization works for symmetric operads, ns operads, shuffle operads and also for associative algebras. In this last case a particular example gives rise to … ns operads, see Sect. 13.14.3.

13.14.2 Morphism Operads

Let $\mathscr{P}$ be an operad. In order to encode the category of morphisms between two $\mathscr{P}$-algebras, we construct a colored operad, called *morphism operad*, as follows. The colors are going to be 0 (light blue) and 1 (dark blue). If the operad $\mathscr{P}$ is presented as $\mathscr{P} = \mathscr{P}(E, R)$, then the morphism operad $\mathscr{P}_{0\to 1}$ admits the following presentation:

$$\mathscr{P}_{0\to 1} = \mathscr{P}(E_0 \oplus E_1 \oplus \mathbb{K}\mathtt{f}, R_0 \oplus R_1 \oplus R_{\mathtt{f}}),$$

where the various ingredients are as follows. The space E_0 (resp. E_1) is a copy of E with input and output colors 0 (resp. 1). The one-dimensional space $\mathbb{K}\mathtt{f}$, of arity 1, has input color 0 and output color 1. The space R_0 (resp. R_1) is a copy of R. The space $R_{\mathtt{f}}$ is spanned by the operations $\mu_0 \circ \underbrace{(\mathtt{f}, \dots, \mathtt{f})}_{n} - \mathtt{f} \circ \mu_1$ where $\mu_0 \in E_0(n)$ and $\mu_1 \in E_1(n)$ are copies of the same operation $\mu \in E(n)$. So, in $\mathscr{P}_{0\to 1}$ we have:

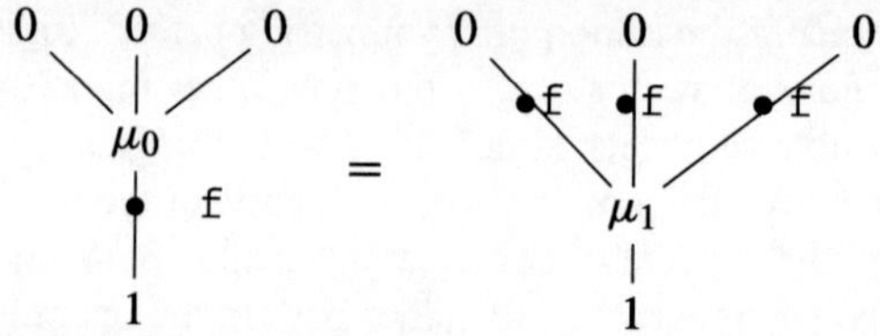

An algebra over $\mathscr{P}_{0\to 1}$ is determined by a $\mathscr{P}$-algebra A_0, a $\mathscr{P}$-algebra A_1, and a morphism of $\mathscr{P}$-algebras $f : A_0 \to A_1$.

When $\mathscr{P}$ is a quadratic operad, the colored operad $\mathscr{P}_{0\to 1}$ admits a presentation involving cubical terms. So Koszul duality theory, as developed in this book, cannot be applied. However, when $\mathscr{P}$ is a Koszul operad, one can apply a generalization of Koszul duality theory given in [MV09a, MV09b], called *homotopy Koszul duality*. This gives the minimal model for the operad $\mathscr{P}_{0\to 1}$. An algebra over this resolution is the data of two $\mathscr{P}_\infty$-algebras A_0 and A_1 together with an ∞-morphism, as defined in Sect. 10.2.2, between them.

In the particular case $\mathscr{P} = As$, the minimal model of $As_{0\to 1}$, like the construction done in Chap. 9 for As, see also [Mar04]. It turns out that, here, the role of Stasheff polytope (alias associahedron) is taken by the *multiplihedron*, cf. [Sta63].

13.14.3 Grafting Algebras vs NS Operads

In [Ron11] M. Ronco introduced the notion of *grafting algebra* (and variations of it). It is in fact a "colored" algebra for which the colors are the natural numbers. By definition a grafting algebra is a graded vector space $A = \bigoplus_{n\geq 0} A_n$, where $A_0 = \mathbb{K}1$, equipped with binary operations

$$\bullet_i : A_m \otimes A_n \to A_{m+n}, \quad \text{for } 0 \leq i \leq m,$$

satisfying:

$$(x \bullet_j y) \bullet_{i+j} z = x \bullet_j (y \bullet_i z), \quad \text{for } 0 \le i \le m, 0 \le j \le l,$$
$$(x \bullet_j y) \bullet_{j+n} z = x \bullet_j (y \bullet_i z), \quad \text{for } 0 \le i < j,$$

for any $x \in A_l$, $y \in A_m$ and $z \in A_n$. It is also supposed that $1 \in A_0$ is a unit on both sides of the operations.

Let us introduce another notation as follows. We put $\mathscr{P}_n := A_{n-1}$ and we let

$$\circ_i : \mathscr{P}_m \otimes \mathscr{P}_n \to \mathscr{P}_{m+n-1}$$

be equal to

$$\bullet_{i-1} : A_{m-1} \otimes A_{n-1} \to A_{m+n-2}$$

for $1 \le i \le m$ under this identification. Strictly speaking there is one operation $\bullet_i$ for each pair of input colors (m, n) whenever $i \le m$. The output color of this operation is $n + m$. The description of grafting algebras given here differs from the one given in [Ron11] by taking the "opposite" operations (our $x \bullet_i y$ corresponds to her $y \bullet_i x$).

From the definition given in Sect. 5.9.4, it follows that a ns operad $\mathscr{P}$ such that $\mathscr{P}_0 = 0$ and $\mathscr{P}_1 = \mathbb{K}$ is equivalent to a grafting algebra (compare with [Ger63]). So the notion of ns operad can be seen as some type of colored algebra.

13.14.4 Tangle Operad, Planar Algebras

The notion of planar algebras was introduced in the late 1990s by Vaughan Jones [Jon99] in connection with its famous invariants for knots. In [Jon10] he informally defines them as follows:

> *A planar algebra consists of vector spaces together with multilinear operations between them indexed by planar tangles–large discs with internal ("input") discs all connected up by non-intersecting curves called strings.*

In fact a planar algebra is an algebra over a colored operad, called the operad of tangles. This operad is a variation of the operad of little discs, where the region outside the little discs has been decorated. We recall here the version given in [Jon10, Definition 2.1.1].

A *tangle* is a disc equipped with (disjoint) little discs with the following decoration, see Fig. 13.1. Each disc (including the outside one) is equipped with an even number of marked points, one of them being distinguished (base-point, pictured as $*$). Each marked point is joined to another one by a path called *string*. Strings without end points are also allowed (see picture), and all the strings are disjoint. The regions formed by the strings are labeled either by $+$ (left blank in the picture) or by $-$ (shaded in the picture), so that adjacent regions are of different labels. These objects are considered up to isotopy, so a tangle is really a combinatorial object (unlike in the case of the little discs operad).

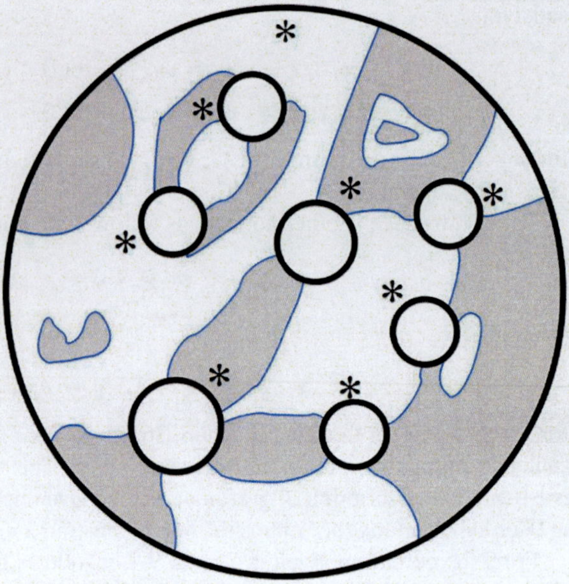

Fig. 13.1 A tangle (Courtesy of V.F.R. Jones)

Composition among tangles is defined as in the case of the little discs operad. More precisely, let T be a tangle and choose an inside circle α with $2k$ marked points on the outside circle. The (partial) composition $T \circ_\alpha S$ is the tangle obtained by plugging S in α while respecting the marked points, the distinguished points and the labels. The fact that S has to have the same number of marked points as α in order to achieve the composition implies that we get a colored operad instead of an operad. It is called the *operad of tangles*. V. Jones defines a *planar algebra* as an algebra over the operad of tangles.

The operad of tangles is a very rich operad which admits lots of possible variations. For instance there is an interesting suboperad made of the tangles with only one interior disc (related to the Temperley–Lieb algebra).

13.14.5 Types of Operads Based on Combinatorial Objects

We have seen in Sect. 5.6, resp. Sect. 5.9.5, resp. Sect. 8.2, that a symmetric operad, resp. nonsymmetric operad, resp. shuffle operad, can be viewed as an algebra over some monad, namely the monad of nonplanar rooted trees, resp. planar rooted trees, resp. shuffle trees. The key ingredient in the construction of these monads is the notion of substitution of a tree at a vertex. A fourth example is given by the ladder trees, which also form a monad under substitution. The algebras over this monad are simply the unital associative algebras. In all these examples the basic tool is a family of graphs (in fact trees here) for which there is a notion of "substitution", see for instance Sects. 5.6 and 8.2. We refer the reader to the paper of D. Borisov and Y.I. Manin [BM08] for a complete study.

There is a lot of examples of this kind, see for instance [Mer10b], and we provide some of them in the following sections. We only describe the set of graphs which is used and the substitution process on them. It is a good exercise to write down the other types of definitions: monoidal, classical, partial, when they exist. One should note that for some of these combinatorial monads, the type of operads that they define (that is the algebras over this monad) are themselves monads, but over a different category. For instance the monad giving rise to symmetric operads is a monad on $\mathbb{S}$-modules, while a symmetric operad is a monad on the category of vector spaces. But it does not always happen that the algebras over a some fixed combinatorial monad are themselves a monad.

There are many other types of operads like dioperads, properads, props, wheeled props, to name but a few.

When a type of operads admits an equivalent definition as a monoid in a monoidal category, one can develop a Koszul duality theory on that level, following the pattern given in this book. The case of colored operad was treated by P. Van der Laan in [vdL03], the case of dioperads was treated by W.L. Gan in [Gan03] and the case of properads was treated in [Val07b].

13.14.6 Cyclic Operads

We consider finite trees (nonplanar nonrooted). Substitution is like in the case of the monad $\mathbb{T}$ defining symmetric operads. We observe that in order to define the monad we need more than an $\mathbb{S}$-module, because among the edges pertaining to a given vertex, there is no pointed one (no root). So, if the number of these edges is $n+1$, then we need an $\mathbb{S}_{n+1}$-module. It is customary, by comparison with symmetric operads, to see this permutation group as generated by $\mathbb{S}_n$ and the cyclic group of order $n+1$, whence the terminology *cyclic operad* for the algebras over this monad. This notion is due to E. Getzler and M.M. Kapranov [GK95a].

It is interesting to look for a presentation of a cyclic operad by means of partial operations. In fact, we first start with a symmetric operad $\mathscr{P}$ and unravel the extra structure needed to make it into a cyclic operad as follows. First, we need an action of the cyclic operator τ, of order $n+1$, on $\mathscr{P}(n)$ so that it becomes an $\mathbb{S}_{n+1}$-module. Second, we have to say how these operators behave with respect to composition in the operad $\mathscr{P}$. The trick is to think of τ as changing the last input into the output and the output into the first input. Then the formulas relating the action of τ with the partial compositions of the operad $\mathscr{P}$

$$\begin{aligned}\tau(\mu \circ_i \nu) &= \tau(\mu) \circ_{i+1} \nu, \quad \text{for } 1 \le i < m,\\ \tau(\mu \circ_m \nu) &= \tau(\nu) \circ_1 \tau(\mu),\end{aligned}$$

follow from Fig. 13.2.

The symmetric operad Ass can be seen as a cyclic operad as follows. We endow $Ass(n) \cong \mathbb{K}[\mathbb{S}_n]$ with the following right action of the symmetric group $\mathbb{S}_{1+n}$. Con-

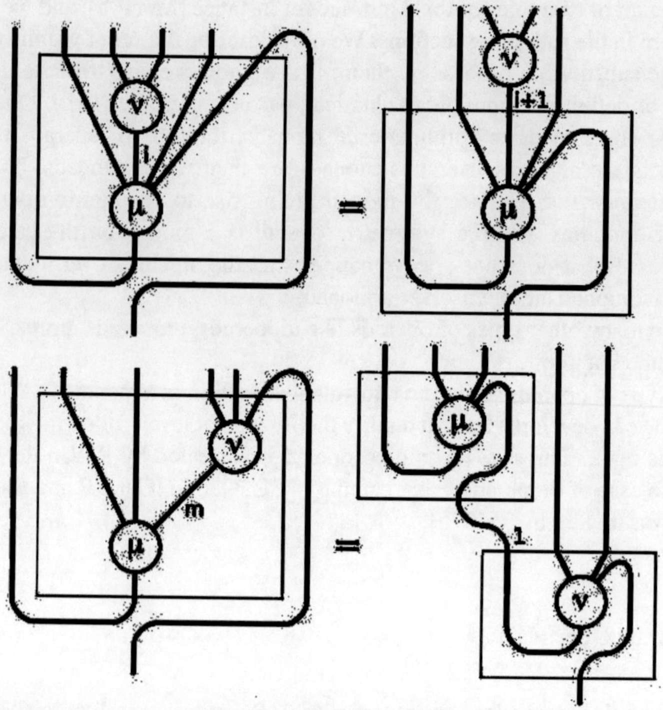

Fig. 13.2 $\tau(\mu \circ_i \nu) = \tau(\mu) \circ_{i+1} \nu$ and $\tau(\mu \circ_m \nu) = \tau(\nu) \circ_1 \tau(\mu)$

sider the set $\mathbb{S}_{1+n}$ as a right $\mathbb{S}_{1+n}$-set given by conjugation: $\omega^g := g^{-1}\omega g$. Let U_{1+n} be the subset of $\mathbb{S}_{1+n}$ made up of the permutations which have only one cycle, that is the orbit of $(0\ 1\ \ldots\ n)$. The bijection $\mathbb{S}_n \to U_{1+n}, \sigma \mapsto (0\ \sigma(1)\ \ldots\ \sigma(n))$ permits us to view $\mathbb{S}_n$ as a $\mathbb{S}_{1+n}$-set. The restriction of this action to the subgroup $\mathbb{S}_n$ of $\mathbb{S}_{1+n}$ is immediately seen to be the right multiplication.

An important case of a cyclic operad is the endomorphism operad. Let A be a finite dimensional vector space, equipped with a nondegenerate pairing $\langle -, - \rangle : A \otimes A \to \mathbb{K}$. This pairing gives rise to an isomorphism $A \cong A^*$, and so we have an isomorphism $\mathrm{End}_A(n) := \mathrm{Hom}(A^{\otimes n}, A) \cong \mathrm{Hom}(A^{\otimes n+1}, \mathbb{K})$. As a consequence, we get an action of $\mathbb{S}_{n+1}$ on $\mathrm{End}_A(n)$, which makes End_A into a cyclic operad.

The data of an algebra over a cyclic operad is given by a morphism of cyclic operad $\mathcal{P} \to \mathrm{End}_A$. For example an algebra over the cyclic operad *Ass* is a finite dimensional associative algebra equipped with a nondegenerate symmetric pairing $\langle -, - \rangle$ such that

$$\langle ab, c \rangle = \langle a, bc \rangle,$$

for any $a, b, c \in A$, see Sect. 13.7.3.

13.14.7 Permutads vs Shuffle Algebras

Let $\underline{n} = \{1, \dots, n\}$. To any surjection $t : \underline{n} \to \underline{k}$, we associate a graph as follows. It is a bipartite graph, the first row is made up of n vertices labeled from 1 to n and the second row is made up of k vertices labeled from 1 to k. There is an edge from vertex i in the first row to vertex j on the second row whenever $t(i) = j$.

EXAMPLE.

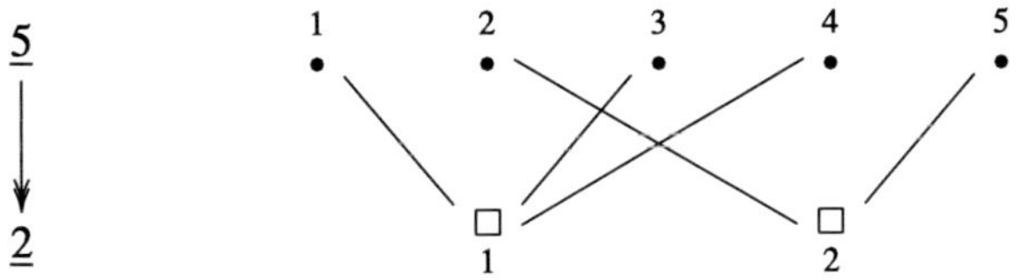

We define a substitution as corresponding to the composition of surjective maps (so we perform substitution only on the second row vertices). This combinatorial data defines a monad $\mathbb{P}$ over the category of arity-graded vector spaces (compare with ns operads, cf. Sect. 5.9.5). An algebra over the monad $\mathbb{P}$ is called a *permutad*, cf. [LR12].

When presenting a permutad in terms of partial operations, one sees that this notion is exactly the same as the notion of "shuffle algebra" introduced by M. Ronco in [Ron11]. Interpreting the bipartite graphs corresponding to surjections as shuffle trees whose underlying tree is a left comb, cf. Sect. 8.2, we get a comparison between permutads and shuffle operads.

There is a permutad pAs similar to the ns operad As, in particular pAs_n is one-dimensional. Its minimal model can be described using the *permutohedron* (in place of the *associahedron*).

13.14.8 Modular Operads

There is an obvious definition of substitution for finite graphs whose edges are labeled by nonnegative integers. A *modular operad* is an algebra over the monad on $\mathbb{S}$-modules defined by the "stable" graphs, see [GK98] for details. One of the motivating examples of a modular operad is $\overline{\mathcal{M}}_{g,n}$, where $\overline{\mathcal{M}}_{g,n}$ is the moduli space of genus g stable curves with n marked points.

It is a higher genus extension of the notion of cyclic operads.

13.14.9 Properads

We consider the automorphism classes of finite connected directed graphs, see Appendix C.4.1. So each vertex has some inputs, let us say $n \geq 1$ and some outputs, let

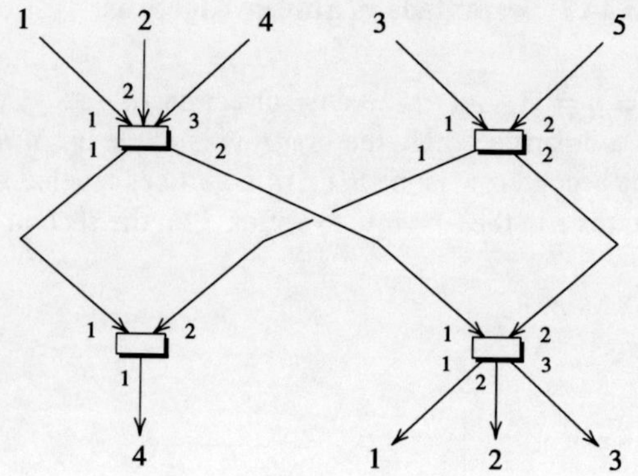

Fig. 13.3 Labeled connected graph

us say $m \geq 1$. Among the outer edges, depending on the orientation, we have leaves, oriented towards the vertex, and roots, oriented outwards the vertex. We suppose that for each vertex the set of inputs (resp. outputs) is endowed with a bijection to $\{1, \ldots, n\}$ (resp. $\{1, \ldots, m\}$). We also suppose that the set of leaves (resp. the set of roots) has a similar labeling, see Fig. 13.3.

We define a substitution of a graph at a vertex like in the general case, but taking the labelings into consideration: the labeling of the set of leaves of the to-be-inserted graph is used to identify them with the inputs of the vertex, and similarly for the roots.

From this data we can construct a monad on $\mathbb{S}$-bimodules, i.e. families of $\mathbb{S}_n \times \mathbb{S}_m^{op}$-modules. A *properad*, as introduced slightly differently in [Val07b], is an algebra over this monad.

Observe that if we restrict ourself to the graphs whose vertices have one and only one output, then we get the notion of symmetric operad (since $\mathbb{S}_n \times \mathbb{S}_1 \cong \mathbb{S}_n$). Properads model operations with several inputs and several outputs. Many types of bialgebras can be encoded by properads. Contrary to modular operads, properads can act on infinite dimensional vector spaces and do not require any pairing.

In [Val07b], the category of $\mathbb{S}$-bimodules is endowed with a monoidal product, which extends the monoidal product $\circ$ of $\mathbb{S}$-modules. In loc. cit., a properad is defined as a monoid in this monoidal category. The Koszul duality for properads is proved following exactly the same pattern as in Chaps. 3 and 7.

Appendix A
The Symmetric Group

Il faut, lorsque vous étudiez les propriétés du cercle, de la sphère et des sections coniques, que vous vous sentiez frères par l'esprit d'Euclide et d'Archimède et que, comme eux vous voyiez avec ravissement se développer le monde idéal des figures et des proportions dont les harmonies enchanteresses se retrouvent ensuite dans le monde réel.

Jean Jaurès, Toulouse (1914)

The purpose of this appendix is essentially to fix the notations and to recall some elementary facts about the representations of the symmetric group that are needed in this book. The main result is the Schur Lemma which says that in characteristic zero any homogeneous polynomial functor is completely determined by a representation of the symmetric group.

For more details the reader can consult any book on the subject, for instance [Mac95, Sag01].

A.1 Action of Groups

A.1.1 Group Algebra, Representation

The group algebra $\mathbb{K}[G]$ of a group G over a commutative algebra $\mathbb{K}$ is the free module with basis the elements of G. The product is induced by the product in the group:

$$\Big(\sum_i a_i g_i\Big)\Big(\sum_j a_j g_j\Big) = \sum_{i,j} (a_i a_j)(g_i g_j).$$

The tensor product of modules over $\mathbb{K}[G]$ is often denoted by $- \otimes_G -$ instead of $- \otimes_{\mathbb{K}[G]} -$.

J.-L. Loday, B. Vallette, *Algebraic Operads*,
Grundlehren der mathematischen Wissenschaften 346,
DOI 10.1007/978-3-642-30362-3, © Springer-Verlag Berlin Heidelberg 2012

A *representation* of the group G is a left module M over $\mathbb{K}[G]$. The action of $g \in G$ on $m \in M$ is denoted either by $g \cdot m$ or by gm if there is no ambiguity. The algebra $\mathbb{K}[G]$ is called the *regular representation* of G when viewed as a left module over itself.

A.1.2 Invariants and Coinvariants

To any representation M of G is associated its space of invariants M^G and its space of coinvariants M_G defined as follows:

$$M^G := \{m \in M \mid g \cdot m = m, \ \forall g \in G\},$$
$$M_G := M/\{g \cdot m - m \mid g \in G, \ m \in M\}.$$

The notation follows the usual convention since M^G is contravariant in G and M_G is covariant in G.

There is a natural map from invariants to coinvariants:

$$M^G \rightarrowtail M \twoheadrightarrow M_G.$$

Whenever G is finite and $\#G$ is invertible in the ground ring, then this map is an isomorphism. Indeed an inverse is given by $[m] \mapsto \frac{1}{\#G} \sum_{g \in G} g \cdot m$, where $[m]$ is the class of $m \in M$ in M_G.

A.1.3 Restriction and Induction

Let H be a subgroup of G. The *restriction* of a representation of G to H is simply the same space but viewed as a module over H. If M is a right H-module, then the *induced representation* is the following representation of G:

$$\mathrm{Ind}_H^G M := M \otimes_H \mathbb{K}[G],$$

where $\mathbb{K}[G]$ is viewed as a left module over $\mathbb{K}[H]$ through the multiplication in $\mathbb{K}[G]$. Observe that the space $\mathrm{Ind}_H^G M$ can be identified with the space $M \otimes_{\mathbb{K}} \mathbb{K}[H \backslash G]$.

A.1.4 Equivariance

Let M and N be two (left) G-modules. A linear map $f : M \to N$ is said to be G-*equivariant*, or simply *equivariant*, if, for any $g \in G$, $m \in M$ one has $f(g \cdot m) = g \cdot f(m)$. In other words f is a morphism in the category of G-modules. The space of

G-equivariant maps is denoted by $\mathrm{Hom}^G(M, N)$ since it is made of the morphisms which are invariant for the action of G by $(g \cdot f)(m) := g^{-1} \cdot f(g \cdot m)$. If L is another G-module, one has a natural isomorphism

$$\mathrm{Hom}^G\big(M, \mathrm{Hom}(N, L)\big) \cong \mathrm{Hom}(M \otimes_G N, L).$$

A.2 Representations of the Symmetric Group $\mathbb{S}_n$

A.2.1 The Symmetric Group $\mathbb{S}_n$

By definition the symmetric group $\mathbb{S}_n$ is the group of automorphisms of the set $\{1, 2, \ldots, n\}$. Its elements are called *permutations*. The image of the permutation σ is denoted either by $(\sigma(1), \ldots, \sigma(n))$ or, sometimes, by $[\sigma(1) \ldots \sigma(n)]$. The neutral element is denoted either by 1 or 1_n or even id_n. Let us denote by s_i the permutation which exchanges the elements i and $i+1$ and leaves the others fixed. It is called a *transposition* and sometimes denoted by $(i\ i+1)$ (cycle notation). The transpositions generate $\mathbb{S}_n$. In fact $\mathbb{S}_n$ is presented by the set of generators $\{s_1, \ldots, s_{n-1}\}$ and the set of relations

$$\begin{cases} s_i^2 = 1, & \text{for } i = 1, \ldots, n-1, \\ s_i s_j = s_j s_i, & \text{for } |i - j| \geq 2, \\ s_i s_{i+1} s_i = s_{i+1} s_i s_{i+1}, & \text{for } i = 1, \ldots, n-2. \end{cases}$$

This is called the Coxeter presentation of the symmetric group. In general there is no preferred way of writing a permutation in terms of the Coxeter generators. However, for any permutation σ there is a minimum number of generators in such a writing. It is called the *length* of the permutation and is denoted by $\ell(\sigma)$. For instance we have $\ell([321]) = 3$ since $[321] = s_1 s_2 s_1$.

A.2.2 Irreducible Representations

A representation of $\mathbb{S}_n$ is said to be *irreducible* if it is not isomorphic to the direct sum of two nonzero representations. It can be shown that the isomorphism classes of representations of the symmetric group $\mathbb{S}_n$ are in one-to-one correspondence with the *partitions* of the integer n, that is the sequence of integers $\lambda = (\lambda_1, \ldots, \lambda_r)$ such that $\lambda_1 \geq \cdots \geq \lambda_r \geq 1$ and $\sum_i \lambda_i = n$. It is often helpful to represent such a sequence by a *Young diagram*. For instance the Young diagram of (4, 2, 1) is

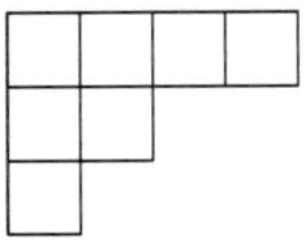

The irreducible representation associated to λ is denoted by S^λ and called the *Specht module*. Any finite dimensional representation M of $\mathbb{S}_n$ is isomorphic to the sum of its *isotypic* components M^λ, one for each irreducible representation:

$$M = \bigoplus_\lambda M^\lambda$$

where M^λ is isomorphic to the sum of a finite number of copies of S^λ. This number is called the multiplicity of S^λ in M. For instance the multiplicity of S^λ in the regular representation is equal to the dimension of S^λ. From this result follows the formula:

$$\sum_\lambda (\dim S^\lambda)^2 = n!.$$

EXAMPLES.

$n = 1$. The unique irreducible representation of $\mathbb{S}_1$ is one-dimensional.

$n = 2$. There are two partitions of 2: $\lambda = (2)$ and $\lambda = (1, 1)$:

The representation S^2 is the one-dimensional trivial representation, so $S^2 = \mathbb{K}x$ and $[21] \cdot x = x$. The representation $S^{1,1}$ is the one-dimensional signature representation, so $S^{1,1} = \mathbb{K}x$ and $[21] \cdot x = -x$.

$n = 3$. There are three partitions of 3: $\lambda = (3)$, $\lambda = (2, 1)$ and $\lambda = (1, 1, 1)$:

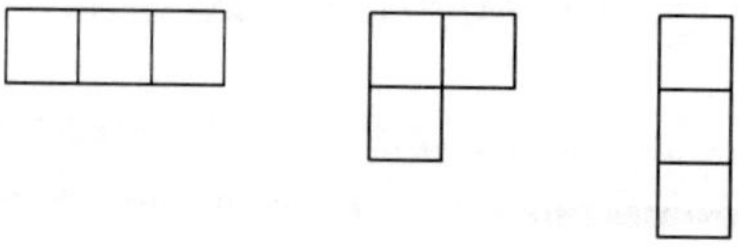

The associated representations are the trivial representation for $\lambda = (3)$ and the signature representation for $\lambda = (1, 1, 1)$. For $\lambda = (2, 1)$ it is the *hook representation* which can be described as follows. Consider the three-dimensional space $\mathbb{K}x_1 \oplus \mathbb{K}x_2 \oplus \mathbb{K}x_3$ on which $\mathbb{S}_3$ acts by permutation of the indices. The kernel of the linear map $\mathbb{K}x_1 \oplus \mathbb{K}x_2 \oplus \mathbb{K}x_3 \to \mathbb{K}$, $x_i \mapsto 1$ is clearly invariant under the action of $\mathbb{S}_3$. This is the two-dimensional hook representation $S^{2,1}$. Observe that the three elements $x_1 - x_2, x_2 - x_3, x_3 - x_1$ lie in the kernel, but are not linearly independent since their sum is zero. Any two of them form a basis of $S^{2,1}$. The vector $x_1 + x_2 + x_3$ is invariant under the action of $\mathbb{S}_3$, hence it is isomorphic to S^3 and there is an isomorphism

$$\mathbb{K}x_1 \oplus \mathbb{K}x_2 \oplus \mathbb{K}x_3 \cong S^{2,1} \oplus S^3.$$

$n=4$. There are five partitions of 4: $\lambda=(4)$, $\lambda=(3,1)$, $\lambda=(2,2)$, $\lambda=(2,1,1)$ and $\lambda=(1,1,1,1)$:

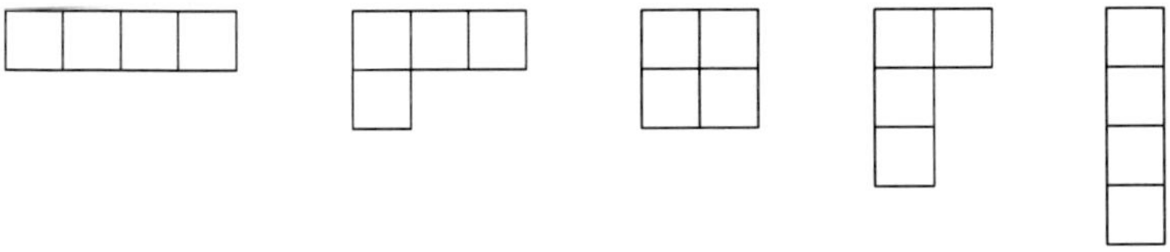

The associated irreducible representations are as follows:

S^4 is the one-dimensional trivial representation,
$S^{3,1}$ is the hook representation, that is the kernel of $\oplus_{i=1}^{i=4}\mathbb{K}x_i \to \mathbb{K}$, $x_i \mapsto 1$,
$S^{2,2}$ is described below,
$S^{2,1,1}$ is the kernel of $\oplus_{i=1}^{i=4}\mathbb{K}x_i \to \mathbb{K}$, $x_i \mapsto 1$, where we let $\mathbb{S}_4$ act on the four-dimensional space by permutation of the indices *and* multiplication by the signature,
$S^{1,1,1,1}$ is the one-dimensional signature representation.

In the set of four elements $\{a,b,c,d\}$ we consider the non-ordered subsets of non-ordered pairs. So we have only three elements:

$$\big\{\{a,b\},\{c,d\}\big\};\ \big\{\{a,c\},\{b,d\}\big\};\ \big\{\{a,d\},\{b,c\}\big\}.$$

We let $\mathbb{S}_4$ act on the ordered set $\{a,b,c,d\}$ as usual, hence the three-dimensional vector space $\mathbb{K}^3$ spanned by the aforementioned subsets of pairs is a representation of $\mathbb{S}_4$. Consider the map $\mathbb{K}^3 \to \mathbb{K}$ which sends each generator to 1. The kernel is a two-dimensional representation of $\mathbb{S}_4$. This is $S^{2,2}$.

A.2.3 The Schur Lemma

The following result, called the *Schur Lemma*, says that, for an infinite field, any homogeneous polynomial functor of degree n is of the form $V \mapsto M(n) \otimes_{\mathbb{S}_n} V^{\otimes n}$ for some $\mathbb{S}_n$-module $M(n)$. It justifies the choice of $\mathbb{S}$-modules to define an operad in characteristic zero.

Lemma A.2.1. *If $\mathbb{K}$ is an infinite field, then any natural transformation $\theta : V^{\otimes n} \to V^{\otimes n}$ is of the form $\theta_n \cdot (v_1, \dots, v_n)$ for some $\theta_n \in \mathbb{K}[\mathbb{S}_n]$.*

Proof. Let $GL(V)$ be the group of isomorphisms of the space V (general linear group). Let $\alpha \in GL(V)$ act on $\operatorname{End}(V)$ by conjugation:

$$\alpha \cdot f := \alpha f \alpha^{-1}.$$

We let $GL(V)$ act on $\mathrm{End}(V^{\otimes n})$ diagonally, and we denote by $\mathrm{End}(V^{\otimes n})^{GL(V)}$ the space of invariants. We first show that, for $r = \dim V \geq n$, the map

$$\xi : \mathbb{K}[\mathbb{S}_n] \to \mathrm{End}\big(V^{\otimes n}\big)^{GL(V)}, \qquad \theta_n \mapsto (\omega \mapsto \theta_n \cdot \omega)$$

is an isomorphism.

Let $e_1, \dots, e_r$ be a basis of V and let $f \in \mathrm{End}(V^{\otimes n})^{GL(V)}$. The element $f(e_1, \dots, e_r)$ can be written $\sum a_{\mathbf{i}}(e_{i_1}, \dots, e_{i_r})$ where $a_{\mathbf{i}} \in \mathbb{K}$ and $\mathbf{i} = (i_1, \dots, i_r)$. Replacing e_1 by λe_1, for $\lambda \in \mathbb{K}$, the element $f(e_1, \dots, e_n)$ is multiplied by λ. Since the field $\mathbb{K}$ is infinite, the index 1 must occur once and only once in each sequence $\mathbf{i}$. Similarly each index $2, \dots, n$ must occur once and only once. It follows from this fact that each sequence $\mathbf{i}$ is a permutation of $(1, \dots, n)$. Therefore we get

$$f(e_1, \dots, e_n) = \sum_\sigma a_\sigma \sigma(e_1, \dots, e_n)$$

and so $f = \theta_n$ for $\theta_n = \sum_\sigma a_\sigma$. Indeed, for any other vector basis of $V^{\otimes n}$ one can find $\alpha \in \mathrm{End}(V)$ such that this vector is $\alpha(e_1, \dots, e_n)$. Since f is $GL(V)$-invariant one has

$$\begin{aligned} f\big(\alpha(e_1, \dots, e_n)\big) &= \alpha\big(f(e_1, \dots, e_n)\big) \\ &= \alpha\big(\theta_n(e_1, \dots, e_n)\big) \\ &= \theta_n\big(\alpha(e_1, \dots, e_n)\big) \end{aligned}$$

as expected.

Since $r = \dim V \geq n$ by hypothesis, the element θ_n is unique because the vectors $\sigma(e_1, \dots, e_n)$ are linearly independent in $V^{\otimes n}$. Finally we have proved that ξ is an isomorphism.

Since $\xi(\theta_n)$ is $GL(V)$-invariant, it follows that any natural transformation of $V^{\otimes n}$ into itself is completely determined by some $\theta_n \in \mathbb{K}[\mathbb{S}_n]$. □

REMARK. In positive characteristic p there are transformations of functors which do not come from morphisms of $\mathbb{S}$-modules. For instance the Frobenius map given by the pth power, is one of them:

$$F : \big(V^{\otimes n}\big)_{\mathbb{S}_n} \to \big(V^{\otimes pn}\big)_{\mathbb{S}_n}, \qquad F(x_1, \dots, x_n) = \big(x_1^p, \dots, x_n^p\big).$$

It is a consequence of the fact that the binomial coefficients are zero in characteristic p.

Appendix B
Categories

... the discovery of ideas as general as these [category and the like] is chiefly the willingness to make a brash or speculative abstraction, in this case supported by the pleasure of purloining words from the philosophers: 'Category' from Aristotle and Kant, 'Functor' from Carnap (Logische Syntax der Sprache) *and 'natural transformation' from then current informal parlance.*

Saunders MacLane in "Categories for the working mathematician"

This appendix is a brief but complete survey on categories: functors, transformations of functors, adjoint functors and monads (a word borrowed from Leibniz). In the last part, we introduce the notions of model categories, homotopy categories and derived functors. The purpose is essentially to fix the terminology and notations used in this book. For more details we refer to the books of Saunders MacLane [ML98] and Daniel Quillen [Qui67].

B.1 Categories and Functors

The idea of the notion of category is to put emphasis on morphisms instead of objects themselves. Category theory is the "algebra of algebra"; it provides a general common framework for many theories.

B.1.1 Categories

A *category* C is made up of objects and, for any two objects C and C' we are given a set of morphisms, also called maps or arrows, denoted either by $\mathrm{Hom}_{\mathsf{C}}(C, C')$, $\mathrm{Hom}(C, C')$ or $\mathsf{C}(C, C')$. For any morphism $f : C \to C'$, the object C is called the

J.-L. Loday, B. Vallette, *Algebraic Operads*,
Grundlehren der mathematischen Wissenschaften 346,
DOI 10.1007/978-3-642-30362-3, © Springer-Verlag Berlin Heidelberg 2012

source of f and the object C' is called the *target* of f. There is a notion of composition of two morphisms $f: C \to C'$ and $g: C' \to C''$ provided that the target of f is the same as the source of g. The composite is denoted as usual $g \circ f: C \to C''$ or gf if there is no ambiguity. Composition of morphisms is supposed to be associative. To any object C, there exists an identity morphism denoted either by Id_C, or Id. Composition with the identity does not change the morphism. By reverting the orientation of all arrows, one defines the *opposite category* denoted C^{op}.

Left cancelable morphisms, i.e. $fg = fg'$ implies $g = g'$, are called *monomorphisms* and denoted by $\rightarrowtail$. Right cancelable morphisms are called *epimorphisms* and denoted by $\twoheadrightarrow$. Invertible morphisms are called *isomorphisms* and denoted by $\xrightarrow{\cong}$ or $\cong$ or simply by $=$ when there is no ambiguity.

B.1.2 Functors

Given two categories C and D there is a notion of *functor* $F: \mathsf{C} \to \mathsf{D}$. It consists in the following data. First to any element C of C, it associates an element $F(C)$ in D. Then to any morphism $f: C \to C'$ in C, we are given a morphism $F(f): F(C) \to F(C')$ in D such that $F(g \circ f) = F(g) \circ F(f)$ and $F(\mathrm{Id}_C) = \mathrm{Id}_{F(C)}$.

A *subcategory* $\mathsf{C} \subset \mathsf{D}$ is a category made up of a subclass of objects of D, and for any pair X, Y of objects of C, the set of morphisms is a subset of that of D: $\mathrm{Hom}_{\mathsf{C}}(X, Y) \subset \mathrm{Hom}_{\mathsf{D}}(X, Y)$. A subcategory $\mathsf{C} \subset \mathsf{D}$ is called *full* when $\mathrm{Hom}_{\mathsf{C}}(X, Y) = \mathrm{Hom}_{\mathsf{D}}(X, Y)$, for any pair X, Y of objects of C.

A functor $F: \mathsf{C} \to \mathsf{D}$ is *faithful* if the set-theoretic maps

$$\mathrm{Hom}_{\mathsf{C}}(X, Y) \rightarrowtail \mathrm{Hom}_{\mathsf{D}}\big(F(X), F(Y)\big)$$

are injective, for any pair X, Y of objects of C. It is the case for a subcategory of a category. A functor $F: \mathsf{C} \to \mathsf{D}$ is *full* if the set-theoretic maps

$$\mathrm{Hom}_{\mathsf{C}}(X, Y) \twoheadrightarrow \mathrm{Hom}_{\mathsf{D}}\big(F(X), F(Y)\big)$$

are surjective, for any pair X, Y of objects of C. It is the case for full subcategories.

B.1.3 First Examples

The category of sets, with set-theoretic maps, is denoted by Set. The category of vector spaces over the field $\mathbb{K}$, with linear morphisms, is denoted by $\mathsf{Vect}_{\mathbb{K}}$ or, more often, simply by Vect. One defines the *forgetful functor*, $\mathscr{U}: \mathsf{Vect} \to \mathsf{Set}$, by associating to a vector space its underlying set. The image of a linear morphism under this functor is simply the underlying set-theoretic map.

In the other way round, to any set X one can associate a vector space denoted either by $\mathbb{K}[X]$ or by $\mathbb{K}X$, which is spanned by the elements of X. Any set-map $f: X \to Y$ can be uniquely extended linearly to define a vector space morphism $f: \mathbb{K}[X] \to \mathbb{K}[Y]$, whence a functor $\mathscr{F}: \mathsf{Set} \to \mathsf{Vect}$.

B.1.4 Natural Transformation

If F and G are two functors $\mathsf{C} \to \mathsf{D}$, there is a notion of *natural transformation of functors* $\alpha : F \to G$ from F to G as follows. For any object C of C, we are given a morphism $\alpha(C) : F(C) \to G(C)$ in D. It is natural in the following way: for any morphism $f : C \to C'$ in C, it is required that the following diagram

$$\begin{array}{ccc} F(C) & \xrightarrow{F(f)} & F(C') \\ {\scriptstyle \alpha(C)}\downarrow & & \downarrow{\scriptstyle \alpha(C')} \\ G(C) & \xrightarrow{G(f)} & G(C') \end{array}$$

is commutative. Notice that one can compose natural transformations of functors. When $\alpha(C)$ is an isomorphism for any C of C, α is called a *natural isomorphism* and denoted by $\alpha : F \cong G$. When such a natural isomorphism exists, one says that the two functors F and G are naturally isomorphic.

B.1.5 Equivalence of Categories

A functor $F : \mathsf{C} \to \mathsf{D}$ is called an *equivalence of categories* if there exists a functor $G : \mathsf{D} \to \mathsf{C}$ such that the two composites $GF \cong \mathrm{Id}_{\mathsf{C}}$ and $FG \cong \mathrm{Id}_{\mathsf{D}}$ are naturally isomorphic to the identity functors. There exists a more strict notion of isomorphism of categories, when a functor $F : \mathsf{C} \to \mathsf{D}$ admits a strict inverse. Such a situation almost never appears in practice since a functor changes the objects and it is not possible to get them back. However, it is sometimes possible to recover an object in the same isomorphism class, whence an equivalence of categories as the following proposition shows. A functor $F : \mathsf{C} \to \mathsf{D}$ is called *essential surjective* if for any D in D, there is a C in C such that $D \cong F(C)$.

Proposition B.1.1. *A functor $F : \mathsf{C} \to \mathsf{D}$ is an equivalence of categories if and only if F is essential surjective, full and faithful.*

For example, the category of vector spaces with a given basis is equivalent to Vect.

A *skeletal subcategory* C of a category D is made up of one and only one object per isomorphism class and full sets of morphisms. In this casc, the inclusion functor $\mathsf{C} \to \mathsf{D}$ is an equivalence of categories.

B.1.6 Limits and Colimits

An object 0 of a category C is called an *initial object* if there exists one and only one map $0 \to X$ associated to any object X of C. An object which satisfies the dual

Table B.1 First examples of limits and colimits

Type of diagram	Limit	Colimit
$\emptyset$	terminal object	initial object
$\bullet \quad \bullet$	products	coproducts
$\bullet \rightrightarrows \bullet$	equalizers	coequalizers
$\bullet \longrightarrow \bullet \longleftarrow \bullet$	pullback (fiber product)	
$\bullet \longleftarrow \bullet \longrightarrow \bullet$		pushout (fibered sum)

property is called a *terminal object*. In the category of sets, the initial object is given by the empty set and every set with one element $\{*\}$ is a terminal object.

The data of a diagram of type D in a category C is equivalent to the data of a functor from a category D, made up of the underlying graph, to the category C:

$$Z \longrightarrow X \longleftarrow Y \iff (F : \bullet \longrightarrow \bullet \longleftarrow \bullet \to \mathsf{C}).$$

A *cone of* F is an element C in C together with maps $g_x : C \to F(x)$ in C, for any object x in D, such that $F(f) \circ g_x = g_y$, for any map $f : x \to y$ in D. A morphism of cones is a map $h : C \to C'$ in C which makes the following diagram commutative

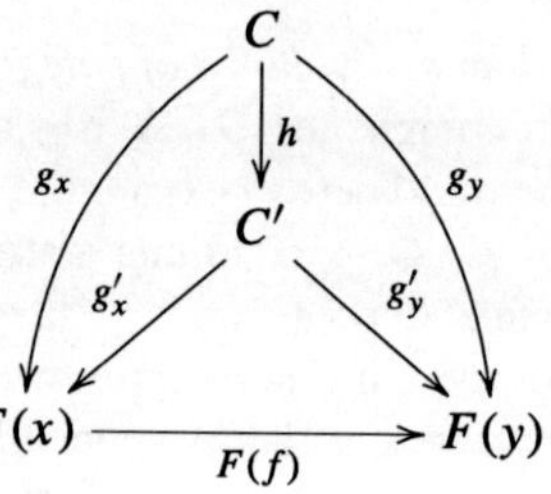

for any $f : x \to y$ in D. This forms the category $\mathsf{cone}(F)$ of cones of the functor F. Dually, by reversing the arrows g, one defines the category $\mathsf{cocone}(F)$ of *cocones* of the functor F.

A *limit* $\lim_F$ of the diagram (respectively functor) F is a terminal object in the category cone_F of cones over F.

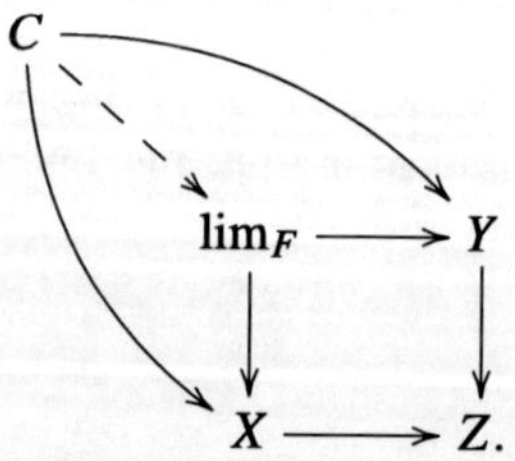

Dually, a *colimit* colim_F of the diagram (respectively functor) F is an initial object in the category cocone_F of cocones over F. Table B.1 gives the main examples of limits and colimits.

B.2 Adjoint Functors, Free Objects

The notion of adjoint functors allows one to compare two categories in both directions. It has been introduced by D.M. Kan in [Kan58].

B.2.1 Adjoint Functors

Let $L : \mathsf{C} \to \mathsf{D}$ and $R : \mathsf{D} \to \mathsf{C}$ be a pair of functors.

$$L : \mathsf{C} \rightleftharpoons \mathsf{D} : R.$$

The functor L is said to be a *left adjoint* to R, and R a *right adjoint* to L if, for any objects C in C and D in D, there is a natural bijection

$$\mathrm{Hom}_{\mathsf{D}}\big(L(C), D\big) \cong \mathrm{Hom}_{\mathsf{C}}\big(C, R(D)\big).$$

The word "natural" means that such a bijection is required to be compatible with the morphisms of each category.

Let C be an object of C. Taking $D = L(C)$, there is a particular element in $\mathrm{Hom}_{\mathsf{D}}(L(C), L(C))$ which is $\mathrm{Id}_{L(C)}$. Its image under the adjunction bijection gives a morphism $\upsilon(C) : C \to R(L(C))$ in C. The resulting transformation of functors

$$\upsilon : \mathrm{Id}_{\mathsf{C}} \to RL$$

is called the *unit of the adjunction*.

Dually, let D be an object of D. Taking $C = R(D)$, there is a particular element in $\mathrm{Hom}_{\mathsf{C}}(R(D), R(D))$ which is $\mathrm{Id}_{R(D)}$. Its image under the adjunction bijection gives a morphism $\varepsilon(D) : L(R(D)) \to D$. The resulting transformation of functors

$$\varepsilon : LR \to \mathrm{Id}_{\mathsf{D}}$$

is called the *counit of the adjunction*.

Conversely, the adjunction bijection is given by the unit as follows

$$\big(f : L(C) \to D\big) \longmapsto \big(R(f) \circ \upsilon(C) : C \to R(D)\big)$$

and by the counit as follows

$$\big(\varepsilon(D) \circ L(g) : L(C) \to D\big) \longleftarrow\!\!| \; \big(g : C \to R(D)\big).$$

Proposition B.2.1.

- *When* $L : \mathsf{C} \rightleftharpoons \mathsf{D} : R$ *form a pair of adjoint functors, the unit* υ *and the counit* ε *of the adjunction satisfy the following relations*

$$\left(L \xrightarrow{L\circ\upsilon} L(RL) = (LR)L \xrightarrow{\varepsilon\circ L} L \right) = \mathrm{Id}_L,$$

$$\left(R \xrightarrow{\upsilon\circ R} RL(R) = R(LR) \xrightarrow{R\circ\varepsilon} R \right) = \mathrm{Id}_R.$$

- *In the other way round, let $L : \mathsf{C} \to \mathsf{D}$ and $R : \mathsf{D} \to \mathsf{C}$ be a pair of functors. If there exist two natural transformations $\upsilon : \mathrm{Id}_{\mathsf{C}} \to RL$ and $\varepsilon : LR \to \mathrm{Id}_{\mathsf{D}}$, whose above composites are equal to the identity, then (L, R) form a pair of adjoint functors.*

EXAMPLES. The two aforementioned functors

$$\mathscr{F} : \mathsf{Set} \rightleftharpoons \mathsf{Vect} : \mathscr{U}$$

are adjoint.

In the category of sets the functors $- \times Y$ and $\mathrm{Hom}(Y, -)$ are adjoint to each other since there is a natural bijection

$$\mathrm{Hom}(X \times Y, Z) \cong \mathrm{Hom}\big(X, \mathrm{Hom}(Y, Z)\big)$$

for any sets X, Y, Z.

Similarly, in the category of vector spaces the functors $- \otimes V$ and $\mathrm{Hom}(V, -)$ are adjoint to each other since there is a natural bijection

$$\mathrm{Hom}(U \otimes V, W) \cong \mathrm{Hom}\big(U, \mathrm{Hom}(V, W)\big)$$

for any vector spaces U, V, W. Of course, here, $\mathrm{Hom}(V, W)$ is equipped with its vector space structure.

Notice that the unit and the counit of an adjunction provide good candidates for natural isomorphisms to prove that L and R form an equivalence of categories.

B.2.2 Free Objects

Let $\mathscr{U} : \mathsf{C} \to \mathsf{D}$ be a functor such that the objects of D are the same as the objects of C except that we do not take into account some of the data. Such a functor is often called a *forgetful functor*. We have already seen an example at Sect. B.1. Another one is when the category C is the category of unital associative algebras, D is the category of vector spaces and $\mathscr{U}$ assigns to the associative algebra A its underlying vector space. So, $\mathscr{U}$ consists in forgetting the algebra structure. Most of the time the image of an object by a forgetful functor is denoted by the same symbol (i.e. C instead of $\mathscr{U}(C)$).

Let $\mathscr{F} : \mathsf{D} \to \mathsf{C}$ be a functor left adjoint to $\mathscr{U}$. The image of an object D of D by $\mathscr{F}$ is called a *free* object. Observe that this notion of freeness is relative to the structure which has been forgotten. In the first example, when $\mathscr{U} : \mathsf{Vect}_{\mathbb{K}} \to \mathsf{Set}$ is the functor which assigns to a vector space its underlying set, we forget the linear space structure. Hence, its left adjoint $X \mapsto \mathbb{K}[X]$ assigns to a set X, the free vector space on X. In the other example, the left adjoint is the tensor algebra functor.

A composite of left adjoint functors is still left adjoint. So the left adjoint of the functor As-alg $\to$ Set, which assigns to a unital associative algebra its underlying set, is $X \mapsto T(\mathbb{K}[X])$.

The free object $\mathscr{F}(V)$ (in C) over the object V of D is characterized by the following property. For any object A in C and any morphism $f : V \to \mathscr{U}(A)$ in D, there exists a unique morphism $\tilde{f} : \mathscr{F}(V) \to A$ in C, which renders the following diagram commutative (in D):

$$\begin{array}{ccc} V & \xrightarrow{\upsilon(V)} & \mathscr{U}(\mathscr{F}(V)) \\ & {}_{f}\searrow & \downarrow{\scriptstyle \mathscr{U}(\tilde{f})} \\ & & \mathscr{U}(A). \end{array}$$

As said before, when $\mathscr{U}$ is a forgetful functor, we usually write A instead of $\mathscr{U}(A)$ and also g instead of $\mathscr{U}(g)$ for morphisms. So the commutative diagram is usually written:

$$\begin{array}{ccc} V & \xrightarrow{i} & \mathscr{F}(V) \\ & {}_{f}\searrow & \downarrow{\scriptstyle \tilde{f}} \\ & & A. \end{array}$$

Under this property, $\mathscr{F}(V)$ is unique up to a unique isomorphism.

Observe that $\mathscr{F}$ stands for "Free" and $\mathscr{U}$ stands for "Underlying".

B.2.3 Representability

A functor $F : \mathsf{C} \to \mathsf{Set}$ is said to be *representable* if it is isomorphic to a functor of the form $\mathrm{Hom}_{\mathsf{C}}(X, -)$, for an object X in C. This means that there are natural bijections $F(C) \cong \mathrm{Hom}_{\mathsf{C}}(X, C)$.

Similarly a contravariant functor, i.e. a functor from the opposite category, is said to be *co-representable* if it is isomorphic to a functor of the form $\mathrm{Hom}_{\mathsf{C}}(-, Y)$, for an object Y in C.

The celebrated Yoneda Lemma says the following.

Lemma B.2.2. *If $F : \mathsf{C} \to \mathsf{Set}$ is a functor and C is an object in* C, *then there is a bijection*

$$\mathrm{Nat}\big(\mathrm{Hom}_{\mathsf{C}}(C, -), F\big) \cong F(C)$$

which sends each natural transformation $\alpha : \mathrm{Hom}_{\mathsf{C}}(C, -) \to F$ to $\alpha(C)(\mathrm{Id}_C)$.

B.3 Monoidal Category

The notion of monoidal category [Bén63, ML63] is the natural extension of the notion of cartesian product of sets and of tensor product of vector spaces.

B.3.1 Definition of a Monoidal Category

A *monoidal category* is a sextuple $(\mathsf{C}, \Box, \alpha, \mathrm{I}, \lambda, \rho)$ where

- C is a category;
- $\Box : \mathsf{C} \times \mathsf{C} \to \mathsf{C}$ is a bifunctor, i.e. a functor from the category defined by the cartesian product of objects and morphisms;
- α is a natural isomorphism

$$\alpha(A, B, C) : (A\Box B)\Box C \xrightarrow{\simeq} A\Box(B\Box C),$$

  satisfying the following pentagon relation

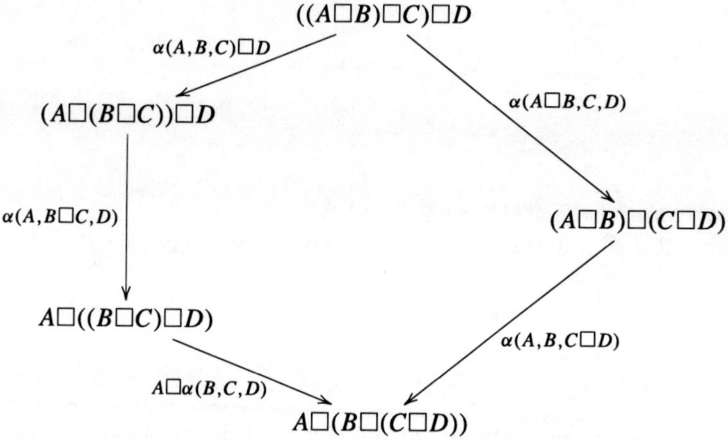

  where notations like $\alpha(A, B, C)\Box D$ stand for $\alpha(A, B, C)\Box\,\mathrm{Id}_D$;
- I is an object of C;
- λ is a natural isomorphism $\lambda(A) : \mathrm{I}\Box A \xrightarrow{\simeq} A$;
- ρ is a natural isomorphism $\rho(A) : A\Box\mathrm{I} \xrightarrow{\simeq} A$, such that

$$\lambda(\mathrm{I}) = \rho(\mathrm{I}) : \mathrm{I}\Box\mathrm{I} \to \mathrm{I} \quad \text{and such that}$$

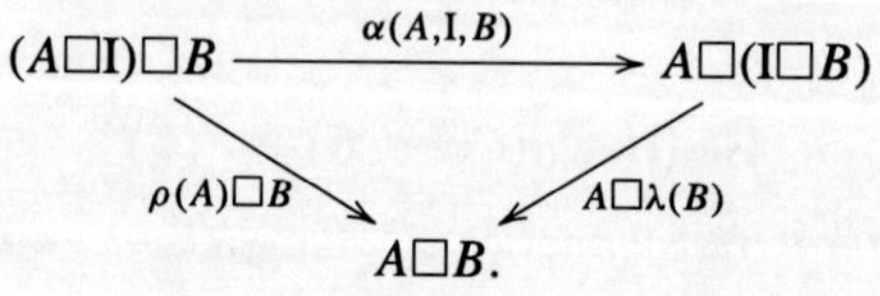

A monoidal category is called *strict* if the three natural isomorphisms α, λ and ρ are identities.

EXAMPLES. The category $(\mathsf{EndoFunct}_{\mathsf{C}}, \circ, \mathrm{Id}_{\mathsf{C}})$ of endofunctors of a category C together with the composite of functors is a strict monoidal category.

The category of sets $(\mathsf{Set}, \times, \{*\})$ and the category of vector spaces $(\mathsf{Vect}, \otimes, \mathbb{K})$ are monoidal categories. Though they are not strict, the morphisms are canonical.

B.3.2 Monoid

A *monoid* (M, γ, η) in a monoidal category $(\mathsf{C}, \square, \alpha, \mathrm{I}, \lambda, \rho)$ is an object M of C endowed with

- a morphism called *multiplication* $\gamma : M \square M \to M$, which is associative

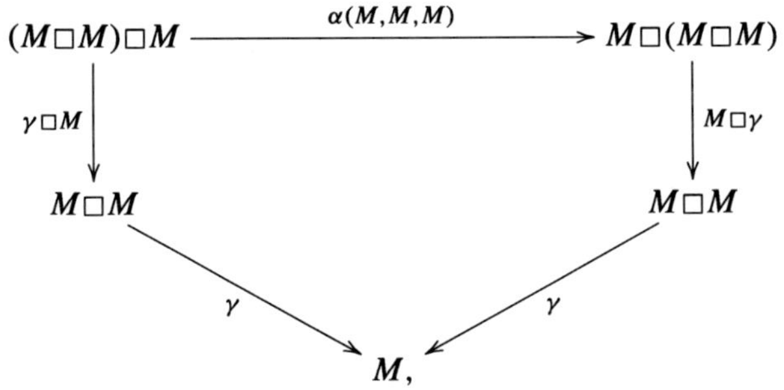

- and a morphism called *unit* $\eta : \mathrm{I} \to M$ such that

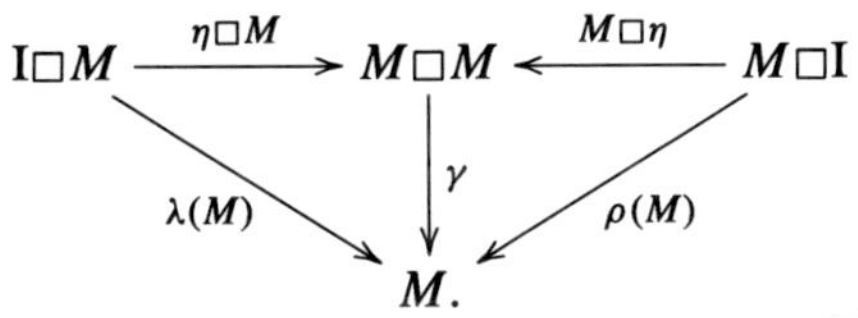

EXAMPLES. A monoid in the monoidal category $(\mathsf{Set}, \times, \{*\})$ is an ordinary monoid or semigroup. A monoid in the monoidal category $(\mathsf{Vect}, \otimes, \mathbb{K})$ is a unital associative algebra. A monoid in the monoidal category of $(\mathsf{EndoFunct}_{\mathsf{C}}, \circ, \mathrm{Id}_{\mathsf{C}})$ is called a monad, see Sect. B.4.

Reversing all the arrows, $\Delta : C \to C \square C$ and $\varepsilon : C \to \mathrm{I}$ for instance, one gets the notion of a *comonoid* (C, Δ, ε). A *(left) module* (N, ζ) over a monoid (M, γ, η), is an object of C equipped with an *action map* $\zeta : M \square N \to N$ satisfying compatibility relations with the multiplication γ and the unit η.

B.3.3 Monoidal Functor

A *(lax) monoidal functor* between two monoidal categories $(\mathsf{C}, \Box, \alpha, \mathrm{I}, \lambda, \rho)$ and $(\mathsf{D}, \boxdot, \beta, \mathrm{J}, \nu, \sigma)$ is a functor $F : \mathsf{C} \to \mathsf{D}$ endowed with

- a natural transformation

$$\phi(A, B) : F(A) \boxdot F(B) \to F(A \Box B),$$

- a morphism $\psi : \mathrm{J} \to F(\mathrm{I})$ in D,

  such that the following properties hold:

- Associativity compatibility:

$$\begin{array}{ccc}
(F(A) \boxdot F(B)) \boxdot F(C) & \xrightarrow{\beta(F(A),F(B),F(C))} & F(A) \boxdot (F(B) \boxdot F(C)) \\
\downarrow{\scriptstyle \phi(A,B)\boxdot F(C)} & & \downarrow{\scriptstyle F(A)\boxdot\phi(B,C)} \\
F(A \Box B) \boxdot F(C) & & F(A) \boxdot F(B \Box C) \\
\downarrow{\scriptstyle \phi(A\Box B,C)} & & \downarrow{\scriptstyle \phi(A,B\Box C)} \\
F((A \Box B) \Box C) & \xrightarrow{F(\alpha(A,B,C))} & F(A \Box (B \Box C)).
\end{array}$$

- Unit compatibility:

$$\begin{array}{ccccc}
\mathrm{J} \boxdot F(A) & \xrightarrow{\psi \boxdot F(A)} & F(\mathrm{I}) \boxdot F(A) & \xrightarrow{\phi(\mathrm{I},A)} & F(\mathrm{I} \Box A) \\
& & \searrow{\scriptstyle \nu(F(A))} & & \downarrow{\scriptstyle F(\lambda(A))} \\
& & & & F(A)
\end{array}$$

$$\begin{array}{ccccc}
F(A) \boxdot \mathrm{J} & \xrightarrow{F(A) \boxdot \psi} & F(A) \boxdot F(\mathrm{I}) & \xrightarrow{\phi(A,\mathrm{I})} & F(A \Box \mathrm{I}) \\
& & \searrow{\scriptstyle \sigma(F(A))} & & \downarrow{\scriptstyle F(\rho(A))} \\
& & & & F(A).
\end{array}$$

The preceding definition is motivated by the following proposition.

Proposition B.3.1. *The image $F(M)$ of a monoid (M, γ, η) of C under a lax monoidal functor $F : \mathsf{C} \to \mathsf{D}$ is a monoid of D with product*

$$F(M) \boxdot F(M) \xrightarrow{\phi(M,M)} F(M \Box M) \xrightarrow{F(\gamma)} F(M)$$

and unit

$$\mathrm{J} \xrightarrow{\psi} F(\mathrm{I}) \xrightarrow{F(\eta)} F(M).$$

A (lax) monoidal functor is called *strong* if ϕ is a natural isomorphism and ψ is an isomorphism. It is called a *strict* monoidal functor if they are identities. Notice that, for a morphism between two monoidal categories, being monoidal amounts to an extra *structure*, whereas being strict is an extra *property*.

EXAMPLE. The forgetful functor $(\mathsf{Vect}, \otimes, \mathbb{K}) \to (\mathsf{Set}, \times, \{*\})$ is a lax, but not strong, monoidal functor.

Two monoidal categories are *monoidally equivalent* if they admit a pair of strong monoidal functors which provide an equivalence of categories.

B.3.4 Coherence Theorem

In practice, any theorem on monoidal categories can be proven by, first, proving it for strict monoidal categories and then, by invoking the following proposition [ML98, Chap. VII].

Proposition B.3.2 (Strictification). *Every monoidal category is monoidally equivalent to a strict monoidal category.*

This proposition is essentially equivalent to the next theorem.

Theorem B.3.3 (Mac Lane's coherence theorem). *In a monoidal category* $(\mathsf{C}, \square, \alpha, \mathrm{I}, \lambda, \rho)$, *every diagram whose vertices come from words in* $\square$ *and* I *and whose edges come from the natural isomorphisms* α, λ *and* ρ *commute.*

This coherence theorem claims that it is enough to check the pentagon and triangle commutative diagram in the definition of a monoidal category to get that "any" diagram commutes.

B.4 Monads

B.4.1 Definition of a Monad

A *monad* $(\mathscr{T}, \gamma, \eta)$ in a category C consists of a functor $\mathscr{T} : \mathsf{C} \to \mathsf{C}$ and two natural transformations

$$\gamma : \mathscr{T} \circ \mathscr{T} \to \mathscr{T}, \quad \text{and} \quad \eta : \mathrm{Id}_{\mathsf{C}} \to \mathscr{T},$$

which make the following diagrams commute

$$
\begin{array}{ccc}
\mathscr{T}\circ\mathscr{T}\circ\mathscr{T} & \xrightarrow{\gamma\circ\mathrm{Id}} & \mathscr{T}\circ\mathscr{T} \\
\downarrow{\scriptstyle \mathrm{Id}\circ\gamma} & & \downarrow{\scriptstyle \gamma} \\
\mathscr{T}\circ\mathscr{T} & \xrightarrow{\gamma} & \mathscr{T}
\end{array}
\qquad
\begin{array}{ccccc}
\mathrm{Id}_{\mathsf{C}}\circ\mathscr{T} & \xrightarrow{\eta\circ\mathrm{Id}} & \mathscr{T}\circ\mathscr{T} & \xleftarrow{\mathrm{Id}\circ\eta} & \mathscr{T}\circ\mathrm{Id}_{\mathsf{C}} \\
& \searrow\!\!\!\searrow & \downarrow{\scriptstyle \gamma} & \swarrow\!\!\!\swarrow & \\
& & \mathscr{T}. & &
\end{array}
$$

So a monad is nothing but a monoid in the strict monoidal category of endofunctors $(\mathsf{EndoFunct}_{\mathsf{C}}, \circ, \mathrm{Id}_{\mathsf{C}})$. Similarly a *comonad* is a comonoid in the monoidal category of endofunctors.

EXAMPLE. When $(\mathsf{C}, \square, \mathrm{I})$ is a monoidal category, any monoid (M, γ, η) gives rise to the monad $\mathscr{T}_M(C) := M \square C$.

Proposition B.4.1. *Any adjunction $(L, R, \upsilon, \varepsilon)$ gives rise to a monad structure on $\mathscr{T} := RL$ by $\gamma := R\varepsilon L$, $\eta := \upsilon$ and dually to a comonad structure on $\mathscr{T}^c := LR$ by $\Delta := L\upsilon R$, $\varepsilon := \varepsilon$. Conversely, every monad and comonad arises in that way.*

B.4.2 Algebra over a Monad

An *algebra* over the monad $(\mathscr{T}, \gamma, \eta)$ is an object A of C equipped with a morphism $\gamma_A : \mathscr{T}(A) \to A$ which makes the following diagrams commute

$$
\begin{array}{ccc}
\mathscr{T}\circ\mathscr{T}(A) & \xrightarrow{\gamma(A)} & \mathscr{T}(A) \\
\downarrow{\scriptstyle \mathscr{T}(\gamma_A)} & & \downarrow{\scriptstyle \gamma_A} \\
\mathscr{T}(A) & \xrightarrow{\gamma_A} & A
\end{array}
\qquad
\begin{array}{ccc}
\mathrm{Id}_{\mathsf{C}}(A) & \xrightarrow{\eta(A)} & \mathscr{T}(A) \\
& \searrow\!\!\!\searrow & \downarrow{\scriptstyle \gamma_A} \\
& & A.
\end{array}
$$

There is an obvious notion of morphism of $\mathscr{T}$-algebras. It can be shown, cf. loc. cit., that any monad $\mathscr{T}$ is completely determined by its category of $\mathscr{T}$-algebras together with the forgetful functor to the underlying category C.

EXAMPLE. Pursuing with previous example of the monad $\mathscr{T}_M$ associated to a monoid M, an algebra over $\mathscr{T}_M$ is a left module over M.

B.4.3 Module over a Monad

There is also a notion of *left $\mathscr{T}$-module*. It is an endofunctor $\mathscr{M} : \mathsf{C} \to \mathsf{C}$ equipped with a natural transformation $\gamma_{\mathscr{M}} : \mathscr{T} \circ \mathscr{M} \to \mathscr{M}$ such that the following diagrams are commutative

$$
\begin{array}{ccc}
\mathscr{T} \circ \mathscr{T} \circ \mathscr{M} & \xrightarrow{\mathscr{T}(\gamma_{\mathscr{M}})} & \mathscr{T} \circ \mathscr{M} \\
{\scriptstyle \gamma(\mathscr{M})} \downarrow & & \downarrow {\scriptstyle \gamma_{\mathscr{M}}} \\
\mathscr{T} \circ \mathscr{M} & \xrightarrow{\gamma_{\mathscr{M}}} & \mathscr{M}
\end{array}
\qquad
\begin{array}{ccc}
\mathrm{Id}_{\mathsf{C}} \circ \mathscr{M} & \xrightarrow{\eta(\mathscr{M})} & \mathscr{T} \circ \mathscr{M} \\
& \searrow\!\!= & \downarrow {\scriptstyle \gamma_{\mathscr{M}}} \\
& & \mathscr{M}.
\end{array}
$$

One recovers the aforementioned definition of an algebra over a monad by considering the constant functor, which sends every element of C to the object A.

B.5 Categories over Finite Sets

B.5.1 Category of Finite Sets

Let Fin be the category of finite sets with any set-maps. We denote by Bij its subcategory made up of finite sets and bijections. Its skeleton is the category denoted Γ which has only one object for any nonnegative integer n, namely the set $[n] := \{0, 1, \ldots, n\}$. So a morphism $f : [n] \to [m]$ is completely determined by the integers $f(i) \in \{0, 1, \ldots, m\}$ for $i = 0, \ldots, n$. Since Γ is subcategory of Fin any functor Fin $\to$ C determines a functor $\Gamma \to$ C by restriction. In the other way round, if the category C admits finite products, then any functor $M : \Gamma \to$ C can be extended to Fin by the following trick:

$$M(X) := \Big(\prod_{\mathsf{Bij}([n],X)} M\big([n]\big) \Big) / \sim$$

where $|X| = n$ and the equivalence relation $\sim$ is as follows. Let $\mu \in M([n])$ and let $f, g \in \mathsf{Bij}([n], X)$. Then we define

$$(f; \mu) \sim \big(g; g_\bullet^{-1} f_\bullet(\mu)\big).$$

B.5.2 Simplicial Category

Putting on $[n] := \{0, 1, \ldots, n\}$ the usual total order $0 < 1 < \cdots < n$, we consider the morphisms $f : [n] \to [m]$ in Γ which are weakly increasing, i.e. $f(i) \le f(j)$ for any $i < j$. Since they are stable under composition, they form a subcategory denoted Δ and called the *simplicial category*.

$$\Delta \rightarrowtail \Gamma.$$

B.5.3 Category of Noncommutative Finite Sets

We now introduce the category of *noncommutative finite sets*, denoted $\Delta\mathbb{S}$, as follows. It has the same objects as Γ, but the morphisms are set-maps enriched with the following data: each fiber is equipped with a total order. It means that for $f:[n]\to[m]$ and any $i\in[m]$ the set $f^{-1}(i)$ is totally ordered. It is clear that, for a composite $g\circ f$ of such maps, the fibers $(g\circ f)^{-1}(i)$ are also totally ordered. So we get a well-defined category. For instance there are two elements in $\mathrm{Hom}_{\Delta\mathbb{S}}([1],[0])$. There is only one set-map, which sends 0 and 1 to 0. The unique fiber is $\{0,1\}$ and there are two different total orders: $0<1$ and $1<0$. Each one of them gives a map in $\Delta\mathbb{S}$.

Forgetting the information about the order of the fibers gives a full functor

$$\Delta\mathbb{S}\twoheadrightarrow\Gamma.$$

Let us denote by $\mathbb{S}$ the groupoid with the same objects as Γ and with isomorphisms $\mathrm{Aut}([n])=\mathbb{S}_{n+1}$. There is an obvious inclusion $\mathbb{S}\rightarrowtail\Delta\mathbb{S}$ since any fiber reduced to one element is totally ordered.

There is also an inclusion $\Delta\rightarrowtail\Delta\mathbb{S}$ since we can identify the morphisms of Δ with the morphisms of $\Delta\mathbb{S}$ such that the total order of the fiber coincides with the total order induced by the natural total order $0<1<\cdots<n$.

Proposition B.5.1. [FL91, Lod98] *Any morphism $f:[n]\to[m]$ in the category of noncommutative finite sets $\Delta\mathbb{S}$ can be written uniquely as the composite $f=\varphi\circ\sigma$:*

$$[n]\xrightarrow{\sigma}[n]\xrightarrow{\varphi}[m]$$

where $\sigma\in\mathrm{Aut}([n])=\mathbb{S}_{n+1}$ and $\varphi\in\mathrm{Hom}_\Delta([n],[m])$.

B.5.4 Comments

From the above proposition any composite $\omega\circ\varphi$ where $\omega\in\mathrm{Hom}_{\mathbb{S}}([m],[m])$ and $\varphi\in\mathrm{Hom}_\Delta([n],[m])$ can be written

$$\omega\circ\varphi=\omega_*(\varphi)\circ\varphi^*(\omega)$$

where $\omega_*(\varphi)\in\mathrm{Hom}_\Delta([n],[m])$ and $\varphi^*(\omega)\in\mathrm{Hom}_{\mathbb{S}}([n],[n])$. The family of maps

$$\mathrm{Hom}_\Delta([n],[m])\times\mathbb{S}_{m+1}\to\mathbb{S}_{n+1}\times\mathrm{Hom}_\Delta\big([n],[m]\big),\qquad(\varphi,\omega)\mapsto\big(\varphi^*(\omega),\omega_*(\varphi)\big)$$

is called a *distributive law*. The pair $(\Delta,\mathbb{S})$ with its distributive law is sometimes called a *matching pair of categories*. The category $\Delta\mathbb{S}$ is sometimes called a *bicrossed product* of the categories Δ and $\mathbb{S}$.

The cyclic permutation $(0\ 1\ \ldots\ n) \in \mathbb{S}_{n+1}$ generates the cyclic group C_{n+1} of order $n+1$. If the permutation ω is cyclic, then so is $\varphi^*(\omega)$:

$$\omega \in C_{m+1} \subset \mathbb{S}_{m+1} \to \varphi^*(\omega) \in C_{n+1} \subset \mathbb{S}_{n+1}.$$

As a consequence, there is a well-defined subcategory of $\Delta\mathbb{S}$ which is made up of the morphisms $\varphi \circ \sigma$ where σ is a cyclic permutation. It is denoted by ΔC and called the *cyclic category*, or *Connes' category* [Con85]:

$$\begin{array}{ccccccc} & & \Gamma & \xrightarrow{\simeq} & \mathsf{Fin} & & \\ & & \uparrow & & & & \\ \Delta & \rightarrowtail & \Delta\mathbb{S} & \leftarrowtail & \mathbb{S} & \xrightarrow{\simeq} & \mathsf{Bij} \\ & & \uparrow & & & & \\ & & \Delta C. & & & & \end{array}$$

The column $\Delta C \rightarrowtail \Delta S \twoheadrightarrow \Gamma$ is not an exact sequence!

B.5.5 Categories of Pointed Sets

The category of *pointed finite sets*, denoted Fin', is such that an object is a finite set together with the choice of one of its elements called the base-point. The morphisms are the set-maps which send the base-point of the source to the base-point of the target. Let us choose the point 0 as the base-point of $[n]$. The skeleton of Fin' is denoted by Γ'. In each of the categories considered in the preceding section, we consider the subcategory made up of morphisms respecting the base-points. This subcategory is denoted by decorating the symbol by $-'$. It is immediate to verify that $\mathrm{Hom}_{\mathbb{S}'}([n],[n]) = \mathrm{Aut}_{(\Delta\mathbb{S})'}([n],[n]) = \mathbb{S}_n$ is viewed as the automorphism group of $\{1,\ldots,n\}$. On the other hand we can check that $(\Delta C)' = \Delta^{\mathrm{op}}$ (the opposite category of Δ):

$$\Delta^{\mathrm{op}} = (\Delta C)' \rightarrowtail (\Delta S)' \twoheadrightarrow \Gamma'.$$

B.6 Model Categories

A homotopical category is usually defined as a category of fractions, à la Gabriel–Zisman, where one localizes with respect to some family of morphisms. Another way of doing, proposed by Dan Quillen, is to distinguish two sub-families of morphisms: the *fibrations* and the *cofibrations*, required to satisfy some axioms. This extra data *models* the homotopy category. Moreover model category structures allow one to do homotopy theory in many other categories than that of topological space.

The papers [GS07, DS95] provides two concise presentations for this topic. For more details, we refer the reader to the original lecture note [Qui67] and to the book [Hov99].

B.6.1 Localization of Categories

One localizes categories in the same way as rings: by adding to a category formal inverse morphisms for some class of morphisms to make them become isomorphisms [GZ67].

Let C be a category with a distinguished subclass W of morphisms. A *localized category* of C at W is a category L, endowed with a functor $\rho : \mathsf{C} \to \mathsf{L}$ such that

1. For any $f \in W$, its image $\rho(f)$ is an isomorphism in L.
2. For any functor $F : \mathsf{C} \to \mathsf{D}$, which sends elements of W to isomorphisms in D, there exists a unique functor $\widetilde{F} : \mathsf{L} \to \mathsf{D}$, which factors F through ρ

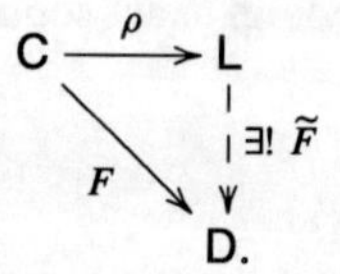

When such a category exists, property (2) implies that it is unique up to a unique isomorphism. We now give a realization of it, denoted by $\mathsf{C}[W^{-1}]$.

- ◇ Objects: Objects of C,
- ◇ Morphisms: To any pair (A, B) of objects of C, we consider the chains $\mathrm{Ch}(A, B)$ of maps either from the morphisms of C or from the *formal inverse* of elements of W between A and B, like

We define the set (when it is actually a set) of morphisms $\mathrm{Hom}_{\mathsf{C}[W^{-1}]}(A, B)$ by the set $\mathrm{Ch}(A, B)/\approx$ of chains quotiented by the following relations

$$C \xrightarrow{f} D \xrightarrow{g} E \approx C \xrightarrow{g \circ f} E,$$

$$C \xrightarrow[f]{\sim} D \dashrightarrow C \approx \mathrm{Id}_C,$$

$$D \dashrightarrow C \xrightarrow[f]{\sim} D \approx \mathrm{Id}_D.$$

Because of the form of this definition, it is difficult to work in practice with such a category. The following results give a method to compare various homotopy categories.

Let C and D be two categories with, respectively, two subclasses W_{C} and W_{D}. Let $L : \mathsf{C} \rightleftharpoons \mathsf{D} : R$ be a pair of adjoint functors, which preserve the two classes W_{C} and W_{D}. This means that the image under L of any map of W_{C} lives in W_{D} and dually for R. In this case, this adjunction induces an adjunction

$$\widetilde{L} : \mathsf{C}[W_{\mathsf{C}}^{-1}] \rightleftharpoons \mathsf{D}[W_{\mathsf{D}}^{-1}] : \widetilde{R}\,.$$

Proposition B.6.1. *When the unit and the counit of an adjunction $L : \mathsf{C} \rightleftharpoons \mathsf{D} : R$ are defined by morphisms in the localizing classes W_{C} and W_{D} or by isomorphisms, the induced adjunction is an equivalence between the localized categories $\mathsf{C}[W_{\mathsf{C}}^{-1}]$ and $\mathsf{D}[W_{\mathsf{D}}^{-1}]$.*

B.6.2 Homotopy Category

We have no choices: when W is the class of quasi-isomorphisms of chain complexes or of dg algebras, the localized category $\mathsf{Ho}(\mathsf{C}) := \mathsf{C}[W^{-1}]$, called the *homotopy category*, or *derived category* for chain complexes, is the good framework for homological algebra. The homotopy category is the universal category for which quasi-isomorphisms become isomorphisms. For instance, such categories are the source and target categories for derived functors, see Sect. B.7.1.

B.6.3 Definition of Model Category

Let C be a category with a distinguished class W of morphisms called the *weak equivalences*, denoted by $\xrightarrow{\sim}$. The category C is endowed with a *model category structure* if it admits two other distinguished classes of morphisms: the *fibrations*, denoted by $\twoheadrightarrow$, and the *cofibrations*, denoted by $\rightarrowtail$, which satisfy the following axioms.

[$\mathcal{MC}$ 1] (Limits–colimits) The category C admits finite limits and finite colimits.

[$\mathcal{MC}$ 2] (2 out of 3) For any pair f, g of composable morphisms, if two of the three morphisms f, g, gf are weak equivalences, then so is the third one.

[$\mathcal{MC}$ 3] (Retracts) These three classes of morphisms are stable under retracts.

[$\mathcal{MC}$ 4] (Lifting property) For any commutative diagram (solid arrows) of the form

$$\begin{array}{ccc} A & \longrightarrow & B \\ \big\downarrow & \nearrow & \big\downarrow \\ X & \longrightarrow & Y, \end{array}$$

there exists a map $X \to B$ if either the cofibration $A \overset{\sim}{\rightarrowtail} X$ is a weak-equivalence or if the fibration $B \overset{\sim}{\twoheadrightarrow} Y$ is a weak-equivalence.

[$\mathcal{MC}$ 5] (Factorization) Any morphism $f : A \to B$ factors into two ways:

$$f = A \overset{\sim}{\rightarrowtail} X \twoheadrightarrow B \quad \text{and} \quad f = A \rightarrowtail Y \overset{\sim}{\twoheadrightarrow} B.$$

Notice that the class of cofibrations (respectively fibrations) is completely characterized by the class of fibrations (respectively cofibrations), which are weak equivalences, under the lifting property [$\mathcal{MC}$ 4].

EXAMPLES. The toy model for this notion is the category of nonnegatively graded chain complexes. It is endowed with a model category structure where

◇ the weak equivalences are the quasi-isomorphisms,
◇ the fibrations are the degree-wise epimorphisms, $C_n \twoheadrightarrow D_n$, for $n \geq 1$,
◇ the cofibrations are the degree-wise monomorphisms, $C_n \rightarrowtail D_n$, with projective cokernel, for $n \geq 0$.

The categories of topological spaces and simplicial sets are two other important examples.

B.6.4 Fibrant and Cofibrant Objects

An object C of a model category C is called *cofibrant* if the map from the initial object $0 \rightarrowtail C$ is a cofibration. Dually an object F is called *fibrant* if the map to the terminal object $F \twoheadrightarrow *$ is a fibration. A *cofibrant resolution*, or *cofibrant replacement*, of an object A of C is a cofibrant object $C \overset{\sim}{\to} A$ weakly equivalent to A. The factorization axiom [$\mathcal{MC}$ 5] ensures that such a resolution always exists. The lifting property [$\mathcal{MC}$ 4] shows that cofibrant replacements are in some sense unique.

Dually, one defines the notion of *fibrant resolution*, or *fibrant replacement*, $A \overset{\sim}{\to} F$.

In the aforementioned example of nonnegatively graded chain complexes, every object is fibrant. A chain complex is cofibrant if and only if it is projective. Therefore, a cofibrant resolution is nothing but a projective resolution.

B.6.5 Main Theorem

We consider the full subcategory C_{cf} of fibrant and cofibrant objects. On the sets $\mathrm{Hom}_{\mathsf{C}_{\mathsf{cf}}}(A, B)$ of morphisms of this category, one defines an equivalence relation $\sim$ called *homotopy equivalence*. Since it is well behaved with respect to composition of maps, it induces a quotient category $\mathsf{C}_{\mathsf{cf}}/\sim$.

Theorem B.6.2. [Qui67] *The following two categories are equivalent*

$$\mathsf{Ho}(\mathsf{C}) = \mathsf{C}\big[W^{-1}\big] \cong \mathsf{C}_{\mathsf{cf}}/\sim .$$

In the example of chain complexes, this notion of homotopy equivalence is the classical notion of homotopy equivalence between morphisms of chain complexes. This theorem asserts that the homotopy category (or derived category) is equivalent to the category of projective complexes with the homotopy classes of morphisms.

B.6.6 Model Category Structure on Operads

The category of dg operads can be endowed with a model category structure [Hin97], which allows one to perform homotopical algebra at that level.

Proposition B.6.3. *Over a field* $\mathbb{K}$ *of characteristic* 0, *the category of dg operads admits a model category structure where*

- ◇ *the weak equivalences are the arity-wise quasi-isomorphisms,*
- ◇ *the fibrations are the arity-wise and degree-wise epimorphisms,*
- ◇ *the cofibrations are characterized by the lifting property.*

When the characteristic of the ground field is not equal to 0, the result still holds but one has to restrict oneself to reduced operads $\mathscr{P}(0) = 0$ [Hin97]. Otherwise, one gets a *semi*-model category structure [Spi01], which a slightly weaker notion.

Under some assumptions, the category of operads over a symmetric monoidal model category admits a model category structure where the weak equivalences (respectively the fibrations) are the arity-wise weak equivalences (respectively the fibrations) [BM03a, Fre09a]. This applies to the categories of topological operads and simplicial operads, for instance.

Proposition B.6.4. *In the category of nonnegatively graded dg operads with the model category structure of Proposition* B.6.3, *any quasi-free operad* $(\mathscr{T}(X), d)$ *is cofibrant.*

A quasi-free operad $(\mathscr{T}(X), d)$ is called *triangulated* after D. Sullivan, when its space of generators X is endowed with a exhaustive filtration $F_0X = \{0\} \subset F_1X \subset \cdots \subset F_nX \subset \cdots$ satisfying $d(F_nX) \subset \mathscr{T}(F_{n-1})$. Over unbounded chain complexes, triangulated quasi-free operads are cofibrant.

B.6.7 Model Category Structure on Algebras over an Operad

The category of dg algebras over an operad can also be endowed with a model category structure [Hin97].

Proposition B.6.5. *Let $\mathbb{K}$ be a field of characteristic 0 and let $\mathscr{P}$ be a dg operad. The category of dg $\mathscr{P}$-algebras, admits a model category structure where*

- ◇ *the weak equivalences are the quasi-isomorphisms,*
- ◇ *the fibrations are the degree-wise epimorphisms,*
- ◇ *the cofibrations are characterized by the lifting property.*

In general, when the characteristic of the ground field is not equal to 0, one has to require that the underlying $\mathbb{K}[\mathbb{S}_n]$-modules $\mathscr{P}(n)$ of the operad $\mathscr{P}$ are projective, for any n. In this case, one says that the dg $\mathbb{S}$-module $\mathscr{P}$ is $\mathbb{S}$-*cofibrant*, because it is cofibrant in the model category of dg $\mathbb{S}$-modules. When the dg operad $\mathscr{P}$ is $\mathbb{S}$-cofibrant, the category of dg $\mathscr{P}$-algebras admits a semi-model category structure, with the same classes of maps as above [Spi01, BM03a, Fre09a].

Proposition B.6.6. *In the category of nonnegatively graded dg $\mathscr{P}$-algebras with the model category structure of Proposition* B.6.5, *any quasi-free $\mathscr{P}$-algebra $(\mathscr{P}(X), d)$ is cofibrant.*

As for operads, a quasi-free $\mathscr{P}$-algebra $(\mathscr{P}(X), d)$ is called *triangulated*, when its space of generators X is endowed with an exhaustive filtration $F_0X = \{0\} \subset F_1X \subset \cdots \subset F_nX \subset \cdots$ satisfying $d(F_nX) \subset \mathscr{P}(F_{n-1})$. Over unbounded chain complexes, triangulated quasi-free $\mathscr{P}$-algebras are cofibrant.

In the literature, one usually defines the notion of *homotopy $\mathscr{P}$-algebra* by a category of algebras over a cofibrant replacement $\mathscr{Q} \xrightarrow{\sim} \mathscr{P}$ of the operad $\mathscr{P}$. It is well defined in the following sense.

Lemma B.6.7.

- ◇ *Any cofibrant operad is $\mathbb{S}$-cofibrant.*
- ◇ *Let $\mathscr{P} \xrightarrow{\sim} \mathscr{P}'$ be two weakly equivalent $\mathbb{S}$-cofibrant operads. The homotopy categories of dg $\mathscr{P}$-algebras and dg $\mathscr{P}'$-algebras are equivalent.*

$$\mathsf{Ho}(\mathsf{dg}\ \mathscr{P}\text{-}\mathsf{alg}) \cong \mathsf{Ho}\big(\mathsf{dg}\ \mathscr{P}'\text{-}\mathsf{alg}\big).$$

This lemma shows that the homotopy category of homotopy $\mathscr{P}$-algebras is independent of the chosen cofibrant resolution. In this book, we provide canonical ones, the ones given by the Koszul duality theory. Since the 1970s, it has been known that the categories of algebras over cofibrant (topological) operads have nice homotopy invariance properties [BV73], the homotopy transfer theorem along weak equivalences for instance.

B.7 Derived Functors and Homology Theories

Cartan–Eilenberg [CE56] and Grothendieck [Gro57] gave a method to define (co)homology theories in abelian categories by using the notion of derived functor. This latter notion was generalized by Quillen beyond the additive case in the context of model categories. It provides a way to show that homotopy categories are equivalent.

B.7.1 Derived Functors

Recall that an *abelian category* is a category whose Hom-sets are abelian groups and which satisfies some assumptions, see Chap. VIII of [ML98]. In an abelian category, one can define the notion of projective (resp. injective) objects and (resp. injective) resolutions, as usual.

Let $F : \mathsf{C} \to \mathsf{D}$ be an additive functor between two abelian categories. If this functor does not preserve short exact sequences but only the right part of them, one would like to measure the default from being left exact. To do so, one defines the notion of *(left) derived functor*, denoted $\mathbb{L}F$. If the abelian category C has enough projective objects, then every object A admits a projective resolution $P_\bullet \xrightarrow{\sim} A$, unique up to homotopy equivalences. The left derived functor of F at A is defined by the homology groups of $F(P_\bullet)$:

$$\mathbb{L}_\bullet F(A) := H_\bullet\big(F(P_\bullet)\big).$$

Its definition does not depend on the chosen projective resolution. Hence, it is well defined. More generally, the derived functor $\mathbb{L}F$ is defined to be a functor between the derived categories of chain complexes of C and of D respectively.

There are two main advantages to define homology theories like this. In the first place, this conceptual approach permits us to prove general results, valid for any homological theory defined via derived functors. Then, it makes it easier to compute homology groups; for a particular object A, there are sometimes ad hoc small projective resolutions of A. Therefore, the left derived functor is computable with this resolution. For more details, we refer the reader to Chap. III of [GM03].

B.7.2 Tor and Ext Functors

We illustrate these ideas on the following example. Let $f : R \to S$ be a morphism between two rings. We consider the categories of left modules over R and S respectively, that we denoted by R-Mod and S-Mod. By pulling back along f, an S-module M becomes an R-module: $r.m := f(r).m$, for $r \in R$ and $m \in M$. We denote this functor by $f^* : S\text{-Mod} \to R\text{-Mod}$. The opposite functor $f_* : R\text{-Mod} \to S\text{-Mod}$ is the extension of scalars given by the relative tensor product $f_!(N) := S \otimes_R N$. These two functors are additive. Since they form a pair of adjoint functors

$$f_! : R\text{-Mod} \rightleftharpoons S\text{-Mod} : f^*,$$

f_* is right exact and f^* is left exact. Therefore, we can consider the left derived functor of f_* denoted $\mathbb{L}(S \otimes_R -)$. More generally, one can derive the functors $M \otimes_R$

$-$ and $-\otimes_R N$ on the left. A classical result states that they give the same homology, that is $\mathbb{L}(M\otimes_R -)(N)\cong\mathbb{L}(-\otimes_R N)(M)$. Therefore, this defines a bifunctor called the *Tor functor* denoted by $\mathrm{Tor}_\bullet^R(M,N)$ or $M\otimes_R^{\mathbb{L}} N$. Under this notation, we have

$$\mathbb{L}f_*(N)=\mathrm{Tor}_\bullet^R(S,N)=S\otimes_R^{\mathbb{L}} N.$$

When R is an augmented algebra over a ring $\mathbb{K}$, with augmentation denoted by $\varepsilon: R\to\mathbb{K}$, the functor $\varepsilon_!(M)=\mathbb{K}\otimes_R M$ is equal to the functor that gives the indecomposable elements of M for the action of R: $\mathrm{Indec}(M):=M/\overline{R}.M=\varepsilon_!(M)$. In order to study the properties of the action of R on M, we derive this functor on the left. We consider the left derived functor

$$\mathbb{L}\,\mathrm{Indec}(M)=\mathbb{L}\varepsilon_*(M)=\mathbb{K}\otimes_R^{\mathbb{L}} M=\mathrm{Tor}_\bullet^R(\mathbb{K},M).$$

The Hochschild homology of an associative algebra A with coefficients in a module M (not a bimodule) is defined in this way, i.e. via $\mathrm{Tor}_\bullet^A(\mathbb{K},M)$. The Chevalley–Eilenberg homology of a Lie algebra $\mathfrak{g}$ with coefficients in a module M is also defined in this way, i.e. via $\mathrm{Tor}_\bullet^{U(\mathfrak{g})}(\mathbb{K},M)$ where $U(\mathfrak{g})$ is the universal enveloping algebra of $\mathfrak{g}$.

Dually, replacing $M\otimes_R N$ by $\mathrm{Hom}_R(M,N)$, one defines the *Ext functor* $\mathrm{Ext}_R^\bullet(M,N)$ as the right derived functor of the Hom functor.

B.7.3 Total Derived Functors

Quillen settled in [Qui67] a way to derive functors in a non-additive setting by using the notion of model categories as follows.

Let C and D be two model categories. A *Quillen functor* is a pair of adjoint functors

$$F:\mathsf{C}\rightleftharpoons\mathsf{D}:G$$

such that F preserves cofibrations and G preserves fibrations. (The functor F is the left adjoint.) Notice that this condition is equivalent to F preserving cofibrations and acyclic cofibrations; it is also equivalent to G preserving fibrations and acyclic fibrations.

In this case, one extends the aforementioned definition of derived functor in the same way. The image of any object A in $\mathscr{C}$ under the *total left derived functor* $\mathbb{L}F$ is given by the image $F(C)$, where C is a cofibrant replacement of A, $C\xrightarrow{\sim}A$. One proceeds dually, with fibrant replacements, to define the *total right derived functor* $\mathbb{R}G$. These two total derived functors form a pair of adjoint functors between the associated homotopy categories

$$\mathbb{L}F:\mathrm{Ho}(\mathsf{C})\rightleftharpoons\mathrm{Ho}(\mathsf{D}):\mathbb{R}G.$$

B.7.4 Equivalence of Homotopy Categories

One of the main uses of the notion of total derived functors is to provide equivalences between the homotopy categories. This is heavily used by Quillen in [Qui69], where he showed that the homotopy theories of various categories are the same.

Proposition B.7.1. *Let*

$$F : \mathsf{C} \rightleftarrows \mathsf{D} : G$$

be a Quillen functor. The induced adjunction

$$\mathbb{L}F : \mathsf{Ho(C)} \rightleftarrows \mathsf{Ho(D)} : \mathbb{R}G$$

is an equivalence of categories if and only if, for any cofibrant object X of C and for any fibrant object Y of D, a map $X \xrightarrow{\sim} G(Y)$ is a weak equivalence in C if and only if the adjoint map $F(X) \xrightarrow{\sim} Y$ is a weak equivalence in D.

In this case, the Quillen functor is called a *Quillen equivalence*.

Appendix C
Trees

Et comme cela m'est arrivé plusieurs fois je compris que le meilleur moyen de m'expliquer à moi-même quelque chose qui me paraissait inexplicable était de la décrire et de l'expliquer à d'autres.

Stephan Zweig in "Magellan"

We introduce the combinatorial objects named "trees" and "graphs". There are various sorts of trees: planar, binary, nonplanar (abstract), circled, ladder, and so on. They play an important role in the operad theory because the free operad can be explicitly described in terms of trees.

Moreover they are closely related to a geometric object called the associahedron (alias Stasheff polytope) for which we give two different cellular realizations. They both play a role in the analysis of operads associated to the operad As.

C.1 Planar Binary Trees

C.1.1 Magma and Planar Binary Trees

A *magma* is a set X equipped with a binary operation, that is a set map $X \times X \to X$ without any further assumption. The free magma on one element, say x, is made of all the parenthesizings of a word of finite length in x:

$$x, (xx), \big((xx)x\big), \big(x(xx)\big), \big(\big((xx)x\big)x\big), \dots$$

It is slightly easier to work with combinatorial objects which are in bijection with the parenthesizings: the *planar binary rooted trees*, abbreviated into *pb trees* (note that graph theorists would say plane instead of planar). We denote by PBT_n the set of pb trees with n leaves (we do not mention the word rooted anymore, unless necessary):

J.-L. Loday, B. Vallette, *Algebraic Operads*,
Grundlehren der mathematischen Wissenschaften 346,
DOI 10.1007/978-3-642-30362-3,

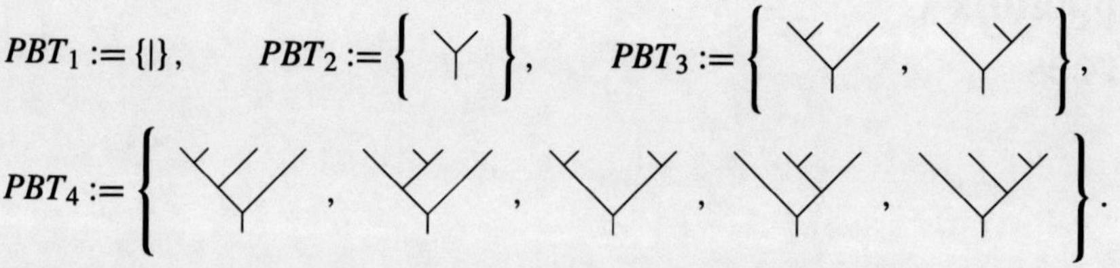

So $t \in PBT_n$ has one root, n leaves, $(n-1)$ vertices, $(n-2)$ edges.

The *grafting* of the trees $r \in PBT_n$ and $s \in PBT_m$ is the tree $r \vee s \in PBT_{n+m}$ obtained by joining the root of r and the root of s to a new vertex and adding a new root. Under the bijection with parenthesizings, grafting corresponds to the product in the magma. For $n > 1$, any tree $t \in PBT_n$ can be written uniquely as the grafting of uniquely determined trees:

$$t = t^l \vee t^r.$$

So there is a bijection $PBT_n = \bigcup_{p+q=n, p\geq 1, q\geq 1} PBT_p \times PBT_q$.

The number of elements in PBT_{n+1} is known to be the *Catalan number* $c_n = \frac{(2n)!}{n!\,(n+1)!}$, that is

n	0	1	2	3	4	5	6	7	…
c_n	1	1	2	5	14	42	132	429	…

The generating series for the Catalan numbers is

$$\sum_{n\geq 1} c_n x^n = \frac{1-\sqrt{1-4x}}{2}.$$

It is helpful to enumerate the leaves from left to right beginning with 1 as in the following example:

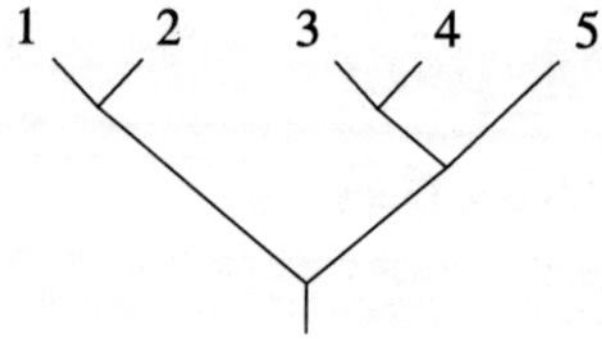

As a consequence the set of vertices gets a total order: the vertex i lies in between the leaves i and $i+1$:

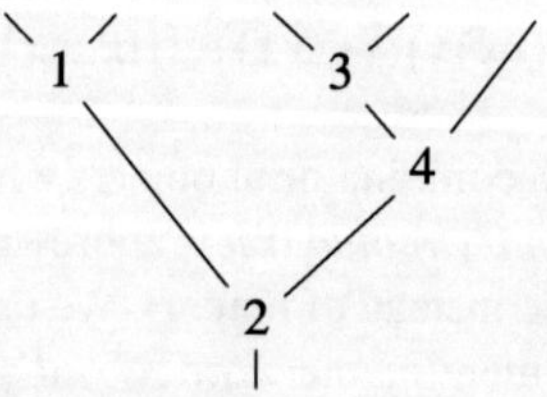

For two trees t and s the *partial composite* $t \circ_i s$ is the tree obtained by identifying the root of s with the ith leaf of t. The result is a tree with $m-1+n$ leaves if the number of leaves of t (resp. s) is m (resp. n). This construction makes sense whenever $1 \le i \le m$.

The partial composition obviously satisfies the following relations for any $r \in PBT_l, s \in PBT_m, t \in PBT_n$:

$$\begin{cases} \text{(I)} \ \ (r \circ_i s) \circ_{i-1+j} t = r \circ_i (s \circ_j t), & \text{for } 1 \le i \le l, 1 \le j \le m, \\ \text{(II)} \ (r \circ_i s) \circ_{k-1+m} t = (r \circ_k t) \circ_i s, & \text{for } 1 \le i < k \le l. \end{cases}$$

These two relations correspond, respectively, to the following two situations:

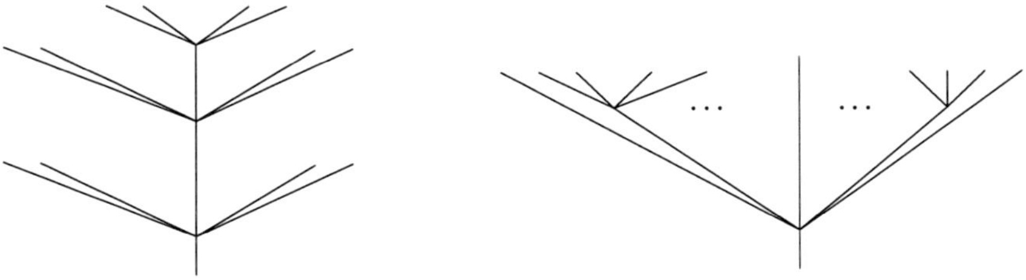

C.1.2 Tamari Poset Structure on PBT_n

On the set of planar binary trees PBT_n there is a poset (partially ordered set) structure, called the *Tamari poset* structure, which has a lot of interesting properties with respect to the algebraic operations on trees. It is defined as follows. Let us suppose that $s \in PBT_n$ can be obtained from $t \in PBT_n$ by moving some edge, or leaf, from left to right over a vertex:

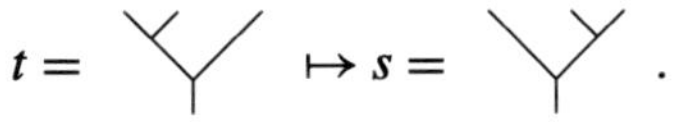

Then we say that t is less than s and we write either $t \to s$ or $t < s$. It is called a covering relation. The poset structure of PBT_n is induced by these covering relations. In low dimension we get

The left (resp. right) comb is the minimum (resp. maximum) of PBT_n for this poset structure.

C.1.3 From Permutations to pb Trees

For any n there is a map $\varphi : \mathbb{S}_n \to PBT_{n+1}$ compatible with several algebraic structures. It is constructed as follows. First, it is helpful to introduce the notion of *leveled*

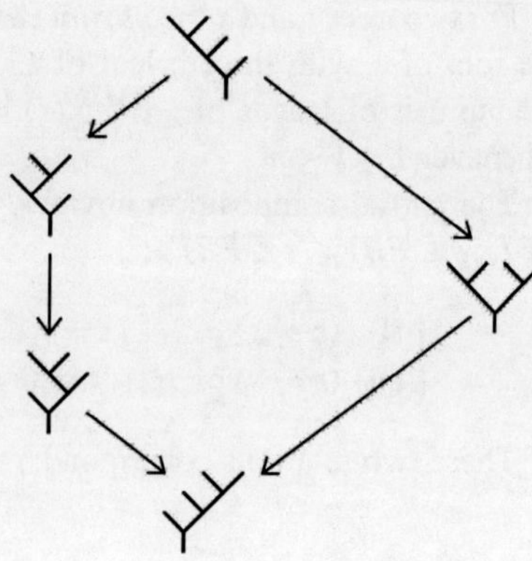

Fig. C.1 The Tamari poset PBT_4

tree. It is a pb tree such that each vertex is assigned a level (from top to bottom, 1 to n), with the requirements that there is only one vertex per level. So the rightmost tree in Fig. C.1 gives rise to two leveled trees. For a given leveled tree, assigning to each vertex its levels determines a permutation of $\{1, \ldots, n\}$. Whence a bijection between $\mathbb{S}_n$ and the set of leveled trees with $n+1$ leaves, denoted by $\widetilde{PBT}_{n+1}$. Forgetting the levels gives the map

$$\varphi : \mathbb{S}_n \cong \widetilde{PBT_{n+1}} \to PBT_{n+1}.$$

See for instance [Lod01, Appendix A].

C.2 Planar Trees and Stasheff Polytope

Orienting the edges of a given pb tree from top to bottom, we see that each vertex has two inputs and one output. We now consider the planar trees for which any vertex has k inputs and one output. We first consider the case when $k \geq 2$. We denote by PT_n the set of planar (rooted) trees with n leaves:

$PT_1 := \{|\}$, $PT_2 := \{$  $\}$, $PT_3 := \{$  $\}$,

$PT_4 := \{$ 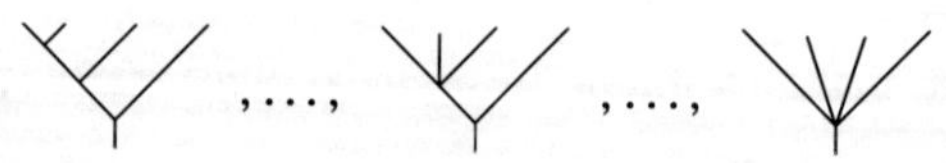 $\}$.

Each set PT_n is graded according to the number of vertices, i.e. $PT_n = \bigcup_{n \geq k} PT_{n,k}$ where $PT_{n,k}$ is the set of planar trees with n leaves and k vertices. For instance $PT_{n,1}$ contains only one element which we call the *n-corolla* (the last element in the above sets). It is clear that $PT_{n,n-1} = PBT_n$.

Any tree $t \in PT_n, n > 1$, is uniquely obtained as the grafting of k trees:

$$t = t^{(1)} \vee \cdots \vee t^{(k)}, \quad \text{with } k \geq 2.$$

The notion of partial composite (grafting $\circ_i$) extends to all the planar trees. The number of elements in PT_{n+1} is the *Schröder number* (also called super Catalan number) and is denoted C_n:

n	0	1	2	3	4	5	6	7	...
C_n	1	1	3	11	45	197	903	4279	...

Let us introduce the generating series in two variables

$$f_y^{\mathcal{K}}(x) := \sum_{n\geq 1}\Big(\sum_{k\geq 0}(-1)^k \# PBT_{n+1,n-k}\, y^k\Big)x^n.$$

One can show (cf. [Sta97a] for instance) that

$$f_y^{\mathcal{K}}(x) = \frac{1-(2+y)x-\sqrt{1-2(2+y)x+y^2x^2}}{2(1+y)x}.$$

C.2.1 The Associahedron, Alias Stasheff Polytope [Sta63]

The *associahedron* is a cell complex $\mathcal{K}^n$ of dimension n which can be realized as a convex polytope as follows. To any tree $t \in PBT_{n+2}$ we associate a point $M(t) \in \mathbb{R}^{n+1}$ by the following algorithm: $M(t) := (a_1b_1, \dots, a_{n+1}b_{n+1})$ where a_i is the number of leaves standing on the left side of the ith vertex, and b_i is the number of leaves standing on the right side of the ith vertex. Equivalently

$$M(t) = M\big(t^l \vee t^r\big) = \big(M\big(t^l\big), pq, M\big(t^r\big)\big)$$

whenever $t^l \in PBT_p$ and $t^r \in PBT_q$.

It can be shown that all the points $M(t)$ lie in the hyperplane $\sum_i a_ib_i = \frac{(n+1)(n+2)}{2}$ and the convex hull of all these points is the Stasheff polytope (cf. [Lod04a]).

From this construction it follows that the Stasheff polytope is a cell complex whose cells are in bijection with the planar trees. For instance the vertices correspond to the planar binary trees and the big cell corresponds to the corolla. More precisely the cells of dimension k of $\mathcal{K}^n$ are in bijection with the planar trees having $n+2$ leaves and $n+1-k$ vertices.

EXAMPLES.

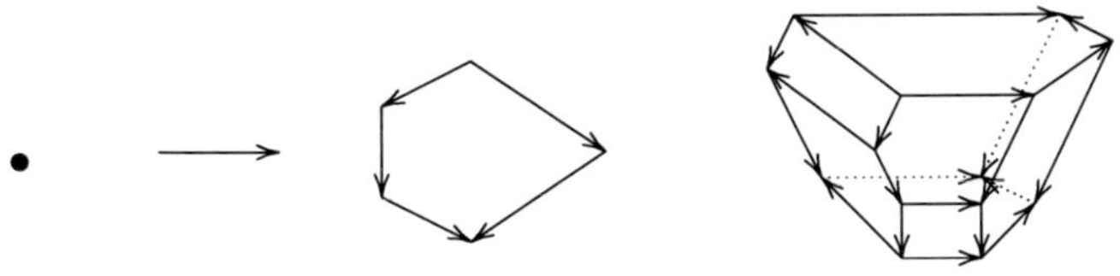

$\mathcal{K}^0$ $\mathcal{K}^1$ $\mathcal{K}^2$ $\mathcal{K}^3$

C.2.2 The Cubical Version of the Associahedron

For any tree $t \in PBT_{n+2}$ and any point $u \in I^n$ (here $I = [0,1] \subset \mathbb{R}$ is the unit interval) we associate a *metric tree* (t, u) by requiring that the edges have a length given by u. Observe that, when some of the coordinates of u are 0, the shape of the metric tree is not binary anymore. We quotient the disjoint union of the cubes $PBT_{n+2} \times I^n$ by the following equivalence relation:

$$(t, u) \sim (s, u)$$

if the associated metric trees are the same. For instance

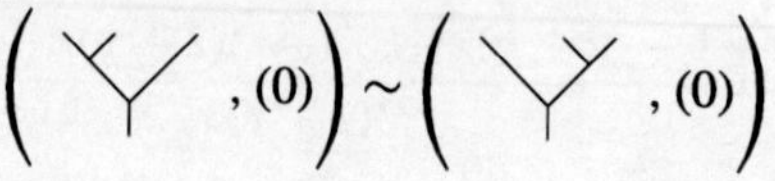

since both metric trees are the tree 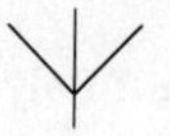 . The quotient $\mathcal{K}^n := PBT_{n+2} \times I^n / \sim$ gives a new cell decomposition of the Stasheff polytope denoted by $\mathcal{K}^n_{cub}$. It is due to Stasheff [Sta63] and was used in [BV73]. Observe that the top-cells are n-cubes and there are c_{n+2} of them:

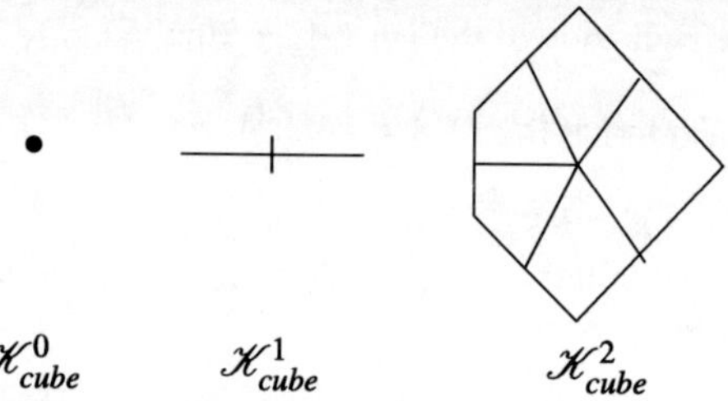

$\mathcal{K}^0_{cube}$ $\mathcal{K}^1_{cube}$ $\mathcal{K}^2_{cube}$

We are going to describe combinatorial objects which encode all the cells of this cellular decomposition.

C.2.3 Circled Trees

By definition a *circled planar tree* is a planar rooted tree in which each vertex has been replaced by a planar tree whose number of leaves is the same as the number of inputs of the vertex. Figure C.2 shows some examples.

The set of circled planar trees with n leaves is denoted by CPT_n.

Proposition C.2.1. *The set of circled planar trees encodes the cells of the cubical cell decomposition of the associahedron.*

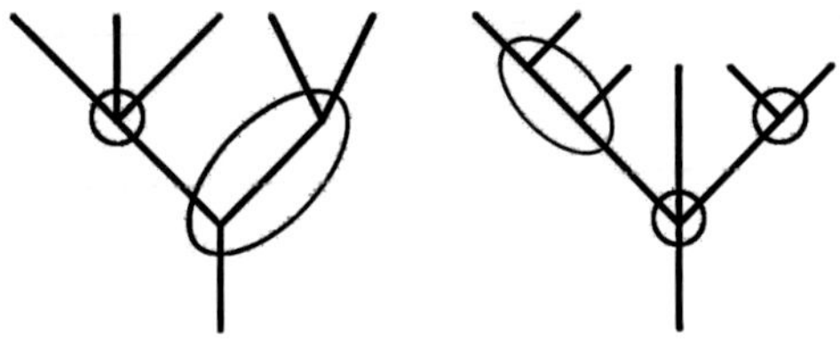

Fig. C.2 Examples of circled trees

Proof. It suffices to encode a given cube and to check that the equivalence relation described in Sect. C.2.1 is coherent with the encoding.

Let t be a planar tree. Choose a bijection between the set of vertices and the directions in the cube. A cell in the cube is determined by the choice of $0, \frac{1}{2}, 1$ in each direction. For instance, in the case of the interval I, the choice 0 encodes for the 0-cell $\{0\}$, the choice $\frac{1}{2}$ encodes for the 1-cell I, the choice 1 encodes for the 0-cell $\{1\}$. We construct a circled tree out of this data and the convention as follows. For a given direction, the choice of 0 shrinks the associated edge to a point. If the choice is $\frac{1}{2}$, then the two adjacent vertices of the edge are going to be encircled in the same circle. If the choice is 1, then the two adjacent vertices are going to be encircled by different circles.

EXAMPLES. See Fig. C.3.

It is immediate to check that the encoding is coherent with the equivalence relation $\sim$ described in Sect. C.2.1. □

From the construction of the cell-complexes $\mathscr{K}^n$ and $\mathscr{K}^n_{cub}$ we see that there is a natural cellular map $\mathscr{K}^n \to \mathscr{K}^n_{cub}$. At the chain complex level it induces a graded map

$$C_\bullet(\mathscr{K}^n) \to C_\bullet(\mathscr{K}^n_{cub}), \qquad \text{big cell} \mapsto \sum \text{top-cells}.$$

C.3 Trees and Reduced Trees

Formally a tree is a connected graph whose geometric realization is acyclic, see Sect. C.4.1. All the trees involved in this book are *rooted trees*: there is one preferred outer edge. As a consequence each vertex has a preferred flag, which we call the *output*. The other flags of the vertex are the *inputs*.

C.3.1 Planar Trees

A rooted tree is planar when there is given a total order on the set of inputs at each vertex. In the preceding sections of this appendix the vertices of the trees were at least trivalent (one output and at least two inputs). But when the operad has unary

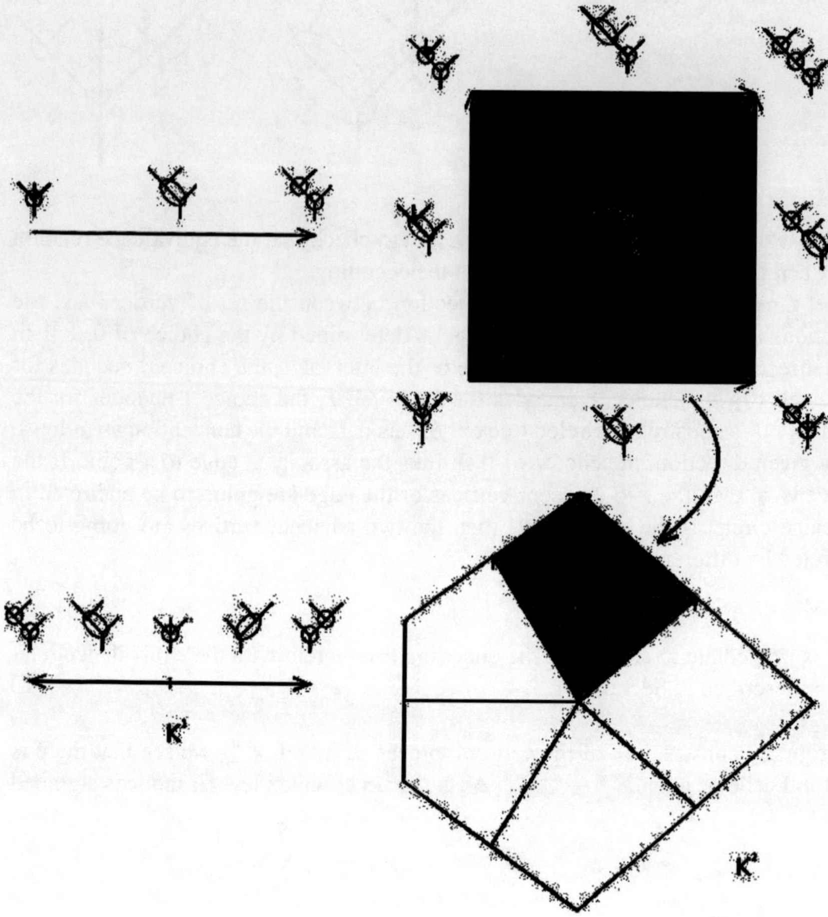

Fig. C.3 $\mathcal{K}^2$ with circled trees

operations (on top of the identity), and/or 0-ary operations, then one needs to allow vertices with one input and/or vertices with no input. Here are examples:

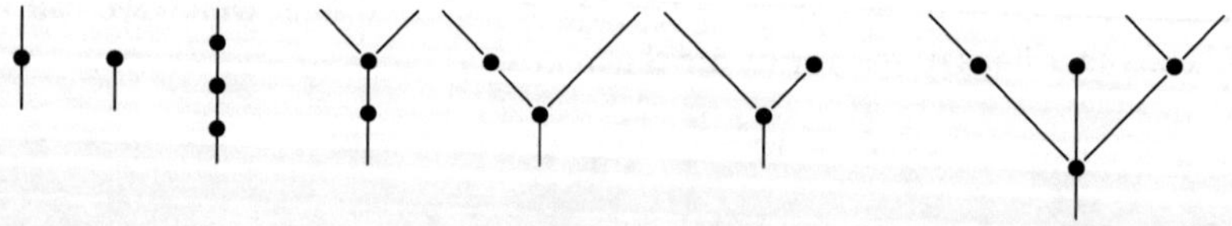

The trees whose vertices have at least one input at each vertex are called *reduced trees* in this book. We still denote the set of reduced planar trees with n-leaves by PT_n when there is no ambiguity.

C.3.2 Trees

Ignoring the order on the set of inputs, we get the notion of *nonplanar rooted trees* or simply *rooted trees*. In this case, in order to keep track of the structure of the tree, one often needs to label the inputs of a given vertex by a given labeling set, which can be $\{1, \dots, k\}$, where k is the number of inputs.

It should be noted that the number of rooted trees (planar or not) with a fixed number of leaves is finite when the valence is greater than or equal to 3. However this is no longer true when the valence is greater than or equal to 2 (resp. 1). For instance there is a countable number of ladders (one leaf):

We denote the set of (nonplanar) rooted trees with n leaves by RT_n. Restriction to trees which are at least trivalent (resp. bivalent, resp. univalent) should be either said or clear from the context.

C.4 Graphs

We give a formal definition of graphs. The definition of trees is a particular case.

C.4.1 Graphs

A *finite graph* Γ is determined by two finite sets: the set of vertices $\mathrm{vert}(\Gamma)$, the set of flags (or half-edges) $\mathrm{flag}(\Gamma)$ and two maps $s : \mathrm{flag}(\Gamma) \to \mathrm{vert}(\Gamma)$ and $\sigma : \mathrm{flag}(\Gamma) \to \mathrm{flag}(\Gamma)$ such that $\sigma^2 = \mathrm{id}$.

An element $u \in \mathrm{flag}(\Gamma)$ which is stable under σ, that is $\sigma(u) = u$, is called an *outer edge*. A geometric realization of the graph Γ is obtained as follows. For each element $x \in \mathrm{vert}(\Gamma)$ we take a point, for each element $u \in \mathrm{flag}(\Gamma)$ we take an interval $[0, 1]$. Then we identify the point 0 of u with the point $s(u)$ and we identify the point 1 of u with the point 1 of $\sigma(u)$ (observe that if u is stable this identification has no effect). The geometric realization of Γ is a 1-CW-complex.

For instance, if $\mathrm{vert}(\Gamma) = \{x, y\}$, $\mathrm{flag}(\Gamma) = \{a, b, c, d, e, f\}$ and, s, σ are given by

$-$	a	b	c	d	e
$s(-)$	x	x	x	y	y
$\sigma(-)$	a	b	d	c	e

then the geometric realization of Γ is

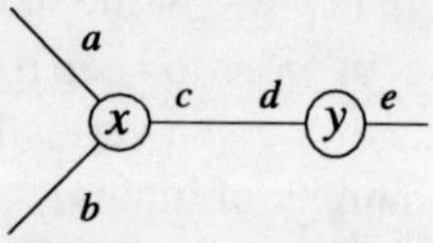

There are obvious notions of *outer edges* (made of only one flag) and *internal edges* (made of two flags). A graph is said to be connected whenever its geometric realization is connected. A morphism of graphs $\Gamma \to \Gamma'$ consists of a pair of maps $\mathrm{vert}(\Gamma) \to \mathrm{vert}(\Gamma')$ and $\mathrm{flag}(\Gamma) \to \mathrm{flag}(\Gamma')$ which commutes to the structure maps s, s' and σ, σ' respectively.

By definition a *tree* (or *abstract tree* or *nonplanar tree*) is a connected graph whose geometric realization has no cycle (acyclic cell complex). A *rooted tree* is a tree with a preferred outer edge which is called the *root*. In this case the other outer edges are called the *leaves*. A rooted tree can be oriented. We choose to put the leaves on top and the root at the bottom (like in nature). This orientation, from top to bottom, permits us to define the set of *inputs* of a given vertex and the output of this vertex, as flags of this vertex. The union of two (different) flags at 1 is called an edge (or sometimes *internal edge*).

If we put a total order on each set of inputs, then we get the aforementioned notion of planar tree.

Notice that an isomorphism of trees is made up of two bijections between the set of vertices and the set of flags respectively. When a tree admits no vertex without incoming edges (leaves or internal edges), then any of its automorphisms is completely characterized by the bijection of its leaves.

Given a set X and a bijection of the set of leaves with X, we have a "tree labeled by X". The set of isomorphism classes of trees labeled by X is denoted by $RT(X)$.

C.4.2 Graph Complexes

A *graph complex* is a chain complex whose module of chains is spanned by some graphs (with no outer edges) and the boundary map is obtained via the contraction of edges (one at a time). Though several classical chain complexes were recognized lately as being examples of graph complexes, the very first example appeared in the work of Kontsevich [Kon93] (see Example 3).

EXAMPLE 1 (The Hochschild complex). Let A be an associative algebra and consider the space of "ladders" whose vertices are decorated by elements of A:

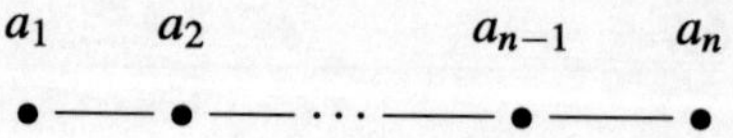

So the space of n-chains is $A^{\otimes n}$. The contraction of the ith edge gives a new ladder whose ith vertex is decorated by the product $a_i a_{i+1}$. The boundary map b' is the alternating sum of these contractions, cf. Sect. 9.1.6.

EXAMPLE 2 (The cochain complex of the Stasheff polytope). Let n be an integer and let C^k be the space spanned by the planar rooted trees with $n+2$ leaves and $n+1-k$ vertices ($0 \leq k \leq n$). Contracting an internal edge gives an element in C^{k+1}. The sum over all the internal edges (with ad hoc signs, see below for the treatment of the signs) gives rise to a boundary map, so, to a cochain complex which can be identified with the cochain complex of the Stasheff polytope:

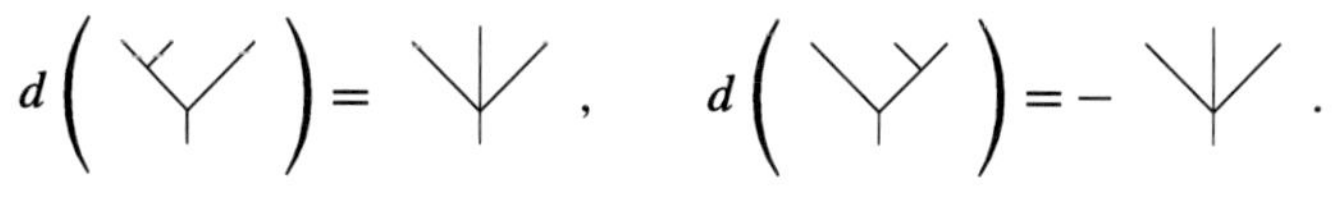

EXAMPLE 3 (Kontsevich graph complex for *Com*). Consider the space spanned by the connected graphs (no outer edges) with n edges. Let C_n be the quotient by the graphs which have at least one loop. Contracting an edge of such a graph gives an element in C_{n-1} (possibly 0 if a loop has been created in the process). The sum over all the edges (with appropriate signs) of these contractions, gives rise to a boundary map, and so to a chain complex. This is Kontsevich's original graph complex [Kon93] appearing in a computation of homology groups related to the operad *Com*.

Handling Signs in Graph Complexes Here is an efficient way to handle the signs in graph complexes. First, we consider the space of graphs with an orientation of each edge and a labeling of the vertices by the integers $\{1, 2, \ldots, k\}$ where k is the number of vertices. Then we mod out by the following equivalence relation:

- reversing an arrow changes the sign in front of the element,
- exchanging the labelings i and $i+1$ changes the sign in front of the element.

Let Γ be such a graph and let e be an internal edge whose end (resp. source) vertex is labeled by the integer j (resp. i). One can always suppose that $i < j$. The boundary map is

$$d(\Gamma) = \sum_e (-1)^j \Gamma/e$$

where Γ/e is the graph obtained by contracting e, keeping the label i, deleting the label j and normalizing the labels (each label u is replaced by $u-1$ for $u > j$). It is a good exercise to check that $d^2 = 0$, see for instance [Bur10].

References

Ada56. J. F. Adams, *On the cobar construction*, Proc. Nat. Acad. Sci. U.S.A. **42** (1956), 409–412. 41, 412

AL04. M. Aguiar and J.-L. Loday, *Quadri-algebras*, J. Pure Appl. Algebra **191** (2004), no. 3, 205–221. 531

AL07. M. Aguiar and M. Livernet, *The associative operad and the weak order on the symmetric groups*, J. Homotopy Relat. Struct. **2** (2007), no. 1, 57–84 (electronic). 190

AM99. G. Arone and M. Mahowald, *The Goodwillie tower of the identity functor and the unstable periodic homotopy of spheres*, Invent. Math. **135** (1999), no. 3, 743–788. 308

AM10. M. Aguiar and S. Mahajan, *Monoidal functors, species and Hopf algebras*, CRM Monograph Series, vol. 29, 2010. 129

And74. M. André, *Homologie des algèbres commutatives*, Springer-Verlag, Berlin, 1974, Die Grundlehren der mathematischen Wissenschaften, Band 206. 428, 447

AP10. M. Ammar and N. Poncin, *Coalgebraic approach to the Loday infinity category, stem differential for $2n$-ary graded and homotopy algebras*, Ann. Inst. Fourier (Grenoble) **60** (2010), no. 1, 355–387. 522

Bén63. Jean Bénabou, *Catégories avec multiplication*, C. R. Acad. Sci. Paris **256** (1963), 1887–1890. 580

Bac76. K. Baclawski, *Homology and combinatorics of ordered sets*, Ph.D. thesis, Harvard Univ., Cambridge, Mass., 1976. 302

Bac83. J. Backelin, *A distributiveness property of augmented algebras, and some related homological results*, Ph.D. thesis, Stockholm Univ., 1983. 90, 110, 111

Bal97. David Balavoine, *Deformations of algebras over a quadratic operad*, Operads: Proceedings of Renaissance Conferences (Hartford, CT/Luminy, 1995), Contemp. Math., vol. 202, Amer. Math. Soc., Providence, RI, 1997, pp. 207–234. 428, 433

Bal98. ———, *Homology and cohomology with coefficients, of an algebra over a quadratic operad*, J. Pure Appl. Algebra **132** (1998), no. 3, 221–258. 428, 448, 475

Bar68. M. Barr, *Harrison homology, Hochschild homology and triples*, J. Algebra **8** (1968), 314–323. 431

Bat98. M. A. Batanin, *Homotopy coherent category theory and A_∞-structures in monoidal categories*, J. Pure Appl. Algebra **123** (1998), no. 1-3, 67–103. 376

BB69. M. Barr and J. Beck, *Homology and standard constructions*, Sem. on Triples and Categorical Homology Theory (ETH, Zürich, 1966/67), Springer, Berlin, 1969, pp. 245–335. 470

BB09. M. A. Batanin and C. Berger, *The lattice path operad and Hochschild cochains*, Alpine perspectives on algebraic topology, Contemp. Math., vol. 504, Amer. Math. Soc., Providence, RI, 2009, pp. 23–52. 510

J.-L. Loday, B. Vallette, *Algebraic Operads*,
Grundlehren der mathematischen Wissenschaften 346,
DOI 10.1007/978-3-642-30362-3,

BC04. J. C. Baez and A. S. Crans, *Higher-dimensional algebra. VI. Lie 2-algebras*, Theory Appl. Categ. **12** (2004), 492–538 (electronic). 276

BCK+66. A. K. Bousfield, E. B. Curtis, D. M. Kan, D. G. Quillen, D. L. Rector, and J. W. Schlesinger, *The* mod $-$ *p lower central series and the Adams spectral sequence*, Topology **5** (1966), 331–342. 80, 82, 87, 117

BD04. A. Beilinson and V. Drinfeld, *Chiral algebras*, American Mathematical Society Colloquium Publications, vol. 51, American Mathematical Society, Providence, RI, 2004. 534

BE74. M. G. Barratt and Peter J. Eccles, *Γ^+-structures. I. A free group functor for stable homotopy theory*, Topology **13** (1974), 25–45. 487

Bec69. Jon Beck, *Distributive laws*, Sem. on Triples and Categorical Homology Theory (ETH, Zürich, 1966/67), Springer, Berlin, 1969, pp. 119–140. 288, 290, 322

Ber78. G. M. Bergman, *The diamond lemma for ring theory*, Adv. in Math. **29** (1978), no. 2, 178–218. 90, 103, 104

Ber96. Clemens Berger, *Opérades cellulaires et espaces de lacets itérés*, Ann. Inst. Fourier (Grenoble) **46** (1996), no. 4, 1125–1157. 488

Ber09. A. Berglund, *Homological perturbation theory for algebras over operads*, arXiv:0909.3485 (2009). 376

BF85. J. Backelin and R. Fröberg, *Koszul algebras, Veronese subrings and rings with linear resolutions*, Rev. Roumaine Math. Pures Appl. **30** (1985), no. 2, 85–97. 112

BF04. C. Berger and B. Fresse, *Combinatorial operad actions on cochains*, Math. Proc. Cambridge Philos. Soc. **137** (2004), no. 1, 135–174. 487, 488, 510

BFLS98. G. Barnich, R. Fulp, T. Lada, and J. Stasheff, *The sh Lie structure of Poisson brackets in field theory*, Comm. Math. Phys. **191** (1998), no. 3, 585–601. 444

BG96. A. Braverman and D. Gaitsgory, *Poincaré-Birkhoff-Witt theorem for quadratic algebras of Koszul type*, J. Algebra **181** (1996), no. 2, 315–328. 79

BGG78. I. N. Bernšteĭn, I. M. Gel'fand, and S. I. Gel'fand, *Algebraic vector bundles on* $\mathbf{P}^n$ *and problems of linear algebra*, Funktsional. Anal. i Prilozhen. **12** (1978), no. 3, 66–67. 551

BGS82. A. Björner, A. M. Garsia, and R. P. Stanley, *An introduction to Cohen-Macaulay partially ordered sets*, Ordered sets (Banff, Alta., 1981), NATO Adv. Study Inst. Ser. C: Math. Phys. Sci., vol. 83, Reidel, Dordrecht, 1982, pp. 583–615. 302

BJT97. H.-J. Baues, M. Jibladze, and A. Tonks, *Cohomology of monoids in monoidal categories*, Operads: Proceedings of Renaissance Conferences (Hartford, CT/Luminy, 1995) (Providence, RI), Contemp. Math., vol. 202, Amer. Math. Soc., 1997, pp. 137–165. 154, 155, 477

BL10. N. Bergeron and M. Livernet, *The non-symmetric operad pre-Lie is free*, JPAA **214** (2010), 1165–1172. 183

BL11. N. Bergeron and J.-L. Loday, *The symmetric operation in a free pre-Lie algebra is magmatic*, Proc. Amer. Math. Soc. **139** (2011), no. 5, 1585–1597. 516

BM03a. C. Berger and I. Moerdijk, *Axiomatic homotopy theory for operads*, Comment. Math. Helv. **78** (2003), no. 4, 805–831. 591, 592, 209, 343, 376

BM03b. D. Blanc and M. Markl, *Higher homotopy operations*, Math. Z. **245** (2003), no. 1, 1–29. 356

BM08. D. V. Borisov and Y. I. Manin, *Generalized operads and their inner cohomomorphisms*, Geometry and dynamics of groups and spaces, Progr. Math., vol. 265, Birkhäuser, Basel, 2008, pp. 247–308. 159, 315, 562

BMR04. H.-J. Baues, E. G. Minian, and B. Richter, *Crossed modules over operads and operadic cohomology*, K-Theory **31** (2004), no. 1, 39–69. 469

BR10. E. Burgunder and M. Ronco, *Tridendriform structure on combinatorial Hopf algebras*, J. Algebra **324** (2010), no. 10, 2860–2883. MR 2725205 (2011m:16079) 531

Bro59. E. H. Brown, Jr., *Twisted tensor products. I*, Ann. of Math. (2) **69** (1959), 223–246. 37, 48, 55

BT82. Raoul Bott and Loring W. Tu, *Differential forms in algebraic topology*, Graduate Texts in Mathematics, vol. 82, Springer-Verlag, New York, 1982. MR 658304 (83i:57016) 28

Buc06. B. Buchberger, *An algorithm for finding the basis elements of the residue class ring of a zero dimensional polynomial ideal*, J. Symbolic Comput. **41** (2006), no. 3-4, 475–511, Translated from the 1965 German original by Michael P. Abramson. 106

Bur06. Dietrich Burde, *Left-symmetric algebras, or pre-Lie algebras in geometry and physics*, Cent. Eur. J. Math. **4** (2006), no. 3, 323–357 (electronic). 21

Bur10. Emily Burgunder, *A symmetric version of Kontsevich graph complex and Leibniz homology*, J. Lie Theory **20** (2010), no. 1, 127–165. 607

BV73. J. M. Boardman and R. M. Vogt, *Homotopy invariant algebraic structures on topological spaces*, Lecture Notes in Mathematics, Vol. 347, Springer-Verlag, Berlin, 1973. 592, 602, 149

BV81. I. A. Batalin and G. A. Vilkovisky, *Gauge algebra and quantization*, Phys. Lett. B **102** (1981), no. 1, 27–31. 534

Car56. Pierre Cartier, *Hyperalgèbres et groupes formels*, Séminaire "Sophus Lie" **2** (1955–56), Exposé No. 2, 1–6. 18

Car58. Henri Cartan, *Théorie des fibrés principaux*, Séminaire Henri Cartan; 9e année: 1956/57. Quelques questions de topologie, Exposé no. 4, Secrétariat mathématique, Paris, 1958, p. 12. 38

Car55. ———, *Dga-modules (suite), notion de construction*, Séminaire Henri Cartan (7) **2** (1954–55), Exposé No. 3. 37, 41, 51, 55

CE48. C. Chevalley and S. Eilenberg, *Cohomology theory of Lie groups and Lie algebras*, Trans. Amer. Math. Soc. **63** (1948), 85–124. 81, 494

CE56. H. Cartan and S. Eilenberg, *Homological algebra*, Princeton University Press, Princeton, N. J., 1956. 592, 80, 81

CFK01. I. Ciocan-Fontanine and M. Kapranov, *Derived Quot schemes*, Ann. Sci. École Norm. Sup. (4) **34** (2001), no. 3, 403–440. 440

CG08. X. Z. Cheng and E. Getzler, *Transferring homotopy commutative algebraic structures*, J. Pure Appl. Algebra **212** (2008), no. 11, 2535–2542. 486

Cha01a. Frédéric Chapoton, *Algèbres pré-Lie et algèbres de Hopf liées à la renormalisation*, C. R. Acad. Sci. Paris Sér. I Math. **332** (2001), no. 8, 681–684. 184

Cha01b. ———, *Un endofoncteur de la catégorie des opérades*, Dialgebras and related operads, Lecture Notes in Math., vol. 1763, Springer, Berlin, 2001, pp. 105–110. 517, 530

Cha02. ———, *Opérades différentielles graduées sur les simplexes et les permutoèdres*, Bull. Soc. Math. France **130** (2002), no. 2, 233–251. 531

Chi05. Michael Ching, *Bar constructions for topological operads and the Goodwillie derivatives of the identity*, Geom. Topol. **9** (2005), 833–933. 308

CK98. A. Connes and D. Kreimer, *Hopf algebras, renormalization and noncommutative geometry*, Comm. Math. Phys. **199** (1998), no. 1, 203–242. 308, 519

CL01. F. Chapoton and M. Livernet, *Pre-Lie algebras and the rooted trees operad*, Internat. Math. Res. Notices (2001), no. 8, 395–408. 515, 517, 519

CL07. ———, *Relating two Hopf algebras built from an operad*, Int. Math. Res. Not. IMRN (2007), no. 24. 308

CL10. J. Chuang and A. Lazarev, *Feynman diagrams and minimal models for operadic algebras*, J. Lond. Math. Soc. (2) **81** (2010), no. 2, 317–337. 376

CLP02. J. M. Casas, J.-L. Loday, and T. Pirashvili, *Leibniz n-algebras*, Forum Math. **14** (2002), no. 2, 189–207. 546

Coh76. F. R. Cohen, *The homology of $\mathscr{C}_{n+1}$-spaces*, The homology of iterated loop spaces (Cohen, Frederick R., Lada, Thomas J. and May, J. Peter), Lecture Notes in Mathematics, Vol. 533, Springer-Verlag, Berlin, 1976, pp. vii+490. 493, 509, 511

Con85. Alain Connes, *Noncommutative differential geometry*, Inst. Hautes Études Sci. Publ. Math. (1985), no. 62, 257–360. 587, 336, 461

Cos07. Kevin Costello, *Topological conformal field theories and Calabi–Yau categories*, Adv. Math. **210** (2007), no. 1, 165–214. 539

Cur12. P.-L. Curien, *Operads, clones and distributive laws*, Proc. Int. Conf., in Nankai Series in Pure, Applied Mathematics and Theoretical Physics **9** (2012), 25–49. 291

DCV11. G. Drummond-Cole and B. Vallette, *Minimal model for the Batalin–Vilkovisky operad*, ArXiv:1105.2008 (2011). 202, 532, 537

Del93. Pierre Deligne, *Letter to Stasheff, Gerstenhaber, May, Schechtman, Drinfeld.* 510

DGMS75. P. Deligne, P. Griffiths, J. Morgan, and D. Sullivan, *Real homotopy theory of Kähler manifolds*, Invent. Math. **29** (1975), no. 3, 245–274. 29

DK07. V. Dotsenko and A. Khoroshkin, *Character formulas for the operad of a pair of compatible brackets and for the bi-Hamiltonian operad*, Funktsional. Anal. i Prilozhen. **41** (2007), no. 1, 1–22, 96. 308

DK09. ———, *Free resolutions via Gröbner bases*, arXiv:0912.4895 (2009). 537

DK10. ———, *Gröbner bases for operads*, Duke Math. J. **153** (2010), no. 2, 363–396. 264, 267, 268, 278, 283, 286, 287

DM69. P. Deligne and D. Mumford, *The irreducibility of the space of curves of given genus*, Inst. Hautes Études Sci. Publ. Math. (1969), no. 36, 75–109. 545

Dok09. Ioannis Dokas, *Zinbiel algebras and commutative algebras with divided powers*, Glasg. Math. J. **51** (2009), no. 10, 1–11. 521

Dot07. Vladimir Dotsenko, *An operadic approach to deformation quantization of compatible Poisson brackets. I*, J. Gen. Lie Theory Appl. **1** (2007), no. 2, 107–115 (electronic). 288

DS95. W. G. Dwyer and J. Spaliński, *Homotopy theories and model categories*, Handbook of algebraic topology, North-Holland, Amsterdam, 1995, pp. 73–126. 588

DTT07. V. Dolgushev, D. Tamarkin, and B. Tsygan, *The homotopy Gerstenhaber algebra of Hochschild cochains of a regular algebra is formal*, J. Noncommut. Geom. **1** (2007), no. 1, 1–25. 405, 510

Dup76. Johan L. Dupont, *Simplicial de Rham cohomology and characteristic classes of flat bundles*, Topology **15** (1976), no. 3, 233–245. 486

Dup78. ———, *Curvature and characteristic classes*, Lecture Notes in Mathematics, Vol. 640, Springer-Verlag, Berlin, 1978. 486

Dzh99. A. Dzhumadil'daev, *Cohomologies and deformations of right-symmetric algebras*, J. Math. Sci. (New York) **93** (1999), no. 6, 836–876, Algebra, 11. 475

EFG05. K. Ebrahimi-Fard and L. Guo, *On products and duality of binary, quadratic, regular operads*, J. Pure Appl. Algebra **200** (2005), no. 3, 293–317. 316, 531

EFG07. ———, *Coherent unit actions on regular operads and Hopf algebras*, Theory Appl. Categ. **18** (2007), 348–371. 553

EML53. S. Eilenberg and S. Mac Lane, *On the groups of $H(\Pi, n)$. I*, Ann. of Math. (2) **58** (1953), 55–106. 41

Far79. F. D. Farmer, *Cellular homology for posets*, Math. Japon. **23** (1978/79), no. 6, 607–613. 302

FBZ04. E. Frenkel and D. Ben-Zvi, *Vertex algebras and algebraic curves*, second ed., Mathematical Surveys and Monographs, vol. 88, American Mathematical Society, Providence, RI, 2004. 538

FHT01. Y. Félix, S. Halperin, and J.-C. Thomas, *Rational homotopy theory*, Graduate Texts in Mathematics, vol. 205, Springer-Verlag, New York, 2001. 417

FL91. Z. Fiedorowicz and J.-L. Loday, *Crossed simplicial groups and their associated homology*, Trans. Amer. Math. Soc. **326** (1991), no. 1, 57–87. 586, 151, 336

FM97. T. F. Fox and M. Markl, *Distributive laws, bialgebras, and cohomology*, Operads: Proceedings of Renaissance Conferences (Hartford, CT/Luminy, 1995), Contemp. Math., vol. 202, Amer. Math. Soc., Providence, RI, 1997, pp. 167–205. 288, 556

FOOO09a. K. Fukaya, Y.-G. Oh, H. Ohta, and K. Ono, *Lagrangian intersection Floer theory: anomaly and obstruction. Part I*, AMS/IP Studies in Advanced Mathematics, vol. 46, American Mathematical Society, Providence, RI, 2009. 552

FOOO09b. ——— , *Lagrangian intersection Floer theory: anomaly and obstruction. Part II*, AMS/IP Studies in Advanced Mathematics, vol. 46, American Mathematical Society, Providence, RI, 2009. 552

Fra01. Alessandra Frabetti, *Dialgebra (co)homology with coefficients*, Dialgebras and related operads, Lecture Notes in Math., vol. 1763, Springer, Berlin, 2001, pp. 67–103. 475

Fra08. ——— , *Groups of tree-expanded series*, J. Algebra **319** (2008), no. 1, 377–413. 184

Fre00. Benoit Fresse, *On the homotopy of simplicial algebras over an operad*, Trans. Amer. Math. Soc. **352** (2000), no. 9, 4113–4141. 130, 136, 137, 138, 139, 489, 502

Fre04. ——— , *Koszul duality of operads and homology of partition posets*, in "Homotopy theory and its applications (Evanston, 2002)", Contemp. Math. **346** (2004), 115–215. 194, 213, 220, 221, 222, 229, 232, 297, 303, 306, 307, 431

Fre06. ——— , *Théorie des opérades de Koszul et homologie des algèbres de Poisson*, Ann. Math. Blaise Pascal **13** (2006), no. 2, 237–312. 475, 505, 506

Fre09a. ——— , *Modules over operads and functors*, Lecture Notes in Mathematics, vol. 1967, Springer-Verlag, Berlin, 2009. 591, 592, xx, 132, 428, 448, 461, 462, 464, 470

Fre09b. ——— , *Operadic cobar constructions, cylinder objects and homotopy morphisms of algebras over operads*, Alpine perspectives on algebraic topology, Contemp. Math., vol. 504, Amer. Math. Soc., Providence, RI, 2009, pp. 125–188. 376

Fre10. ——— , *Koszul duality complexes for the cohomology of iterated loop spaces of spheres*, ArXiv e-prints (2010). 513

Fre11a. ——— , *Iterated bar complexes of E-infinity algebras and homology theories*, Algebr. Geom. Topol. **11** (2011), no. 2, 747–838. 513

Fre11b. ——— , *Koszul duality of E_n-operads*, Selecta Math. (N.S.) **17** (2011), no. 2, 363–434. MR 2803847 488, 512

Frö99. R. Fröberg, *Koszul algebras*, Advances in commutative ring theory (Fez, 1997), Lecture Notes in Pure and Appl. Math., vol. 205, Dekker, New York, 1999, pp. 337–350. 62, 66, 72

Gan03. Wee Liang Gan, *Koszul duality for dioperads*, Math. Res. Lett. **10** (2003), no. 1, 109–124. 563

GCTV09. I. Galvez-Carrillo, A. Tonks, and B. Vallette, *Homotopy Batalin-Vilkovisky algebras*, Journal Noncommutative Geometry (2009), arXiv:0907.2246, 49 pp. 253, 257, 444, 445, 474, 506, 510, 532, 537, 538, 539

Ger63. M. Gerstenhaber, *The cohomology structure of an associative ring*, Ann. of Math. (2) **78** (1963), 267–288. 21, 178, 183, 184, 364, 428, 433, 436, 506, 507, 547, 561

Ger64. ——— , *On the deformation of rings and algebras*, Ann. of Math. (2) **79** (1964), 59–103. 433

Get94. E. Getzler, *Batalin-Vilkovisky algebras and two-dimensional topological field theories*, Comm. Math. Phys. **159** (1994), no. 2, 265–285. 257, 507, 509, 532, 533, 536, 537, 546

Get95. ——— , *Operads and moduli spaces of genus* 0 *Riemann surfaces*, The moduli space of curves (Texel Island, 1994). Progr. Math., vol. 129, Birkhäuser Boston, Boston, MA, 1995, pp. 199–230. 229, 232, 545, 546

Get09. ——— , *Lie theory for nilpotent L_∞-algebras*, Ann. of Math. (2) **170** (2009), no. 1, 271–301. 486

GH00. P. G. Goerss and M. J. Hopkins, *André-Quillen (co)-homology for simplicial algebras over simplicial operads*, Une dégustation topologique [Topological morsels]: homotopy theory in the Swiss Alps (Arolla, 1999), Contemp. Math., vol. 265, Amer. Math. Soc., Providence, RI, 2000, pp. 41–85. 428, 448, 456

Gin06. V. Ginzburg, *Calabi-Yau algebras*, arXiv:math/0612139 (2006). 535

GJ94. E. Getzler and J. D. S. Jones, *Operads, homotopy algebra and iterated integrals for double loop spaces*, hep-th/9403055 (1994). 161, 194, 214, 220, 229, 360, 405, 427, 428, 509, 511, 548

GK94. V. Ginzburg and M. Kapranov, *Koszul duality for operads*, Duke Math. J. **76** (1994), no. 1, 203–272. 194, 214, 229, 264, 308, 360, 427, 428, 448, 497

GK95a. E. Getzler and M. M. Kapranov, *Cyclic operads and cyclic homology*, Geometry, topology, & physics, Conf. Proc. Lecture Notes Geom. Topology, IV, Int. Press, Cambridge, MA, 1995, pp. 167–201. 159, 232, 563

GK95b. V. Ginzburg and M. Kapranov, *Erratum to: "Koszul duality for operads"*, Duke Math. J. **80** (1995), no. 1, 293. 264, 308

GK98. E. Getzler and M. M. Kapranov, *Modular operads*, Compositio Math. **110** (1998), no. 1, 65–126. 565

GL89. R. Grossman and R. G. Larson, *Hopf-algebraic structure of families of trees*, J. Algebra **126** (1989), no. 1, 184–210. 519

GM88. W. M. Goldman and J. J. Millson, *The deformation theory of representations of fundamental groups of compact Kähler manifolds*, Inst. Hautes Études Sci. Publ. Math. (1988), no. 67, 43–96. 496, 497, 500

GM03. S. I. Gelfand and Y. I. Manin, *Methods of homological algebra*, second ed., Springer Monographs in Mathematics, Springer-Verlag, Berlin, 2003. 593

Gne97. A. Victor Gnedbaye, *Opérades des algèbres* $(k+1)$*-aires*, Operads: Proceedings of Renaissance Conferences (Hartford, CT/Luminy, 1995), Contemp. Math., vol. 202, 1997, pp. 83–113. 545

Gri04. P.-P. Grivel, *Une histoire du théorème de Poincaré-Birkhoff-Witt*, Expo. Math. **22** (2004), no. 2, 145–184. 18

Gro57. A. Grothendieck, *Sur quelques points d'algèbre homologique*, Tôhoku Math. J. (2) **9** (1957), 119–221. 592

GS07. P. Goerss and K. Schemmerhorn, *Model categories and simplicial methods*, Interactions between homotopy theory and algebra, Contemp. Math., vol. 436, Amer. Math. Soc., Providence, RI, 2007, pp. 3–49. 588

GS10. J. Giansiracusa and P. Salvatore, *Formality of the framed little 2-discs operad and semidirect products*, Homotopy theory of function spaces and related topics, Contemp. Math., vol. 519, Amer. Math. Soc., Providence, RI, 2010, pp. 115–121. 538

GV95. M. Gerstenhaber and A. A. Voronov, *Homotopy G-algebras and moduli space operad*, Internat. Math. Res. Notices (1995), no. 3, 141–153 (electronic). 447, 509, 547, 548

GZ67. P. Gabriel and M. Zisman, *Calculus of fractions and homotopy theory*, Ergebnisse der Mathematik und ihrer Grenzgebiete, Band 35, Springer-Verlag, New York, Inc., New York, 1967. 588

Har62. D. K. Harrison, *Commutative algebras and cohomology*, Trans. Amer. Math. Soc. **104** (1962), 191–204. 431, 484

Hin97. Vladimir Hinich, *Homological algebra of homotopy algebras*, Comm. Algebra **25** (1997), no. 10, 3291–3323, see also Erratum arXiv:math/0309453. 591, 428, 448, 462

Hin03. ———, *Tamarkin's proof of Kontsevich formality theorem*, Forum Math. **15** (2003), no. 4, 591–614. 510

HK91. J. Huebschmann and T. Kadeishvili, *Small models for chain algebras*, Math. Z. **207** (1991), no. 2, 245–280. 352

HKR62. G. Hochschild, B. Kostant, and A. Rosenberg, *Differential forms on regular affine algebras*, Trans. Amer. Math. Soc. **102** (1962), 383–408. 508

HM10. J. Hirsch and J. Millès, *Curved Koszul duality theory*, arXiv:1008.5368 (2010). 252, 405, 445, 552

HMS74. D. Husemoller, J. C. Moore, and J. Stasheff, *Differential homological algebra and homogeneous spaces*, J. Pure Appl. Algebra **5** (1974), 113–185. 37, 38, 49

Hof10a. Eric Hoffbeck, *Γ-homology of algebras over an operad*, Algebr. Geom. Topol. **10** (2010), no. 3, 1781–1806. MR 2683753 (2011f:16022) 470

Hof10b. ———, *Obstruction theory for algebras over an operad*, ArXiv e-prints (2010). 470

Hof10c. ———, *A Poincaré-Birkhoff-Witt criterion for Koszul operads*, Manuscripta Math. **131** (2010), no. 1-2, 87–110. 264, 267, 278, 283, 286

Hol06. Ralf Holtkamp, *On Hopf algebra structures over free operads*, Adv. Math. **207** (2006), no. 2, 544–565. 553

Hov99. Mark Hovey, *Model categories*, Mathematical Surveys and Monographs, vol. 63, American Mathematical Society, Providence, RI, 1999. 588

HS10. K. Hess and J. Scott, *Twisting structures and strongly homotopy morphisms*, arXiv:1007.1934 (2010). 402, 403

HS93. V. Hinich and V. Schechtman, *Homotopy Lie algebras*, I. M. Gel'fand Seminar, Adv. Soviet Math., vol. 16, 93, pp. 1–28. 496

Hua97. Yi-Zhi Huang, *Two-dimensional conformal geometry and vertex operator algebras*, Progress in Mathematics, vol. 148, Birkhäuser Boston Inc., Boston, MA, 1997. 538

HW95. P. Hanlon and M. Wachs, *On Lie k-algebras*, Adv. Math. **113** (1995), no. 2, 206–236. 546

Jac62. Nathan Jacobson, *Lie algebras*, Interscience Tracts in Pure and Applied Mathematics, No. 10, Interscience Publishers (a division of John Wiley & Sons), New York–London, 1962. 139, 501

Jon99. Vaughan F. R. Jones, *Planar algebras, i*, arXiv:math/9909027 (1999). 561

Jon10. ———, *Quadratic tangles in planar algebras*, arXiv:1007.1158 (2010). 561

Joy86. A. Joyal, *Foncteurs analytiques et espèces de structures*, Combinatoire énumérative (Montreal, Que., 1985/Quebec, Que., 1985), Lecture Notes in Math., vol. 1234, Springer, Berlin, 1986, pp. 126–159. 129

JR82. S. A. Joni and G.-C. Rota, *Coalgebras and bialgebras in combinatorics*, Umbral calculus and Hopf algebras (Norman, Okla., 1978), Contemp. Math., vol. 6, Amer. Math. Soc., Providence, RI, 1982, pp. 1–47. 22

Kac98. Victor Kac, *Vertex algebras for beginners*, second ed., University Lecture Series, vol. 10, American Mathematical Society, Providence, RI, 1998. 538

Kad80. T. V. Kadeišvili, *On the theory of homology of fiber spaces*, Uspekhi Mat. Nauk **35** (1980), no. 3(213), 183–188. International Topology Conference (Moscow State Univ., Moscow, 1979). 347, 348

Kad82. ———, *The algebraic structure in the homology of an $A(\infty)$-algebra*, Soobshch. Akad. Nauk Gruzin. SSR **108** (1982), no. 2, 249–252 (1983). 352, 480

Kad88. ———, *The structure of the $A(\infty)$-algebra, and the Hochschild and Harrison cohomologies*, Trudy Tbiliss. Mat. Inst. Razmadze Akad. Nauk Gruzin. SSR **91** (1988), 19–27. 485

Kan58. Daniel M. Kan, *Adjoint functors*, Trans. Amer. Math. Soc. **87** (1958), 294–329. 577

Kap93. Mikhail M. Kapranov, *The permutoassociahedron, Mac Lane's coherence theorem and asymptotic zones for the KZ equation*, J. Pure Appl. Algebra **85** (1993), no. 2, 119–142. 247

Kar87. Max Karoubi, *Homologie cyclique et K-théorie*, Astérisque (1987), no. 149, 147. 461

Kas93. Takuji Kashiwabara, *On the homotopy type of configuration complexes*, Algebraic topology (Oaxtepec, 1991), Contemp. Math., vol. 146, 1993, pp. 159–170. 488

Kau07. R. M. Kaufmann, *On spineless cacti, Deligne's conjecture and Connes-Kreimer's Hopf algebra*, Topology **46** (2007), no. 1, 39–88. 510

Kau08. ———, *A proof of a cyclic version of Deligne's conjecture via Cacti*, Mathematical Research Letters **15** (2008), no. 5, 901–921. 539

Kel01. Bernhard Keller, *Introduction to A-infinity algebras and modules*, Homology Homotopy Appl. **3** (2001), no. 1, 1–35 (electronic). 340

Kel05. ———, *Deformation quantization after Kontsevich and Tamarkin*, Déformation, quantification, théorie de Lie, Panor. Synthèses, vol. 20, Soc. Math. France, Paris, 2005, pp. 19–62. 433, 501

Klj74. A. A. Kljačko, *Lie elements in a tensor algebra*, Sibirsk. Mat. Ž. **15** (1974), 1296–1304, 1430. MR 0371961 (51 #8178) 491

KM94. M. Kontsevich and Yu. Manin, *Gromov-Witten classes, quantum cohomology, and enumerative geometry*, Comm. Math. Phys. **164** (1994), no. 3, 525–562. 545

Knu83. Finn F. Knudsen, *The projectivity of the moduli space of stable curves. II. The stacks $M_{g,n}$*, Math. Scand. **52** (1983), no. 2, 161–199. 545

Kon93. Maxim Kontsevich, *Formal (non)commutative symplectic geometry*, The Gel'fand Mathematical Seminars, 1990–1992, Birkhäuser Boston, Boston, MA, 1993, pp. 173–187. 606, 607

Kon99. ———, *Operads and motives in deformation quantization*, Lett. Math. Phys. **48** (1999), no. 1, 35–72, Moshé Flato (1937–1998). 203

Kon03. ———, *Deformation quantization of Poisson manifolds*, Lett. Math. Phys. **66** (2003), no. 3, 157–216. 215, 490, 495, 497, 500, 501

Kos47. Jean-Louis Koszul, *Sur les opérateurs de dérivation dans un anneau*, C. R. Acad. Sci. Paris **225** (1947), 217–219. MR 0022345 (9,196h) 28

Kos50. ———, *Homologie et cohomologie des algèbres de Lie*, Bull. Soc. Math. France **78** (1950), 65–127. 81

Kos85. ———, *Crochet de Schouten-Nijenhuis et cohomologie*, Astérisque (1985), no. Numero Hors Serie, 257–271, The mathematical heritage of Élie Cartan (Lyon, 1984). MR 837203 (88m:17013) 532

KP11. O. Kravchenko and M. Polyak, *Diassociative algebras and Milnor's invariants for tangles*, Lett. Math. Phys. **95** (2011), no. 3, 297–316. MR 2775128 (2012a:57018) 526

Kra66. David Kraines, *Massey higher products*, Trans. Amer. Math. Soc. **124** (1966), 431–449. 350, 356

Kra00. Olga Kravchenko, *Deformations of Batalin-Vilkovisky algebras*, Poisson geometry (Warsaw, 1998), Banach Center Publ., vol. 51, Polish Acad. Sci., Warsaw, 2000, pp. 131–139. 537

KS96. Yvette Kosmann-Schwarzbach, *From Poisson algebras to Gerstenhaber algebras*, Ann. Inst. Fourier (Grenoble) **46** (1996), no. 5, 1243–1274. 519

KS00. M. Kontsevich and Y. Soibelman, *Deformations of algebras over operads and the Deligne conjecture*, Conférence Moshé Flato 1999, Vol. I (Dijon), Math. Phys. Stud., vol. 21, Kluwer Acad. Publ., Dordrecht, 2000, pp. 255–307. 352, 510

KS04. Yvette Kosmann-Schwarzbach, *Derived brackets*, Lett. Math. Phys. **69** (2004), 61–87. 519

KS09. M. Kontsevich and Y. Soibelman, *Notes on A_∞-algebras, A_∞-categories and noncommutative geometry*, Homological mirror symmetry, Lecture Notes in Phys., vol. 757, Springer, Berlin, 2009, pp. 153–219. 539

KS10. ———, *Deformation theory. I [Draft]*, http://www.math.ksu.edu/~soibel/Book-vol1.ps, 2010. 433, 447, 501

Lad76. T. Lada, *Strong homotopy algebras over monads*, The homology of iterated loop spaces (Cohen, Frederick R., Lada, Thomas J. and May, J. Peter), Lecture Notes in Mathematics, Vol. 533, Springer-Verlag, Berlin, 1976, pp. vii+490. 398

Lap01. S. V. Lapin, *Differential perturbations and D_∞-differential modules*, Mat. Sb. **192** (2001), no. 11, 55–76. 386

Laz55. Michel Lazard, *Lois de groupes et analyseurs*, Ann. Sci. Ecole Norm. Sup. (3) **72** (1955), 299–400. 121

Lei04. Tom Leinster, *Higher operads, higher categories*, London Mathematical Society Lecture Note Series, vol. 298, Cambridge University Press, Cambridge, 2004. MR 2094071 (2005h:18030) xx, 159

Ler46. Jean Leray, *Structure de l'anneau d'homologie d'une représentation*, C. R. Acad. Sci. Paris **222** (1946), 1419–1422. 28

LH03. K. Lefevre-Hasegawa, *Sur les A-infini catégories*, arXiv:math/0310337 (2003). 37, 49

Lis52. William G. Lister, *A structure theory of Lie triple systems*, Trans. Amer. Math. Soc. **72** (1952), 217–242. 546

Liv98a. Muriel Livernet, *Homotopie rationnelle des algèbres sur une opérade*, Prépublication de l'Institut de Recherche Mathématique Avancée [Prepublication of the Institute of Advanced Mathematical Research], 1998/32, Université Louis Pasteur Département de Mathématique Institut de Recherche Mathématique Avancée, Strasbourg, 1998, Thèse, Université Louis Pasteur (Strasbourg I), Strasbourg, 1998. 428

Liv98b. ———, *Rational homotopy of Leibniz algebras*, Manuscripta Math. **96** (1998), no. 3, 295–315. 405, 417

Liv99. ———, *On a plus-construction for algebras over an operad*, K-Theory **18** (1999), no. 4, 317–337. 470

Liv06. ———, *A rigidity theorem for pre-Lie algebras*, J. Pure Appl. Algebra **207** (2006), no. 1, 1–18. 308

LM95. T. Lada and M. Markl, *Strongly homotopy Lie algebras*, Comm. Algebra **23** (1995), no. 6, 2147–2161. 401

LM05. ———, *Symmetric brace algebras*, Appl. Categ. Structures **13** (2005), no. 4, 351–370. 517

LN12. J.-L. Loday and N. M. Nikolov, *Operadic construction of the renormalization group*, Proceedings of international conference held in Varna June 2011, Springer (2012). 152, 184

Lod06. Jean-Louis Loday, *Inversion of integral series enumerating planar trees*, Sém. Lothar. Combin. **53** (2004/06), Art. B53d, 16 pp. (electronic). MR MR2180779 (2006k:05016) 550

Lod82. ———, *Spaces with finitely many nontrivial homotopy groups*, J. Pure Appl. Algebra **24** (1982), no. 2, 179–202. 349

Lod89. ———, *Opérations sur l'homologie cyclique des algèbres commutatives*, Invent. Math. **96** (1989), no. 1, 205–230. 484, 505

Lod93. ———, *Une version non commutative des algèbres de Lie: les algèbres de Leibniz*, Enseign. Math. (2) **39** (1993), no. 3-4, 269–293. 8, 520, 521

Lod94. ———, *Série de Hausdorff, idempotents eulériens et algèbres de Hopf*, Exposition. Math. **12** (1994), no. 2, 165–178. 20, 36

Lod95. ———, *Cup-product for Leibniz cohomology and dual Leibniz algebras*, Math. Scand. **77** (1995), no. 2, 189–196. 520, 521, 522, 524, 526

Lod97. ———, *Overview on Leibniz algebras, dialgebras and their homology*, Cyclic cohomology and noncommutative geometry (Waterloo, ON, 1995), Fields Inst. Commun., vol. 17, Amer. Math. Soc., Providence, RI, 1997, pp. 91–102. 522

Lod98. ———, *Cyclic homology*, second ed., Grundlehren der Mathematischen Wissenschaften [Fundamental Principles of Mathematical Sciences], vol. 301, Springer-Verlag, Berlin, 1998, Appendix E by María O. Ronco, Chap. 13 by the author in collaboration with Teimuraz Pirashvili. 586, 28, 151, 335, 336, 385, 461, 483, 535

Lod01. ———, *Dialgebras*, Dialgebras and related operads, Lecture Notes in Math., vol. 1763, Springer, Berlin, 2001, pp. 7–66. 600, 520, 522, 524, 526, 527, 529

Lod02. ———, *Arithmetree*, J. Algebra **258** (2002), no. 1, 275–309, Special issue in celebration of Claudio Procesi's 60th birthday. 522, 525

Lod04a. ———, *Realization of the Stasheff polytope*, Arch. Math. (Basel) **83** (2004), no. 3, 267–278. 601

Lod04b. ———, *Scindement d'associativité et algèbres de Hopf*, Actes des Journées Mathématiques à la Mémoire de Jean Leray, Sémin. Congr., vol. 9, Soc. Math. France, Paris, 2004, pp. 155–172. 553

Lod06. ———, *Completing the operadic butterfly*, Georgian Math. J. **13** (2006), no. 4, 741–749. 530

Lod07. ———, *On the algebra of quasi-shuffles*, Manuscripta Math. **123** (2007), no. 1, 79–93. 531

Lod08. ———, *Generalized bialgebras and triples of operads*, Astérisque (2008), no. 320, x+116. 18, 151, 527, 553, 554, 557, 558

Lod10. ———, *On the operad of associative algebras with derivation*, Georgian Math. J. **17** (2010), no. 2, 347–372. 552

Lod12. ———, *Geometric diagonals for the Stasheff associahedron and products of A-infinity algebras*. 355

Löf86. Clas Löfwall, *On the subalgebra generated by the one-dimensional elements in the Yoneda Ext-algebra*, Algebra, algebraic topology and their interactions (Stockholm, 1983), Lecture Notes in Math., vol. 1183, Springer, Berlin, 1986, pp. 291–338. 72

LP93. J.-L. Loday and T. Pirashvili, *Universal enveloping algebras of Leibniz algebras and (co)homology*, Math. Ann. **296** (1993), no. 1, 139–158. 464, 475, 521, 522

LQ84. J.-L. Loday and D. Quillen, *Cyclic homology and the Lie algebra homology of matrices*, Comment. Math. Helv. **59** (1984), no. 4, 569–591. 335, 336, 385

LR04. J.-L. Loday and M. O. Ronco, *Trialgebras and families of polytopes*, Homotopy theory: relations with algebraic geometry, group cohomology, and algebraic K-theory, Contemp. Math., vol. 346, Amer. Math. Soc., Providence, RI, 2004, pp. 369–398. MR 2066507 (2006e:18016) 323, 531

LR06. ———, *On the structure of cofree Hopf algebras*, J. Reine Angew. Math. **592** (2006), 123–155. 548, 549, 553, 555

LR10. ———, *Combinatorial Hopf algebras*, Clay Mathematics Proceedings **12** (2010), 347–384. 154, 185, 519, 548, 549, 559

LR12. ———, *Permutads*, Journal of Combinatorial Theory A (2012). 336, 565

LS93. T. Lada and J. Stasheff, *Introduction to SH Lie algebras for physicists*, Internat. J. Theoret. Phys. **32** (1993), no. 7, 1087–1103. 366, 496

LTV10. P. Lambrechts, V. Turchin, and I. Volić, *The rational homology of the space of long knots in codimension* >2, Geom. Topol. **14** (2010), no. 4, 2151–2187. 447

LV12. J.-L. Loday and B. Vallette, *Multicomplex, spectral sequence and Koszul duality*. 386

Mac95. I. G. Macdonald, *Symmetric functions and Hall polynomials*, second ed., Oxford Mathematical Monographs, The Clarendon Press Oxford University Press, New York, 1995, With contributions by A. Zelevinsky, Oxford Science Publications. 567, 128, 329, 494

Man87. Yu. I. Manin, *Some remarks on Koszul algebras and quantum groups*, Ann. Inst. Fourier (Grenoble) **37** (1987), no. 4, 191–205. 62, 66, 90, 111, 114

Man88. ———, *Quantum groups and noncommutative geometry*, Université de Montréal Centre de Recherches Mathématiques, Montreal, QC, 1988. 62, 66, 90, 111, 113, 114

Man99a. Marco Manetti, *Deformation theory via differential graded Lie algebras*, Algebraic Geometry Seminars, 1998–1999 (Italian) (Pisa), Scuola Norm. Sup., Pisa, 1999, pp. 21–48. 501

Man99b. Yu. I. Manin, *Frobenius manifolds, quantum cohomology, and moduli spaces*, American Mathematical Society Colloquium Publications, vol. 47, American Mathematical Society, Providence, RI, 1999. 545

Man01. M. A. Mandell, *E_∞ algebras and p-adic homotopy theory*, Topology **40** (2001), no. 1, 43–94. 487

Man06. ———, *Cochains and homotopy type*, Publ. Math. Inst. Hautes Études Sci. (2006), no. 103, 213–246. 487

Mar92. Martin Markl, *A cohomology theory for $A(m)$-algebras and applications*, J. Pure Appl. Algebra **83** (1992), no. 2, 141–175. 469

Mar96a. ———, *Distributive laws and Koszulness*, Ann. Inst. Fourier (Grenoble) **46** (1996), no. 2, 307–323. 229, 264, 288

Mar96b. ———, *Models for operads*, Comm. Algebra **24** (1996), no. 4, 1471–1500. 203

Mar04. ———, *Homotopy algebras are homotopy algebras*, Forum Math. **16** (2004), no. 1, 129–160. 560

Mar06. ———, *Transferring A_∞ (strongly homotopy associative) structures*, Rend. Circ. Mat. Palermo (2) Suppl. (2006), no. 79, 139–151. 352

Mas58. W. S. Massey, *Some higher order cohomology operations*, Symposium internacional de topología algebraica International symposium on algebraic topology, Universidad Nacional Autónoma de México and UNESCO, Mexico City, 1958, pp. 145–154. 339, 350, 384

May66. J. P. May, *The cohomology of restricted Lie algebras and of Hopf algebras*, J. Algebra **3** (1966), 123–146. 80

May69. ———, *Matric Massey products*, J. Algebra **12** (1969), 533–568. 350, 356

May72. ———, *The geometry of iterated loop spaces*, Springer-Verlag, Berlin, 1972, Lectures Notes in Mathematics, Vol. 271. 142, 148, 149

Mén89. M. Méndez, *Monoides, c-monoides, especies de Möbius y coàlgebras*, Ph.D. thesis, Universidad Central de Venezuela, http://www.ivic.gob.ve/matematicas/?mod=mmendez.htm, 1989. 308

Men09. Luc Menichi, *Batalin-Vilkovisky algebra structures on Hochschild cohomology*, Bulletin de la Société Mathématique de France **137** (2009), no. 2, 277–295. 535

Mer99. S. A. Merkulov, *Strong homotopy algebras of a Kähler manifold*, Internat. Math. Res. Notices (1999), no. 3, 153–164. 352

Mer04. ———, *Operads, deformation theory and F-manifolds*, Frobenius manifolds, Aspects Math., E36, Vieweg, Wiesbaden, 2004, pp. 213–251. 295

Mer05. ———, *Nijenhuis infinity and contractible differential graded manifolds*, Compos. Math. **141** (2005), no. 5, 1238–1254. 323, 518

Mer08. ———, *Exotic automorphisms of the Schouten algebra of polyvector fields*, arXiv:0809.2385 (2008). 522

Mer10a. ———, *Wheeled Pro(p)file of Batalin-Vilkovisky formalism*, Comm. Math. Phys. **295** (2010), no. 3, 585–638. 376

Mer10b. ———, *Wheeled props in algebra, geometry and quantization*, European Congress of Mathematics, Eur. Math. Soc., Zürich, 2010, pp. 83–114. 534, 563

Mey78. Jean-Pierre Meyer, *Acyclic models for multicomplexes*, Duke Math. J. **45** (1978), no. 1, 67–85. MR 486489 (80b:55012) 386

Mil09. D. V. Millionshchikov, *The algebra of formal vector fields on the line and Buchstaber's conjecture*, Funktsional. Anal. i Prilozhen. **43** (2009), no. 4, 26–44. 485

Mil10. Joan Millès, *The Koszul complex is the cotangent complex*, arXiv:1004.0096 (2010). 417, 469, 470, 475

Mil11. ———, *André-Quillen cohomology of algebras over an operad*, Adv. Math. **226** (2011), no. 6, 5120–5164. MR 2775896 405, 428, 448, 475

ML63. Saunders Mac Lane, *Natural associativity and commutativity*, Rice Univ. Studies **49** (1963), no. 4, 28–46. 580

ML95. ———, *Homology*, Classics in Mathematics, Springer-Verlag, Berlin, 1995, Reprint of the 1975 edition. xxii, 26, 28, 51, 78, 80, 266, 349

ML98. ———, *Categories for the working mathematician*, second ed., Graduate Texts in Mathematics, vol. 5, Springer-Verlag, New York, 1998. 573, 583, 593, xxii, 456

MM65. J. W. Milnor and J. C. Moore, *On the structure of Hopf algebras*, Ann. of Math. (2) **81** (1965), 211–264. 18

Mnë09. P. Mnëv, *Notes on simplicial BF theory*, Mosc. Math. J. **9** (2009), no. 2, 371–410. 486

Moe01. Ieke Moerdijk, *On the Connes-Kreimer construction of Hopf algebras*, Homotopy methods in algebraic topology (Boulder, CO, 1999), Contemp. Math., vol. 271, Amer. Math. Soc., Providence, RI, 2001, pp. 311–321. 154

Moo71. J. C. Moore, *Differential homological algebra*, Actes du Congrès International des Mathématiciens (Nice, 1970), Tome 1, Gauthier-Villars, Paris, 1971, pp. 335–339. 37, 38

MR96. G. Melançon and C. Reutenauer, *Free Lie superalgebras, trees and chains of partitions*, J. Algebraic Combin. **5** (1996), no. 4, 337–351. 492

MR09. M. Markl and E. Remm, *(non-)Koszulness of operads for n-ary algebras, gagalim and other curiosities*, arXiv:0907.1505v2 (2009). 545

MS02. J. E. McClure and J. H. Smith, *A solution of Deligne's Hochschild cohomology conjecture*, Recent progress in homotopy theory (Baltimore, MD, 2000), Contemp. Math., vol. 293, Amer. Math. Soc., Providence, RI, 2002, pp. 153–193. 447, 510

MSS02. Martin Markl, Steve Shnider, and Jim Stasheff, *Operads in algebra, topology and physics*, Mathematical Surveys and Monographs, vol. 96, American Mathematical Society, Providence, RI, 2002. vi, xx, 121, 148, 232, 360

MT78. J. P. May and R. Thomason, *The uniqueness of infinite loop space machines*, Topology **17** (1978), 205–224 (English). 151

MV09a. S. A. Merkulov and B. Vallette, *Deformation theory of representations of prop(erad)s. I*, J. Reine Angew. Math. **634** (2009), 51–106. 447, 549, 560

MV09b. ———, *Deformation theory of representations of prop(erad)s. II*, J. Reine Angew. Math. **636** (2009), 123–174. 447, 477, 560

MY91. M. Méndez and J. Yang, *Möbius species*, Adv. Math. **85** (1991), no. 1, 83–128. 297

New42. M. H. A. Newman, *On theories with a combinatorial definition of "equivalence"*, Ann. of Math. (2) **43** (1942), 223–243. 103

NR66. A. Nijenhuis and R. W. Richardson, Jr., *Cohomology and deformations in graded Lie algebras*, Bull. Amer. Math. Soc. **72** (1966), 1–29. 366, 433, 473

NR67. ———, *Deformations of Lie algebra structures*, J. Math. Mech. **17** (1967), 89–105. 366, 433, 473

Nto94. Patricia Ntolo, *Homologie de Leibniz d'algèbres de Lie semi-simples*, C. R. Acad. Sci. Paris Sér. I Math. **318** (1994), no. 8, 707–710. 522

OG08. J.-M. Oudom and D. Guin, *On the Lie enveloping algebra of a pre-Lie algebra*, J. K-Theory **2** (2008), no. 1, 147–167. 517

Pio01. D. I. Piontkovskiĭ, *On Hilbert series of Koszul algebras*, Funktsional. Anal. i Prilozhen. **35** (2001), no. 2, 64–69, 96. 74

Pir94. Teimuraz Pirashvili, *On Leibniz homology*, Ann. Inst. Fourier (Grenoble) **44** (1994), no. 2, 401–411. 522

Pir02a. ———, *On the PROP corresponding to bialgebras*, Cah. Topol. Géom. Différ. Catég. **43** (2002), no. 3, 221–239. 151, 183, 190, 336

Pir02b. ———, *Polynomial functors over finite fields (after Franjou, Friedlander, Henn, Lannes, Schwartz, Suslin)*, Astérisque (2002), no. 276, 369–388, Séminaire Bourbaki, Vol. 1999/2000. 85

Pir03. ———, *Sets with two associative operations*, Cent. Eur. J. Math. **1** (2003), no. 2, 169–183 (electronic). 190

Pos95. L. E. Positsel'skiĭ, *The correspondence between Hilbert series of quadratically dual algebras does not imply their having the Koszul property*, Funktsional. Anal. i Prilozhen. **29** (1995), no. 3, 83–87. 74

PP05. A. Polishchuk and L. Positselski, *Quadratic algebras*, University Lecture Series, vol. 37, American Mathematical Society, Providence, RI, 2005. xxi, 62, 66, 74, 75, 90, 100, 102, 111, 117

Pri70. S. B. Priddy, *Koszul resolutions*, Trans. Amer. Math. Soc. **152** (1970), 39–60. 61, 62, 66, 75, 80, 82, 90, 100, 106, 117

Pro86. A. Prouté, *A_∞-structures, modèle minimal de Baues-Lemaire et homologie des fibrations*, Reprints in Theory and Applications of Categories [2011] **21** (1986), 1–99. 37, 405

Qui67. Daniel G. Quillen, *Homotopical algebra*, Lecture Notes in Mathematics, No. 43, Springer-Verlag, Berlin, 1967. 573, 588, 591, 594

Qui69. ———, *Rational homotopy theory*, Ann. of Math. (2) **90** (1969), 205–295. 595, 17, 18, 405, 417, 485, 487

Qui70. ———, *On the (co-) homology of commutative rings*, Applications of Categorical Algebra (Proc. Sympos. Pure Math., Vol. XVII, New York, 1968), Amer. Math. Soc., Providence, RI, 1970, pp. 65–87. 428, 447

Ree58. R. Ree, *Lie elements and an algebra associated with shuffles*, Ann. of Math. (2) **68** (1958), 210–220. 19

Ret93. Vladimir S. Retakh, *Lie-Massey brackets and n-homotopically multiplicative maps of differential graded Lie algebras*, J. Pure Appl. Algebra **89** (1993), no. 1-2, 217–229. 384, 497

Reu93. Ch. Reutenauer, *Free Lie algebras*, xviii+269, Oxford Science Publications. 20, 490, 492

Rez96. C. W. Rezk, *Spaces of algebra structures and cohomology of operads*, Ph.D. thesis, MIT, 1996. 154, 376, 477

Rob03. Alan Robinson, *Gamma homology, Lie representations and E_∞ multiplications*, Invent. Math. **152** (2003), no. 2, 331–348. 470

Ron00. María Ronco, *Primitive elements in a free dendriform algebra*, New trends in Hopf algebra theory (La Falda, 1999), Contemp. Math., vol. 267, Amer. Math. Soc., Providence, RI, 2000, pp. 245–263. 548

Ron02. ———, *Eulerian idempotents and Milnor-Moore theorem for certain non-cocommutative Hopf algebras*, J. Algebra **254** (2002), no. 1, 152–172. 516, 548, 559

Ron11. ———, *Shuffle algebras*, Annales Instit. Fourier **61** (2011), no. 1, 799–850. 183, 274, 560, 561, 565

Roo95. Jan-Erik Roos, *On the characterisation of Koszul algebras. Four counterexamples*, C. R. Acad. Sci. Paris Sér. I Math. **321** (1995), no. 1, 15–20. 74

RW02. A. Robinson and S. Whitehouse, *Operads and Γ-homology of commutative rings*, Math. Proc. Cambridge Philos. Soc. **132** (2002), no. 2, 197–234. 470, 493

Sag01. B. E. Sagan, *The symmetric group*, second ed., Graduate Texts in Mathematics, vol. 203, Springer-Verlag, New York, 2001, Representations, combinatorial algorithms, and symmetric functions. 567

Sam53. Hans Samelson, *A connection between the Whitehead and the Pontryagin product*, Amer. J. Math. **75** (1953), 744–752. 497

Sch68. Michael Schlessinger, *Functors of Artin rings*, Trans. Amer. Math. Soc. **130** (1968), 208–222. 497

Sch93. Albert Schwarz, *Geometry of Batalin-Vilkovisky quantization*, Comm. Math. Phys. **155** (1993), no. 2, 249–260. 534

Seg04. G. Segal, *The definition of conformal field theory*, Topology, geometry and quantum field theory, London Math. Soc. Lecture Note Ser., vol. 308, Cambridge Univ. Press, Cambridge, 2004, pp. 421–577. 537

Sei08. Paul Seidel, *Fukaya categories and Picard-Lefschetz theory*, Zurich Lectures in Advanced Mathematics, European Mathematical Society (EMS), Zürich, 2008. 552

Ser06. Jean-Pierre Serre, *Lie algebras and Lie groups*, Lecture Notes in Mathematics, vol. 1500, Springer-Verlag, Berlin, 2006, 1964 lectures given at Harvard University, Corrected fifth printing of the second (1992) edition. 499

Šev10. Pavol Ševera, *Formality of the chain operad of framed little disks*, Lett. Math. Phys. **93** (2010), no. 1, 29–35. 538

Sin06. D. P. Sinha, *Operads and knot spaces*, J. Amer. Math. Soc. **19** (2006), no. 2, 461–486 (electronic). 447

Sin09. ———, *The topology of spaces of knots: cosimplicial models*, Amer. J. Math. **131** (2009), no. 4, 945–980. 447

Smi89. J. H. Smith, *Simplicial group models for $\Omega^n S^n X$*, Israel J. Math. **66** (1989), no. 1-3, 330–350. 488

Smi01. V. A. Smirnov, *Simplicial and operad methods in algebraic topology*, Translations of Mathematical Monographs, vol. 198, American Mathematical Society, Providence, RI, 2001, Translated from the Russian manuscript by G. L. Rybnikov. xx

Spi01. Markus Spitzweck, *Operads, algebras and modules in general model categories*, arXiv:math/0101102 (2001). 591, 592

SS85. M. Schlessinger and J. Stasheff, *The Lie algebra structure of tangent cohomology and deformation theory*, J. Pure Appl. Algebra **38** (1985), no. 2-3, 313–322. 417, 435, 496, 497

ST09. P. Salvatore and R. Tauraso, *The operad Lie is free*, J. Pure Appl. Algebra **213** (2009), no. 2, 224–230. 183, 493

Sta63. Jim Stasheff, *Homotopy associativity of H-spaces. I, II*, Trans. Amer. Math. Soc. 108 (1963), 275–292; ibid. **108** (1963), 293–312. 601, 602, 336, 337, 339, 560

Sta93. ———, *The intrinsic bracket on the deformation complex of an associative algebra*, J. Pure Appl. Algebra **89** (1993), no. 1-2, 231–235. 435

Sta97a. R. P. Stanley, *Enumerative combinatorics. Vol. 1*, Cambridge Studies in Advanced Mathematics, vol. 49, Cambridge University Press, Cambridge, 1997, With a foreword by Gian-Carlo Rota, Corrected reprint of the 1986 original. 601, 109, 301, 550

Sta97b. Jim Stasheff, *From operads to "physically" inspired theories*, Operads: Proceedings of Renaissance Conferences (Hartford, CT/Luminy, 1995), Contemp. Math., vol. 202, 1997, pp. 53–81. 350

Sta10. ———, *A twisted tale of cochains and connections*, Georgian Math. J. **17** (2010), no. 1, 203–215. 325, 376

Sto93. C. R. Stover, *The equivalence of certain categories of twisted Lie and Hopf algebras over a commutative ring*, J. Pure Appl. Algebra **86** (1993), no. 3, 289–326. MR 1218107 (94e:16031) 130

Str08. Henrik Strohmayer, *Operads of compatible structures and weighted partitions*, J. Pure Appl. Algebra **212** (2008), no. 11, 2522–2534. 308

Str09. ———, *Operad profiles of Nijenhuis structures*, Differential Geom. Appl. **27** (2009), no. 6, 780–792. 323

Str10. ———, *Prop profile of bi-Hamiltonian structures*, J. Noncommut. Geom. **4** (2010), no. 2, 189–235. 308

Sul77. Dennis Sullivan, *Infinitesimal computations in topology*, Inst. Hautes Études Sci. Publ. Math. (1977), no. 47, 269–331 (1978). 405, 417, 487, 496, 497

SV12. H. Strohmayer and B. Vallette, *Homotopy theory of homotopy operads*, Preprint (2012). 228

Tak94. L. A. Takhtajan, *Higher order analog of Chevalley-Eilenberg complex and deformation theory of n-gebras*, Algebra i Analiz **6** (1994), no. 2, 262–272. 546

Tam99. D. E. Tamarkin, *Operadic proof of M. Kontsevich's formality theorem*, 51 p., Thesis (Ph.D.)–The Pennsylvania State University. 510, 539

Tam03. ———, *Formality of chain operad of little discs*, Lett. Math. Phys. **66** (2003), no. 1-2, 65–72. 510

Tam07. ———, *What do dg-categories form?*, Compos. Math. **143** (2007), no. 5, 1335–1358. 510

Tou04. V. Tourtchine, *On the homology of the spaces of long knots*, Advances in topological quantum field theory, NATO Sci. Ser. II Math. Phys. Chem., vol. 179, Kluwer Acad. Publ., Dordrecht, 2004, pp. 23–52. 447

Tou06. Victor Tourtchine, *Dyer-Lashof-Cohen operations in Hochschild cohomology*, Algebr. Geom. Topol. **6** (2006), 875–894 (electronic). 493

Tra08. T. Tradler, *The Batalin-Vilkovisky algebra on Hochschild cohomology induced by infinity inner products*, Annales de l'institut Fourier **58** (2008), no. 7, 2351–2379. 535

Tsy83. B. L. Tsygan, *Homology of matrix Lie algebras over rings and the Hochschild homology*, Uspekhi Mat. Nauk **38** (1983), no. 2(230), 217–218. 336

TT00. D. Tamarkin and B. Tsygan, *Noncommutative differential calculus, homotopy BV algebras and formality conjectures*, Methods Funct. Anal. Topology **6** (2000), no. 2, 85–100. 507, 510, 533, 537

TZ06. T. Tradler and M. Zeinalian, *On the cyclic Deligne conjecture*, J. Pure Appl. Algebra **204** (2006), no. 2, 280–299. 539

Uch10. Kyousuke Uchino, *Derived bracket construction and Manin products*, Lett. Math. Phys. **93** (2010), no. 1, 37–53. 519

Ufn95. V. A. Ufnarovski, *Combinatorial and asymptotic methods in algebra*, Algebra VI, Encyclopedia Math. Sci., vol. 57, Springer, Berlin, 1995, pp. 1–196. 74, 107

Val07a. Bruno Vallette, *Homology of generalized partition posets*, J. Pure Appl. Algebra **208** (2007), no. 2, 699–725. 297, 303, 323

Val07b. ———, *A Koszul duality for props*, Trans. of Amer. Math. Soc. **359** (2007), 4865–4993. 159, 220, 242, 264, 288, 550, 559, 563, 566

Val08. ———, *Manin products, Koszul duality, Loday algebras and Deligne conjecture*, J. Reine Angew. Math. **620** (2008), 105–164. 158, 159, 242, 250, 264, 308, 518, 529, 531, 541

Val09. ——— , *Free monoid in monoidal abelian categories*, Appl. Categ. Structures **17** (2009), no. 1, 43–61. 154

vdL02. P. van der Laan, *Operads up to Homotopy and Deformations of Operad Maps*, arXiv:math.QA/0208041 (2002). 184, 391, 447

vdL03. ——— , *Coloured Koszul duality and strongly homotopy operads*, arXiv:math.QA/0312147 (2003). 183, 378, 391, 398, 559, 563

vdLM02. P. van der Laan and I. Moerdijk, *The bitensor algebra through operads*, hep-th/0210226 (2002). 185

vdLM06. ——— , *Families of Hopf algebras of trees and pre-Lie algebras*, Homology, Homotopy Appl. **8** (2006), no. 1, 243–256. 154

Vin63. È. B. Vinberg, *The theory of homogeneous convex cones*, Trudy Moskov. Mat. Obšč. **12** (1963), 303–358. 21

Vor00. Alexander A. Voronov, *Homotopy Gerstenhaber algebras*, Conférence Moshé Flato 1999, Vol. II (Dijon), Math. Phys. Stud., vol. 22, Kluwer Acad. Publ., Dordrecht, 2000, pp. 307–331. 510

VV97. A. Vinogradov and M. Vinogradov, *On multiple generalizations of Lie algebras and Poisson manifolds*, Secondary calculus and cohomological physics (Moscow, 1997), Contemp. Math., vol. 219, 1997, pp. 273–287. 546

Wan67. J. S. P. Wang, *On the cohomology of the* mod − 2 *Steenrod algebra and the non-existence of elements of Hopf invariant one*, Illinois J. Math. **11** (1967), 480–490. 82, 117

Wei94. C. A. Weibel, *An introduction to homological algebra*, Cambridge Studies in Advanced Mathematics, vol. 38, Cambridge University Press, Cambridge, 1994. 28, 86, 199

Whi41. J. H. C. Whitehead, *On adding relations to homotopy groups*, Ann. of Math. (2) **42** (1941), 409–428. 497

Wig89. D. Wigner, *An identity in the free Lie algebra*, Proc. Amer. Math. Soc. **106** (1989), no. 3, 639–640. 17

Wil07. S. O. Wilson, *Free Frobenius algebra on the differential forms of a manifold*, arXiv:0710.3550 (2007). 445

Zee57. E. C. Zeeman, *A proof of the comparison theorem for spectral sequences*, Proc. Cambridge Philos. Soc. **53** (1957), 57–62. 55

Zha12. Young Zhang, *Homotopy transfer theorem for linearly compatible di-algebras*, preprint (2012). 356

Zin12. G. W. Zinbiel, *Encyclopedia of types of algebras 2010*, Proc. Int. Conf., in Nankai Series in Pure, Applied Mathematics and Theoretical Physics, vol. 9, World Scientific, Singapore, 2012. 479

ZSSS82. K. A. Zhevlakov, A. M. Slin'ko, I. P. Shestakov, and A. I. Shirshov, *Rings that are nearly associative*, Pure and Applied Mathematics, vol. 104, Academic Press Inc. [Harcourt Brace Jovanovich Publishers], New York, 1982, Translated from the Russian by Harry F. Smith. 543

Zwi93. Barton Zwiebach, *Closed string field theory: quantum action and the Batalin-Vilkovisky master equation*, Nuclear Phys. B **390** (1993), no. 1, 33–152. 496

Index

J.-L. Loday, B. Vallette, *Algebraic Operads*,
Grundlehren der mathematischen Wissenschaften 346,
DOI 10.1007/978-3-642-30362-3, © Springer-Verlag Berlin Heidelberg 2012

W

Z

List of Notations

J.-L. Loday, B. Vallette, *Algebraic Operads*,
Grundlehren der mathematischen Wissenschaften 346,
DOI 10.1007/978-3-642-30362-3, © Springer-Verlag Berlin Heidelberg 2012